HERBAL HARVEST

HERBAL HARVEST

Commercial Organic Production of Quality Dried Herbs

THIRD EDITION

Greg Whitten

BLOOMINGS BOOKS

First published in Australia in 1997 by Agmedia
Second edition published by Bloomings Books Pty Ltd in 1999
This third edition published in 2004 by

Bloomings Books Pty Ltd
Melbourne, Australia
Phone +61 (0)3 9427 1234
Facsimile +61 (0)3 9427 9066
sales@bloomings.com.au
www.bloomings.com.au

Bloomings Books is a specialist publisher and distributor of horticultural, agricultural and natural history books.

Distributed in North America by
CHELSEA GREEN PUBLISHING COMPANY
P.O. Box 428
White River Junction, Vermont 05001
800-639-4099
www.chelseagreen.com

© Greg Whitten

All rights reserved. No part of this publication may be reproduced, stored in a retrieval system, or transmitted in any form or by any means, electronic, mechanical, photocopying, recording or otherwise without the prior writen permission of the Publisher.

The publisher would welcome readers' comments: warwick@bloomings.com.au

Cover design by Watts Graphic Design, Melbourne
Typeset by Desktop Concepts Pty. Ltd, Melbourne
Photographs by Greg Whitten, Jenny Grinlington
Illustrations by Julia McLeish
Printed in China through Bookbuilders
Publisher Warwick Forge

National Library of Australia Cataloguing-in-Publication entry

Whitten, Greg, 1946– .
Herbal harvest: commercial organic production of quality dried herbs.

3rd ed.
Bibliography.
Includes index.
ISBN 1 876473 47 9.

1. Herbs. 2. Herbs – Processing. 3. Herb farming.
4. Herbs – Marketing. I. Title.

635.7

While the information contained in this publication has been formulated in good faith, the contents may not take into account all those factors which need to be considered before putting information into practice. Accordingly, reliance should not be placed on anything contained herein as a substitute for specific advice.

Front cover: Broad-leaf Echinacea (*Echinacea purpurea*), by Jenny Grinlington.

Foreword

During the 20th century, the use of traditional herbal medicine has been eclipsed by Western scientific medicine. The term 'medicine' itself has come to mean only medicine which conforms with the Western scientific model. However, in the past couple of decades, disillusionment with that branch of medicine has meant that more and more people have begun to look for alternatives. In Australia in the mid-1990s, it is estimated that half of the population is using some form of alternative medicine[1]: for many this is herbal treatment.

Growing, harvesting, drying and processing conditions have an enormous impact on the effectiveness of herbs as medicines. Each of the plants has its own particular likes and dislikes in terms of optimum conditions, as well as differences in time of harvest and the part of the plant (root, leaves, flowers) to be used medicinally. Herbs that are used as medicines should be grown without the use of chemical pesticides and fertilisers. One does not want the medicines adding to health problems. As Hippocrates advised the physicians of his day: 'First, do no harm'.

Pharmaceuticals are purchased in bottles and boxes, and as consumers we rarely think about their quality. It is a different story regarding fruit and vegetables: consumers instinctively know the difference between good and poor quality. Good quality foods look better and they taste better.

It is the same with herbal medicines: good quality herbs look better and taste better than poor quality ones. However, there is an additional factor: good quality herbs will work well, while inferior ones will work poorly or not at all. It is, therefore, false economy to buy poor quality herbs, even if they are cheap, as they will not do the job required.

Many professional herbalists use between one and two hundred different plants in their practice: some in large amounts, some in small amounts and rarely. Until relatively recently, most medicinal herbs used in the western world were grown in Europe, especially Eastern Europe. The environmental problems faced there, particularly since the Chernobyl disaster, make that area unsuitable as a source of plant medicines.

Australia, by contrast, enjoys a cleaner environment and a variety of climates which allow for the growing needs of many different medicinals. It can be argued that wherever possible it is preferable to use locally grown herbs in medicines. Apart from the economic advantages of buying Australian, these herbs will be fresher and often more effective than imported ones, especially if they are grown according to the methods described in this book.

There is the opportunity for Australia to become a major producer of medicinal herbs, both for the domestic and overseas markets. The use of herbs as medicines is not going to go away, but its development will to a large extent be determined by the availability of good quality medicinals. The enormous increase in demand for those herbal medicines which have become so popular in recent years – herbs such as Echinacea, Ginkgo and Ginseng – is indicative of the potential of the market. Less 'glamorous' herbs are also needed, such as Peppermint and Thyme.

Greg Whitten has pioneered the development of techniques for producing premium quality medicinal herbs in this country. He has inspired and supported many growers starting out in this field. For more years than I care to remember, people have been asking 'When's Greg going to write a book?' His knowledge about optimal growing conditions for medicinals and the best ways of harvesting and drying the plants to retain their maximum therapeutic effectiveness is grounded in years of trial and error.

By encouraging growers to produce good quality medicinals, this book makes an important contribution to herbal medicine.

Sue Evans,
Lecturer in Herbal Medicine, School of Natural and Complementary Medicine

Southern Cross University, Lismore, New South Wales

1 MacLennan A.H., Wilson, D.H., Taylor, A.W., 'Prevalence and cost of alternative medicine in Australia', *The Lancet*, 1996, 347: 369-73.

Acknowledgements

I would like to extend my gratitude to all those who have in one way or another contributed to this book.

First I would like to thank my parents, Beth and Wes Whitten, for their loving encouragement and support over the years, my partner Liz who has helped me enormously with her keen proof-reading eye and constructive criticism and my daughter Lucie for her help with the phrasing of some passages I found difficult.

Jenny Grinlington has generously contributed a number of superb photos. Sue Evans was instrumental in encouraging me to write this book and has kindly helped with details of medicinal usage of herbs. I would also like to thank Mike Brouwer and Natalie Greenwood of Southern Light Herbs, Rob and Julie Carolane, Jim Newton, Martin and Ann Joyce, my son Forrest, Greg and Libby Maulder of Highland Herbs Tasmania, Michael and Linda Gardiner, Peter Purbrick, Linda Bradbury, Dave Marks, Carol Lewis-Shaw, Iain Larner, Assunta Hunter, Chris Read of Diemen Pepper, and Col and Pam Maple.

I am also greatly indebted to all those wonderful people in many places and at different times who have shared their knowledge, skills, understanding and inspiration in the fields of organic and biodynamic gardening and farming, the use of tools and machinery, the appreciation of plants and their usage as herbs.

I would also like to thank all those whose words of encouragement and positive feedback have helped make it possible for me to complete what has proved to be a much bigger undertaking than I had ever envisaged. My sincere thanks to my family for their phenomenal patience and my apologies for all the duties, tasks and social participation I have neglected during the years I have been working on this book. My apologies as well to the herbs in my gardens and the other aspects of the farm which have had to endure an extended period of less than optimum care while my attentions have been devoted elsewhere.

Photographs by Jenny Grinlington are on text pages 1, 9, 44 (top left), 58 (right), 257, 258, 259 (top), 269, 309, 345, 356, 369, 395, 408, 429, 449, 497, 499; and colour plates 1.6, 2.3, 2.5, 3.3, 3.6, 4.1, 4.2, 5.2, 5.4, 6.2, 6.3, 6.7, 7.2.

Location of Colour Plates

Facing page 166
1.1	Alfalfa	*Medicago sativa*
1.2	Angelica	*Angelica archangelica*
1.3	Bergamot	*Monarda didyma*
1.4	Burdock	*Arctium lappa*
1.5	Calendula	*Calendula officinalis*
1.6	Catnip	*Nepeta cataria*

Facing page 167
2.1	Chamomile, German	*Chamomilla recutita*
2.2	Comfrey, English	*Symphytum officinale*
2.3	Chaste Tree	*Vitex agnus-castus*
2.4	Coriander	*Coriandrum sativum*
2.5	Crampbark	*Viburnum opulus*
2.6	Dandelion	*Taraxacum officinale*
2.7	Dandelion root rot	

Facing page 182
3.1	Dock, Yellow	*Rumex crispus*
3.2	Echinacea, Broad-leaf	*Echinacea purpurea*
3.3	Echinacea, Narrow-leaf	*Echinacea angustifolia*
3.4	Elder, Black	*Sambucus nigra*
3.5	Fennel	*Foeniculum vulgare*
3.6	Ginseng, American	*Panax quinquefolius*
3.7	Hawthorn	*Crataegus monogyna*

Facing page 183
4.1	Hops	*Humulus lupulus*
4.2	Horsetail, Field	*Equisetum arvense*
4.3	Hyssop	*Hyssopus officinalis*
4.4	Lemon Thyme	*Thymus x citriodorus*
4.5	Lady's Mantle	*Alchemilla vulgaris*
4.6	Marshmallow	*Althaea officinalis*

Facing page 358
5.1	Melissa Balm	*Melissa officinalis*
5.2	Pasque Flower	*Anemone pulsatilla*
5.3	Mountain Pepper	*Tasmannia lanceolata*
5.4	Passionflower	*Passiflora incarnata*
5.5	Peppermint	*Mentha x piperita*
5.6	Nettle, Greater	*Urtica dioica*

Facing page 359
6.1	Raspberry	*Rubus idaeus*
6.2	Red Clover	*Trifolium pratense*
6.3	Rose, Dog	*Rosa canina*
6.4	Rue	*Ruta graveolens*
6.5	Rust on Peppermint	
6.6	Rose, Sweet Briar	*Rosa rubiginosa*
6.7	Sage	*Salvia officinalis*

Facing page 374
7.1	Spearmint	*Mentha spicata*
7.2	Valerian	*Valeriana officinalis*
7.3	St John's Wort	*Hypericum perforatum*
7.4	Wood Betony	*Stachys officinalis*
7.5	Thyme	*Thymus vulgaris*
7.6	Yarrow	*Achillea millefolium*

Facing page 375
8.1	Red Clover - trade & premium grade
8.2	German Chamomile - premium grade & trade
8.3	Peppermint - premium grade & trade
8.4	Spearmint - premium grade & trade
8.5	Culinary Grade & Tea Grade
8.6	Decoction - particle size
8.7	Manufacturing Grade - aerial parts

Contents

Foreword	v
Acknowledgements	vi

1. An Introduction — 1
Growing Herbs in Australia — 2
How I Came to be Growing Herbs — 2
The Current Situation in the General Herb Trade in Australia — 4

SECTION 1 – General Information

2. A Brief Outline — 7
The Importance of Growing Organically — 7
General Cultivation Requirements — 8
Choice of Crops — 9
Harvesting and Drying — 9
Marketing — 10
Capital Requirements — 11
Herb Fever — 11

3. Making a Start — 13
Climate — 13
Temperature — 14
Rainfall — 15
Wind — 15
Frost — 15
Evaluating the Suitability of Climate — 15
Site — 15
Microclimate — 15
Soil — 17
Irrigation — 19
Slope — 19
Pollution and Contamination — 19
Previous Management — 20
Pests and Diseases — 21
Area — 21
Choice of Crops — 21
Other Requirements for Getting Started — 21
Drying and Processing Facilities — 21
Cultivating and Harvesting Equipment — 22
Propagation Material — 22
A Source of Compost — 22
Organic Certification — 22
What Else? — 22

4. Weed Management and Control — 23
Overview — 23
What Is a Weed? — 23
The Impact of Weeds — 23
Laissez-faire Approaches — 23
Managing Weeds in Different Types of Crops — 23
Choice of Crop According to Weed Control Needs — 25
A Systematic Approach to Weed Control — 25
Strategies for Effective Weed Control — 25
Good Initial Soil Preparation — 25
Initial Bare Fallow — 27
Layout — 29
Effective and Efficient Tools and Equipment — 32
A Plan of Action — 37
Killing Weeds Thoroughly — 39
Action at Critical Times — 39
Dealing with Problem Weeds — 40
Mulching — 41
Avoid Letting Herbs Become Weeds — 43
Quarantine — 43
Flame Weeding — 44
Hot Water Weeding — 45
Tickle Weeding — 45
Selective Grazing — 45
Mowing — 45
What about the Benefits of Weeds? — 46
Making Weeding a Priority — 46

5. Propagation and Planting — 47
The Trial Plot — 47
Identification — 47
Propagation Methods — 48
Vegetative Propagation — 48
Growing Herbs from Seed — 51
Propagation Facilities — 55
Planting Out — 56
Timing — 56
Layout — 56
Rotation — 60

6. Herb Growth Types — 65
Spreading Herbs — 65
Expanding Clump Herbs — 66
Perennial Crown Herbs — 67
Woody Perennials — 67
Trees and Shrubs — 68
Annuals, Biennials and Short-lived Perennials — 69

7.	**Compost**	**71**
	Desirable Characteristics of Compost	71
	Materials for Making Compost	71
	Gathering Compost Materials	74
	Factors for Successful Compost Making	78
	Turning Compost	79
	Spreading Compost	80
	Alternatives to Making Compost	85
8.	**Irrigation**	**87**
	Water Supply	**87**
	Types of Irrigation Systems	**88**
	Flood Irrigation	88
	Furrow Irrigation	88
	Sprinklers	89
	Trickle Irrigation	90
	Management of Water	**91**
	Setting Up the Irrigation System	91
	Managing Sprinkler Irrigation	92
	Managing Trickle Irrigation	95
9.	**Pests and Diseases**	**97**
	Factors Reducing Pest and Disease Problems	97
	Managing Pests and Diseases	97
	Strategies for Pest and Disease Control	98
	Strategies for Individual Pests	**100**
	Grasshoppers	100
	Aphids	101
	Snails	102
	Slugs	103
	Two-spotted Mite (Red Spider)	103
	Red-legged Earth Mite	104
	Strategies for Individual Diseases	**104**
10.	**Biodynamic Aspects**	**105**
	A Biodynamic Approach to Dried Herb Production	**105**
	The Biodynamic Preparations	105
	The Compost Preparations	108
	Using the Compost Preparations	112
	Moon and Planet Rhythms	113
	Pest and Disease Control	116
	The Farm as an Organism	117
	Associations Involved with Biodynamics	118
	Further Reading	118
11.	**Harvesting**	**119**
	Harvesting Leaf, Leaf and Flower, Aerial Parts	**119**
	Timing	119
	Equipment	120
	Making a Catching Scythe	122
	Other Possibilities for Harvesting	123
	Technique	127
	Harvesting Flowers	**133**
	Timing	134
	Equipment	135
	Technique	137
	Harvesting Roots	**140**
	Timing	140
	Equipment	140
	Technique	141
	Harvesting Fruit	**144**
	Timing	144
	Equipment	144
	Technique	144
	Harvesting Seed	**145**
	Timing	145
	Equipment	145
	Technique	146
	Harvesting Bark	**147**
	Timing	147
	Equipment	147
	Technique	148
12.	**Drying**	**151**
	Drying for Quality	**151**
	Factors Affecting Quality	151
	Selecting a Drying System	155
	Drying on Portable Screens	156
	Making Drying Screens	156
	The Drying Shed	**160**
	Ambient Air Drying	160
	Procedures for Ambient Air Drying	**167**
	Drying Leaf Herbs with Ambient Air	167
	Drying Flower Herbs with Ambient Air	171
	Drying Fruit Herbs with Ambient Air	173
	Drying with Heated Air	**174**
	Design	174
	Building a Small Cabinet Dryer	176
	Operating the Cabinet Dryer	179
	Drying Leaf Herbs in the Cabinet Dryer	181
	Drying Flower Herbs in the Cabinet Dryer	182
	Drying Fruit Herbs in the Cabinet Dryer	182
	Drying Root Herbs in the Cabinet Dryer	182
	Drying Bark Herbs in the Cabinet Dryer	184
	A Wood-Fired Cabinet Dryer	185
	Solar Heat Options	185
	Circulation Drying with Heated Air	188
	Other Drying Systems	**190**
	Outside Drying	190
	Drying in Bunches in Shade	190
	Inside Floor Drying	190
	Recirculation of Used Air	191
	Drying with a Heat Pump Dehumidifier	193

Sun Drying under Cover	193		**SECTION 2 - Individual Crops**	
A Greenhouse Covered with Black Plastic	194		Species Covered	243
Freeze Drying	194		Explanatory Notes on the Individual Crop Entries	243
Comparison of Drying Costs of Various Systems	194			
Sharing a Central Drying Facility	194	**15.**	**Spreading Herbs**	**247**

13. Processing 197

Methods of Processing 197
Categories of Herbs According to Use 197
Processing Facilities 198
Processing Leaf Herbs 199
Equipment for Preparing Tea and Culinary Grade Leaf Herbs 199
Technique 205
Soil Contamination 213
Processing Flower Herbs 214
Processing Root Herbs 214
Milling 214
Rubbing 214
Removing Soil Contamination 214
Processing Fruit Herbs 214
Cleaning Fruits 214
Processing Bark Herbs 214
Quality Assessment 214
Important Aspects of Quality 215
Packing and Storage 216
Storage Containers 216
Labelling 219
Recording 221
Insect and Vermin Control in Storage 222
Shipping 223

14. Marketing and the Economics of Herb Growing 225

Marketing 225
An Overview of the Australian Herb Market 225
Buyers of Herbs Grown in Australia 227
Marketing Organic Herbs 228
Price Stability 230
The Therapeutic Goods Act 1991 232
Financial Viability 235
Expenditure 235
Comparisons of Gross Returns 236
Establishing and Maintaining a Market 240
Maintaining Your Market 241
Where to From Here? 241
Group Marketing 242

15. Spreading Herbs 247

Coltsfoot 253
Couch Grass, English 256
Gipsywort 257
Golden Seal 258
Ground Ivy 258
Horseradish 259
Horsetail, Field 261
Licorice 265
Passionflower 269
Pennyroyal 272
Peppermint 273
Raspberry 286
Spearmint 291
St. John's Wort 292
Tansy 293
Yarrow 294

16. Expanding Clump Herbs 297

Bergamot 298
Chamomile, Roman 300
Comfrey, English 301
Dandelion 302
Hops 309
Melissa Balm 312
Mugwort 313
Nettle, Greater 314
Oregano 319
Scullcap 321
Tarragon, French 323
Valerian 326
Violet, Sweet 332

17. Perennial Crown Herbs 335

Agrimony 336
Alfalfa 337
Artichoke, Globe 341
Celandine, Greater 342
Chives 343
Echinacea, Narrow-leaf 345
Elecampane 349
Fennel 350
Figwort 350
Garlic 351
Ginseng 354
Goat's Rue 356
Horehound, Black 357
Lady's Mantle 358

Lemon Grass	358
Marshmallow	361
Meadowsweet	364
Motherwort	366
Pasque Flower	367
Plantain, Greater	367
Poke	368
Red Clover	369
Stoneroot	375
Vervain	376
Wood Betony	378
Wormwood	379

18. Woody Perennials — 381
Feverfew	382
Gum Plant	384
Hyssop	385
Lavender, English	387
Lemon Thyme	388
Marjoram, Sweet	391
Rosemary	392
Rue	394
Sage	395
Thyme	397

19. Trees and Shrubs — 401
Balm of Gilead Poplar	403
Bay	404
Cascara Sagrada	407
Chaste Tree	408
Crampbark	409
Elder, Black	410
Ginkgo	410
Hawthorn	412
Horse Chestnut	413
Lemon Verbena	414
Linden	416
Rose, Dog	418
Willow	419
Witch Hazel	420

20. Annuals, Biennials and Short-lived Perennials — 421
Angelica	422
Aniseed	424
Basil	425
Blessed Thistle	427
Burdock	428
Calendula	432
Caraway	435
Catnip	436
Celery	438
Chicory	439
Coriander	450
Dill	451
Echinacea, Broad-leaf	451
Mullein, Great	457
Oats	457
Parsley	459

21. Wildcrafting and Weed Harvesting — 463
Broom, English	469
Centaury	470
Chickweed	471
Cleavers	472
Couch Grass, English	474
Dandelion	474
Dock, Yellow	475
Elder, Black	477
Eucalyptus (Blue Gum)	480
Fennel	481
Hawthorn	483
Horehound, White	487
Melissa Balm	488
Mountain Pepper	489
Mullein, Great	490
Nettle, Lesser	492
Pennyroyal	494
Periwinkle, Greater	495
Plantain, Greater	495
Plantain, Narrow-leaf	495
Poke	496
Red Clover	498
Rose, Wild	498
Shepherd's Purse	501
Sorrel, Sheep	502
St John's Wort	504
Variegated Thistle	506
Yarrow	508

22. Information Charts — 509
Information Chart 1 – Growing	509
Information Chart 2 – Harvesting, Drying, Prices and Marketing	519

Appendixes — 527
1. Leaf Shapes and Arrangements	529
2. Diagram of a Flower and Illustrations of Flower Arrangements	530
3. Suppliers, Buyers and Organisations	531

Bibliography — 533

Glossary — 535

General Index — 543

Index of Individual Herb Crops — 555

Metric Conversion Tables — 556

1

An Introduction

Herbs are plants that have acquired a significance for humankind because of their particular flavours, aromas, medicinal qualities or other attributes. With their diversity of forms, habits and botanical affinities they have become woven into human culture.

Originally used in the regions to which they were indigenous, many herbs have been embraced by cultures far from their places of origin through the cultural interaction and mobility of our era. Others have ceased to be used, even in their areas of origin, as cultures originating from other parts of the world become dominant in those regions. This is very much the case in Australia, where traditional Aboriginal knowledge of indigenous native herbs has been overlooked and virtually lost in many areas, although some good work is now being done to preserve and foster that knowledge where it remains intact.

In Australia the distance between the cultures has been so great, and the European attitude towards the Aborigines so narrow and prejudiced, that very little knowledge of Australian indigenous herbs has been absorbed into our culture. In this way we are quite alienated from our native flora. This is in contrast to North America where native Americans' herbal tradition has been able to contribute greatly to current European–American knowledge and use of herbs.

Consequently Australian herbal usage is based almost entirely on northern hemisphere plants, generally from temperate climates somewhat different from our own. Hopefully with time and our increasing awareness of the environment around us, we will see greater use made of our native flora. This is beginning to happen on a significant scale with some native culinary herbs used in bush tucker cuisine; nevertheless, we are still a long way from the development of a herbal tradition based on Australian native flora, so for a long while yet we will continue to rely on species whose origins are elsewhere. In Australia we predominantly follow the Western herbal tradition which mostly relies on species of European and North American origin.

Whatever the tradition, herbs need to be something more than just bags of shrivelled faded parts of plants or bottles of dark liquids. Quality is a vital aspect of herbalism, and

Local herb growers have found that it is possible to produce herbs of outstanding quality here in Australia.

this involves more than just an assessment of a herb's known active constituents.

A number of local herb growers have found that it is possible to produce dried herbs of outstanding quality here in Australia, in terms not only of known active constituents, but also of colour, aroma, flavour and effectiveness. This is borne out time and time again by the many people who tell us that they have never tasted teas like ours and by the numbers of practitioners who tell us that they get much better results from using our premium quality herbs.

Growing Herbs in Australia

There is currently very little available information on the commercial production of herbs in Australia, particularly with regard to the harvesting, drying and processing to produce a really good quality dried herb that contains all the plant's vital qualities.

Indeed, the popular literature embraces a lot of misinformation. For instance many books advise the reader to 'strip all the leaves off the stems before spreading them out to dry'! I wonder how many would-be growers have given up after producing a crop of blisters and a few handfuls of bruised leaves.

Unfortunately, the need for information on herb growing has spawned a number of people with a smattering of knowledge who, for various reasons, have taken up roles as consultants and trainers. However, lacking any depth of knowledge on the subject, they are unable to give would-be herb growers adequate guidance.

This lack of good information has resulted in poor quality herbs coming onto the market and a large number of failed ventures. On the other hand, some growers, who have had the opportunity to see how it can be done, have been able to develop successful herb growing enterprises.

How I Came to be Growing Herbs

To place this book and my approach to herb growing in perspective, I will explain how I came to be growing herbs.

Growing up in suburban Australia, I developed an interest in plants in my teens and seemed headed for an academic career in that field. However, that was the 1960s when the interaction of the established old and the emerging new brought about a tumultuous upheaval for many of us entering adult society in that period. For me it meant a growing disinclination towards the academic sphere, culminating in dropping out and heading off down the road in search of experience and fulfilment.

I discovered that I enjoyed manual work and took jobs digging ditches, tuna-fishing, water-boring, and working in factories, saw-mills and on railways. I learned new skills and discovered new strengths through this experience.

My search led me overseas and to a commune in Northern California. Here I rediscovered my interest in plants and got involved in the growing of vegetables and a number of other crops.

Here also a rather fortuitous small event occurred that later greatly shaped the course of my life. The previous owners, who had farmed and lived and survived in those mountains for 50 years or so, were in the process of moving out as we moved in. As they cleared their sheds of the accumulations of decades, they were consigning their discards onto a bonfire. Among these was a number of old agriculture pamphlets. I happened to spot these and salvaged them before the flames caught them.

One pamphlet, published in 1921 by the US Department of Agriculture, was entitled 'Drying Crude Drugs'. Well it turned out that the pamphlet was not quite on the subject that such a title might lead one to expect today. Seventy years ago the term 'drugs' referred to medicines and most of these were derived from plants. So this pamphlet was about drying medicinal herbs, using methods and technology available in the early part of this century.

My coming across the pamphlet coincided with an era when some of us became rather passionate about self-sufficiency.

The author, Greg Whitten, with young Elder trees.

An Introduction

In 1970 it seemed that the only way to survive, or at least to maintain health and sanity, was to equip ourselves to provide for all our needs: grow our own food, and provide our own shelter, clothing and other necessities.

Meanwhile the US Army was making a great endeavour to transform part of South-East Asia into a wasteland and I was asked to join them in their efforts. To decline the invitation it was necessary for me to leave the country.

An opportunity came, thanks to the support and financial assistance of my parents, to buy a small farm in Quebec, Canada: 40 hectares with woods, sugar maples, 4 hectares of cleared land, a house and a barn. So with my first wife I set out to live a life of complete self-sufficiency, independent of modern technology.

I can't say we managed to achieve our high goals, but we did manage to live without our own car, grew more than 90% of our own food and produced a small income from maple syrup that provided for the things we needed to purchase.

Those 5 years were spent among wonderful French Canadian people who had only 30 years previously followed a lifestyle similar to that which we sought. They still remembered and practised many of their traditional ways and were more than happy to share their knowledge and skills. This equipped me with some understanding and proficiency in the use of manual tools, and that proved enormously useful later when I became involved in growing herbs.

In that climate, with snow on the ground 7 months of the year, we converted every spare blade of grass, or anything else edible, into hay to feed our animals through winter. I learned quite a bit about the cutting and drying of plants in those years. I learned to use the scythe, the reaping hook and the cradle – a large scythe with long wooden fingers mounted above the blade – for harvesting grain.

This experience with the scythe and cradle later led me to develop the 'catching scythe' for harvesting herbs. This tool has proved to be the backbone of intermediate-scale herb growing, for not only is it an efficient method of harvesting that preserves quality, it also saves your back from the strain of long hours bending down. More about it later.

Growing herbs was something that I had thought about doing for a number of years, but it wasn't until 1977 that I had the opportunity to try it out.

By this time we had moved across the world to a property in southern Tasmania, and were experimenting with a few crops to grow for an income. I tried a few organic vegies, but in those days they were not easy to sell. Garlic seemed a possibility and my first crop met an enthusiastic response to its quality and strong flavour.

I was interested in trying a few other crops and, after digging out and re-reading that old agricultural pamphlet on drying, I realised it was within my scope to produce dried herbs, so I gave a few a try: Peppermint, Spearmint, Chamomile and Comfrey. These plants seemed to strike a chord within me: there was something enthusing about them. When I took the dried herbs, with their vibrant fragrance and colour, into the shops the response was inspiring and they sold well.

The Peppermint was the most popular. There was such a tremendous difference between the quality I was able to produce and the imported debris that was (and often still is) being sold in the shops as Peppermint. Marketing was incredibly easy: the herbs spoke for themselves. They looked and smelled good and people enjoyed the flavours instead of having to grimly gulp them down.

Thus Southern Light Herbs was born. It wasn't long before I was growing an increasing range of culinary, tea and medicinal herbs, and marketing on a wholesale basis around Hobart and elsewhere in Tasmania. Bass Strait limited expansion northward, but my herbs gradually became known on the mainland by a few herbal practitioners who appreciated the quality of the herbs and found that their use in preparations was giving much better results.

This eventually led to a new development. Green Pharm Health Products started buying large quantities of my herbs for use in liquid medicines. They were producing tinctures and extracts from organically grown herbs and they needed high quality raw materials. This was something new in the field, as until then all the manufacturers of herbal medicines had been only using imported materials.

These cheap imported herbs are often of questionable quality and origin, frequently bearing little resemblance to the plants they claim to represent, instead being stale, discoloured, with little flavour or aroma, doubtful levels of active constituents, and who knows what residues? At least one manufacturer used to beef up the tinctures with caramel if the extraction process resulted in a liquid that looked a bit pale.

Unless you have had some experience in the herb trade, these may seem like rather outlandish accusations. But their actuality has meant a good market for high quality organically grown herbs in Australia among herb users who are looking for something better.

So with a view to meeting this need I went into a joint venture with Green Pharm. Together we bought a property in north-east Victoria, rather aptly named 'Twin Creeks'. I set out to grow a wide range of herbs there, and they set up to manufacture herbal medicines using herbs grown on the farm and from other reputable sources.

At 'Twin Creeks' we grew about 50 species of herbs that went into Green Pharm herbal medicines or were distributed as dried herbs through Southern Light Herbs, which developed a clientele all over the country. For about 5 years the farm was something of a focal point for people interested in herbal medicine and growing herbs.

While the herb production side was very successful, unfortunately the 'Twin Creeks' joint venture was not able to continue, due substantially to the impact on the manufacturing side of the *Therapeutic Goods Act 1991*. It was a choice of get really big or get out and we felt we would be getting out of our depth and sacrificing too much to join in racing all the other rats.

After trying for more than two years to sell 'Twin Creeks' as a going herb farm, it was finally bought by some hobby farmers and the herb beds were abandoned.

Nevertheless the spirit and influence of 'Twin Creeks' lives on among those who were involved in it, and propagation stock has been distributed to herb growers all over the country.

Southern Light Herbs continues in full stride, having been taken over by friends in central Victoria, and is the largest distributor of premium quality organically grown herbs in Australia.

The Current Situation in the General Herb Trade in Australia

Today there is a great interest in the use of herbs for medicine and for food. Indeed the two uses are not easily separable. Many herbs with therapeutic effects have become incorporated in our diet and are so widely consumed as to be no longer regarded as herbs: witness Tea and Coffee (although some might argue that they are more harmful than therapeutic). Other herbs with undisputed medicinal properties, such as Garlic, Sage and Cloves, are widely used as condiments, but usually without any conscious awareness of their therapeutic effects.

Not all popular herbs are beneficial. One widely used herb from the Nightshade family is responsible for thousands of deaths annually, yet it is legal and sold everywhere. A fair proportion of the proceeds from it go into government coffers, which may be why the *Therapeutic Goods Act* and Regulations don't give tobacco a mention, while they strive to ensure the public is safe from imaginary perils that might be lurking among the vast array of herbs that challenge bureaucratic understanding.

Imported Herbs

Millions of dollars worth of herbs are being imported into Australia every year. The problem is that most of them are of dubious quality. Grown predominantly in third-world countries or in Eastern Europe, much of the imported product comes from areas of heavy industrial pollution or is subject to unregulated use of pesticides and other chemicals.

It is usually harvested and dried without much concern for the quality of the end product, which is often over-mature, adulterated with other species and handled in large volumes before drying, with consequent fermentation and deterioration. Commonly it is artificially dried at excessive temperatures or else field cured where it is exposed to sun and rain. Often it is not fully dry when put into storage. Contamination with soil, dung or other even less savoury things is not uncommon – for example, just the other day I came across a cigarette butt in a bulk jar of imported Chamomile in a local health food shop.

Storage is often in inadequate containers allowing re-absorption of moisture. Fumigation with methyl bromide or ethylene oxide is done as a matter of course. Irradiation of herbs is legal in many countries and quite acceptable to Australian authorities for reducing the bacterial count in herbs for therapeutic use.

Even if they haven't been irradiated there may still be radioactive residues thanks to Chernobyl fallout, which still affects large areas of Europe. Testing is only done sporadically on herbs coming into Australia for general consumption. If they are found to be contaminated, the distributor is allowed to dilute the contaminated product with uncontaminated herb from another source to bring it down to the permissible level!

Quality

Of course everybody involved in the trade says they only deal in herbs of the highest quality. However, when you compare the difference between dried herbs produced using the methods outlined in this book, and the imported herbs you will find on the shelves of the less discerning health food shops, you will see, smell and taste a remarkable difference. The trouble is that most people in the herb trade have never seen good quality herbs so they are under the impression that what they are handling is as good as any other. Much herb is sold as liquid extracts, in capsules or in tea bags, where the difference in quality is not clearly apparent until the consumer comes to use the product.

Section 1

General Information

The object of this book is to make it possible for more people to produce good quality herbs here in Australia.

Section 1 outlines the general principles and practices of organic and biodynamic herb production. Section 2 provides specific information on the cultivation of individual herb species.

The book is primarily written for the small- to intermediate-scale commercial growing operation with about 0.4 hectare in herbs with individual plots of herbs ranging from 30 to 1000 square metres. However, much of the information and many of the techniques are applicable for smaller and larger undertakings: for home gardeners or herbal practitioners growing their own raw materials, or for growers contemplating farming on a much larger broad-acre scale.

Essentially the principles are the same, but the operations involved can be quite different for these various scales of production.

Basically the methods described are those that I and a few other growers have found suit us and our lifestyles. We prefer to grow on a scale sufficient to provide a reasonable return for our efforts, but one that doesn't require an enormous investment or a large work force. We use machinery where it is effective at reasonable cost, but use hand methods where these are most suited to our quality standards and level of production. We are also geared to producing a marketable range of high quality herbs. This involves producing a sufficient supply of each herb in the range rather than excessive quantities of any one herb.

Geographically and climatically the book is written with temperate Australia in mind, but generally the information would be relevant for other countries with similar climates. Much of it would be also be relevant for other climates in Australia and elsewhere, but allowances would need to be made for differing local conditions.

It is also important for the reader to bear in mind that in some areas the information presented here is in the process of evolution, and it is offered to the reader in that light. These methods for growing, drying and processing are so far the best that we have been able to develop to produce herbs of high quality on a scale suited to our operation and in the style we want to do it. But rest assured, they are not the ultimate. We are learning all the time and improving technique and technology as more people turn their attention and resources to the production of high quality herbs.

My hope is that herb growers will be able to use the information and ideas presented, either directly or to draw on as a source in developing methods, systems and technology that suit their own particular situations.

Map showing the approximate climatic area of temperate Australia covered by this book.

For Readers Outside Southern Australia

The techniques and systems described in this book are generally applicable to temperate Australia (see map on page 5) which is characterised by mild to hot summers and cool to mild winters. Annual temperature range is from minus 10 to 0°C (15 to 32°F) to 35 to 45°C (95 to 112°F). Rainfall ranges from 250mm to 2000mm (10 to 80 inches) per annum.

Nevertheless much of the information is quite relevant for readers in other parts of Australia and the rest of the world. It is basically a matter of being aware of how your climate and growing conditions differ from southern Australia and how that will affect your own production system with regard to growing, harvesting and drying. And of course, you will need to evaluate your own market situation and economies of production, both of which may vary according to the country you live in.

Regions with a similar climate

Many parts of the world enjoy a similar temperate climate. One general indication of climatic similarity is whether Eucalyptus trees from southern Australia such as *Eucalyptus globulus* (Southern Blue Gum) grow well in your region.

Southern Europe, most of southern USA and especially all of the west coast, Chile, Argentina, New Zealand, South Africa etc are climatically very similar to southern Australia and should suit the growing of most or all of the species described in this book, bearing in mind each plant's particular requirements and local climatic variations.

Regions with a colder climate

Many of the herb species described in this book will thrive in colder climates, especially if they originate from northern Europe or North America where winters are much colder than in Australia.

Winter extremes are not the only factor. You need to also consider whether summer temperatures are warm enough for a particular plant's growing requirements.

Climate can also affect other aspects of production such as weed control. A colder climate may mean less winter weed growth, but it may also mean less opportunity for effective weed control during other times of the year if conditions are cool and moist.

Where the growing season is shorter, yields may be lower. On the other hand if the plant is growing in its preferred climate it should suffer less from stress and disease so that quality and perhaps even yields may be improved.

Drying may be more of a challenge in colder climates, particularly those with summer rainfall. There may be a need for greater reliance on artificial heat or a dehumidifier to produce good drying conditions.

Regions with a warmer climate.

Where climates are warmer than southern Australia you will find that many cool climate species do not thrive or may not even survive unless a modified, shaded environment is provided. Even if they grow well, they may lack the cool winters or the cool nights needed to stimulate development of desirable qualities such as essential oils.

Rather than trying to grow temperate species in tropical regions it is probably more realistic to focus on other species which come from hotter climates, are suited to your region and cannot be grown in cooler regions.

In some cases, annual crops such as German Chamomile which are grown as summer crops in temperate regions can be grown as winter crops in hotter regions where the mild winters resemble a cool temperate summer.

Harvesting and drying may present particular challenges in a hotter climate if a premium quality final product is to be achieved. Careful attention is needed to avoid bulk quantities of harvested herb generating heat and composting (see page 120). Slow drying and rehydration in hot humid conditions can also cause rapid deterioration.

Additional heat may be an option to assist drying but this can be at the expense of quality if volatile constituents are lost at higher temperatures over 35°C.

In this situation, a heat pump dehumidifier may be a more viable option because of its ability to lower humidity without raising temperature, though capital outlay and access to this technology may be limiting factors.

Weed Control

Every region and climate will have its own particular problem weeds. Each climate will favour certain weeds but it will also provide opportunities for their control. Readers will need to observe and discover these, but most of the weed control techniques described in this book should work in a wide range of climates if applied appropriately.

Northern Hemisphere Readers

While Southern Hemisphere readers are quite used to reading books written in the other hemisphere and transposing the months of the year in relation to the seasons, this may be a new experience for many in the Northern Hemisphere.

To translate Southern Hemisphere months to the equivalent month, season-wise, in the Northern Hemisphere, substitute as follows:

January = July	May = November	September = March
February = August	June = December	October = April
March = September	July = January	November = May
April = October	August = February	December = June

Metric conversion tables can be found on page 556.

Marketing

In spite of its Asia-Pacific location, Australia has a "western" economy similar to Canada, USA and western Europe. Prices and costs quoted in this book are relevant to Australia and are quoted in Australian dollars.

Countries with lower labour costs, lower prices and lower average levels of disposable income will have to assess the financial feasibility of herb production in the light of their own particular economic situation and local and export market opportunities. Local demand for individual herbs will also vary according to cultural traditions and tastes.

2

A Brief Outline

The Importance of Growing Organically

The benefits of herbs are all due to natural processes at work in the plants themselves, in their effects on our senses, in their actions on our physiology, in our perception of them, and in the practices and traditions that have grown up around them. For their full potential to be realised, herbs need to be grown in conditions that foster these processes.

Like all plants, herbs reflect the vitality of the situation they are growing in. If this is a clean and healthy environment, where natural materials are used to maintain the fertility and structure of the soil, where they have clean water and air, where pests and diseases are kept in balance by natural forces or, failing that, where they are controlled by the use of biological or other non-chemical methods, then the benefits that we are seeking from these herbs will be developed to their fullest extent.

Unless you are prepared to follow organic methods, much of this book will be wasted on you and your herbs will lack vitality due to the upsetting of their environment. To think that somehow the use of a few chemicals won't affect the subtle forces that work through the medium of these plants is very short-sighted. Apart from the harm it may do to your land, yourself, your family and your employees, any use of chemicals compromises the quality of your produce as it disturbs the delicate biochemistry of the herbs, consequently interfering with the benefits to the end-user, and possibly even having harmful effects.

The term 'organic' embraces quite a broad range of farming practices. At one end of the spectrum it is basically a conventional approach where concentrated organic fertilisers, such as chicken manure, with a high level of freely soluble nutrients, are substituted for chemical fertilisers and plant-derived pesticides are used to control pests. At the other end of the spectrum is a more sophisticated approach where the farmer works with natural processes and forces to aid the development and maintenance of a fertile productive system that fosters the healthy balanced growth of plants and animals. Biodynamics is at this end of the spectrum, but quite a few farmers who would describe themselves simply as organic are to be found here as well.

Many of us come into organic farming at the conventional end of the spectrum, because with our cultural background we tend to see things in analytical terms, and to see our role as that of controlling nature. Then as we develop more understanding of how natural systems work and more confidence in relying on them for successful outcomes, we tend to move towards the biodynamic end of the scale.

Biodynamics

Biodynamics can be defined as a sophisticated organic system. In my own herb growing operation I have been following this approach for the past 9 years, with good results. Biodynamics and herb growing make a natural combination, as a number of herbal preparations are essential elements of biodynamic practice and biodynamic methods produce herbs of excellent quality. But I will concede that for some growers biodynamics does not mean very much and they don't feel comfortable with some of its concepts. This is not really a problem as far as I am concerned, as there is scope for a diversity of views and different ways of understanding how we do things. If you are using good organic methods – making plenty of compost, staying away from fertilisers in a bag and raw animal manures, building your soil fertility and the vitality of your land – then you will be able to produce herbs of comparable quality.

While some people do draw a strong distinction between biodynamic and organic, for the purposes of this book, where I have used the term 'organically grown' it includes 'biodynamically grown'.

The Market Outlook

Market trends are reflecting the growing realisation that herbs should be organically grown. People are becoming aware that this is an important aspect of quality and they are prepared to pay significant premiums for it. This is the fastest growing sector of the herb trade and there is good demand for a wide range of organically grown herbs.

On the other hand, if you are producing chemically grown herbs, you can't expect to get any more for them than for the cheap inferior quality herbs that are plentifully available on the market. For most species, the costs of production here (even taking chemical shortcuts) are too high to be sustained by such low prices unless you can grow them on a very large scale, which will require massive capital inputs and could be risky. Witness the plight of the Hop and Tobacco industries in Australia, where many growers are in dire straits, trying to compete with low world market prices.

Certification

In order to give consumers an assurance that the organically grown produce they purchase is genuine, certification schemes have been set up by a number of organisations. These inspect the farming operation, examine the producer's understanding of organic techniques and test the soil or the produce for chemical residues.

In order to have produce accepted as organically grown, it is necessary to obtain certification from one of the recognised certifying bodies. This involves some cost to the grower, but this is soon recouped with the premium prices obtained.

Why Not Just Residue Free?

A faction of the conventional agricultural sector is promoting a minimal chemical residue approach to farming. While this can be seen as an improvement on some past practices, it is based on a limited view that it is the chemical residues that are the sole concern and if you cut down on the use of organochlorines and other pesticides that may leave detectable residues in produce, then it will be as good as organically grown. However, this overlooks the problems created by the use of the other synthetic chemical inputs used in such farming systems. The natural balance of the growing system is upset by their impact on soil flora and fauna, by their effects on the biochemistry of the plants. The uptake of freely soluble nutrients by plants causes unbalanced growth that lacks the vitality of an organically grown plant. This can be manifested as a susceptibility to pests and diseases, and a disruption of the biochemical make-up of the plant. In the case of herbs, whose properties and actions are often due to the complex interaction of various constituents and other factors, plants grown in a regime of synthetic chemical inputs are not going to be on a par with those grown organically.

Hydroponics

I hope some people will forgive me for even bringing up the subject, but it is surprising how many people suggest growing herbs hydroponically under the impression that it is somehow compatible with or equivalent to organics. Hydroponics involves growing plants in an almost totally artificial environment, their roots sitting in a liquid medium of dissolved nutrients, so there is no way they are going to have the vitality of an organically grown plant. While the size of hydroponically grown plants may be impressive, they won't have the same inner goodness, the quality of the dried product is likely to be very disappointing and the cost of production prohibitive.

General Cultivation Requirements

Most of the herbs we are dealing with here are perennial row crops, or at least they are initially established as such and then allowed to expand and fill in the spaces between the rows. A large proportion of them go dormant each winter, dying back to the ground. This sets them apart from most other crops that are in common cultivation. They can't be just put in like a crop of wheat, left to grow unattended and then harvested a few months later.

Generally a plot of a perennial herb can expect to be in the ground for at least 5 years and take up to 2 years to reach full production. An appropriate management needs to be developed, particularly with regard to site selection, initial preparation, propagation, weed control, fertilisation and crop rotation.

As mentioned before, the majority of herbs we use come from a temperate (and often a humid) northern hemisphere climate. This generally means they are adapted to mild to warm humid summers with ample rainfall, and cold to very cold winters. There are very few parts of Australia with such a climate.

While our milder winters may be cold enough to induce dormancy in most herbs, there is usually enough warmth to foster a rampant growth of weeds, which can cause major problems in some herb crops.

Weed control is perhaps the biggest task the organic herb grower has to face. While a conventional grower would resort to an array of chemicals to keep the weeds down, the biological grower has to find more skilful means to manage them. This we will deal with in greater detail further on. With some herbs, weed control needs a lot of attention and hard work, but while a challenge, it is not insurmountable.

In southern Australia, irrigation is generally needed for most herbs, as summers are drier here than in the places the herbs originated.

To build up and maintain the high levels of fertility generally required, the organic herb grower needs a good compost-making system. Fertilising with raw manures is not recommended, as they can promote unbalanced growth, can often contribute to weed problems and can end up as contamination in the dried herb. Organic fertilisers out of a bag come a very poor second to good compost. They can even lead to soil degradation because they contribute too much in the way of soluble nutrients, and too little organic matter.

A diversity of crops has advantages for plant health, spreads labour, better utilises drying facilities, and can have marketing advantages.

Choice of Crops

Most growers find it best to maintain diversity and grow a number of herb crops: I would recommend at least five or six. The advantages are that the use of drying facilities and available labour can be spread through the season as harvests and other operations will fall at different times for different crops. A mixture of crops, or mosaic polyculture, creates a healthier, more balanced growing environment.

With diversity, seasonal and market fluctuations are less disastrous. A large volume of one herb alone may be difficult to sell in the specialised organic market, as complementary volumes of all the other herbs are required to make up the range the market requires. A number of growers each producing a number of different herbs makes for a more stable market situation.

The herb grower will have to make a choice as to which crops to grow. This will depend on markets, suitability of conditions, growing techniques, the plants the grower feels an affinity with and what other growers are concentrating on.

Harvesting and Drying

It is in the area of harvesting and drying that specialised equipment and techniques are needed to ensure a high quality product. Growers generally need to do their own harvesting and drying. Few herbs are required in the fresh state, apart from those for the fresh culinary market.

Harvesting

On a small to intermediate scale, harvesting is mostly done by hand. Leaf crops growing in very small plots are harvested with a sickle or reaping hook, while those in plots which we can get a swing at are harvested with a modified scythe. On a larger scale, specialised equipment would need to be devised that does not bruise or damage the crop. Ordinary forage harvesters or hay-balers simply will not do, as they are too destructive to the quality of the herb.

Roots are dug with a fork or a spade, though on a larger scale they could be done with equipment used for harvesting vegetable root crops such as carrots or potatoes.

Flowers and fruits are picked with the fingers or a comb, though mechanical harvesters have been developed for large-scale operations.

Drying

Drying is a critical aspect of herb production. The conventional drying systems used for crops such as hay, dried fruits, Hops, grains etc. will give a very inferior product in the case of most herbs. For best results, herbs need to be spread in thin layers on screens and dried under cover in the shade, without any remoistening during the drying process. Where small quantities are involved, herbs can be tied and hung in bundles, but this soon becomes tedious when any volume is involved.

Over most of southern Australia, ambient air drying is possible much of the time. This uses the drying capacity of the air in prevailing conditions, but almost everywhere some additional heat will be required at times when conditions are less favourable.

For large-scale drying, it is something of a challenge to develop a mechanised system that will handle large volumes of throughput without compromising quality. No doubt it is possible, but I don't know of anyone who has done it yet. All the large-scale drying set-ups I know of produce nothing better than second-grade herbs: there is always a significant loss of flavour and/or colour.

Smaller operations may be more labour intensive, but – particularly for leaves and flowers – they can consistently produce a much higher quality dried herb.

Processing

Once dry, the herb needs to be processed into a form suitable for the purchaser. With most leaf herbs for tea and culinary use, the leaves need to be broken up and the stalks removed by rubbing the dried herb through screens. Herbs intended for manufacturers of herbal medicines are generally sold in a coarser form and chopping with a chaff-cutter is usually adequate. The grower needs to be equipped to carry out these operations.

Roots that have been dried but which are still in large pieces can generally be sold to manufacturers and wholesalers.

While a centralised drying and processing facility serving a number of growers is a possibility, these people would need to be in close proximity and communication, and would need to be able to co-ordinate their activities well.

Because of the amount of work involved in drying and processing, and because problems encountered in this area are often caused by inadequacies in the growing and harvesting operations, it is best that the grower also be responsible for the drying and processing. This way growers can ensure that their herbs get optimum treatment, assuring the quality of the final products.

Storage

One of the advantages of dried herbs is that they keep. If there is no immediate market, most can be held for a year or two without any great deterioration in quality. This requires adequate storage facilities that are proof against moisture, vermin and weather, such as plastic bags inside 200 L drums with tight fitting lids, and the drums stored in a shed.

Marketing

In Australia the herb trade seems to be settling down to more or less three streams: trade herbs, manufacturing-grade organic herbs, and premium-grade organic herbs.

Trade Herbs

'Trade herbs' refers to the quality of herb used by the general trade for culinary, tea and medicinal purposes. Trade herbs are almost all imported, cheap and of poor to mediocre quality, but they still dominate the market because of the lack of availability of better quality herbs and because those in the distribution network make their decisions on price, as they have not yet come to understand the importance of quality.

Manufacturing-grade Organic Herbs

Manufacturing-grade organic herbs are mostly used in the manufacturing of extracts, tinctures, ointments, capsules, tablets and some herb teas by a few manufacturers in Australia specialising in organically grown herbs. Most imported organic herbs fall into this category. While some quality has been sacrificed in the production of a cheaper product, it is still generally much better than that of trade herbs. Some manufacturers are adopting standards for these herbs – such as minimum levels of known active constituents – so the trend will probably be towards increasingly better quality herbs to meet market requirements.

The price of manufacturing-grade organic herbs falls somewhere between trade and premium-grade organic, though it is not consistently halfway. With some herbs it tends to be closer to the trade price, while with others it is closer to the premium-grade organic price. The determining factor is usually the price of imported organic herbs.

Premium-grade Organic Herbs

Premium-grade organic herbs are grown, harvested, dried and processed under optimum conditions and using methods that ensure a very high quality product which attracts a much higher price than the trade herbs. This category is mainly used as herb teas, because this is where aspects of quality such as flavour, colour, and freedom from stalk and contamination are most obvious to the consumer.

Premium-grade organic herbs are mostly handled by a small number of wholesalers and are sought by better quality health-food and whole-food shops, some practitioners, and a few small manufacturers. Most of these herbs are sold as tea grade (the herb is in reasonably large pieces), but some are a finer culinary grade.

High quality organically grown herbs can be used in manufacturing a premium-grade range of liquid herbal medicines, but at present this is not being done on any scale. There are signs of some interest in this being shown by the manufacturing sector, so this market might open up if sufficient volume becomes available.

Price

As stressed before, there is not much potential for competing with the price of trade herbs, so to be viable, the grower with the small- to medium-sized operation needs to concentrate on the two sectors of the organic market where better prices can be obtained.

The price for premium-grade organic herbs is generally determined according to the local cost of production. For most herbs this means the small- to intermediate-scale grower can make a reasonable return producing for this category.

The viability of producing manufacturing-grade organic herbs is not as straightforward. For some herbs, the returns can be comparable even though the prices are lower because of higher yields due to the inclusion of the stalks, larger-scale production, and lower costs of drying and processing: quality being not so critical. However, this does not apply to all species, the price offered for some being way too low for production of manufacturing-grade herb to be a viable proposition for local growers.

Export Markets

While there must be a potential world-wide market for premium-grade organically and biodynamically grown herbs, the chicken that will lay this egg is not yet hatched. These markets cannot materialise until the product is available in sufficient volumes. But nobody would dare invest in an export-orientated herb growing operation without an assurance that the market is there.

With local marketing, a start can be made on a small scale, feeding the market and increasing production as it develops, but this is more difficult to do in exporting because of the distances involved.

There already exists a potential for exporting some manufacturing-grade organic herbs if they can be produced at prices competitive with those overseas, and on a scale that would make freight and distribution economical.

Until recently, the general world market has been orientated towards the distribution of cheap herbs and, as already stated, most people in the trade have no concept of quality. They tend to be quite happy with this situation as it ensures very comfortable profit margins on low-cost herbs from third-world countries and Eastern Europe.

This situation is starting to change, and around the world there is a trend towards better quality and organic production methods, because this is what many herb users are looking for.

Within Australia, people are becoming aware that there is an incredible difference between cheap imported products and high quality locally grown herbs, and the market is opening up for us. This is happening at a grass-roots level now and I believe it will increase as more growers come into the field, provided there is continuity of supply and consistent high quality.

At present there is a shortage of most organic herbs, but the situation is delicate and growers need to co-ordinate with each other to ensure complementary production of a range of herbs to facilitate marketing and avoid overproduction: as far as this is possible. With consistent availability, the market for premium quality herbs will grow as more people become acquainted with them.

Capital Requirements

Small- to intermediate-scale herb growing (up to 0.4 ha) is labour intensive rather than capital intensive. Apart from a few major items – such as irrigation system, tractor, rotary hoe, mower, and forage harvester or some other mechanised compost-making system – the equipment required is mostly of a nature that you can construct yourself at moderate cost: in fact you can't buy it off the shelf.

While it does help to have some capital available to get the operation off to a good start, it is possible to start off on a small scale and expand gradually as the operation pays for itself.

If you are starting off with a small plot, you can get by without the tractor and, if you enjoy a bit of physical exertion, you may even get by without the rotary hoe. There are a few hand tools all growers will need to invest in. The ones I recommend cost a bit more than average because they are of good quality and design, but they are good value as their efficiency enables me to make the best use of my time and effort.

Herb Fever

The idea of growing herbs has a strong romantic appeal to many people who see it as an avenue of escape from many of the contentions of modern life. There is something idyllic about the prospect of making an income while immersing oneself in aromatic fields of herbs.

Sometimes this starry-eyed vision leads to outbreaks of high-pitched enthusiasm about herb growing – a phenomenon that might be termed 'herb fever' – with resulting 'herb rushes' as people charge blindly into herb growing in search of fulfilment and fortunes.

Fortunately herb fever is not as virulent as the gold fever that brought many of our ancestors halfway around the world in search of their fortune and a new life. Outbreaks of herb fever seem to be more self-limiting, probably because nobody has yet struck it rich, though a few individuals have done quite well firing up people's expectations of making a lot of money for very little outlay.

Realities

The best treatment for herb fever is a dose of reality. Producing dried herbs does offer the smallholder the opportunity of making an income from a small area of land, but a few points do need to be taken into consideration.

First of all you need to be happy on a moderately low income. You can't expect to get rich growing herbs, but you can make a living or a part-income from it, provided you enjoy working hard, have a bit of ingenuity, and stubbornly refuse to give up when problems arise.

You also need to be prepared to make it a high priority on your personal agenda and let a few other things slide. A passion for herbs probably helps, but it needs to be balanced with a solid methodical approach to the operation. You also need to be in good physical condition and to hold a certain pride and joy in having an orderly, well-managed and productive garden.

Generally it is best to start off on a small scale and learn as you grow. You can expect it to take at least 2 years before you are making any income out of it and probably a further 2 or 3 years before you reach your full potential. Meanwhile you need to be able to survive financially while getting established.

Most of the successful herb growers I know find it suits them because it fits in with their lifestyle, which doesn't require a lot of income to sustain. Others rely on it to supplement income from other sources.

Most of us choose to grow herbs because it is an enterprise based on the land, doing healthy work we enjoy, and which creates a wholesome product we can feel proud of.

3
Making a Start

At this point we will look at what the aspiring grower has to come up with or take into consideration when making a start in dried herb production.

The first thing you will need is a place to grow the herbs. You may already own or have access to some land or you may be looking to buy land. Either way you will need to *assess* and *select* a location. This means considering a number of factors, in particular, climate, soil, site, irrigation water, contamination, pests, diseases, weeds and previous management.

In essence, herb production is about developing an understanding of the cycles and interactions of these factors with each other and with the plants, and working in the light of this knowledge.

CLIMATE

Climate involves the cycling and interacting of water, air, the warmth of the sun and even the earth, which has as much of a role in determining climate as the other factors, though a less obvious one. Earth acts as a reservoir for warmth, retaining heat for a while, but cooling and heating faster than water. Where there is a large expanse of land some distance from the moderating influence of the sea, more extremes of temperature will occur. And where earth rises higher into the air in the form of hills and mountains, this modifies climate very significantly.

Each given place on the Earth is subject to the continual changes and interactions of water, air and warmth, which tend to follow cycles and patterns that have regional characteristics: this is climate.

You may be in a position to make some choice about the region and hence the climate you are going to grow your herbs in, but even if you are already established on a piece of land, you will no doubt want to assess how suitable it will be for particular herbs.

While a species of herb may have an optimum climate that suits it best, most will grow reasonably well and give adequate yields of good quality in a range of climates. However, this range varies from species to species and it is important to ensure that what you intend to grow is suited to your particular climate.

The term microclimate is used to refer to local variations in climate according to the topography, nearby vegetation, buildings etc. It is another aspect of climate that needs to be taken into consideration, and can be used to advantage in placing herbs in situations that best suit them.

Temperature

As mentioned earlier, most of the herbs we will be considering come from the northern hemisphere temperate zone. These tend to do better and yield better, both in quantity and quality, in climates where there are distinct seasons: where winter is cold enough to induce dormancy and the warmer seasons bring on a flush of growth.

Essential Oil Levels

For many aromatic herbs of temperate origin, the southernmost regions of Australia with long summer days and weather and temperature fluctuations seem to foster a good essential oil content. Tasmania is renowned for this, but other less southerly regions can be quite good if there is a contrast between day and night temperatures. Areas with hot days and cool nights in summer – such as in parts of north-east Victoria where the valleys intersect the mountain ranges – can produce outstanding levels of essential oil in Mints. But most of Victoria, southern New South Wales and the ranges and tablelands of New South Wales (and probably the higher regions of southern Queensland) are suitable for producing a reasonably good quality of these aromatic herbs.

On the other hand, the steady mellow warmth of subtropical coastal regions of northern New South Wales and Queensland produces a significantly lower oil content in many temperate herbs. Even so, these herbs will still be better than the very mediocre material that is imported from overseas.

Effects of Australian Temperatures on Herbs and Their Management

Herbs of southern European origin usually do well in southern Australia, but it is often not well understood that the majority of herbs (as opposed to spices) used in our culture come from colder climates than ours. This means that there are very few places in Australia too cold for herb growing and in warmer regions many species can suffer stress in hot weather. Even much of Tasmania, which many mainlanders seem to have the impression is undergoing its own little private ice age, can be a bit on the warm side for some herbs.

Many of the northern temperate-zone herbs will do very well in our cooler higher altitudes, though drying will need more assistance.

Winter Weeds

One advantage of a cold winter is that it reduces winter weed growth. A major problem in warmer climates is that most of these temperate herbs go dormant, being adapted to much colder winters than ours, but many of our local weeds grow actively throughout winter and can totally smother a patch of dormant herbs, or at least create a lot of work getting the patch free of weeds again.

Winter Grass can quickly take over a plot when the herb crop is dormant.

Rainfall

In dried herb production, it is important to understand and work with the dynamics of water: the movements, the various cycles and the interactions with the plant and its environment. Balanced water absorption by the crop is fostered through good growing practices and then this water is removed in the harvesting and drying process. Weeds are destroyed as a result of interfering with their uptake of water: damaging their root systems at a time when they are losing moisture causes them to desiccate.

Optimum Rainfall

Rainfall affects herb growing in a variety of ways. It is always useful to get rain at the right time and water from the sky is preferable to having to put it on yourself, except that you can't turn it off when you have enough. There is nothing like real rain for making things grow, but untimely rain can make harvesting and drying difficult, and it can interfere with cultivation and weed control.

A happy medium would range from around 750 mm per year in Tasmania to around 1000 mm in northern Victoria and New South Wales. There are very few places in Australia where some irrigation won't be necessary: even where there is 2000 mm a year, the rain doesn't always come at the right time.

If sufficient good irrigation water is available, a low rainfall is no great problem. If your soil is light and well drained, then a high rainfall can be coped with.

Good availability of water during the growing season is essential for good yield and good quality. Many herbs from cooler climates will do quite well in our hotter conditions provided they have plenty of water.

Soil moisture must be sufficient to foster good growth, but it needs to be accompanied with dry sunny weather so this growth is balanced. Warm drizzly rain that goes on for days and weeks can encourage lush growth that lacks vitality.

On the other hand, if plants are deprived of moisture, growth slows and stops, leaves start to fade and die, and the quality and quantity of the final product is diminished.

Distribution through the Year

Rainfall distribution is also of significance. Southern Australia generally has a winter maximum with usually little effective rainfall during the warmest part of the summer. This has both advantages and disadvantages. The drier conditions make harvesting and drying easier and assist weed control, but a good reliable water supply and irrigation system is needed to provide optimum conditions for growing.

Further north, summer rainfall areas will usually receive more moisture during the growing season. However, there are inevitable dry spells and droughts as this region is subject to the now familiar *El Niño* effect. In moist summer conditions, weeds tend to be more of a problem as they germinate more readily and are harder to kill.

Humidity

Drying involves the removal of water from the harvested plant. The manner and conditions in which this water is lost greatly influence the quality of the final product. If the herb can be dried reasonably quickly at temperatures of not more than 35°C (for aromatic herbs) or 45°C (for others), without delays or periods of remoistening, then the quality will be better.

Humidity, or the amount of water vapour suspended in the air, has a very significant influence on herb drying. If the prevailing humidity is fairly high during the harvest period, it will be necessary to employ an artificial drying system of some type.

In most parts of Tasmania, Victoria and inland New South Wales, ambient air conditions during summer are usually adequate for drying, though a back-up artificial drying system will probably be needed at times. In humid summer rainfall regions, ambient air drying is rather unreliable.

Wind

Wind is another aspect of climate that can have a significant impact on many plants. Some regions are naturally windier than others, but the effect of wind varies enormously according to the actual site, as local topography may reduce or accentuate the force of the wind.

The most destructive effects are those of cold winds on heat-loving plants and hot dry winds on plants that need cool moist conditions, but a number of herbs just don't tolerate being blown about by any kind of wind. This may be evidenced by poor growth, burnt leaves, and more serious physical damage. My one and only attempt to grow a crop of Summer Savory in Tasmania ended when a blustery westerly snapped the plants off at ground level and blew them all away!

Even the welcome sea breeze that brings cool relief on a hot day can cause problems by remoistening herbs during drying, making them too damp to process and put into storage.

Frost

Most herbs are not adversely affected by frost, though there are a few species of tropical origin that are frost tender, such as Basil and Lemon Grass.

Evaluating the Suitability of Climate

It is not possible to have a climate that is perfect for all herbs. First, they are very diverse in their requirements. Secondly, many of them originate in climates that are quite different from our own and will only thrive if we provide a modified environment for them. In terms of climate, this means irrigation and, for some, protection from extremes of heat or cold or from wind.

If anything, my own preference is for a climate with a reliable winter rainfall and drier summer weather that facilitates drying and weed control. There is also a less tangible aspect of the southern climate that seems to bring out the vitality and aromatic qualities of so many herbs.

Within every region there can be enormous variations in microclimate, especially where the topography is varied, so climate cannot be considered alone without looking at how it expresses itself in a particular site.

SITE

In assessing a potential site for its suitability for herb growing, there are a number of additional factors to take into consideration.

Microclimate

The general prevailing climate will be influenced to a greater or lesser extent by the nature of your specific site. In most regions and even on most farms there will be microclimatic variations due to aspect, exposure to wind and sun, proximity to water, the influence of surrounding vegetation etc

For example, in regions with a hot summer, this will be moderated in sites with protection from hot winds, or those with some afternoon shade. On the other hand, the summer heat will be accentuated on sites with a northern slope, but this can be an advantage in cooler regions. Local topography can greatly increase or reduce the effect of wind.

These factors all need to be taken into consideration when choosing a site for herb growing. If it is possible to make use of a variety of microclimates, it is an advantage if you are looking to grow a range of species with different preferences. For instance at 'Twin Creeks' I found that species such as Peppermint and Spearmint did well in the hotter more exposed sites, while species such as Melissa Balm and Meadowsweet preferred a sheltered site near the creek where it was a little cooler with some afternoon shade. Here they grew taller and lusher and didn't suffer so much from leaf burn or pests.

Consideration of microclimate is particularly relevant, because while you can't do anything much about the regional climate, you can with microclimate. You can decided which piece of land you are going to buy or where on it you are going to grow things. You can also have an influence on the microclimate of a particular site by planting trees and constructing shelters or shade.

In hotter conditions some herbs from cooler climates, such as Coltsfoot and Violet, will need all-day shade, while others will be happy in a semi-shade situation. A situation with morning sun and afternoon shade is good for heat-sensitive plants. The morning sun is cooler and the plants are fresh from recuperation overnight and the stress doesn't really hit them until mid-afternoon, so if shade comes over them at this time, they are spared.

On the other hand, morning shade and afternoon sun is less beneficial as the afternoon sun is hotter, the air is hotter and drier, and often there is more wind.

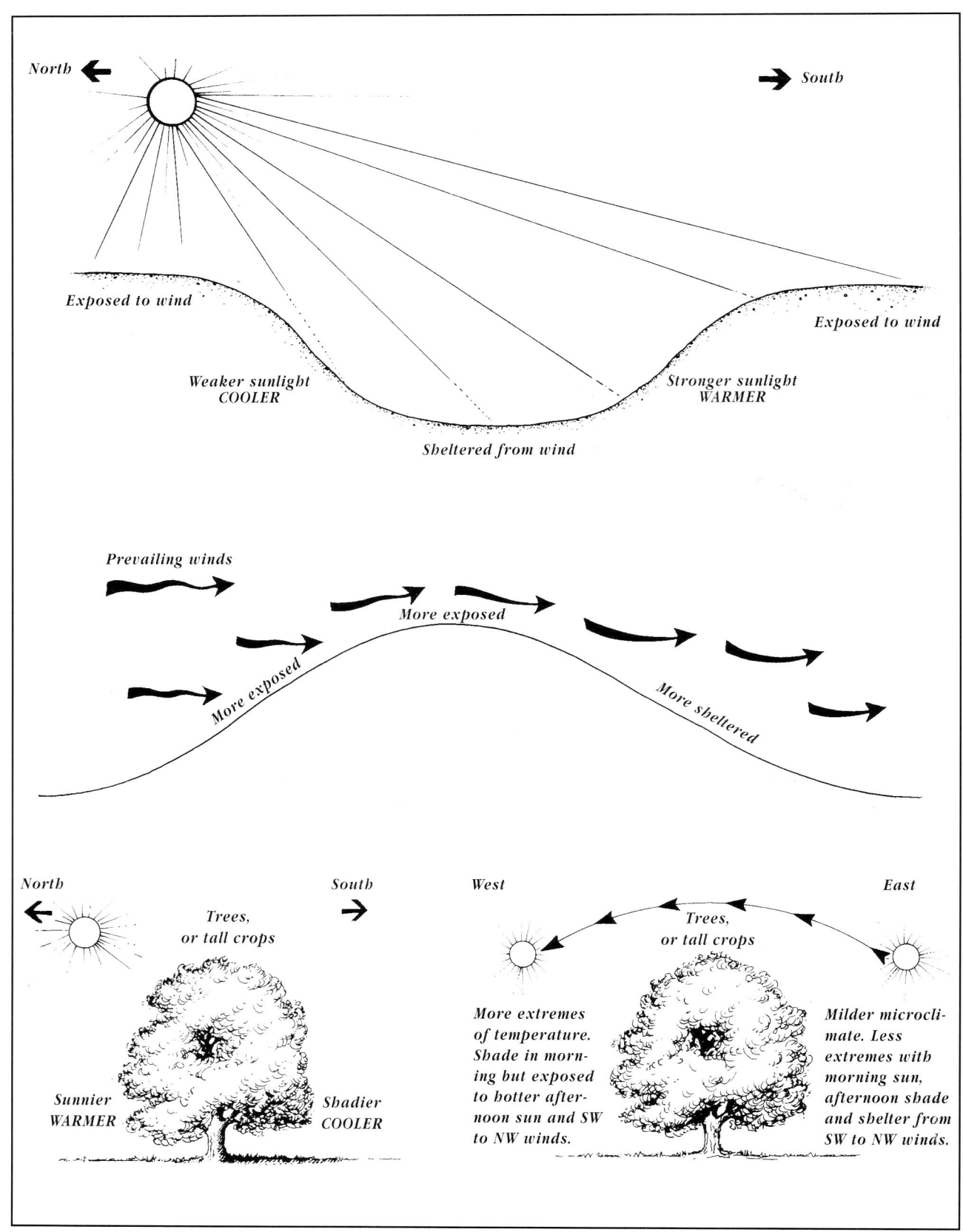

Variations in microclimate due to local influences.

Soil

In many ways the success of your herb growing venture is rooted in the soil. While various other factors will have a vital effect on the outcome of your efforts, many of these will influence your plants through the soil, so it is important that it be a suitable medium. While much can be done to improve a soil, and a less-than-perfect soil can often suffice, the better the soil is to begin with, the easier your work in the garden will be.

Naturally there is some diversity in the growing requirements of various herbs, but it is possible to make a number of fairly safe generalisations for a soil that will suit the great majority of herbs you are likely to want to grow.

The ideal soil is light enough for ease of cultivation but with good moisture (and nutrient) retention, and it is reasonably fertile with plenty of organic matter. It needs to be well drained with no waterlogging occurring during wet periods at any time of the year.

Soil Texture

The red volcanic soils that occur in parts of the eastern states are excellent for herb growing, with their light friable texture, their fertility and their ability to hold moisture while allowing any excess to pass through quickly, but other soils can be quite suitable. For instance, many alluvial soils have similar qualities.

The lighter friable soils are preferable because of the frequent cultivation necessary for weed control. Their structure is less easily damaged by cultivation and they are ready to work sooner after rain or irrigation, making effective weed control more achievable. If the soil is too sandy though, it will be subject to rapid drying with the attendant risk of moisture stress for your plants. Also, the leaching away of nutrients can be a problem in light sandy soil.

While many herbs will do well on heavier soil, cultivation and other activities are more liable to damage the soil structure. Heavy soils can only be worked within a narrow range of moisture content, so often they are too wet or too dry at critical times, which can make planting or weed control difficult.

Stones

Stones in the soil can range from being of mere nuisance value to a major problem, depending on their size and frequency.

Some people like to see a few stones in their soil as a source of minerals that will be released as the stones gradually break down, and stones can help keep the soil open.

On the negative side, stones can be an obstruction to easy working of the land, impeding operations such as cultivation, digging and even mowing. They can blunt and damage tools and equipment. Hoeing in stony soil requires more effort and it is hard to keep an edge on the blade.

With crops like Valerian, tiny stones get trapped among the fine roots and are hard to wash out.

If there aren't too many of them, stones can be gathered and removed: fortunately they don't grow back, so the situation can be improved over time. Stony areas can be used for crops that don't require much cultivation or those to be used as compost-making materials.

Fertility

It is good to start with a reasonably fertile soil. There is a popular misconception that herbs need to be grown on poor soils for the best quality: this certainly has not been my experience.

Most herbs need at least moderate fertility for good growth and quality. Some, such as Peppermint and Nettle, need very fertile conditions for optimum growth. Of course rank unbalanced growth, forced with chemical fertilisers, or dubious 'organic' practices using things like chicken manure, is not going to produce the best quality herbs, but good organic methods using balanced compost will promote healthy vigorous growth in these species.

Poorer soils can be built up, but this can take a while and a fair bit of work. They probably will need outside sources of materials for compost and soil building. This can mean a year or two will pass before optimum fertility levels are reached, but then good management will sustain them.

Earthworms

It is always wonderful to see earthworms in the soil and while most organic growers try to encourage them, some good soils don't seem to have many. Conversely, there can be quite significant numbers in poor soil, if conditions suit them.

Soil Drainage

Another very important aspect of the soil is how it behaves in response to an excess of water.

The term used to describe this is 'drainage' and refers to the soil's internal drainage: its ability to allow excess water to percolate down through the soil profile and away from the root zone.

Good drainage is critical for a majority of herbs. Only a few of the commonly grown herbs will tolerate waterlogging – prolonged saturation of the root zone with water – and most of these will do better where they have good drainage.

I feel it is worth while here to go into some detail about drainage because a good understanding of it is essential to avoid some sticky problems.

Often people are under the impression that if a piece of ground is on a slope and water can run off it, then that land is well drained. However, this same piece of sloping ground may have severe waterlogging problems. It depends on the porosity of the soil, especially the subsoil. It can also depend on what is going on further up the slope. Water that is percolating down through the soil from further up the slope may come to the surface lower down.

Observing Soil Drainage

It is often difficult to predict how a potential herb-growing site is going to behave under conditions of excess water, so preferably the site should be observed through a prolonged wet period when the soil has become fully saturated with water. Then, by digging a hole about 300 mm deep, you can observe how long it takes the excess water to percolate down through the soil or whether it is retained in the root zone by an impervious layer of rock, clay or hardpan in the soil profile, causing waterlogging.

Effects of Waterlogging

When all the spaces between the soil particles become filled with water, air is excluded from the soil and this causes the drowning of the root system of those species that cannot tolerate this condition.

Most herbs cannot withstand waterlogging. Perennial herbs need a situation that is going to be suitable for them over at least several years. A few days of waterlogging in a wet winter can do severe damage from which a crop may not recover.

The damage is often not evident straight away. The tops of the plants may appear normal for a while, but when warm weather hits them their damaged root systems cannot cope and they yellow or die quite suddenly.

Other Indications of Poor Drainage

Often a decision on the selection of a site has to be made when there has not been an opportunity to observe it through a wet season, in which case you have to look out for various signs of waterlogging.

Some plants are indicators of poor drainage: notably rushes, buttercups and some species of sedge and dock. Native vegetation can be a good indicator, too, if you are familiar with your local species. The success and failure of previous crops can also give you a fair idea if you know their tolerances. If Lucerne does well, then the drainage is probably good enough for most herbs.

In an area in pasture, look out for a hard lumpy surface caused by cattle pugging during waterlogged periods. Smoother surfaces tend to be better drained, but not necessarily so.

It is a good idea to dig down to a depth of at least 900 mm to see what the subsoil is like. A narrow soil auger (hand-driven) makes this task easier if there aren't too many rocks. If the subsoil is clayey, is it a solid clay or is it a bit crumbly or gravelly? The solid clay is likely to be impermeable, while the crumbly and gravelly material should be better draining. A layer of hardpan or bedrock can be impermeable, too.

Colour of the soil is an indicator at times. Grey soils or subsoils with little orange streaks (often along roots) or orange mottling can be an indication of periods of waterlogging.

Duplex Soils

Duplex soils can be misleading. These have one type of soil on the surface and another type of subsoil. The extreme is light sandy soil underlain by a heavy clay at a depth of up to 600 mm, not uncommon in southern Tasmania.

Rushes are an indication of poor drainage.

If you have had experience with this type of soil, you will understand the problems it can generate. When I started growing herbs in Huonville, Tasmania, I selected a piece of ground with a light, friable sandy soil, worked it up, put crops in and watched them burgeon and flourish. Then, after a series of heavy rains the first winter, I found the rows awash as a copious spring surfaced and continued to flow prolifically for a couple of months after the rain stopped.

On investigation with an auger, I found that underneath about 450 mm of light sandy soil was a solid layer of heavy clay, quite impermeable. So, while the water could percolate easily down through the topsoil, repeated winter rains soon saturated this top layer. As the water couldn't go past the clay subsoil, it simply flowed over it, underground, coming to the surface in the lower part of my garden and running down the rows between the plants.

The result was that half my garlic died from waterlogging, in spite of a 10° slope and at least a 20 m fall below the garden to the bottom of the valley.

One can endeavour to alleviate such drainage problems with agricultural drains, raised beds etc. and the situation can be significantly improved, but in my experience these measures are often not entirely satisfactory and sensitive plants may still suffer. If you are obliged to use such areas, it may be advisable to concentrate on crops that will tolerate wet feet, or on annual crops grown during the drier time of the year.

Soil pH

While pH is a major point of concern in conventional agriculture, in the organic field it is not given as much attention. The maintenance of high levels of organic matter tends to act as a buffer against the effects of acidity. Repeated applications of compost will gradually bring up the pH of an acid soil where more organic material is being brought in than taken away.

Probably of more concern is an area where compost-making materials are being harvested and taken away on a regular basis, as this will gradually bring about an increase in soil acidity.

Soil pH is a difficult thing to pin down and a complex matter: it varies during the season, there are different ways of measuring it that give different figures, and most of the cheaper soil test kits available will not give accurate readings.

Generally, if the pH is between 5.5 and 6.5 it will be adequate for most herbs. A few need a higher pH, while some do well in more acid conditions.

Applications of crushed dolomite or limestone can be used to improve an acid soil, though a high pH level – above 7 – may cause problems with some species.

Irrigation

In most parts of southern Australia irrigation is essential, even in areas of high summer rainfall, as there are not many herbs that can be grown satisfactorily without it. Being shallow rooted, many herbs suffer easily from moisture stress. Without irrigation, the disruptions to growth during dry spells will result in poor yields or even total losses. So, in selecting a location for growing herbs, you need a reliable supply of good clean non-saline irrigation water that won't let you down – even in the most severe drought – and a reliable system for getting the water onto the crop in a manner that best suits the herbs.

Bear in mind that in summer, herbs such as the Mints require up to 50 mm of water (rainfall plus irrigation) per week, while most others need at least 25 mm. After investing your valuable time, labour and money, you can't afford to run short of water in the peak of the growing season.

Slope

A level or gently sloping area is best for crops that are going to be maintained in rows, as the continual cultivation involved brings the likelihood of erosion on steeper ground. The amount of slope tolerable depends on the nature of the soil and the local climate, such as whether heavy downpours are likely to be a frequent occurrence.

In some situations a degree of slope is difficult to avoid, so erosion control measures will need to be implemented in layout and management.

Slope can sometimes be used to advantage by working uphill for jobs where a lot of bending is involved, as in harvesting flowers.

Pollution and Contamination

In selecting your herb-growing site, it is important to avoid contaminated areas that are the legacy of human activity. Busy roads and industrial areas are surrounded by heavy metal fallout. Urban regions are suspect with their smog and industry, as are regions of intensive agriculture with their widespread use of chemicals, many of which are inclined to stray from their intended targets or linger in the soil for many years after they are used.

Places with a history of growing crops that attracted heavy usage of pesticides, such as Tobacco, Apples and Potatoes, need to be avoided with their likely residues of lead (which it seems will never go away), DDT and Dieldrin (half-lives of 7–20 years). Even small domestic orchards can be a problem unless you are certain of their history as they are notorious for having even higher residues than commercial orchards (people tended to spray pesticides on thickly to make sure of doing a good job). The ground around old sheep-dips should be avoided too.

Water supplies can be a source of contamination. Upstream activities can be the origin of unsavoury substances such as sewage effluent, industrial wastes, pesticide residues, and heavy metals from roadsides or from mining tailings. These can prove difficult to monitor or regulate. Flood waters can carry large volumes of contaminated soil from neighbouring land and deposit it liberally over your certified ground.

So the catchment of your potential water supply needs to be checked out and possibly the origin of your alluvial soil if you are growing on river flats.

Find out all you can about the previous cropping history of the property and if there is any doubt have a soil test done by a reliable analyst. This may save you much chagrin when you later want to obtain certification for your produce.

Also find out what you can about the habits of your neighbours and what sort of things might come floating in the air and onto your crops. Some regions of intensive agriculture are so heavily into the use of chemicals that it is best to avoid them in preference for a place amongst grazing land or a little pocket surrounded by bush: as long as that bush is not likely to be made into a pine plantation. The latter should be given a wide berth because of the aerial spraying of heavy numbers like 2,4,D, 2,4,5,T (agent orange) and Atrazine used for controlling broadleaf weeds among the pines.

All these factors can affect your gaining and retaining biodynamic or organic certification and for a level 'A' certification there needs to be several years history of good management, developing the full vitality of the farm and its produce.

Previous Management

Previous management or mismanagement can affect your operation for quite a few years. A piece of land that has been managed well, along good biodynamic or organic lines can be a real asset, saving you many years of work correcting errors of the past and building up a high level of vitality.

On the other hand, ground that has been flogged hard with overgrazing, pelted with high analysis fertilisers, or subject to years of row-cropping without rotation, is going to take a while to recover. It may be better to look elsewhere unless you feel strongly driven by a desire to heal that piece of the Earth.

Weed Legacies

Weeds are commonly inherited from previous use or misuse of the land. Remember the old adage: 'One year's seeding – seven years weeding'. Weeds may not have been a major problem in the previous crops or pasture but, as we shall come to elaborate later in this book, weeds are generally a much bigger problem in organic herb growing than they are in most other forms of agriculture. A level of weed infestation that is quite tolerable among vegetables or in an orchard is enough to write off some herb crops like the Mints.

Weeds are marvellously adapted to surviving and proliferating. A botanical diversity of seeds, roots and runners are just lying waiting for the sort of conditions you will be creating for your beloved herbs to grow in. Then like mighty hordes they will rise and lay claim to the ground you have prepared.

Old orchards (remember we were going to avoid these anyway) and cropping areas can be thick with perennial weeds and a big reservoir of weed seeds. Old Hop-gardens are said to be the worst. Pastures are not so bad if they have been well managed and not invaded by bracken, blackberries or other tenacious vegetation, but they will still have their fair share of weeds and the seeds of those weeds.

Land that has been freshly cleared from bush is usually fairly free from weeds of cultivation, though these days most of us would find great difficulty with the idea of cutting down any more of our greatly diminished forests.

Most growers will have to contend with a weed legacy. Fortunately, all weeds can be controlled, or at least managed, given sufficient time, skill, effort and dedication, but it is good to know what you are in for, as weeds can create a lot of work and they are probably the biggest single cause of failure of organic herb crops. So it is worth looking over your prospective plot to ascertain what weed problems you are likely to encounter.

Dealing with heavy infestations of large perennial root weeds like Dock, Bracken and Blackberry is a long-term project that may take 2–3 years. Spreading root and rhizome weeds such as Couch, Sorrel, Kikuyu and other stalwarts of the perennial ranks will require a summer bare fallow to get rid of them. What may be harder to ascertain is the volume and variety of other weeds whose seeds lie dormant in the soil you are thinking of stirring.

Ask questions and look around at other areas under cultivation nearby with similar histories. Don't be afraid of tackling a weedy patch – you may have no choice – but be prepared for what you are getting into.

Pests and Diseases

Some regions and some locations will be more subject to pests or diseases than others. Grasshoppers have proved to be a major problem at times in northern Victoria, though their depredations seem to have a definite pattern. Red-legged earth mites can attack establishing crops in some circumstances. Snails are best avoided, if possible, but in many established areas they are fairly well entrenched. In Tasmania, slugs can be a problem at times.

In general, healthy plants are less susceptible to the ravages of pests and diseases, so in an organic regime they are less of a problem than weeds. However, some furry four-legged and two-legged creatures do show a predilection for healthy plants. In bush situations it is often not possible to grow anything at all without a major fortification and a high-voltage wire serpent around the lush temptation of your garden.

Area

In growing high quality herbs using the methods described in this book, a production area of about 0.4 ha of herbs per person seems to be the most that can be managed (and this probably with some extra help at peak times of the year). Possibly larger areas could be grown once the many hurdles of developing the appropriate technology have been overcome. A lower standard of quality could be produced on a larger scale using mechanised harvesting and bulk drying systems.

In the first year or so, it is best to limit yourself to a smaller area, between 400 and 1000 m^2 so you can get the hang of it and develop a working knowledge of all aspects of the operations involved in producing dried herbs.

In addition to the area planted in herbs, you will probably need at least three times that area of reasonably fertile land to produce material for compost. Ideally the herb-growing operation should be part of a larger farming operation involving pasturing of livestock and, perhaps, the growing of crops other than herbs. But beware of spreading your energy too thinly over too many areas that need your close attention, as some of them are liable to suffer unless other people are available to focus on these other aspects of the farm.

Choice of Crops

As there are some hundreds of herbs that could be grown, you will need to make a decision at some point as to which ones you are going to concentrate on. There are a number of factors to take into consideration. Choose crops that are suitable, marketable, profitable, complementary to your other crops and operations, and for which you have some affinity.

Suitability: Your crops need to be suited to your growing situation, giving good yields and quality.

Marketability: There needs to be a reliable demand for your produce. Crops that have a broader market, that is, those that have several different markets, are a safer proposition than those that have a single specialised market.

Profitablility: The price obtained for the product needs to be worth your cost, time and effort.

Complementarity: The herbs you choose need to fit in with your other crops and operations. Crops that complement each other well will enable you to spread your workload more evenly through the year, particularly with regard to harvesting and drying operations, so you can make optimum use of your time and drying facilities.

Affinity: Choose crops that you feel an affinity for. Herbs tend to elicit fairly strong likes and dislikes among people. By choosing species that you like, you will enjoy your work more, the plants will respond to you, and the results will be better.

Number of Crops

Some degree of diversity in the herbs you choose is an advantage as it gives you a buffer against crop failures or marketing problems. Growing a number of different crops provides some protection against pests and diseases and enables you to spread your workload and better utilise your facilities, as mentioned above.

On the other hand, too great a diversity can lead to a loss of efficiency, as smaller plots and harvests are more fiddly and time consuming, and volumes may be too small to command a market.

In general, the best balance is to concentrate on 5 to 10 crops. Of course some growers may find their situation requires that they focus on more than this if they need to produce a broader range to fulfil their particular market requirements. Others may need to grow fewer if they need to achieve a large enough economy of scale or make adequate use of specialised equipment for a particular crop.

OTHER REQUIREMENTS FOR GETTING STARTED

A number of other items that will be required for the operation will need to be given some thought at this stage.

Drying and Processing Facilities

No matter what sort of drying process you envisage, you will need covered space. This should be reasonably close to your garden, vermin proof, and able to be closed up fairly tight. You will probably have to build the shed or modify an existing building. As for your drying system, you will almost certainly have to count on constructing that yourself, unless you're inclined to spend a lot of money on a high-tech dryer designed for another purpose and which quite likely won't give you as good a result drying herbs as a home-built dryer.

Depending on the crops grown, you will also need some processing equipment, such as wire screens for rubbing and sifting, and perhaps a chaff-cutter and a winnower.

You won't need to have all this up and running before you can make a start, but you will need to have them ready for when you begin harvesting significant volumes of herbs. Drying and processing equipment and facilities are all covered in detail in later chapters.

Cultivating and Harvesting Equipment

For an operation of around 0.4 ha, the following equipment will probably be found essential or at least very useful. You may find it is possible to get by with hiring, borrowing or sharing the larger items when needed. On smaller operations, more can be done by hand or with smaller machinery.

Tractor

A tractor in the 25–40 hp range, is useful for breaking new ground and carrying out initial cultivations. It is also handy for large-scale mowing and compost making. Together with a forage-harvester it can be used to mechanise a large part of the gathering of compost materials. A front-end loader would be handy, if you can afford it, otherwise a rear silage fork and a fair bit of twisting in the seat will do the job of stacking, turning and loading compost.

Small Machinery

A medium-sized rotary hoe (5–8 hp) will be needed for general cultivation and for inter-row and perimeter cultivation. A largish mower or self-propelled slasher is necessary for cleaning up stubble after harvest and for keeping weeds and grass down around the garden.

Hand Tools

For harvesting on a small to intermediate scale, most growers use hand tools: a modified scythe and a sickle or reaping hook for leaf crops, a fork or spade for various roots, and a comb for harvesting flowers.

On a larger scale, harvesting can be done mechanically, though for many herbs this results in a loss of quality due to problems that arise when large volumes are handled.

Hand tools are also used a great deal for weed control: various hoes, a wheel hoe, and small implements for weeding close in among the plants. A good set of fingernails helps too.

Propagation Material

You will need sufficient good propagation material for the area you are planting. Many herbs are best propagated vegetatively. A reliable source for these can be your own multiplication plots or another herb grower. Other herbs must be grown from seed. More about this later.

A Source of Compost

Compost will be needed to build up and maintain the level of fertility required by most herb crops. Ideally this compost should be made on the farm from materials grown there, but in some situations growers may need to rely on outside sources.

Organic Certification

Most of the markets for quality herbs require those to be certified organically (or biodynamically) grown. The prospects for selling non-certified produce as organic are becoming very limited as people realise the need to give consumers a bona fide guarantee as to how produce has been grown.

Certification involves some initial expense, in the range of $300–600, and you need to have made some sort of a start so that an inspector can see how you are going about things. If you are just starting out growing on a trial basis, you may want to delay the expense of certification until you are ready to make a commitment. Initially you can join one of the certifying bodies as an associate member, get a copy of the standards and follow them in your operations. This way you will have your systems up and running when it comes to inspecting and assessing your farm for organic certification: something which you should plan on obtaining by the time you have any significant volume of herbs ready to sell.

What Else?

Other useful ingredients for success:
- Some way of supporting yourself/yourselves and your herb-growing operation for at least 2 years until it starts making enough to support you.
- A constant supply of patience, stubborn dedication, ingenuity, energy and time.
- An enjoyment of working with plants, and a green thumb.
- And, if possible, regular contact with someone who is already producing dried herbs successfully.

4

Weed Management and Control

OVERVIEW

What Is a Weed?

From a farmer or gardener's point of view, a good working definition of a weed is that it is a plant that is growing in the wrong place and is able to keep propagating itself, so that some effort is required to restrict or eliminate it.

The Impact of Weeds

Effectively limiting weeds is one of the keys to success in most farming ventures. This is especially so in herb growing because of the problems that weeds can generate.

Contamination

Contamination with weeds is unacceptable in a high quality dried herb. For herb teas and culinary herbs the tolerable limit is virtually none. It doesn't take very much grass – much less than 1% – to make something like Peppermint tea look really awful. Weeds are impossible to separate from the herb once it has reached the processed stage and are very tedious to remove from the harvested crop.

The odd weed can be picked off the drying screens but if there is any quantity, this becomes very time consuming. Trying to harvest some crops from among weeds can be very slow and tiresome, and is often not worth the time and labour.

In herbs destined for manufacturing herbal medicines, the amount of extraneous plant material must be kept to an absolute minimum.

While it usually looks as if there is more contamination than there really is, buyers won't realise this and they won't be too impressed. How do they know what is mixed in with the herb they are buying? It might be something that interferes with the herb's action or even something toxic. There have been cases of poisoning caused by herbs adulterated with toxic weeds such as Belladonna.

One area the new Therapeutic Goods regulations are very strict on is the identity of the herbs that are being used therapeutically. Contamination with extraneous plant material could make your product unacceptable.

Competition

Competition from weeds can greatly reduce yields of many herbs. Some major crops, such as Mints, are particularly susceptible to weed infestation and it is easy to lose them entirely.

As mentioned before, in most parts of Australia we have mild winters when many weeds can grow unchecked while most herbs are dormant. So it is very easy to lose control of the situation during this period. What may have been a productive patch of vigorously growing Mint with the odd weed here and there can, over winter, become a patch of vigorously growing weeds with the odd Mint trying to emerge here and there.

While a small proportion of herbs can succeed in the face of weed competition, most will not do too well. I have seen more herb-growing failures (and experienced a few myself) as a result of weeds than anything else. Regular attention is needed if it is going to be your herbs that succeed and not the weeds. And as most of the herbs we are considering are perennial crops, weed control is a perennial problem.

Laissez-faire Approaches

There tends to be a popular view that weed problems will just disappear when you 'get the soil right' with organic methods, liming, applications of 500, weed peppers or whatever. These approaches may be quite adequate in some fields of agriculture where factors such as grazing influence weed dominance and where a certain level of weeds is tolerable and may even be beneficial. However, these approaches are simply not going to effectively keep the weeds out of most herb crops.

If you try to rely solely on these methods, which may be quite appropriate in a pasture situation, you will most likely find that is just where you end up: in a pasture situation, with a lovely stand of grasses, clovers and other plants plus a few nourishing herbs in the sward. A very healthy situation for a cow to be in, but not such a good one for your herb venture.

Managing Weeds in Different Types of Crops

To develop an understanding of the methods and labour required to maintain adequate weed control, it is worth looking at the natural growth habits of the plants we are dealing with. Being such a diverse group, they include a wide range of growth habits, so weed control methods and the labour input required vary considerably.

Effective management and control of weeds is an important aspect of herb growing.

Crops Requiring Intensive Weed Control

At one end of the spectrum are a number of major crops, such as Peppermint and Spearmint, that are spreading herbs with a natural affinity for growing in mixed stands with other plants. Their open spreading habit leaves plenty of spaces for weeds to get established. Consequently they cannot be expected to remain weed free without considerable help. But unless we can maintain a pure stand of these herbs, harvesting becomes too much of a problem.

Staying on top of the weeds in this situation is a bit like maintaining control of a motor vehicle: it requires constant awareness and attention. If you momentarily lose control of the car you are driving, you can very rapidly find yourself in an inextricable situation. With weeds it is much the same, except that the process of losing control occurs slowly and insidiously. You may not realise that you have let your herb plot go careering out of control until it is starting to disappear in the undergrowth.

While a conventional grower would then resort to a herbicide to take control of the situation, the biological grower has to find more skilful means to keep the weeds down, which can be very difficult when they have already taken over the crop. The best approach is to avoid letting things get to this stage in the first place.

To achieve this requires a methodical approach that can be summed up in two rather well-known adages:

'One year's seeding – seven years weeding.'

'A stitch in time will save you nine.'

Kill the weeds while they are small and don't allow any seeds to fall (with my apologies for not writing the rest of this chapter in verse).

Successful control of weeds is really a matter of staying ahead of them. You need to keep knocking them over soon after they emerge, not letting any go to seed or send out runners.

However, if you are always behind the weeds, things will get progressively worse. You will be fighting bigger and tougher weeds that are harder to kill, continually dropping seeds, and spreading themselves underground through your herb plots.

Of course, keeping ahead of the weeds is a never-ending task, but if you devote yourself to it diligently and consistently, using effective techniques and strategies, it gradually gets a lot easier as you develop a few skills, as the regenerative weeds are mostly eliminated and the reservoir of weed seeds in your soil is gradually depleted.

Crops Requiring Minimal Weed Control

At the other end of the spectrum there are a few herbs whose growth is so vigorous under the right conditions, that weeds cannot compete with them.

One example is Lucerne, or Alfalfa as it is known in the herb trade. Once established, it regenerates so quickly between harvests that the weeds are left way behind. The first cut in spring will be weedy, but this can be discarded (it makes excellent compost) and the later harvests are normally clean enough without any additional weed control. The weeds are suppressed by regular close mowing and the vigorous regrowth of the Alfalfa and remain as an under-storey below the level of harvest.

A number of trees and shrubs can be maintained simply by mowing. Nevertheless, virtually all of this group need some attention to weeds during their establishment phase, until they are strong enough to dominate their surroundings.

Crops Requiring Regular Weed Control

In the middle of the spectrum is a range of herbs which, once established, is relatively easy to maintain free of weeds by regular cultivation and hoeing around the plants. These herbs grow in more or less solid clumps or crowns, so they can hold their own against the weeds to a degree, though if neglected most will sooner or later be engulfed by the competition.

Some annual and biennial flower and root crops need a period of regular cultivation until they are established, but then a few weeds can be tolerated as long as they don't compete with the crop or interfere with harvesting.

Choice of Crop According to Weed Control Needs

Growers need to assess their capacity to effectively manage weeds, and only take on as much as they can successfully deal with in terms of area and choice of crops.

While there is the option of concentrating only on those herbs that require minimal weed control, the turnover of these crops alone is not large enough to sustain very many growers. There is a big demand for crops like Peppermint and Spearmint that require intensive weed control, and for other crops – more to the middle of the weed control spectrum – that still require a fair amount of regular attention.

If the grower can focus on a number of different herb crops needing various intensities of weed control, the labour required can be kept to a manageable level. This diversity also fits in with optimum usage of drying facilities, and marketing requirements.

A Systematic Approach to Weed Control

In essence what we are endeavouring to do in establishing a herb crop is to replace the existing vegetation with our chosen plants. Usually the ground will have been occupied by a variety of plants that are well adapted to growing and surviving on a continuous basis in that particular place. Naturally they will have provided for any untoward events by ensuring that their progeny are waiting as seeds, roots and runners, ready to take their place in great numbers over a prolonged period.

Into this situation we are introducing herb species not necessarily chosen for their ability to survive unassisted. So it would be rather optimistic to expect them not only to survive but to dominate the scene to the extent that no other plants are present. Not only that, but we put our herbs at a disadvantage by regularly hacking bits off them, cutting them down to the ground or digging them up.

Consequently the viability of most herb crops depends on us intervening on their behalf to give them sufficient advantage to hold their own, often as a pure stand.

Most growers will need a number of strategies to restrict or control a variety of weeds in a range of different crops. Fortunately every weed seems to have at least one or two weaknesses that allow the observant and skilled organic grower to find a means of overcoming it or at least a way of limiting its proliferation.

Weed Consciousness

To achieve and maintain an effective weed management and control program, you need to develop an understanding of the nature of your weeds: their life cycles, their preferences and dislikes, their strengths and weaknesses. You need to foster an attitude of 'weed consciousness'. By this I don't mean a state of torpor induced by an excess of some kind of weed, but a conscious awareness of the weeds you are dealing with and an unbearable feeling that gnaws away at the back of your mind not giving you any peace until the weeding has been done.

On the other hand, if you approach weeds in a sporadic manner, not worrying about them until they are out of hand, they will never cease to be a major problem until you decide it is worth putting in the effort to get ahead of them.

STRATEGIES FOR EFFECTIVE WEED CONTROL

Good Initial Soil Preparation

When starting out to grow herbs, most growers will not be so fortunate as to have a piece of ground that is ready to plant into straight away. Usually the soil will need to be ploughed and/or cultivated in some way to remove existing vegetation, be it bushy growth, pasture, or weeds remaining from a previous crop.

Soil preparation is the initial step in the management and control of weeds that might interfere with your intended crop. How you go about this will have a major impact on how successful and how demanding ongoing weed management and control will be.

The aim is to achieve a thorough kill of the existing growth, carrying the operation out in a manner and at a time that deprives the plants of any opportunity to recover and proliferate.

If there is much length of growth, it should be mowed first as this will make ploughing or cultivation easier.

To avoid damage to the soil structure, ploughing and cultivation should be carried out when the soil has a suitable moisture content and is of a friable nature. The soil should crumble in your hand rather than smear and run together. One test is to take a piece of soil in your closed hand and squeeze it really hard: if small droplets of water start to appear between your fingers, the soil is too wet.

Ploughing

Ploughing is the initial upturning or deep loosening of the soil. There are different types of implements available that can do the job in various ways.

Ploughing Established Growth or Pasture: Where there is a thick turf or mat of surface growth, this can to be turned over (not too deeply) with a mouldboard plough, or cut up and partially turned with a disc plough or a set of heavy off-set discs.

Sometimes the initial break can be made with a chisel-tined implement, but often this tears up the turf in large lumps that choke the tines and are difficult to work down later.

Another option is to do the initial breaking of the soil with a rotary hoe that is powerful enough for the job. This has the advantage of cutting up the turf, but it can be hard on soil texture and leave a pan just below the depth of the tines. The latter problem could be overcome by following the rotary hoe with a tined implement to cultivate to a greater depth.

Ploughing is seldom sufficient to kill off the existing plant cover, so it needs to be followed promptly with a series of cultivations to complete the job. Avoid just ploughing up ground and leaving it to grow weeds, though in colder climates, a late autumn ploughing can enable an earlier start in spring, and lets the frost work on the upturned soil.

Ploughing Previously Cultivated Ground: Ploughing may or may not be necessary or advisable. If the soil is rather compacted, chisel ploughing or ripping may be of benefit, though this can bring up weed seeds that have been lying dormant at some depth in the soil.

Where soil is able to be simply worked up with a disc or tined harrow, there is certainly no need to turn it over with a mouldboard first.

Cultivation

Ploughing is followed by cultivation or harrowing with an implement designed to further break up the soil, kill any regrowth, and create a finer tilth suitable for planting into.

Disc Harrows: Disc harrows are invaluable for early cultivations as they cut up tough clumps of turf, pushing it down rather than pulling more up onto the surface.

Spring-tined Harrows and Tillers: There are various implements with tines that drag through the soil with a spring action and tend to bring objects such as stones, roots, turf etc. up to the surface. These implements can be good for later cultivations, bringing weed roots up where they can dry in the sun, and stones so they can be gathered up. Where there is not a thick layer of turf, they can also be used for an initial cultivation, instead of ploughing.

Fine-toothed Harrows: Fine-toothed harrows are good for doing the final cultivations, as they don't penetrate very deeply into the soil and they leave a smooth level finish.

Rotary Hoe: The rotary hoe is another option for doing the later cultivations. It leaves a fine tilth and achieves a good weed kill, but in many soils repeated passes can damage soil structure and create a pan just below the depth of cultivation.

Green Manuring

Green manuring involves the planting of a crop specifically for cultivating back into the soil. The advantages are that available nutrients that might be leached away in bare soil will be taken up by this crop and, when worked in it contributes some organic matter to the soil. Legumes that fix nitrogen can contribute this to the soil also.

Another valuable role that green manures can play over winter is to reduce erosion.

However, if good weed control is a prime objective, some caution needs to exercised in green manuring.

The main problem is that while the green manure crop is in place, weeds can establish, proliferate and form seeds and runners in a situation where you can't do anything about them.

In order to avoid this, the green manure must be sown fairly late in autumn and worked in very early in spring, before the weeds have too much chance to get away. This means that you don't get much benefit of organic matter or nitrogen fixation from the crop.

The other problem that can arise is that the green manure may be hard to kill out or work in, creating problems for the next crop.

Where the ground has just been worked up after being in grass for a number of years, green manuring is redundant. In this situation the grass itself has been a green manure: it is a wonderful builder of organic matter in the soil and the fibre of its roots and leaves will remain in the soil for a few months, helping hold it against erosion.

Adding Compost and Animal Manures during Cultivation

Adding compost and animal manures during cultivation is sometimes advocated but can lead to a loss of nutrients if there is going to be a long delay before planting.

Raw animal manures normally have high levels of weed seeds in them, so they can significantly contribute to subsequent weed problems. Some certification guidelines require that animal manures be composted before use.

Initial Bare Fallow

If you just work up a patch of ground and plant your herbs out in it straightaway, you will find that any regenerative weeds present, such as Couch or Sorrel, will soon burst forth in vigorous regrowth. Those botanical Hydras are able to sprout several new heads every time you cut one off and are just waiting for something to come along and chop up their roots and disperse them, while at the same time destroying their competition. Then they can set about rapidly claiming a stronger foothold on the ground they have been growing in and you will be faced with the Herculean task of getting them out of your herbs.

As for the seeds of all the weeds that grew on that ground over the previous 40 years or more, a fair measure of them will be triggered into germination by the cultivation. The surface of the soil will be garnished with a green multitude of cotyledons emerging to assert their life forms.

If you are lucky enough to be able to make out your emerging herb shoots among this rampant verdure and you have enough time on your hands, you may be able to salvage the situation with some diligent hoeing and fingernail work. This can be achieved on a small plot of a few square metres, but if you are looking at growing 0.1 ha or more of herbs, you need a better plan.

Before you curse the day you ever took on biological farming and reach for the agent orange, just pause a moment. Before the days of chemical warfare there were methods used successfully by growers to combat these problems. One such technique is known as 'bare fallow' or, in its more refined form, 'stale seed-bed'.

Bare fallowing involves continued cultivation to keep the ground bare of vegetation for at least 3 months during the growing season, until the roots of all the regenerative weeds are exhausted and they die. It also germinates and kills a lot of the seeds that are waiting in the soil ready to invade your crops. To be effective, it needs to be done thoroughly, otherwise you will just end up with more weeds than before.

Starting a Bare Fallow

When starting a bare fallow patch, first plough and harrow the ground until you have thoroughly killed all the green parts of the plants that are growing on your plot. You don't need a fine seed-bed at this stage, just a thorough kill.

It is best to start a bare fallow in mid- to late spring while there is still plenty of moisture in the soil to encourage the weed roots to sprout and the seeds to germinate. If the soil is too dry, seeds may lie dormant in the soil, waiting for suitable conditions. Alternatively, irrigation could be used.

Starting the fallow in October–November gives you the summer months for killing weeds, and then often the ground can be ready for an autumn planting.

Killing Regrowth

Of course the regenerative roots and rhizomes of your foes will still have plenty of life in them and it won't be long – about a week – before you see green shoots appearing. These first shoots draw on the plant's reserves to grow. The aim is to allow this to occur, then to hit them again before they have begun to replenish their reserves. Some plants such as Sorrel will begin to send out new runners in as little as 2 weeks, so don't wait any longer than this before giving them another intensive going over.

Later on you may be able to allow longer periods between cultivations as the recovery of the weakening weeds slows down, but keep a close eye on them and don't allow any leaves to grow for more than a week or they will start replenishing their reserves. In the event of an untimely rain enabling the weeds to recover, the cultivation will need to be repeated as soon as possible.

It is a good idea to go over the patch by hand with a hoe a day or two after each cultivation and carve into any weeds that may have had the fortune to survive your harrow or rotary hoe.

Suitable Implements

Cultivation implements that cut up the roots and rhizomes, such as rotary hoe or disc harrows, are good to use at the beginning of the fallow. They will hasten the exhaustion process because the smaller pieces of root will each send up a shoot. Larger pieces can hold out longer because they still send out only one shoot and can keep more in reserve for another go later.

Spring-tooth harrows are good for pulling roots and rhizomes up to the surface and letting them bake dry in the sun. This is a good way to deal with things like bracken and blackberry roots.

Disc harrows.

The Stages of a Bare Fallow

Basically a bare fallow consists of two stages.

First Stage: The first 3 months or so of bare fallow, carried out during the warmer, drier part of the year, it is essential to get rid of the roots and rhizomes of regenerative weeds that are very hard to deal with after a crop has been put in.

If this first stage of the bare fallow is carried out thoroughly and consistently, you should find that virtually all the perennial roots and rhizomes are exhausted and dead. Large-rooted perennials such as dock, bracken and blackberry may take somewhat longer. If there aren't too many regenerative weeds and the soil conditions allow it, then it is advantageous to dig them out with a fork before commencing the fallow. Alternatively, if you can drag them out with spring-tined implements and leave them on the surface, they will dry out in the sun. If you can at least get out the main rootstock, the smaller pieces won't have the same endurance.

A variation on the bare fallow theme is to enlist the help of a pig to take care of these tenacious roots.

The demise of their roots and rhizomes does not mean you have seen the last of these regenerative weeds, as most will still have seeds in the soil waiting for the opportunity to try again.

Second Stage: During the second stage of the bare fallow, the aim is to deplete the reservoir of weed seeds that has built up in your soil over the years. The length of this second stage can be varied according to how soon you need to plant, how much weed germination you can cope with in your particular crops and your situation. If you are keen to get your crops in and don't mind the extra workload of intensive weeding, or if the planned crops are easily maintained, this second stage can be cut short or eliminated. On the other hand, if you want to do a thorough job and make subsequent weed control easier, you can extend the bare fallow for as much as a year, though bear in mind that this will be depleting organic matter and increasing erosion risk, especially on a sloping site.

Stale Seed-Bed

Towards the end of the period of fallow, the cultivations should become progressively shallower as the weed growth becomes weaker. This kills the weeds that have germinated without bringing fresh seeds to the surface to germinate after your crop has gone in.

This progressively shallower cultivation develops what is known as a stale seed-bed, and is a very useful technique if you intend direct sowing seeds, or establishing spreading herbs like the Mints. The idea is to deplete the weed seeds in the top 20–50 mm of soil so they won't be germinating among your crop. The final cultivations should be quite shallow and the crop only planted when it is evident that weed germination is going to be sparse.

Promoting Germination of Weed Seed

To promote germination of weed seed during the fallow period, especially if you are developing a stale seed-bed, periods of moistening are required, as the weed seeds won't sprout if the soil is too dry. Instead, they will wait until you put your crop in and water it, and then up they will come. So if it doesn't rain much during the time your plot lies fallow, it is worth irrigating to encourage more weed germination between cultivations.

This watering is best timed so that weed tops are left on the surface to dry for a couple of days after cultivating. Once they are thoroughly dead, follow with an irrigation which will germinate further weeds at about the right time for them to be killed at the next cultivation.

If it is possible to roll the plot before watering, this will encourage more germination as weeds seem to be inhibited from germinating if the soil is too fluffy.

Weed Survival Mechanisms

Most weeds have seeds with variable dormancy so that they won't all germinate at once. Consequently it takes a long time to exhaust the reservoir of weed seeds in the soil. In fact it is virtually impossible to do so because there are always some seeds lurking down in the depths of the soil, waiting for some fortuitous event to bring them up near the surface where germination might be worth while.

After all, the weeds we have around us today are the progeny of hundreds or even thousands of generations that one way or another have survived human efforts to annihilate them. During that time they have become very adept at surviving.

Some, like Sorrel, have very short establishment periods and quickly put out multiplying roots. Any piece of these roots is capable of regenerating. Other weeds, such as Lesser Nettle, flower and start forming seeds when they are only a few weeks old. The seeds are very inconspicuously located at the base of the second or third pair of true leaves.

Depletion of Organic Matter

It should be borne in mind that a bare fallow is a destructive process, as the production of organic matter on that bit of ground is stopped during the warm part of the year when all the breakdown processes are at their height. So it does deplete organic matter and you can see this going on over the period of the fallow as the fibrous plant material decays into finer particles and the soil loses its structure a bit. Heavy rain is not so much a problem at the beginning of the fallow if the ground was formerly in pasture, as there will be a lot of coarse organic matter to hold on to the soil. However, after 3 months or so, the soil will become more susceptible to washing away, as this coarse material breaks down.

It is important to leave a contour pattern on the surface after each cultivation to discourage erosion. Slopes of more than 10% (6°) are particularly vulnerable.

Larger cultivated areas are more susceptible to erosion than smaller ones because of the volumes of water that build up and flow across them in heavy rain.

Because it is a destructive process, when you undertake a fallow, you need to make sure it is done well so it doesn't have to be done again. If the ground has been a few years under grass, there will normally have been a good build-up of organic matter during this period and you can afford to spend a bit of it in the interest of effective weed control.

On Completion

While a bare fallow is an invaluable tool for reducing your weed problems, it is never going to get rid of them all. There will still be some seeds remaining in the soil, and new weed seeds, roots and rhizomes will find their way into your garden from surrounding areas. So a period of fallow needs to be followed up by a consistent, ongoing weed-control program.

Layout

The layout of your garden determines not only its visual and aesthetic impact, but also the effectiveness and efficiency of the various operations involved in growing. As weed management and control is one of the cornerstones of successful herb growing, it should be one of the prime considerations in designing the layout of the herb garden.

Placement of Crops

In planning the layout of your crops, it is important to bear in mind the weed control implications of where each species is placed. Herbs such as Alfalfa, in which an understorey of weeds is tolerated, or those managed with mowed lanes of Clover and grass, should be kept separate from the main garden, as the weeds among them can be a continual source of problems if they spread to neighbouring crops.

Herbs that tend to become weeds among other herbs should have their own contained area with a buffer zone to prevent them spreading. Particularly bad are Coltsfoot and

A mowed and cultivated buffer zone can effectively protect crops from invading weeds.

Horsetail. Their runners can spread several metres in a year given the right conditions, and can be a real problem in the rest of your garden. Chamomile self-sows very freely and also tends to harbour weeds. This can be a problem for adjacent crops or for herbs that follow it in rotation. Dandelion, especially the wild form, can send seeds floating on the wind over the rest of your garden.

Problem herbs like these can be grown separately, or else adjacent to crops that are able to resist invasion or that are easily cultivated and maintained free of weeds.

Where possible, crops that are vulnerable to weed invasion are best placed some distance in from the perimeter of the garden so they are more protected from weeds coming in from around the outside edge.

Buffer Zones

There is a need to protect the outside perimeter of an area where weed control is being maintained.

One option is to build a wall deep enough to prevent weed roots from growing under it and high enough to prevent seeds from blowing over it. This is a long-term project, however, and few growers would ever find the time or money to do it.

A practical alternative is to cultivate a buffer zone around the perimeter of the weed-control area. A cultivated strip about a metre wide should be maintained between the crops and adjacent grass. The rotary hoe is eminently suitable for this, as in the buffer zone it doesn't really matter if soil structure is affected and the action of the rotating blades does more damage to invading roots and stems than just chopping them off with some other implement.

If this bare soil is cultivated every few weeks and a further 3 m wide strip of grass beyond the cultivated zone is kept mown to prevent seeds forming and blowing into the garden, then the incursion of new weeds into your hard-won ground will be greatly reduced.

This mowed strip also facilitates access to the garden as it can allow a vehicle to drive around the perimeter for harvesting, compost-spreading etc.

The buffer zone can be augmented by planting strong-growing herbs such as comfrey, mugwort, rue, rosemary etc. in the beds adjacent to the perimeter. The shade and competition these herbs present will help defend your borders against invasion. Alternatively, crops that are easily cultivated and maintained free of weeds can serve a similar purpose.

Avoid too many grassed paths through your garden as they will be a continual source of invasive weeds. Any such paths need to be maintained in the same way as the perimeter. Also avoid creating or including islands of weeds within the cultivated area of your garden: these can arise in spaces between crops and around stone outcrops or other obstacles such as irrigation equipment, sheds and greenhouses.

Fences are also problem areas as they shelter weeds that form seeds, which blow into your garden. Fences need to be kept back behind the buffer zone.

Row Crops

While a few crops are best sown down as a solid stand or meadow-crop, most species should be laid out in rows, at least initially, as this greatly facilitates efficient cultivation of the ground around the plants.

Of course some crops will sooner or later expand and take over the spaces between the rows to form a meadow-crop, but laying them out in rows enables you to control weeds during the critical establishment phase.

Accurate Placement of Rows It is worth going to some trouble to set out rows accurately, because plants that are out of line are likely to get bowled over when you cultivate between the rows. Otherwise you will be obliged to swerve here and there to miss your herb plants and you'll miss a lot of weeds, too.

The rows don't have to follow a dead straight line: they can be laid out accurately in sweeping curves which follow the contours or other patterns, as long as the spacing is consistent and the rows can be followed easily with whatever implements you are using so that cultivation is efficient and effective.

Row crops that are well laid out can make weed control much easier.

For straight rows, I like to use a string-line pulled tight about 150 mm above the ground between carefully measured out stakes at the ends of the rows. If it is windy, if the rows are long, or the ground is undulating, one or more stakes may be needed along the line. The string-line is kept in place until the plants or seeds are set in the ground to ensure they are accurately placed, and then it is moved on for laying out the next row.

Curved rows can be set out using a string line with extra stakes to establish the general curve, or a hose can be laid on the ground as a guide.

Distance between Rows: The distance between rows can be critical. The aim is to create a spacing for each species that will provide for optimum growth while allowing access between the plants for as long as necessary for cultivation. This spacing may vary somewhat according to your weed situation, the area of land available, your choice of implements, and your own tastes.

Sometimes it makes sense to trade space for time and plant rows sufficiently far apart to enable cultivation between them with larger, faster implements. This will allow a longer period of cultivation before the herbs grow to the stage where there is no longer enough room between the rows. Wider spacing can help reduce hand-labour during the early stages of establishing a garden when there is still a lot of weed growth.

On the other hand, some crops are better planted closer so that a solid canopy is formed sooner to suppress weeds. Where weeds are less of a problem, rows can be planted closer to obtain greater yields in the first season.

Distance between Plants: The spacing of the plants or clumps within the rows is important, too, to allow room for optimum development and optimum density, and to allow enough space between the plants for ease of access with cultivation tools. Weeding plants that are too close for a hoe to pass easily between them is much more time consuming.

The time and effort involved in setting up the rows and spacing the plants accurately will be amply repaid by the amount of time and effort saved in cultivation and hand-hoeing, as these jobs are so much easier when the plants are precisely placed. It means that more of the weeds are accessible to the larger, more efficient implements, and less fiddly hand-weeding is required.

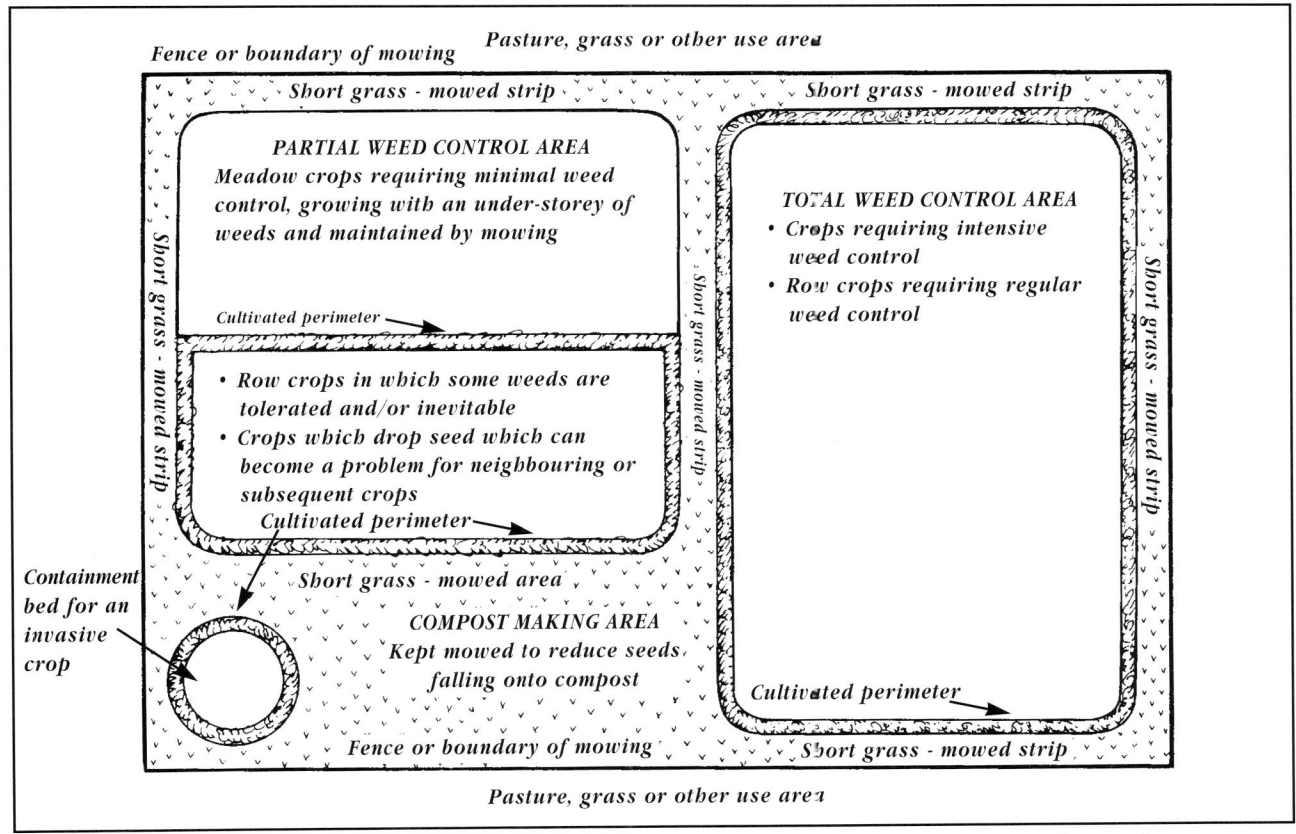

A herb-growing layout designed for effectiveness and efficiency of weed management and control.

Dense Planting

A very dense planting is sometimes advocated as a solution to weed problems, but it is often only partially effective and cannot be relied on as a sole means of weed control for most crops.

While a crop with a dense canopy will suppress weed growth, there is always a critical development period when weeds have an opportunity to establish. There will also be periods after a leaf harvest or when the crop goes dormant in winter when weeds can get away. In a densely planted stand, these weeds may be difficult to get at because of inadequate space for hoeing or cultivation.

Some crops do not thrive in a dense stand as the individual plants become weak due to competition for moisture and nutrients. This can lead to thinning of the canopy, allowing weed development.

A better arrangement for many species is to go for a fairly dense planting in the rows, while leaving plenty of space between the rows for cultivation, application of compost, and as a reservoir of moisture and nutrients. This will foster vigorous growth that will hold its own against weeds while still allowing ease of cultivation for weed control.

Effective and Efficient Tools and Equipment

Getting the right tools for the job and using them effectively is important if weed control is going to be successful and manageable. Naturally your needs are going to vary according to the scale of your operation, the crops you are growing, your budget, your physical capacity and your preferred style of doing things.

My own choice of tools and equipment is something of a balance between the enjoyment and appreciation of using good hand tools efficiently and the need to mechanise some operations for greater productivity, allowing more time and energy to devote to other aspects of the operation.

Rotary Hoe

For cultivation between the rows, a rotary hoe is useful where the space allows, in fact it is almost indispensable for a number of tasks, but I may need to justify my advocacy of this controversial implement.

Initially I attempted to set up my herb garden so I could cultivate between the rows with a draught horse. This soon proved impractical because most of my plots were too small so the rows were too short for ease of working. The variation in the growth habits and rates of development among herbs made it difficult to combine different species to make up longer rows. Another problem was that Nugget

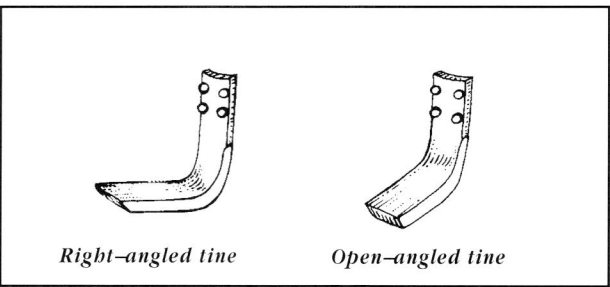

Comparison of rotary hoe tines. The right-angled tine is more damaging to soil structure than the open-angled tine.

did not ever get enough steady row work to get the hang of it. A good horse can be excellent for inter-row cultivation and if I'd had a couple of acres of Peppermint, it might have been a more practical proposition, as Nugget would have had more regular work and come to understand that he was supposed to be stepping on the bare soil between the rows, rather than right on the plants.

Consequently inter-row cultivation with the draught horse was always a rather harrowing experience and eventually I came to the point that I felt I could tolerate the sound of a rotary hoe better than the sound of my own voice shouting

The broad-fork can be used to loosen up the soil to some depth and overcome the pan formation effect of the rotary hoe.

and cursing at my big-hoofed friend. I don't think he minded being superseded all that much: he was quite happy just eating grass in the paddock.

A lot of deprecating things have been said about rotary hoes. While at least some of them are more or less true, a bit of caution in selecting and using the rotary hoe can avoid or reduce many of the problems.

Pan Formation: The rotary action of the tines tends to form a pan or compacted layer of soil just below their operating depth. Any cultivation implement will do this to some extent, but perhaps not as much as the rotary hoe. This is aggravated when the soil is worked in too moist a state, so it is important to wait until soil has dried to a good friable texture, especially at the depth your rotary hoe reaches. The shape of the tines also affects this pan formation. Some makes of rotary hoe have tines bent at a right angle so that the whole width of every blade is polishing at the same depth. These will form a hardpan faster than types that have a more open angle so that only the tip is going down to the full depth.

Even so, over a period of time a harder layer will be formed and this will need to be loosened up by cultivation with a chisel plough or by working over the garden with a broadfork: not actually turning the soil over but sticking the fork in and rocking it back and forth every 300 mm or so. I have thought about the possibility of making a small ripper to attach to the rotary hoe to perform this task, but haven't got around to trying it yet.

Damage to Soil Structure: Another problem with the rotary hoe is that the speed of the tines tends to be too fast and they break down the little crumbs of soil, destroying the structure. In dry conditions the crumbs of soil are shattered into dust, while in wet conditions they are smeared together. The result is that the soil becomes hard and cloddy. This can be reduced by rotary hoeing only when soil moisture is suitable and by keeping tine speeds down. You won't give the weeds as good a blistering, but you will do less harm to your soil. There seems to be some difference between makes of rotary hoe as to the operating speed of the tines.

Other Objections: There is a lot of popular antagonism to rotary hoes and you will sometimes hear rash statements like: 'Rotary hoes only cut the weeds off at ground level and they just grow back again'. Of course if the weeds are things like Sorrel or Couch, they certainly will grow back again, no matter what you cultivate them with, but for most of the non-regenerative weeds, the rotary hoe is more effective at killing them than other implements. This is because the action of the tines cuts and bruises the weeds so they don't recover so easily.

Using the Rotary Hoe: Where the rotary hoe becomes almost indispensable is for cultivating thick stands of weeds because it works them into the soil rather than pushing them aside into the rows. It is excellent for working compost into the soil, for rejuvenating beds of spreading herbs and for dealing with the stubble of annuals like Chamomile. It is also invaluable for maintaining a buffer zone around the garden.

There is quite a wide range of sizes and styles of rotary hoe. The cheapest is the rotary tiller: this has tines at the front that churn away, dragging you up and down the rows, all the while veering from side to side and flying out of your hands when you hit a rock or a solid clump of weeds. After chasing around on the end of one of these for a while, I invested in an 8 hp rear-tine rotary hoe. This has proved to be a valuable ally: large enough to work reasonable areas, but still manoeuvrable in tight spots.

Nevertheless it is important to use the rotary hoe with discretion. To minimise damage to soil structure I try to use it only when soil and weed conditions oblige me to, or for tilling in applications of compost. Wherever possible I prefer to use a wheel hoe for inter-row cultivation.

An alternative to the rotary hoe is becoming popular in Europe. It is known as the reciprocating spader and is now available here. It works with an up and down spading action that is supposed to be less inclined to form a hard pan, though it leaves a rougher surface. I have yet to see one in action.

Wheel Hoe

The wheel hoe is a marvellous tool with a wheel in front and different types of blades and attachments. Two long handles with grips enable you to push it along with the blade(s) skimming just below the surface cutting the weeds off as it goes. It can cut a swathe up to 300 mm wide, though usually it is more effective when set a bit narrower.

You can push the wheel hoe at a steady pace up and down the rows. Sometimes there has to be a fair bit of back and forth action to clear the blade(s) or to scalp the tough patches.

It isn't quite as fast as the rotary hoe because it takes two or more passes to cover the same width, but it is quieter and cheaper and doesn't have the same compacting effect. You get a bit of a sweat up though, so you may prefer to use the wheel hoe in the cooler part of the day. Morning cultivation is usually more effective because the damaged weeds are then subjected to the heat of the day and are more likely to succumb.

Using the wheel hoe is practical for inter-row cultivation on a scale up to about 0.4 ha, where cultivations are frequent, the soil is friable and the weeds are kept small. If you are handling a larger area on your own, you may have to resort to the rotary hoe and tractor in order to cover it all. Even so, there still would be a place for the wheel hoe to follow afterwards and clean up close to the plants. Later, as the plants get bigger and the space between the rows becomes too tight for machinery, the wheel hoe becomes very handy.

Hand Hoes

When it comes to hand-hoeing, having a good weapon is half the battle. The hoe must be balanced, comfortable to use, and the blade must be sharp and strike the ground at the right angle, so that it is drawn into the weeds rather than skimming over them. The implements that you are likely to find in the average hardware store are next to useless, blunt and clumsy. No wonder a lot of people feel they have to resort to herbicides.

A comfortable and effective manner of holding and using hand hoes is important. Note that the hoe is held thumbs upward.

Take the ordinary garden hoe. As my friend and mentor Eliot Coleman points out, this is a relic of Stone Age design. The original hoe was probably made from a flat piece of stone bound at right angles to a stick. With the advent of metal this same original shape was emulated and it has continued in use to this day, despite its great limitations. Somehow people have been slow to explore the possibilities of new materials and techniques available to improve the humble hoe; instead we have focused on chemical wizardry and brute force liberated from fossil fuels.

There are, however, a number of hand hoes available today that make use of developments in design and metallurgy. These tools are easier and more pleasant to use, and are more productive as they are designed to make optimum use of human effort.

The hoes I use the most are two versions of the shallow-blade hoe.

Sneeboer Gooseneck Hoe: One type of shallow-blade hoe is referred to by some growers as the gooseneck hoe, though really the gooseneck between the blade and the handle is only a minor feature. This hoe is made by the Sneeboer firm in Holland. The blade is of stainless steel, which has the advantage of being tougher and holding an edge longer than most metals. It also does not rust, so the surface stays smooth and the soil tends to cling to it less. This hoe is skilfully designed and handcrafted to hang balanced in the hand, so it doesn't try to roll over as the blade strikes the ground. The blade comes in a number of widths: I find the 200 mm one the most useful.

Holding and Using the Hoe: Because the blade is shallow, you can direct your effort to it more effectively and with more control. It can be used in a chipping action, working forwards along the row, or with a sweeping action, holding the hoe with your thumbs upward, drawing it towards you as you step backwards. This technique takes a little practice until you learn to 'see with your feet' and not tread on your plants as you work your way backwards along the row. If you are ahead of the weeds and are just cleaning up the recently germinated seedlings and the odd bigger one that managed to survive your previous weeding, then this second stroke is the one you will use the most as it is much faster.

Another stroke is to turn the hoe on its side for chipping between plants set closely in the rows. And the point of the corner can be used to hook out weeds that are coming up in the dense mats of young regrowth of some herbs. Altogether it is a wonderful tool and a joy to use: I have spent many tranquil hours swinging it among the herbs.

Coleman Gung-Hoe: Another version of shallow-blade hoe has recently become available here and is known as the Coleman Gung-Hoe. Developed by organic gardener, teacher and writer, Eliot Coleman, in conjunction with a Swiss manufacturer, it approaches the ultimate in hand hoes, a real state of the art tool. It is designed on similar principles to the gooseneck hoe, but it is much lighter and even shallower. It is the ideal tool for the grower who likes to keep ahead of the weeds, killing them while they are small. Being light and shallow bladed, it is easy and fast to use as it skims just under the surface and does not move as much soil around. But if you have bigger weeds to tackle, it is a bit light for the job, though there is usually little justification for letting the weeds get that big.

My preference now is to use the Coleman hoe whenever I can because it is so light and efficient and I only resort to the other hoes when it can't do the job. It can be used in the same ways as the gooseneck hoe, though it is not so good for hooking weeds out by the corner of the blade unless you grind the shoulder back a bit to make the point of the corner a bit sharper.

Dutch Hoe: The Dutch hoe is another useful tool. This type of hoe has a flat horizontal blade, with sharpened edges in front, behind and on the sides. The blade slides along, slicing the weeds off just below the surface. It is used with a sort of back and forth scuffling action as you walk forwards along the row. The style I prefer is made by the Dutch firm Sneeboer, the same manufacturer as the gooseneck hoe. This Sneeboer hoe is also of stainless steel and has a curved blade which meets less resistance when pushed through the soil. Unless your soil is very light, a 150 mm blade is ample: a wider blade offers too much resistance.

The Dutch hoe is preferred above all others by some growers, though I don't use it as much as the shallow-blade hoes. It is excellent for working along under overhanging foliage and over ground covered with young weeds. It is a good summer hoe when the soil is a little on the dry side and just slicing the weeds below the surface is enough to kill them. In the moist conditions of late autumn, winter and early spring it is not so effective as it does not lift the weeds out as much as the shallow-blade hoe and they tend to just set down again and keep growing.

Stirrup Hoe: The stirrup hoe is a bit out of the ordinary and has its devotees. It has a thin blade bent in the shape of a flattened 'U' so it looks a bit like a stirrup. It is hinged at the ends so it swings back and forth as you use it in a forward and backward motion. This swinging action means that the angle the blade strikes the ground is at the optimum for cutting into the weeds.

This hoe is handy for sweeping up and down rows and patches that are too narrow or short for the wheel hoe. It is probably a little faster and more effective for this than the gooseneck and the Dutch hoe, but it lacks the accuracy of these because the blade moves about and has rounded corners. It is not so effective for working right up close to some plants and may need following up with one of the other tools.

Small Hand Tools: Having got as many weeds as you can using the larger tools, there will still be a few that have managed to elude you by being too close to the plants or too deeply rooted. To get at these you will have to get down on your hands and knees and bend.

Fingernails are quite useful for getting little weeds out but a few simple tools can make it easier to get the more difficult ones and improve your efficiency.

Weeding Claw: The weeding claw is one tool I wish I had found or thought of 20 years ago as it would have saved me a lot of trouble. It looks a bit like Captain Hook's legendary claw, only beaten out flat for agricultural use. Wolf's Tools make one, but most of their outlets don't stock it, you have to order it in. Alternatively, it wouldn't be hard to make your own from a piece of steel bent into shape and sharpened.

It is like a miniature hoe and it can be used for chipping, sweeping or hooking the weeds at close quarters. It is particularly useful for cleaning up weeds close in around small plants where you wouldn't dare strike with a hoe for fear of injuring your friends. And it is useful for scratching out the root systems of such offenders as Sorrel and Couch.

Tweaker: Another tool useful for getting things out by the roots is what I call a 'tweaker'. It used to be in action every weekend on many a suburban lawn, to get everything out that wasn't grass, but it has waned in popularity since the advent of agent orange (sold under other names, of course) which people think does the job so much quicker and easier! (Too bad about that annoying rash that never seems to go away.)

Anyway the tweaker has two little fingers like a curved 'V' and can be used to lever out weeds that have a bit of a grip on the ground.

The Humble Screwdriver: For deep-rooted things like Dock, a big, heavy duty screwdriver is the best bet. Push it down on a few sides of the entrenched culprit and lever the weed out. It is the most reliable way of getting the big weed roots out whole, with a minimum of disturbance to adjacent herb plants.

Rotary hoe.

Wheel hoe.

Gooseneck hoe.

Coleman Gung hoe.

Dutch hoe.

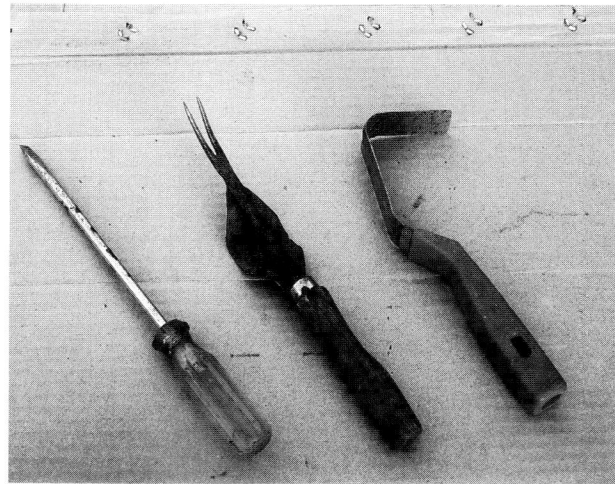
Screwdriver, tweaker and weeding claw.

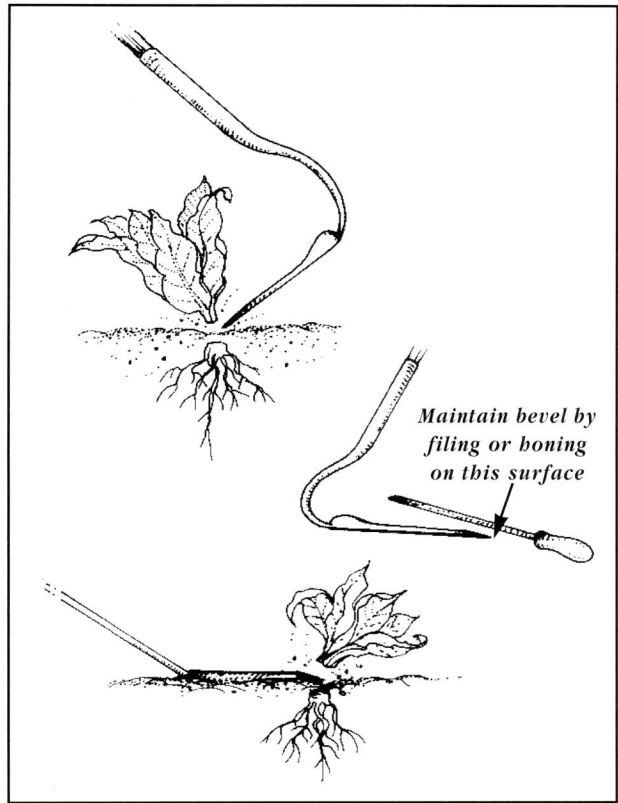

The bevel and sharpening of hoe blades.

Maintaining Weeding Tools

Hand tools such as the wheel hoe and other hoes with blades need to be kept sharp to be effective. While the swinging blades tend to be self-sharpening, fixed blades need regular touching up. Most blades are sharpened with a bevel on the upper side of the blade as it strikes the ground. This helps draw the blade down into the object of your attention. This bevel should be maintained by sharpening it with a file or a stone every few hours. It does not have to be deadly sharp, but a reasonably good edge will make your work easier and more effective. It is important to maintain the acute angle of the bevel: don't allow it to gradually become square.

Blades should be scraped clean after use, to prevent soil caking hard on them as it dries, and to prevent rusting on non-stainless blades. Not that they will rust away in a hurry, but soil clings to rusty surfaces and creates friction as you work.

Handles should be oiled occasionally and the tools kept in a dry place. This will help protect wooden handles from cracking, and keep blades shiny.

Sources of Hand Tools

Some of the high quality hand tools described cannot be obtained in ordinary hardware stores. The two suppliers listed below both do mail orders.

Gundaroo Tiller
Allsun Farm, Gundaroo, NSW 2620
Telephone (02) 6236 8173

Suppliers of a number of useful high quality hand tools, including a wheel hoe and Coleman Gung-Hoe.

Hollander Imports
87 Brooker Highway, Hobart, TAS 7000
Telephone (03) 6234 5111

Suppliers of a range of agricultural machinery and tools, including Sneeboer stainless steel gooseneck and Dutch hoes, other hand tools and a wheel hoe.

A Plan of Action

For the optimum use of your time and labour to achieve effective weed control, a planned approach is necessary. For inter-row cultivation it is usually best to start at one side of the garden and work progressively across it to ensure that no areas are overlooked. This might be varied a little if there are sections that need urgent attention, or ones that have been recently irrigated and need to be skipped over and come back to when they have dried out a little.

The important thing is to view the garden as a whole system that needs to be managed as such. Weeds that are out of control in one part will very likely sooner or later affect the rest of the garden.

Aim to go over the garden every 2 to 4 weeks (according to the season) in order to keep ahead of the weeds, so that most of what you are dealing with are small young weeds and the few that are always missed will not have had time to go to seed or spread runners before the next time around. Avoid getting behind in your weeding schedule as it is only too easy for a spell of wet weather or some other event to keep you out of the garden and enable the weeds to get so far advanced that it is a monumental task to catch up with them.

From the point of view of efficiency and effectiveness, it helps to follow a sequence of operations.

Step 1: First go over the area about to be cultivated with a bucket (or a big pocket) and collect up any weeds that are forming seed or about to flower. If there are any regenerative weeds like Couch or Sorrel that might survive your cultivation and maybe proliferate as a result, they should be carefully dug out while they are few in number and before they manage to spread. Watch for inconspicuous little flowers and seed heads. By collecting this propagative material and disposing of it elsewhere, you will be progressively reducing the weed regeneration in your cultivated area.

(If seed heads and flowers are few and far between, you might prefer to collect them as you come across them during subsequent operations, as long as they are not being overlooked and tilled into the soil.)

Step 2: Commence the cultivation with the largest implement(s) you are going to use. Possibly the rotary hoe first or maybe just the wheel hoe on its own, to rip into the weeds that are occupying the spaces between the rows. Timing is important here: the soil must be dry and friable enough for you not to damage the soil structure, and there must be a good chance of killing the weeds. It is best if a cultivation can be followed by a few days of dry conditions that will kill off the weeds that have been damaged and unearthed. In cultivating just before a rain or an irrigation, the weeds are more likely to revive again.

Step 3: A few hours later (or a day later) when the dying weeds have at least started to wilt and the ones that were missed are obviously still thriving, follow up with the next largest implement: a hoe of some type to clean up the weeds that survived your first onslaught and also to get those that are in closer to the herb plants.

Step 4: Having done as much of the operation as possible in an upright position, go among the plants with a weeding claw, tweaker or whatever, and get the weeds that were not accessible to the larger tools. By doing this operation last, you will ensure that you have minimised the amount of kneeling and bending work, which is the slowest way of weeding and the hardest on the back.

Taking Care of the Back

Many people's backs will stand only so much continuous bending before they start to give trouble. If there is a lot of bending work that can't be avoided, it is a good idea to alternate it with some standing work, which allows the back to straighten out for a while.

Consequently you may want to stage Step 4 in conjunction with Step 3. Rather than finish Step 3 completely before starting on Step 4, you can do Step 3 on part of the garden and then start Step 4 on that section, then Step 3 on another section etc., alternating them as you progress across the garden.

A few other points can help minimise the strain on the back of this type of bending work. Where you are working on sloping ground, always try to face uphill rather than downhill, so you don't have to bend so far. It is amazing what a difference a few centimetres can make.

Don't be afraid to get down on your knees, as this reduces your height considerably and the distance you have to bend.

If working with one hand, rest one knee on the ground, like a person about to be knighted (some republicans may wish to find some other analogy for this posture) and then lean your free forearm on the other knee to relieve your back of some of the weight of your upper torso.

Disposal of Chopped Weeds

What is to be done with the weeds once you've got them out? Some growers prefer to meticulously collect up all the weeds and consign them to the compost heap. While this is a good use for them, it is usually not necessary to go to all that trouble. Generally they can be left where they fall and return to the dust whence they came, as long as they are reasonably sure of dying where they are left, without dropping any seeds or striking any roots. This requires a bit of judgement as to the likely weather conditions and the stage of development of the particular weed. Bear in mind that once the plant has been beheaded, it is not dead straightaway. It may start to desiccate, but it is only dead when that process has reached an irreversible stage, which varies from species to species.

Most weeds are able to continue their flowering and seed forming process for some time after being cut off from the ground, so while a weed may only have a few flowers on it when you chop it down, by the time it finally dies it may have succeeded in forming several hundred viable seeds. This can mean several hundred more weeds you'll have to chop down over the next few years. So if you would rather eliminate one weed now than spend the next 7 years battling with its progeny, you might prefer to stick that weed (or just its flower heads) in your pocket and drop it off somewhere outside your garden, or into a good hot compost pile. Generally if a weed has not yet begun to form flowers when it is chopped, it is safe to leave where it falls: it will normally die before it can form any seeds, though a few juicy-stemmed weeds such as Purslane and sometimes Capeweed are able to continue for a long time. In fact Purslane can bake in the sun for weeks until it looks quite shrivelled and then come to life again with the next rain.

It takes a bit of practice to judge whether to leave or remove a weed, and often it is bit of a gamble. If rain or irrigation revive your recently hoed weeds, you may need to cultivate again sooner than you had intended. I usually try to flick the weeds out into the bare ground between the rows where the sun and wind can get at them better. Then, if by chance they manage to revive, they are out in the open where they can be dealt with easily next time around.

In the moist conditions of late autumn, winter and early spring, it is very likely that chopped weeds won't die at all. If they are a problem, then it may be best to rake them up and cart them away or rake them into little piles between the rows, turning the piles over regularly so the top weeds can smother the bottom ones.

Killing Weeds Thoroughly

It may seem a little trite to include this section, but it is very important to understand just how to go about tackling a weed to ensure it is down and unable to get up again.

Basically, the aim is to arrest the weed's life processes. In the heat of summer this is relatively easy: all you have to do is disturb the roots a bit and the plant won't be able to obtain enough water to replace what is being lost. In less extreme conditions, though, you have to be more thorough to ensure your efforts have the desired effect. This means you have to do a bit more damage to the plant than just stir its roots or poke it with the blade. Fortunately most weeds will die if the top part is removed from the bottom part. In other words, if the roots are separated from the aerial parts each will die without the other.

Most of our weeding implements are designed to do just this. Make sure that each weed is properly cut off at the base below the crown or the lowest point on the stem capable of sending out new leaves. This may necessitate a few roots being left on the top part but if conditions are dry enough, these won't be enough to sustain the plant. Otherwise, up-end the aerial part so that any roots can't take hold very easily. Hoes with blades that strike the ground at an angle tend to lift the cut weeds a little, but in moist conditions the flat blades on some wheel hoes and the Dutch hoe are inclined to leave weeds flat and allow them to re-root.

Action at Critical Times

There are a number of critical times through the season when a window of opportunity exists to destroy weeds at a relatively easy stage which, if missed, may result in weeds proliferating in conditions where it is difficult to control them.

Newly Emerged Weeds

The younger the weeds are, the easier it is to kill them. If you can organise your cultivations so that you are hitting the weeds when they have just emerged, most of them can be irreparably damaged just by touching them. Their young stems have little fibre in them so they snap easily and a quick pass with a blade will do in minutes what would take hours to do a few weeks later.

The Spring Flush

One of the most critical times in the weeding calendar is that period in the flush of spring when everything is growing so rapidly, particularly the weeds, and moist conditions make it difficult to kill them before they start to go to seed.

Often a wet winter has made it difficult to get into the garden and many weeds may be quite advanced as spring comes on. It is a race to get weeds under control again before they become inaccessible amidst the burgeoning herbs (or the herbs are lost among the weeds). If these weeds start dropping seeds there will be a price to pay later on with even heavier weed infestations.

Mid- to Late Autumn

Mid- to late autumn often brings the germination of a crop of weeds that must be cleaned up while conditions are still dry enough. If these autumn weeds are allowed to survive into winter, they will prove very hard to kill and may be dropping seed (or spreading runners) by early spring.

Newly Planted Crops

During the establishment phase of new crops, it is important to ensure that weeds are not given the opportunity to get a start among them. The disturbance, compaction and watering associated with planting tends to germinate a crop of new weeds. Young plants that are too small to compete with the weeds can easily be overshadowed and lost. Those that are vigorous enough to survive can end up in a mixed stand with the weeds, and it is a major task to sort out this kind of situation later on.

In the early stages of emergence or just before then, there is an opportunity to deal with newly germinating weeds before they become too tough.

Where the emerging shoots of a planted crop are robust enough, it is sometimes possible to save a lot of work by going over the rows with a rake, stiff broom or tickle weeder and giving the surface a good stirring. This will destroy the majority of the young weeds at a vulnerable stage while doing minimal damage to the tougher emerging shoots of your crop.

Early Regrowth after Harvest

Often an under-storey of young weeds develops under the canopy of a crop approaching harvest. When the aerial parts of the plant are cut, these weeds become accessible for a period before the regrowth gets too tall. If left, they can grow up among the next harvest and become much more difficult to deal with.

Sheep Sorrel (*Rumex acetosella*), a regenerative weed that spreads by means of its extensive roots, which send up new shoots along their length.

Dealing with Problem Weeds

A few weeds are particularly difficult to manage, so you need to watch out for them and deal with them appropriately as far as possible.

Regenerative Weeds

Some weeds have the ability to regenerate from pieces of root, rhizome or stem and require special attention. We need to be on continual guard against these tenacious little hangers-on which have found a niche in our agriculture. We will continue to have them around us as long as we provide the right conditions for them. If an initial bare fallow is carried out thoroughly, you should only have their seedlings to deal with over most of your plot, though very likely you will have to cope with continuous incursions from around the perimeter of your garden.

You need to keep an eye out for these regenerative weeds while at work in your garden. If they are still very small, you can safely knock them over and leave them as you do any other weed, but once they have formed their regenerative roots or rhizomes, if you just chop them it can simply cause them to multiply.

When you find a larger one of them among your herbs, it is worth spending a little time making sure you have dug out the regenerative parts before they spread and the problem becomes unmanageable.

Winter-active Weeds

A few species of weeds are well adapted to growing and multiplying during the colder, wetter part of the year when it is difficult to control them using the methods already described.

Winter Grass (*Poa annua*) is one weed that is often a problem. I have spent years battling it, with only limited successes. Generally it has managed to spread itself throughout the crops in spite of my efforts. Growing in cool conditions – autumn, winter and spring in Victoria and even through summer in Tasmania – it is the only weed I know of that can form viable seeds during moist winter conditions when no amount of cultivation will kill it out.

After a few years of this I realised I was going to have to live with winter grass. Fortunately it is an annual and not very tall, so it is not a great problem in herbs that form dense stands or clumps or among those that are harvested well above the ground.

Where winter grass creates havoc is among the spreading herbs, such as the Mints, that go dormant down to the ground during winter while it grows prolifically to form a dense sward. When the herbs try to emerge in spring, their shoots are partially smothered by the winter grass and what does manage to grow through can't be harvested without picking up substantial amounts of yellow straw and seed, which look terrible in the finished product. Even a very small proportion of winter grass in the harvest can seriously downgrade quality because it shows up so clearly against the green leaf.

There are a couple of techniques that work quite well for coping with this and other winter-active weeds. One is to cut back the Mint beds into narrow rows each autumn so only a thin strip needs to be hand-weeded and the rest can be mechanically cultivated when conditions allow. There is more detail on this strategy in first part of Chapter 15, 'Spreading Herbs' in Section 2.

The other technique that has proved successful has been the use of a temporary black plastic mulch over the herb beds while they are dormant. This is described below under 'Mulching: Winter-dormancy Plastic Mulch'.

Mulching

Mulching can be of much assistance in controlling weeds in some situations. It also helps conserve moisture and is often a source of nutrients, but it can have major limitations. Having once naively followed some of the many exhortations about how wonderful mulch is and consequently experienced some disasters with it, I tend to be rather cautious about relying on it for weed control.

Permanent Plastic Mulch

Apart from its use as a temporary mulch in winter, I have grave doubts about the use of plastic as a mulch. Used on a permanent basis, it restricts access to the soil so no compost can be worked in and it must surely affect the soil fauna and flora, especially in summer, when it creates abnormally high temperatures near the surface. These temperatures would also risk burning the more heat-sensitive herbs. Many plastics perish under ultraviolet radiation, and breakdown products may wash into the soil or be absorbed by the plants. And when the mulch finally deteriorates, the pieces of shredded plastic are liable to end up contaminating harvested crops.

Apart from all this, permanent plastic mulch simply won't work for those herbs with a spreading habit: the very ones that suffer the greatest weed problems.

Winter-dormancy Plastic Mulch

Using black plastic over beds for only a short period of 4 to 8 weeks when the herbs are dormant in midwinter avoids many of the problems described above. The aim is to cut the light off from winter-active weeds so they turn yellow and exhaust themselves while the herbs are still more or less dormant. The mulch must go on in the coldest part of the year – around the end of June or early July – so there is enough time to smother the weeds before the herbs start to move in mid-August.

Not all herbs will take to being under such a mulch but those spreading and expanding clump herbs that go dormant in winter seem to tolerate it fairly well. I have used this method successfully with plots of Spearmint, Peppermint, Coltsfoot, Horsetail, Valerian, Greater Nettle and a few others (but it will kill winter-active herbs such as Pennyroyal). It works well for many winter-active weeds, particularly annual weeds, most grasses and clovers. It will not kill regenerative weeds such as Couch or Sorrel, as they have enough reserves in their roots and rhizomes to see them through.

A black plastic dormancy mulch can be used to smother winter-active weeds in meadow-crops of herbs that go dormant in winter.

Make sure you have completely opaque black plastic as some turns out to be a semi-transparent dark grey. Hold it up and look at the sun through it. If the sun shines through dimly looking like nuclear winter, the plastic might be sufficiently transparent to allow your weeds to linger on, so put it on double or use a darker plastic. There is no need to go for woven weed mat, a medium to heavy sheet plastic will do.

It is best to mow your stubble down low first to reduce damage to the plastic. Hold the plastic down with spadefuls of soil along the edges and here and there over the sheet so the wind doesn't lift it up and carry it away. Alternatively, old tyres can be used to weight it down.

Keep an eye on what is going on under the plastic. Once the weeds have faded to a yellowish colour after 4–8 weeks you can lift it off. Even if the weeds still look a little green after 8 weeks, don't leave the plastic on any longer as your herbs may shoot excessively. The weeds should soon burn off in the sunlight if they have become soft enough.

When lifting the plastic off, try to avoid dropping any soil that is on top of it onto your now clean patch, because it will probably introduce more weeds. If your herbs have started to put out long anaemic shoots, it is best to mow these off. Like those long white shoots stored potatoes put out, they aren't much good for growing on. If cut off, the herb will soon put out more vigorous, new healthy green shoots.

The plastic can be folded carefully and stored for use next winter, and can be reused many times.

Mulches of Biological Origin

Mulches of organic material can be quite effective for suppressing weed growth in a good many crops. They have the added benefit of conserving moisture and building up the soil as they break down and they also help prevent soil from splashing up onto the leaves of the herbs. Many materials can and may be used for mulching, but there are a number of things to bear in mind when considering using them on herbs.

Fabric-type Mulches: Fabric-type mulches made of paper or wool have some advantages over plastic in that they break down to become organic matter, though they still restrict access to the soil. Avoid using old carpet or underlay as these have often been found to contain high levels of pesticides and their use is not acceptable for organic certification. Black and white newsprint is regarded as acceptable, but coloured inks can contain heavy metals (personally I don't trust either of them).

Beware of Introducing Weed Seeds: Any mulch to be used in your weed-control zone must be free of weed seed otherwise you are just introducing problems for later on. This eliminates hay, and even straw commonly has a lot of viable seed in it. Seed-containing materials can be used for mulching trees and shrubs and those crops where a few weeds won't be a problem.

Avoid Getting Mulch Mixed with Parts To Be Harvested: Mulch must not end up among the part of the plant you are harvesting. This means that for most leaf crops it must sit close to the ground; consequently fluffy materials are not suitable and in windy situations anything light enough to blow around is likely to cause problems.

Avoid Chemical Contamination: Some products from conventional agriculture or industry may contain chemicals that could contaminate your crop or soil.

Adequate Thickness: To be effective a mulch needs to go on thickly enough to smother any emerging weeds. However if Couch, Sorrel or other regenerative weeds are present, no amount of a loose mulch will be thick enough to smother them: they will just grow up through it and thrive.

Texture: The nature of the mulch needs to be such as to allow hoeing or cultivation if any weeds do come through. In other words, it needs to be in small enough pieces.

Avoid Nitrogen Starvation: In order to break down into the soil without starving the plants for nitrogen, the mulch needs to have a good nitrogen carbon ratio. Low nitrogen, high carbon materials like sawdust need to be mixed with nitrogen-rich material to balance them if they are incorporated into the soil.

Fermentation: Some mulches high in nitrogen can ferment, releasing substances antagonistic to plant growth. These materials, such as lawn clippings, leaf waste, seaweed etc., should first go through a composting process with other materials to balance them.

Slugs and Snails: Mulches can provide a wonderful haven for slugs and snails, which can be a real problem for some crops.

Soil Temperature and Moisture: Mulches tend to keep soil temperatures down, which can be a disadvantage in spring, especially in cooler climates or for crops that need warmer temperatures. In these situations it may be preferable to delay mulching until the soil has warmed up.

Mulches can also make it difficult for an excessively moist soil to dry out sufficiently to promote optimum growth. In heavier soils this can be a real problem. On the other hand, mulching in dry conditions can impede water penetration.

Compatibility with Irrigation: Some mulches shed water and some absorb quite a lot, so there can be problems getting irrigation water through the soil and to the plants' roots. Drip lines can be laid under mulch. Overhead irrigations must be sufficient to saturate the mulch and soak into the soil.

Adequate Supply: For most growers, finding sufficient volume of materials that satisfy these requirements is difficult, so at best they will only be able to rely on mulch for weed control for quite small areas.

The stems that are sifted out in the rubbing when processing tea grade leaf can be chopped in a chaff-cutter and make excellent mulch, but I have only ever had enough of this to use on small areas. My whole annual production of stems at 'Twin Creeks' was only enough to mulch a small bed of Coltsfoot.

A few growers may be fortunate enough to have a large supply of suitable material for mulching but for the majority, mulching will not be much of an option.

Avoid Letting Herbs Become Weeds

Noxious Weeds

A number of valuable herbs are potential noxious weeds and several are declared as such in some states. St John's Wort and Poke are aggressive weeds in many areas, best harvested in the wild and kept out of your garden as they can spread.

Poke could present toxicity problems if it comes up among other herbs and is accidentally harvested with them. I grew it as a crop for a while in Victoria, but it was starting to come up in all sorts of unexpected places as a result of birds eating the ripe berries and passing them out elsewhere. Fortunately I caught the problem in time. I was able to destroy the escapees and as it was impossible to prevent the Poke from forming at least some berries, I discontinued growing it.

Not so lucky was a Mr Paterson of Chiltern in north-east Victoria. Quite some time ago he planted *Echium lycopsis* in his garden for its attractive purple flowers. It did very well and the next year he noticed a few plants had come up on the other side of the fence. The year after they were across the paddock. Since then they have spread all over southern Australia, carrying his name with them: Paterson's Curse.

Precautions with Other Herbs

A number of species of herbs can become weeds of more or less serious proportions within your garden. The more serious are those that form aggressively spreading runners or inevitably drop a lot of seed. These herbs need special measures to contain them that are described under 'Containment' in Chapter 5, 'Propagation and Planting' and in Chapter 15, 'Spreading Herbs'.

Quite a few other species will form viable seed if given the opportunity, so if for some reason you are not going to harvest a crop, or even an individual plant, it is best to cut it off before it forms seeds and these are scattered all around among other crops where they could be a nuisance to get rid of once they germinate.

Quarantine

To avoid bringing weeds into places where they are not wanted, it is worth instituting your own quarantine procedures. There are two aspects to this: quarantine within the farm, and the quarantining of material from outside it.

Within the Farm

Quarantine on the farm involves adopting practices that reduce or prevent weeds and seeds coming from elsewhere on the farm into the weed-control area of your garden. To be effective, the following areas need to be addressed:

Buffer Zones: Good buffer zones need to be maintained around the perimeter of your cultivated area as described under 'Layout'.

Mulches and Fertilisers: These should be free of weed seeds or should be put through a hot composting process to destroy any weeds.

Composting System: The composting system needs to produce enough heat to kill weeds and seeds. Only compost that has been through the hot part of the heap should be used in the weed control area. Either turn the compost so it all goes through the centre of the heap, or use the outside part of the heap in other areas.

Buried Weed Seeds: Deep cultivation can bring up a new lot of weed seeds that have been lying dormant below the normal level of cultivation. Avoid doing this in crops where the weeds could be a problem to deal with.

From Outside the Farm

Some weed problems can be prevented by not having the weeds in the first place. If there are some serious weeds, such as Couch or Winter Grass, that you don't already have on your farm it is probably worth instituting quarantine

Due to its open spreading nature, Horsetail growing as a meadow-crop requires intensive weed control.

Meadowsweet requires regular inter-row cultivation and hoeing around the plants.

The input required for weed control is minimal where Alfalfa is growing as a meadow-crop. This is because of Alfalfa's dense canopy, its vigorous regrowth after mowing, and because it is harvested above the level of the weeds.

procedures to reduce the risk of bringing new species of weeds onto your farm.

Attention should be paid to possible avenues by which new weeds are likely to find their way onto your farm.

Mulches, Organic Fertilisers and Compost Materials: These should be free of weed seeds or should be put through a hot composting process to destroy anything that might grow. Beware of weed seeds left on the bed of your vehicle or falling on the ground when unloading. Keep an eye out for new weeds starting to grow in these areas.

Livestock: Livestock coming onto the farm are liable to bring new weeds in their manure. Stock feed and hay brought in from outside could bring in new weed seeds.

Propagation Material: New weeds can easily hitch a ride right into the middle of your garden with propagation material you bring in from elsewhere. Seed can be contaminated with weed seeds, and soil on plants and propagation material can contain seeds or viable pieces of weeds.

Propagation material should be checked carefully or washed before bringing them onto the farm. After planting, watch out for new weeds emerging around new plantings; you may be able to nip them in the bud.

Flame Weeding

Flame weeding involves the use of an LP gas or kerosene flamer that is directed onto the weeds and kills them by cooking them.

Shields are used to protect the crop from damage as the flamer travels up the row.

While at first this may seem an easy answer for weed control, in fact many weeds are very difficult to kill with a flame once they are established. Their leaves are designed to cover the growing point in the crown, which is in close contact with the soil and thus protected from the high temperatures. Consequently flaming is only effective on freshly germinated weeds, which are normally quite easy to kill with cultivation anyway.

The main advantage of flame weeding is that it can be used in situations where cultivation is not possible or where it is undesirable.

With careful timing it can be used to kill weed seedlings along newly planted rows before your crop has emerged.

Because it does not disturb the soil, it does not encourage the germination of a fresh batch of weeds in the way that cultivation tends to, and this can be an advantage in some situations.

On the minus side it is noisy, a bit dangerous, and burns up fossil fuels and some of your organic matter. It probably affects soil flora and fauna and it indiscriminately kills insects and other small animals in its path.

Hot-water Weeding

Hot-water weeding is probably not an option for most small growers due to the cost of the equipment. Not having had any experience with it, I can't speculate as to how useful it would be for weed control with herbs.

Tickle Weeding

Tickle weeding involves the use of light springy tines, which are designed to disturb and flick out young weeds while doing minimal damage to the more robust crop plants.

Its main use in organic agriculture is during the establishment phase of plants that are flexible and tough enough to endure the tickling when the young weeds they are infested with are in the fragile post-emergent stage.

Most commercial models are tractor mounted, but a hand version is made by one supplier. On the same principle I have found a spring grass rake (with steel, plastic or bamboo tines) or even an ordinary garden rake quite effective for dealing with young weed seedlings coming up quite thickly among newly planted rows of Mint and Nettle.

The young shoots of these herbs are tough enough to stand a fair bit of knocking around. The odd piece does get broken off, but the rhizomes underground have plenty of reserves to send up new shoots to replace any that have been damaged. It is important to perform this operation while the weeds are still small and fragile. If they are too advanced they will be tougher and able to survive.

With good timing a high percentage of weed kill can be achieved for a small outlay of time and effort. A few weeds will survive but this method can save hours of tedious back bending work.

Selective Grazing

Most grazing animals are rather selective in their diets, and many herbs are unpalatable to them, so this is a possibility worth exploring.

A large organic farm in the USA reportedly uses geese to weed its Peppermint, but my own trials with sheep and geese were not all that successful.

While the sheep did do a good job with the White Horehound, they devastated the Mint. After they had eaten the grass and other weeds down a bit, they turned on the Peppermint and ate it right down too.

As for the geese, in the Raspberry they left the crop alone and ate some of the weeds such as the Clovers and the finer grasses, but they left the Plantain, Dock and Cocksfoot, which soon filled up the spaces. In the Peppermint they did graze down most of the weeds, but they also took to nipping the Mint plants off with their beaks and just left them on the ground.

Apparently the Chinese train their young geese to eat the weeds they want controlled, but I don't know of anyone here who has done this.

The other problem with the sheep and geese was that their droppings did not all drop on the ground, some landed on the Mint. This was not a problem with the Raspberry and the Horehound, for being taller plants they were harvested at a safe height.

For herbs that are being grown for therapeutic use, contamination with animal dung could result in problems with microbial counts in the finished product exceeding maximum acceptable levels.

Mowing

With a few crops that are tall and/or vigorous enough with a dense canopy of leaf, the weeds can be adequately managed mostly by mowing.

Generally initial cultivation is needed before planting and, in some cases, during early development to enable the plants to get established.

Trees, Shrubs and Hedges

Most trees and shrubs can be maintained with initial cultivation and mowing, as well as some herbs that can be grown as a hedge. Often the clippings can be used as a mulch up close to the plants.

Rapid Regrowth Species

Some vigorous meadow-crop species such as Alfalfa and Red Clover can be maintained with an under-storey of weeds. Periodic mowing gives these species an advantage over the weeds because their regrowth is so rapid that it outstrips them. Because the crop quickly forms a dense canopy, the weeds are suppressed and they mostly remain as an under-storey. This enables the herb to be harvested above the level of the weeds. After a leaf harvest, the stubble and under-storey of weeds are mowed down close to the ground, which sets the weeds back and encourages the regrowth of the crop. With Red Clover the mowing is less frequent and is usually only carried out once or twice a year, as the need arises. These crops tend to get weedy over winter when their growth is slower, but they can be cleaned up again in spring by mowing. It is important to allow them a sufficient recovery period between mowings, around 6 weeks (or more in the cooler part of the year), as these plants pour all their reserves into their regrowth. If they are cut again before they have the opportunity to replenish, they can be severely set back or killed.

What about the Benefits of Weeds?

Many of the plants we have been referring to as weeds do unquestionably bring benefits to our agricultural systems.

Grasses with their fibrous roots are wonderful soil builders. Clovers and other legumes fix nitrogen from the air, bringing it into the system. Deep-rooted weeds help open up the soil and bring up nutrients and minerals from below. Some weeds have the ability to extract and concentrate minerals that are deficient in the soil and all of them contribute organic matter to some degree.

The question that is sometimes raised is whether it would be better to let some weeds grow among the herbs so the system can gain from these benefits. Unfortunately for most species we cannot afford to do this because of the problems that the weeds create in the management and harvesting of a commercial crop. However, we can still gain the wonderful benefits of these weeds by giving them a place to grow outside the cropping area and harvesting them on a regular basis for making the compost that is used on the herb crops.

Making Weeding a Priority

Success in weed management depends on doing the work at the time it can be performed most efficiently and effectively. Usually this is while the weeds are very small. This requires a shift in attitude for many of us because at this early stage the tiny weeds are not affecting our crops and we are not yet experiencing any of the consequences of the weed infestation. In the pressure of other work, it is easy to let weeds go until we finally start to experience the consequences of neglecting them. By this stage it may be too late as the task will by now be one of much greater magnitude. With a head start, the weeds are likely to race away to seed or proliferate with their rhizomes and roots while we are battling to catch up with them, and thus the stage is set for a never-diminishing struggle with generations and generations of multiplying weeds.

Most of our day-to-day problems do not behave this way. If we don't get around to doing the washing-up, the dirty dishes remain where they are until we experience the consequence of no more clean dishes. But the unwashed dishes don't multiply and spread all over the house so that by the time we run out of clean dishes we can't move because we are completely smothered by rapidly multiplying dirty ones. While it might take a few hours to get all the washing-up done, because the problem would not be expanding exponentially the situation would still be salvageable. I don't mean you should necessarily leave the washing up until the weeding is all done, but if you are going to stay ahead of the weeds, you need to avoid getting diverted onto other activities while the weeds quietly run amok.

The secret to making weed control a manageable affair is to make it a high priority. Try to cultivate a state of weed consciousness so you are constantly aware in of what is happening with the weeds and how your management practices and decisions affect or influence the development and control of weeds.

But don't let yourself just be guilt driven. Instead focus on some of the joys of weeding: the pleasure of looking over a well-maintained garden, the reward of bountiful harvests of weed-free produce and the pleasant quiet hours tending your crops. Indeed, the personal transformation experienced as a result of many quiet hours spent immersed in weeding your herbs can prove as rewarding as the more tangible results of the work.

One of the great rewards of the effort put into weed control is being able to look over a bountiful well-maintained patch of herbs.

5

Propagation and Planting

THE TRIAL PLOT

Before diving head first into a full-scale herb plantation, it is both wise and practical to set up a trial plot. This will give you an opportunity to become familiar with the herbs you are interested in growing, to assess their suitability for your conditions and to assess how suited you feel to growing them.

Rather than spend a lot of energy theorising about which species are suitable crops for your region, the best thing is to just plant a small amount of a few species. Then you can see how they grow and learn a bit about the plants and the operations involved in growing, harvesting, drying and processing them.

The production from your trial plot can serve for market testing. You need to show manufacturers and distributors that you can produce the goods: they get so many enquiries from would-be herb growers that they get tired of wasting their time on false leads. This way you can get valuable feedback on quality and marketability.

Provided you don't mind doing a bit of intensive weed control, your trial plot can go in at the same time that you start your bare fallow or even a little before. In a small trial plot, weed control without first doing a bare fallow can be manageable, whereas it wouldn't be so on a bigger scale.

Depending on the species, the weed situation, and the labour available, a trial plot would normally be 100–400 m².

Another big benefit derived from a trial plot is the propagation material it will generate. This can save you a lot of money, especially for species that can't be grown from seed. Many crops can be established faster from divisions or runners than by growing them from seed.

After a season's growth, most species will yield good quantities of vegetative propagation material. Typically there is a 20–50 times increase, enough to plant out a much bigger plot.

By the time your bare fallow is completed, your trial plot will have given you a good idea of what grows best for you and you can set about establishing your main garden. Autumn is a good time to plant many perennial species if you have a relatively weed-free area to put them into: they can get themselves established before going dormant for winter and will then take off early next spring, giving good production that season.

A trial plot provides the opportunity to assess the suitability of a species to your situation and the chance to multiply propagation material.

IDENTIFICATION

The trial plot will also give you the opportunity to properly check the identification of your plants and ascertain whether you have a suitable variety. Among herbs there can often be difficulty with accurate identification. Many common names refer to more than one species, which are sometimes quite unrelated. Botanical names are usually more precise but unfortunately they can at times be confusing as they get changed from time to time and some species have two or three botanical names in common usage.

There is no easy solution to this problem of names, and in many ways I prefer to use common names, provided they are generally recognised standard names. They often convey something of the folk tradition and the popular connection with the plant. On the other hand the botanical name is usually more academic, more universal in that it traverses language and cultural boundaries, and it often reflects ancient traditional associations with the plant.

The reality is that we need to be familiar with both nomenclatures. If both the common name and the botanical name are used to refer to a plant, there is usually no problem as to its identity.

It is most important to ensure that the herb you are growing or wild harvesting is the correct species and the preferred variety. I cannot stress this too highly.

Unfortunately, in the early days of the revival of interest in herbs, one of the foundation herb nurseries got the identifications wrong on a number of their plants. They subsequently supplied starting material to a great many other herb nurseries around the country and so these errors have been perpetuated and multiplied. This I came to know about because some friends bought out the original nursery and it took them a while to sort everything out.

I also know from my own experience how many times I have obtained propagation material and found it to be either incorrectly identified or an inferior variety not worth growing. Some of it has taken a while to get rid of too, as by the time I found out it was the wrong one, it was well established.

So how does one cope with this situation? First, try to find a reliable supplier and then check the identification with at least two books to confirm it. Unfortunately even the best herb books often contain a few inaccuracies or are hard to comprehend, so if there is any doubt you should consult a reliable authority or send a specimen to your buyer.

The National Herbariums in Melbourne and Sydney can be extremely helpful in sorting out some identification problems.

Material obtained from friends or growing wild also needs checking thoroughly, as does anything grown from seed. Seed is very easily mixed up so it needs to be grown to maturity to confirm its identity.

For example there is a phenomenon referred to as the 'next bin syndrome', where seed is accidentally taken from the next bin and you get a herb that is alphabetically adjacent to the one you thought you were getting.

Unfortunately almost every seed house sells useless things such as Tarragon and Peppermint seed. The 'Tarragon' seed will be nothing but flavourless Russian Tarragon, as the preferred French Tarragon does not form seed. The so-called Peppermint seed is usually a rank-flavoured form of Spearmint, for Peppermint is sterile also. Both Peppermint and French Tarragon need to be grown from rhizomes.

Many of the essential-oil containing herbs vary greatly in their oil content and consequently their flavour and effectiveness. To a large degree this variation is genetic, so it is important to start off with good varieties. This is much easier to assess if you are dealing with plants rather than seeds.

PROPAGATION METHODS

Propagation is a vast subject, so I will primarily focus on aspects of propagation and planting that are relevant to this scale and style of herb growing.

Vegetative Propagation

Vegetative propagation is the growing of new plants from parts of an established plant rather than raising them from seed. Many species of herbs can be propagated vegetatively from divisions, rhizomes or cuttings. Of course, there are a number of species which are not suited to this, particularly annuals, biennials and a few perennials that grow more vigorously from seed.

Advantages of Vegetative Propagation

Faster Establishment: Vegetative propagation generally produces an established crop more quickly as the young plants are larger and more vigorous than seedlings. This means harvesting can begin sooner, generally the first season, and full production is achieved earlier.

This earlier production may justify the greater expense of purchasing vegetative propagation material.

Simpler Procedures: Vegetative propagation of most herbs is often simpler as it enables direct planting, which is easier than raising seedlings and transplanting.

Cloning: It ensures that the new plants have the same desirable characteristics as the parent.

Identification: It is usually easier to confirm the identification of vegetative propagation material, as one can generally see the parent plant.

Disadvantages of Vegetative Propagation

Quantity: It may be difficult or expensive to obtain sufficient quantities of propagation material for a large planting. It can take a while to build up enough stock.

Narrow Genetic Base: Clones have a narrow genetic base. This lack of diversity means there is less, or sometimes no opportunity for selective breeding and natural selection for characteristics such as pest and disease resistance

Availability: Vegetative propagation material for some species or varieties may simply not be available.

Division

Division is essentially the dividing up of a single large plant into pieces that will each grow into a new plant. Division is suited to many species. The method of dividing the plant varies according to the species and the nature of its growth. Each piece needs to have viable roots and leaves, or buds ready to form these.

Crown Division

Many clump-forming herbs can be cut or torn apart. The pieces of stem or crown, with roots attached, are then replanted. It is important to bear in mind that these small root systems will only support a small amount of leaf until they recover from the shock of transplanting and re-establishing themselves. Depending on the plant and the time of year, the leaves need to be cut back, sometimes drastically, otherwise they can transpire too much water and desiccate the plant. This is often not well understood: we tend to relate more to the attractive above-ground parts of the plant than to the hidden root system. People often feel they would be committing an act of barbaric cruelty to take a knife to those beautiful leaves and flowers. But really it is much more cruel to dig up a plant, reduce its root system and then replant it with all its leaves transpiring in the sun and wind, but with insufficient roots to support that amount of water loss.

There needs to be an equilibrium between the root system and the leaf system in the newly separated plant, with the balance in favour of the root system so the new plant is not stressed for lack of moisture or nutrients.

Root Division

A number of species will form new plants from any piece of root. After digging up the plant and shaking the soil out of the root system, the roots can be broken into suitable pieces, usually 50–100 mm long, and planted where they are to grow.

Runners

Some herbs form runners – horizontal stems or roots – that extend out from the parent plant for varying distances and form new plants with their own root and leaf systems. The term 'runner' can refer to a stolon or horizontal stem, but generally it is used more loosely to include rhizomes, extensive roots and stolons.

Rhizomes: Technically a rhizome is a stem that extends more or less horizontally underground.

Stolons: A stolon is of similar form to the rhizome, but is above the surface.

Extensive Roots: Some species form a horizontal root that performs a similar function, extending some distance from the parent plant and giving rise to buds that send up new crowns and roots.

Plants that form runners can be dug up and the runners separated off and broken into manageable lengths for replanting. Often these pieces don't have any leaves or fine roots yet formed. The rows need to be clearly marked so weeding can proceed without excavating the newly planted crop.

Cuttings

Many woody species can be grown from cuttings, though some more readily than others. Softer stemmed species are not as suitable unless they form above-ground runners that strike easily. It is a useful method for establishing a plot when division is difficult or not an option. It is often much easier to prevail upon someone to let you take some cuttings than to dig up their plants and divide them.

Some species strike very easily while others require a green thumb, with careful attention to timing, soil type, warmth and moisture conditions.

Species vary as to whether cuttings should be hardwood, softwood, green hardened wood or tips, and vary as to the suitable time of year. Most require protection from drying out until their roots are formed.

As the success rate can be variable, cuttings are best started in a nursery bed, or in a tray at least 70 mm deep with provision for drainage. Sandy soil can be used or a mixture of equal parts river sand and peat moss may be preferred.

The usual procedure is take sections of stem 60–200 mm, or longer for some trees, and ideally there should be four or five nodes on each piece, but there need to be at least two.

If the wood is hardened enough, heeled cuttings can be taken. Short side shoots are torn away from a main stem so there is a sort of pointed tip and heel at the base. These tend to be more successful.

At least one-third and preferably two-thirds of the length of the cutting should be stripped of any leaves and inserted into the soil so that at least two nodes are covered.

Many authorities advocate the use of hormone rooting powder but this is not essential and is really not compatible with biodynamic and organic methods. It is usually possible to get very good results without it.

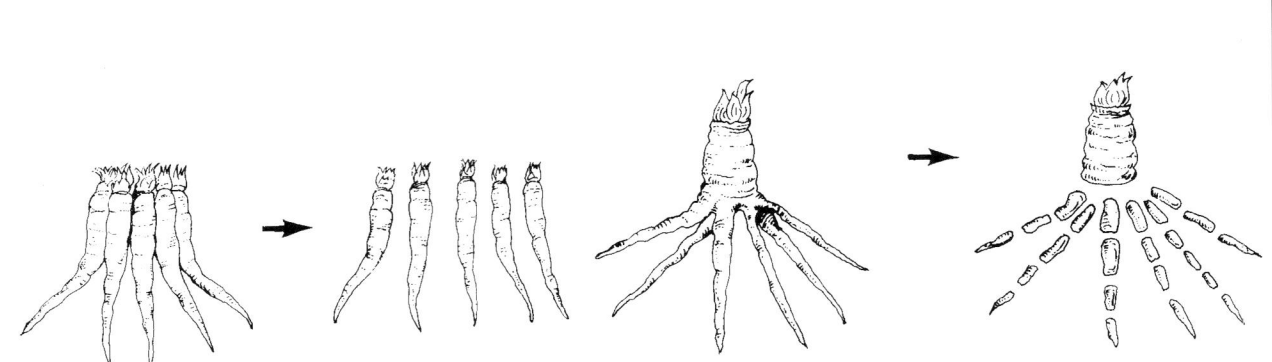

Crown division is the separation of a crown or clump into pieces which will then each form a new plant.

Root division is possible for some species that are able to regenerate from separated pieces of root.

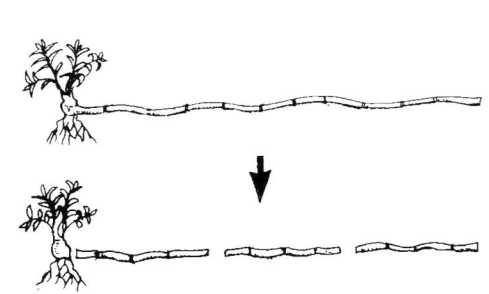

Some herbs form runners – rhizomes, stolons or extensive roots – that give rise to new plants.

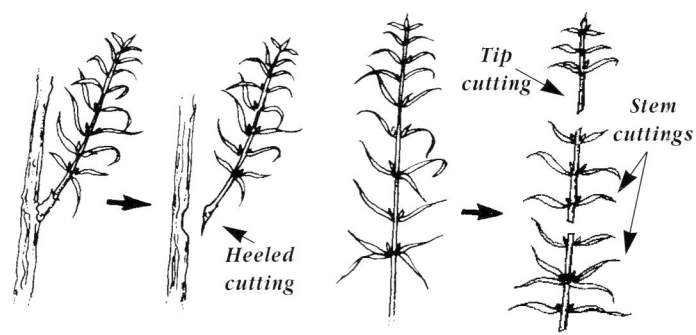

Cuttings can be used for propagating many woody species.

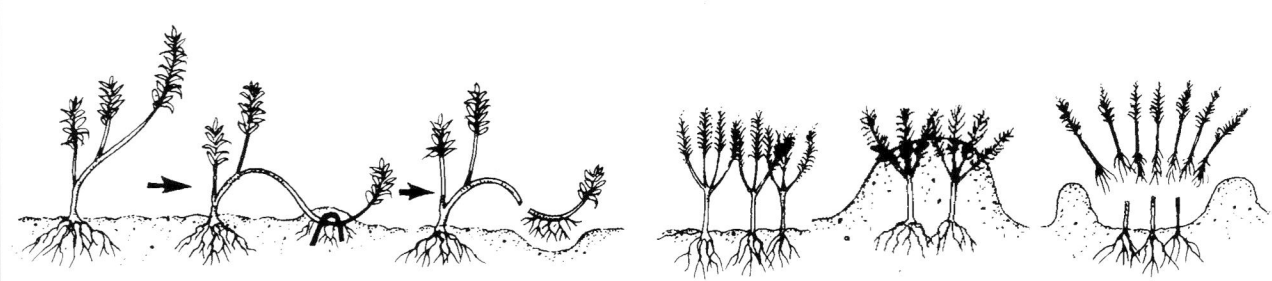

Layering is the development of roots on a woody stem before it is separated from the parent plant.

Propagation methods.

Provision needs to be made to prevent dehydration. Leaves should be cut back and a humid atmosphere maintained or a misting system used if prevailing air conditions are too dry. The soil or rooting medium should be kept moist but not waterlogged.

Once they have struck roots, the cuttings can be potted up or planted out. It may be necessary to grow them on for a while until the optimum time for setting them in their permanent positions: usually late autumn to early spring.

Layering

Layering is really just striking a cutting before separating it from the parent plant. Many plants will readily form roots from their stems where they are in firm contact with the soil. This can be encouraged by pegging a stem down and covering it with soil or simply by pulling soil up around the stems of a plant. This latter method works well for many woody perennials and shrubs.

Once the roots are well formed, the stem can be cut from the parent and you have the beginnings of an independent plant that can be moved to a new site. It may be necessary to cut back the leaf system a bit so it is in balance with the roots.

Plants grown from layered material will usually do quite well as they aren't subjected to as much shock as cuttings or crown divisions.

Direct Planting

Direct planting is practical where the propagation material has a good prospect of surviving and thriving. Generally this is so with larger pieces, except for cuttings which are usually best held until they have established an adequate root system.

Of course it is important that direct planting be done at a time when the conditions are suitable: usually from mid-autumn to spring when it is cool enough to avoid stressing the establishing plants.

Pots and Trays

Smaller pieces of propagation material may need to be grown on for a while in trays, pots or soil blocks until they are large enough for planting out in the open. The procedure is much the same as for growing from seed.

If large numbers of plants are required from a limited amount of propagation material, dividing it up into small pieces and growing these on to a viable size before planting them out is an option.

Each piece of crown must have some root and a bud or shoot. Rhizomes and stolons can be cut into quite small sections, but they must have at least two nodes. Roots without nodes should be a minimum of 25 mm long.

Nursery Beds

Where in-ground conditions are suitable, nursery beds are another option for some easily transplanted species that don't require a closely controlled propagating environment. Here the propagation material can be grown on in a garden situation until it is ready to plant out.

A nursery bed has the advantage of not requiring such close management as growing in pots and trays. Often it can be set up to be irrigated and weeded as part of the main herb growing area, avoiding the need for a separate system.

Where stronger plants are needed for direct planting, or where there are normally some losses in propagation, nursery beds can be a simple set-up for growing stock on.

Growing Herbs from Seed

Some herbs, particularly annuals and biennials, can only be grown from seed. For many of the other species there is a choice between propagating vegetatively or from seed.

Advantages of Growing from Seed

Quantity: For larger plantings it is usually easier and cheaper to obtain seed in sufficient quantity.

Transportability: Seed is easier and cheaper to ship, especially from overseas.

Availability: A greater range of species and varieties may be available as seed.

Genetic Diversity: Growing from seed should ensure a broader genetic base.

Vigour: This applies only to a few species that are more vigorous when grown from seed.

Disadvantages of Growing from Seed

Longer Lead Time: Growing from seed, especially small seeded species, means a longer period before production begins and little, if any, harvest the first season.

Identification and Variety Problems: There is a greater risk of getting the wrong species or an inferior variety when buying seed as it often difficult to determine the origin. These problems may not become apparent until a year or more down the track, when the plant can be positively identified.

More Elaborate Procedures: Some species need special techniques and facilities for good results.

Starting Very Small Seeds: Many species of herbs have very small seeds that require careful tending to get them started. These need to be raised in seed-trays or flats until they are large enough to transplant out into the garden.

A soil that combines the qualities of moisture retention and permeability is preferable. On a small scale there is often no need to go in for fancy soil mixes. I usually just get some good soil from as weed-free a section of the garden as I can find and add some fine mature compost, and maybe a little sand if the soil needs lightening to facilitate watering and easy emergence of germinating seeds.

Some people like to heat-sterilise the soil to prevent damping off. I usually don't bother with this any more as I find most seedlings will grow all right without it. A healthy soil flora and good air circulation will prevent the development of damping off. In fact, wiping out the soil flora destroys all the beneficial bacteria and leaves the way open for the possible invasion of pathogens.

Having filled the seed-tray with soil, levelled and firmed it, a sieve can be used to sprinkle the seeds evenly over the surface. Fine soil can then be sieved over the seeds until they are just (and only just) covered by a thin layer.

When watering, the seed-tray should be set in a shallower tub of water (not too cold) to soak it up from underneath, as any watering from above is likely to wash the small seeds away or break off newly emerged seedlings.

Once fully moistened, the seed-tray should be set somewhere warm and sheltered, with a sheet of glass or plastic over the top. Leave a small gap of 5–10 mm along the edge of the tray to allow for a little air circulation.

The surface must not be allowed to dry out. The temperature should be in the 15–25°C range, but this can vary according to the needs of a species.

Step 1: Seed tray is filled with soil, levelled and firmed.

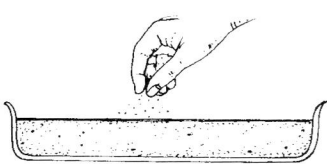

Step 2: Seed is sprinkled evenly over surface.

Step 3: Fine soil is sieved over the seeds until they are just covered by a thin layer.

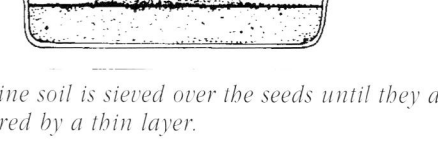

Step 4: Water by setting the tray in a shallow tub of water to soak up from underneath.

Step 5: Cover with a sheet of glass or plastic and sheet of newspaper if in direct sunlight.

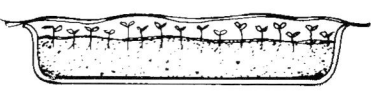

Step 6: When germination commences remove the glass, but leave the newspaper on for one more day.

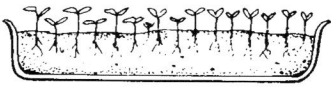

Step 7: The next day the seedlings should be strong enough to withstand direct sun.

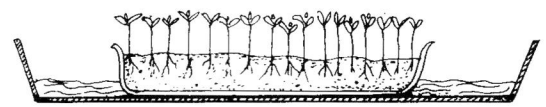

Step 8: Until they are robust enough to water from above, continue to water by soaking in a shallow tub.

Very small seeds need special procedures to achieve good results.

If the tray is in direct sunlight, a sheet of newspaper should be placed over the glass. Each day the glass should be turned over. When germination is evident the glass should be removed and the newspaper left on alone for one more day, by which time the seedlings should be ready to withstand direct sunlight.

Not So Small Seeds

There are a number of options for planting seeds that are not so small.

Direct Seeding

In general, the larger the seed the less amenable the young seedling is to transplanting and the stronger and faster its growth will be if direct-sown in its final position. Species with large seeds are usually able to compete with weeds adequately to get established and are easier to see and cultivate around for good weed control.

Direct seeding is also the only option for some species that are sown down as a meadow-crop, such as Red Clover and Alfalfa.

Direct seeding should be carried out at a time when soil moisture and temperature are suitable. Additional irrigation may be necessary to keep the surface moist until germination occurs.

Sowing Row Crops

Seed should be sown in as straight a row as possible and covered with soil to two to four times the seed's diameter, depending on soil type and moisture. Bury seed deeper in lighter soils or drier conditions, shallower in heavier soil or moister conditions.

Usually it is advisable to sow seed somewhat more densely than the desired final spacing in order to ensure sufficient plants survive. Unwanted plants can be thinned out or maybe transplanted into gaps in the row.

For smaller seeds, direct seeding should only be considered if weed germination is light. Small seedlings can easily get smothered and lost in vigorous weed growth. Generally this means that, for at least the first year or two, most growers will need to plant these seeds in trays or nursery beds.

Sowing Meadow-crops

A few species are grown as annual or perennial meadow-crops and need to be direct-sown because they will not transplant. These will need fairly good prior cultivation and bare fallow before sowing.

The aim is to get a good even coverage. Soil conditions are important, especially for small seed which will require that the surface be kept moist until germination.

Depending on scale and equipment available, there are two methods of sowing: broadcasting and drilling.

Broadcasting: The simplest method is to broadcast the seed by hand or with a spinner: a hand-held device which distributes the seed evenly over the surface.

Large seeds, such as oats can then be disc-harrowed or cultivated into the soil.

Smaller seeds can be lightly covered by dragging a chain or a branch with a lot of fine twigs across the ground. Sometimes smaller seeds will give adequate germination just by watering them in with a good irrigation or fall of rain.

Drilling: This is the planting of seed in closely spaced rows, 150–200 mm apart, and is only practical if you have suitable seeding equipment.

The advantage is that the seed can be sown at its optimum depth, giving better results and economising on seed.

Seed-trays

The most basic of seed trays is an ordinary flat tray about 50 mm deep with drainage holes in the bottom. The same soil mix can be used as described for starting small seeds. It should be filled to about 10 mm from the top and pressed firmly down.

The seeds should be sprinkled evenly about the surface at a density that will ensure sufficient numbers without overcrowding. You may prefer to place the seeds more accurately, 10–25 mm apart.

Cover them with about twice their diameter of soil. When watering, avoid heavy droplets that might uncover or disturb the seeds.

Compartmentalised seed-trays are available. One style, moulded from styrofoam, is quite popular but I have never been impressed with its results. It has individual compartments for the seedlings, of an inverted pyramid shape which tapers down to a point at the base but there is really not enough space for sufficient soil for root development and a reserve of moisture. They appear to have been designed with chemical nutrients and automatic watering systems in mind.

The other problem is that bits of styrofoam inevitably break off and blow around the place – not the best thing to find its way into your finished product.

A slightly more expensive option is a more durable black polythene tray with larger compartments that provide for more root development and give better results.

Soil blocks, a recent development that reduces transplant shock when raising seedlings.

Soil Blocks

Soil blocks are another option that has just become available in this country. This system has been used very successfully by organic vegetable growers overseas and would undoubtedly be very suitable for herbs.

Cube-shaped blocks are pressed out of a mixture of peat, compost and soil. A special block press makes a number of these at a time (these are available from Gundaroo Tiller, see Appendix). They are then placed beside each other in trays and the seeds are sown direct into the blocks. Because the blocks are separate, the roots stay within them as the plants develop. The blocks can then be placed directly into the garden when the plants are ready to transplant. Because this involves very little disturbance of the root system, transplant shock is minimised.

Nursery Beds

The use of nursery beds is another option for most species, except those with the smallest and largest of seeds.

If you have a reasonably weed-free situation in your garden, you can direct-sow in small nursery beds for transplanting later. Provided you choose the right time of year, generally spring or autumn, moist conditions can be adequately maintained to foster germination. Once established, the seedlings will be watered by your normal routine of irrigation.

This frees you from the tedium of daily watering-can rituals and reduces the possibility of moisture stress, as in the open garden there are deeper reserves of moist soil to sustain the little plants through hot spells or if you go away for a few days. Though growth out in the open may be a bit slower, the plants tend to be healthier.

One disadvantage is that it may be harder to protect the very young seedlings from snails and slugs if these are prevalent.

Making soil blocks.

Nursery beds can be an easy way of raising seedlings for some species.

Seed Treatments

The seeds of a few species of herbs have a built-in dormancy that prevents or reduces germination. There are various treatments that can help overcome this, including stratification and priming.

Stratification

Some herbs need a period of stratification or moist cold treatment to break dormancy in their seeds. This is no doubt a natural mechanism in plants that originate from cold climates. It prevents them from germinating in the moist warm conditions of autumn as the small seedlings would be unable to survive the extremes of winter. In the wild state, this dormancy is broken by the seeds lying in the soil over winter, after which they will germinate freely the following spring. (Note: A few species have a 2-year dormancy.)

This natural process can be imitated by placing the seeds in cold moist conditions for at least 4 weeks prior to planting. Some species require temperatures below freezing, while others need temperatures just above freezing. Mix the seed with moist sand or peat moss and put it in a plastic bag in the fridge or freezer for a month. (Moisture is an important part of the process so just putting a dry packet of seed into the fridge or freezer won't be adequate). Plant immediately afterwards.

Alternatively, the seeds can be planted in a seed-tray in autumn and left outside in a location where they will be kept moist and exposed to frosts over winter. They will then germinate when conditions are suitable the following spring.

Priming

Priming often just consists of soaking the seeds in water for a period of time, usually 24 hours, before planting. Often warm water can be more effective.

Rain water tends to stimulate germination, too. A fall of rain can sometimes bring on germination that weeks of watering has failed to prompt.

Some seeds may need special treatment, like passing through a digestive tract, soaking in weak acid, fermenting in a mash, soaking in hot water, or nicking the outer seed coat.

Propagation Facilities

A greenhouse and shadehouse can be handy for propagation, though unless you have a need for raising large quantities of seedlings or other propagation that cannot be done in the open soil, you can probably get by quite well without them. In over 15 years of herb growing I have hardly ever used either. In our mild climate with our long growing season, I have always managed with direct planting, nursery beds in the open soil, or setting a few seed-trays on the windowsill.

The main advantage of the greenhouse is in getting propagation material to an advanced stage for early planting, to obtain an extended growing season, or to profit from an early crop. While this can be useful for some annuals and biennials, in general our growing seasons are long enough for most herbs and for a dried product there is no price advantage in an early harvest.

One problem with greenhouses is that they create an artificial environment, which can result in stressed or soft, unbalanced growth and consequent problems with pests and diseases, which tend to build up in the greenhouse.

Another option is the use of cloches to help start direct sown crops like Basil which require warm conditions.

Cloches are clear polythene tunnels about 500 mm high that can be removed or folded back as the season gets warmer and the plants get bigger.

In warmer regions a shadehouse may be useful for moderating the excessively hot and dry conditions that often come on strongly in spring, making it difficult to keep seed-trays moist enough for good germination and growth.

Cloches can be useful for starting plants that need warm conditions in spring.

PLANTING OUT
Timing

Autumn Planting
Especially in regions with hot summers, perennial herb plots planted out in autumn will usually yield much better crops the following season than spring plantings. This is because they have had the opportunity to get well established in the mild conditions of autumn and early spring, developing a good root system before hot weather hits them in late spring and early summer.

However, weed control over the first winter can be a problem with species that are vulnerable to weed invasion and smothering.

Spring Planting
Spring-planted herbs will usually be lower yielding the first season but will then catch up. The advantage is that weed control is easier if the plants are set out just prior to spring growth commencing. This is usually around mid- to late August in southern regions. This way the herbs will grow strongly from the start and will be more competitive against the weeds. Conditions for effective weed control will be better as spring progresses and the herb rows will be easier to follow.

Prevailing Conditions
For best success, transplanting should be carried out when the prevailing weather and soil conditions will favour the establishment of the new plants. This depends to a large extent on the nature of the material you are planting out. Plants in soil blocks or pots, provided they are not too advanced or root bound, will suffer less shock in transplanting than, say, crown divisions, which have to re-establish their root systems after the disturbance of being separated.

Hardening Off
If plants have been in an artificial environment, such as a greenhouse or shade-house, they should be hardened off by being placed in the open for a few days before transplanting.

Soil Moisture
The plants themselves need to be well watered before transplanting and the soil they are going into should be moist but not sticky. If soil does need additional watering prior to planting, it is best to do this some time beforehand to let it even out and dry a little, or else put the water on after transplanting.

Planting into excessively wet soil will cause compaction around the plants and wherever you step.

Weather
Ideally the weather should be cool for several days after transplanting. A good time to do it is just before a rain, so hopefully the weather will be cool and moist for a few days afterwards. This is not always easy to arrange, so you may have to resort to supportive measures to reduce the shock of transplanting.

If the weather is rather warm, then transplanting is best carried out late in the day so the plants are not subject to as much stress. Their roots should not be allowed to dry out at all.

Supportive Measures
Removing some leaf will help reduce moisture stress. Water your plants thoroughly before transplanting them. Additional waterings after transplanting, so the soil is kept near saturation point for a few days, will help too.

If you are obliged to transplant in hot dry conditions and your new plants are likely to need additional help, placing shadecloth over them for a few days will make an enormous difference.

Transplants should not be allowed to wilt, or even approach the point of wilting, as this means they are under stress and their survival or at least their continued good growth is threatened.

Layout
There is plenty of scope in laying out a commercial herb garden for individual and artistic expression, but also for errors and learning experiences as you find out which herbs would have been better placed elsewhere or managed in different ways.

Planning the layout of your garden is a rather complex operation involving the consideration of a number of factors.

Flat Profile Beds v Raised Beds
There seem to be two schools of thought when it comes to laying out a garden: those who like to grow on a flat surface and those who like to grow on raised beds.

Flat Profile Beds
My own preference is for a flat profile, provided soil conditions and climate allow it. Many aspects of the operation are easier if the surface of the ground is more or

less flat. Inter-row cultivation is simpler, so weed control is more efficiently and effectively carried out. Access for other operations, such as harvesting and spreading compost, are easier. Water soaks in better and more evenly into the flat surface and it holds moisture better. Harvesting with the scythe is also much easier.

Raised Beds

Raised beds are popularly constructed as narrow elevated beds about 1 m wide with steep sides and deeply recessed paths some 200 mm lower than the bed.

Raised beds' main advantage is in heavy soil and poorly drained situations where aeration in the root zone needs improving. In situations where the soil would lie cold and wet for a long time in spring, raised beds allow it to warm up earlier in the season and this gives better growth. Also there tends to be less compaction around the plants as people and machinery stay mostly on the paths between the beds.

Weighed against this are a number of disadvantages. Getting adequate water onto raised beds during the peak of the growing season can be a real problem as it tends to run off and the raised bed dries out more in the sun and wind. Weeds are inclined to proliferate in the paths and on the steep sides of the beds and continually invade them. Weed control in these areas is labour intensive as they are inaccessible to the rotary hoe and wheel hoe, so weeding has to be done by hand. Consequently weeding tends to be left undone or is done too late, after the weeds have spread their seeds and runners into the beds.

Raised beds can create problems when harvesting with the scythe, as it is hampered by the shape of the bed: the blade tends to cut into the soil on the steep edges and pick up trash.

Unless your soil has a drainage problem, these disadvantages generally outweigh the advantages. A light well-drained soil will warm up quite adequately in spring without being raised up. It will also be more productive and easier to manage as a flat profile.

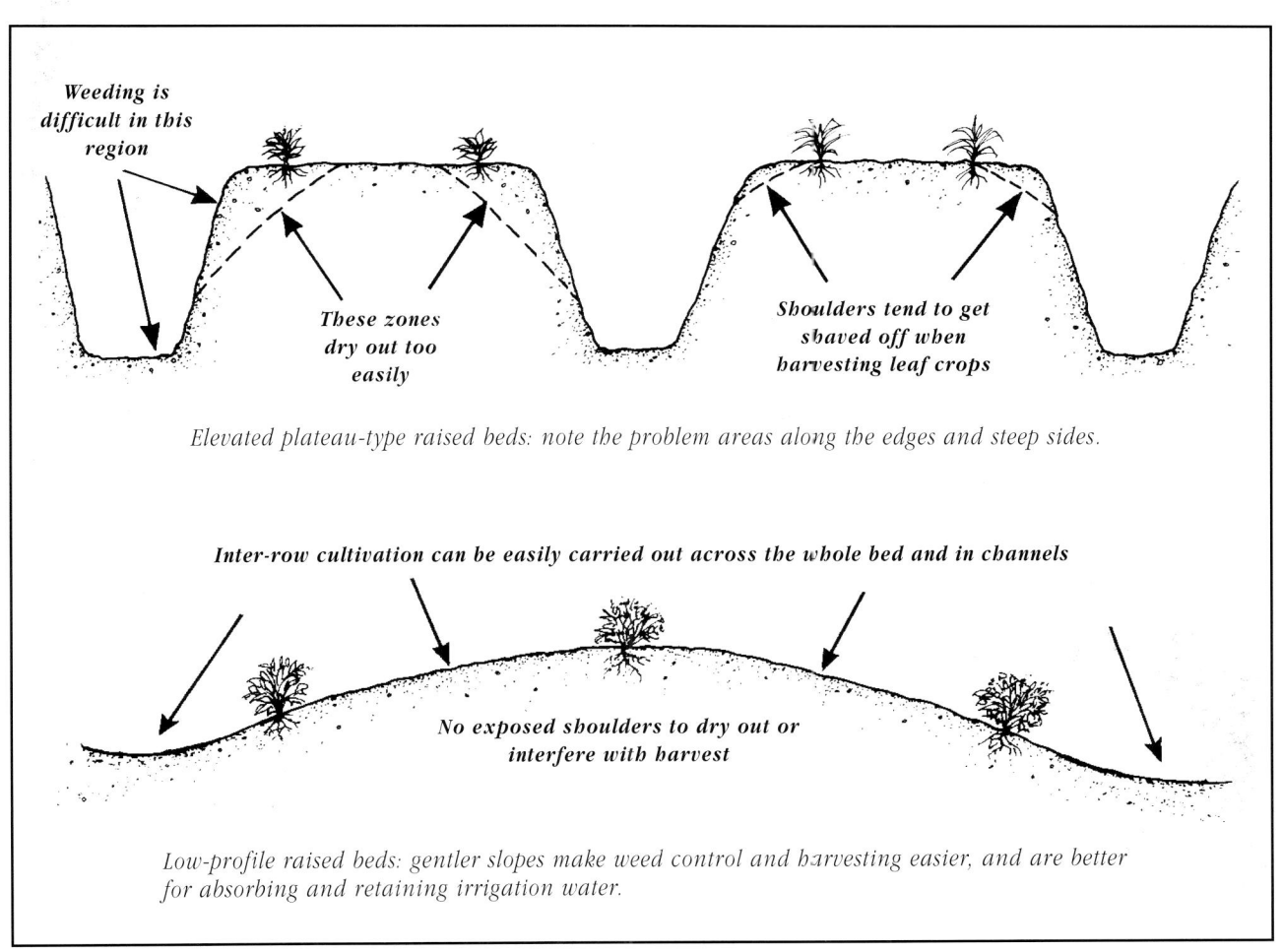

Elevated plateau-type raised beds: note the problem areas along the edges and steep sides.

Low-profile raised beds: gentler slopes make weed control and harvesting easier, and are better for absorbing and retaining irrigation water.

Elevated-plateau and low-profile raised beds.

Low-profile Raised Beds

If your soil situation does necessitate raised beds, then rather than create the narrow elevated tablelands one usually sees, much broader lower profile raised beds will be more manageable and will also be easier to set up and maintain. This broad gentle profile is better from the point of view of irrigation, weed control and harvesting.

I like to set the beds up about 3 m wide with a gentle rounded profile and a fall of 100–150 mm from the highest point in the middle of the bed to the bottom of the valley at the edge.

Of course, provision needs to be made for water to flow away from the valleys between the raised beds, so it is not just lying there.

Weed Control

Weed control is one of the most important factors to take into consideration in deciding on layout, as the arrangement of the garden has a big impact on how effective and efficient weed control can be. This aspect is covered in detail in the previous chapter.

Erosion Control

Preventing the loss of soil is vital and unfortunately it is an aspect of garden planning that is often overlooked. There is no point in spending all your efforts in building up the soil only to have it gradually washed away. And in some regions it may not be such a gradual a process: a few good thunderstorms sending water ripping down your rows can carry the most valuable portion of your garden off on its way to the sea.

First, avoid steep ground. If cultivating on a significant slope, make your rows follow the contours. This may take a little more care and time in setting up, but it does look very attractive on a curved contour and greatly reduces soil loss. Avoid long downhill sections. Cultivated strips across the slope – each no wider than 20 m with strips of grass in between – will reduce run-off and consequent erosion.

Ease of Harvest

Where harvesting of leaf is going to be done with a scythe, plots of each species need to be at least 2.7 m wide to allow enough space to get a good swing. This usually means three rows at least 0.9 m wide, though often I prefer the wider spacing of 1 m or even 1.2 m for spreading herbs. Consequently a mixed garden will tend to be made up of herbs planted in blocks rather than long single rows of individual herbs.

Different microclimates can be used to locate plants in situations that suit them best. A creek lined with willows in the middle ground provides a cooler, sheltered, partly shaded microclimate for the beds in the foreground, which are planted with species that would be stressed if growing in the open area in the background where more heat-tolerant crops are fully exposed to sun and wind. ('Twin Creeks' in north-east Victoria)

Microclimate

Inevitably there are going to be variations in microclimate in different parts of your garden, even if they are only the variations that have been created by the plants themselves. Tall plants create a warmer microclimate on their north side and a cooler one on their south side. Cooler sites may be found in areas of your garden with some afternoon shade or where there is shelter from hot northerly winds. Warmer

Curved rows following the contour to reduce erosion can be very pleasing to the eye.

sites may be found in areas receiving full sun or those sheltered from cold south-westerly winds.

It is worth the trouble of trying to site plants in the microclimate that suits them best, as it can mean the difference between a plant struggling to maintain itself and a plant yielding an abundant harvest.

Proximity to Trees

In trying to locate plants close to the shelter and shade of trees, you will probably find your herbs have to compete with vigorous tree roots attracted by the moisture and fertility of your garden. It can be surprising how far trees will extend their lateral roots. Eucalypt and Willow root systems can easily reach out 20 m or more from their trunks: you can get an idea of this in late spring as soil moisture is dropping in a pasture with trees in it. Around the trees the grass will wilt while the grass out in the open continues to grow. There is usually a distinct boundary that defines the extent of the tree root systems. You may see the same thing in the plants in your garden between irrigations in summer or just in an area where the growth is poor.

At 'Twin Creeks' the opportunity came to dig around the tree-affected gardens with a trenching machine. I cut a trench 900 mm deep and discovered Willow and Eucalypt roots up to 100 mm thick that had been pumping water and nutrients out of the garden. No wonder the herbs in their proximity hadn't been doing well.

Just about all the roots I found were in the top 600 mm, so deep ripping (with a 200 L drum full of water attached to the ripper to help it bite in) beside the gardens every year or two proved sufficient to keep the tree roots back. So if parts of your garden are close to trees, leave enough room around the perimeter to get through with machinery to cut the tree roots on a regular basis

Leaves falling from trees can be a problem in some crops as they can contaminate those with close foliage that tends to hold onto fallen leaves. Avoid trees – particularly deciduous trees – when planting crops such as Parsley, Mint or Thyme that might be harvested around the time of leaf fall.

Soil Types and Drainage

It is important to identify variations in soil types and drainage so plants can go into situations that suit them. Plants that will tolerate waterlogging should be sited in any wet spots so the drier sites are available for those plants that need them.

Irrigation System

Your layout needs to be compatible with your irrigation system. Inevitably there will be areas with extra water where sprinklers overlap and drier areas at the outer edge of a sprinkler's throw. Try to make sure these don't cause problems and use them to advantage where possible. If you are using sprinklers, plan your garden within the limits of their reach.

Tall plants need to be sited at the edge of a sprinkler's throw so they are not creating a 'sprinkler shadow' beyond them. Some herbs use large amounts of water and have aggressive root systems that can detrimentally affect plants nearby with shallow root systems and high moisture requirements.

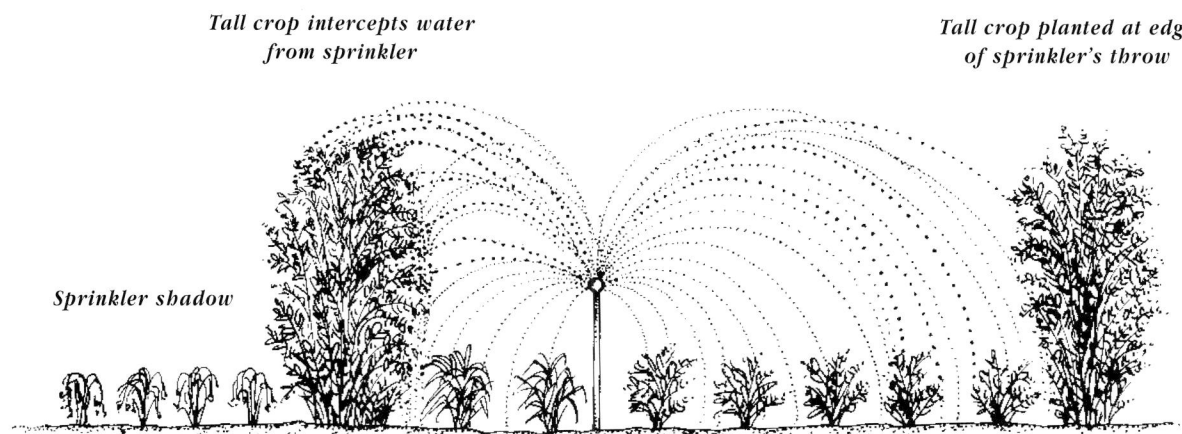

Sprinkler shadows can be caused by tall crops, so these should be located at the edge of the sprinkler's throw.

Containment

Quite a few herbs have a tendency to spread beyond the location in which they are placed. The extent and rate of this expansion varies according to the species. Mint rhizomes will advance about 500 mm a year, Coltsfoot about 2 m and Horsetail up to 9 m a year. Some herbs, such as Plantain, Vervain and Chamomile, even when harvested regularly, drop viable seeds, which come up prolifically around them, while wind-dispersed seeds like those of Dandelion will tend to come up here and there all over your garden.

The design of your garden must take into account the need to contain these venturesome herbs. The most intrepid of them need their own plot separated from others by a safe distance of unwatered, regularly mown grass or a regularly cultivated buffer zone. In small gardens you may prefer physical barriers, such as solid walls or tubs, to prevent runners moving out into the wider world. (For more detail on containment see Chapter 15, 'Spreading Herbs'.)

Moderate spreaders, such as the Mints, are manageable in a mixed garden, provided they are located next to non-spreading species with distinctly different foliage. This will make it much easier to spot any wayfaring plants and deal with them before they become too entrenched.

It is a good principle to arrange the layout so that species in adjacent beds are of sufficiently dissimilar appearance to reduce the risk of mixing up species in harvest, especially if you sometimes have inexperienced people helping you.

It is important to avoid very similar plants such as Spearmint and Peppermint becoming mixed together as they would be very difficult to sort out. They need to be separated by a safe distance, with at least one block of a quite different herb between them.

Companion Planting

Companion planting is an area that greatly interests some people and there are companion planting guides of compatible and incompatible combinations of herbs.

Personally I do not doubt that plants have an influence on each other and there are some clear examples of plants having strong antagonistic effects on other species. In my own observations, most of this influence can be explained by gross factors, such as competition for nutrients and moisture and sunlight. My impression is that generally the positive influences often alluded to require that the companion plants be growing right next to each other, intermingled as it were. Unfortunately this is mostly not practical in a commercial operation: we need to keep the species sufficiently separate to facilitate harvesting.

There are many subtle and complex interactions at work in the associations of plants and it may not be all as simple and straightforward as most companion planting guides would indicate. However, if companion planting strikes a chord with you, it is important to follow it as this is the sort of guiding inspiration that will lead to the development and sustaining of your own expression in the form of a bountiful herb garden that involves you as part of its whole organisation.

In many ways I do follow a system of companion planting myself, but the factors I take into account in determining whether to place plants near each other, or well separated, are things like competition, shading, shelter, sprinkler shadowing, invasiveness, similarity of appearance, harbouring of pests and diseases, ease of access etc.: generally factors of a physical nature that influence the growth of the plants and the ease and efficiency of management.

Rotation

With perennial herbs there is not generally a lot of scope for rotation. Most of these plants seem to be quite happy where they are for at least 5 years. Because many take some getting rid of, it is sometimes easier to rejuvenate an established plot than to change it over to something else.

But if production is declining after a number of years of the same perennial crop, and rejuvenation is not possible or fails to improve productivity, then it is good practice to rotate to another crop.

There are a number of annual herbs that need to be rotated because generally they do not do well if grown in the same location year after year. There can be a build-up of disease or pests, and yields will deteriorate. Non-proliferating root crops such as Elecampane, Marshmallow and even Valerian can be included in these rotations, because, although they are perennials, they are dug up and replanted every year.

The general principle is to rotate with crops of a different type. For example, you can follow a root crop with a leaf crop, then a seed crop and maybe also a flower crop. It need not necessarily be in that order, but by growing crops of a different nature, the demands they place on the soil seem to balance out and it keeps the pests and diseases guessing.

This may not always be possible if there is a predominance of root crops, but you can try to rotate among the root crops so that a crop of one botanical family is followed by one belonging to another family.

Crops that self-sow heavily or that result in heavy weed infestations will need to be followed by crops that can cope with this problem. Strong growing, widely spaced, large root crops are usually suitable for this as they enable regular cultivation and any residual weeds will not interfere greatly with harvesting.

Green Manures in Rotations

It is good practice to include green manure crops in a rotation, particularly when the soil might otherwise be lying bare until it is time to plant the next crop.

However, green manure crops must be fitted into a place in the rotation where the following crops can cope with the trash, surviving green manure plants and often a legacy of weeds that remain after the green manure crop has been worked in.

Oats and Rye Corn are good crops to put in over winter as they are active in cold conditions. Peas and Tic Beans also make good growth over winter and can add nitrogen to the system, though they may require inoculation with the right bacteria to form nodules on their roots.

Summer green manures can result in an even greater proliferation of weeds going to seed, so they need to used with caution.

Pests and Diseases

There may be ways that the layout of the garden will affect the impact of pests and diseases and the effectiveness of any control measures. Many of these only become problems because we create a favourable habitat for them and their impact can be reduced by creating a habitat that does not encourage their build-up to problem proportions.

In some regions grasshoppers can be a real problem and it may be necessary to plan a fenced walkway around your garden and allow a few domesticated friends with eager beaks to patrol it. Grasshoppers sometimes seem to be selective as to which parts of the garden they attack, so vulnerable plants should be located away from these areas.

Snails need to be planned for, if you have them, though it would be best to plan not to have them. Initially at 'Twin Creeks' there were only a few snails around the house, but after a number of years they had managed to slither their way into most of the herb gardens. Among the taller growing species their presence can generally be tolerated, but in the lower growing herbs they are more of a problem in that they scalp off new regrowth and get picked up in harvesting. The problem was that I had planted plots of woody perennials right next to herbaceous species like the Mints, so there were always good hiding places for snails where the birds couldn't find them. We had a busy flock of magpies – sometimes helped by a few ibises – but they couldn't get into the hiding places deep in thick bushes where the snails sheltered.

If a snail problem is likely to arise in your garden, I would recommend planting all your woody herbs in one area. If the snails get among them, they will find plenty of shelter but generally they won't be too much of a problem in these plants. In another area, preferably separated by strips of mown grass and cultivated soil, place all those herbs that get harvested down to the ground, plus any annuals and other herbs that are vulnerable to attack by snails.

Because these crops are regularly harvested to ground level and die back to the ground in winter, there will not be the cover available for snails to survive in any great number as the birds will keep them cleaned up.

There may be other hungry-mouthed creatures with which you have to deal. Many marsupials are quite fond of new flavours, so in bush areas you might have to construct quite substantial fences to keep them away. Rabbits can be a problem for a few species, particularly Chamomile.

If you do have to fence, it is best to leave a grassed area at least 2 m wide, but preferably 3 m wide, between the edge of your garden and the fence. This area needs to be mowed to provide part of your buffer zone against seed falling or blowing into your garden, as fences are inevitably a haven for seed-forming weeds (see 'Buffer Zones' in the previous chapter).

Spacing

In setting out your herb plots, it is worth going to the trouble of measuring out the spacing of the rows accurately. This way you can avoid tapered plots and problems in later operations. It is a nuisance getting stuck halfway down the row with a rotary hoe because the plants were set 1 m apart at one end but only 0.9 m at the other.

For those who are not in the habit of carrying a tape measure into the garden, sufficiently accurate spacing can be achieved using an ancient measuring system that most of us still carry wherever we go. This is based on lengths of various parts of the body (see diagram next page).

Of course you will need to calibrate this for your own dimensions, and if more than one person is measuring out spacings, you will need to calibrate with each other for consistency.

There is an optimum spacing for each herb so the plants of that species have adequate room for development and you have room to get at the weeds, work in compost etc. If you can afford the space, leaving larger distances between the rows of some crops will make weed control easier during the establishment period. On the other hand, excessive empty space means lower densities, lower yields and more ground you have to cultivate. With some crops a denser

Provision of Shade

Some crops need partial or total shade. This can sometimes be found near tall vegetation, but often it is necessary to provide shade structures.

Shadecloth is quite strong and very durable: some of mine has now spent nearly 15 years in the sun. Shadecloth comes in a range of densities.

A permanent structure can be used that is tall enough to work inside, with shadecloth attached to the roof and walls.

Another option is a low frame over which the shadecloth is slung on a temporary basis, so it can be removed for weeding, harvesting etc., and also during the cooler part of the year when it is not needed.

The shadecloth should come down close to ground level on at least the northern and western sides. It may need some attention to ensure that irrigation is not obstructed. Water will penetrate more or less level shadecloth, but shadecloth that is sloping too much may conduct much of the water away so there will wetter and drier spots.

Advantages of the temporary shade system are that it can be moved easily for rotations and the cloth can be lifted off when conditions are cooler to allow plants to benefit from the extra light.

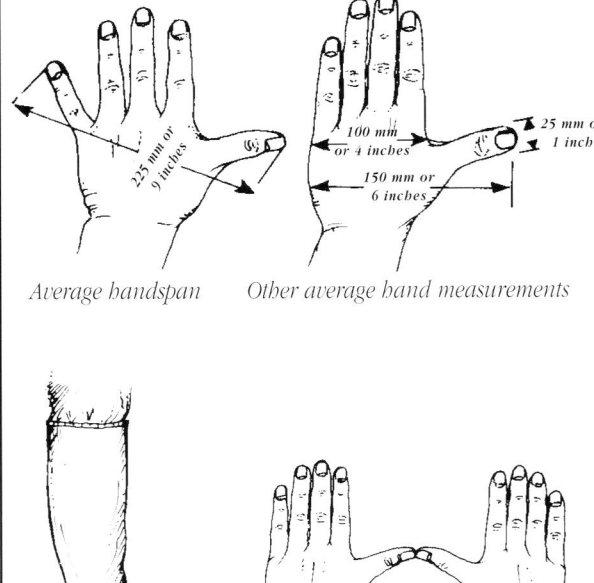

Average handspan   *Other average hand measurements*

Length of average boot or average double hand width (thumbs outstretched with tips touching, fingers together)

Average pace length (less accurate than the other approximations as an individual's pace length will vary somewhat)

An ancient but convenient measuring system that is always at hand.

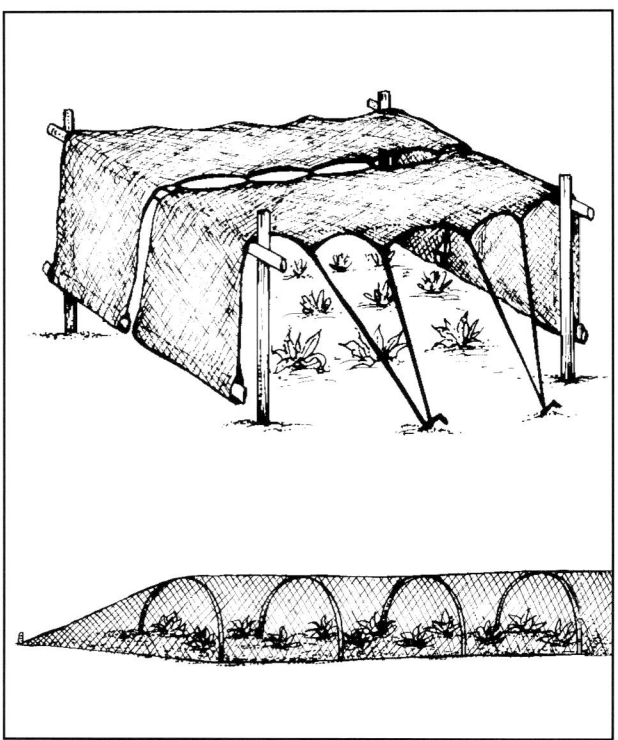

Temporary shadecloth structures for shade-loving plants. The shadecloth can be easily removed for harvesting and weeding, and in winter when it is not needed.

planting enables them to form a canopy that smothers out weeds and facilitates harvesting.

When starting out, there is a tendency to plant things too close together: the tiny plants you are setting out don't give you much of an idea how big they will grow. If possible, first take a look at some established plants that have reached their full development.

Vehicle Access

In planning layout, consideration should be given to the need to get close enough to all parts of the cropping area with a vehicle for spreading compost, carrying harvested herbs etc.

If a mowed perimeter is maintained, this can serve as vehicle access around the garden. There may need to be additional grassed strips through the crops or some crops may be of a nature that they can tolerate being driven over occasionally to allow access to the central part of the garden.

Small Herb Gardens

This section on layout has been written with particular reference to intermediate-scale herb growing operations cultivating a diversity of species. Of course, in smaller gardens, where only a few plants of each species are grown, most of these factors are still relevant, but the dimensions won't be the same. Some of the options mentioned here won't be open to you, but some other options that are appropriate to a smaller operation will be. For instance, it will be feasible to be more artistically expressive in the layout and construction of your garden and some problems, like weeds, can be coped with using more labour intensive methods.

Broad-acre Operations

On a larger broad-acre scale there won't be the range of options that are open to the intermediate- and small-scale herb operations, as everything will have to be organised for efficient large-scale management. Some of the inevitable insoluble problems of modern agriculture start to creep in and economic necessity becomes a major driving force in many decisions.

Nevertheless, many of the principles mentioned here still hold, and a skilful approach to broad-acre cropping can be ecologically and economically sustainable, as many Australian organic and bio-dynamic farmers have shown.

6

Herb Growth Types

There are a number of different ways of classifying herbs, and in most books one finds them arranged in the alphabetical order of their names, by their botanical families, by their medicinal actions, or by the part of the plant that is used.

From the point of view of growing herbs, I find it helps to understand the needs of each species if they are classified according to their growth types, that is, their habits of growth and manner of regeneration.

This way herbs fall reasonably well into six groups of plants:

Spreading Herbs

Expanding Clump Herbs

Perennial Crown Herbs

Woody Perennials

Trees and Shrubs

Annuals, Biennials and Short-lived Perennials.

Thus the various aspects of management, such as propagation, cultivation and harvesting, are generally quite similar for the plants within each group.

If the grower is familiar with each of these groups and their habits and needs, it becomes easier to work out the requirements of a new herb by recognising which of the groups it belongs to.

The nature of a plant's growth and regeneration is largely determined by where the vital force is centred in the plant. It is from this region that the plant's regeneration and most vigorous growth comes forth. It is important to know where this focus of regeneration is for each species of plant, because it determines the manner in which that species is propagated, harvested, and managed for optimum growth.

Some plants regenerate more strongly at ground level – from buds in their crowns, or from rhizomes – while others will only send out new growth from higher up the plant, and if cut off too low will die or recover only very slowly.

Of course a number of plants do not fit neatly into these categories. There are some that are exceptions to this pattern and some that seem to fall in between two groups in their characteristics. This is just a reminder that we are dealing with the natural disorder of things, with its subtle complexities that cannot be just slotted into a simplified view of the world.

Spreading Herbs

Species in the group of 'Spreading Herbs' include a number of economically significant herbs and present the herb grower with a significant share of problems. Because of their wandering habit, they are often difficult to contain and weed control can be quite a challenge. Typical plants are Peppermint, Spearmint, Coltsfoot, Yarrow, Horseradish, Passionflower and Horsetail.

Focus of Regeneration

This group has its most vigorous regeneration at or below ground level from runners that give rise to new growth extending some distance from the parent plant. Runners can reach from 200 mm for some species like Peppermint to 5 m or more for Horsetail and Passionflower.

Propagation

The usual method of propagation for this group is by runners, though some can also be grown from seed, and one or two by root division. The runners are normally broken into pieces, laid out end-to-end in a shallow furrow and then covered with soil.

Cultivation and Management

Good initial weed control prior to planting is essential for most of this group as their open spreading nature tends to allow plenty of space for weeds to establish themselves where they are difficult to control. Most of these herbs are dormant in winter, which allows weeds to over-run them. A black plastic mulch during winter dormancy works well in this situation.

Generally these crops are best established in rows, but will soon spread to form a solid block or meadow. After a while yields may decline as the matted growth becomes too tight. Often the herbs can then be rejuvenated by rotary hoeing strips through the patch, leaving narrow rows from which new growth can spread.

Special precautions need to be taken with many of this group to ensure they don't invade other herb plots or get into places where they are not wanted.

Harvesting

Where leaf or aerial parts are required, these are usually harvested as low as good leaf extends and the stubble then

A 'Spreading Herb' – Spearmint – regenerating after harvest and mowing. Note the open growth spreading out from the original row.

mowed to stimulate a more vigorous regeneration from ground level rather than allowing the remaining stems to send out the new growth. This top growth is usually poorer quality and less vigorous. Unless the stems are cut right back, the top growth tends to suppress the more desirable ground-level regeneration.

Expanding Clump Herbs

The 'Expanding Clump' group of herbs is rather similar to the spreading herbs group in their habit, but their expansion goes on at a slower rate, so the result is denser growth in the form of a clump that gets bigger every season. Consequently weed control and containment are not such a problem. Examples are Nettle, Melissa Balm, Dandelion, Comfrey and Oregano.

Focus of Regeneration

This group regenerates most vigorously at or below ground level but the plants usually need to be cut back hard to stimulate them to shoot from the base. Some form runners, while others can regenerate from any piece of root as well as from the crowns. A few regenerate only from crowns that are continually expanding and dividing.

Propagation

Those forming large enough runners can be planted in furrows and covered. For the others, small clumps, pieces of crown or root can be planted in rows.

Cultivation and Management

Good weed control is still essential, but this group is able to better resist invasion by weeds as the clumps do not provide as many opportunities for them to get a hold. The growth of these herbs in spring is generally stronger and more focused, so they are better able to compete with weeds that have grown up through the winter.

These crops are laid out in rows and will gradually fill up the spaces over a number of seasons, so enough room – the amount varies for different species – needs to be left to allow for this.

Herbs in this group that are harvested for the roots, like Dandelion and Comfrey, will usually regenerate from root pieces that are left behind after harvest. After a season or two this more or less results in a meadow-crop that may require hand-weeding between the clumps or, if the weeds are too prolific, can be cut back to row culture.

An 'Expanding Clump' herb – Nettle – in mid-regrowth. Note the solid clump and canopy.

Harvesting

Leaf and aerial parts are normally harvested low on the stems. Leaf harvests are usually followed with a close mowing to clean up the stubble and stimulate regrowth from the base.

Perennial Crown Herbs

Perennial crown herbs are characterised by the formation of a crown at ground level, which may divide and slowly expand, but the plant basically stays in the same place and only gradually enlarges in size. Some species have crowns that do not divide. Most of this group go dormant in winter, dying back to the crown. Examples are Wormwood, Ladies Mantle, Marshmallow, Vervain, Elecampane, Garlic, Alfalfa and Red Clover.

Focus of Regeneration

The crown is the focus of regeneration with these plants.

Propagation

Many of these herbs are propagated by dividing their crowns: sometimes these will break apart easily but with some species they need to be cut apart. The crown divisions can be planted directly in the ground from late autumn to early spring. Spacing varies from species to species depending on their size. Normally the roots alone do not regenerate, only the crown.

Non-dividing crown herbs like Alfalfa and Red Clover need to be grown from seed.

Cultivation and Management

These herbs are quite easy to keep in their place and most do not make aggressive demands on the space around them. Generally they are grown in rows, though some are direct-sown as a meadow-crop. Weed control (with the exception of Garlic) is not a big problem with them as they grow strongly from the single crown. In fact some are good plants to grow in locations where weeds are inclined to be a problem. The root crops in this group are amenable to rotation with other crops, as they do not regenerate from pieces of root that break off, only from the crown itself.

Harvesting

Most crown herbs regrow best when cut right back to the ground. Root crops, such as Elecampane and Marshmallow, tend to make a large clump of roots below the crown so digging proceeds more rapidly than for roots in the other groups, though it may take a bit of a heave to get them out of the ground.

Figwort is a typical 'Perennial Crown' species of the dividing type. Note the dense growth around a tight cluster of crowns.

Woody Perennials

Woody perennials are woody-stemmed bushes that usually retain their leaves year round and grow to a height of up to 2 m, though many never get as big as this. Most have multiple stems and many tend to expand sideways through natural layering. Examples are Thyme, Sage, Hyssop and Rosemary.

Focus of Regeneration

Woody perennials usually have their focus of regeneration among the leaves fairly high on the plant and regenerate only feebly or not at all from the lower portion.

Propagation

Most of these herbs can be grown from cuttings or by layering. Some can be divided by pulling apart clumps of stems and separating them into pieces, each with roots and leaves, provided they are not too old and woody.

Many can be grown from seed as well, but their seed tends to be small and seedlings slow to establish.

Cultivation and Management

After some attention to ensure they are free of weeds in their initial growth, woody perennials are generally not very demanding as far as weed control goes. They do tend to provide a haven for snails, so this can affect surrounding plants.

In planning their layout, make sure you are not obstructing access to other parts of your garden for machinery or loads of compost as they don't tolerate being run over. Some, like Thyme, need to be encouraged to form a solid bed to facilitate harvesting.

Harvesting

Harvesting must be done with some care to ensure that sufficient leaf is left for regeneration: generally you can take two-thirds of the leaf and leave one-third. Regular harvests (one to three cuts a year) help maintain the shape of these plants and prevent them from becoming too straggly. Usually the first year it is best to trim them only lightly.

Hyssop, a typical 'Woody Perennial' with bushy growth and woody stems.

Trees and Shrubs

The 'Trees and Shrubs' group of herbs is somewhat similar to the woody perennials, but these plants are taller and larger, they form trunks and branches, and they live longer. They grow from 2 to 60 m.

Focus of Regeneration

Trees vary as to their focus of regeneration, some are capable of coppicing from the base, others from higher on the tree, while yet others have their focus of regeneration only in the branches or even closer to the growing tips and so need to be harvested by careful pruning.

Propagation

This group of herbs is usually grown from seed or cuttings, depending on the species.

Cultivation and Management

The 'Trees and Shrubs' group is rather diverse in terms of management. Because of their dominating nature they usually need to be grown apart from smaller herbs and careful consideration needs to be given to choice of site, as most do not take kindly to being moved.

Once a tree or shrub is established, weeds are usually not a great problem, but some trees and shrubs may need help to overcome competition initially, after which periodic mowing or mulching around the base is generally adequate. If harvesting entails the death of the tree, provision needs to be made for regular planting to ensure a continuous supply.

Harvesting

It may be some years before you can expect much of a harvest from trees. Leaves are harvested by pruning or, in the case of large trees, by felling the whole tree. Harvesting bark usually requires the sacrifice of the tree trunk. A bit drastic perhaps, but some trees will coppice and provide a continual source of harvest. In some cases bark can be obtained from branches or even twigs.

Fruit and buds are picked by hand or combed. Flowers are picked by hand or, in some cases, are harvested with the leaves. The harvest of fruit and flowers may be limited to the lower part of taller trees, depending on how versatile you are when perched at the top of a ladder.

In a forest situation, harvesting can be part of a thinning operation in an ongoing sustainable management of the forest as a whole. Leaves or bark might be collected from trees being cut for timber.

Hawthorn: a small tree harvested for its leaves and flowers in spring, and its fruit in autumn.

Annuals, Biennials and Short-lived Perennials

The species in the annuals, biennials and short-lived perennials group are characterised by their short lifespan and their need to be grown from seed. Often they will self-sow once established.

Annuals normally flower, go to seed and die in their first season, though some will live on into their second year. Their life cycle ranges from 3 to 15 months, but they will normally flower in their first season. Examples are Basil, German Chamomile and Calendula.

Biennials normally form a crown of leaves and make an enlarged root in their first season and then in the second season they flower, form seed and die, though this varies a bit according the plant's development. Sometimes they may take more than two seasons to complete the cycle, or sometimes a cold spell can trigger the plant into flowering in the first season. Examples are Curly-leaf Parsley, Burdock and Caraway.

A few short-lived perennial herbs are included in this group, as they are best treated as annuals. While they may live beyond 2 years, they are often not very productive if kept going. They are also best grown from seed because their seedlings are more vigorous than plants grown from divisions. Examples are Broad-leaf Echinacea and Catnip.

Focus of Regeneration

Annuals

The regeneration capacity of annuals is usually not very strong and is mostly centred fairly high on the plant.

Biennials

Biennials generally regenerate from the crown during their first season and until they have advanced to the flowering and seeding stage.

Short-lived Perennials

Short-lived perennials generally come back from the lower stem or the crown.

Propagation

Annuals and Biennials

Annuals and biennials all need to be grown from seed: some can be sown direct, while others with small seed may need to be grown in trays or nursery beds and then transplanted. Keep your eye out for volunteers – seedlings that come up themselves around previous annual crops – as they can save you a lot of propagation time.

Short-lived Perennials

While propagation by crown division may be a possibility for these short-lived perennial species, plants grown from seed are generally more vigorous and live longer.

Cultivation and Management

Some attention will need to be paid to timing to ensure that these crops go in at the optimum time and that seedlings are available for sequential plantings, where needed. Weed control is, as always, important, but with most of this group, development of the crop is usually rapid and, except for autumn sown crops, there isn't so much problem with winter weeds.

Herbs of this group should be rotated with other crops. Some care needs to be taken to ensure that sequences are compatible. For example, crops like Chamomile that are inclined to re-seed heavily need to be followed by strong growing plants that are easy to control weeds in, like Elecampane, Marshmallow or Burdock.

Annuals

Annual crops are well worth considering for some early returns when getting established in dried herb production, as they can be in full production a couple of months or so after planting.

For the most part they are sown in spring and harvested during that season, though some can be autumn sown for an earlier harvest the following season. Sometimes annuals will carry on into a second season of production, but this is usually diminished.

Biennials

Biennial crops are usually sown in spring. Leaf can be harvested later that same season. Root harvest is at the end of the first season and seed is harvested the following season.

If the weed situation is good enough, some biennials can be sown in late autumn to give them a better start the following spring, but still will not flower until the spring after that. However if they go in too early in autumn, they may (depending on the species) behave as if that was their first season and then flower the first spring.

For continuous production, biennials will need to be sown every year.

Short-lived Perennials

Short-lived perennials normally reach their peak production in about 12 months.

Harvesting

Annuals

When annuals are harvested for leaf, cut them as they come into flower or a little before. It is important to leave enough leaf to enable good regrowth and to remove all the flowers to stimulate more leaf growth. Species vary a bit as to how they respond to this: some are rather insistent on flowering, so if you cut off the flowers you just get more flowers and few leaves.

Biennials

Leaves of biennials normally can be harvested down to the ground during their first season, but once the plant starts to go to flower there is no stopping it, and no amount of cutting back will make it revert to leafy growth.

The roots of biennial herbs need to be harvested at the end of their first season. If allowed to continue they will go to flower and the root won't have much value – all the plant's energy will have gone up into the flowering and seeding process.

Short-lived Perennials

The tops of short-lived perennials can be cut quite low in harvesting, as their regenerative capacity is quite good while they are young and vigorous. Roots are best harvested after one or two seasons, before the plant goes into decline.

An annual, Oats normally goes through its life cycle in 4–8 months.

A biennial, Angelica normally completes its life cycle in 2 years.

A short-lived perennial, Catnip tends to become leggy after two to three seasons and dies.

7

Compost

The compost heap is the focal point of any biological farming operation. It can be seen as the centre of a vortex where materials from all over the farm are gathered together to go through an intense process of decay and enlivenment to then be dispersed outwardly to various parts of the farm. It is the means by which growers can focus the biological processes of the farm onto specific areas they wish to raise to a higher state of activity.

In order to achieve and maintain a level of fertility that will produce good yields, most herbs will require regular dressings of compost. Some will require more than others, but there are very few which do not require it at some stage in their growth.

Those species that require a high level of fertility come from habitats where organic materials naturally accumulate, such as alluvial soils and other places where they are continually being brought from elsewhere and deposited. Where this biological activity is occurring, there is a continual release of nutrients, some of which are taken up by the plants and some of which are 'lost' through leaching or through release into the atmosphere. The level of activity is sustained, however, through the continual deposition of more organic matter brought in by floods, wind, gravity or animals.

In a cultivated crop situation, there are additional depleting factors. Part of the soil surface is kept bare and moist, with additional water added during the warm season. This speeds up the breakdown processes and makes more moisture and nutrients available to the crops, but it consumes organic matter and releases some nutrients in soluble and volatile forms that are dispersed elsewhere.

Further depletion occurs when crops are harvested. Three or four cuts of leaf a year carry away significant quantities of organic matter and nutrients. Together these losses are too much for the soil to sustain. Most herb crops are not soil building: their roots tend to be shallow, they do not return a surplus of organic matter to the soil, and only a few are able to fix nitrogen from the atmosphere.

Consequently these losses must be replenished from an outside source. Compost made from materials gathered from elsewhere is the best way to sustain the herb garden. Preferably all of this, or at least the bulk of it, should come from the farm itself, but this ideal may not always be attainable. Where the herb production is part of a larger farm system that includes pasture and animals and perhaps other crops, there will be materials available which can be used for compost, such as surplus hay, manure, grass clippings, straw etc. Of course, the farm needs to be organised so that this material is available for compost making and isn't needed for other purposes, such as feeding starving animals in order to keep them alive.

Desirable Characteristics of Compost

From the point of view of herb growing, finished compost should have the following characteristics:

Free of Weeds and Seeds: In order to avoid introducing a whole lot of extra weeds into your weed control area, finished compost needs to be free of weeds and seeds. Unless the starting materials are weed-free, it will be necessary for the compost to have attained sufficient heat to destroy all weeds and seeds.

Free of Unacceptable Levels of Toxic Residues: Some materials which might otherwise be good additions to the compost can contain pesticide residues, heavy metal residues and other nasties. Certification bodies have guidelines for the residue levels acceptable in materials for use as inputs in compost making. A good reason for basing compost making on materials from your own farm is that you know their history.

Able to Provide Nutrients and Contribute to Humus Levels in the Soil: Nutrients should not be in a freely soluble form, but instead should be available to the plant through interaction with the soil flora and colloidal humus. Try to get away from the idea of fertilising the plant, and focus on feeding the biological activity of the soil which the plant is a part of.

To make a contribution to this activity in a way that will benefit the growth of your crops, the materials in compost need to be transformed by a period of intense breakdown activity by bacteria and fungi, which generates a lot of heat. This needs to proceed to a stage when the heap has cooled down somewhat and a good proportion of the organic matter has been converted to colloidal humus, which is dark brown to black and like a thick slime.

Materials for Making Compost

The word compost comes from the Latin *compositum*, meaning composite, and refers to the use of a balanced mixture of materials to get a good composting activity going.

For a favourable outcome to this process, there needs to be a good balance of materials, and sufficient moisture and air to feed and fuel the biological activity.

Carbon:Nitrogen Ratio

In the materials used for the compost, there needs to be a good ratio of carbon to nitrogen: the C:N ratio. At one end of the scale are woody materials like sawdust, which are high in carbon and low in nitrogen, while at the other end are things such as straight chicken manure, which are high in nitrogen and low in carbon. There needs to be an overall balance between carbon and nitrogen in the soil and in the compost. The C:N ratio of finished compost should be about 15:1, but the starting materials for making the compost should average around 30:1. This provides sufficient carbon to fuel the composting process without losing nitrogen.

If the carbon level is too high in the finished product, nutrients will be tied up by the soil flora as they work on decomposing the carbonaceous substances and draw on nitrogen in the soil to do this. The result is a short-term lack of nitrogen available for plant growth.

If the nitrogen level in the compost is too high, then nitrogen in the form of ammonia will be lost and there will be too much soluble nitrogen added to the soil. This will be taken up by the plants and result in lush unbalanced growth. In some conditions, toxic levels of nitrate can build up in the leaves of plants. Meanwhile, in the soil, the high levels of nitrogen speed up the breakdown and loss of organic matter, with a consequent deterioration of the soil structure.

A balanced ratio of carbon to nitrogen in the compost will result in a healthy soil with good levels of nutrients available to the plants through biological activity and a good residue of organic matter that will contribute to the humus level of the soil with sustained benefits to fertility and structure.

Some Good Mixtures for Compost Making (in approximate percentages)

Hay 60%, Seaweed 30%, Sawdust 10%,

Foraged Grass (half green, half dead) 50–80%, Cow manure 20–50%

Foraged Grass (half green, half dead) 50–80%, Herb Marc (milled extracted herb) 20–50%

Foraged Grass (green) 80–90%, Sawdust 10–20%

While this is not intended to be an exhaustive list, it gives an indication of how nitrogenous and carbonaceous materials can be combined in the making of compost.

Value of Various Compost Ingredients

Bark

Bark is faster to break down than sawdust as it contains more nutrients, but some species contain high levels of tannins or resins that are antagonistic to plant growth.

Cutting or shredding bark to a manageable form may be a problem.

Bracken Fern

Bracken seems to be reluctant to break down in the compost heap and even in the soil afterwards, especially older material, though it eventually will. As an ingredient in the compost heap, it should be used in moderation.

There is some concern about the risk to health posed by Bracken's air-borne spores, so it should not be gathered when sporulating.

Chicken Manure

Because chickens are fed on concentrated high protein food without much roughage in it, their manure is high in nitrogen and low in carbon. It does contain a good balance of nutrients, so if it is composted with sufficient carbonaceous material, it can be a good contribution to the compost.

It is important that only a small proportion of chicken manure go into the compost, preferably not more than 30%, or the result will be too high in available nitrogen, and will foster rank unbalanced growth and even deplete soil organic matter.

It is also important when chicken manure is used in the compost that the heap is carefully turned so it all goes through a good heating process. This is because contamination with chicken manure can create microbial levels in herbs exceeding Therapeutic Goods Regulations limits.

Care must be taken when sourcing chicken manure. Very few chickens are raised organically so there is always the possibility of contamination. The feed they are raised on could contain pesticide residues that pass into the manure or there could be contamination from some treatment applied to the chickens or the shed: these are commonly sprayed out with a pesticide to control mites and lice. Approval should be obtained from your organic certifying body before using chicken manure from a non-organic source.

Cow Manure

Cow manure is highly regarded as an ingredient of compost, as the cow has a beneficial influence on the overall fertility of the farm. Cow manure usually has a nice balance of carbon to nitrogen, but it is a 'cold' manure in that it does not generate much heat in a heap.

Animal manure from non-organic sources could be contaminated, so approval should be sought from your certifying body.

Bear in mind that cow manure will very likely have a lot of weed seeds in it.

Fish Waste, Blood and Bone

Fish waste, and blood and bone contribute a lot of nitrogen and phosphorous, but virtually no carbon, so they need balancing with carbonaceous material.

They need to be used in thin layers to prevent putrefaction. Some fish might contain a bit of mercury as they are very high on the food chain, so some moderation is recommended in using them.

Grass

By grass I mean the grass itself and whatever weeds, clovers etc. are growing with it. The behaviour of grass in the compost heap is very much dependent on its stage of maturity and how much old dead material is standing with it.

Young leafy material is the most nitrogenous, and highest in moisture content. In a compost heap this young leaf tends to mat down and go slimy, sealing off the air and making the heap anaerobic. It needs to be mixed with coarser carbonaceous material. With sufficient aeration, grass will generate a lot of heat.

Grass in the flowering stage is more balanced, containing a good proportion of leaf, and enough fibre to keep it aerated.

Old grass which has seeded and turned brown tends to be rather dry and coarse, and higher in carbon, so it needs to be mixed with moister nitrogenous material. A stand of grass that has not been mown or grazed for several years will contain a lot of old dead material in it that does not heat or show much activity in the compost heap, so it needs to be mixed with generous quantities of more active material.

Hay

Hay is a good balanced ingredient once it has been adequately moistened (see section below on 'Making Compost from Bales'). Old hay, swept from the floor of a shed or lying around outside in bales, may not have much heating ability left in it.

Horse Manure

Even though horses eat the same sort of diet as cows, their manure behaves quite differently in the compost heap, generating a lot of heat. Some care needs to be taken with it as it can cause overheating, driving off nitrogen as volatile ammonia and drying out the heap to the point that activity stops. Consequently horse manure should be mixed with other ingredients to temper it somewhat.

Horse manure from non-organic sources could be contaminated, and approval should be sought from your certifying body.

Bear in mind also that it will very likely have a lot of weed seeds in it.

Sawdust

Some people may feel that sawdust is not a worthwhile ingredient for good compost, and on paper it would appear to be a negligible perhaps even detrimental contribution. It has an extremely high carbon to nitrogen ratio: in other words it has virtually no nutrients. It has a reputation for being acid and retarding growth, even killing things when used on its own. A pile of sawdust is an almost sterile environment, taking decades to break down.

However, a small proportion of sawdust in a compost heap has a sort of catalytic effect, quite out of proportion to its meagre contribution of nutrients. It stimulates activity by retaining moisture, facilitating water percolation into the heap, insulating and aerating (especially things like seaweed, cow manure and short grass clippings which are likely to go anaerobic). The lignin in sawdust is slow to break down, providing long-term organic matter.

The important thing is not to get too enthusiastic and use too much of it or you will cause nitrogen starvation and other problems. Sawdust that has had the tannins leached out of it by standing for a few years is best. Hardwood sawdust is preferable, and it is best to avoid sawdust from timbers which have high levels of natural preservatives in them, such as Huon Pine, Jarrah and Red Cedar, as these may adversely affect plant growth.

Composts containing sawdust tend to give a slower release of nutrients, particularly nitrogen, so you will need to take this into account when using it by giving higher doses to heavy feeders such as the Mints and Nettle or giving them a compost based on materials with a lower overall carbon to nitrogen ratio.

Seagrass

Seagrass is quite different to seaweed and is mostly found washed up along the shores of sheltered bays. It does not seem to contain a lot of nutrients. Being a fairly fibrous material, it takes a long time to break down. It would appear to have a high carbon to nitrogen ratio.

Seaweed

Seaweed is more commonly found along shores exposed to the open sea and there is quite a variety of types. They behave differently from seagrass in that they don't contain any cellulose, so as they break down they turn into a slimy sticky mass. This can put a damper on composting activity unless the seaweed is mixed in thin layers with other materials that can absorb it, preventing it from sealing off airflow through the heap.

From a handling point of view, the frond and feathery seaweeds are easier to cope with than bull kelp, which tends to come ashore in great masses.

Seaweeds contribute nitrogen and potassium to the compost, plus many valuable trace elements.

Sheep Manure

Sheep manure can sometimes be obtained in good quantities from under shearing sheds. As it will be very dry, it will need to be moistened or mixed with other material with sufficient excess moisture.

Sheep manure from non-organic sources could be contaminated, and approval should be sought from your certifying body. Bear in mind that it will very likely have a lot of weed seeds in it.

Soil

While a small proportion of soil is beneficial in the compost heap, it does not take much to really damp down the activity, so it should be added with caution.

Straw

Straw needs to be moistened in a similar manner to hay (see section below on 'Making Compost from Bales'). Straw is usually more towards the carbonaceous end of the spectrum.

Urine and Slurry

Urine and slurry are excellent when mixed and absorbed into carbonaceous material. Urine complements the dung of animals, as it tends to contain the potassium and the soluble nitrogen component of the excrement.

Wood Ash

Wood ash is a good source of potassium, calcium and other minerals. A very small proportion of wood ash is beneficial in the heap, but too much can lead to a loss of nitrogen as ammonia from nitrogenous material.

Comparison of carbon to nitrogen ratios of various compost materials (dry basis, approximate figures)

Material	C:N
Urine	0.8:1
Blood	3:1
Fish Waste	4:1
Abattoir Wastes	4:1
Chicken Manure	5:1
Night Soil	6–10:1
Purslane	8:1
Pig Manure	8:1
Sheep Manure	10:1
Grass Clippings (young)	12:1
Horse Manure	14–25:1
Cow Manure	14–25:1
Seaweed	19:1
Lucerne Hay	16–20:1
Red Clover Hay	20:1
Grass Hay	19–40:1
Straw, Oat	48:1
Bracken Fern	50:1
Straw, Wheat	128:1
Sawdust, Rotted	208:1
Sawdust, Raw	511:1

Gathering Compost Materials

If you are making enough compost to sustain 0.4 ha or so of herbs, the gathering of sufficient materials from around the farm and putting them together can be quite a laborious procedure when it is all done by hand. If it is possible to mechanise some of the operations of compost making, then the manufacture of adequate volumes becomes more easily achievable.

The Feed-out Shed

One system is to feed livestock in a shed with absorbent bedding, such as straw, old hay or sawdust, laid down on the floor. Periodically the bedding, manure and wasted hay are scraped up by hand or with a front-end loader and piled up for composting. Other available materials could be added to the heaps in layers or could be spread on the floor to be trampled in with the manure.

Making Compost from Bales

Often spoiled or surplus baled hay or straw is available for making into compost. If you have to bring compost materials in from outside the farm, this can be a practical way to do it. It is also a possibility for gathering on-farm materials if the machinery is available.

The secret to making good compost from dry bales is to get the material thoroughly moistened. The trouble is that the dry hay or straw tends to resist wetting and sheds water, so it is difficult to get good moisture penetration.

One method of overcoming this with small square bales is to have an old bathtub full of water with a hose running into it. As you pull the strings off each bale, immerse it in the bathtub before spreading it on the heap. The bale only has to stay under long enough for the water to penetrate. The 'cakes' of hay will act as a sponge, and will go onto the heap dripping water, which will then slowly soak into the material and achieve a good moisture level.

Some sort of variation on this system would have to be used for wetting hay in round bales. It may be possible to roll them out flat, side by side and run a sprinkler on them and then pick them up with a front-end loader, a rear fork or a buck rake.

The wetted hay should be built up in layers – alternating with other materials that complement it – to get good activity going.

Making Compost with a Forage Harvester

A forage harvester is one of the best systems for gathering and making compost from on farm materials, particularly if you have a lot of grass, weeds or other standing growth that needs to be chopped and gathered.

The most practical and affordable type of forage harvester for this mounts on the 3-point linkage of a tractor and has 1.2 m cut. The cutting action is a horizontally rotating shaft with rubber-mounted blades that chop and whoosh the grass up and out through a chute that is directed towards a cart attached behind the forager. When full, the cart is taken to the composting area and emptied by a dumping mechanism, depositing a load about the size and shape of a round hay bale.

Several brands of this type of forage harvester were in production at one stage: Lilly was the most common and parts are still obtainable for it. The availability of these machines seems to be regional, being more frequent in the traditional silage-making areas such as southern Victoria or coastal New South Wales. The cost of a forage harvester varies. The front part – that is, the cutter and chute – sells second hand for between $400 and $1500, depending on the region. The back part – the silage cart – is a bit harder to find, but would be worth about the same amount. Alternatively a tipping trailer can be fitted with high sides to do the same job.

The forage harvester does two things that an ordinary mower or slasher won't do: it chops the material into short pieces that are ideal for composting and handling, and it picks them up and loads them, which saves a lot of effort and time. It is excellent for converting grass, weeds, even blackberries and ferns, into compost material.

The piles of material can be dumped for later building into a heap, as described below.

On-field Mixing of Materials

Alternatively, if you have a good supply of other materials to go into the compost, these can be spread out onto the grass before it is foraged. Then the machine will pick them up with the grass, chop and mix them up, and blow them into the cart. The mixture can be then be dumped and formed into heaps with minimal further handling.

With this system you need to shut a paddock up until there is a good 300 mm or more of solid grass, then drive around the perimeter ahead of the forage harvester with a load of herb waste, manure, sawdust, seaweed or whatever, throwing a few good shovelfuls every few metres.

The trick is to throw the stuff so that it lies on the grass as much as possible rather than falling through to the ground. If it is lying on the grass, the forage harvester will pick most of it up, but if it is on the ground, much more will be left behind.

The proportions are not terribly critical and depend on what you have available. If your grass has a good proportion of stem and dead leaf in it, it won't need much other material to provide sufficient aeration. Lush green grass will be inclined to mat down and go anaerobic, forming silage rather than compost, so it needs more aerating material such as sawdust or herb waste mixed with it.

The forage harvester is then taken around the paddock, cutting the grass and picking up the mixture. If the loads of mixed material are dumped in two rows right next to each other, a front-end loader or a rear silage fork, can be used to pick up one row and pile it on top of the other so a long windrow about 1.8 m wide and 1.5 m high is built up.

The forage harvester mows, picks up and blows the grass and manure mixture into the cart behind.

When full, the load is dumped in the compost-making area.

The loads are dumped in two parallel rows, as close to each other as possible.

The rear silage fork is an inexpensive implement for moving volumes of compost around.

One row of dumped loads is stacked on top of the other to create a long continuous heap.

Finishing and final shaping of the heap is done by hand. Then it is covered with a layer of straw or other coarse material.

Building the Compost Heap

There are a number of different styles of building compost heaps, some with quite sophisticated aeration structures.

The general principle for all of these is to put the heap together with your grass, hay, straw, or whatever, in alternating layers with other materials, such as manure, seaweed, thin layers of sawdust, kitchen scraps etc. Of course if these materials were mixed together in the foraging operation described in the previous section, this layering will not be necessary.

For optimum development of activity, the compost heap should be built to around 1.5 m high and 1.8 m wide, in a windrow of whatever length suits you. As the activity gets going, the heap will soon begin to settle.

A covering of straw or other material free of weed seed will help greatly by retaining heat and moisture so the composting process reaches the outer layers of the heap.

At this point the biodynamic grower will add the compost preparations to the heap. These six preparations are made from parts of herbs: Yarrow flowers, Chamomile flowers, Nettle leaf, Oak bark, Dandelion flowers and Valerian flowers, which have each been through a special decay process and stored ready for use. These preparations can be obtained through a number of biodynamic organisations in Australia.

Factors for Successful Compost Making

Optimum Size

To reach and maintain a good level of activity in the compost heap, a critical mass needs to be achieved so there is a sufficient build-up of heat and sustained biological activity.

A small heap may heat for a while, but it will cool off more quickly. Its centre is smaller and there will be a greater proportion of exterior that does not heat. This means that the desirable conditions for composting and destruction of weed seeds exist in only a small part of the heap.

Too large a heap may result in too large a region of anaerobic conditions in its centre, while in some cases excessive heat can be generated.

The optimum dimensions for a compost heap (before being covered with an outer layer of coarse material) are around 1.8 m wide, 1.5 m high and a minimum of 1.8 m long. Length can be extended indefinitely, as a large heap is best built as a long windrow rather than as a huge round pile.

Balance of Ingredients

A good balance of ingredients is important to provide nutrients and fuel for the biological activity (this is covered extensively in the section on 'Making Compost', see above).

Adequate Moisture

An adequate but not excessive level of moisture is critical for good progress in composting.

Starting with a good mixture of materials that complement each other in their content and retention of moisture is basic for good composting.

A heap that is too dry initially will heat for a while until it drives off all its moisture. In extreme circumstances in a very large heap, this can even lead to spontaneous combustion, but usually it just cools off once it dries out.

On the other hand, excess moisture can also stop the composting activity and the heap will stay cold.

A handful of compost should feel quite wet, but not dripping. It should not be possible to squeeze water out of it.

Judging moisture content is part of the skill of composting. One has to estimate how much moisture is in the various materials that are going into the heap, how much they will heat and dry out, and then guess what the prevailing weather conditions will be. In dry weather a compost heap will lose moisture, while in wet conditions soaking rains can penetrate the upper part of the heap.

The shape of the heap can make a difference in that a pointed or rounded top will shed water, while a flat top will encourage it to soak in. Long-cut material will tend to shed water more than short pieces.

It is rare for a compost heap not to end up too dry or too wet in places. Even if moisture levels are good to start with, the heating process will dry out some sections, particularly in the centre, while other sections will get additional water from rain or condensation around the outside of the heap.

While it may be possible to spear holes in the compost heap and pour water into them to moisten it, usually it is more effective to add water while turning the heap.

Aeration

The biological activity in the compost heap is dependent on a supply of air to sustain it. If this is not available the activity will stop or go anaerobic. To ensure adequate aeration of the heap there needs to be enough porous material for air to penetrate to the centre. Stemmy grass, straw, sawdust etc. all help keep the structure open and complement other materials that are good ingredients but which can restrict air flow if they are in thick layers on their own.

Even so, as the heap settles, the centre can often lack air and is inclined to become anaerobic. Some growers use special techniques to foster better airflow into the centre of the heap. Sections of bales can be leant against each other in a row on the ground, to make a tunnel around which the heap is then built. Another option is to tie a few long poles together in a bundle on the ground and build the pile over them or else lay down a base of coarse woody stems that will hopefully encourage air circulation from underneath.

I usually don't worry about these special constructions, and just cope with a centre that goes a bit like silage. It sweetens up when it gets some air, so if the heap is turned, it can be placed nearer the outside.

Monitoring

It is important to keep an eye on the activity and progress of the compost heap. Heat is the best gauge of activity and should peak a few days after the heap has been put together.

If you check the centre of the heap at this time, it should be too hot for the hand to endure, around 65–70°C. Temperatures will then slowly decline, but when a heap is turned they should peak again shortly afterwards.

If this level of heat is not achieved, something is wrong. This is most likely to be insufficient or excess moisture, or too high a C:N ratio, but there are other possible causes.

A Word of Caution

Once when I was showing a friend my compost, I plunged my hand into the heap to see how warm it still was inside. He was rather horrified, exclaiming that he would never do such a thing because he had sometimes had snakes living in his compost heaps.

I replied that I wasn't too worried as I had never seen any in mine, but I was eating my words a few weeks later when I went to load compost from that heap.

Only inches from where I had plunged my hand in, I found a recently shed snake skin. Further down in the pile I discovered its former wearer: a healthy-sized red-bellied black snake. It was rather reluctant to move on its way, so it went for a trip in a bag to the nearby State Park.

Since then I have been a little more cautious about poking my hand into the compost.

Turning Compost

Turning compost provides the opportunity to examine the condition of the heap and correct anything that is inhibiting the processes of composting. It also stimulates the biological activity of the heap and accelerates the composting process.

Using Unturned Compost

If the original ingredients contained adequate moisture and enough rain fell on the heap (or you watered it) while it was composting and conditions were not too dry or too wet, it may not be absolutely necessary to turn the heap to get a useable compost.

The exterior and base will probably still contain some weed seeds, so these parts of the heap should only be used where they won't cause problems, such as around established trees or on crops that are managed with a weed under-storey. Most herbs will need compost from the middle of the heap, where the heat has killed all weeds and seeds, and the material is more broken down.

However, if the heap can be turned, much better compost will be made as it will be more evenly broken down and virtually all weed seeds will be eliminated.

Timing

Usually the optimum time to turn the compost heap is after about 6 weeks, when its activity has slowed down and its temperature has dropped. This allows the initial composting and heating process to progress about as far as it is going to go. Turning the heap will then stimulate a second wave of activity, and by the time this has settled down, the compost should be fairly mature.

If finished compost is needed urgently, it can be turned much sooner and at frequent intervals to keep maximum activity going. Turning the heap every few days will speed up the process so the compost can be ready in a few weeks. Normally it takes around 3 months if the heap is turned once.

Technique

During turning close attention should be paid to moisture content. Sometimes a desirable level of moisture can be obtained by mixing wetter sections with drier sections. But if the heap is too dry overall, it will be necessary to add water.

This is best achieved by enlisting someone to hold a hose capable of putting out a good jet of water and having them wet down all the dry material as it becomes exposed in the turning process.

If the heap is too wet, some suitable dry material may need to be added. If the weather is good, it may be possible to spread the heap out to dry it somewhat.

Aeration is another important aspect of turning, so lifting and dropping forkfuls of compost in the rebuilding process will produce an open structure in the new heap.

From the point of view of eliminating weeds, it is important to ensure that all the material which has not yet been through a good heating process goes into the middle of the heap and that which has already been through the middle goes onto the outside or the base. Essentially the aim is to turn the heap inside out.

This turning operation may seem a bit complex, but is worth the extra trouble. It ensures that all the material goes through the hot part of the heap, producing a better more evenly broken down compost that is free of weeds and seeds: so that when you spread the compost, you won't also be sowing a prolific crop of weeds and undoing all your efforts in getting your crops clean.

Spreading Compost

Rates of Application

The amount of compost required annually by different herbs varies from virtually none to about 75 t/ha. Because we are generally talking about plots of herb of much less than a hectare, it is easier if these rates are converted into kilograms per square metre. This conversion is easily done by dividing the t/ha rate by 10 to arrive at a kg/m² rate, that is, 10 t/ha = 1 kg/m², 50 t/ha = 5 kg/m².

Heavy feeders like Nettle and Mint will need compost at a higher rate, 5–7.5 kg/m² per annum while the majority of herbs will require around 2–4 kg/m² per annum. A few, like German Chamomile, will normally need little or no compost, provided the soil is at least moderately fertile or has had compost on a previous crop. Of course, these figures will vary significantly according to the nature of your soil, and how long it has been getting good applications of compost.

Approximate rates for individual herbs are given in the second part of this book.

Estimating How Much Compost To Apply

Compost does vary greatly in volume and weight, depending on its stage of decomposition and its moisture content, so you will have to allow for this. Coarser, less mature compost will be much bulkier and needs to be put on much more liberally for the same effect.

You may not be able weigh the compost, but you can estimate how much a load weighs and work out what area it needs to cover.

Example – A 1 tonne ute or a well-loaded 1.8 x 1.2 m trailer will carry about 750 kg of compost. If this is to be spread at 5 kg/m², the load will cover 150 m², or a strip 30 m long and 5 m wide.

If this amounts to roughly a cubic metre of compost and it is spread evenly over 150 m² of meadow-crop, then it will be an average of about 6.6 mm thick.

With a row-crop being given 5 kg/m² of compost, if this is spread on the bare inter-row space which is one-third of the total area occupied by the crop, then it will need to be 3 times as thick (an average of about 20 mm).

Mature v Immature Compost

I usually like to put the more mature compost on to herbs that are heavy feeders, as it is faster in its action. Where large amounts need to go on, denser, crumblier mature compost is easier to work with. The less mature compost is more appropriate for lighter feeders that don't need such fertile conditions for optimum growth.

Personally, I don't mind using a relatively immature compost and letting it complete its decomposition in the soil. There are those who would disagree with this, but others feel it involves the mineral aspect of the soil in the decay process and some nutrients are more likely to be taken up by the soil as they are released. You do have to put on larger quantities, though, to get the same response, because it is bulkier and less concentrated.

New Ground

In reasonably fertile soils, when a long established pasture is initially ploughed up and planted, there is a burst of activity as the old turf, manure, grass roots etc. break down. Often in this situation not much additional compost is required the first season. This can lull the grower into complacency. Such a burst of growth will only last for about a season, after which productivity and soil structure will deteriorate unless they are sustained with increased quantities of compost.

Poorer Soils

On the other hand, in poorer soils the normal application of compost may seem to bring little response. Sometimes it takes a heavy dressing to build up the soil to an adequate level of activity, which can then be sustained.

Forest soils sometimes contain a lot of dark organic matter that looks wonderful but does nothing. This is because they are high in carbon and low in nitrogen and so need a lot of compost with a good nitrogen content to activate them.

Timing of Compost Application

Most of the perennial herbs will need a dressing of compost each year. Generally the best time to get this on is in early spring, normally September in Victoria and Tasmania, just

Compost

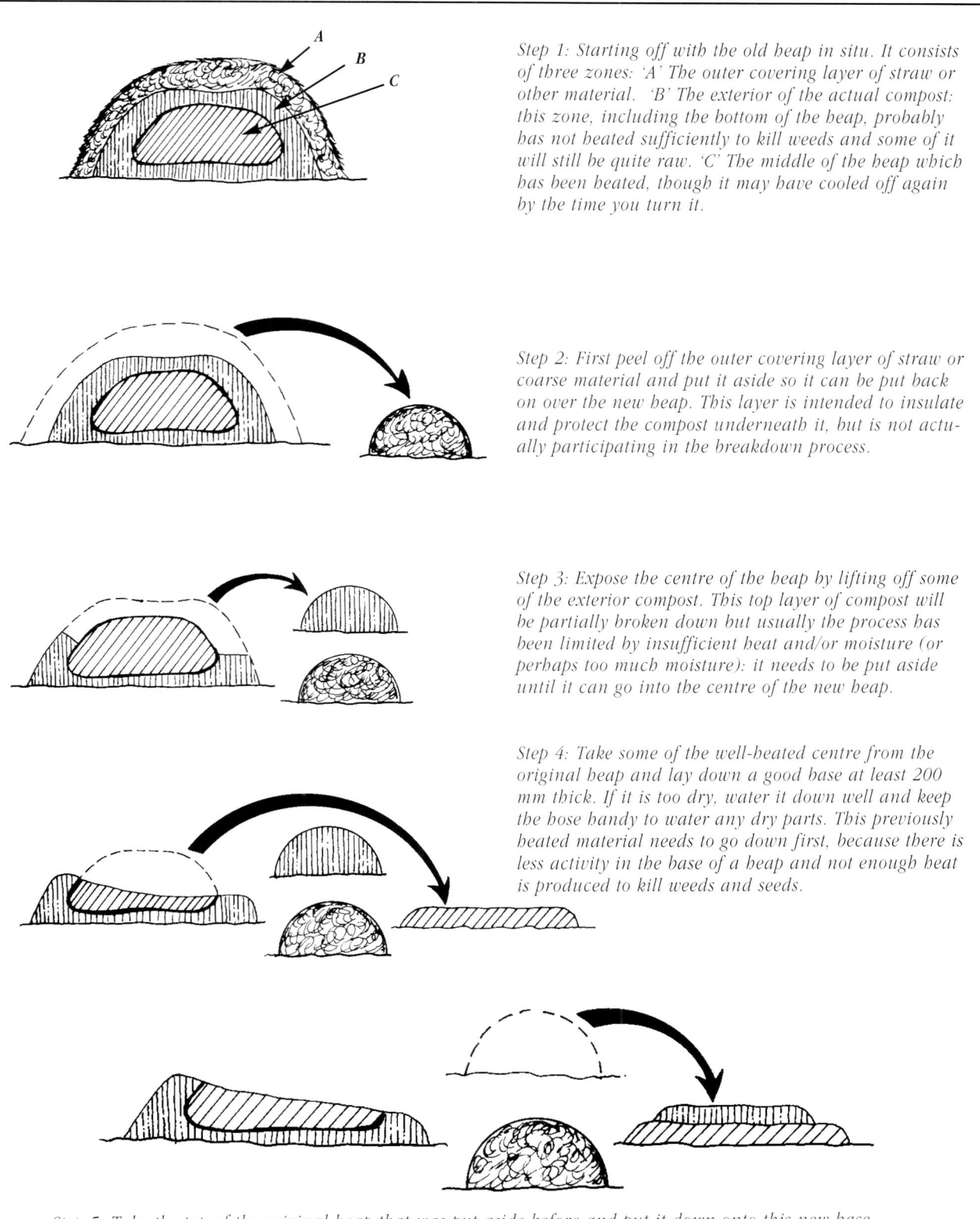

Step 1: Starting off with the old heap in situ. It consists of three zones: 'A' The outer covering layer of straw or other material. 'B' The exterior of the actual compost: this zone, including the bottom of the heap, probably has not heated sufficiently to kill weeds and some of it will still be quite raw. 'C' The middle of the heap which has been heated, though it may have cooled off again by the time you turn it.

Step 2: First peel off the outer covering layer of straw or coarse material and put it aside so it can be put back on over the new heap. This layer is intended to insulate and protect the compost underneath it, but is not actually participating in the breakdown process.

Step 3: Expose the centre of the heap by lifting off some of the exterior compost. This top layer of compost will be partially broken down but usually the process has been limited by insufficient heat and/or moisture (or perhaps too much moisture): it needs to be put aside until it can go into the centre of the new heap.

Step 4: Take some of the well-heated centre from the original heap and lay down a good base at least 200 mm thick. If it is too dry, water it down well and keep the hose handy to water any dry parts. This previously heated material needs to go down first, because there is less activity in the base of a heap and not enough heat is produced to kill weeds and seeds.

Step 5: Take the top of the original heap that was put aside before and put it down onto this new base, leaving an uncovered margin of about 150 mm around the outside edge of the new base.

Turning the compost heap.

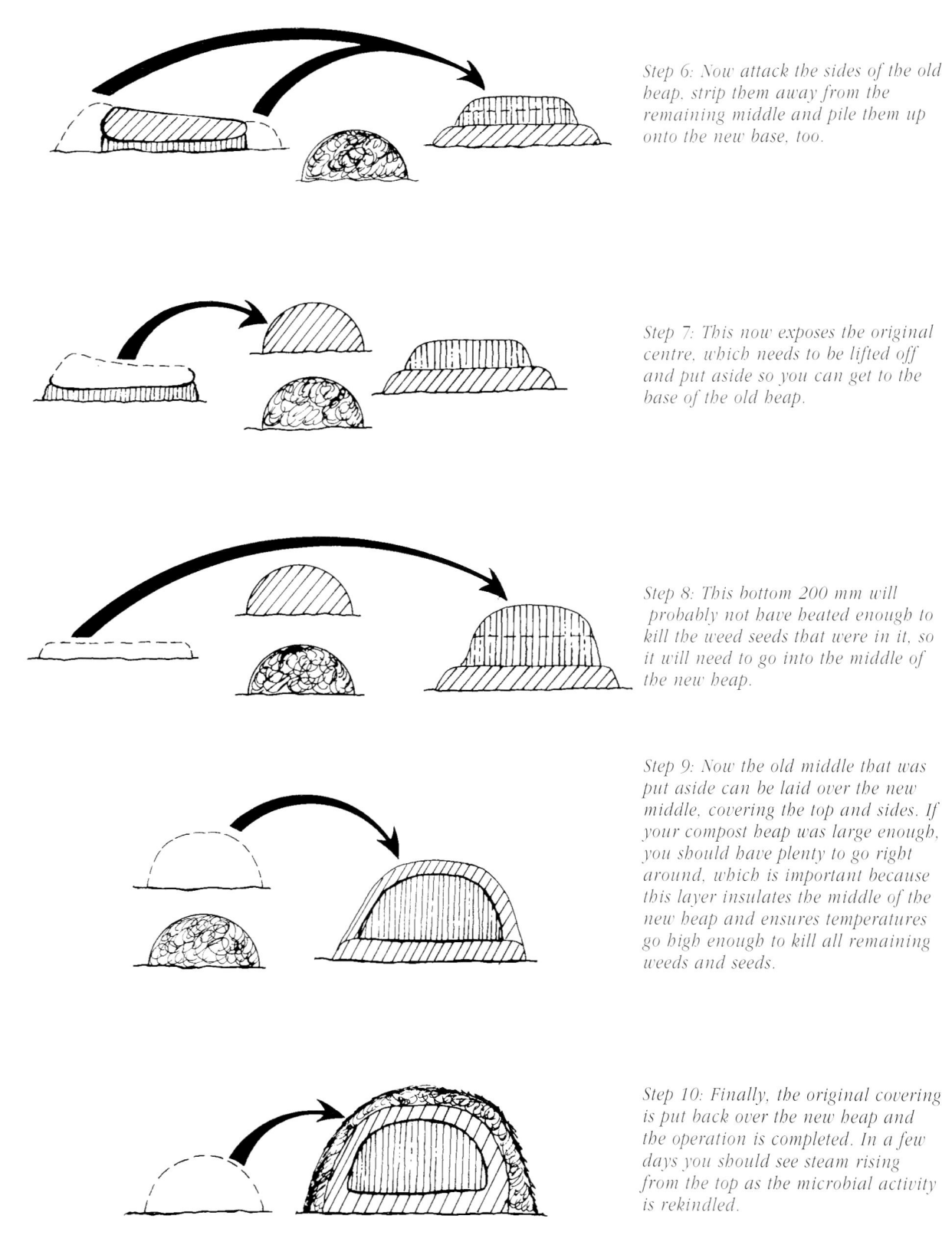

Step 6: Now attack the sides of the old heap, strip them away from the remaining middle and pile them up onto the new base, too.

Step 7: This now exposes the original centre, which needs to be lifted off and put aside so you can get to the base of the old heap.

Step 8: This bottom 200 mm will probably not have heated enough to kill the weed seeds that were in it, so it will need to go into the middle of the new heap.

Step 9: Now the old middle that was put aside can be laid over the new middle, covering the top and sides. If your compost heap was large enough, you should have plenty to go right around, which is important because this layer insulates the middle of the new heap and ensures temperatures go high enough to kill all remaining weeds and seeds.

Step 10: Finally, the original covering is put back over the new heap and the operation is completed. In a few days you should see steam rising from the top as the microbial activity is rekindled.

as growth is commencing, and the soil has dried out sufficiently to allow the compost to be worked in. This ensures the plants have a good level of nutrients available for optimum growth through the season. Heavy feeders may sometimes need one or more additional application(s) later but generally a liberal dressing of compost in spring will sustain steady growth right through the season.

The compost needs to go on early, before emerging herbs have reached any size, so the compost is not lodging on their leaves as this can result in contaminated harvests.

Tools for Handling Compost

Once again, efficiency in a labour-intensive operation depends on having suitable tools. The coarser materials are best handled with light pronged forks: a three-pronged hayfork for the lighter hay and straw, and a four- or five-pronged manure fork for heavier stuff and coarser finished compost. I prefer a five-pronged short-handled manure fork for turning and spreading, but these are difficult to find in this country: mine came back with me from Canada. A four-pronged long-handled fork is easier to find here and will do for most jobs.

Hand tools useful for handling compost: four- and five-pronged forks and a large aluminium shovel.

For handling fine materials like sawdust and for spreading crumbly compost, a large aluminium shovel is handy. The size and lightness of these shovels makes handling bulky materials like compost much easier, but you need to avoid excessive loading or jarring as the aluminium is not as strong as steel.

Technique

Where possible, the compost should be placed in the space between the rows and rotary-hoed or otherwise worked in. This mixes it into the soil where biological activity can make its nutrients available to the plants. Left on the surface it is inclined to dry out, and volatile constituents are lost into the air, while further breakdown and release of nutrients is slower because they are bound up in the colloidal humus. For meadow-crops there is not much choice but to leave it on the surface.

Often there is quite a volume of compost to work into the soil between the rows. I find it is best to work the ground up first with a couple of passes with the rotary hoe and then spread the compost onto this loosened soil. A couple more passes will then work it in quite well. If you try to rotary-hoe compost into firm soil, the blades tend to churn around in the compost and don't get down into the soil much, but if the soil is loose to start with, the compost will work in better.

Initial Plantings

It is to be hoped that you will have been making compost so it is ready to go on at the same time as you plant your herbs in the ground. Putting the compost on and working it in before planting is probably easier, but I generally prefer to put it on after the herbs go in for several reasons.

If the compost is a bit coarse, as mine often is, this can interfere with the contact the small plants need to make with moist soil about 50 mm below the surface. The coarseness can cause excessive aeration and drying out around their root systems. If the compost is a bit raw, it may burn the young plants' roots where it contacts them.

Similarly, coarse compost can interfere with the germination of seed and early growth of seedlings. It makes it difficult to get a good seed-bed.

I often don't know how far my propagation material will go, so the final layout of plots is unclear until after they are planted. By waiting before applying the compost, I can ensure that each species gets its optimum rate of application. Species vary enormously in their requirements and for a few of them an excessive dose can be as harmful as an inadequate one.

If these factors are not a problem in your situation, then putting the compost on before planting is certainly more efficient. Throwing compost out over clear ground from a vehicle is much quicker than walking up and down the rows with a shovelful at a time, then carefully pouring it so it lies beside and not on top of the plants.

Generally I put the compost on as soon after planting as is practical. Where a crop is planted in autumn and will go dormant over winter, I tend to give it a light dressing in autumn to help it get established, and then a full dressing at the usual time in spring.

Problem Crops

Meadow-Crops

Getting compost effectively onto heavy feeders like Mint and Nettle that have become a meadow-crop can present a problem, as it is not possible to work the compost into the soil. Spreading it on top of the newly emerging shoots in early spring is not entirely satisfactory. If the compost is at all coarse, lighter pieces of it will simply sit on top of the shoots and be lifted up by them as they grow, so a good amount of the compost can still be up there when you make the first harvest in late spring, and inevitably some will be gathered with the herb.

The best thing is to reserve your crumbliest compost for these crops, making sure you get it on very early. If it is too lumpy it may be necessary to rub it through a 40 mm mesh (rabbit wire) screen to break it up a bit.

It should go on while the plants are still very small: around 50 mm and not more than 100 mm tall. When you spread the compost on, rake it over so that it works in around the shoots. They can then push up through and around the lumps. Immediately follow spreading with a good watering to wash any compost off the plants and into the soil.

If frequent light waterings are possible over the next few weeks, this will keep weighting and washing the compost down so the crop will grow up past it. Keep a check on it for several weeks to make sure the pieces of compost are staying down and not getting lifted up as the shoots grow. You may need to intervene with a broom or a rake if lumps of compost are still riding on top of the leaves.

A special compost using crumbly materials can be made specifically for meadow-crops like Peppermint and Nettle.

Growing these heavy feeders as a meadow-crop may work for 2 or 3 years, but eventually production will decline as growth becomes very tight and stunted. At this point the patch will need to be rejuvenated by tearing strips out of it with the rotary hoe, leaving narrow rows about 900 mm apart. A good dressing of compost can be spread between these rows and worked in. The resulting boom in growth can seem little short of miraculous.

Woody Perennials

The other herbs that present problems when applying compost are the small woody perennials like Thyme and Lemon Thyme. To facilitate harvesting, these are maintained as a solid canopy with no space between the plants to put compost on. As the tops never die down, compost can't be spread over the crop.

I have not yet found a satisfactory solution to this one: to date I have just let the yields gradually run down. Fortunately the woody perennials aren't heavy feeders and if the soil is reasonably fertile to start with, adequate yields will be sustained for several years before they decline. If means of getting some compost onto them could be devised, production might be sustained much longer.

It may be possible to deliver a compost slurry via a pipe or funnel poked into the bushes below the canopy level.

Sustainable Compost Production

Reliable sources are needed for your compost materials. Where these are coming from the farm, the production of them needs to be sustainable.

On a larger farm where livestock are grazing, it may be possible to harvest grass and manure from areas where fertility is concentrated, such as around stockyards, dairies, and places where stock camp at night.

Other growers will have to look at setting aside an area specifically for growing materials for compost production. If this is to be sustainable, some attention will need to be paid to following practices that will ensure this.

Planting deep-rooted soil-building plants such as Phalaris, Cocksfoot and Lucerne will draw on reserves of moisture and minerals in the deeper layers of the soil. Legumes will bring nitrogen into the cycle. Many weeds have a significant contribution to make, too. If you are sowing down an area for production of compost materials, it might be worth including a number of deep-rooted weeds and herbs in the mixture to help get a good balance of plants for a healthy vibrant compost.

Whether it is possible to sustain production on an area that is regularly being harvested and taken away is a point of some argument among organic, bio-dynamic and conventional farmers.

At one end the 'nutrients in/nutrients out' school says that sooner or later the nutrients in the soil will be depleted and production will decline and eventually cease. On the other hand, biodynamic advocates say that plants don't necessarily follow chemical and physical laws, and in a biodynamic system adequate minerals will continue to be available to the plants on a long-term basis, even though these may appear to be lower in soil analyses.

At this stage there is not enough long-term evidence to support or refute either view, though certainly biodynamic farmers have been shown to be able to sustain production despite falling levels of measurable phosphorus in their soil.

Nevertheless, most people agree that if soil fertility is low initially, it will need to be built up with inputs from outside sources until it reaches a productive level. Personally I feel that if you have to bring in minerals and manures, it is best to apply them to a grass-based crop that will use them to create a surplus of organic matter, which can then be used as a basis for compost to sustain herb production.

Whatever system is followed, a watchful eye needs to be kept on the productivity of the area that is being harvested for compost making and appropriate steps need to be taken to ensure it is sustained.

Alternatives To Making Compost

Commercial Organic Fertilisers

For those who are unable to consider making compost or who can't make enough, it is possible to purchase organic compost and fertilisers that meet certification requirements, though some certifying bodies require that inputs from outside sources be phased out after a conversion period.

Most of the commercially available organic fertilisers are fairly concentrated, containing a high level of nutrients and relatively little carbon, so their direct contribution to humus levels in the soil is limited. They may be quite suitable on meadow-crops that don't require cultivation and produce a net gain of organic matter.

However, for cultivated row crops where soil humus levels need to be sustained by inputs from elsewhere, these concentrated organic fertilisers may actually result in a decline in organic matter as the activity stimulated by the added nitrogen consumes it.

It may be possible to purchase something more balanced, such as spent mushroom compost (but beware of residues of pesticides used to control mushroom flies), or compost made from bark and fish waste, that contains higher levels of carbon and makes a long-term contribution to soil humus levels. If these materials don't already have certification as approved inputs, permission should be obtained from your certifying body before using them.

Animal Manures

Although traditionally used directly on the land, animal manures can cause problems with unbalanced growth. Organic certification guidelines require that these only be used after they have been through a heat composting process. This should involve a mixture of other materials to create a more balanced compost, rather than just being left to heat in a heap of pure manure.

The other problem with non-heat composted animal manures is that they can be full of weed seeds, even when they have been sitting around for a long time.

Liquid Manures

While liquid manures may have a role in an organic system as a supplement for use on soil, foliar feeding is restricted by certification guidelines as it is against natural principles to feed plants through their leaves.

I am inclined to question liquid manures as they cannot contribute anything in the way of organic matter to the soil, and growers tend to use them to provide freely soluble nutrients to prop up plants growing in an impoverished soil.

Green Manures

The use of green manures in herb growing is generally limited to those crops that are managed in annual rotations. Most short-term green manure crops do not result in a significant contribution to soil organic matter levels, though they can help reduce losses between crops when the soil would otherwise be bare. Leguminous green manures can bring nitrogen into the system.

To make a substantial contribution to the soil, the grower needs to consider long-term green manure crops, such as deep rooting grasses, together with legumes such as Red Clover and Lucerne, which need to occupy the ground for a period of 2 or more years.

From the point of view of weed control, green manures can lead to problems because they tend to allow development of weeds as an under-storey. Very few are capable of totally smothering weeds. It may be possible to reduce this somewhat by sowing the crop late enough in autumn so weeds can't seed before winter and then work the crop in early enough in spring before any weeds have a chance to develop to the seeding stage, but this is going to limit the development of the green manure crop itself, and it won't provide its full benefit.

Summer green manure crops will be particularly prone to weed proliferation.

Mulches

With some crops, mulches that contribute nutrients and organic matter to the soil can be used to sustain fertility.

However, as noted in chapter 4, 'Weed Management and Control', mulches have their limitations.

8
Irrigation

Water is essential for plant growth. The majority of herbs come from moist habitats, many are shallow rooted or have foliage that transpires a lot of water. Consequently they need a constant supply of moisture for good growth. Interruptions in the availability of water can result in stunted development or even dying back, affecting both yields and quality.

The erratic nature of our climate in Australia means that even in high rainfall areas at least some irrigation will be necessary to avoid growth setbacks due to lack of moisture. In most regions, there is only a limited number of species of herbs which could viably grow commercially without irrigation.

An irrigation system is rather expensive to set up, but you can't afford not to have it. There is a great variety of options available and your requirements need to be considered carefully to ensure that your money is well spent. An inadequate irrigation system is money wasted if it lets you down when you need it most.

WATER SUPPLY

Your irrigation water may come from your own dam, spring, bore, permanent creek or river, or from an irrigation or town water scheme.

The things to consider with a water supply are: Can sufficient volume be drawn from it on a daily and a seasonal basis? How reliable is the supply? And how good is the water?

Volume and Flow

During summer most herbs require 25–50 mm of water a week, rainfall and irrigation together. Is your spring, creek or bore able to supply this much? Can your pump deliver this? Does your dam hold enough?

Calculate the area you are intending to irrigate, multiply this by the average amount (in millimetres) of water you estimate you will need to put on each week, multiply by the number of weeks a year you expect to irrigate, and you will have an estimate of the annual volume of water you will need.

It is best to err on the safe side, and also to allow for 300–600 mm of evaporation from the surface of a dam over summer.

Example: A dam with an average width of 20 m, average length of 50 m and average depth of 2 m will hold 2000 m^3 (2 ML) of water.

In a 750 mm/annum rainfall area in Tasmania, a grower might expect to irrigate for 12 weeks of the year, at an average of 25 mm/week for a total of 300 mm over the season. After allowing for 300 mm evaporation, the dam above will contain 1700 m^3 (1.7 ML). This would irrigate an area of approximately 5660 m^2 (0.56 ha) for one season.

In a 750 mm/annum rainfall area in northern Victoria, a grower might expect to irrigate for 20 weeks at an average of 25 mm/week for a total of 500 mm over the season. After allowing for evaporation of 600 mm over the season, the dam above will contain 1400 m^3. This would irrigate around 2800 m^2 (0.28 ha) for one season.

Dam volumes are expressed these days in megalitres or millions of litres.

 1 million gallons = 4.546 ML
 1 megalitre (ML) = 1000 m^3

Reliability

One critical question is will there be sufficient water to supply you through the worst foreseeable drought? Will your spring or creek continue to flow, or will your dam hold enough to supply you through a second season if there is no run-off to refill it during a drought?

For added insurance, dam capacity should be sufficient for two seasons irrigation if there is any possibility that some years there won't be enough run-off to fill it.

In planning your storage capacity, make sure you have enough to supply your crops through the longest foreseeable drought.

Quality

Quality is another aspect that is difficult to assess, as water quality will vary from one time to another. Salinity will increase during dry conditions. Floods can stir up pesticide residues and other unsavoury substances and neighbours can cause sporadic contamination of your water supply with pesticide use and misuse, septic tank outflows, dairy effluent etc. Livestock often go down in water: a weak animal tends to gravitate to the lowest point and can end up dying and decomposing in your creek or dam.

Irrigation schemes and town water have their own problems: they are often contaminated with the results of human activity and also substances like chlorine and herbicides that the authorities, in their wisdom, see fit to introduce to the water.

Those drawing from the Murray–Darling irrigation schemes take in from a large catchment area and the water is the diluted effluent of all the industries and other human activities upstream. Not only that, herbicide is periodically added to the water to kill weed growth that might choke the channels. Town water almost inevitably has chlorine, fluoride and sometimes clarifying agents such as alum in it.

So unless you have access to water that flows from a pristine source, you are going to have to cope with the possibility of less than perfect water. Irrigation water does not quite have to be of drinking quality, but it should be reasonably clean and certainly should not contain anything that might leave measurable residues of anything toxic or infectious. Your certifying body should be able to give you guidelines to follow for water for use on leafy vegetables, and herbs would fall into the same category.

Salinity

Salinity is becoming an increasing problem in inland areas. Some of the causes are related to irrigation, but there are other causes as well. The effects of salinity can be reduced or increased by different systems of irrigation management, depending on whether they flush the salt through the soil, or allow it to build up near the surface.

The tolerance of various herbs to salinity and saline irrigation water is an area that needs further research. Until more information is available, to be on the safe side, it is best to follow the guidelines for other crops that are susceptible to salinity.

There has been one instance of a herb grower using saline water for irrigation that resulted in the dried herbs having a peculiar salty flavour.

Salinity is mostly a problem of drier regions, but it can also occur in some localities with higher rainfall. If there is any doubt, you should have your water supply tested.

TYPES OF IRRIGATION SYSTEMS

Water can be applied in a number of different ways. It can be flowed on as a flood or in furrows, or it can be brought through pipes and delivered as a trickle or by sprinklers.

For most growers it will be a choice between these latter two systems.

Flood Irrigation

I have no personal experience with flood irrigation of herbs, except accidentally when a control timer failed to turn off some sprinklers and the area around and below the them was flooded. My impression is that flood irrigation for perennial herbs would have its limitations.

Soils would need to be light and well drained so that the flooding was not likely to damage soil structure or create waterlogging. When soils are flooded, the saturation can cause the soil structure to collapse. Any impediment to free drainage in the subsoil will result in anaerobic conditions, which will damage the root systems of many herbs.

The other difficulty with flood irrigation is that slight variations in ground level can result in an uneven distribution of water. With plants that are shallow rooted, as many herbs are, this will result in loss of production from under-watered areas and possibly problems with timing of harvest as the development of the crop could be uneven.

Row cultivation leaves furrows and lines on the surface of the soil. These and the rows themselves will tend to divert the water from its intended course, so flood irrigation would generally only be considered in crops that are managed as a meadow-crop.

Furrow Irrigation

Flowing water onto the crop along furrows between the rows or beds is another possibility. The soil needs to be of a structure that gives good lateral movement of water. If the structure is too open, the water flow is more downward into the soil profile, rather than horizontally towards the plants. Again shallow rooted plants are liable to receive inadequate watering. Furrow irrigation needs careful management as to the layout and height of the beds or rows and careful attention as to the flow of water to ensure an adequate but not excessive quantity reaches all the plants.

I must say there is something very fulfilling about flowing water down furrows to moisten the soil around the crops.

This method of irrigation has an ancient tradition that dates back to the birth of civilisation, but for most situations it is not very practical.

Sprinklers

There are a number of different types of sprinklers, but the general principle is the same: the water is projected through the air as droplets and lands on the surface of the soil.

Advantages

Availability to Shallow Root Systems: Because the water is applied to the surface of the soil, this moistening of the upper layer ensures that all those herbs with shallow roots have water available at a level they can reach.

Biological Activity: Keeping the upper layer of soil moist fosters the biological activity in this region, ensuring continued availability of nutrients to the plants.

Moderation of Soil Temperatures: When there is moisture on or near the surface, its evaporation has a cooling effect in hot weather, and plants that are being grown in climates hotter than those they originated from will feel less stress. Their roots will be cooler and the leaves will suffer less from the baking effect of heat rising from a hot soil surface. Many of them can stand being baked from above, as long as they are not being baked from below as well.

Wetting of Leaves: The wetting of the leaves by the droplets from the sprinklers has a beneficial effect on some species. It also washes off some of the dust that tends to accumulate during dry weather.

Disadvantages

Higher Water Consumption: Because a proportion of the water applied to the surface is lost through evaporation, sprinkler irrigation requires more water than trickle irrigation.

Higher Pressure: Sprinklers need high pressure to function properly.

Wind: This can be a problem in exposed sites, especially with fine droplets, as the wind can blow the water away from where it is intended to fall.

Leaf Burn: In the heat of the day, water from sprinklers can burn the leaves of some species.

Soil Splash: Sprinkler irrigation can cause soil to splash onto the leaves of low-growing crops.

Crops Falling Over: Some crops can become top-heavy close to harvest and fall over with the weight of water on their leaves.

Choice of Sprinklers

There is a huge range of choice in sprinklers. It seems that the technology of projecting water into the air has attracted more than its fair share of innovative attention: no doubt Sigmund Freud would have had some enlightening insights on the reasons for this.

Every grower seems to pick out a different sprinkler to suit their particular situation and taste. The main factors to take into consideration are outlined below.

Droplet Size

The spray needs to be fine enough to minimise splash. Large droplets hitting the surface will splash soil up onto the lower leaves and this will go in with the harvested crop. While some soil splash may be inevitable, a finer sized droplet will avoid much of the problem. The force of heavy droplets is more likely to flatten some crops.

Area of Coverage

Each individual sprinkler needs to give a reasonably broad coverage to minimise the number of hoses around the garden.

Evenness of Distribution

There needs to be a fairly even coverage over the garden. While some areas of overlapping are inevitable and are usually not a big problem (provided drainage is good), patches of under-watering can result in poor growth where plants can't obtain adequate moisture.

Evenness of water distribution varies from one type of sprinkler to another: some will throw more water to the centre while others will throw more to the perimeter. Some types of sprinkler need to be overlapped 50% to compensate for this: that is, they are placed so that each sprinkler throws to the next one and every point in the area is watered by at least two sprinklers.

Wind is another factor here: on windy sites very fine sprays will be blown away from their intended target, which may result in under-watering.

Pressure Requirement

If water pressure is too high or too low for a sprinkler's design, uneven distribution and other problems result.

Compatibility with the Dimensions of your Garden

To economise on water usage, you will need sprinklers that more or less fit in with the measurements of your garden. For beds 5 m wide, a sprinkler with a 10 m diameter coverage would be wasteful, unless you have plenty of water and you don't mind growing extra grass.

Choice of sprinklers depends on a number of factors.

Rate of Application

The rate of application needs to be suitable for your irrigation regime. If you are using electricity for pumping, it is more economical to buy a small pump and irrigate for a longer period. Then if you wish to saturate your soil to a good depth, you can use sprinklers that apply the water at a slower rate, around 25 mm in 8–10 hours, so that all of it soaks in, especially on heavier soils, slopes or raised beds.

On the other hand, if you are using a petrol or diesel pump, you may want to put the water on at a higher rate, say 25 mm in 2–4 hours, so you don't have to get up in the night and refuel.

Trickle Irrigation

Trickle irrigation involves delivering water to the vicinity of the plants by means of drippers, perforated pipes or mini-sprays, so that it soaks into the soil in a limited region around these.

Advantages

Economisation of Water: Because a greater proportion of the water soaks into the deeper layers of the soil, there is reduced loss from surface evaporation.

Low Pressure: These systems operate at low pressure, so a smaller pump or lower fall will be adequate, and cheaper pipes and fittings can be used.

Less Weed Development: Because the surface of the soil is drier, there is less weed germination and growth during the growing season.

Better Penetration of Heavy Mulch: Because the water is delivered at a single point rather than over the surface, more will penetrate a heavy mulch and soak into the soil.

Less Problem with Salinity: There is less surface evaporation and concentration of salts. If some of the water applied soaks down into the water table, it can carry salts away with it so they are not building up in the soil. This means it is possible to safely use water that is too saline for other irrigation systems.

Less Weight on Crops Prone to Falling Over: Trickle irrigation avoids the added weight of water on the leaves and flowers of crops that are liable to fall over just prior to harvest.

Disadvantages

Limitations for Closely Spaced Plants: Water tends to flow from the drippers in a downward direction through the soil. Lateral spreading of the water depends on the nature of the soil: in lighter soils it tends to soak straight down without much spread to the side. This can result in very little water being available to shallow-rooted plants more than 150 mm or so from the drippers.

Trickle irrigation of closely spaced crops and meadow-crop herbs requires a lot of lines and frequent drippers, which tend to get in the way of cultivation and harvesting. Trickle systems are perhaps more practical for the wider-spaced larger plants and trees, as these can be serviced with fewer lines.

Blockages: A blocked dripper means that a plant is going without water. Blockages tend to go unnoticed unless you spend a lot of time checking the system.

The water needs to be clean and finely filtered to prevent blockages. Sometimes bacterial growth in the lines causes blocked drippers. This may or may not be a problem in your situation, but the chlorine flushing normally recommended to dissolve build-up in lines would be a no-no under an organic or biodynamic system: you would have to find some other means of coping with it.

Limited Soil Activity: Trickle irrigation delivers water to a small volume of soil in the vicinity of the dripper, so only a portion of the root system of larger plants is watered. This means that biological activity and root function may be reduced in the rest of the soil when it is too dry. In a biological system this does not seem desirable.

Less Flexibility: Plants that have developed under a trickle system concentrate their root systems in the vicinity of the drippers. Because only a small zone is being watered, there is less reserve of moisture in the soil, and because the plants lack an extensive root system, if there is an interruption to supply, the plants can run out of water more rapidly.

Another thing that must be avoided with trickle systems is overwatering, which can cause even more damage than under-watering. If these regions around the drippers receive excessive water, they become saturated and anaerobic. If the water still continues to drip, then plants with sensitive roots systems will be damaged and may die.

MANAGEMENT OF WATER

A good irrigation system allows the herb grower a measure of control over that vital but sometimes elusive element – water.

Most plants consist of between 60% and 95% water, which is continually being transpired into the air and replaced via the root system. Water is needed as a raw material for the photosynthesis process. Free circulation of air through the pores of the leaves is needed to provide the carbon dioxide essential for photosynthesis, and to dispose of the oxygen produced. Consequently some water is going to be transpired and lost as this air circulates. If the plant is unable to support this water loss by drawing on water in the soil, the leaf pores will close down and photosynthesis will be slowed or stopped.

If there is further water loss, then the plant will attempt to reduce this by wilting or even by shedding leaves. It is important to realise that photosynthesis slows or ceases some time before wilting commences. This means that growth will be affected well before the plant shows overt signs of moisture stress. As your eye becomes attuned to the appearance of the plants, you will probably be able to detect the early signs of moisture stress: the leaves start to look as if they have hardened, and if you are regularly among the plants you will notice that their growth is slower.

Water is also needed to provide the medium for the biological processes going on in the soil which the plant depends upon and participates in to obtain nutrients through its roots. Most of the soil flora and fauna need a moist environment for optimum activity.

However, this does not mean that it is necessary to continually soak crops with copious volumes of water. As in most things, moderation will give the best results.

Setting Up the Irrigation System

Layout of Pipelines

Growers tend to disagree among themselves as to whether pipes should be buried or left on the surface. If you are absolutely certain that your pipes are where you want

One system for setting up sprinklers, where the surface lines are supplied by a main line buried at the end of the garden. The crop rows will be laid out parallel to the surface lines to minimise interference with inter-row cultivation. For even coverage, this type of sprinkler is spaced so that each throws just past the base of the next.

them, then you might consider burying them, but you may regret it later if you want to deep rip your garden, move it or change the configuration of your sprinklers, or if your standpipes keep getting in your way.

I like the idea of burying the mains up to the garden, but in the garden I prefer to have all the pipes and hoses above ground. Perhaps part of this is because I have never been sure that I wasn't going to move in the not too distant future. Four times in my herb-growing career I have rolled up my pipes, hoses and fittings and moved the irrigation system to a new location. If the pipes had been buried, I would have been forced to leave them and buy new ones.

The nuisance of having pipes and hoses in the way can be minimised if the pipes are laid out along the rows rather than across them. That way you can have most of your rows clear of pipes, and hoses can be moved one way or the other as necessary.

An Organic Alternative to Teflon Tape

Have you ever worried about what happens to the bits of Teflon tape that flake off your pipe fittings and float off into the garden? Apparently Teflon's breakdown products are extremely toxic if it is burned; who knows what the pieces of Teflon might do if they end up in your herbs.

A cheap reliable alternative to Teflon tape is good old beeswax. After using it for nearly 20 years now, I can attest that it has proved totally dependable on various metal and plastic fittings, even in extremely hot conditions when sometimes the metal fittings would be too hot to touch.

If you carry the wax in your pocket, it is usually warm enough to smear onto the thread. The thread does have to be dry and not oily. Work the wax around with your thumb to get it reasonably bonded to the thread (it only needs to go on the male thread), then screw the joint together.

Filters

For most finer sprinklers and for trickle irrigation, there needs to be an adequate filtration system. This helps eliminate problems with blockages.

Pressure Variation

Variations in pressure from one part of a system to another can occur as a result of differences in height or friction due to the distance from the pump or source.

It may be possible to balance these two factors against each other to compensate for height differences by running the pipes downhill and reducing pipe diameters on the distant lower sections to create more friction and lower pressures at the bottom of the hill.

Use of greater diameters for pipes to more distant sprinklers will help maintain adequate pressure.

Alternatively outputs can be equalised by using gate valves, pressure regulating valves or, in the case of trickle irrigation, pressure compensating drippers.

Automatic Irrigation Control

Initially I had one sprinkler to each four or five posts and used to move them to a new position every night. Over the years I finally got tired of running around with a torch moving sprinklers and getting a soaking at the same time. Eventually I invested in an automatic controller, a few solenoids and a sprinkler for every post. This set-up certainly makes life easier, but it has also had its drawbacks.

One advantage is that irrigation can be entirely at off-peak rates. In addition it can be set up on a 4-day irrigation cycle watering a quarter of the garden each night. In hot, dry summer conditions, the majority of herbs produce significantly better crops under this regime, because the more frequent watering avoids moisture stress and keeps soil surface temperatures down. However, because the system is not being operated manually, if a sprinkler malfunctions or has been moved during weeding or harvesting and has not been replaced, it might be several days before it is noticed.

To reduce these problems one should institute periodic inspections of the gardens and flushing of the lines to keep the irrigation system running smoothly. At 'Twin Creeks' I was always amazed at the black geysers of muck we used to flush out of the lines once a month. Although the creek water looked quite clean, a fine sediment used to elude the filters and would deposit on the inside of the pipes. If allowed to build up, occasionally bits of muck would break free and block a sprinkler.

Managing Sprinkler Irrigation

There are a number of factors to take into consideration in assessing irrigation requirements and determining the optimum frequency and rate of irrigation. Every situation will be different according to the climate, soil, site, species and stage of development of the crop.

Arrangement of Sprinklers

The system I have used for about 15 years is based on little two-arm brass sprinklers that spin around and deliver a fine spray that gives a fairly even 25 mm of water over a diameter of 8–9 m in 8–10 hours. These are on the end of flexible hoses with a piece of wire twisted around the hose below the sprinkler, with two tails about 150 mm long sticking down from the sprinkler. These can then be stuck into the tops of sections of old 13 mm or 20 mm galvanised pipe about a metre long, which are driven into the ground as posts, and placed so that the sprinklers can cover the whole garden. The wires hold firmly in the tops of these posts, but can be easily slipped out to move the sprinkler to another post or to shift a hose out of the way of the wheel hoe.

Another system for movable posts uses 9 mm steel rods 1 m or so long. These have a movable rubber stop on them and 150 mm length of 13 mm poly pipe slipped over the rod and resting on the movable stop. The wire tails of the sprinkler can be slipped down the gap between the polypipe sleeve and the rod. The advantage of this system is that usually the rods can be inserted into the ground with hand pressure, which makes moving them easier, and the height of the sprinkler can be easily adjusted by sliding the rubber stop up or down the rod.

There are many other ways of mounting sprinklers and every grower seems to come up with a different arrangement.

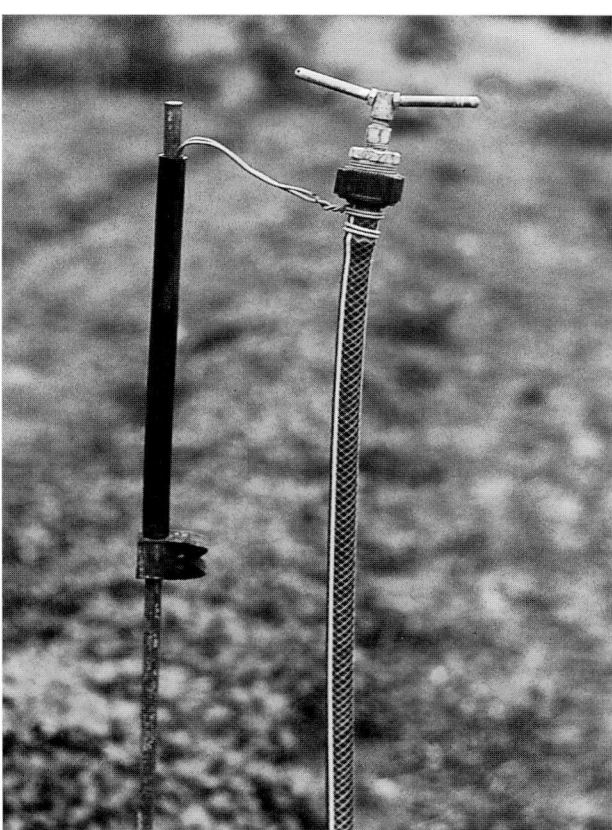

Brass two-arm sprinkler with wire tails attached for fitting onto posts. A 9 mm steel rod with 13 mm poly sleeve and rubber stop allow height adjustment of the sprinkler and make for easy moving to a different spot.

Assessing the Full Depth of the Soil

Always look deeper than the surface in assessing your garden's moisture status. The top 25 mm of the cultivated soil between the plants may be dry, but there can still be plenty of water below that for herbs with decent root systems.

On the other hand, the surface may have been nicely moistened by a light irrigation or shower of rain, but underneath that the soil could be bone dry. If it is a hot day, as the sun approaches its zenith this surface moisture will be quickly burned off and your plants will be gasping for water for the rest of the day.

Good Penetration

The first 6 mm or so of water doesn't count for much as it just wets the surface and is soon dried off. To be effective, rain or irrigation needs to be in amounts of at least 12 mm at a time. If there is a light fall of rain, you can follow it up with sufficient irrigation to make it count.

One irrigation of 24 mm will be worth more than two of 12 mm irrigations a couple of days apart, as less of the large single irrigation will be lost to surface evaporation and more will soak into the deeper part of the soil.

Effects on Other Operations

Organise your watering schedule so that it fits in with your weed control and harvesting.

Cultivation usually needs to be done 1–3 days after watering, depending on the soil, the weather, and the amount of water. After a cultivation, the weeds will die more readily if you can refrain from watering them until they have succumbed to the damage you have done to them, so try to carry out your weeding after an irrigation or fall of rain, rather than just before.

Harvesting needs to wait at least a day after irrigation or significant rain to allow the plants to dry off underneath and the soil to firm up a little, so the activity is not compacting it.

Try to time irrigation so crops prone to falling over are not watered at a vulnerable stage just prior to harvest when the extra water on their leaves or flowers would make them top heavy.

Prevailing Conditions

Naturally plants' moisture requirements are going to vary according to the prevailing conditions. In irrigating, you need to anticipate what the weather conditions are likely to be over the next few days to ensure that there will be sufficient moisture available.

Keep a regular check on soil moisture to ensure that you are putting enough water on each time to maintain adequate moisture levels until the next watering.

Soil Type

Take account of the moisture-holding capacity of your soil. Light sandy soils hold less water and lose it more easily than heavier soils. A good level of organic matter also increases a soil's water-holding capacity.

Wind

Wind can greatly increase moisture losses by evaporation and transpiration. It can wreak havoc with sprinkler irrigation, especially fine droplets, making it difficult to place the water where you want it to fall. If wind is likely to be a problem at times in your situation, this can be alleviated with movable sprinklers, watering at night when there is usually less wind, using larger droplets, and keeping ahead in the watering schedule so there is always sufficient moisture for an extra day or two in the event an application of water has to be delayed.

Crop Development and Spacing

A crop in its early stages of development will not need the same amount of irrigation as one that is approaching harvest and carrying a lot of leaf that is transpiring much more water.

Plants that are closely spaced will need more additional moisture than those that have larger areas to draw on.

Beware of Overwatering

Excessive irrigation can cause a number of problems:

- Waterlogging in heavy soils or those with impeded drainage, with consequent damage to root systems.
- Leaching of soluble nutrients in lighter soils, damage to soil structure in heavier soils, and harm to vital soil micro-organisms.
- Soil that is continually moist on the surface will encourage greater germination and growth of weeds, while making it more difficult to kill weeds.
- In some situations it can also foster lush growth that is soft and lacking in vitality, with a low dry-matter content and a tendency to become top-heavy and fall over.

In particular don't water too heavily in spring, and ease up as you approach the end of the irrigation season in autumn. Soil that is a little on the dry side is warmer and growth will be better at these cooler times of the year.

One year in late March at 'Twin Creeks' I gave the garden an extra good soaking because I was going away for a few days. Virtually next day the 'autumn break' came, early and in earnest. Weeks of repeated heavy rains delighted the local graziers whose pastures responded with lush growth, but my poor garden was in a soggy state and never did dry out properly again before winter. Digging root herbs that year was difficult and some species suffered badly from rotting as a result of the excessive water.

Leaf Burn

Quite a few species of plants are susceptible to leaf damage if water hits their leaves in hot conditions or when they are wilted. Having seen the damage this can do to corn and beans, I haven't risked testing out all the herbs to see which ones are susceptible.

This problem can be avoided by watering at night, early morning or late in the day. If plants are wilting in the heat of the day, unless you know they are able to tolerate water on their wilted leaves, it is best to wait until conditions have cooled off before irrigating, or else apply the water to the surface of the soil, keeping the leaves dry.

Time of Day for Watering

There may not always be a choice, but watering in the daytime does have a number of disadvantages. Evaporation is greater, so more of your precious water just goes to humidifying the atmosphere (perhaps to the benefit of some desert island thousands of miles downwind). As this water evaporates, if there is some clay or minerals in it, these are deposited on the leaves, making them look rather murky. Daytime tends to be windier, so the water is more likely to get blown off course and end up where it isn't needed.

These problems can be largely avoided by watering at night or, failing that, in the early morning and evening.

My preference is for night watering, particularly as I usually have a few potatoes and tomatoes being watered by the same system as the herbs and these are liable to suffer from blight if watered during the day.

At 'Twin Creeks' I was able to set up an automatic watering system that watered a quarter of the gardens each night for up to 8 hours at off-peak electricity rates. This worked out very cheaply and it was also the best time to water everything. A relatively small 1.5 hp electric pump drawing on a small permanent creek easily supplied enough water for 0.4 ha of herbs with this system. It also ensured there was adequate pressure, as during the day quite a bit of water was being used by the manufacturing operation.

Frequency

Optimum frequency of irrigation will normally range from 7 days to 4 days, depending on prevailing conditions.

In cooler or more humid conditions, a 6- or 7-day irrigation cycle may give best results as it allows more opportunity for the soil to dry out a little between waterings, so weeding and other operations can be carried out.

In hotter drier conditions, a watering regime based on a shorter cycle seems to benefit most herbs: with a 4-day cycle the surface does not get as dry and hot as with a longer cycle and in hot spells there is less moisture stress

In some circumstances – such as for germination of shallow-planted seed or the establishment of transplants – more frequent, even daily or twice-daily irrigation may be necessary.

Rate

During the growing season, the total water requirement (rainfall plus irrigation) will normally be in the range of 10–50 mm/week. If significant rain falls, the irrigation should be reduced or delayed.

The amount of each application will vary according to the frequency: more frequent applications can be smaller, though not in direct proportion, as more is lost to evaporation. For example in north-east Victoria I was applying around 30 mm once every 6 days through summer. On a shorter cycle in the same conditions, applications needed to be 25 mm once every 4 days.

Saving Water

Water consumption can be reduced by a number of strategies:

Less Frequent Heavier Irrigations: This achieves deeper penetration of moisture where plants can draw on it for longer and there is less loss from evaporation.

Following Up Falls of Rain: If there is a significant fall of rain of 25 mm or more, irrigation can be postponed. If there is a fall of rain that amounts to somewhat less than your normal irrigation, then it will save water to immediately top this up with a light irrigation as there will be proportionately less loss from surface evaporation if the rain and the irrigation are combined.

Regular Cultivation: Maintaining a loose layer of dry soil or dust mulch on the surface will greatly reduce loss of moisture by evaporation. This breaks the upward movement of water by capillary action to the soil surface that occurs when the soil is firm. Depending on the soil, this may need to be done after every irrigation or rain to be effective.

Managing Trickle Irrigation

Management of trickle irrigation differs from sprinkler irrigation in a number of ways:

Monitoring: This needs to be done more frequently and precisely because of the narrower margin of safety with a trickle system.

Surface Evaporation: This is of much less significance as only a small proportion of the surface is irrigated.

Wind: Provided the water is being delivered at or close to ground level, the system will be immune to the effects of wind.

Overwatering: This can be particularly damaging in a trickle system, as the roots are concentrated in the watered area: if these are damaged by waterlogging, then the plant has few other roots to fall back on.

Time of Day for Watering: This is more flexible as leaf burn, leaf stain, evaporation and wind are not a problem with trickle irrigation.

Frequency: Because the volume of soil being irrigated is smaller, there is a smaller reservoir of moisture for the plants to draw on, so irrigations have to be more frequent: every 1–3 days in hot weather.

Rate: It is harder to measure how much water is going on as it depends on the spacing of the drippers or perforations and how much each delivers. With careful monitoring the optimum rate of application can be established for a particular site.

9
Pests and Diseases

In general, dealing with pests and diseases among herbs requires much less of the organic grower's time, effort and ingenuity than dealing with weeds.

With weeds one is striving to replace an established flora that has a strong hold on the location, but with pests and diseases there are a number of factors that work or can be made to work in the grower's favour.

Factors Reducing Pest and Disease Problems

Natural Resistance

While there are a number of pest and disease problems that can arise among herbs, these are generally fewer than for most other cultivated crops. The reason many herbs are of interest to us is because of the active constituents they contain, which are in fact part of the plants' defence systems, so the herbs we are choosing to grow usually have quite effective means of resisting most pests without our assistance.

The majority of herbs have the overall resistance of wild plants, because they are closer to the wild form than most of our food crops, which have been selectively bred for such attributes as nutritional value, high yield etc. rather than for resistance to pests and disease.

Isolation

Because herbs are grown on a limited scale in Australia, in situations isolated from each other and their native habitats, there is a quarantine situation in effect. Most species have made it here without the specific pests and diseases that affect them in their native habitats.

Vitality of the Organically Grown Plant

To a large degree, the health of a plant is a reflection of the health and vitality of the environment that it is growing in. Pests and diseases have a strong disposition to attack plants that are weakened by some adverse factor.

This is a guiding principle followed by organic and bio-dynamic growers alike. Problems with pests and diseases are greatly reduced by focusing on providing a healthy environment for the plant, using organic substances, primarily compost, to feed the soil rather than the crop. Plants grown organically have a vitality and resilience that enables them to resist many pests and diseases that will seriously affect plants suffering from the imbalanced growth induced by chemical fertilisers or other inputs. The evidence of the validity of this is the fact that thousands of organic and biodynamic farms around the world are operating viably with minimal pest and disease problems.

Natural Controls and Predators in an Organic System

One of the principles of organic farming is the fostering of an ecological diversity and vitality that will provide a suitable habitat for beneficial predators and other natural controls to exist.

Managing Pests and Diseases

Nevertheless there are some little critters whose metabolisms can cope with the essential oils, alkaloids, stinging hairs etc. and which don't seem to mind chewing on an organically grown plant or two. If their numbers build up and significant damage begins to appear in a herb crop, you need to look at how to deal with the situation.

In many cases the first thing to do is to wait. While this is perhaps the hardest thing to do, it is often the best course of action. Rather than mount a pre-emptive attack in an attempt to nip in the bud what may appear to be a growing infestation, hold off for a while and see how serious the problem really is. If you come in with a heavy hand and try to bludgeon pests out with poisons, then you will quite possibly find yourself locked into an ever-increasing level of noxious and expensive inputs.

By waiting and watching, you are not just doing nothing, you are in fact giving the vitality that you have endeavoured to foster in your crops the opportunity to work and restore the balance. In perhaps 90% of cases, natural controls will develop and correct the situation if you can be patient enough.

Equilibrium is not necessarily a steady state: it often exists as cyclic fluctuations around a norm. Many pests and diseases are self-limiting. Their incidence may increase for a while until it declines again in response to an increase in a controlling predator, parasite or disease, a change in the weather, or even a response from the plant itself.

Frequently leaf damage does not affect growth as much as one might expect, as more light will strike the lower leaves which photosynthesise more and compensate for much of the loss. If less than 20% of leaf surface gets eaten, this usually does not significantly reduce overall growth.

Fortunately most dried herbs are broken or cut as part of their processing after drying. This means that holes in the leaves and other minor blemishes don't show up in the finished product.

It is also important to bear in mind that a low incidence of a 'pest' may be necessary to sustain populations of its controlling predators. Attempting to wipe out a pest entirely may instead succeed in wiping out its controls while allowing small numbers of the pest to survive. These can then multiply without any limitation.

By observing the development and progress of the remaining 10% or so of pests and diseases that do become a problem in an organic system, you should be able to gain a better knowledge of their nature. This knowledge and understanding of a pest or disease's habits, patterns and life cycles can be the key to a suitable means of coping with it.

Try to work out the conditions under which a pest or disease proliferates and what its life cycles are. You may have created the habitat that favours it and there may be something you can do to change this so as to discourage it. There may be a way of breaking its life cycle at some point that reduces the incidence of the pest or disease to a tolerable level. Alternatively, you may be able to obtain what you want from the crop before the pest or disease takes its share.

If you can restrain your anxieties and hold back until you have a good comprehension of the situation, you will be able to minimise the amount of input needed to maintain production.

Strategies for Pest and Disease Control

Quite likely there will be a few pests and diseases that will need some action on your part to prevent unacceptable losses.

'Control' is perhaps not quite the right word because it carries with it the implication of mastery and domination, whereas here we are talking more in terms of influencing outcomes rather than governing them.

To be acceptable, a pest or disease control strategy must be within the guidelines of organic certification standards, provide a reasonably reliable means of dealing with the problem economically, and not have significant harmful effects on the product or the local and wider environment.

Alleviating Stress

Many of the herbs you are growing will be outside their natural habitat, outside the range of optimum conditions for which they are adapted, consequently they will be under a certain amount of stress, perhaps only at certain times of the year: maybe during the heat of summer, or during periods of excessive rain or humidity. It is during these times of stress these plants will be most susceptible to attack.

If you can do something to alleviate this stress, such as provide the plant with more moisture, shade or wind protection, additional compost, relocate it to a more favourable site or in some other way improve the plant's environment, often the pest or disease will disappear or its effects will diminish.

My own experience with grasshoppers in north-east Victoria certainly seemed to bear this out. In a number of cases I had two plots of the same herb, one plot in the main garden that was exposed to the full heat of summer, and one in a cooler shadier creek-side garden.

With several species that have a preference for cool mild conditions, the grasshoppers would voraciously attack the plots in the main garden where the plants were under stress, while the plots near the creek would be left virtually untouched. When I changed over to more frequent watering, which reduced moisture stress and lowered temperatures around the plants, I found that the grasshopper damage was significantly diminished in most species growing in the hotter sites.

Interrupting Life Cycles

Many pests and diseases follow a cyclic pattern and if you can find a way to interrupt these cycles, you can obtain satisfactory yields with very little intervention.

Often the development of a disease or pest can be broken by harvesting as there is nothing for it to live on until the plant regenerates. This effect can be heightened by close mowing and even by flaming the stubble after harvesting.

This approach works well for Mint rust and for two-spotted mite (red spider).

Polyculture

This principle aims to create conditions of diversity that simulate a natural ecosystem. An assortment of species growing together tend to have a protective effect on each other.

A large-scale monoculture, where a vast area of a single crop is grown, is an inherently unhealthy situation that is out of balance. Few plants in the wild grow as large pure stands of the one species, because most need the presence of other species for one reason or other.

In an extensive monoculture, conditions are more conducive to pests and diseases as they constitute a large target for them to find in the first place, and then a large area for them to spread without interruption. There are also fewer niches for predators as there is little diversity of flora and fauna.

With herbs in commercial production, it is not possible to have mixed stands, but it is quite practical on a small scale to maintain a mosaic culture, so that the different species are in close proximity and exerting a beneficial influence on each other.

It is also good to have trees, including natives, in the vicinity as they provide a habitat for bird and insect life that can have a beneficial effect on the garden.

Layout and Management

Some pests and diseases are favoured by particular layouts of crops that create a habitat for them to thrive in from where they can move onto vulnerable crops nearby. Some weeds harbour undesirables such as red-legged earth mite, vegetable weevil or slugs and snails. These latter can find wonderful havens in woody perennial herbs and other cultivated plants or in mulch that give them permanent shelter from birds and other predators.

By modifying the layout or the management regime, these sources of food or shelter for such pests can be removed and their numbers can be reduced to less damaging levels. This has to be done at the appropriate stage in the pest's life cycle as it can drive hordes of hungry critters off in the direction of a vulnerable crop.

A variation of this strategy is to leave enough of some food source for a pest to feed on elsewhere so it is not driven into your garden in search of food.

Crop Rotation

Crop rotation is another form of polyculture. With herbs it is primarily an option for annuals, biennials and those perennials that are dug up and replanted each season. If the same or similar crops are grown repeatedly in the same location, there tends to be a build-up of pests and diseases associated with the species grown.

Rotating dissimilar crops that do not share pests or diseases can interrupt this build-up and will often eliminate the problem. Complex rotations involving at least three species tend to be most effective. Some crops can be safely grown two seasons running, as long as they are then rotated with several years in other crops.

Quarantine

It is worth paying some attention to quarantine as it is often a very effective form of prevention that saves a lot of time and effort.

With diligent care and surveillance, you can probably avoid bringing in pests that are not mobile enough to make it on their own. These usually come onto the farm in propagation material or soil, or on equipment and clothing, particularly footwear.

Peppers

Using peppers is a biodynamic pest control technique practised and advocated by some. The principle of it is similar to homeopathy.

While some spectacular responses have been reported, there appears to be a general inconsistency of results. For more detail see Chapter 10 'Biodynamic Aspects.'

Natural Pesticides

While there are a number of plant-derived insecticides that are approved by certifying organisations, many organic growers prefer not to use them.

Even natural pesticides are mostly non-selective so they will kill predators, not only of the pests you are trying to control, but also predators that are keeping the numbers of other potential pests in check. So your attempt to intervene and eliminate one problem may unleash a number of other pest problems that are perhaps even more serious than the original one.

Some natural pesticides also kill frogs and fish and could have other side effects that are detrimental to the healthy vitality of your garden or which affect the subtle biochemistry of some herbs' active constituents.

Other Pest Control Options

There is quite a variety of approved substances and techniques used on vegetables and other crops, some of which may be of value for pest and disease control in herbs. The reader is recommended to consult books on the subject of organic pest and disease control, and organic certification standards for allowed inputs.

Changing Crops

Changing crops may be the best solution when you have not been able to find a suitable effective strategy for a crop that is afflicted by a pest or disease.

There is such a range of herb crops to choose from that it doesn't make much sense to persist with species that require elaborate life-support systems to keep them in production. In most cases they can be grown more easily in some other location that suits them better, and there are alternative crops that will do much better on your place.

Trial Plots

Trial plots will bring to light many of the pests and diseases you will have to contend with, enabling you to assess which pest problems are manageable and which ones are too difficult.

However, not everything will emerge in the trial plot situation and there will always be challenges arising that are a test for one's comprehension and ingenuity. In the dynamics of the farm, one can view these as opportunities to gain some insights into the workings of nature and the cosmos.

STRATEGIES FOR INDIVIDUAL PESTS

Covered here are a few pests that can affect a fairly wide range of herbs. Pests and diseases that are specific to one or only a few species are covered under the herbs concerned, in Section 2.

While this list of pests and the strategies for dealing with them are not exhaustive, they do give examples of the sort of problems that can be encountered and how organic approaches can provide solutions.

Grasshoppers

There are a number of species of grasshopper.

Locusts

Locusts are large voracious grasshoppers that can reach plague proportions, travelling across the country and eating everything in their path. Fortunately most small growers won't ever encounter them, as they occur mostly in inland areas.

While locusts can be devastating to annual crops, for most perennial herb crops a plague of locusts would only be a temporary setback, as the herbs will soon regenerate.

Wingless Grasshopper

The wingless grasshopper is the common, small brown grasshopper. The term 'wingless' is something of a misnomer as the adult grasshoppers do have wings, but they don't travel far with them. These grasshoppers are fairly widespread and can be a problem in warmer drier regions. While they may be present in cooler and moister regions, their numbers don't seem to build up to the same degree.

Much of the damage is done by the young grasshoppers before they develop wings. The eggs tend to be heavily concentrated in certain places and, in the spring, dense mobs of the young hooligans hang around together close to where they hatched, gradually eating away the leaves of the plants that take their fancy.

If their numbers aren't so great, they tend to just eat the lower older leaves, but if there is a swarm of them, they will strip all the leaves and then move on to adjacent plants. As they get bigger they eat more, of course, and can strip some herbs, leaving just stalky skeletons.

Their intensity seems to vary from one year to another, and sometimes they will just suddenly disappear, usually towards the end of summer, after a rain.

Apparently there is a disease affecting their gills that can spread through them as an epidemic in humid conditions, which is probably the cause of their sudden disappearances.

Susceptible Herb Species

Wingless grasshoppers will attack a wide range of herbs, including some that you would not expect anything to enjoy, such as Wormwood.

They seem to be rather fickle in their habits and while I am sure there is definitely a method in their madness, I can't say I completely understand it.

Particularly vulnerable are plants that are under stress, such as those species that are hard pressed to adjust to hot, dry summer conditions.

As mentioned earlier, my experience in north-east Victoria has been that where a susceptible species was also growing in a cooler site nearby, the grasshopper damage on the less stressed plants was quite limited, while those out in the open heat were devoured. However, there were sections of the garden where grasshopper damage was always more severe, irrespective of the species. This may have been because these parts of the garden were close to the grasshoppers' breeding sites, or because of some factor that was stressing the plants in those areas.

I frequently used to have problems with grasshoppers attacking a patch of Peppermint that was next to a bed of Agrimony, which the hoppers just adored. Large numbers would descend on the Agrimony and when it had been stripped, they would move onto the Peppermint. Most of the damage was along the border adjacent to the Agrimony and at the end of the crop where it fronted onto grass.

Management Strategies

Poultry

The best solution to grasshopper control seems to be a few hungry ducks or chooks that are given the range of the herb garden for a while and then allowed to patrol the perimeter. A pair of Khaki Campbell ducks are able to maintain good grasshopper control over more than 0.4 ha.

My own erratic lifestyle and the prevalence of foxes in Victoria, precluded the keeping of poultry as I could not be sure that they would always be safely locked up at night. Consequently I had to battle on with other approaches to the problem.

Scoop Netting

One way of reducing grasshopper numbers without spraying is to sweep them up with a large scoop net like a giant square-mouthed butterfly net. With regular repeated passes of this net, and tipping the caught grasshoppers into a large plastic bag lining a 200 L drum they couldn't hop out of, I found I could remove significant numbers from the garden. I used to let them go a good distance away in another happy hopping ground (no, not the big one up in the sky) where they wouldn't cause any problems.

Naturally it was not possible to eliminate them as some would always evade the net and more would hop in from elsewhere, but it did help reduce numbers to a level where the damage was more tolerable.

Encouraging Plant Vigour

Plants that are growing more vigorously tend to be less attractive to grasshoppers and can also endure a bit more damage. Providing good conditions for strong, healthy balanced growth will help ensure this.

Increasing the rate and frequency of watering during the heat of summer reduces the damage by grasshoppers. The additional moisture on the soil surface keeps the temperature around the plants lower, reducing heat stress in vulnerable species and encouraging more growth.

The grasshoppers seem to prefer older leaves, so if the plants are growing strongly enough, there will be enough of these older leaves to feed them all and yields won't be affected greatly.

Meadow-crops that have become very dense tend to lose vigour and suffer more at the hands of pests like grasshoppers. Replanting or rejuvenation will promote more vigorous growth that can cope better with the ravages of pests.

Keeping Them Fed Elsewhere

Keeping grasshoppers fed elsewhere won't get rid of them, but it will help reduce the numbers moving onto your crops. If the grasshoppers run out of food in one area, they will naturally move next door in search of something to eat. If this happens to be your garden, you will have a lot of extra hungry mouths to deal with.

At 'Twin Creeks' there was initially pasture alongside the main garden. One summer, cows were allowed to graze this pasture off quite hard, creating hordes of refugee grasshoppers that then moved over into the lush growth of my garden. Consequently, that year was the worst ever for grasshopper damage.

After that I moved the fence of the adjoining pasture back, so that there was an ungrazed strip 30 m wide between the pasture and the garden. This provided more food for the grasshoppers in that area and reduced the numbers migrating into the garden.

Aphids

At times the number of aphids builds up – usually during mild, humid spring weather – and this situation might last for about a month until conditions become drier and ladybird numbers increase.

Susceptible Herb Species

Most soft-stemmed herbs seem to be susceptible to aphids. The majority can withstand a fairly heavy infestation, though Dill may succumb and die.

Management Strategies

Ladybirds

Given half a chance, ladybird numbers will build up and take care of the aphids. If you want to speed up this process, it may be possible to purchase ladybirds and introduce them into the garden.

Avoiding the use of pesticides, including natural ones, will help maintain ladybird numbers, as they are affected by them.

Avoiding High Nitrogen Levels

In my early herb-growing days, I used to get a big build-up of aphids every year, but I am convinced that the causative factor was my use of deep litter chicken manure. Although this contained quite a lot of sawdust and had been standing heating in a heap for some time, it wasn't real compost as it didn't have a balance of other ingredients. There was a lot of soluble nitrogen in it, as evidenced by the ammonia coming off it.

The aphids would appear shortly after this chicken manure was applied. My feeling is that the consequent lush growth of the plants – pushed into imbalance by the excessive availability of nitrogen in the manure – would attract and foster the build-up of aphids.

Since adopting biodynamic methods which avoid excessive nitrogen levels, I have seen very few aphids on my crops. I no longer use chicken manure, except for very small quantities as part of a complete compost. The biodynamic compost I make has less freely available nitrogen in it and no smell of ammonia. While the growth of the plants may be less rapid, it is more robust and of a healthy, balanced nature.

Snails

Snails can be a problem in many crops, not only from the point of view of how much they eat, but also because they are not quick enough to get away when the crop is harvested. Consequently they end up going onto the drying screens with the harvested leaf, where they just go to sleep in their shells and slowly desiccate.

Small snails and fragments of the larger ones can go through the rubbing screens, and become a significant contamination problem in the finished product.

The common snail is an introduced creature and if you are fortunate you may not have them on your land. It is quite easy to inadvertently introduce them in propagation material, though, or in any item that has been sitting outside if these come from an infested location. This can be prevented by careful inspection and by washing the soil off any suspect material (and disposing of this soil in a place where any hatching snails won't be a problem).

Many growers on established properties may find that the snails are already well entrenched around their homesteads and ways will have to be found to prevent snails becoming a major problem.

Susceptible Herb Species

Many crops are susceptible to snails. They are most vulnerable at the seedling and transplanting stages or when emerging in early spring.

Snails like to take cover in bushes, under rocks and debris, in crevices around buildings etc., and when conditions are moist enough to suit them, they come gliding out in search of food and adventure. Usually this is at night or when it is raining. Many of the bushy perennial herbs offer snails the sort of year-round shelter they delight in and they can build up in big numbers while feeding on surrounding plants.

Crops that are harvested down to ground level, like Spearmint and Peppermint, only offer snails good cover once they have formed a canopy. If there are places nearby where the snails can take refuge after harvest and during winter, then they are susceptible to snail infestation.

However if the crop is growing in a more open environment, away from the sort of places where snails can abide during winter and between harvests, they will not have the opportunity to ever build up, as birds and other predators will be easily able to find them and keep them from proliferating.

Management Strategies

Poultry

Running poultry in and around the garden at appropriate times can help prevent or alleviate the problem. Ducks and chooks are quite fond of snails, however they may not be able to find them unless it is raining enough to encourage the snails to come out during the day.

Snail Bait

I have never been in favour of snail baits as there is a serious risk of non-target animals taking them or of predators eating dead or dying snails. Certification guidelines do allow the use of methaldehyde as a snail bait, but require that it be used in a trap that does not allow entry to non-target species and which is supposed to hold the poisoned snails. However, by good planning, layout and management, with many crops it is possible to avoid the need to use snail bait.

Layout

If snails are likely to be a problem, susceptible crops should be sited away from any bushy perennials and other havens that are likely to provide year-round cover for them. This way any snails will be accessible to birds, which will usually keep the numbers down. Magpies are very keen snail patrollers, or else poultry could be introduced at appropriate times when they are not likely to inflict too much collateral damage.

Delaying Cultivation at Critical Times

If freshly germinating weeds around the crop seedlings or transplants are left uncultivated for a short period during their most vulnerable stage, they can help reduce damage to the crop by providing other food for the snails.

The risk with this strategy is that if left too long, the weeds may be difficult to control and will provide more snail habitat.

Timing of Harvest

If you do find you have an infestation in a leaf crop that is approaching harvesting, you can minimise the number of snails you pick up by harvesting several days after rain or irrigation so the conditions inside the plant canopy are as dry as possible (without causing moisture stress to the crop, of course).

If conditions are moist, the snails will be all through the leaves, but as it becomes drier they will tend to migrate to the lower parts of the plant and onto the ground to avoid drying out. By harvesting after it has been dry for several days, most of the snails will be below the level of the harvest. Then invite the magpies in for a feast among the stubble, or call in your chooks.

Drying Snail-infested Herb

You should carefully pick out any snails as you spread the herb on the drying screens. When it has finished drying, it is a good idea to lift it from the screen with the hands rather than dumping it off, as usually any snails among the leaves will have migrated down onto the screens as the herb dries out. By carefully lifting the dried herb from the screen, most of the hidden snails will be revealed.

Winnowing could be another option for removing snails from rubbed leaf herbs, though lighter shell fragments may not separate easily.

Slugs

In cool humid regions, slugs can be a problem for young seedlings and transplanted crops. They tend to be fairly ubiquitous, but usually they only build up to problem proportions if they have conditions that suit them. They proliferate in a weedy garden which provides abundant food and shelter for them.

The worst periods tend to be in late autumn and early spring when moist conditions prevail and the slugs can get around easily without desiccating. They need suitable cover to shelter in and stay moist during the day and in dry spells.

They are a greater problem on heavier soils whose surface stays moist more of the time. They also thrive where they can obtain shelter in mulches, particularly those of plant origin, or in coarse compost lying on the surface.

Management Strategies

Regular Cultivation

A regime of loose bare soil between plants that is maintained by frequent cultivations creates a habitat that does not favour the build-up of slugs.

Less Frequent Watering

Allowing the surface to be dry as much as possible will discourage slugs.

Delaying Cultivation at Critical Times

As for snails, see above.

Predators

While birds, frogs and even a native carnivorous snail will eat slugs, they usually don't manage to make much of an impression on them. Chooks might be useful, as long as they don't damage other things in the garden.

Slug Bait

As for Snail Bait, see above.

Problems with Harvest

Slugs are less inclined to be picked up in harvesting leaf crops than snails, as they normally don't remain in the canopy during the day. However, if the leaves are moist and low growing – as for a final harvest of low regrowth in May – there could be slugs among the leaves.

The problem is that slugs are difficult to spot and when dried and shrivelled up, they pass easily through the screen. Desiccated slugs are hard to find in the dried herb as they look like a dark piece of rolled leaf. If they remain in the herb though, there could be serious consequences. If the herb is used to make tea, the slugs will take up water and make a gruesome sight, expanded to full size and floating around in the top of the teapot.

Avoid harvesting low regrowth when moist, if slugs are prevalent. Winnowing is probably the only practical way to get rid of them from tea-grade leaf, but it is not an option for aerial parts.

Two-spotted Mite (Red Spider)

A mite and not a spider, though rather spider-like in appearance, the two-spotted mite is barely large enough to see with the naked eye. These pests are so small, you may not realise they are there and attribute the damage to some other cause.

The symptoms are a yellowing of the leaves, which progress to turning brown and falling off. The mites are red and live on the underside of the leaves, sucking the sap.

Susceptible Herb Species

Melissa Balm is particularly susceptible, but a few other species, such as Hops, Greater Célandine and Meadowsweet, may also suffer at times.

Infestations usually occur from late spring to autumn in warm to hot dry conditions. Plants that are stressed during this period are more vulnerable.

Management Strategies

Improved Growing Conditions

Providing a cooler moister growing environment will usually reduce the incidence of the problem.

I had two patches of Melissa Balm at 'Twin Creeks': one was in a hot exposed site and it always was badly hit by mites; the other patch was in a cooler, more sheltered plot, closer to the creek, with trees around, though it received almost full sun. This second plot was hardly ever affected by mites, nor was the Melissa Balm that grew wild alongside the creeks in the district.

Interruption of Life Cycle

The life cycle of mites can be interrupted fairly easily where they are affecting herbs that are able to regenerate from ground level. Simply harvest the crop (or if it is too badly damaged, just cut the plants off at the base) and mow the stubble off with a lawn mower, set as close to the ground as possible. As there will be virtually no leaves remaining, the mites will be deprived of their nourishment and die off. When the new growth emerges in a week or so, it should be free of the mites and will normally remain free of them until harvest.

The clippings can remain on the ground beside the plants, they don't seem to reinfest the new growth.

Nearby crops that are also infested should be cut at the same time, if possible, to reduce the risk of a new infestation.

Another option is to introduce predatory mites, which are commercially available and are used for mite control in a number of crops such as apples.

Red-legged Earth Mite

Red-legged earth mites are similar in appearance to two-spotted mites (see above), but they are somewhat larger and hungrier. They have black bodies with bright red legs and live in the soil, swarming up onto the plants to feed.

They usually peak during mild conditions in spring and autumn and seem to prefer lighter, sandier soils.

Susceptible Herb Species

Red-legged earth mites will attack a range of herbs, doing most damage to emerging plants and seedlings, especially if these are under some sort of stress, such as transplanting.

Infestations seem to be worse in the first year after ploughing up pasture, or if large numbers of weeds that harbour them, such as Capeweed, are allowed to persist in the garden.

Management Strategies

Because there are always a good proportion of red-legged earth mites underground and plenty of eggs in the soil, there is not much point in spraying them with anything, even natural pesticides like Pyrethrum, as it won't make much difference to numbers: indeed it is more likely to result in an increase because it will kill any predators.

Fortunately the infestations are normally short-lived and usually doing nothing will see a decline in numbers as hotter and drier or colder and wetter weather comes on. This is a good example of a pest that at first looks as if it will not be stopped until· it has denuded the face of the Earth, but turns out to be self-limiting.

Plant Vigour

Usually if the plants are strong and healthy, they can survive the period of peak infestation and will recover once the earth mites decline. Vigorously growing plants will normally manage to outgrow the damage. This means providing adequate moisture and a good dressing of compost to promote strong growth.

Crop Management

By observing affected crops, you may be able to work out which factors, such as weather conditions, soil type, weed control practices, planting times, irrigation and fertilisation regimes, crop rotations etc., favour or discourage the development of the earth mites and devise strategies that will reduce their impact.

In one case a grower had two soil types, one a light sandy soil derived from sandstone, and the other a medium to heavy loam derived from basalt. The red-legged earth mite infestation was entirely limited to the sandy soil. No damage occurred on the basalt soil even though it was only a few metres away and well within range of the mites' spread. The solution was simply to grow vulnerable crops on the basalt soil.

STRATEGIES FOR INDIVIDUAL DISEASES

As diseases are generally specific to one or a very few species, they are dealt with under the herbs concerned in Section 2.

10

Biodynamic Aspects

A BIODYNAMIC APPROACH TO DRIED HERB PRODUCTION

Most of the methods and systems outlined in the other chapters of this book are compatible with biodynamics, though they describe an organic regime. To be biodynamic involves the adoption of a number of practices, such as the use of special preparations and the working towards an understanding of the principles of biodynamic agriculture. The biodynamic approach to agriculture was originally outlined by Rudolf Steiner more than 70 years ago and has been going through a process of evolution and development since then. It is now practised by thousands of farmers around the world, including many in Australia.

Although I am a biodynamic grower myself, I have tried to refrain from presenting the production of dried herbs solely in that context. Some organic growers find biodynamics difficult to relate to, so I have, for the most part, sought to stay on common ground.

This chapter is intended for readers who are interested in going more deeply into the biodynamic side of herb growing. Those who find biodynamics a bit too airy fairy or out of this world can skip this section, though they are also invited to read it, as perhaps I may be able to convey some understanding of this method of farming.

First I should emphasise that biodynamics involves seeing the farm as a whole system: an organism. From a biodynamic point of view, the separating of practices into organic and biodynamic categories, as I have done here, does not convey this conception particularly well, as the division is an artificial one.

In a biodynamic system, all the practices are woven together as part of a whole. Biodynamic practices are not just something that can be tacked on to an organic system to make it biodynamic. The grower's understanding and outlook has to undergo a transformation, which affects the way they see the farm and how they go about doing things. How big these changes will be depends on the individual.

I had been farming organically for some 15 years before I was introduced to biodynamics. For me it has added a dimension missing in organics. It hasn't been difficult to change over, as many of my attitudes and practices were basically biodynamic anyway, but it has provided me with an overall understanding of the farm that ties everything together. Biodynamics has helped me improve my farming practices, made my plants healthier and my products a better quality.

Biodynamics is a vast and dynamic subject, and there is some diversity of views on most aspects of it. This can lead to communication difficulties, but for the most part biodynamic growers find they have a lot to gain by sharing and listening to each other's outlook and understanding.

The basis of this approach to farming was put together in a series of lectures given in 1924 by Rudolf Steiner. A group of German farmers had approached him seeking answers to the increasing problems they were encountering as agriculture was becoming more reliant on chemical inputs.

People around the world are still striving to fathom the great depths of what Steiner indicated in these lectures on agriculture. This can become something of a major intellectual exercise. Perhaps for many of us the intellect needs something to grab hold of and chew on for a while, keeping it occupied and allowing a more subtle understanding of biodynamics to develop on other levels.

The Biodynamic Preparations

The fundamental basis of biodynamics is the use of a number of preparations indicated by Steiner as important for regenerating and sustaining the soil.

Horn Manure – Preparation 500

Horn manure (Preparation 500) is effectively the cornerstone of biodynamics. It is made by filling cow horns with cow manure, and burying them underground over winter. In late spring they are dug up and the transformed manure, which now has a sweet earthy smell, is put into storage.

This preparation is normally put on the land twice a year, in spring and autumn. The best time is when the moon's path through the sky is descending (see 'Moon and Planet Rhythms' below), the soil is moist, the weather sunny, mild and not too windy or threatening heavy rain. When conditions are suitable, a small amount of 500, usually 35 g per 13.5 L of clean lukewarm water per 0.4 ha, is stirred in a round container: one preferably made of copper, wood or stainless steel. Up to 540 L can be stirred at a time by a mechanical stirrer, but 135 L is about the limit for hand-stirring.

Stirring the 500

In this land of vast acreages, many farmers have to use mechanical stirrers, but on a small farm of up to around 8 ha, hand-stirring is quite manageable.

Many growers prefer machine-stirring as they feel the machine can do the job better. On the other hand, Rudolf Steiner was quite emphatic when asked whether the stirring operation could be mechanised. His reply was that while the spraying of the 500 could be done mechanically, the stirring should be done by hand because of the importance of conveying human energy and spirit to the liquid that is to be sprayed over the farm. In Europe most 500 is still stirred by hand.

Personally, on a small property of 5 ha, I could not afford to spend $1000 or more for equipment that would only get used three times a year and might save me 6 hours work in that time. On larger properties, though, machine-stirring becomes an economic necessity.

One of the most important things is the personal benefit I feel from stirring a batch of 500 by hand for an hour. It can be a transforming experience, bringing one into a more intimate and meaningful connection with the land. Through participating in the process in this way, my senses seem to be opened so that I can see more, feel more and understand more of what is happening on the land around me.

An old copper, like the one our grandmothers and great-grandmothers used to boil the clothes in, is an ideal vessel for stirring smaller volumes: enough for up to about 1 ha.

For larger areas, 1.2–2.8 ha, a container fashioned out of an old 180 L copper hot-water heater with the top cut off works well. A convex bottom helps develop a good vortex when stirring. I sit mine on an iron stand with three legs, which enables a fire to be lit underneath it to warm the water. When the water is lukewarm, the fire is put out, the horn manure is added, and stirring commences.

A large vessel can be stirred using a pole with a bunch of twigs attached to the stirring end and looking a bit like a witch's broom. This stirrer is suspended from a swivel on a branch or a beam above. Alternatively, one can roll the sleeves up, reach down into the vessel and stir with the hand alone. It does take a bit of effort to get the mixture moving, but I rather enjoy being close to the swirling water.

The aim in the first phase of the stirring cycle is to create a deep vortex that extends as close to the bottom as possible. The circular stirring organises the water into a state of order, all rotating in the one direction.

Then the arm or stirrer is withdrawn, and the water is allowed to continue its rotation for a while. How long this pause is depends on the individual. Some like to keep a vigorous activity going and start stirring in the reverse direction after only a few seconds, to begin the second or 'chaos' phase.

Others like to watch the gradual transformation from a mighty whirlpool to a gently revolving smooth-dished surface, before then starting to stir in the opposite direction. And then some of us like to get the water moving vigorously for the first 5–10 minutes and then take it a bit more slowly.

The second phase is known as the chaos phase. The organised momentum of the water is disrupted and thrown into turbulent chaos when the direction of the stirring is reversed. Then as stirring continues in this new direction, gradually order is established again and the water begins to form a vortex in the opposite direction.

This cycle of stirring in reversing directions is continuously repeated and the water is kept in motion for an hour.

There is no set way to stir, and each individual will approach it differently, though the basic principle is always the same. It is probably good for everyone who is involved with the growing operation to participate in the stirring, though I find doing a whole hour by myself to be quite a profound experience (a comparison of stirring in Europe came to the conclusion that the 500 was more effective if it was stirred by the one person for the whole hour).

When an hour has elapsed, the stirring is completed. It is now time to spread the 500 over the land.

Spreading the 500

When spraying the 500, the aim is to scatter the liquid in raindrop-sized droplets that make contact with the soil. If the vegetation is thick, the droplets may need to be larger so some of the 500 gets down into the soil.

The liquid should be strained before spreading, to avoid blockages and wear on the pump, as there is usually a bit of grit in it.

For spreading the 500 on a few acres, a fire-fighting type knapsack spray is good, provided it has never been used for anything other than water. I prefer to use an old brass one. With the nozzle slightly closed, this can throw a shower of droplets about 6 m either side. Walking briskly while pumping steadily will put out around 13.5 L/0.4 ha, but the rate of consumption should be checked after 0.4 ha, to make sure it is going on at about right the right rate. If I go flat out, this way I can spread 2.8 ha on my own in 2 hours: this is the maximum period that is recommended to keep the 500 after it is stirred. But it is nice to have some

Fresh cow manure is packed into cow horns and buried in the ground over winter. The transformed horn manure, or 500, is ready for use by late spring and can be stored indefinitely.

Clean water is heated to lukewarm and the 500 is dissolved and added. Various methods of stirring can be used, including machine stirrers.

The stirred and strained 500 is sprayed out onto the land.

help or to be doing a smaller area and be able to go at a more leisurely pace.

Alternatively 500 can be spread by hand using a hearth brush or a bracken frond dipped into a bowl and flung out. This is somewhat slower, but quite suitable for small areas.

For larger areas, a tractor-mounted sprayer can be used. This should be set up specifically for biodynamic sprays, not for anything else, and nozzles should be used that give a large enough droplet. On some very large holdings, aircraft are employed for spreading.

I like to put the 500 on all the areas that are in some sort of cultivation, all grazing and compost collecting areas, orchards and introduced ornamentals, but generally not on areas where the growth of native species is being encouraged.

If the 500 is bringing in the forces that foster the growth of European-type introduced plants, these are going to compete more strongly with native species, which are mostly quite different in their relationship to the soil.

From the point of view of herb production, as most are akin to European-type ecosystems, the 500 should be used around them.

Unfortunately nobody was there to ask Steiner what we should be putting on our Australian natives to help them withstand the influences of the European-type environment that we have been creating in their vicinity.

Perhaps some inspiration will come forth with a suitable preparation, drawing on native plants and animals, that can be used to help bring in those forces that nurture our native ecosystems.

Horn Silica – Preparation 501

While 500 works in conjunction with the cosmic forces that foster fertility and growth to increase the vegetative process of plants, these need to be complemented with the forces of light and warmth to foster development of the flowering and fruiting process, which is where the horn silica is used.

This is made by grinding quartz, preferably clear well-formed crystals, to a fine powder and then burying it in a cow's horn over the summer. When dug up it usually shows little apparent physical change, which is not surprising as silica is a fairly stable substance.

A quantity of 2.5g/ha is used and 501 is stirred in the same way as 500, but it is usually applied at different times: for leaf, when growth is at its most lush, and for flowers, fruit and roots, when the plants are just starting to develop those parts that are to be harvested, and maybe again later when they are ripening. It is also best to spray the 501 when the moon's path through the sky is ascending (see 'Moon and Planet Rhythms' below).

The 501 is put on in the early morning when it is clear weather and the sun is still fairly low in the sky. Avoid windy conditions. With 501 the aim is to make contact with the leaves and other parts of the plants. Use as fine a spray as you are able to with the equipment you have available and cast it upwards so that it drifts down onto the plants. Naturally the grower gets quite a dose of it, too. As the sun shines through the spray, it creates little rainbows.

Two schools of thought have emerged in Australia regarding the use of this preparation. Some approach the use of 501 very cautiously from the point of view that the forces of light and warmth are already very strong in our climate and there is a widespread concern that 501 might burn the plants.

Other growers feel that it is important to use 501 as its influence is needed in the balanced development of the plant. My own experience using 501, both in the hot conditions of Victoria and the cooler climate of Tasmania, is that it has not burned any plants and this has also been the experience of other growers in different parts of Australia.

I am inclined to wonder why 501 should burn things. Biodynamic growers continue to use 500 on fertile soil and it does not cause the plants to bloat and fall over, even though the vegetative forces are already strong, so why would 501 have the extreme effect of burning them?

The Compost Preparations

Rudolf Steiner spoke of the need to make compost, and his directions for building a compost heap would be hard to improve upon. In addition, he indicated a series of six herbal preparations to be added to the compost heap once it is built. These are each prepared in a specific way. They help the compost to become receptive to the cosmic forces and regulate various mineral processes.

As a biodynamic herb grower, I was soon drawn into the making of these preparations. At 'Twin Creeks' I was growing most of the herbs that are used in the preparations and the gardens there were a suitable place for making them.

The processes and transformations involved are quite fascinating. Participating in the making of the compost preparations helps bring about a deeper understanding of the dynamics of the biological processes at work on the farm. I think much of this happens on an unconscious level, permeating the biodynamic grower's perceptions of the world.

Of course, just using 500 and 501 will have this effect to a degree, but getting involved with the other preparations seems to heighten this and add more dimension to it.

The making of the compost preparations is best undertaken by an organisation responsible for the fostering of biodynamics. While individuals can make their own, because each farm will only require very small quantities and many farmers do not have the herbs available, it makes sense for the preparations to be made and stored centrally, and then distributed from there, and this way the quality can be assured.

As biodynamic herb growers are likely to get involved in the making of preparations, it is important that they have some understanding of the processes involved. The actual filling, burying, digging and storage is normally done by a group, under the supervision of someone with some experience in making the preparations to ensure their quality, so I have not gone into fine detail on these aspects.

Individual growers will often be called to do the gathering and drying of the herbs beforehand on their own, so I have tried to cover this more completely.

Yarrow Flowers – Preparation 502

The flowers of Yarrow (*Achillea millefolium*) are in umbels that bear some resemblance to the antlers of a stag. To make this preparation, the flowers of Yarrow are packed into the bladder of a stag, hung in a sunny place over summer and buried in the ground over winter until the following spring. The buried bladders may need to be contained and protected in a manner that will allow the soil influence to reach them.

This preparation helps regulate the potassium and sulphur processes.

If you are collecting Yarrow flowers for this preparation, it often has to be done a season in advance, as in many regions the plant does not flower until after the summer solstice, and the bladders need to be filled before this date. The flowers can be gathered once they have opened, but it doesn't matter if they have advanced towards the fruiting stage.

It is the flowers themselves that are required, with the stalks and leaves removed. There are a few options for doing this.

Some prefer to harvest the individual flower heads with a pair of small scissors, snipping them off at the base so no stem is included. The flowers can be used fresh if the time is right, or else dried and stored until the next spring.

Another method is to harvest the whole umbels, dry them in the shade, and later, at your leisure on a rainy day or sitting around the fire at night in winter, the flowers can be carefully snipped off.

A quicker alternative is to carefully harvest the umbels, avoiding or removing any leaves as far as possible, and then dry them. When they are brittle dry, the umbels be rubbed very carefully through a 4 to 5 dent screen (mesh aperture around 5 mm) so as to separate the flowers without breaking up the stalks at all. Rescreening a couple of times may be necessary to separate the odd stalk that goes through. Yarrow flowers produced by this method may contain a tiny bit of stalk and leaf, though all but perhaps the most discerning will find them quite acceptable.

If you have a good stand of Yarrow, a portion of the flowers can be picked for this compost preparation from the first harvest of the season and the rest of the plant, which will still include plenty of flowers, can be harvested for other purposes.

Yarrow tends to become crowded after a season or two and stops flowering, so if you are growing it specifically for making the preparation, it will need continual rejuvenation. Some strains seem to maintain flowering better than others (see under 'Yarrow' in Chapter 15 'Spreading Herbs').

Chamomile Flowers – Preparation 503

German Chamomile (*Chamomilla recutita*, also known as *Matricaria recutita* or *M. chamomilla*) is used for Preparation 503. Again it is the flowers that are used from this herb. They are stuffed into the small intestine of a cow to make a string of sausages.

A small thin-walled cow's horn, cut off at the tip, makes a good funnel to feed the flowers into the intestine. An interesting mingling of fragrances arises from this activity, with the perfume of the flowers, the smell of the digested diet of the cow arising from the intestines, plus maybe the stink of decay if the intestines have been sitting around for a few days before being filled. This helps bring about an awareness of the interconnectedness of the decay processes and the growth processes. Somehow the aroma of the flowers seems to sweeten the other smells and they don't seem as nauseating.

This preparation is made in autumn and buried in the ground over winter. Usually the sausages are best stuffed into an unglazed earthenware pipe to contain and protect them from little underground marauders. By spring they are well broken down and ready to use.

The Chamomile preparation helps to regulate the calcium process.

The flowers should be picked when they are around the middle of the flowering process. Some authorities insist that the flowers for this must be picked by hand with absolutely no stalk, but this is very time consuming. An alternative is to harvest them with a comb. Excessive stalk among the flowers can be picked out prior to stuffing. Another option could be to rub the dried flowers through a 4 or 5 dent screen (mesh aperture around 5 mm) to separate the stalks. Otherwise the stalks can be easily picked out or sifted out of the finished preparation when it is dug up in spring.

Fresh Chamomile flowers can be used, but dried ones are probably preferable as a greater bulk of dry matter will be packed into the sausages.

Nettle – Preparation 504

Preparation 504 is made from Greater Nettle (*Urtica dioica*). The aerial parts are used, harvested and dried when the plant is in flower and then buried in autumn to be left in the ground for a whole year. They can be packed directly in the soil, though they are usually surrounded by something that will contain and protect them without preventing the influence of the soil from reaching them. Commonly they are packed tightly into an unglazed earthenware pipe.

Nettle helps the humus-forming process in the compost.

The first or second harvest of the season is probably the most suitable as these have a good proportion of flowers, while the later growth tends to flower less or not at all. Make sure you have Greater Nettle (*Urtica dioica*) and not Lesser Nettle (*U. urens*) nor Native Nettle (*U. incisa*) (see under 'Identification' in the entry for 'Nettle, Greater').

Normally the herb is dried and then wetted before packing it into the pipes, but fresh Nettle can be used. Drying conditions are not critical, as it doesn't really matter if the herb goes dark.

While underground, the leaves of the Nettle are transformed into a black humus, but the stalks don't break down much and need to be sifted out of the preparation when it is dug up.

Oak Bark – Preparation 505

Preparation 505 is made from the bark of English Oak (*Quercus robur*), the same species that is used in herbal medicine. But for the biodynamic preparation it is the dead outside bark that is used, not the live inner bark. This is milled to a medium coffee grind and packed into the skull of a domestic animal, usually a sheep or a cow. Then it is buried in the ground over winter in a wet spot where the soil will be waterlogged. Sometimes precautions have to be taken to prevent its being dug up, particularly in Tasmania where devils will drag the skull out and devour it.

The Oak bark preparation helps to regulate the calcium process.

In harvesting the bark, care should be taken not to damage the tree. If the tree is valued as an ornamental, the bark should only be taken from a place where it will not be noticed, or from small portions of the trunk.

There are a number of techniques used, depending on your preferences. Flakes of bark can be removed with a screwdriver. This method is suited to situations where the appearance of the tree is important as a flake can be removed here and there, and it will look more or less like natural wear and tear.

Alternatively, an axe or a drawknife can be used to chip or shave a layer of outside bark from the trunk. If you have somebody to help (who has enough confidence in your ability to control the blade) they can hold a container underneath to catch the bark as it falls.

Another method, which saves having to grind the bark later, is to rasp the bark off the trunk. This method is good for younger trees where the bark is thinner and there is a risk of cutting into live bark. If there is a lot of lichen or moss on the trunk, it should be cleaned off with a wire brush first, as with the rasp you could end up with mostly lichen and not much bark.

Whichever method you use, take care not to cut into the inner bark. The dead outer bark is dry, hard, grey on the outside and dark red where it hasn't been exposed to light and weather. The live inner bark is moist, softer, with a pinkish brown colour.

Dandelion Flowers – Preparation 506

Preparation 506 is made by wrapping Dandelion flowers in the mesentery of a cow. The mesentery is the membrane that surrounds the intestines and is commonly known as the caul fat. This is one of the more difficult preparations to make, first to get enough Dandelion flowers and then to find a cow's mesentery that is not too fatty.

The wrapped Dandelion flowers are buried in autumn and dug up in mid-to-late spring. If the process has worked, they should have turned to a dark sticky humus.

This preparation helps to regulate the silica and potassium processes.

In gathering the flowers for the making of 506, you first need to make sure you have the right species. There are other related plants with similar looking flowers, but these are on branched or solid stems. The true Dandelion has a single flower head at the top of a hollow unbranched stem and a leaf that is smooth with no hairs on it.

The next challenge is finding enough Dandelion flowers. Usually in the wild there will only be a few flowers at a time, so you need to come back each suitable day and pick them over a period of several weeks.

If you are growing Dandelion as a crop, you will very likely find yourself called into service picking flowers for making this preparation, as biodynamic groups all over the country never seem to have enough of it.

For this you will need to produce dried Dandelion flowers that retain their yellow petals and have not developed to the fluffy white pappus or 'puff' stage. To successfully do this you need to understand how the Dandelion flowering process works.

First a bud is formed, which is then raised up on a stalk. Most of the flowers open for the first time in the morning, the actual time depending on the season, the weather and the prevailing temperature. In the heat of midsummer this may be as early as 6 or 7 a.m., but in cooler temperatures it will be later.

The rate of opening progresses quite quickly in warmer conditions, so to catch these new flowers in their early stage of development, you need to be out there soon after the flowers begin to open. If you are too early, some may not have opened enough, but if you are too late, many flowers will be too advanced and will go to puff before they are dry.

The Dandelion flower head is actually a composite of little individual florets that open progressively from the perimeter. The outside florets open first and gradually the opening process moves towards the centre. The unopened florets are clustered in the centre of the flower, curled inwards. This cluster becomes progressively smaller as the petals unfold until finally the flower consists entirely of fully opened petals.

If the flower is picked before the central cluster of unopened florets disappears, the flower head will dry to a shrivelled yellow and green bundle, which will be suitable for making the preparation. If the flower is picked after the centre is fully open, it will progress to the 'puff' stage before it dries.

In warmer weather, the opening process only takes a few hours. By mid- to late morning the process will have finished and it will be too late to pick the flower heads. In cooler weather, the Dandelion flower takes longer to open: up to a day or more. Fully opened flowers will close at the end of the day and reopen the next morning, but the

Dandelion – Preparation 506: the flowers are used for this compost preparation. The flower head on the left has not yet fully opened, as seen by the central cluster of unopened florets, and is suitable to pick for this preparation. The one on the right is fully opened and is too advanced to pick.

second time they will open fully straight away without any central clusters. At the end of the second day they close again and usually then form seed.

The flowers need to be dried rapidly. The first day is critical, so it is not really worth picking them if the sun is not shining, unless you have some artificially heated situation to put them in.

To speed up the initial drying, they can be placed in direct sunlight on a screen for the first day. If this can be set against a reflective wall, such as a corrugated iron shed, they will get some additional heat. You may have to move the screen around to keep it in an optimum drying situation all day. At the end of this first day, place the screen on the top rung of your drying shed, over the wood stove in your kitchen or some place where it can continue in a good drying situation for a few days.

Once the flowers are fully dry, put them in a paper bag and store them in a dry place.

If you go out and pick the flowers every suitable day, you will find the flowering seems to fluctuate according to the season, the weather and other factors, such as the moon and the constellations. It is all rather complex and I can't say I have ever managed to unravel it, but some days there will be four or five times as many flowers as other days.

In southern Australia, Dandelions growing in the wild will normally only flower when the right conditions prevail in spring and will cease flowering when conditions become too dry.

In warmer regions, hot summer weather seems to be detrimental to the flowering of Dandelions: they don't seem to function properly and may not follow the normal pattern of opening. Heat can close the flowers, and sometimes they get infested with little crawling things. In north-east Victoria it is not worth picking the flowers until the hot weather eases off a bit around March. In a cultivated irrigated situation, wild-strain Dandelions will start to flower strongly in autumn. This may be the best opportunity to pick good quantities of flowers. If the Dandelion's roots have been harvested that winter, the regenerating plants tend to be too young to flower much in spring.

The broad-leaf strain (a cultivated variety of Dandelion) does not produce as many flowers: it typically has a burst of flowering in spring and not much after that.

Valerian Flowers – Preparation 507

Preparation 507 is made from the flowers of *Valeriana officinalis*, the same species grown for its roots and used medicinally. The flowers are picked when they are just opening and the juice is extracted from them in water.

This preparation helps regulate the phosphorus process.

There are two or three different methods of making this preparation, but they all start off with picking the flowers in the early opening stage. This will be late spring to early summer.

Valerian flowers open sequentially, that is, one after another, so when you go to pick them you will find them at different stages of development. The best stage is when some of them have opened and there are still a lot of buds.

One method is to carefully cut the flowers off with scissors to avoid including any stalk. This can be very slow if you are trying to harvest any quantity. A quicker method is to pinch the flowers off between the thumbnail and the flat of the forefinger (you need a reasonably long thumbnail for this). If a bowl is placed on the ground, the flower heads can be bent over it for the operation. With a bit of dexterity, it is possible to nip off several flowers with each pinch, and very little stalk is included.

The juice of these flowers is extracted with water. One method is to fill a glass bottle with the flowers and then add just enough water so all the flowers are submerged. The bottle is then left in the sun for about a week, after which the mixture is poured out and the liquid is squeezed out and filtered.

A second method is to crush the flowers with a mortar and pestle and then soak them in clean lukewarm water (25 g flowers per 100 mL water). This mixture is allowed to stand for 24 hours and then strained through filter paper (in one variation on this method, the mixture is allowed to stand in the sun for a week as in the first method).

The Valerian extract is quite slow to go through the filter. Once when we were making this preparation in very hot weather, I put a flagon of the filtered liquid in the fridge while waiting for the rest to pass through. It looked like very weak tea, in fact not much darker than our tank water at the time, which was somewhat tinted by a lot of gum leaves which had fallen in the gutters.

In the heat of the day, a friend came in from working outside and went to the fridge looking for a cold drink. Seeing a flagon of what she took to be water, she upended it and drank a good portion of it before the rather bitter flavour took hold. She must have swallowed enough for several hundred tons of compost. We rather expected her to turn luminous as a result of a massive release of phosphorous, but it didn't seem to have any obvious effects.

The filtered liquid is poured into dark glass bottles, preferably small enough so that, once opened, the contents will be used within a few weeks. When filling the bottles, the liquid should poured in so they are absolutely full, with a meniscus sitting above the brim, and then the lid screwed on tightly. This way they will not contain any air bubbles, and should keep indefinitely. It is important to ensure the bottles don't get too warm, as they can burst: they should be stored in a cool dark place, preferably buried in dry peat moss.

Using the Compost Preparations

The use of the compost preparations is an important aspect of biodynamic farming that has been overlooked in Australia where most farms are broad-acre operations. However, there is a growing awareness of the need to bring the influence of these preparations onto the land by some means. There are a number of methods used but in an intensively cultivated operation such as herb growing, the application of biodynamic compost is the most appropriate.

Making Biodynamic Compost

Biodynamic compost making does not differ greatly from other methods, the main point is the addition of small portions of the preparations 502–507.

A series of six holes is made with a crowbar into the sides of the compost heap. These holes should be more or less evenly spaced around the heap. The first five preparations, 502–506, are each added separately to the heap, a 1–2 g portion of each of the preparations being placed in the centre of a small lump of mature colloidal compost. This lump of compost and preparation is pushed down a hole into the active part of the compost heap so it makes firm contact, then the hole is filled in and covered.

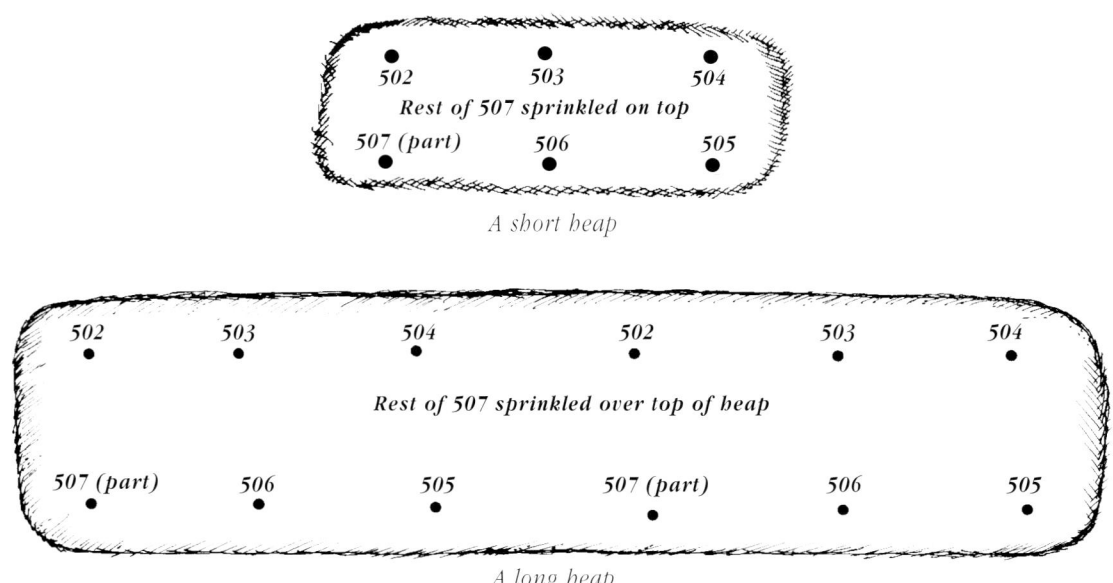

Placing compost preparations into the compost heap. The six compost preparations, 502–507 are each inserted into a separate hole in the compost, and part of the 507 is sprinkled over the top of the heap.

112

Next, 5 mL of the liquid preparation 507 is added to about 4 L of clean lukewarm water and stirred in a bucket or a bowl for 15 minutes (some say 10 minutes, others say 20). Part of this diluted preparation is then poured down the sixth hole and the rest is then sprinkled over the top of the heap.

The only other significant difference between biodynamic and many other methods of making compost is that the biodynamic compost heap is covered with straw or some other available coarse plant material that has similar breathing and insulating qualities (and is preferably free of weed seeds). Coarse grass, pine needles, bracken fern or some coarse but previously heated compost will do. This outside layer helps to contain the heat and moisture, so the composting activity extends further towards the outside of the heap.

Other Methods of Using the Compost Preparations

On many Australian broad-acre farms where compost is not part of the normal regime, some other methods are used to get the influence of the compost preparations onto the land. One option is to stir them in with the 500, but many growers feel this is not appropriate, except perhaps for 507, which is a liquid preparation and is sometimes added to the 500 for the last 15 minutes of stirring.

Some growers use horn manure which has had the compost preparations added to it during storage, but others don't accept this method.

Another approach is to put them into cow manure in what is known as a cow pat pit. After a while this becomes something similar in appearance to Preparation 500, and is put on in the same manner, though its action is different, as it is primarily a vehicle for the compost preparations.

For the small- to intermediate-scale herb grower whose operation is based on the use of compost, these other options are probably not needed, except perhaps to put onto land used for grazing or for the production of compost materials. Even so, a light dressing of compost for these areas would be preferable, if sufficient is available.

Moon and Planet Rhythms

The role of the moon and planets in biodynamic farming is a complex subject and one that merits a whole book of its own. In addition to following the commonly perceived rhythms of the sun, and the vagaries of the weather, many biodynamic farmers organise their activities according to the rhythms of the moon and the planets so as to bring various cosmic influences onto their crops.

The Four Main Moon Rhythms

The moon exerts an influence on plants which changes according to the various rhythms (see overleaf).

Lunar Nodes

Because the moon goes through its ascension and descension rhythm every 27.2 days, while the sun takes a year to complete the same rhythm, the moon must pass through the sun's path twice a month. Biodynamic growers try to avoid farming activities for several hours surrounding these nodes as they tend to exert a negative influence. Emotional crises seem to arise at these times, also accidents and mechanical breakdowns.

Working with Cosmic Influences

These cosmic influences may be more qualitative than quantitative and the effects may not be obvious on a tangible level. Actual increases in yields may not always occur, but there is certainly an effect on the consciousness of the grower and the consumer.

Of course there must be flexibility, and while biodynamic growers endeavour to plan their activities according to moon and planetary cycles, the prevailing weather and season are still major, perhaps *the* major, factors. For example, there is no point in rushing out to harvest in the right moon sign if it is pouring rain.

There is also the reality of modern society, with its many demands on one's time that do not always coincide with the farmer's calendar. Just because a task cannot be performed on the most suitable day, does not mean that one should not do it at all or that the biodynamic calendar should be cast aside. Instead one can use it as an opportunity to observe the outcome and in that way learn more about how these influences work.

Becoming aware of the cosmic rhythms and cycles helps biodynamic growers to see themselves as participating in a complex interacting of forces and influences, working in conjunction with them to contribute to the vitality of the farm.

If readers wish to go into further depth on this subject of moon and planetary rhythms, they are referred to the excellent 'Antipodean Astro Calendar' put out each year by Brian Keats and Sue Pearson. This is available through the various biodynamic associations, which can also be contacted for further information on biodynamics and on obtaining the compost preparations.

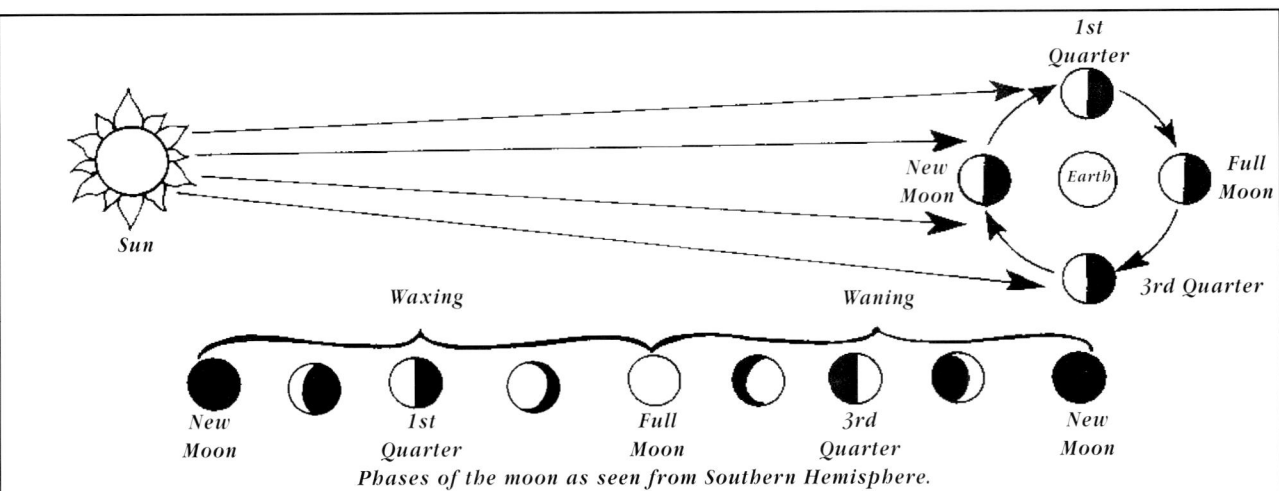

Phases of the moon as seen from Southern Hemisphere.

1. Waxing and waning rhythm: Everybody is familiar with the moon's waxing and waning rhythm over a period of 29.5 days from new moon to first quarter to full moon to last quarter to new moon. Many people notice that their mental and, maybe, physical state is affected by this moon cycle, which also has a strong influence on plant and animal realms. The waxing moon fosters leaf qualities, with lusher growth and moisture uptake, while the waning moon brings on root qualities, declining moisture content and better keeping produce.

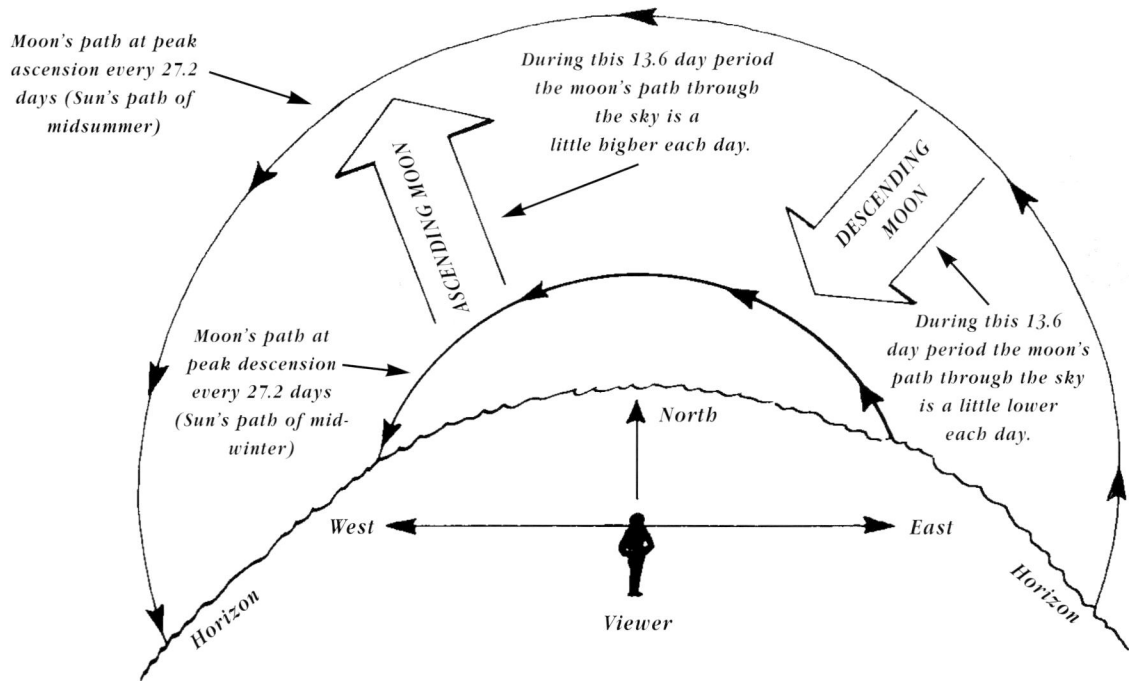

2. Ascending and descending rhythm: The ascending and descending rhythm of the moon does not refer to its daily rising and setting, but rather to the path followed by the moon as it rides through the sky, which is higher and lower in a rhythm of ascending and descending over a period of 27.2 days. This mini-cycle imitates the sun's annual cycle of riding higher in the sky in summer and lower in winter. The ascending moon imparts a spring-like influence, while the descending moon's influence is autumn-like.

The four main moon rhythms. The moon has an influence on the growth and qualities of plants that changes according to its various rhythms.

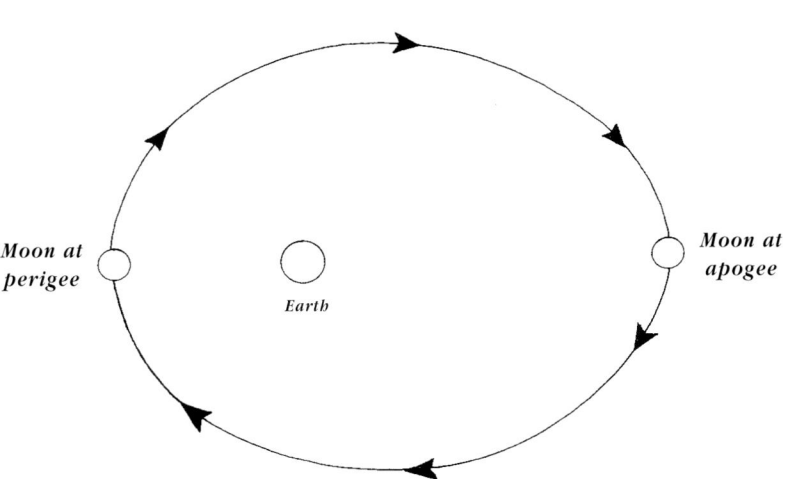

3. Apogee and perigee rhythm: The distance of the moon from the Earth varies in a cycle of 27.55 days as the moon follows an elliptical orbit. When it is closer to the Earth, the moon's influence is stronger, and it is weaker when further away.

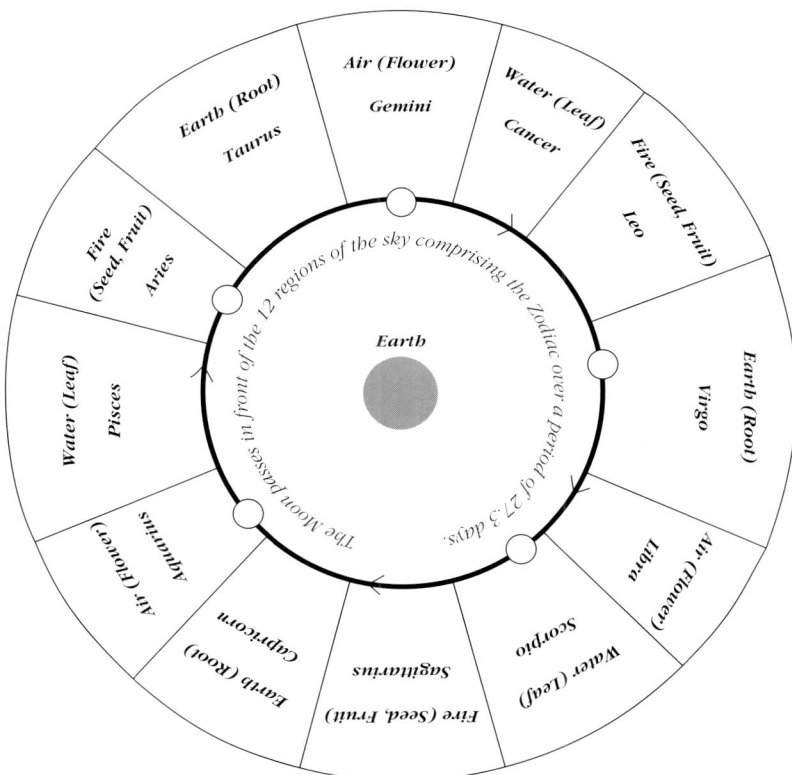

4. Sidereal rhythm: In a 27.3 day cycle, the moon passes in front of the 12 regions of the sky known as the Zodiac. Each of these regions or signs is associated with one of the four elements in nature: Earth, Water, Air and Fire (or Warmth) and the formative forces emanating from these regions have an influence on correlated parts of the plant, which is heightened when the moon is in that region. Thus Earth correlates to root, Water to leaf, Air to flower, and Fire to seed. When the moon is in a particular sign, farming activities will tend to accentuate the related aspect of the plant. Thus if we are growing a plant for its root, then planting, cultivating and harvesting that plant on 'root days', when the moon is in an Earth sign, will further develop its root qualities.

Timing of Operations

In growing herbs under a biodynamic regime, it is important to keep in mind the cycles of the herb itself. Often it is possible to follow the moon and planetary rhythms in performing the various operations and fit this in with the herb's own cycles, but sometimes they just don't synchronise.

Generally the grower tries to follow the moon signs appropriate to the part of the plant that is to be harvested. However, some herbs are harvested for both leaf and root, so some compromise has to be made. In a mixed garden situation, it is usually not possible to time cultivations so they are carried out in the optimum sign for all the plants.

For example, biodynamic growers prefer to harvest flowers on 'flower days', when the moon is in those constellations that exert an influence on the flower aspect of the plant. This means harvesting every 7–9 days or multiples of this period. While this may sometimes fit in very neatly with the flowering cycle, at times it will be too short or too long. The optimum interval between harvests varies according to the species and time of year: in the vigorous growing period of late spring and early summer, flowers develop and mature rapidly, while in the coldest part of the year, this process is much slower. There are also the vagaries of the weather to take into consideration. Consequently it is not practical to rigidly follow the moon calendar throughout the year for flower harvesting, though preference can often be given to harvesting on flower days.

Sometimes a grower will depart from the usual pattern if they wish to accentuate a certain quality in a plant. For instance, where possible, herbs whose leaves contain essential oils are generally planted, cultivated and harvested on flower days as something of a flower quality is desired in the aromatic leaves of these species.

Pest and Disease Control

One of the principles of the biodynamic approach is that a plant does not become diseased in the same sense that humans and animals do. To quote Pfeiffer's translation of Rudolf Steiner's explanation: 'Properly speaking there can be no such thing as sick plants, for the etheric is always healthy. If disturbances occur in spite of this, it is a sign that something is wrong with the environment of the plant, especially the soil'. In other words, the plant reflects the nourishment and vitality that is provided by the situation in which it is growing.

When a pest or disease problem arises, the biodynamic grower will wait and watch, to see if the imbalance will correct itself or if something can be done that will help to restore balance. If some action is called for, they will prefer to approach the problem using skilful means or maybe a biodynamic remedy rather than resort to drastic control measures or even natural pesticides.

Horsetail Tea – Preparation 508

Fungal diseases are sometimes treated with Horsetail tea. This is made from the dried Field Horsetail (*Equisetum arvense*), by boiling approximately 21 g per 1200 mL of water for 20 minutes. When the mixture is cool, it should be strained and diluted, usually in 4.5 L of water and stirred for 10 minutes in the same manner as Preparation 500. The resulting Horsetail tea is then normally applied to the soil, but some farmers apply it directly onto the plant.

One herb grower has obtained reasonable control of Mint rust with weekly applications of a mixture of 508 and 501. I must admit I have never tried it as a single flaming at the appropriate time has always worked well and the prospect of spending 2 hours or more each week spraying seems a bit daunting.

Some growers use leaves of the River Sheoak (*Allocasuarina cunninghamiana*) and other fine-leafed Sheoaks as a substitute for Horsetail for fungus control.

Peppering for Pests and Weeds

Peppering is a method of control indicated by Rudolf Steiner in his lectures, but there is still no consensus among biodynamic growers and farmers as to its effectiveness. The principle is to take the skin of larger animals, the whole of smaller pests like insects, or the reproductive portion of plants, and burn it at particular alignments of the moon and the planets. The ashes are then scattered like pepper in the areas where the infestation is to be discouraged. Sometimes the ashes are potentiated in the same manner as homeopathic preparations. The intention of peppering is to reduce the reproductive fertility of the pest or weed and cause its numbers to decline.

The main problem seems to be that results are unreliable. Perhaps we don't really understand what Steiner meant or we are placing unrealistic expectations on peppering's effectiveness.

Steiner's words were that the pests or weeds would gradually withdraw their energies over a period of years. If peppers do work in this way, then they would be quite useful for a grazing or a broad-acre situation, where the immediate short-term consequences of the infestation are not too serious, and the farmer can afford to wait a while for the peppering to take effect. But in a herb-growing situation where a very high level of weed control is necessary, it would be risky to rely on peppering alone. Of course there would be no harm in making the pepper and putting it on, but the usual weed control methods should be continued.

Even if the peppers don't have a direct tangible effect, the process of making and using them may help the grower form a closer connection with those pests and weeds, leading to a better understanding of them and, perhaps, a means of coping with the problem.

One concern I do have is that growers are turning to peppers because they are looking for something to 'put on' for an easy 'quick fix' solution. This same approach has led to conventional agriculture's heavy reliance on pesticides, herbicides and chemical fertilisers, rather than looking for the causes of problems and what can be done to modify or alleviate the circumstances that have brought them about.

The Farm as an Organism

The concept of the farm functioning as an organism is a fundamental aspect of the biodynamic approach. This involves a diversity of production and the interaction and interdependence of the plant and animal and human realms within the farm, to the point that the farm becomes a whole system, taking on a life of its own.

A biodynamic farm should have at least some domestic animals whose manure is part of the growth-decay-growth cycle, but not all the plant material produced on the farm should be passing through the digestive systems of animals. There should be a surplus available for the making of compost, and some should be passing directly back to the soil to help maintain the levels of organic matter.

Managing a diverse farming operation can be a challenge, particularly when one of the operations is herb growing, which can be fairly demanding if it is to be done well. Other aspects of the farm can easily become neglected or the herbs suffer at critical times because of the pressure of other work.

My own approach is to focus on the herb growing and to limit other aspects of the farm's operation to only those things that will complement the herb operation, or at least not be to its detriment.

For instance, a few cattle can contribute their manure to the compost making, a few sheep can help keep weeds down around the farm (outside the herb garden), ducks and chooks can be effective for controlling insects and snails, geese can be used to graze around trees and bushes. Some vegetable and grain crops can fit in well with herbs that are grown in rotation with other crops.

There do need to be some animals in the system, even if they are native animals, though cows are preferred because their influence contributes to the fertility of the farm more than other animals do. The temperament and digestive processes of the cow and the nature of its manure foster beneficial microbial activity and the building up of fertility.

Another biodynamic principle is that farms should be self-sustaining as far as fertility goes. The ideal biodynamic farm should not need to bring in fertilisers.

Of course everyone agrees that some farms may need to bring in organic or natural mineral fertiliser to initially raise the vitality of their soil, but once a good productive level is reached, it should be possible to maintain this without additional inputs. The regenerative forces in action on a biodynamic farm should be able to sustain the biological activity even though some nutrients are continually being removed in the form of produce sold.

This seems to fly in the face of conventional thinking on the need to replace what is being taken off the farm, if soil nutrients are not going to be depleted. The biodynamic response is that biological systems are not bound by chemical laws and that there are other factors involved.

In comparisons of biodynamic and conventional farms, the measurable levels of soil nutrients in a biodynamic system tend to be lower. Productivity is usually comparable but a little lower, while soil structure, organic matter levels and plant and animal health are superior. Nutrient run-off is lower. Lower operating and input costs generally mean that net returns for the biodynamic farm are very similar to those for conventional farms.

So far, there have been no long-term trials to assess whether biodynamic production can be sustained over a period of 100 years or more without any inputs from outside. However, in an intensively cultivated, high fertility, cropping situation, such as most herbs are grown in, there are too many losses from the cycle for production to be maintained. This requires a high input of compost made from material brought in from outside the garden.

In a biodynamic system this is predominantly or entirely made from plant material and animal manures from the rest of the farm. The herb garden should not be seen as a separate entity, but as part of the whole farm organism. While the rest of the farm may be contributing to the physical maintenance of the herb garden, the herbs will be contributing to the vitality of the farm in less tangible ways. The farm overall can still be essentially self-sustaining, provided the intensive production is only carried out on a limited portion of it: probably no more than 25%.

A biodynamic farm should be looked at as a whole system: the individual aspects of it may be hard to comprehend on their own. A well-managed bio-dynamic farm has a healthy vitality about it that speaks for itself.

Bio-dynamic plants and animals tend to be leaner but healthier. With herbs, the fresh weight yields may be a little lower, but there is more dry matter in the plants and essential oil contents seem to be higher, so the quality is better.

On a biodynamic farm you won't see that lush blue-green growth, symptomatic of plants pushed into imbalance with excess nitrogen, so yields may not be as high as those that can be obtained under conventional farming and some organic systems, but they are sustainable, both ecologically and economically.

Associations Involved with Biodynamics

Bio-Dynamic Farming & Gardening
Association in Australia
PO Box 54, Bellingen NSW 2454
Telephone (066) 55 0404, after August 1997 –
(02) 6655 0404

This is a national organisation with regional contacts throughout most of Australia. Members receive a quarterly newsletter and can obtain biodynamic preparations.

Bio-Dynamic Farming and Gardening
Association in New Zealand
PO Box 306 Napier
New Zealand

This is a similarly structured organisation, which also carries out certification of biodynamic growers in New Zealand.

Biological Farmers of Australia Co-op Ltd
PO Box 2577
Canberra City ACT 2601

This is a national association of organic and biodynamic farmers that carries out organic and biodynamic certification.

Bio-dynamics Tasmania
PO Box 543
Ulverstone TAS 7315

This is a state organisation which produces a newsletter and supplies biodynamic preparations to members.

Biodynamic Agriculture Association of Australia
Powelltown VIC 3797

This is a national association of biodynamic farmers that operates under the guidance of Alex Podolinsky. It supplies biodynamic preparations to members and carries out biodynamic certification.

Further Reading

The following books are available by mail order from:

The Rudolf Steiner Book Centre
307 Sussex St
Sydney NSW 2000
Telephone (02) 9264 5169, fax (02) 9267 1225

or from some of the organisations involved with biodynamics that are listed above.

Agriculture by Rudolf Steiner: A translation of his original lectures. Heavy going in parts, but it is interesting and relevant to go to the source.

Bio-dynamics: A Modern Sustainable Agriculture by Terry Forman – Terry has a wonderful ability to explain biodynamics in ways that make sense and inspire.

Bio-Dynamics – New Directions – New Zealand: A compilation of studies and stories of a number of different biodynamic farms. Very practical and down to earth.

Culture and Horticulture by Wolf Storl: An outline of biodynamic philosophy that most people will find quite readable.

Antipodean Astro Calendar – Biodynamic planting guide & planetary rhythms, compiled and published by Brian Keats and Sue Pearson: an excellent calendar showing cosmic influences, with detailed explanations.

Grasp the Nettle – making biodynamic farming and gardening work by Peter Proctor. An easy-to-read and practical outline of biodynamics. Peter Proctor has 30 years experience in biodynamic management and as a field advisor both in New Zealand and internationally.

11
Harvesting

Harvesting is a vital link in the production of herbs. The quality of the final product can only ever be as good as the material that emerges from the harvesting operation. Harvesting needs to be carried out with considerable care to ensure that the appropriate part is taken from the living plant at the optimum time; that it is vibrant and healthy; and that it is not adversely affected by the harvesting process.

HARVESTING LEAF, LEAF AND FLOWER, AERIAL PARTS

The term 'leaf' is used when the finished product is predominantly leaf, though it may contain a few flowers. Many leaf herbs are at the flowering stage when harvested, so a significant proportion of the dried herb consists of flowers included with the leaves: this is termed 'leaf and flower'. When the finished product comprises the stems, leaves, and often also the flowers, this is referred to as 'aerial parts' or 'herb'.

With few exceptions, leaf, leaf and flower, and aerial parts are harvested and dried in the same manner: by cutting the stems at the appropriate level and drying them intact. Where the stems need to be removed, this is best done in the processing after drying.

Unfortunately, a piece of misinformation frequently repeated in herb books advises readers to strip the leaves from the stems before drying. This not only entails a lot of extra work, but it may bruise the leaves. For almost all herbs, it is easier to get the leaves off the stems by rubbing them through a screen after they have been dried.

Timing

There are a number of factors that need to be taken into consideration in determining when to harvest, and these are outlined below.

Stage of Growth

As a general rule of thumb, leaf, leaf and flower, and aerial parts are harvested when the plants reach the peak of their development and vitality, which is usually at the early flowering stage. If allowed to go beyond this, they start to put their energy into seed formation and the quality of the leaves progressively deteriorates.

With most species, it is primarily the leaves the grower is interested in, so if they reach their prime before flowering commences, harvest is best taken then. Some species start to go into decline before they flower. Some only flower once, early in the season, and their second or third harvests are mostly leaf.

The best guide is too look at the leaves near the base of the plant. As the plant reaches the optimum stage, these start to fade, die and fall off. A decline then ensues as the growth of new leaves slows while the deterioration of older leaf increases. This decline can result in a loss of yield and/or a loss of quality.

There is usually no advantage in having plants mark time, not gaining much and possibly deteriorating, when they could be harvested and getting on with growing a second or third cut.

Active Constituents

The end use of the herb should also be taken into consideration. In some cases the levels of desirable constituents build up later in the plant's development. For example, with Peppermint the concentration of desirable essential oils peaks at flowering, so if an especially high level is required it may be necessary to hold the plants until this stage, even though the appearance and yield of the leaf is deteriorating.

Ginkgo leaves are an exception to general rules: the active constituents are actually higher in the leaf when it falls from the tree in autumn.

Interval between Harvests

The interval between harvests depends on a number of factors: the achievement of optimum yields, the quality of the leaf and the need to allow the plants to build up their reserves again.

For most crops, a 6–8 week interval allows enough time for the plant to replenish its reserves and develop optimum yields, but the interval varies somewhat according to the species, the location and the time of year.

Time of Day

Harvesting in the Morning

The tradition is that leaf should be harvested in the morning when the dew has lifted: this is when the essential oils or other active constituents are said to be at their highest. There is some doubt as to whether this is actually the case, but there is evidence that harvesting early in the day reduces losses of essential oils and there are a few other good reasons for harvesting at that time.

Avoiding Wilting and Bruising: If the plants are harvested while it is still cool, their moisture content and turgidity (or inflation with water) is greater: this makes them easier to cut and also minimises bruising. If plants or cut material are handled in a wilted state, they will bruise to a greater or lesser degree, according to the species.

Bruising causes discoloration of the dried herb: some will go quite black. While this condition is deliberately created in harvesting and processing Oriental Tea and Tobacco, it is not sought in any of the herbs with which we are dealing. If herbs are harvested and handled before the heat of the day develops, there will be less wilting and less bruising.

Avoiding Heating: Heating is another consideration in choosing a time of day to harvest. A pile of any fresh-cut plant material generates heat, just as a compost heap does. So a heap of harvested herb will get hot inside: how long this takes and how hot it gets depends on prevailing conditions, the nature of the herb, the size of the pile and the initial temperature of the herb as it goes into the pile. If the herb is harvested in the cool of the morning, it will take much longer to build up to composting temperature than if it is cut in the heat of the day, because the composting process accelerates at higher temperatures.

Getting Drying Started: If the herb is cut early in the morning, it means that it will be spread out earlier in the day and will get a good start drying.

Keeping Cool: If the days are inclined to be hot, then the heavy work of harvesting and carting will be finished before the heat comes on, and the hot part of the day can be spent doing lighter work, like spreading the herb onto screens and processing.

Harvesting Later in the Day

The above considerations do not mean that harvesting must never be carried out later in the day. If the weather is cool, it is not so critical to harvest early, and even in warmer weather, harvesting can be done as long as measures are taken to reduce the adverse effects of the heat.

Availability of Drying Space

Timing of harvesting can be affected by how much drying space is available. To get optimum usage of drying facilities, it is good to have them operating close to capacity for as much of the time as possible. On the other hand, it may be difficult to arrange a harvesting schedule so that availability of drying space coincides with optimum harvest times.

There will tend to be 'bottlenecks' at certain times of the year, particularly in late spring and early summer when a large number of species are ready at the same time. Consequently it is best to anticipate this and harvest some crops a little on the early side if drying space is waiting to be utilised. There may be a small sacrifice in yield, but the quality will be better than if crops have been held some time past their optimum harvesting stage.

By staggering the first cuts of the season in this way, the timing of the second and third cuts tend to follow suit, as the period of regeneration is approximately the same for many herbs.

There are also a few tricks one can use to shorten the drying time if there is a backlog of crops waiting to be harvested, and we will go into these in the chapter on 'Drying'.

Weather and Irrigation Cycles

Other points to consider are the weather and your irrigation cycle. If the plants have been rained on or watered the night before, the moisture on the leaves in the dense growth of a crop ready to be harvested is often slow to dry off. If the soil is soggy, walking on it will compact it, so it is best to wait a day or so to let the plants dry out and the soil to firm up.

Where ambient air is used for drying, it is best to harvest leaf when the weather offers prospects of reasonable drying conditions ahead, though with most species a day or two of sitting on the screens waiting for better conditions will not harm them greatly, as long as it is not too humid for too long.

Pests and Diseases

It may be necessary to harvest early to save a crop if a pest or disease is building up in it and also to break the life cycle of whatever is attacking it.

Snails (and to some extent, slugs) tend to move up the plants during rain or irrigation, and move down to the ground in dry conditions. If they are a problem, the best time to harvest is after a dry spell of 3 or 4 days, when they will be mostly on the ground below the level of harvest.

Equipment

The leaves of many species need to be handled carefully to avoid bruising. Consequently harvesting equipment should cut leaves off cleanly and handle them reasonably gently.

There are a number of hand tools that can be used for harvesting leaf herbs.

Shears

Garden shears or sheep shears can be used for cutting small quantities of aerial parts. Sheep shears are quite a handy tool to have around the garden. They are made of very good steel that keeps its edge well in spite of sand and grit among the material being cut, and they can be used with one hand. If there is any doubt as to their previous use, it is probably best to give them a good cleanup to remove any unsavoury leftovers from the sheep.

Shears are not a very quick way of harvesting, but their advantage is that they can cut the plant without pulling it at all. This is good for delicate plants, particularly newly established herbs, which any pulling might uproot or otherwise damage. After harvesting with other tools, shears are also useful for trimming plants to get a neat finish that will regenerate evenly.

The Reaping Hook and The Sickle

The reaping hook and sickle are similar in appearance. Most people are familiar with the sickle, if only as the component of the emblem that had many of our compatriots looking under their beds, back in the 1950s. Actually, the implement displayed with the hammer on the Soviet flag looks more like a reaping hook, which is used by farm workers for harvesting grain. While the two implements look similar, there are significant differences between them.

The sickle usually has a rounder curved blade and its blade is honed to a sharp cutting edge, while the blade of the reaping hook is made with a curve that is somewhat elongated towards the tip and has a toothed cutting edge. The sickle is normally used for cutting grass or weeds.

Small harvesting tools: from left to right, shears, sickle and reaping hook.

The reaping hook is said to be one of the most (if not the most) ancient of agricultural tools still in use. It dates from an era when knowledge of steel was limited and it was not possible to make a blade that would keep a sharp edge, so to maintain a usable cutting tool, the reaping hook was made with small teeth.

Traditionally it was used for harvesting grain crops: even until early this century in some Western countries. It still is used occasionally by a farmer who wants to save a few heads of grass for seed. You might find a good reaping hook in a second-hand shop, or one that could be made serviceable with a bit of touching up.

New reaping hooks can still be bought in hardware shops, but often they are not well made. The ones I have seen are a bit small and round bladed and the teeth usually need going over with a small triangular file to get a good sharp point on them. Nevertheless they will do the job well enough and are not expensive.

If you are unable to obtain a reaping hook, a sharp sickle will do the job almost as well.

The Catching Scythe

Since the rediscovery of the usefulness of the scythe, it has been adopted by quite a number of small- and medium-scale herb growers. It makes harvesting on areas of up to 0.4 ha or so quite feasible without having to resort to mechanical harvesting methods that would not be cost effective on this scale, or that might affect quality.

Essentially the catching scythe is a normal scythe fitted with a wire-mesh catcher, which gathers the herb as it is cut. This is the same principle as the 'grain cradle', which is a traditional scythe with four long wooden fingers above the blade that catch the standing oats or wheat as it cuts them. The main difference is that for catching most herbs, the wire mesh works better. The gathered herb is then tipped onto a sheet spread on the ground beside the crop being harvested.

While you can still buy a new scythe in a hardware shop, you will have to devise the catcher yourself. Before attaching it though, you may wish to develop some proficiency with the normal scythe (see the section below on 'Technique').

Harvesting Sheets

Sheets are the most practical way of holding and transporting volumes of cut herb. Bundled in a sheet, the herbs can be kept clean, lifted and carried, and they will heat less than if held in larger volumes, as air can circulate around the bundles.

The catching scythe: this tool greatly increases harvesting efficiency on a small to intermediate scale.

The sheet is placed on the ground to receive the herb as it is harvested. When full, the diagonal corners are tied together to make a bundle for carrying to the drying shed.

A number of sheets will be needed, sufficient to hold all that is going to be harvested at one time. They can be made of sturdy cotton, though polyester is stronger. The optimum size is around 2.4 m x 2.4 m. A larger sheet would be too heavy to lift when full and its contents would heat more. If it is much smaller, too much of the cut herb will miss the sheet and land on the ground.

Making a Catching Scythe

A catching scythe can be made fairly easily by modifying an ordinary scythe. This needs to be in reasonable condition, and not too heavy, bearing in mind that you will be carrying the extra weight of the catcher plus a swipeful of cut herb.

The Ordinary Scythe

At first sight the scythe can appear a rather strange and clumsy prehistoric device. But with a bit of practice, a sharp well-adjusted scythe can function as a wonderful extension of the body, giving a 10-fold increase in efficiency over smaller cutting implements.

Because few people have had the opportunity to become familiar with this tool, and because the majority of scythes available need some adjustment and fine tuning to work properly, it is worth going into some detail about it here.

The scythe basically consists of two parts: the blade and the snath, as the long curved handle of the scythe is called.

The Snath or Handle

A light wooden snath may be okay, or else go for an aluminium one, which will be lighter.

A snath made of aluminium may seem something of an anachronism – an ancient tool like the scythe should really deserve a wooden handle – but while the aluminium is not as strong or elegant as the wood, it will make a catching scythe that is light enough for most people to use.

My first catching scythe had a thick wooden snath, but this made it a bit too heavy. My arms would be quite tired after an hour or so and I couldn't convince other growers to take it on, it felt so weighty and cumbersome. Now with an aluminium snath, my scythe is easier to handle and has won quite a few followers.

The Nibs or Hand-grips

Scythes usually have two wooden hand-grips or nibs, to use the traditional term for them. These can be adjusted to different angles and heights to suit the individual.

Traditionally they are attached by a left-hand threaded bolt, so to loosen them for adjusting, turn the nib clockwise and tighten anti-clockwise. This prevents them from coming undone in use, but modern manufacturers often overlook this and make the nibs with a right-hand thread.

The Blade

The blade needs to be at least 60 mm wide and about 500–750 mm long, measured along the cutting edge, from the tip to the heel, but not including the shank. Avoid a blade longer than this as it will be very cumbersome and of no advantage for catching herbs, not even in the hands of the most proficient user. My own scythe blade is 725 mm long and that is plenty long enough: I often feel a shorter one would be lighter and easier for people to get the hang of using.

There are two types of blade you may come across. In Australia the most common is the English scythe blade. This consists of two pieces riveted together: a broad flat piece of thin steel that forms the face of the blade and a narrow thicker piece that forms the backbone and shank.

The European or Austrian scythe blade is forged out of a single piece of steel. This type is more common in Europe and North America.

I don't think my knowledge of scythes is sufficient to go into the merits of one type of blade against the other, but from the point of view of attaching the wire mesh catcher, the two-piece English blade has some advantage. The catching mesh can be attached by sandwiching it between the two pieces of the blade in a manner that protects the mesh from scuffing on the ground and wearing out.

Adjusting the Angle of the Cutting Edge

The first thing to do is look at the angle the cutting edge of the blade makes with the ground when in action. The preferred angle depends on the use of the scythe. Where it is intended for cutting short grass, the closer the edge skims to the ground the better. Most scythes have a twist in the shank at the base of the blade that brings the cutting edge close to the ground.

For harvesting herbs, however, a cutting edge as low as this will result in problems. First of all it will tend to scoop up soil, stones and lumps of compost. It will also tend to cut the herbs too low: usually one aims to cut a little above the ground to avoid dead and yellow leaves.

Consequently you will probably need to straighten the shank of the blade so the cutting edge is more or less in the same plane as the handle. This will make the back of the blade a little lower than the cutting edge as you swing the scythe, so if it strikes an undesirable item, such as a lump of compost, it will tend to bounce over it rather than scoop it up and throw it in with your harvested herb.

Other Possibilities for Harvesting

Modified Hedge Trimmer

A modified hedge trimmer is another possibility for harvesting some crops where for various reasons a scythe can't be used.

The hedge trimmer is a power tool with a reciprocating toothed blade, which looks a bit like a sawfish's upper jaw. There are 240 volt, 12 volt and two-stroke petrol models available.

At times I have used a 12 volt hedge trimmer fitted with a metal catching tray for harvesting herbs.

I used this for harvesting low growing woody perennials, such as Thyme and Hyssop, for a number of seasons. While it did the job, I was never particularly happy with the set-up.

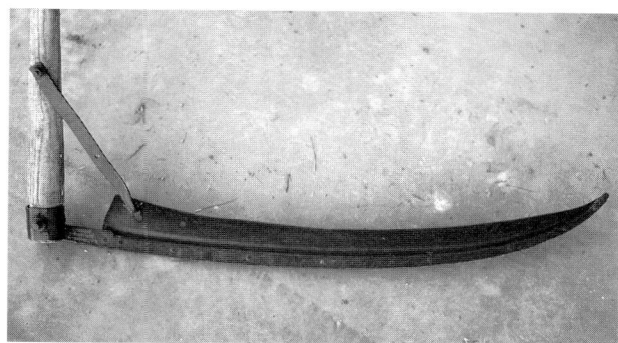

The English scythe blade is made of two pieces riveted together.

With a twist in the shank this scythe blade has its cutting edge set close to the ground.

After straightening the shank, the blade edge is raised. At this angle the blade will catch less debris during harvesting. To straighten the shank of a scythe blade, place one end in a vice and bend the other with a shifting spanner.

Attaching the wire-mesh catcher to the scythe.

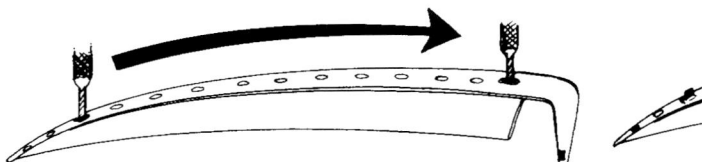

Step 1. First drill out all but the first one or two rivets (from the tip) on the backbone side of the blade and tap them through. This will leave the tip intact. The catcher is attached about 50 mm from the tip of the blade.

Step 2. Now gently prise the pieces slightly apart and fit 3/16 inch x 1 inch (5 mm x 25 mm) roofing bolts (these have low-profile, rounded screw heads) through the holes, with the heads on the lower side (towards the ground). Loosely thread the nuts on, but don't tighten them yet. The first bolt, 50 mm from the tip of the blade will need a flat washer to hold the eye of the wire that supports the upper edge of the mesh catcher.

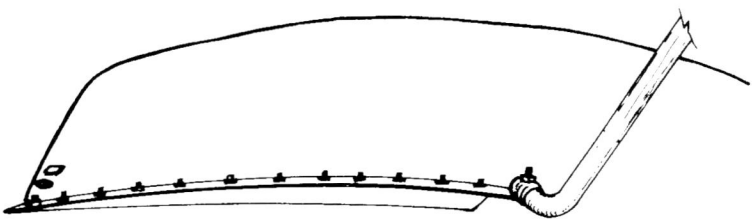

Step 3. This wire supports the leading edge of the catcher so it needs to be reasonably stiff: use 4 mm (8-gauge) wire or else lighter high tensile fencing wire, 2.85 mm (11-gauge). Take a piece of this wire about 1350 mm long. Bend one end of it around to form a small eye at approximately right angles to the wire. This is attached to the blade about 50 mm from the tip by the first bolt and held down by the flat washer. Now shape the wire so that it rises at right angles to the blade for 200 mm, then bends and goes in a slight curve (similar to that of the blade) and meets the snath (scythe handle) at a point 450 mm up the from the base (of the snath itself). Don't attach the wire to the snath yet.

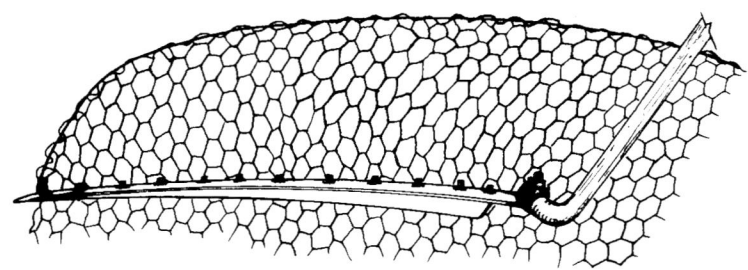

Step 4. For the catcher, 40 mm hexagonal mesh galvanised chicken wire will be strong and durable. Other materials could be used, especially if weight is a major consideration. Heavy-duty knitted shadecloth (70–80% rating) with a few supporting wires threaded through it should work. A lighter gauge hexagonal mesh, such as bird wire, will do for a while but breaks down with repeated use. Whatever material is used, it needs to be something that can be shaped into a compound curve. The material needs to be large enough to sandwich between the two pieces of the blade and wrap around the snath. Actual dimensions will vary according to the length of the blade, but 1200 x 600 mm should be ample There needs to be a smooth uncut edge along one of the long sides. Thread the support wire (which was shaped in Step 3) through this long uncut edge of the mesh.

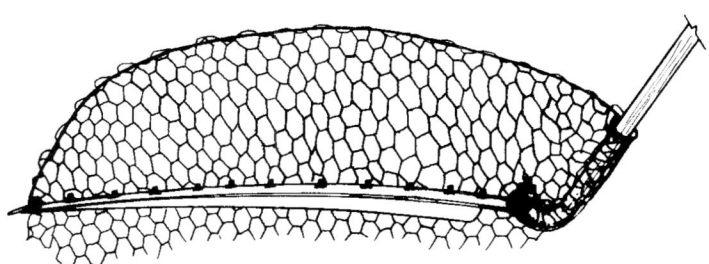

Step 5. Now finalise the shape of the support wire. The 200 mm leading edge above the tip of the blade needs to be at right angles to the ground when the scythe is held in the normal mowing position, so this leading edge of the catcher passes between the cut and the uncut standing herb. If this alignment is incorrect, either it will miss some of the cut herb or else snag on stems that are still standing at the rear of the blade as it passes through. When you think you have the shape right, attach the wire temporarily but firmly to the snath. You may want to adjust its length again after attaching the mesh to the blade.

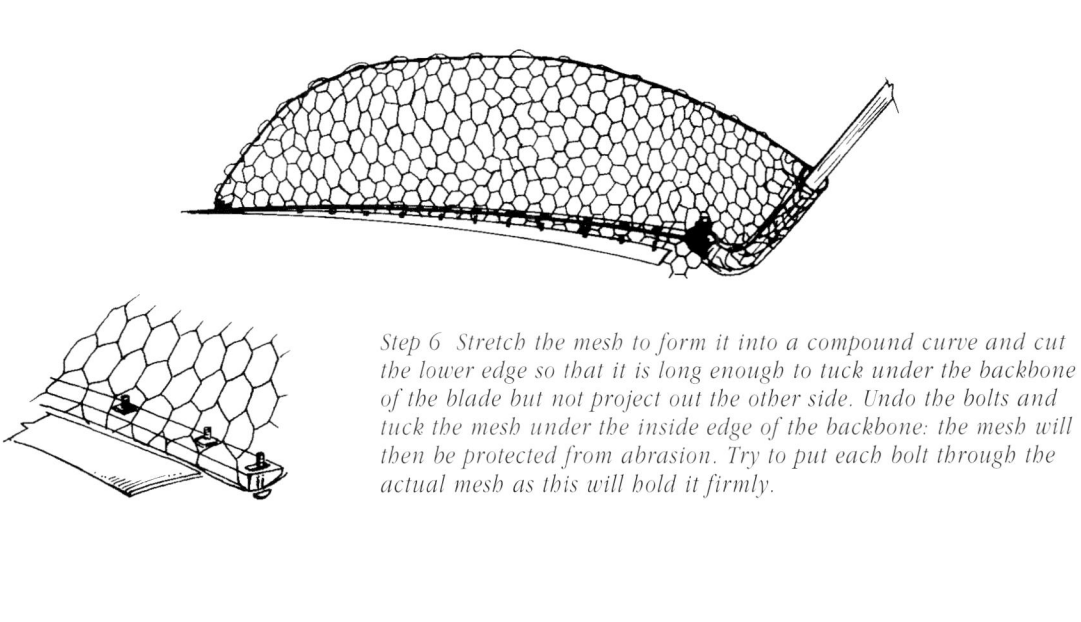

Step 6 Stretch the mesh to form it into a compound curve and cut the lower edge so that it is long enough to tuck under the backbone of the blade but not project out the other side. Undo the bolts and tuck the mesh under the inside edge of the backbone: the mesh will then be protected from abrasion. Try to put each bolt through the actual mesh as this will hold it firmly.

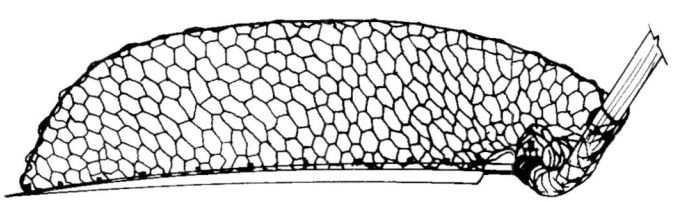

Step 7. When you are happy with the shape of the mesh, tighten the bolts and permanently attach the loose edge around the snath. Finally, permanently tie off the upper support wire where it is attached to the snath.

One problem is that the cut herb tends to be pushed forward by the action of the blades, so it is tricky to get it to fall onto the tray. It works best when cutting uphill, pushing the blade into thick standing herb so that as stems are cut the herb falls back onto the tray. A basket or sheet needs to be close at hand as the tray has to be emptied frequently. The work is awkward as the tool is heavy and has to be operated from a kneeling or crouching position.

I had been reluctant to attempt to scythe woody perennials, because only a short length of stem is harvested: about 25% of the leaf has to be left on the bushes. I was particularly afraid of scalping the plants, so I persisted with the hedge trimmer, but when I tried the scythe I found that it cut well provided the blade was really sharp. With a bit of care, the damage to the plants was not great: only an odd piece here and there was cut a bit low. So the catching scythe has turned out to be much quicker and easier for this job than the hedge trimmer, which has rested on the shelf for years now.

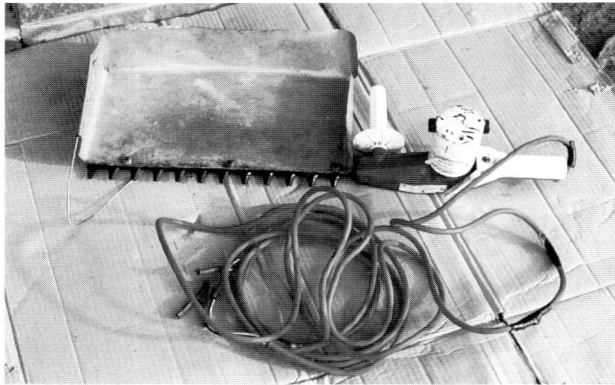

Hedge trimmer with catching tray attached: this tool has proved to be unsuitable for most crops.

I understand some growers use the hedge trimmer (without a catching tray) for harvesting a range of crops, manoeuvring it so cut herb is projected forward onto a sheet laid beside the plants. While this system would work for those crops that are tall and dense enough, I can't see that it would be really any quicker than a skilfully handled reaping hook, as the sheet would need to be moved more frequently and a lot of herb would fall the wrong way: the hedge trimmer requires two hands to hold it. Alternatively, two people could work together, one person cutting while the other holds the herb and throws it onto a sheet.

Just mowing the herb off with a hedge trimmer or any other device, letting it fall on the ground and then raking it up is risky. Soil, compost, mulch and other trash will inevitably be gathered up with the harvest.

The other limitation for the hedge trimmer is the power source. A battery-powered 12 volt unit is not practical for any significant scale of operation. The 240 volt unit is a bit lighter, but needs an extremely long cord (with resulting voltage drop) or a portable generator. The 2-stroke model is unsuitable as it would blow oily smoke over the herb being harvested.

Despite the disadvantages mentioned, this type of cutting blade may have potential for development into a practical small-scale mechanised harvester. It would need some sort of sweeping device to carry the cut herb past the blades and into a bin. The extra weight would probably require the unit to be mounted on motor-driven wheels.

I doubt it could be much faster than harvesting with a scythe, or as versatile and economical, but for those who feel that scything is a bit too prehistoric or too difficult, it may be an option.

Large-scale Mechanical Harvesting of Leaf and Aerial Parts

Large-scale mechanical harvesting of leaf and aerial parts is being done in other parts of the world and even in Australia on a few broad-acre herb growing operations. The main hurdles are the volume of turnover needed to justify the cost of developing and building the machinery, and the difficulty in maintaining quality when handling large volumes of herb. I have yet to see mechanically harvested leaf herb that is premium grade.

Most of it is third grade or trade quality, but I have seen some that is a reasonable second-grade product, and I believe it may be possible to produce a premium-grade herb if close attention were paid to certain critical aspects of harvesting and drying.

Cutting Action

The cutting action of the harvester must be clean and must not damage the rest of the plant. A reciprocating cutter bar is probably the most suitable as it cuts the plant cleanly off at the base without bruising it.

Height of Harvest

The height of harvest needs to be adjustable and controlled by the operator so the cutter bar can be raised or lowered depending on the condition of the crop. This enables the exclusion of deteriorated leaf or excess stem from the harvest.

Gathering

The herb must be gathered as it is cut and not allowed to touch the ground.

A system where the herb falls on the ground when it is cut and is then gathered up may enable the use of conventional hay-making machinery, but the herb will be contaminated and bruised in the process.

Handling

The mechanism for moving the cut herb from the cutter bar to a transporting bin must be gentle, so bruising and crushing is minimal, as must all handling of the fresh herb.

Holding

The size and dimensions of containers for holding and carrying should be designed to minimise compression and heat build-up. They should be broad rather than high, and the depth of herb should be limited. The maximum depth would vary according to the species, as they differ markedly in their robustness and in heat generation.

Timing

The timing of the operation must be arranged to minimise any holding periods between harvesting and drying. In other words, the drying system must be able to take the volume of material as soon as it is cut. There is no point in a harvesting system that delivers large quantities of herb that sit around in piles for long periods composting while waiting to be dried.

The larger the volumes of herb being handled, the greater and more rapid is the generation of heat, and the shorter any holding times must be. In warm weather, an hour or two sitting in a large pile can be enough to ruin the colour of many herbs and destroy or volatilise active constituents resulting in a significant loss of quality.

The Reaper and Binder

For large-scale harvesting of leaf herbs, one possibility would be to modify an old reaper and binder. This is a traditional grain harvesting machine, originally drawn by horses, but these days usually towed by a tractor. There are still a few around in regular use for harvesting oats for chaff.

The reaper and binder has a cutter bar, which cuts the crop at a level adjustable from just above the ground to some distance up the stems. A rotating reel sweeps the crop gently against the cutter bar, and onto a moving canvas platform behind it. As it is cut, the crop falls onto this and is carried up to the side where it is bundled and tied into sheaves, which are dropped gently to the ground.

The bundling and tying part of the process would have to be replaced with some sort of bin which is periodically emptied as the harvested crop is delivered into it by the canvas elevator. The operation of this machine is quite gentle as it is designed to avoid shattering and losing the grain as it is harvested.

Other Possibilities

Alternatively a suitable harvester could be built up around a reciprocating cutter bar mower by adding a sweeping mechanism to carry the cut herb into a bin or a sheet behind the cutter bar.

Questionable Equipment – the Forage Harvester

Avoid the temptation to tear into a crop with equipment that will damage it in the harvesting process. Some years ago I saw a magazine article about an aspiring Australian herb grower who was harvesting herbs with a forage harvester. This must have bruised the crop horribly and picked up a lot of trash and soil in the process. A forage harvester is a difficult piece of machinery to clean properly, so there would be choice lumps of pulverised material sticking to the inside of it and coming loose in the next crop, after it had fermented and dried a little.

I did get to see some samples of the product of this operation later and it looked a bit like compost that hadn't progressed to completion because it had dried out. This was being put into tea bags and marketed in an attractive package. Needless to say the product has not made a great impression on the market.

Conventional Hay-making Equipment

Sometimes hay-making equipment is advocated as an option for harvesting herbs, but this would also be courting disaster in terms of quality.

Apart from the traditional reciprocating cutter bar mower, most hay mowers knock the grass around a bit. This speeds up the drying by damaging the stems and fluffing up the grass, which is fine for making hay as grass is tough and fibrous, not easily bruised, and cows aren't too fussy anyway. But most herbs will suffer terribly if put through the same treatment.

The other problem with all hay-cutting equipment is that it will leave the cut herb on the ground. It would then need to be raked up, which would probably result in contamination with soil, compost, trash etc. Raking and baling will cause severe bruising of many herbs if they are wilted, or shattering and loss of leaf if they are brittle dry. Unless the herb is fairly dry, stems and all, it is liable to heat and go mouldy in the bales.

Technique

Using the Reaping Hook and the Sickle

The action of the reaping hook is somewhat the reverse of the usual manner of using a sickle to cut grass. The teeth are aligned to cut with a drawing motion towards you rather than a hacking motion away from you.

The blade is pushed tip first into the standing crop and hooked around a number of stems, drawing them together with a partial anti-clockwise motion of the blade. The left hand (apologies left-handers, please read the opposite) then grasps these and the right hand draws the blade to the right, cutting the stems as the teeth saw across them. Care needs to be taken to avoid cutting off parts of your left hand: it may be wise to wear a glove on it.

The left hand then casts the cut herb into a low basket or onto a sheet spread on the ground. It is more efficient if one can work progressively forwards, with the crop on the right, and the basket or sheet on the left.

The cutting motion needs to be practised for a while to get it flowing smoothly. Try to focus on cutting the stems by sawing the teeth across them with a moderate pressure, rather than just pulling the blade hard against them. This requires a partial clockwise rotation of the blade as you draw it to the right.

While many plants won't be too fussed how you cut them off, some are more delicate and, if tugged too hard, they may break off or come out by the roots. The cutting motion I am trying to describe here will minimise tugging at the plant.

If you are unable to obtain a reaping hook, a sharp sickle will do the job almost as well: simply use it with the same motion, again taking care to avoid cutting your left hand. It is a matter of some debate which heals more quickly – the clean straight cut of the sickle or the jagged gash of the reaping hook – but if you are careful with both you won't have the opportunity to find out.

Reaping hooks and sickles are suitable for harvesting quite small plots, for young plants that are not dense enough or robust enough to use a scythe on, for tough-stemmed and tall plants that need to be cut fairly high above the ground, and for people who find the scythe a bit formidable.

They are good for harvesting a weedy crop where care needs to be taken to separate the herb you are seeking from other plants. This also makes them handy for wild-crafting or harvesting weeds. While they are significantly quicker than shears, they are still much slower than a scythe.

Using the Ordinary Scythe

The ordinary scythe, without a catcher, is not a very satisfactory tool for harvesting herbs, as it leaves the cut material scattered on the ground. But it is useful for trimming grass and other growth around the place, and it is probably a good idea to get the hang of using one by

Harvesting with the reaping hook. This ancient tool can be useful for harvesting some crops.

cutting things that don't matter too much, so you can develop some confidence before you tackle harvesting your valuable herb crops with a catcher attached.

The best thing would be to find an old codger who has some skill with the scythe and get him (or her, women used the scythe, too) to show you how it is done. Then practice as much as you can until you get the hang of it. In our grandparents' day, people used to scythe their lawns, scythe grass for feeding livestock, and even for making hay. Where I lived in Quebec in the 1970s, it was common for farmers to go around with a scythe and clean up all the corners the tractor couldn't get into: winters are long there and every blade of grass is valuable.

Someone could write a book on scything, as it is an art in itself. Some of my neighbours in Quebec grew up with a scythe in their hands and it was simply amazing to watch the skill with which they could handle it. Nevertheless, you do not have to be a master with the scythe to be able to do a reasonable job with it.

Before starting, there are few aspects of the scythe that need attention to ensure it will cut effectively and make it easier to use.

The Blade

To cut well, the blade of the scythe must be sharp and straight. By 'straight' I mean that while it has an intended curve in it, this should be all in one plane. If the scythe is new there should be no problem, but a used one may have been speared into the ground and bent. If this is the case, it can probably be straightened fairly easily, as long as it isn't cracked.

Sharpening: To sharpen the blade, a scythe-stone is needed. This is a fairly coarse, abrasive stone, usually round or oval in cross-section, 200–300 mm long, 15–40 mm thick. But first the blade may need grinding to remove excessive shoulder on the cutting edge. Getting a fine edge makes mowing with the scythe much easier: it will almost feel as if it is pulling itself through the grass if it is well sharpened.

Continual sharpening with the scythe-stone tends to gradually make the cutting edge coarser – the taper becomes too fat – so the shoulders need to be ground back to make the taper finer. Ideally this should be done on a traditional sandstone grindstone turned by hand, but how many of us have one of these? Otherwise a good file held at a fine angle will eventually do the job or it can be done very cautiously with a bench grinder or an angle grinder, taking care not to overheat the blade.

Some people sharpen both sides of the blade, while others prefer to just sharpen the upper side to get a finer bevelled edge.

The blade should then be touched up by stroking the scythe-stone along the length of the blade, drawing it downwards at the same time. The stone should be held at a very acute angle to the blade to ensure a fine edge is maintained.

The strokes should start at the heel (handle end) of the blade, working towards the tip and slightly down. This action creates microscopic teeth on the edge that are aligned forwards to cut into the grass as the blade passes across it.

If the scythe is being sharpened both sides, then the strokes can alternate from one side to the other, working from the heel to the tip. With a bit of confidence this can be done rapidly in the same style as butchers sharpen their knives.

Developing the knack of getting a good cutting edge on the blade is half the battle won in scything, particularly for tough plants like grass. Fortunately most herbs are quite easy to cut.

Adjusting the Angle of Openness: Having obtained a good edge on the blade, the openness, or angle between the length of the blade and the length of the snath may need adjusting. This is done by undoing the eyebolt – at the base of the snath – enough so the shank of the blade can be lifted out of one retaining hole on the guide plate and placed in another. Then retighten the eye-bolt. Some snaths have an adjustable guide plate that can be moved by loosening a second bolt.

The angle of openness of the blade needs to be set so it will lead into the grass, strike it at the optimum angle and cut off a nice neat bundle. If the angle of the blade is too open, it will snag on too much grass and will probably stall, or will leave a tuft of uncut grass. If the angle is too closed, it won't catch much grass and won't cut much of it either.

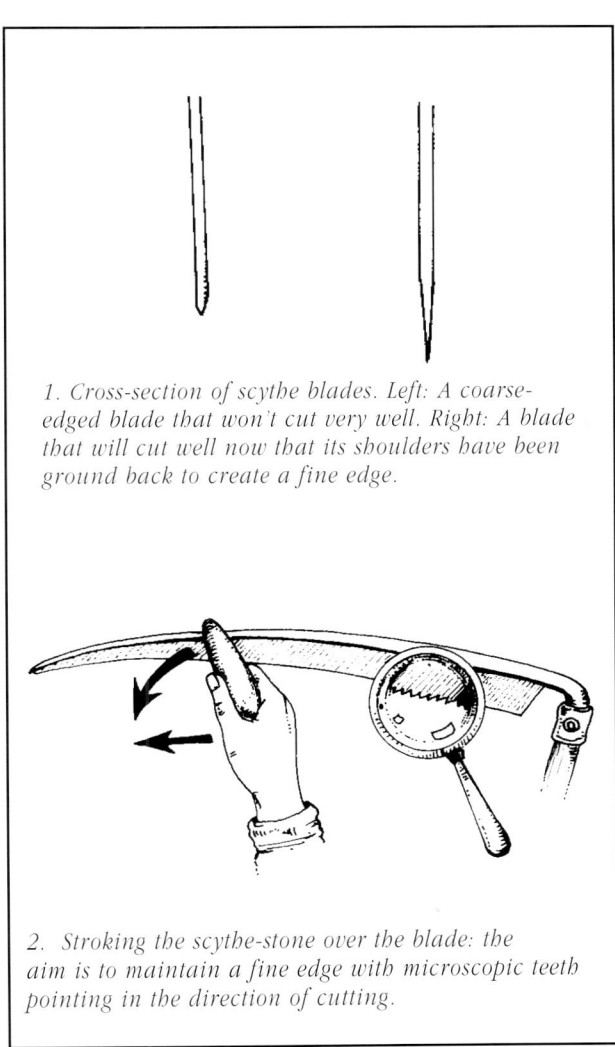

1. Cross-section of scythe blades. Left: A coarse-edged blade that won't cut very well. Right: A blade that will cut well now that its shoulders have been ground back to create a fine edge.

2. Stroking the scythe-stone over the blade: the aim is to maintain a fine edge with microscopic teeth pointing in the direction of cutting.

Sharpening the scythe blade.

The Nibs

The next points of adjustment are the hand-grips or nibs. Usually these can be loosened by turning them clockwise as they are normally left-hand thread (though these days some are right-hand thread). On older tools the threads are often seized – probably from the salty sweat soaking into them through long hours of scything on sultry summer mornings – and may need some careful persuasion to loosen.

Everyone seems to have their own positioning for the nibs. Basically, the weight of the scythe hangs on the right hand (unfortunately, left-handers will have to adapt themselves, as no left-handed scythes seem to exist). The blade should swing comfortably just above the surface of the ground.

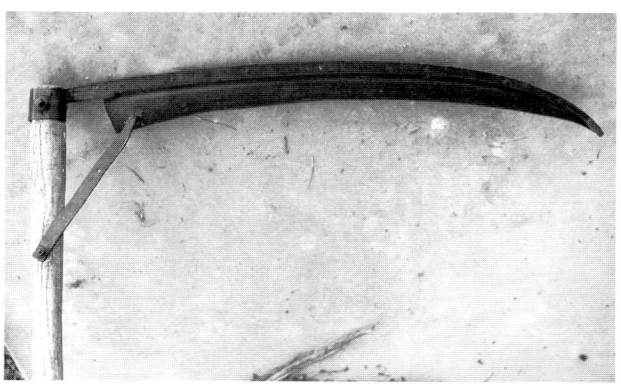

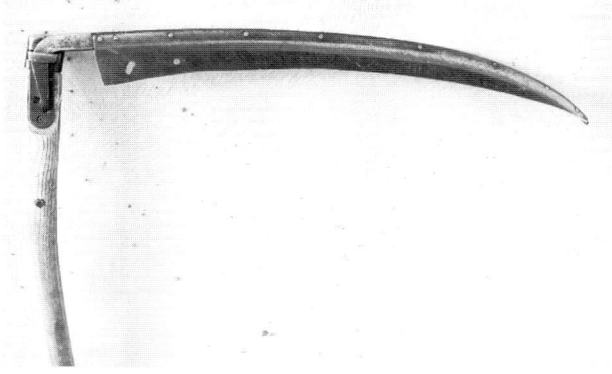

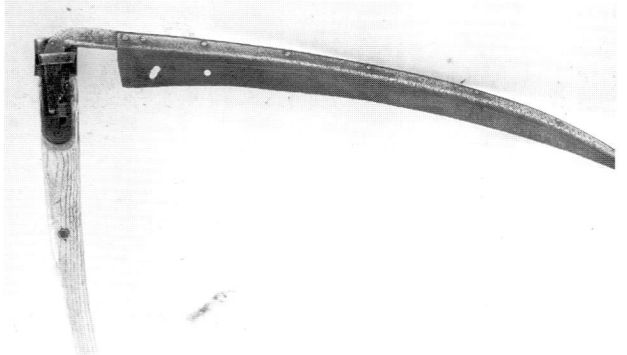

Adjusting the angle of openness on the scythe.

The left hand's task is more of a guiding role, while the right hand delivers most of the force and acts as a fulcrum for the left hand to control the height and angle of the blade.

Now To Cut Some Grass

The Stroke: First, stand before the grass you wish to cut, with the scythe hanging comfortably just above the ground. A good scything action is somewhat like a swinging pendulum: swing the scythe a little back and up to the right, then swing it down to the left. As the blade comes down to ground level, slip the tip into the base of the standing grass and swing it through, cutting with a slicing action. As the blade moves through the grass, try to keep the heel (base) of the blade down as this will give you a cleaner cut and keep the tip from spearing into the turf.

As the blade comes clear of the grass to the end of the stroke, allow it to swing to the left and up a bit until its momentum is spent, then swing back to the right, ready to take another cutting stroke. By stepping forward slightly each stroke, you can move progressively along a line, cutting a swathe as you go.

A Systematic Approach: Successful scything depends very much on a systematic approach to the grass, or whatever it is you are cutting. This should be from a position where you cut from right to left and the cut grass falls into an open space to the left. Where this is not possible, you may need to cut an initial pass or swathe laying the cut grass on top of standing grass. This will then create an opened line that you can then work into. It is much easier to cut grass that has not been laid over or trodden on, so keep the kids and the dogs out of the grass out until it has been mowed.

You can scythe in a series of swathes starting at one side of the patch you are mowing. When you get to the end of that side, return to the beginning and start another swathe parallel and to the right of the first, working your way across the patch. Alternatively, you can go right around the patch in a clockwise direction, continuing in a spiral until you reach the centre.

Traditionally, teams of mowers would work in this manner, the leader starting out, followed by the others one by one, each commencing their own swathe and racing each other to stay ahead or cut the leader off at the corners. Tolstoy gives a beautiful description of this in one of his short stories.

Variation in Scythe-ability: Plants vary enormously in their 'scythe-ability'. The easiest would be a tall, standing grain crop like Oats, still in the green stage, with straight clean stems and no undergrowth. The scytheperson's nightmare would be something in the order of a lush tangled growth of weeds and vines, some of them prickly, that has fallen over, dried off, and grown up again, having been romped over by stock, kids and motorbikes, so that it lies every which way, with sticks and twigs all through it, on a steep slope with rocks, stumps, fencing wire and possibly a snake hidden here and there.

Normally one would avoid taking on such a task, but these are some of the challenges you may encounter in scything long growth around the place. Thankfully most herbs are much easier to cut than grass, as they are not as tough; and in the garden one has most of the other hazards under control, although sometimes herb crops will fall over.

Scything Fallen Stands: Usually plants fall over all in one direction, domino style, under the weight of wind or rain. Sometimes this direction will vary a bit from one part of the patch to another,

The way to approach this situation is to place yourself facing in the same direction the plants have fallen and cut across the fallen stems at right angles, right to left. You will need to stand up a little straighter and hold the scythe closer to you, so the cutting edge of the blade is tipped up a bit, and take a shorter, more hooked sort of stroke.

Using the Catching Scythe

The principles of using the catching scythe are basically the same as outlined above for the ordinary scythe. The main difference is that whereas before you just had to concentrate on the scythe cutting neatly, you now have to also concentrate on it catching what it cuts and then dropping the harvest onto a sheet spread on the ground beside you.

Mowing and Catching: Fortunately most herbs are relatively easy to scythe, so this frees some of your concentration for the catching and dropping part. Your scything action probably needs to be modified a little to get this to work. Try to watch the leading edge of the catcher to make sure you have the alignment right, and don't try to cut too much at a time, as you will tend to spill more.

Place yourself and the sheet so that it is on your left and a little behind you. As you reach the end of each cut with the scythe, lift the tip a little more to hold the catch and keep swinging to the left. To dump the caught herb, just drop the tip down as you reverse swing back to the right for the next cut. With a bit of practice you will manage to land most of the cut herb on the sheet and after a while you won't drop much at all, while maintaining a nice, free swinging rhythm.

Try to start from one end of a plot, placing yourself so you can reach across three to six rows and land the harvest on the sheet. Work your way down the plot, pulling the sheet along when it gets too far back. When the sheet is full enough – the limit is usually what can be reasonably lifted – tie the diagonally opposite corners together to make a tight bundle and spread a new sheet out.

Watching out for Extraneous Material: While you are working, watch out for bits of trash you might have accidentally flicked onto the sheet with your feet, and brush them off before they get included with the herb. If you snag any soil or trash with the scythe and accidentally drop it onto the pile, stop and pick it out straight away: once it is buried it is difficult to find and separate.

Floppy Short Growth: Sometimes you will find that some of the shorter growth that is not vertical will bend over, rather than be cut by the blade. If this is happening, you can usually pick it up by scything from the opposite direction. After the sheet has passed it, cut it from the other side, or else, after first cutting towards the sheet, clean up the remainder by scything towards the uncut herbs and dropping the catch on top of them. This latter method works well if the standing rows are dense and broad for the cut herb to land on top of them and stay there. It is then gathered by the scythe as you cut that section.

Efficiency and Versatility: For small- to moderate-scale harvesting, the catching scythe is a very efficient hand tool. In good going, for example in a nice straight stand of Alfalfa, in about an hour one person can harvest a patch 30 x 10 m, yielding up to 200 kg fresh herb or about 50 kg dried aerial parts.

Harvesting with the catching scythe: with a little practice this can be an efficient versatile harvesting method.

One of the beauties of the catching scythe is its versatility: you can harvest higher or lower according to the herb or its state of maturity. If there is an under-storey of weeds, as with Alfalfa, you can cut above them, or if the lower leaves of a crop have turned yellow, you can raise your blade a little to skim over them. Often the levels of these vary within a crop but altering the cutting height is not difficult with the catching scythe.

As you become skilled with the scythe, you will be able to use it to harvest some of the trickier herbs such as the woody perennials like Thyme, Sage and Hyssop.

For the grower producing high quality herbs who has a market for a diversity of crops, the catching scythe offers an inexpensive, efficient method of harvesting small- to moderate-sized plots. In this situation mechanised harvesting would not only be an expensive investment, but the amount of time saved would be insignificant.

In the above example of Alfalfa, while it takes about an hour to harvest the herb, bundling, lifting and carting take about half an hour, loading drying screens about $4^{1}/_{2}$ hours and chaff-cutting and bagging about 4 hours. So the harvesting part of the operation does not represent an area where time needs to be saved. On this scale of production, mechanical harvesting of leaf would not be significantly quicker than a catching scythe, and the quality is likely to be compromised.

Could Herbs be Field Cured Like Hay?

Field curing is sometimes advocated for drying herbs. It might be a possibility in a reliably dry climate, but it is a risky way to dry a valuable crop and there are a number of problems inherent with this method. At best it will only result in a product of second or third grade quality.

The simplest method might be to just mow the crop like hay and let it lie in the sun until it is dry, and then rake and bale it. For most species, however, the results will be disappointing. The leaves of most herbs shatter very easily when dry, much more than grass and even more than Lucerne or clovers, so a lot of leaf – the most important part of the herb – will be lost in the raking and baling.

Other problems include fading due to exposure to the sun, discoloration of leaf due to contact with water in the form of dew or rain (this will turn some herbs almost black), debris picked up from the ground, and moisture still in the herb stems when baled, which could result in mould growth or make the herb difficult to process.

Drying in windrows might reduce the shattering, sun-fading and some of the water staining problem (though it would still be at risk from a significant fall of rain). If the herb were cut with a good length of stubble, the windrow might sit on top of this, minimising trash pick up.

For aerial parts, field curing could be more difficult with herbs that are slow to release the moisture in their stems. They would have to wait in the field until they were thoroughly dry, or would need to be brought in partly dry for further drying inside, which would involve additional facilities, plus handling and costs.

The risk of rain is something one can't do much about. In southern Australia we generally have dry summers, but sometimes weather conditions prevail that would be disastrous for anyone trying to field cure herb crops, especially in the past few seasons when there has been a trend to wetter summers. One thunderstorm could ruin a crop at a critical stage of drying.

To sum up, field curing is risky, with a number of problems that lack reliable solutions. Quality would be adversely affected to an inconsistent degree, so it would be difficult to build a reputation and a market. The big attraction of field curing is that it doesn't cost much, but if you consider how much the loss of quality is costing, then it can be a very expensive way to dry herbs.

Parts to Harvest and Things to Avoid

When harvesting aerial parts, it is important to take into consideration the herb's focus of regeneration to ensure optimum regrowth and continued production. This has been dealt with in Chapter 6, Herb Growth Types, and there is more detail in the entries for individual herbs.

As far as practicable, during harvest avoid including faded and dead leaves, excessive stem, weeds, soil, compost, mulch or any other extraneous material that might contaminate the end product. Birds often perch on sprinklers, so sometimes there can be quite a concentration of droppings on the leaves below.

Of course nobody expects herbs to be growing in a totally sterile environment, but with a little care, contamination can be kept to an absolute minimum. Generally it is easier and preferable to avoid contamination at the harvesting stage rather than try to find and remove it later.

Soil Splash

Sometimes the lower leaves of herbs have significant amounts of soil adhering to them as a result of splashing caused by heavy rain or irrigation. It may be possible to shake soil off the occasional sprig by whacking it against your leg (this is not advised with Nettle).

Washing may be a possibility, but it is time consuming, difficult, and may damage the herb and impede drying.

If much of the leaf has soil adhering to it, it is probably easier to remove this after drying and processing, by sifting out the fines and discarding them.

Post-harvest Treatment: Mowing Stubble

Most spreading, expanding clump, and crown herbs (and some biennials in their first year) regenerate more vigorously if all the stubble is mowed back hard to the ground. This fosters regeneration from below or at ground level, and can help in the control of a number of pests and diseases, as it effectively removes their food source for a period.

It also ensures that the next harvest is not contaminated with the stubble of the previous one. Most herbs' first growth of the season is taller than their later regrowth. The second growth tends to be shorter, but broader and denser. Overall there may be as much in the second cut, but it usually needs to be harvested at a lower height than the first. The third harvest is usually lower again.

If there is stubble remaining from a previous cut, this can make harvesting difficult, as the scythe tends to ride along on the tough old stalks of the stubble, and when it does cut them it picks up short pieces of dead stem that look trashy and are hard to sift out of the dried herb. These problems can be avoided if the stubble is mowed off low to the ground after each harvest. Then, as each regrowth gets progressively shorter through the season, harvesting is easier and the herb is of better quality.

Timing
The stubble must be mowed off within 3 days to avoid setting back any new regrowth.

Equipment
The best implement for mowing the stubble of most herbs is a reasonably powerful ordinary lawn-mower that can cut almost to ground level. This helps with rust and other pest control. A lawn-mower can usually be manoeuvred anywhere in the garden and can go up and down undulations. Otherwise a slasher will do the job, if you can get access, or a weed trimmer can be used on small beds. Failing that, go over the patch with a normal scythe, well sharpened: it will usually cut lower than the catching scythe.

Technique
Care should be taken to avoid throwing debris and soil onto other crops. Approach the crop by mowing around its perimeter and always going in the direction that throws the cut pieces in towards the centre.

Crops Whose Stubble Should Not Be Mown
Woody perennials, annuals and a few other herbs that need to retain some leaf because their focus of regeneration is above ground, should not be mown. Mowing them off to ground level is likely to kill them, or at least set them back terribly.

It is often advisable to leave the stubble of the last harvest of the season standing. This provides a little cover and shade over the ground and helps suppress weed germination and growth going into winter. Often the last harvest is such a low cut that its stubble will not be a problem to the higher growth of the next spring, but if it is likely to be so, it can be mowed off just before the new growth comes away after winter.

Mowing off Unharvested Crops
The majority of cultivated herbs thrive on, and may even depend upon, repeated harvesting to maintain their vigour. If for some reason you are unable to harvest a plant when it is ready – maybe there is no requirement for it at the time – it is usually best to just prune it back as if it were being harvested and leave the trimmings on the ground or, if that may cause problems, compost them.

Mowing off the stubble after harvesting to stimulate more vigorous and cleaner regrowth.

HARVESTING FLOWERS

The purpose flowers are designed to serve – to facilitate the fertilisation and formation of the seed – is only transitory so they are usually not very robust and are at their optimum for only a short period before they deteriorate. Because of this delicate nature, their harvesting and subsequent handling require special care to avoid loss of quality.

There is a concentration of potency in flowers which, with care, can be retained in the dried herb but which can easily cause problems in the harvesting process. Small volumes of flowers develop heat more rapidly than other herb parts, so care needs to be taken to avoid deterioration as a result of this.

Timing

Stage of Development

With most flowers, the optimum time to harvest is when they have just fully opened. However, it may not be practical to pick over them every day, in which case there is a need to strike a happy medium. Adequate quality can usually be achieved by harvesting at regular intervals so that the flowers being picked range from newly opened to older flowers that are still in good condition.

Often the flower development is very uneven at the time of the first picking, but it is usually best to go over all the bushes in the section or row you are harvesting. This helps even up the flower development of subsequent harvests, otherwise there will always be some plants out of synchrony.

Interval between Harvests

Most herbs flower sequentially: they don't open all at once, but one after the other. Consequently there are usually several harvests necessary at intervals of 3–20 days, depending on the species, growing conditions, weather, season etc.

Ideally the interval should be as short as practical, but there are a number of factors to take into consideration.

In order for there to be enough flowers to make it worth while harvesting, the interval between harvests needs to be adjusted to the length of time between a flower opening and beginning to deteriorate. This period will allow the maximum number of flowers within an optimum range of maturity and usually varies from 3 days to 2 weeks. The interval between harvests should not greatly exceed this period, otherwise there will be an excessive proportion of overmature flowers.

Another approach for some crops is to harvest when 80–90% of the flowers are within the optimum range of flowering, 5–15% are large buds ready to open, and a small proportion (not more than 5% and preferably less) have reached the completion of flowering and the petals are beginning to fall.

Another factor to consider is whether there is a big proportion of large buds just about to open, in which case you might delay picking a day or two to get them at a better stage.

On the other hand, if flowering is very dense, and the weight of the flowers is beginning to collapse the bushes, it is best to harvest before this happens, as collapsing results in a tangled mass that is difficult to harvest.

Biodynamic growers may prefer to harvest on 'flower days' (for more detail see 'Moon and Planet Rhythms' in Chapter 10, 'Biodynamic Aspects').

Inevitably there will be occasions when wet weather interferes with harvesting. Ideally flowers should not be harvested when they are wet, as this can cause some discoloration, but in extended periods of rainy weather it may be best to harvest the flowers wet rather than leave them to become overmature on the bushes. If provision can be made to dry them rapidly, the deterioration will be less than if they stay on the bushes.

Harvesting and Drying Capacity

Harvesting of flowers should be carried out when good drying conditions can be provided for 2–3 days after harvest. This way the flowers are quickly 'fixed' in their prime state and don't continue maturing. If you are relying on ambient air drying, harvesting should take place in the early morning on a sunny day.

Also, you need to consider your picking and drying capacity in light of the quantity of flowers that are approaching the optimum stage. If a glut of harvest is coming up, it may be best to start on the early side, in order to stagger the harvesting. If this is established for the first picking of a plot, the later harvests will mature in a similar sequence.

Most people will find there is a limit to how long they feel like sustaining this sort of work, as the back and the plucking hand feel the strain after a while. I prefer to harvest smaller quantities every day or so, picking for 2 or 3 hours at a time, rather than spending a whole day at it.

This regime fits in well with morning harvesting and sequential loading of a cabinet dryer, so optimum drying at maximum capacity can be maintained. With some crops, sequential plantings can enable harvesting to be spread over a period of 6 months or more, where the climate is suitable. If 1 kg or so of dried flowers is produced each day or two, this can amount to a significant quantity over the season.

Time of Day

If the weather is inclined to be warm, it is usually best to harvest flowers in the morning, once the dew has dried off them.

At this time of day, the heads seem to snap off more easily. Later, as the day warms up, the plants will wilt slightly and the stems will become a little more leathery and tougher. This makes the flowers harder to pluck and more come away with stems or branchlets attached.

Other good reasons for picking in the early part of the day are that the flowers then get some good initial drying for the rest of the day, and they will heat less rapidly if they are cool when picked. In the heat of the day, even a bushel (36 L) basket of flowers can get very hot in half an hour or so. However, if the situation necessitates it, you can harvest later in the day, provided you take measures to prevent heating and ensure rapid drying.

Equipment

Larger flowers can be harvested by hand. Smaller flowers can be more efficiently picked with a comb. There are a number of variations of design used in different parts of the world.

The Harvesting Comb

The design I started off trying to harvest flowers with was a Blueberry rake: a device used in North America (and perhaps other places) to comb through wild Blueberry bushes and pluck the berries from among the leaves and twigs. The Blueberry rake is more like a large comb than a rake, with an open box attached to catch and hold the berries as they are plucked.

I soon found this needed modifying, because flowers tend to fly when plucked, so a guard overhanging the comb was added. After experimenting with a couple of prototypes I finally settled on a design that has worked well for many years, though the next one will have a few additional modifications.

Making a Flower Comb

A comb can be 100–250 mm wide: 200 mm seems to be good width for general purposes.

Wire Teeth

The teeth of the comb can be 90–120 mm long: the shorter length is good for surface combing, while the longer teeth can penetrate deeper into the bushes and are preferable if the comb is to also be used for harvesting fruits like Rose hips.

The teeth are made from high tensile galvanised fencing wire, 2.5 mm (12-gauge) is adequate for teeth 90 mm long but for a longer comb (particularly if it is going to be used for harvesting Rose hips as well) then 2.85 mm (11-gauge) high tensile wire will be stronger and the teeth less inclined to bend in use. (Two grades of high tensile fencing wire are available: the stiffer grade should be used.)

Combs can be used for harvesting some species of flowers.

The pieces of wire should be cut 25 mm longer than the comb length and carefully straightened. One end of each wire is ground to a rounded point (for the leading end of the comb), while the other end is left square.

The base of the comb is a piece of reasonably hard wood – about 40 x 15 mm cross-section – that is not inclined to split. If there is any concern about splitting, the wood can be supported with small bolts. Accurately spaced holes, which are slightly smaller than the thickness of the wire, are drilled into the wood. The square ends of the wire teeth are driven into these holes.

An alternative approach devised by one grower is to use a couple of store-bought 'Afro' combs mounted side by side. These have a space of 3.5 mm between teeth that are about 75 mm long.

Teeth Spacing

The optimum space between the teeth will vary according to the size of the flowers you intend to harvest: 3.5 mm seems to be good for the smaller German Chamomile and Red Clover varieties, but a 4.5 mm spacing is good for larger flowered strains of these species, as it harvests fewer buds. It is also a better spacing for Rose hips. This means that using 2.5 mm gauge wire, the holes will be drilled on 6–7 mm centres.

I find it useful to have two combs – one with 3.5 mm and one with 4.5 mm spaced teeth – for use on different-sized flowers or in different conditions. Sometimes I will start off with the larger spaced comb and change over to the 3.5 mm comb if the flowers are starting to wilt a little and are slipping through the 4.5 mm spacing.

With a bit of ingenuity it would be possible to design a flower comb with easily interchangeable sets of teeth of different spacings, for use on different species or in different conditions.

The Handle

It helps if the handle of the comb has an adjustable strap that sits across the back of the right hand to help relieve some weight from the fingers.

The Catcher

The woven wire mesh for the catcher is somewhat difficult to obtain, but is probably the best material to use. It is available from the same suppliers as the screen for rubbing tea and culinary-grade herbs. I used the same mesh as for culinary grade – 8 dent (0.9 mm wire, 2.24) mm aperture, but a lighter gauge is adequate if the comb is only used for flowers.

Alternative materials could be worth considering. Shade-cloth (70–80%) is strong enough (but not for Rose hips) if it is supported with wire in a few places. A metal tin will work as a catcher, but tends to have sharp edges and goes rusty after a while.

What about a Blade on the Comb?

One suggestion sometimes offered is to attach a sharp blade to the base of the comb to cut the heads off more easily. I did attempt this on one of my prototypes, but it didn't seem to do much: the blade was soon choked up with flowers.

Basket

The harvesting comb has the capacity to hold 1–2 L of flowers, so a suitable basket is needed to empty it into. A basket is better than a bucket, as it breathes and reduces heating somewhat. It needs to be strong, but not too heavy or bulky, with a weave close enough to hold the flowers and to prevent the picking up of soil through its base.

It needs to be large enough to hold at least an hour's picking, but small enough to set down between the rows and wide enough not to fall over too easily. A good size is around 250 mm across at base, 450 mm at the top and 400 mm high. This will normally hold as much as can picked in 3 hours. A smaller one could be used, but if it were larger it would be difficult to find places to set it down among the flowering crop.

An alternative is to wear a basket on your back, tea-picker style. In my younger days, when I believed my back to be indestructible, I used to do this, but these days I try to lessen the strain on that vulnerable part of the human anatomy and sit the basket on the ground. Wearing a basket might not be so bad if less leaning were required, but the bushes are usually not tall enough to enable one to maintain a vertical posture.

Mechanical Harvesting

Mechanical harvesting is possible and is being done with Chamomile in Europe and New Zealand, and with Boronia for essential oil extraction in Tasmania, but there can be problems of contamination with stem, leaf and weeds, as well as heating, bruising, discoloration and delays in drying.

A Czech Chamomile harvester looks a bit like a grain header: the flowers are plucked by rotating combs and picked up by vacuum. The whole machine is as much as 7 m wide. To be viable, production would need to be on a large scale, as the cost of the machinery and associated drying facility would be substantial.

While mechanical harvesting can produce a cheaper product on a broad-acre scale, the delicate transitory nature of flowers makes it difficult to ever obtain a premium-grade product comparable to hand-harvested flowers.

Because of the very mediocre quality of most mechanically harvested flowers, the production of a premium-grade hand-harvested product is a viable proposition where the price obtainable is sufficient to justify the extra labour.

Technique

Combing

The Combing Action

The combing action is difficult to describe without being able to demonstrate. Assuming you are right handed (left-handers please substitute right for left and left for right), hold the handle in your right hand, with the strap over the back of your hand to take some of the weight. The aim is to slide the teeth under the flowers so as to catch them just below the heads and then pluck them with a lifting or pulling action, while foliage and smaller buds (it is hoped) slip through the spaces between the teeth.

The technique varies according to the form of the bushes and their manner of flowering. Upright growth with the flowers nicely arrayed at the top, but not terribly thick, can often be combed with a long sweeping action. Denser flowers require a shorter stroke, while growth that has collapsed and tangled requires a jabbing sort of stroke, poking into the flowers and lifting them clear.

In very thick flowering you may need to comb the upper layer first and then work deeper into the bush. There is a limit to how deep it is worth combing, as this tends to pick up a lot of foliage and stem and can result in some frustrating tangles.

The left hand can be used to help guide the flowers into the teeth of the comb, and sometimes to hold the bush if the branches are tearing away. I also use the left hand a lot to help lift the comb, so the right hand does not have to do all the work. Normally I place the fingers of the left hand under the end of the teeth when lifting, but if the flowers are such that they can be harvested with long sweeping strokes, I take hold of the top of the mesh cage at the front with the left hand, supporting and guiding the comb as it skims the bushes.

Clearing the Comb

With every few strokes, or sometimes every stroke, you need to pass your left hand over the underside of the comb, and pull off the stalks that are hanging through, as there are always some flowers that do not snap off cleanly. The upper side of the comb needs checking, too, as some stem and leaf will be gathered there as well. If this is done diligently all the while, there will be less stalk in your final product. Also watch out for weed flowers, seed heads and other extraneous matter. After clearing the comb, tip it upwards to allow the picked flowers to fall into the bottom of the catcher.

When you drop pieces of stalk and trash, make sure they fall to the ground rather than land on the bushes where they might be gathered up next time you harvest and then have to be picked out again.

Combing Strategy

Try to comb in the direction the stalks are leaning. This is not always possible, but even when you have to insert the comb across or against the angle of the stems, it is usually possible to pluck the flowers by lifting or pulling in the direction of the lean. This will result in cleaner, easier harvesting and will leave the bushes in a tidier arrangement for subsequent harvests.

Work progressively along the row, and try to pick clean. Of course there will inevitably be some flowers left on the bushes, but these should be mostly those that are too small to be caught in the comb or too deep in the bushes for easy harvesting. Tangled portions are best just skimmed over.

The main thing is to avoid leaving flowers near the top and outside of the bushes, where they will be picked up in a state of deteriorating over-maturity in the next harvest. With small-flowered species, the occasional one will not matter, but they should not be so numerous as to stand out in the finished product.

Posture

If the bushes are tall, you may be able to harvest them from a standing position, but lower growing bushes are best harvested kneeling, provided your knees can stand it, as this is much easier on your back. You need to be careful not to tangle the bushes as you shuffle along the rows: it is best to maintain the same pathways each harvest, so as to minimise the disturbance.

Minimising Leaf and Stem

It is always frustrating how much leaf and stem gets picked up when combing flowers like Chamomile and Red Clover, but you can minimise this when you develop a little skill with the comb. Try to avoid snagging tangles, and don't try too hard to get flowers that are deep inside the bush.

With some flowers, the inclusion of a proportion of leaf can be acceptable, but standards depend on the species and the intended use. For example, with Red Clover there are usually three to six leaflets attached at the base of the flower that are difficult to separate and their inclusion is usually acceptable.

Optimum timing of harvest helps, too. If the flowers are harvested a bit too early in their development, more leaf tends to be gathered, while a delayed harvesting can be difficult because of multiple layers of flowers, and tangles where bushes collapse with the excessive weight.

Getting your crop to grow in an easily 'comb-able' manner is more than half the battle. Lush leafy growth is more prone to tangling and tends to have a lower proportion of flowers, which is why high nitrogen levels need to be avoided for most flower crops.

Careful combing and regular clearing of the comb will reduce the amount of stalk and leaf included among the flowers. The rest will need to be picked out as they go onto and come off the drying screens. With Red Clover, winnowing with a large fan gets a lot of the unattached leaf out.

If some suitable separation method could be developed for removing stalk and leaf from comb-harvested flowers, combing could proceed at a faster rate: often about half the time is spent clearing the comb of stalk and leaf.

The Holding Basket

When the comb starts to feel a bit heavy, empty it into the basket, which can be on the ground alongside you or carried on your back.

Try to avoid holding a volume of harvested flowers for any length of time before spreading it on the drying screens. The 'critical mass' of flowers is less than for most other herbs and even a small basket full of flowers can begin to generate excessive heat in a short time in warm conditions. If there is going to be any delay in getting the flowers onto the drying screens, they should be occasionally stirred around in the basket, and pulled away from the middle and piled up at the sides to help the dissipation of heat. Better still, they can be poured out onto a sheet in a layer thin enough to allow any heat to dissipate.

Take care when emptying the contents of your basket, as it may have soil or other debris clinging to the base that can accidentally drop off into the flowers as you tip them out.

Rate of Harvest

In very good going, I have harvested enough to yield more than 1 kg dried flowers (5–6 kg fresh) in an hour's picking.

The comb is inserted into the flowers, with the left hand used to guide them in.

The flowers are plucked by a lifting or pulling action. Note the strap over the right hand to reduce the weight carried by the fingers.

The comb is cleared by running the hand under the teeth to break off any longer stems hanging through.

but my average is more like 0.5 kg/hr (2.5–3 kg fresh) and I usually quit when yields drop much below 0.25 kg/hr (1.25–1.5 kg fresh), depending on the price of the dried product.

Whether hand-combing of flowers is going to be worth your while depends on being able to get a good premium for a high quality product, because at trade or even manufacturing-grade prices, you won't be making much of an hourly rate for your time.

Hand-picking

Larger flowers and tree flowers are usually harvested with bare hands. Harvesting some smaller flowered species by hand-picking may also be an option as a much higher quality can be produced. Hand-picking can ensure that each individual flower is picked at its optimum stage of development, but this is only financially viable if a good price can be obtained.

As hand-picking tends to be a fairly slow operation, it is necessary to be as efficient as possible, otherwise it won't be a viable proposition.

Having long thumbnails can be handy for nipping some flowers off below the head. If the stems are a bit tough, it may help to file your nails a bit sharper. Other flowers may be easier to pick by clasping the stalk just below the flower head between the index finger and forefinger.

If a basket or bucket can be worn in front at a suitable height, this will speed up picking as it allows both hands to work without interruption. To speed up the picking motion, rather than putting individual flowers into the container as you pick them, tuck them into your hand until you have a handful and then drop them in together.

Trees

Picking tree flowers from a ladder is rather slow work. Where possible, the use of a wire hook from ground level is quicker. If this is made with a hook at one end and a loop in the other, it can be hooked over a branch to pull it down to a reachable height and held there by placing a foot through the loop, leaving both hands free. A little caution needs to be exercised when bending down strong branches with this technique, to avoid finding yourself hanging upside down from a branch.

Rates of Hand-picking

With most species of flowers, hand-picking a good stand typically produces about 0.33 kg/hr dried weight.

Other Methods

Another method of harvesting sometimes used, but one which I can't recommend, is to skim over the top of the plants with a hedge trimmer or some other device that will cut and catch the tops. The problem with this method is that you end up getting a lot more than just flowers – you get a lot of stalk and leaf as well – which dilutes the herb and makes it unacceptable for many purposes.

When a herb is supposed to be flowers only, but has a high proportion of leaf and stem in it, there are problems. To get a suitable product for packaging and tea-making, the herb ends up having to be milled or rubbed, so the attractiveness of the flowers is destroyed. Apart from which, the stalk and leaf might dilute the herb as they usually don't have the same properties or flavour as the flowers.

Things to Avoid in Harvesting Flowers

Insects

Keep your eye out for insects, particularly for beetles revelling in rapture among the flowers, and gently shake them off or pick them out. Then they won't end up hidden

Hand-picking of some flowers can produce a product of premium quality.

in your final product, dried and dead, to reappear floating in someone's cup. They are not such a problem if you are drying in the open air, as they will use their wings to gain their freedom when it suits them, but in a closed drying situation like a cabinet dryer, there may be no escape for them. Most customers don't appreciate insects in their herbs and they are not amused when you assure them that this is a clear indication of the product's pesticide-free status.

Over-mature Flowers

If practical, flowers that are too advanced to harvest should be picked and discarded. This will get them out of way and encourage plants to produce more flowers. Of course this is not realistic with Chamomile, but in that case a few older flowers won't affect the appearance much, while they will stand out badly in Red Clover or Calendula.

Deterioration

Harvested flowers should not be held for any length of time before they commence drying, as they will deteriorate rapidly.

If there is any delay in getting the harvested flowers onto the drying screens, precautions should be taken to prevent heat build-up. Spreading them out on a sheet in a thin layer – no more than 100 mm thick – will help dissipate any heat. Even so, holding times should be minimal as flowers will continue to mature and fade unless they are dried quickly.

HARVESTING ROOTS

Harvesting roots by hand is work that can be quite enjoyable and interesting. It brings one in touch with the garden in its dormant state. The various forms of different roots are a fascination in themselves, and digging and washing them is an opportunity to delve into the hidden realms of life beneath the surface.

Timing

Most herb roots are harvested when the plant is dormant. As growth slows and/or the tops die down in late autumn or early winter, the plant stores its reserves and active constituents in its root system, so this is normally the optimum time to harvest that part of the plant.

Harvesting is usually easiest when the tops have completely died off and before the new shoots emerge, as this reduces the trimming required. The period can be extended a little either way. Some species do not lose their leaves in winter.

The harvesting of roots can be affected by adverse soil conditions. It is best if the soil is dry enough to be friable so it will shake off the roots, then washing them is easier and not so much precious topsoil is taken away. Harvesting in drier conditions also avoids damage to soil structure.

Consequently when the ground is dry enough for digging, the opportunity should be taken to harvest sufficient roots to keep the drying facility operating to capacity for some time in the event of inclement weather. In winter, most roots will store quite well for several weeks in woven polypropylene bags, provided they are not overly damaged in harvest and they are kept in the shade and out of the wind. If they are looking a little dry, a sprinkling of water will help keep them moist. They can then be washed, chopped and put into the dryer as space becomes available. By harvesting enough to keep the dryer going for several weeks, if you are beset by inclement weather – as is common in winter in southern Australia – you won't get behind in your drying schedule.

Equipment

The Garden Fork

For hand digging, most root crops can be effectively dug with a robust garden fork. However, it is not easy to find one strong enough to withstand the force that needs to be applied to lever the more tenacious roots out of the ground.

When assessing a fork's strength, pick it up and place the shank of the fork – where it goes into the base of the handle – against your knee. Taking the tip of one tine in one hand and the handle in the other, pull it against your knee as if you were trying to break a piece of firewood. If the tine bends, reverse the procedure to straighten it again, put the fork back on the shelf, and take a look at another brand.

If the fork passes the strength test, the next points of weakness are the handle above the ferrule and the shank where it joins the fork. On these you will just have to use your best judgement as to whether they look strong enough. Unfortunately if they do break, it is usually without warning and you may find yourself suddenly flying backwards into the mire.

It is difficult to find a good, strong harvesting fork. Another option is the broad-fork normally used for breaking up compaction layers in the soil. Being very robust, it can be used for digging large roots. There are 3 prong and 5 prong models available (see Appendix for suppliers).

The Spade

Deep-rooted crops such as Burdock, Licorice and Narrow-leaf Echinacea will require a strong narrow spade to excavate around the root and dig it up. My own preference is a stainless steel spade made in Holland by Sneeboer (see Appendix for suppliers), which is a skilfully crafted tool and a joy to dig with. The round cutting edge gives better penetration and the face of the spade is dished, which gives a little more control over the soil that is picked up, and stiffens the blade. The blade is also slightly tapered so it is a little broader at the cutting edge, which makes working in a deep trench easier.

The Vine Hoe

Valerian, with its fine root system, is best dug with a pronged hoe, variously called a Canterbury hoe or a vine hoe, or sometimes referred to as a potato fork. This same tool works well for digging propagation material for the Mints, Nettle etc.

Washing Facility

All herb roots will need thorough washing, so if any quantity is being grown, a good washing set up is needed.

On a scale with an annual production of up to around 500 kg dried roots, a system that can be used for washing involves spreading the roots out on a couple of sturdy 1500 x 940 mm chicken-wire screens and systematically directing water from two 20 mm jet nozzles all over them. An extra screen of the same construction is helpful for the purpose of turning the roots. The water pressure needs to be quite strong to lift the soil off efficiently, but not so great as to damage the roots.

Usually it is most efficient to work with two screens set up side by side. These need to be raised a little above the ground, sitting on logs or some such support, to allow the water to flow away underneath. Some support across the middle of the screens may be needed to keep the mesh from sagging to the ground under the weight of heavy clumps of roots. Shadecloth screens on the ground underneath will reduce muddy back-splash and catch a lot of the smaller roots that fall through the chicken-wire.

It is good to set this all up close to a source of high-pressure water in a place where the large volume of water running off won't cause problems (that is, not just above your garden) and where the soil and debris that washes off the roots can be salvaged. It should also be a place where a conglomeration of herbs, which may strike from the various viable pieces that fall into the grass, won't go invading where you would rather they didn't. It is to be hoped that the spot chosen will also be a little out of the wind and where the sun will keep you a bit warmer and more cheerful on wintry days, as washing roots can be miserable work if you are cold and wet.

Some tools for harvesting roots: garden fork, vine hoe and spade (see also broad-fork on page 32).

Technique

Digging

The technique for digging most roots is fairly straightforward. Start at one end of the patch and dig them up as if you were digging a garden. Some of the larger ones may take a bit of leverage and loosening before they come away. If two people with forks can get on opposite sides of big plants like Elecampane and Comfrey, it makes the work a lot easier, and you are less likely to break implements.

As you get each clump out, thump it down on the ground, or on a couple of sticks laid across the hollow where the previous roots came out. Provided the soil is dry enough to be friable, this will knock it loose from around the roots (and fill in the hollow at the same time). If your soil is too wet and sticky, this is going to be a very frustrating and messy job: best to wait a few days to let it dry out and then go for it and get as much done as you can while conditions are good.

A simple but efficient set-up that makes washing of roots easier.

Replacing Soil

Because digging roots involves shifting a fair bit of earth, it is important to avoid covering undug roots. If subsoil is brought up, try to put it back in the bottom of the hole rather than bury good topsoil down there. The aim is to return the patch, as best you can, to a level plot with the topsoil still on top. This is most easily done as you finish each plot, or even each day, by raking it over while it is still loose and light. Once the soil gets a soaking it settles and is harder to move around.

Trimming

If the roots have tops that need trimming off, they can be thrown into a wheelbarrow as they come out of the ground. When the barrow gets full, the crowns can be trimmed or one person can be doing this while the other(s) dig(s).

Some herbs can be trimmed with a knife, others need a butcher's cleaver and a solid board for chopping on: it depends how tough they are. For most herbs, aim to cut the crown off around the base of the leaves or the stem.

Trimming can be fairly time consuming with some of the smaller roots with multiple crowns, like Dandelion, but on the larger roots, like Marshmallow and Elecampane, it proceeds very quickly.

Another option is to trim the tops before digging. This may be preferable, provided you can still find the roots after their tops are gone, and provided the top is not needed to help lift the root out. Trimming this way can be done with a mower set low, a weed trimmer, a sharp hoe, or a sickle swung close to the ground. On a larger scale, a slasher or a forage harvester could be used for trimming some crops, provided the height can be set accurately.

Storing Dug Roots

Once trimmed, the roots can be dropped into a woven polypropylene bag (a device for holding the bag open is handy) and can then be held until they are ready to be washed.

It is best to only wash as much root as you can commence drying in the following few days. A fair amount of bruising and breaking can occur in the washing process, and this makes the roots susceptible to mould and rotting if they are not dried soon afterwards. It is usually best to hold dug roots in an unwashed state: they will deteriorate less, provided they can breathe but not dry out, and provided they are kept cool but not allowed to freeze. Usually it is not difficult to find a place in the shade, out of the wind, where the rain or an occasional watering will keep them moist. Most roots will keep for several weeks this way.

However, if your soil is at all heavy, holding unwashed roots may not be advisable. Heavy soil can dry hard and become very difficult to wash off. In this case the roots should be washed straight after digging. It may be necessary to set up some sort of preliminary drying situation so the roots can start drying and deterioration of the cut and damaged surfaces will be arrested. More on this in the next chapter.

Washing

Using the set-up described in the previous section (see 'Washing Facility'), both screens are loaded with a single layer of roots or clumps. If a nozzle is held in each hand, the jets can be directed so the water strikes each root from two angles. Proceed to hose all the roots from one side of the screens, then move around to a new position 90°, or 180° from the first, depending on the shape and soil retention of the roots you are washing. Give them the same treatment from this angle and then move again until you are back to your starting position.

Now you are ready to turn the roots over and wash the other sides. The quickest way to turn most roots is to flip the screen by placing another chicken-wire screen on top, lifting them both together by one side, dragging them back a little more than the width of the screen and quickly flipping them over so they land back in their original place, still together, but with the new screen underneath and the old screen on top.

If this is performed successfully, the roots will be sitting on top of the new screen with their dirty sides facing up. However, if you hesitate a little at the wrong moment or if the roots are very round and smooth, like Burdock, then you may end up with an empty screen and a pile of roots on the ground beside it. This flipping technique will probably take a few goes to get the hang of, but it is much quicker than turning roots individually. Two people holding one end each can do it more easily, but this extra pair of hands may not always be available.

Having turned the roots over, repeat the same washing process. When this is finished you may need to break the crowns apart to get access to the soil trapped between the roots – this depends on the form of the root and how effective your first washing was at getting into the tight spots and 'armpits'.

The washing of each herb varies according to its requirements. Some roots only need a single hosing on each side. Others always need two separate washings: a light first washing to remove the superficial soil from the roots, and then a thorough second washing after they have been broken apart.

Valerian is the most time-consuming root to wash, as it is inevitably a tangled mass of crowns and rootlets with soil trapped within. There is more on the specific requirements of this and other roots in the section on individual herbs.

It is important that washing be done thoroughly, as contamination of herbs with soil, little stones, sticks, compost and other debris is not acceptable in a high quality product.

Soil contamination can result in serious problems for the end user. Its microbial content can also cause major headaches for manufacturers of herbal medicines who are trying to comply with regulations.

Mechanisation of Root Harvesting

Mechanisation of root harvesting is certainly worth looking into, as hand digging and washing methods are fairly labour intensive and time consuming.

Mechanical Harvesters

Herb roots are somewhat more robust than leaves or flowers, so the problems of heating, bruising and discoloration are not as formidable for the harvesting and handling of large volumes. And roots can be safely held for several weeks before drying. Consequently they would be more amenable to large-scale mechanical harvesting without losing quality.

There is some rather sophisticated machinery available for digging conventional root crops such as potatoes and carrots, and this may be suitable, possibly with some modification, for harvesting the roots of a number of species.

Most herb roots are much less uniform in shape and size than crops like carrots and potatoes, and some form awkward clumps, which may cause problems for conventional machinery.

Washing procedure for roots.

Step 1. The roots are initially hosed off to remove superficial soil. Hosing from a number of angles gets rid of more soil.

Step 2. Turning the screenful of roots over.

Step 3. After hosing off the second side, the roots are pulled apart to expose any trapped soil. The broken roots are then hosed off on both sides until they are thoroughly clean.

This technology is not cheap, so growers would need to know that they had the capacity to produce and sell sufficient volume to justify the capital outlay. They would also need to ascertain whether their soil and climate were going to be reliably suitable for harvesting with this machinery: it is no good having the crop waiting in the ground and the machinery sitting in the shed while you wait in vain for the soil to dry out.

Contract harvesting is another possibility worth exploring if there are contract root harvesters nearby, but beware of possible contamination with pesticide residues in soil from other properties. Contact your certifying organisation for guidelines on preventing this.

A cheaper alternative might be to simply plough the roots out. There is a simple implement like a double-sided plough with rods extending out from the mouldboards that is sometimes used for digging potatoes. It ploughs them out, bringing them to the surface as the soil falls through the rods attached to mouldboards. The potatoes are then picked up from the ground. I haven't tried it on herb roots, but it should work for large shallow roots as long as it doesn't break them up too much.

Mechanical Washers

Washing could also be mechanised: there are carrot washers available, but again growers would have to ascertain whether this machinery would do an adequate job on knobbly, twisted and tangled herb roots, and whether production and market volume warrants the expenditure.

Viability of Hand-digging

Root crops have provided a minor portion of the production of most growers of premium quality herbs in Australia. Most root herbs are not called for in large quantities in the retail market on which we focus, they are grown to complement the range of herbs we are offering. Even though they are labour intensive crops, the winter harvest comes when there is not a lot of pressure of other work, so labour is available in a self-employment situation, and the drying facilities are otherwise empty.

HARVESTING FRUIT

There are number of fruits commonly used as herbs. Many of what we generally regard as seeds are technically fruits, but from the point of view of harvesting, if the fruit is of a fleshy nature, it is regarded as a fruit, but if it is harder and drier, it is dealt with as a seed.

The main difference from the grower's point of view is that fleshy fruits require specific harvesting and drying techniques, as they are generally picked and dried at the prime fleshy stage, whereas seeds need to be harvested more towards the fully mature dry stage.

Timing

Generally fruit is harvested from when it develops full colour until it starts to get soft. The earlier in this period that fruit is harvested, the better the quality, but you may have to contend with more leaf in the harvest. Chaste Berry is an exception: it more or less dries on the bush and is most easily harvested after the leaves have fallen.

Equipment

Some fruits are picked by hand, so you just need a basket. For picking from trees, it helps if the basket can be held in front, suspended from your shoulders. Better still would be an apple-picking bag that opens at the base for unloading.

Gloves are needed for harvesting species armed with thorns.

An alternative for harvesting some fruits is to use a comb similar to that described for harvesting flowers. For Rose hips this is more efficient than trying to pick with gloves on. The spacing of the teeth of the comb may need to be varied according to the different sized fruits.

Clean woven polypropylene bags are good for holding and carting fresh fruits, provided they have not been used for anything that might contaminate the product.

Technique

The harvesting technique varies according to the species, but the general principle for most is to get the fruit off by whichever method you can devise that is quickest, not worrying too much about avoiding leaves and other trash in the process, as these can be fairly easily removed by screening and/or winnowing once the fruit has been dried.

If the dried fruit is required to be whole for use in tea blends, its attractiveness is an aspect of its quality, and care must be taken not to damage it in harvesting and handling. This is more likely to happen if fruit is being harvested late in the season when it is getting soft.

If fruit is going to be used in a milled form, minor damage is inconsequential, provided the fruit is dried before any fermentation sets in.

Fruits are less subject to deterioration after harvest than leaves or flowers, but excessive delays before drying should be avoided as quality can be affected if fruits start to ferment. Usually they can be held a few days, provided they are kept cool.

Take some care in handling to avoid squashing them: how robust they are depends on the species and time of harvest. They are not inclined to heat, so they can be safely held in woven polypropylene bags with spaces between the bags for air circulation.

HARVESTING SEED

There is a range of herbs whose seed is the part used: this includes those seeds that are technically fruits, but because these are not fleshy, they are regarded as seeds for the purpose of harvesting and drying. Although I have dabbled in growing a number of them on a small scale, they have generally not been suited to the facilities and systems used for other types of herbs. They also tend to attract mice into the drying facility with resulting problems.

Threshing and winnowing by hand are time-consuming, labour-intensive operations, and the price of most seed herbs is too low to justify them. Mechanical harvesting methods using conventional machinery are suitable for most seed crops and can produce as good or better quality products than hand methods. Consequently, it is generally not worth considering producing seed herbs on a small scale, except for your own use or interest, or if they are not available elsewhere.

If a threshing machine or a header that can be operated as a stationary machine is available, the seed crop could be hand-harvested, then threshed and cleaned mechanically. This may be viable with some seed crops on an intermediate scale where special handling or curing is required.

Harvesting fruits: in this case Hawthorn berries are stripped from the bushes with gloved hands. The leaf is easily removed from the berries after drying.

Timing

Timing of harvest is critical for most seed crops, as the seed usually does not hang around once it is formed, but continues with the task the plant intended for it: to make contact with the soil and begin the formation of a new plant.

There are two approaches for harvesting seeds.

Field-Ripening: The seed is allowed to fully ripen on the plant and is then harvested before it falls to the ground. This usually involves a piece of high-tech machinery known in this country as a header, and elsewhere as a combine harvester. Harvesting can be done by hand on a small scale, though.

At this stage, the seed should be quite hard, with perhaps a few starting to fall. Some species that are inclined to shatter may have to be harvested a little earlier and the seed further dried after harvest.

Curing in Pooks, Stooks and Indoors: The other technique is to harvest the whole plant at a more immature stage and allow it to finish ripening slowly – or cure – in small stacks known as stooks or pooks, or on screens inside your drying shed. Using this method, the harvest is timed so the plant is cut or pulled when the seed has reached the hard dough stage: usually around 2 weeks before it would be thoroughly ripe. The advantage of this method is that seeds that fall out easily when mature or that are attractive to birds and vermin can be saved and allowed to mature in a more secure situation.

Equipment

The equipment required depends on the herb and the quantity. On a small scale, some herbs are pulled up by the roots or else the harvesting can be done with a reaping hook, catching scythe, or grain cradle. This latter is similar in principle to the catching scythe, but instead of a chicken-wire catcher, it has a catcher made of four thin wooden arms that run parallel to the blade, but above it.

For hand threshing you will probably need a flail. This you will have to make yourself, for although the flail was still in use within living memory in Tasmania for threshing peas, there aren't many around. The flail, or 'stick and a half', consists of two round pieces of wood: a longer thinner one for the handle and a shorter fatter one for the swingle, or swinging part. The two pieces are attached by a leather thong and swivel, or sometimes two swivels.

The swingle should be made of a reasonably heavy wood, such as Blue Gum, that is not prone to splitting or splintering.

On a larger scale, mechanical harvesting is carried out with a header for ripe seed, or at an earlier stage with a reaper and binder that ties the crop into sheaves for stooking, the sheaves being put through a threshing machine when cured.

Technique

Harvesting Field-Ripened Seed

When harvesting mature seed by hand, to reduce shattering or seed falling, it may be best to cut the crop when it is still damp early in the morning of a sunny day. This will depend on the crop. The cut crop can be gently laid in windrows to dry off in the sun, or it can be laid on sheets if it shatters very easily.

Later in the day, when the seed heads have dried out and become brittle, you can prepare a threshing floor on a hard piece of fairly level ground nearby, and lay down a large sheet of canvas, at least 3.6 x 3.6 m. Put a pile of the crop to be threshed, about 300 mm thick, in the middle of the canvas.

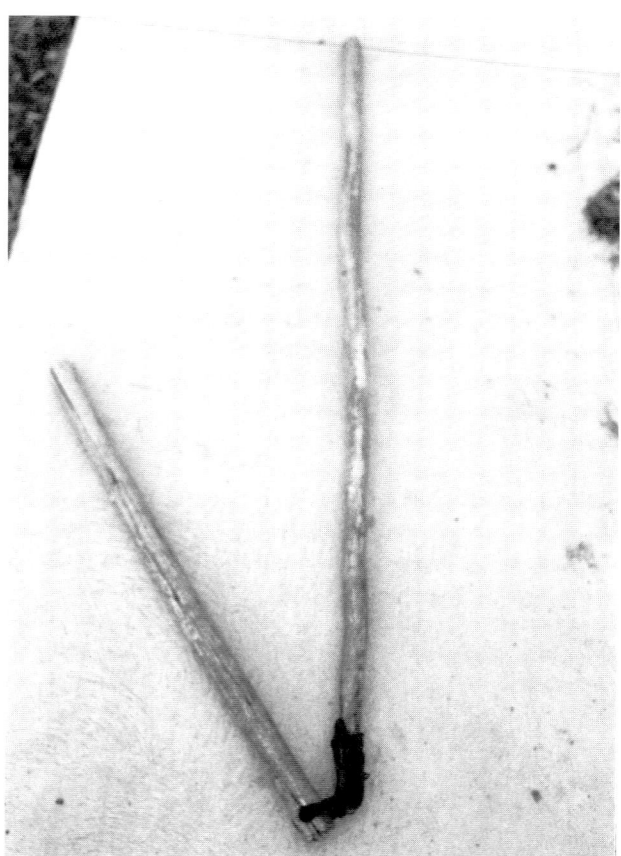

The flail can be used for hand-threshing of seeds. The handle (right) is attached to the swingle (left) by means of a swivel.

Now take up the flail and, swinging it in a circular motion, whack the swingle down flat onto the pile of crop. Continue with the circular motion so that it lifts the flail again and again, bringing the swingle down repeatedly onto the seed heads. The aim is to maintain a steady rhythm of blows that systematically work over the crop on the floor, beating the seeds out of the heads or pods. It will take a little while to get the knack of making the swingle fall hard and flat, but you soon will.

Once the heap has been thoroughly worked over on one side, turn it over and give it a flailing on the other side until examination shows that just about all the seed has been released. Then skim off the straw and put a new layer down to thresh.

Continue until you have finished threshing or until your layer of seed gets too thick. The seed next needs to be screened and winnowed. Screening it once to separate off most of the coarser straw and chaff will makes it easier to pour for winnowing.

Crude winnowing can be done by pouring the seed from a dish into a box or tub with a steady wind blowing, or failing that, in the draught of a large fan. It will take you a little while to get the hang of pouring the seed at the right rate and the right height, while holding it at the right place for the size of the seed and the strength of the wind so the seeds fall into the tub and the lighter chaff blows clear. It usually takes 4 or 5 passes to get seed reasonably clean.

Curing in Stooks and Pooks

If you have only small quantities of harvested crop, you can cure it in a thick layer on drying screens, but an alternative is to cure it in the field in stooks or pooks. The principle of these is to stack the plants in a manner that allows the air to slowly dry them while protecting them from the rain. Being in a close pile, the drying process is slow enough to allow the sap to continue to move up the stems and fill out the seeds. This is known as curing.

Stooks

Stooks (or shocks, as they are sometimes called) are mostly used for curing cereal grains. The nearly mature plants are tied together in sheaves with the seed heads all at one end. These are then leaned against each other with the heads upwards, usually 8–12 together, to make a peaked stook that sheds rain reasonably well.

Pooks

A pook is used for those seed crops, usually broadleaf species, whose form does not lend itself to tying in sheaves and stooking. It is like a paper spike, as found on bureaucrats' desks, but it is made of wood and is about 1200 mm high. It can be easily made from a thin hardwood pole, or a piece of timber around 50 x 50 mm, sharpened at both ends, with one or two shorter cross bars nailed on about 300 mm from the lower end. The main stake is then driven into the ground until firm. Bundles of the harvested crop are 'spiked' onto the pole, with the stems in alternating directions so they hold together as they are built up to the height of the stake.

The purpose of the cross bars is to keep the crop off the ground. It is amazing how well these little pooks shed water. It doesn't matter if the seeds on the outside are moistened during periods of rain as long as they dry out quickly afterwards, which they normally will if they are up in the air.

It takes most crops 2 weeks to cure in stooks or pooks. Then on a good dry day, after a dry spell, the cured crop can be carefully carried to the threshing floor and threshed by hand with a flail, or else fed into a threshing machine.

The main problem with field curing is that the ripening seed is quite attractive to birds and vermin, which – if they take a liking to its flavour – can do a fair bit of damage to a crop, not only by eating it, but also by leaving their little droppings all through it, so it is difficult to clean.

It is virtually impossible to stop this happening to some extent out in the field. Inside the drying shed, it takes a thorough vermin proofing of your facilities to prevent it. One mouse can do quite a lot of damage.

Mechanical Harvesting

Much of our southern Australian climate is suitable for seed and grain production, so harvesting machinery is available in many areas. The growth habits of many seed herbs makes them suitable for broad-acre farming, and mechanical harvesting can produce high quality seed.

Consequently, producing seed herbs is generally more appropriate for broad-acre mechanised systems, provided the markets can be found for the larger volumes produced.

On an intermediate scale, harvesting and curing seed herbs using the methods described above may be a viable proposition if the seeds can be threshed and cleaned by feeding them into a stationary threshing machine or a header set up to operate as such.

In mechanical harvesting it is important to avoid contamination with other seeds that the header may have been harvesting. If previous crops were conventionally grown, contamination with chemical residues could occur unless the machinery is adequately cleaned. Consult your certification body for their recommendations before starting.

HARVESTING BARK

There are a few species of plant whose bark is used as a herb. Harvesting bark tends to be a fairly destructive process, in that it kills the part from which it is taken. Consideration has to be given to the impact on the tree or shrub, the species and the environment. Sustainable practices need to be adopted.

Timing

For most species the best time to harvest bark is in early spring when the sap is flowing. This causes a loosening of the bark, so it comes off very easily.

It is possible to harvest bark at other times of the year, but the removal usually involves more work. If bark has to be peeled with a drawknife or chipped off, a certain proportion of wood is likely to be included, affecting quality.

Equipment

A Brush

A stiff bristle brush may be needed to remove any moss or lichen that is growing on the surface of the bark.

Stooks and pooks: traditional methods for field curing seed and grain crops.

The Peeling Spud

The peeling spud is another tool that I brought back with me from Quebec, where it is used for peeling small logs and billets destined for pulpwood. It has a flat blade, a bit like a piece of leaf spring but with a slight curve down at the tip.

This is the implement of choice for most species being peeled in the spring when the bark slips off easily. If the bark is very thick, an axe or a larger spud may be needed.

For peeling very thin stems and branchlets, a miniature peeling spud may need to be devised. On one occasion I found a thin-edged spoon did a reasonable job.

Axe and Hatchet

In the absence of a peeling spud, an axe can be used on thicker bark in spring, and a hatchet on thinner bark. These two might be able to be used to some effect in peeling tight bark at other times of the year.

Drawknife

A drawknife can be used for shaving off bark that is too tight to peel with the spud. It may also be needed for shaving off dead outer bark.

Sheet

Where the bark is coming off in small pieces, it is best to work over a sheet so the harvest can be gathered easily.

Axe, hatchet and peeling spud.

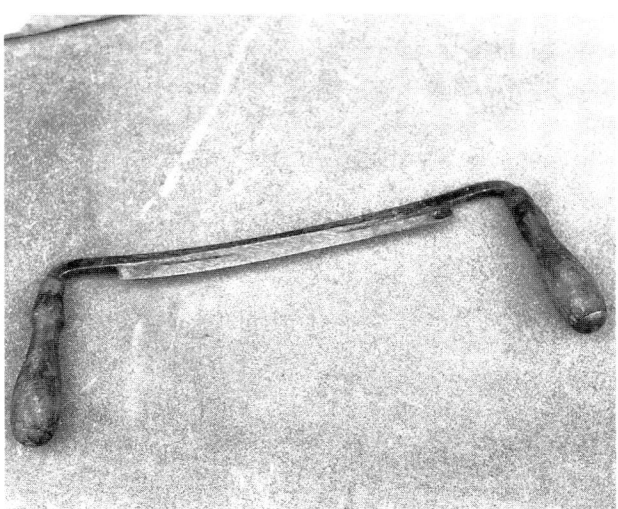

Drawknife.

Technique

Sustainable Production

The first thing to consider in harvesting is the regeneration of the tree or bush from which the bark is being taken.

Coppicing and Pollarding

Some species will coppice or send up new shoots from the base. With these, the main stem can be cut down, leaving a stump that will regenerate.

Some species grow as a clump of stems, so a few of the larger ones can be harvested, leaving the younger ones room to fill out. New shoots will normally emerge at the base of these species.

Some species can be pollarded, or cut off 2–3 m above the ground, where stock will not harm the new shoots.

With all these methods, the important thing is to cut the trunk or stem off before peeling it, taking care to leave a clean stump with a slope on it that will shed water. It is also very important not to damage or lift the bark on the stump, because this will cause it to die back.

Patch Stripping

One questionable method that is often advocated is to strip a patch of bark from the trunk without girdling the tree. The idea seems to be that the tree will not be killed by this operation.

While the bark may eventually grow back over the wound, this injury will sooner or later lead to the downfall of many trees treated in this way. Disease and borers are likely to enter the wound, rotting the wood and weakening the tree, so it ultimately dies or is blown over.

Lopping and Pruning

In some situations, lopping or pruning branches will yield adequate bark. With one species, a method of salvaging the bark from the twigs has been developed which avoids the need to cause any significant damage to the tree.

Selective Cutting and Replanting

Some species will not regenerate when cut down and these have to be replaced with seedlings or cuttings. Selective cutting accompanied by the planting of replacement trees will provide sustainable production.

Preparing for Peeling

Old thick bark may need to have the dead outer layer shaved off down to the live inner bark before peeling. Any moss or lichen should also be removed with a stiff bristle or wire brush

Spring Peeling

When peeling bark during the spring, insert the peeling spud under the bark and lever it away from the wood, working your way around the stem on both sides and up and down its length. It often helps to make some cuts in the bark around the stem with the hatchet so you are not trying to lift too long a piece of bark.

Some barks come away in long strips while others tend to break off in flakes. How the bark behaves will vary from species to species and according to the thickness of the bark.

Very thick bark will probably need an axe or an oversize peeling spud to lift it off. The top corner of the axe blade can be slipped under the bark to lift it in a manner similar to the one described for the peeling spud.

Peeling at Other Times

At other times of the year, if the bark is tight, you may have to resort to shaving it off with a drawknife. It is best to set the drawknife up at a comfortable height on some sort of stand with a sheet underneath to catch the bark as it is shaved off.

Take care not to cut into the wood with the drawknife, because it is difficult to separate wood shavings from bark.

Another option is an axe. Sometimes tight bark can be loosened by whacking it with the back of the axe at regular intervals over the surface.

A sharp axe or hatchet could be used to chip off thick bark that is too tight to peel, but it is likely to chip off a fair bit of wood as well if you are not very careful.

Harvesting bark with a peeling spud.

12
Drying

DRYING FOR QUALITY

Drying is the transformation of the herb from a dynamic, watery, living plant to a more stable dry state. If the drying process is carried out skilfully, then many of the herb's living qualities will be preserved.

The dried herb should retain most of the colour, aroma and therapeutic properties possessed by the fresh plant. However, the quality of the final product depends on more than just getting the herb down to 10% moisture content. It also depends on what has happened between harvesting the living plant and putting the dried herb into storage.

Factors Affecting Quality

Holding Time before Drying

It is important to minimise holding times and volumes of fresh herbs, to avoid build-up of heat and consequent deterioration. It is essential that harvesting and drying be closely co-ordinated to avoid delays in the drying of fresh herb, particularly aerial parts, leaves, flowers and fruits. Roots are usually not as critical and can be held for longer periods before drying, if handled and stored correctly (see Chapter 11, 'Harvesting').

Handling

While most freshly harvested leaves and flowers are reasonably robust, the majority of them are very susceptible to bruising once they begin to wilt. This can sometimes result in severe darkening and discolouration as the herbs dry. Consequently they should be transported and spread out for drying while still in a fresh state – before they have started to wilt – and should then left untouched until they are dry or close to it.

A drying system that does not involve handling the herbs during this critical wilted period will produce a much brighter dried product, which is an important aspect of quality.

Most root herbs and a small number of leaf herbs are not as susceptible to bruising and can be safely handled in a wilted state.

Temperature

The temperature during drying is also critical. If it goes too high, volatile constituents, such as essential oils, will be depleted or lost. Aromatic herbs should be dried at temperatures under 35°C, with the optimum around 30°C, provided humidity is sufficiently low. Other herbs can be dried at temperatures as high as 45°C.

Low temperatures are not harmful in themselves, but they usually mean slower drying. In humid conditions, if it is cool, less deterioration will occur than if it is warm.

If prevailing conditions are too humid, it may be preferable to raise the temperature of the air flowing over the herb by 10–15°C to lower the humidity sufficiently for good drying. Even if this somewhat exceeds the recommended temperature limits, it will be better than having the herbs sitting in warm humid conditions for a prolonged period.

Exposure to Sunlight

Sunlight fades the green colour of most herbs. One day in the sun will make the herb noticeably paler. Indirect light is not as harmful as sunlight, so, unless the herb is going to be exposed to it for a long period of time, it will not cause problems. It won't matter if the sun strikes the drying screens for a short period each day, but the longer the exposure, the greater the fading.

A few herbs are particularly sensitive to fading if exposed to light: they include Alfalfa, Parsley and Chives. On the other hand, not being green, most root herbs can be dried in the sun without affecting their colour. However, sun drying of roots is not practical in most winter rainfall regions.

As for artificial light, while ordinary incandescent globes don't cause any significant fading, fluorescent or neon light fade herbs badly and should be avoided in the drying facility.

Drying Time

The faster the drying process, the less opportunity there is for deterioration, provided the conditions are within the optimum range. However, for most leaf, root, bark, seed and fruit herbs if the drying takes up to 14 days, the result can still be satisfactory, provided suitable conditions prevailed during that time. Flowers need to be dried more rapidly – within 4–7 days – before they deteriorate.

The problem with many commercial drying setups is that there is a lot of pressure to complete the drying rapidly because of the large volume of throughput. The turnover of a drying facility can be increased phenomenally if it is cranked up to 60°C or 70°C, but the quality of most herbs will suffer phenomenally.

Airflow

Herbs dry by losing their moisture to the air they are in contact with. To maintain drying at a steady rate, there needs to be a constant exchange of air around the herbs, so it doesn't become too humid for continued drying.

This exchange of air can be achieved by various means, such as spreading the herbs in thin layers with airspace between, or using convection or fans to move the air through the herb. In a passive system the natural movement of air in a drying shed can be enhanced by well-located vents and doors that allow a greater flow of air.

This need for continual changing of air is often not well understood by people who are first setting up their drying systems and they tend to provide an airflow inadequate for good drying.

On the other hand excessive airflow can cause problems. A nice dry wind blowing through the drying shed is not going to be a great blessing if it lifts the herbs off the screens and scatters them all over the place. In a heated-air situation, too great an airflow can reduce temperatures and the rate of drying. Doors and vents need to be controllable and the layout of the shed needs to be designed to minimise such problems.

Some herbs, particularly those without any weight in their stems, such as Dandelion Leaf, Coltsfoot or Calendula petals, are especially vulnerable to being blown off the screens. A bright coloured herb like Calendula is a real nuisance if it gets blown around the drying shed as the orange petals show up very distinctly in everything else.

While the aim is to encourage the flow of dry air, provision needs to be made to prevent the flow of humid air onto dried or partially dried herb, as this can cause problems. The drying facility needs to be closed up when humid conditions prevail. This is particularly important with ambient air systems as the herb reaches the completion of its drying. This is often in the late afternoon, and the dry herb may need to wait for several hours until it can all be processed and safely stored.

Humidity

The drying of plant material is not quite as simple as evaporating water from a bare surface. The plant has some attraction for moisture and hence a reluctance to let it go and some propensity to reabsorb it. This attraction varies a bit according to the species.

Consequently, drying will proceed until a level of equilibrium is reached between the plant's attraction for water and the evaporative capacity of the air. This is reflected in the amount of moisture retained by the drying plant at a particular relative humidity.

When the relative humidity is very high, 90–100%, there will be little or no drying at all.

When the relative humidity is between about 40% and 90%, drying will proceed to a certain point, but cannot be completed: the moisture content of the plant material cannot be reduced enough for processing and safe storage. If dry or partially dried herb is exposed to air over 40% relative humidity, a degree of remoistening will take place to the point where it is in equilibrium with the moisture in the air.

While a partial level of dryness can be achieved in air of over 40% relative humidity, to complete the final stage of drying and bring the moisture content of the plant down to 10%, air of less than 40% relative humidity is needed for at least this final stage.

Drier air will also speed up the drying process and this generally means better quality.

As the temperature of air increases, so does its capacity to take up water: its saturation level rises. If the air is too humid to adequately dry plant material, this can be overcome by increasing its temperature by 10–15°C. This is usually enough to change a poor drying situation into a good one, as ambient air rarely reaches 100% relative humidity (see graph).

At around 60% relative humidity, the air will eventually dry the herb to a borderline 13% moisture content. However, to obtain a safe level of 10% moisture, the relative humidity needs to be less than 40%.

To get a good quality dried herb, it is preferable that the air has a prevailing humidity low enough to enable steady drying to a 10% moisture level within 5–14 days, depending on the herb.

This may all sound very technical, but while a hygrometer and a moisture meter might be quite useful in assessing the effectiveness and adequacy of your drying system, they are not essential. I have dried herbs for some 15 years now and never once used either: it can all be done by feel and observation.

There are simple tests that can be used on the dried herb to determine if it is adequately dry for storage.

If the leaves are dry enough to break up easily when rubbed in the hand or through a screen, they are dry enough to go into storage safely. Judging the dryness of stems and roots is a bit trickier: when dry they are usually brittle enough to snap with a 'crack' but sometimes there can be hidden moisture and you need to know where to look for it.

You will develop a feel for the prevailing conditions as you notice how they affect the drying of your herbs. Nevertheless, if you have an inclination to do moisture and humidity readings, don't let me talk you out of setting yourself up with the necessary instrumentation.

Re-moistening During Drying

Because dried plant material attracts moisture, if the relative humidity of the air increases above a certain level, dry or partly dry material will actually reabsorb water from the air.

The degree to which this occurs depends on the prevailing conditions, but it can have a disastrous effect on the quality of the herb, particularly its colour. In extreme cases where very high humidity prevails for a number of days, the re-moistened herb might go mouldy instead of drying, but the usual problem is a darkening of the leaves. Nettle and Basil are among the most sensitive – they will turn almost black – but most herbs are affected to some degree.

A small degree of remoistening usually won't affect the herb's quality significantly, but it can cause problems in processing and storage if the herb has not been dried out again. Half an hour in a damp atmosphere can be enough to bring processing to a standstill because the herb won't pass through the rubbing screen and it can raise the herb's moisture content enough for it go mouldy in storage.

Precautions need to be taken to reduce remoistening where possible and when it does occur, to ensure that the herb is redried before processing and storage.

Inadequate Drying

Inadequate drying is a common problem, especially for inexperienced growers. It usually results from assessing the superficial feel of the herb without being aware of residual moisture hidden inside parts of the plant. The problem is

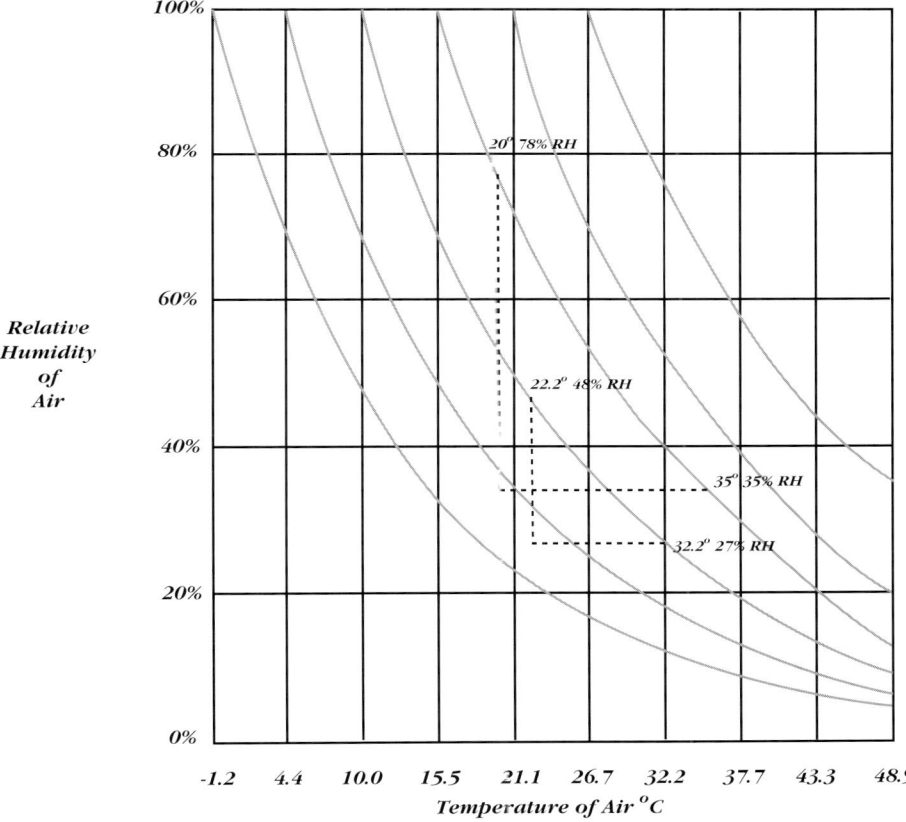

Temperature–humidity graph showing the decrease in relative humidity as saturated air, at various temperatures, is subjected to an increase in temperature. The two examples show the effect on the relative humidity of a 10°C and a 15°C rise in temperature: air of 48% relative humidity at 22.2°C drops to only 27% relative humidity when it is warmed to 32.2°C, and air of 78% relative humidity at 20°C drops to 35% relative humidity when warmed to 35°C.

usually with aerial parts or flowers, where the stalks, buds or fruiting bodies are slower to dry than the leaves or petals. Roots can be still quite moist inside. This hidden moisture can cause the stored herb to go mouldy.

Leaf rubbed through a screen to produce a tea or culinary grade is much less at risk because if it is not dry enough it will not break up and pass through the rubbing screen.

Moisture Migration

Moisture migration is really a problem of storage, but as it is the consequence of inadequate drying, it is important to be aware of the phenomenon.

For anyone with any understanding of the nature of water, it is probably no surprise that it can move from place to place. What can be a surprise though, is to open a bag of herb that has been in storage for a while and find that at one end the herb is sweet and dry, while at the other end it is damp and mouldy.

What has happened? If herb goes into storage with a borderline moisture content of around 13%, then while temperatures are moderate and don't fluctuate, it probably won't deteriorate. However, if there are temperature fluctuations, or if the herb is stored in a place where one end of the container is significantly warmer than the other (for instance, against the outside wall of a heated room in winter), moisture will move from the warm part of the container and condense in the cold part. The result is inevitable: mould growth in the damp portion.

The herb may keep all right for several months until conditions arise that cause moisture migration. To prevent it you need to make sure that your herb is adequately dry before it goes into storage. If its water content is less than 10%, moisture migration will not occur. Even though there is still some residual water in the material, there is not enough to cause problems. Evidently if the herb is sufficiently dry, it does not let go of its remaining water readily and if there is a bit of moisture movement, at 10% there is a capacity to absorb a little without harm.

Vermin

The herb grower needs to be aware that dried or partially dried herbs can be attractive to vermin, and measures need to be taken to protect herbs against them.

Rats and Mice

Rats are the worst problem because they like to eat many kinds of dried herbs, particularly Nettle and Alfalfa. Mice can be a problem, too, but being predominantly seed eaters, they are drawn to fewer crops.

The only really effective way of preventing rats and mice is to make your drying and storage facilities vermin proof by blocking up any holes with concrete or metal flashing (see 'Constructing a Drying Facility' later in this chapter). Trapping and baiting alone are not really adequate measures, because to be poisoned or caught, the little beasts need to be present, and if they are, some damage will have been done before you get the offenders.

This is an important aspect of quality because nobody wants to find a little vermin dropping floating around in their cup of tea. Manufacturers of herbal medicines can have big problems with vermin droppings as they contain bacteria like *E. coli* that are inclined to get regulatory authorities' blood pressure rising.

Possums can be a problem, too, because they are nimble climbers and like to make a day nest among the herbs on your screens if you give them the opportunity.

Moths

Another common but unwelcome visitor, which is very difficult to keep away, is the small grey-brown warehouse moth, which lays its eggs on many species of drying herbs. There is not much that can be done about them until after processing and bagging, when the herbs can be put in a freezer for 3–4 days before going into storage. This destroys any eggs or larvae.

Carbon dioxide fumigation is another safe treatment used successfully on organic and biodynamic grain for export.

Contamination, Intermixing and Misidentification

Various forms of contamination can occur during drying, such as dust and debris blowing onto the screens or undesirable material coming from your building or racks and screens. Problems can occur from paint flaking from old walls or second-hand timber, asbestos sheets deteriorating, plastic disintegrating etc. Flaky paint could be very serious if it is old lead-based paint. Even modern paint is an undesirable contaminant.

The direct exhaust from diesel burners used in some drying systems can result in contamination with combustion products so the use of these is not permitted by some certifying bodies.

Herbs coming in to be dried need to be checked for contamination with weeds, other herbs, fallen tree leaves, soil, snails etc. This is where screen drying is an advantage, because the herb all passes under the scrutiny of the person engaged in spreading it out on the screens. Being in a thin layer, this sort of contamination will be noticed and can be removed. If the crop has too much foreign material to be worth the time to pick it out, the lot should be consigned to the compost.

Intermixing and confusion can be a bit more insidious, happening when you're not aware of it. Wind can blow the herbs off one screen onto another, or a screenful of one herb may get dumped in with a batch of another, or somebody can simply mistake the identity of a herb.

If you are drying several different herbs at once, you need to institute systems that prevent them getting mixed up. Clean, or at least shake or brush down equipment after each use.

Keep the different herbs separate while drying, with a space between them, and keep the screens of each herb all together so you don't lose track of some of them. Try to keep herbs of similar appearance well apart during drying to reduce the possibility of errors.

Take care where you place herbs that are likely to blow around as they dry.

A good recording system will help eliminate errors. Make sure everyone working in the area is familiar with the appearance of all the herbs you are handling. If there is any possibility of confusion, then labels should be used to maintain identification of each herb through the drying process.

Careful Monitoring

Someone needs to take responsibility for ensuring that optimum conditions prevail throughout the drying process and that the herb going into storage is properly dry, of consistent quality, and true to label.

While good drying is not the only factor in achieving a high quality product, it is a vital link in the chain. Although good drying procedures may cost a bit more to set up and follow, they are usually justified by the higher price obtainable for high quality dried herbs.

Selecting a Drying System

A herb-drying system essentially consists of two components: an arrangement for holding the herbs and a supply of air capable of carrying away the moisture in them.

How the herbs are placed and held will influence the options available for air, but there are a number of different combinations possible.

Placement of Herbs

There are quite a few possible arrangements for placing or holding herbs during drying:

Portable Screens: The herbs are spread in thin layers on screens that can be moved around by one person.

Bunches: The herbs are tied in bunches and hung up to dry.

Suspended Screen Floor: A thick layer of herb is spread on a screen floor and air is forced through from underneath.

Screen Shelves: A layer of herb is placed on fixed shelves in a shed.

Shed Floor: Drying is carried out in a layer simply spread out on the floor of a shed.

Outside Floor: A bitumen or concrete surface outside in the open is used to spread the herbs on to dry.

Field Curing: Herbs are dried on the ground in the field where they were grown.

Air

Several options exist for a supply of air for drying. These systems can rely on natural air movement, or circulation can be assisted by convection or a fan:

Ambient Air: This is where drying relies solely on the evaporative capacity of the air under whatever weather conditions prevail at the time.

Ambient Air with Passive Solar Heat Gain: This system relies on ambient air but the design and materials used for the drying shed provide some solar heat gain to raise the temperature of the air inside the shed during sunny weather.

Solar-heated Air: Air is heated by flowing through a solar collector and then flows through the drying facility.

Artificially Heated Air: Air is heated by flowing over an electric heater, heat pump, or a heat exchange burning wood or fossil fuels.

Direct Burner: The exhaust and hot air from a diesel or gas burner, mixed with additional air from outside, is directed through the herbs.

Freeze Drying: Frozen plant material is dried by a vacuum process which volatilises its water content.

Dehumidifier: A condensation process removes the moisture from the air circulating over the herbs.

Designing a System

While there would appear to be a multitude of possible combinations for a drying system, when one considers the need for an economical system that produces a quality product, the choice is narrowed down considerably.

Generally growers will require a drying system that does not involve a large capital outlay, has a low running cost, and can be used efficiently for drying a variety of crops at the same time or for varying quantities.

A number of producers of high quality herbs rely on a system based on the use of portable screens and a combination of ambient air drying with solar heat gain and a back-up system using heat from a solar, electric, wood or gas source. In some situations where ambient air is inadequate for drying, artificial heat is used for the whole drying process.

Drying on Portable Screens

Drying herbs in a shallow layer on portable screens has a number of advantages. It is a reliable and versatile system that makes it easier for the small- and intermediate-scale grower to produce a consistent high quality for a modest outlay and running cost.

The screens can be slid onto racks that hold them one above the other with an air space between them of around 150 mm. If screens need to be moved to a new position during the drying process, this can be done without having to touch the partly dry herbs.

While there are other methods of drying herbs, the general experience is that screen drying in shallow layers produces the best quality dried herb and has other advantages. There are several reasons for this:

Avoids Bruising: Because the herb is spread in thin layers, the whole batch begins drying as soon as it is spread out and the drying can proceed to completion without any need to stir or turn the herb. This minimises bruising and shattering.

Facilitates Removal of Contamination: Spreading the herb by hand onto the screens means that any weeds, discoloured leaf or other contaminants usually show up readily and can be removed.

Avoids Heating and Sweating: If there is an interruption to drying, there is no tendency for the herb to heat or sweat when it is spread in a thin layer. Airflow around the herb helps prevent deterioration.

Ease of Monitoring: The herb dries relatively evenly and it is easy to monitor dryness throughout the crop.

Versatility: Screens can be easily moved about without disturbing and damaging the herb. This makes it easier for transferring them to a backup dryer or dumping them into the processing tray.

Economy and Efficiency: The grower can make the screens in spare time from shadecloth and timber, so capital costs are kept down. The screens help reduce running costs by making it possible to utilise good ambient drying conditions while they prevail, but if it is necessary to resort to a back-up system, the screens can be easily transferred into a cabinet dryer.

The concurrent drying of several different crops is facilitated by the use of screens as they can easily be kept separate.

A further advantage of a screen drying system is that it can be started off on a small scale and expanded gradually by making more screens and racks as production increases.

Making Drying Screens

If you are making the screens yourself, you will need to plan ahead in order to allow sufficient time to season green timber for 2–3 months before assembling the screens.

Screen Dimensions

The size of the screen does not really affect the drying, but it does affect the operator. For efficiency the screens need to be as large as practicable while still being easily managed by one person when they are loaded. They also need to be able to fit into the space you are using for drying, with plenty of room for manoeuvring.

It makes life much easier if they are all the same size, as they can be interchangeable in all positions.

What is referred to here as a standard-sized drying screen is one most growers have found suits best: 1500 x 940 mm. This is a comfortable width to pick up, and two screens can be made side by side from standard 1829 mm shadecloth without any wastage. A loaded 1500 mm screen is about as long as the average person can handle without straining.

Sometimes an 1800 mm long screen is advocated, but this would be too heavy when loaded with dense herbs, such as roots and fruits, and could not be manoeuvred single-handed.

If you have only a very small drying operation and limited space, you may wish to construct smaller 940 x 940 mm screens. Being the same width as the longer screens, these will still fit on the same racks.

Construction

Sandwiching the edges of the fabric between two layers of timber that make up the frame produces a strong durable drying screen. The edges are protected, the fastenings can't come loose and are supported by the pressure of the sandwiched construction.

Materials

Frames

Timber is normally used for the frames as this is easy to fasten the screen material to. Each of the four sides of the frame is made from two pieces of timber with the fabric sandwiched between.

Dimensions for Hardwood Frames: Hardwood is preferable as it is stronger and can be used in smaller dimensions. This makes the screens less bulky for stacking or doubling or tripling up on the racks. The grain of hardwood is easier to see, so you can make sure it is straight enough to provide the necessary strength. But one does need to watch out for brittle heart and avoid sapwood, which will be attacked by little borers.

If only lightweight herbs are being dried, such as leaves and flowers, then 38 x 13 mm is a suitable dimension (fully dried) which gives a screen with a total thickness of about 26 mm. However, if roots or fruits are to be dried, the frames need to be somewhat heavier – at least 38 x 16 mm – to reduce flexing and breaking.

Usually this timber has to be ordered green from a sawmill as these are not standard sizes. Be prepared for a few raised eyebrows when you order a kilometre or two of it for 100–200 screens. If ordering green timber, remember to allow for shrinkage and wastage. Inevitably there will be some unsuitable pieces due to sapwood, brittle heart, warping etc., perhaps as much as 25% wastage. If the timber is green it should be racked out for 2–3 months to allow it to dry and shrink before use.

Naturally these dimensions are not rigid, and you may find some timber of slightly different dimensions at a good price. One thing to bear in mind is that the sides need to be strong enough to bear the weight of the herb without breaking or excessive bending.

They also need to be stiff enough in a lateral direction to keep the long sides straight. If the sides are pulled in too much by the tension of the shadecloth and the weight of the herb, the screen will become too narrow. This can cause a minor disaster if the screen slips off the slats, dumping its contents onto lower screens or the floor below.

On the other hand, excessively wide or thick sides will encroach on the capacity of the screen, or make it unnecessarily heavy and bulky.

If the timbers of the frames are made of thicker material, the screens are going to occupy more space when stacked directly on top of each other for storage or other purposes. This can create a bit of inconvenience, but this is generally endurable.

Softwood Frames: Softwood frames are a possibility but they will need to be thicker to obtain the necessary strength. If you are using radiata pine, you will need to select it very carefully as it has a lot of knots and it is difficult to judge its grain and strength at a glance.

Importance of Using Dry Timber: Whatever timber you are using, it will need to be thoroughly dry otherwise shrinkage will cause looseness: the fabric may pull through and the frames will become wobbly at the corners.

Fabric

There are a number of possibilities for the fabric of the screen. It needs to be porous enough to allow air circulation from underneath and strong enough to bear the weight and handling. It also needs to be non-toxic, durable, and not affected by moisture. It must be of a texture that allows the dried herb to fall easily from it.

There are a number of materials I have tried:

Chicken Wire: My first screens were made of chicken wire (it had not been used for chickens), but while this is strong and durable, the leaves have a tendency to curl around the wires while drying and have to be physically scraped off. This can be rather tedious and time consuming.

Fly Screen: My next batch of screens was made with galvanised fly screen. This was much more expensive, but proved to be much better from the point of view of the leaf clinging problem. However, flywire is not very strong and tends to tear away from the nails in the frames. It also deteriorates in contact with moisture: after a while it starts to go rusty. (Aluminium or fibreglass fly screen are not strong enough for the purpose.)

Shadecloth: Then I discovered knitted shadecloth, which proved to be about twice the strength and half the price of fly screen. It is the material preferred by most growers. The only real objection I have to it is that it is made of plastic.

Hessian: Hessian is sometimes advocated as a natural material, but it is treated with something to facilitate its manufacture and to protect it from insects and mould. It will also go slack on the screens as it stretches with use.

Other Possible Fabrics: Another possibility is light cotton or polyester over chicken wire or shadecloth to give it support. For small seeds and other fine herbs that fall through ordinary shadecloth, a finer screen is very useful.

There are other possible fabrics but most of them would be prohibitively expensive to make the number of screens you will need for a drying operation or else would not be strong enough.

Types of Shadecloth

Knitted Shadecloth: The shadecloth used must be the knitted type: the woven type will pull away from the fastenings and is difficult to tension evenly.

Thickness: Shadecloth comes in different thicknesses described as a percentage of the light they block out, with 50% being the grade most of us have used. Lighter weight grades are too stretchy and will sag after a while.

If you are drying heavier herbs, such as fruits and roots, 70% shadecloth is recommended for its extra strength.

Colour: There is a choice of colour, too. While the colour doesn't seem to affect the drying performance, if you can obtain the white shadecloth it makes loading herb onto the screens a little easier because the white shows through more and any thin or thick areas of herb can be seen more readily.

Assembling Standard-sized Screens

Selecting and Cutting Timber for Frames

Selected straight-grained timber free of knots, sap wood and other defects should be used for the main frame of the screen. This consists of four pieces, two of 1500 mm and two of 940 mm, bolted together at the corners with ¼ inch (6 mm) 'roofing bolts'. (These are small bolts with flattish rounded screw heads, which need to be located on the under side of the screen.)

The screens are designed to slide onto horizontal runners fixed one above the other on vertical posts. The long members on the sides of the screen need to be able to slide freely on the runners. They are made with the 1500 mm lengths underneath and the 940 mm cross-members fastened over them with the screw heads on the underside and the nuts and washers above. This is because the rounded heads won't snag on the ends of the runners as much as the nuts would.

Selecting and Cutting Battens

Four shorter pieces of the same cross-section as the main frame (two of 1420 mm and two of 860 mm) are battened down onto the frame to hold the shadecloth. (Timber rejected as unsuitable for the main-frame members is often adequate for these battens as their strength is not as critical.)

Setting up a Jig

A jig for two screens side by side can be made to hold all the pieces in place in the right positions while they are nailed and bolted together. A reasonable degree of accuracy is needed to ensure the screens come out all the same size. Variations in width of more than 2–3 mm can cause problems in getting the right spacings for the racks they are to slide into.

Setting up the Shadecloth

The shadecloth comes in rolls 1829 mm wide. When the width is stretched tight across two 940 mm screens, a good tension is achieved.

If the shadecloth is suspended so it can be unrolled from above one end of the jig, it can be easily pulled across for fastening down.

Fastening the End Battens

The frame sides and end battens are laid on the jig first. The shadecloth is then laid over and the cross-members are placed on top of the end battens and nailed and bolted to the runners at the corners. Using a washer under the nut will help avoid splitting.

The free end of the shadecloth is battened down first. Then the uncut end is carefully pulled to the right length and fastened down evenly at the other ends with battens. It may take a bit of experimenting at first to get the right length, because shadecloth, being somewhat stretchy, is difficult to measure accurately. Don't cut the shadecloth off yet as it may still need adjusting.

Fastening the Outer Battens

Having fastened the end battens down, the next step is to fasten one outer side batten. By this stage it is starting to get a little tricky. If you drive the nails (30 mm flat-head) through the batten first, these can be hooked into the shadecloth to pull it over and will ensure it is securely fastened at each nail, rather than just being squeezed between the pieces of wood. The nails should be angled out slightly so the cloth does not slip off when pulling the batten into place under tension.

Repeat this on the other side. This should bring the shadecloth to a good even tightness over the two screens. If you can't manage to stretch it across without putting a terrible bow in the sides, or if it ends up too loose or with sags in it, you will need to make an adjustment to the length of the shadecloth at one or both ends.

Fastening Inner Battens and Separating the Screens

When you are satisfied with the tension of the shadecloth, the two inner battens can be nailed down tight and the two finished screens separated by running a sharp knife between them and cutting the shadecloth free at the ends.

Clinching Nails and Rasping Edges

The points of the nails should then all be clinched over, which makes them secure against pulling loose, and protects your hands from being gashed by protruding points. It is probably worth while running over the edges with a rasp to trim off any splinters, many of which may otherwise end up embedded in your flesh over the years, as you shuffle the screens around in your drying shed.

Angling Corners

About 20 mm should be diagonally sawn off the points of the corners as this helps when positioning the screens on the racks by making them less inclined to catch on the posts.

Trimming and Burring Bolts

The protruding ends of the bolts should be trimmed off with a hacksaw so they are flush with the nuts, but this operation could wait a while until all shrinkage is finished. It is probably worth tapping the end of the bolt with a hammer to burr it enough to lock the nut on as nuts can occasionally come off and disappear into the herbs.

The corner of a drying screen, showing the sandwich construction.

Construction Time

With a good jig set-up and the timber already cut to length, one person should be able to assemble a pair of screens in 40–60 minutes once a good routine is established.

Number of Drying Screens Required

The number of drying screens required depends on the system adopted.

Ambient Air Drying of Leaf Crops

If you are relying on ambient air conditions with a solar-oriented shed, then a very rough estimate is that for drying a leaf crop you will need a total screen area of approximately 50% of the surface (including inter-row spaces) it took to grow the crop. In other words, a patch of herbs 3 x 15 m will need, very roughly, 22.5 m^2 of drying surface or somewhere around 16 screens of 940 x 1500 mm.

As the herbs will dry in 1–2 weeks and the new growth reaches harvest stage in about 6 weeks, these screens can be used for 3–5 crops. Provided you are able to stagger your harvesting, you will need only $1/5$ to $1/3$ of 50% of your total leaf crop area, that is 10–17%. Some situations might feasibly need a little more, say up to 20% of total crop area.

At 'Twin Creeks' in north-east Victoria I managed with 180 screens for drying about 2500 m^2 in leaf crops. Drying conditions are usually very good there. This amounted to a drying area of 10% of the crop area, but it did include some medicinal herbs that were not very high yielding. In Tasmania I had roughly the same area in leaf crops, but needed 200 screens as drying conditions were not as favourable.

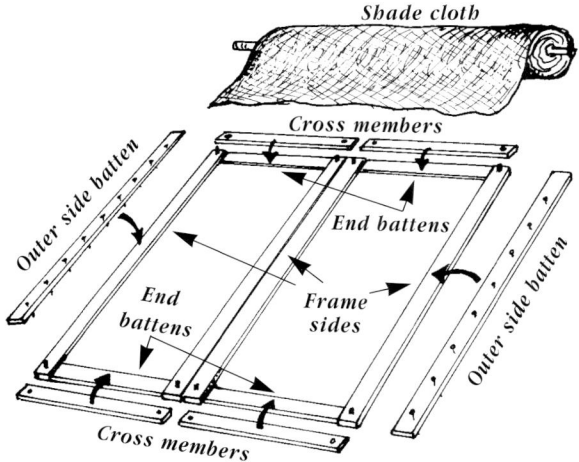

Making portable drying screens in pairs in a jig.
The main frames have been drilled through at the corners and the bolts fitted from underneath the long sides, with the thread sticking up. The cross-members have been lifted off the bolts to allow the shadecloth to be pulled over the screens. They will then be nailed down on to the end battens, sandwiching the shadecloth tightly, and the nuts screwed onto the bolts. The outer side battens will be hooked onto the shadecloth, stretching it tight, and will then be nailed down. When this is done, the two inner side battens will be nailed down and the shadecloth will be cut off at the end and down the middle between the two screens.

Flower Crops

Flower crops need a smaller drying area: 5–10% of crop area, or less if additional heat is used in drying.

Root, Fruit and Bark Crops

Roots and fruits are normally dried with artificial heat and are loaded much more heavily onto the screens. Roots can be harvested and dried over a longer period. This way a smaller drying facility and fewer screens would be needed: around 5% of crop area should be an adequate drying surface if the root crops are harvested and dried sequentially over a 3-month period using artificial heat. About ten times this capacity would be needed if they were all dried at once.

Heated Drying Systems

Where supplementary heating is being used for leaf or flower crops, fewer drying screens will be needed. If constant optimum drying conditions can be provided, the number of screens required can be reduced by as much as half. Depending on the system, the herb can be loaded more heavily on the screens and the drying time will be shorter.

An operation growing 2500 m^2 in leaf crops, 500 m^2 in flower crops and 1000 m^2 in root crops will need around 200 screens of 940 x 1500 mm if using ambient air with solar heat gain in summer and a good heated back-up system for inclement weather and for winter drying of roots.

Of course these are all rough estimations and depend on prevailing weather conditions, the efficiency of the drying facility, the crops being grown, the yields obtained, and how much the harvests can be staggered so the screens are as fully utilised as possible.

Cost and Life Span of Screens

Materials

For 1500 x 940 mm screens made with 50% shadecloth, hardwood (purchased green) and bolts, the cost is $7–10 per screen. The breakdown of this is $3–4 for shadecloth, $3–5 for timber and $1 for bolts and sundries.

Labour

Allow 10–15 minutes per screen for organising materials and 20–30 minutes for assembling the screens: a total of 30–45 minutes per screen. Most growers should be able to make their own screens in quiet times as it is a good inside job for winter or rainy weather. If labour were hired to make them at, say, $12/hr, then the labour component would cost $6–9 a screen.

Working Life

I am still using screens I made 15 years ago, though some are in need of minor repair. With a little care and a few repairs, they should last at least 30 years.

THE DRYING SHED

Those who are embarking on a commercial scale of production will probably need a building, or part of a building, allocated specifically for the purpose of drying. On a smaller scale or on a temporary basis, it may be possible to dry herbs in the attic in your house, a partitioned off section of verandah, or some other such arrangement.

Most herbs need to be dried in the shade in order to produce good quality. There are a number of options for shade drying, but in most parts of southern Australia ambient air can be used quite effectively for much of the year.

Ambient Air Drying

Ambient air drying uses prevailing air conditions to dry herbs. It can be passive, relying on natural ventilation, or it can be assisted with a fan to increase air circulation.

Solar Heat Gain

If the drying facility is well sited and well constructed, it can achieve a degree of solar assistance for no extra cost by way of sunlight on the roof and walls increasing temperatures inside.

With the additional heat gained, drying conditions inside the shed can be improved, making the facility more effective and functional in otherwise unsuitable weather conditions.

This system works well for aerial parts and leaf when the prevailing relative humidity of the air is below 50% most of the time, and below 40% for a good part of the time. The air is then able to progressively take up moisture from the herb at a sufficient rate to dry it fast enough to prevent deterioration and preserve good colour.

For drying flowers, ambient air can be adequate when conditions are very good, but otherwise flowers will need additional heat to dry quickly enough.

In warmer, drier regions, the drying of fruits can sometimes be considered using this system if they are harvested early enough in autumn while drying conditions are still good, otherwise they will need artificial heat.

Drying roots with ambient air is normally not feasible. Roots are usually harvested in winter and, being thick and fleshy, need additional heat to dry well.

The general principle of ambient air drying is to provide good air circulation around or through the herb. This keeps the drying process going as the air often has only a limited capacity to absorb additional moisture. Good air circulation ensures that the air around the herbs is continually being replaced.

Advantages

An ambient air drying system has the following advantages:

Cost: It is relatively inexpensive to set up and operate.

Flexibility of Size: It can be expanded progressively as your operation increases in size.

Simplicity: It does not depend on complex technology: you can build, operate and repair it yourself.

Temperature: It dries the herbs within the optimum temperature range.

Environmentally Friendly: It does not depend on burning fossil fuels, so it does not contribute to carbon dioxide emissions.

Disadvantages

The system has the following disadvantages:

The Vagaries of the Weather: The main problems with an ambient air drying system stem from its being very much subject to the prevailing weather conditions. Your drying program and daily work schedule need to be able to fit in with this.

This problem can be largely overcome with an adequate back-up system using an additional heat source of some type.

Throughput: Volumes of throughput will be smaller than for a dryer that is maintained at a constant temperature and airflow.

Skill: Proper functioning of the system depends on a certain level of skill on the part of the operator in judging prevailing drying conditions and understanding the drying process, though this is not difficult to develop.

Manual Operation: This system does not lend itself to automation as it depends on the surveillance and skill of the operator.

Structure

External Cladding

For ambient air drying in most parts of southern Australia, a drying shed should be clad with material that will collect some supplementary heat from the sun, resist invasion by vermin, not contaminate the herb, be easy to clean and not be too expensive. From these points of view it is hard to go past that great bastion of Australian bush architecture: corrugated iron.

The heat-collecting quality of galvanised iron makes it preferable to zincalume, which is shinier and reflects more heat. These days it may be difficult to buy new galvanised iron, though it is still made. Alternatively second-hand iron (unpainted) could be used – this is cheaper and having weathered a bit, it is usually darker and absorbs more of the sun's heat than shiny new iron.

Other options are painted metal sheeting (some councils will insist on this) using a dark colour, or including some clear panels in the roof. This latter option may need to include some precautions to protect the herbs from fading. Clear panels which are opaque to ultraviolet and shade-cloth suspended over the racks seems to be sufficient for the average leaf herb, but there could be problems with easily faded herbs such as Alfalfa.

In order to transmit heat into the building, the roof and those walls that catch the sun should be single skin with no insulation or lining on them.

Orientation and Solar Efficiency

The shed should be located and aligned to receive as much sun as possible on the walls and roof. If the roof can slant towards north, this will help. If possible the shed should be on an east-west axis with a long windowless heat-absorbent wall on the north side so a line of drying racks can be constructed against it on the inside.

Galvanised iron is a preferred material for external cladding as it improves solar heat gain, makes it easier to exclude vermin, and is easy to clean. Note the rear sliding vent, which allows through ventilation when used in conjuction with a roller door in the opposite wall.

Good location, good alignment and good utilisation of the upper levels can make the difference between a drying shed working adequately or not, particularly during the cooler parts of the season or in cooler regions. The sun's heat on the roof and walls during the day can often raise the temperature and lower the humidity enough to enable herbs to be dried in these less favourable conditions.

In warmer drier regions, the configuration of the drying shed for heat absorption is not as critical. In fact, measures may be needed to keep temperatures down in the hotter part of the year.

Height

Because warm air rises, the herbs will dry more quickly in the upper part of the shed, so it makes sense to fully utilise this space for drying. You will want to be able to get the screens up into this upper region and down again, preferably without having to do too many risky acrobatics. If the racks extend much more than 3 m from the ground, placing and removing screens becomes significantly trickier and slower.

Consequently, there is no point in building a shed much over 3 m internal height. A skillion roof (flat and low pitched) is the most practical from the point of view of

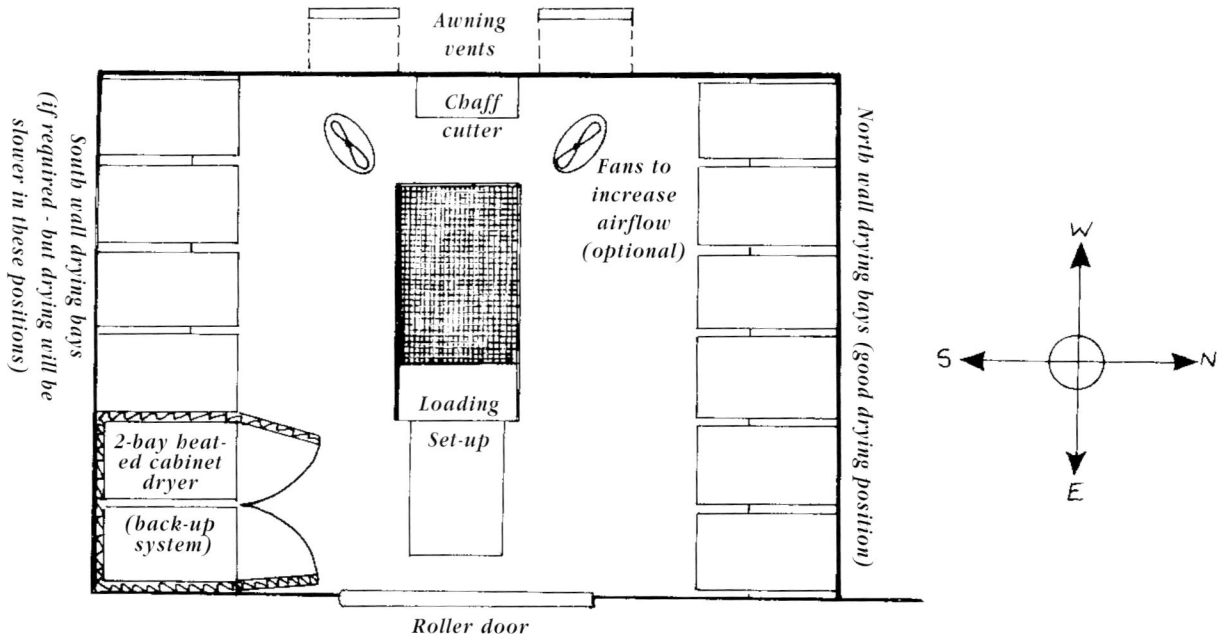

A shed designed for ambient air drying: the blank north wall and skillion roof serve to collect heat from the sun, while the east-west ventilation makes use of prevailing winds.

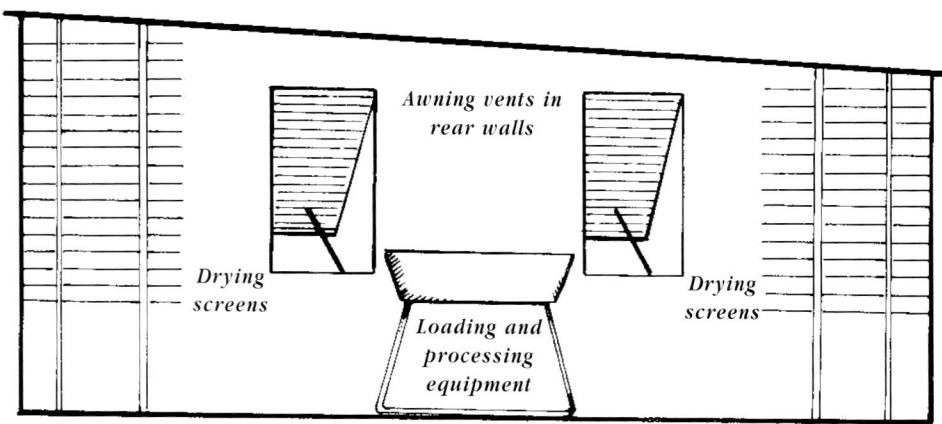

Cross-section of the same drying shed.

utilising the upper levels. Of course, if you already have a shed higher than this, you can still make use of the upper levels by climbing further up a ladder, but if you are going to build a drying shed, keep it lower.

Dimensions and Layout

The dimensions and layout of the drying shed need to be planned around the dimensions of the screens to ensure optimum use of the space and adequate room for the operations carried out in the building.

Generally it makes sense to be able to do the processing in the room in which the herbs are dried. If the herb can be tipped off the drying screens onto the rubbing screen or chaff-cutter tray without taking too many steps, the operation will be much more efficient. It also means the vents and doors can be closed down if the humidity has risen outside and processing can continue for some hours inside before the herbs start to get damp.

In addition to drying space, there needs to be enough space in the drying shed for all the processing equipment as well as space for loading the fresh herb onto the screens.

The greater size of the shed required to accommodate these operations actually increases its effectiveness for ambient air drying as there is more heat absorbed and more air flow for the same volume of herb being dried.

Ventilation

A drying shed relying on ambient air needs maximum air flow when drying conditions are good, though not so much as to blow the herbs off the screens. It also needs to be closed up tight when the humidity gets too high.

Closable Vents

It is quite easy to make a closable vent in a corrugated iron wall by cutting an opening one or more sheets wide in the wall, about 900 mm from the floor and up to the roof. A section made up of corrugated iron a little longer and wider than the space can be hinged from the top. When closed, it covers the opening with a 75–100 mm overlap all around, or it can be propped open to allow the breeze to flow through.

Vents like these are preferable because they keep rain out if you aren't around to shut them and they keep the sun out to a large extent. Being well above ground level, there is no access for vermin as they cannot scale the corrugated iron walls.

Placement of Vents

The best position for the vents is opposite the main door or opposite each other, allowing a cross-flow of air through the shed. Vents and doors should be positioned so they open onto the alley between the racks rather than directly onto the drying herbs. This reduces possible damage from sun and rain and reduces the amount of wind striking the screens where it might lift the herbs up and scatter them.

East-west ventilation is often the more effective and fits in with having a blank northern wall. The aim is to let the steady drying breezes come through, moderate the wild blustery ones, and shut out the moisture-laden ones.

Supplementing Natural Ventilation

A good system for through ventilation should prove adequate most of the time, but in some situations you may need to supplement natural ventilation with fans. A large fan or fans pointed in the general direction of a the loaded drying screens can make a significant difference in their rate of drying during a mild breezeless period.

Avoid Pockets of Dead Air

In designing the drying shed, try to avoid pockets of dead air where air movement will be too restricted for good drying. Corners and screens placed up against cold walls can be a problem, so some provision may need to be made to ensure they have adequate ventilation

Vermin Control

Vermin control is an important aspect of the drying shed and best provided for before a problem arises. Rats and mice have followed Europeans around the globe and there will be few places you might choose to set up a drying operation where you won't find them soon trying to move in. If you happen to be fortunate enough not to share your dwelling place with our ancestral comrades, *Rattus rattus* and *Mus musculus*, you will nevertheless find that native rats and possums – endearing as they may be – can pose a considerable problem.

The best way to deal with all these little critters is to not let them get into the drying shed in the first place. I am continually amazed at the way we construct buildings with complex council regulations applying to so many aspects of the structure, but hardly a thought is given to what can be done to prevent the ravages of rats and mice by simply blocking their access.

Careful attention to keeping them out will prevent vermin problems. I find this a better solution than trying to rely on trapping or poisoning.

Main Points of Entry for Rats and Mice

Rats and mice are able to find any number of possible entry points:

- At the foot of the walls.
- Under and around doors (or through them if left open at night).
- Through small gaps in wooden floors and walls.
- Up climbable outside walls and under the eaves. If they can climb onto the roof via trees, they can get into the gutter and under the eaves.

Sealing Off Gaps

Rats and mice can squeeze through surprisingly small holes and if a gap is too small they will gnaw away with their little teeth until it is big enough to let them through.

Sheet metal and cement mortar are the best materials to seal off entry points. With corrugated iron walls, mortar can be used to fill up the corrugations at ground level. It takes a bit of observation and skill to find all the possible entry points: you need to get down and take a rat's eye look at everything.

Doors

Doors are vulnerable if they don't seal tightly or if they get left open at night. You need to be consistent about closing them at the end of each day or else organise your ventilation so air circulation is good enough with the doors closed.

Some types of doors are virtually impossible to seal adequately and should be avoided. Roller doors are good.

Walls

Corrugated iron is the best material for walls as vermin can't chew through it or climb it. Attention needs to be paid to where it meets the floor and where the sheets lap, particularly at the corners, to ensure there are no entry points.

Trapping and Poisoning

Trapping and poisoning tend to be indiscriminate and can affect predators. They can never be relied on to prevent the problem entirely. Cunning rats become trap and bait shy, and can do a lot of damage in the drying shed before you finally manage to catch them.

Ultrasonic Repellents

I haven't tried ultrasonic transmitters because of a concern about possible harmful effects on people and domestic animals.

Old Buildings

Old buildings are particularly difficult to vermin proof, especially if the walls are of timber or other material that allows rats and mice to climb.

An option in this situation could be to enclose the drying racks in a fine mesh metal cage to prevent vermin access. I had to do this in my first drying set-up in a large old wooden shed, but it was not very satisfactory.

Invasion Time

The worst time of the year for vermin is in autumn when they start to seek shelter and food for winter. You can be lulled into a false sense of security over summer and think that they are not around and then suddenly find you have a major problem on your hands when the weather starts to get cooler and they move in.

Other Creatures

Apart from vermin, there are some other creatures that need to be discouraged from taking up residence in the drying shed.

Starlings will come in under the eaves and make large messy nests if they get a chance.

Swallows may like to nest up near the roof. This may seem delightful at first, but inevitably a growing pile of droppings appears on whatever is below the nest. Keep an eye out for them in spring when they first start nest building and encourage them to look elsewhere: you may need to put screening over your vents for a few weeks.

Possums can be a nuisance, too, but will be kept out by a rat-proof set-up.

Placement of the Drying Screens

A system of racks is needed to hold the screens while the herbs are drying. These are best constructed so the screens are held horizontally, 150 mm above one another, in bays extending from around 900 mm from the ground to the roof. They could extend right down to the floor but these lower positions will not dry as quickly, especially in marginal conditions. This area might better be used for storage: there are plenty of things used in the drying and processing operation that need a place out of the way.

One system for holding the screens has a series of runners nailed onto two 75 x 50 mm vertical posts on each side of the bay, as in a chest of drawers.

Posts

The posts must be very straight and stiff enough to be set 950 mm apart or even 945 mm (if you can work within these tolerances) to accommodate 940 mm wide screens. It is best to place these posts with the 75 mm face at right angles to the screens, as this is the direction greater rigidity is required.

Radiata pine is adequate for the racks and is easier to nail than hardwood. Avoid green timber on account of shrinkage and its tendency to move – you could find that your spaces become too wide or too narrow for the screens – a real problem.

Before erecting the posts, you may need to bevel the leading edges (see below).

Runners

The runners that the screens slide on need to be at least 38 mm wide to hold the screens with enough free play and no risk of slipping off. As 38 x 38 mm is an odd size, it may be difficult to find. Alternatively 75 x 38 mm ripped in half will give 35 x 38 mm pieces which can be used with the wide face horizontal.

The runners don't need to be particularly long, in fact it is easier to slide the screens on and off if they project only 50–75 mm in front of the posts. They just need to be long enough to give the screens reasonable stability when pushed right in.

Bevelling Wooden Posts

With wooden runners, the screens tend to strike the posts as you slide them in. The leading face of the posts needs to be bevelled to reduce this problem: a job most easily done on the ground, before putting them up and attaching the runners. As the screens have their corners cut at a corresponding angle, if they strike the posts they will tend to centre themselves and slide past.

Alternatively, guide rails could be attached to the runners.

Steel Racks

Another alternative is to construct the racks of steel using square section for the verticals and 35 x 35 mm angle for the runners. By setting the angle iron for the runners with the vertical side up, it serves as an outer guide that makes it easier to slide the screens in and out as they can't strike the posts.

Movable Racks

The configuration of some sheds may lend itself to the use of one or more movable racks. Set on wheels, these can be rolled over close to the loading set-up and when full of screens, can be rolled into place. When space is needed to access other bays, they can be rolled out of the way. For

Drying shed set up with wooden racks for holding the drying screens. The lower screens in foreground are empty and are stacked two or three per pair of runners.

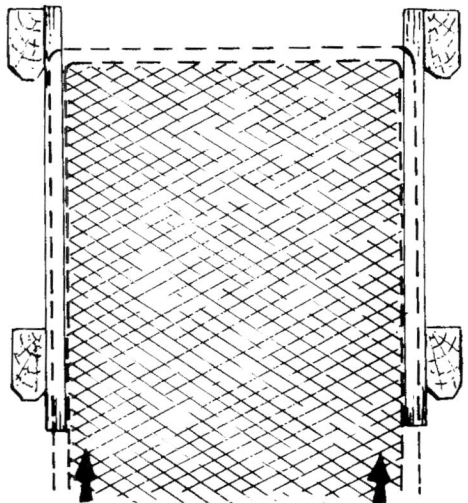

Bevelling of vertical posts makes it easier to slide screens past them.

adequate rigidity, a movable rack will probably have to be made of steel as diagonal bracing of a wooden structure would be in the way of the screens.

Heated Back-up System

Naturally the drying capacity of ambient air is subject to variation according to the time of day, the weather and the season. In warmer drier regions, it can be relied upon for most of the leaf harvesting season – October to May – while in other parts the weather is less dependable. Nevertheless, in all regions some sort of heated back-up system will need to be available to dry or complete the drying of the herbs, especially in late autumn when prevailing conditions are usually unsuitable for satisfactory drying.

Problems arise if the relative humidity of the air is too high to complete the drying process, as the herb will only lose a certain amount of its moisture and no more. If these conditions continue for too long, deterioration will occur. The worst situation is when the herb is dry or almost dry and then gets remoistened in very humid conditions. This needs to be avoided where possible.

Usually for aerial parts and leaves, it will only be necessary to provide additional heat for the last part of the drying process. This is generally the case in most of southern Australia, including most of Tasmania, where summers tend to be dry. In these regions, a heated back-up system of 10–20% of the total drying capacity will be adequate as it will be mostly used for finishing off partially dried leaf herbs and sometimes for drying flowers.

Further north, especially along the coast and ranges, summers tend to be humid, and 'sloppy' weather – rain and drizzle – can set in for days or weeks at a time, which can be disastrous for ambient air drying. In regions with humid summers, this sort of drying set-up will need to have a large heated back-up system capable of handling the complete drying process in damp weather.

The back-up system is best located inside the drying shed, where it is close to the processing facilities and main drying racks so that it is easy to move screens of herbs in and out of it. Its actual location in the shed will depend on the heat source used. If only fuel or electricity is used, it is better located on a southern wall, so it is not taking up a warm spot in the shed. If solar heat is envisaged as part of the back-up system, the back-up system may need to be on the northern wall.

It will need to be contained within its own walls to work efficiently and to avoid loss of heated air. One option is to have removable walls that are put in place when the heated

A heated back-up system is needed for times when conditions are not favourable for the completion of drying with ambient air.

back-up system is in use, but the rest of the time the bay(s) can serve for ambient air drying.

A back-up system contained within the drying shed will need to be vented to the outside so warm moist air coming from it does not condense within the shed and cause problems.

If it is adequately insulated, the heated back-up system will operate more efficiently and can also be used for drying root crops in winter.

Loading Set-up

As you will be spending a fair bit of time loading herbs onto screens, you need an efficient and convenient set-up for loading. This should be close to the drying racks so you don't have far to walk with each loaded screen.

For this a large plywood tray – a little wider and longer than the screens, with sides about 225 mm high – works

1.1 Alfalfa *(Medicago sativa)*

1.2 Angelica *(Angelica archangelica)*

1.3 Burdock *(Arctium lappa)*

1.4 Bergamot *(Monarda didyma)*

1.5 Calendula *(Calendula officinalis)*

1.6 Catnip *(Nepeta cataria)*

Plate 1

2.1 Chamomile, German *(Chamomilla recutita)*

2.2 Comfrey, English *(Symphytum officinale)*

2.3 Chaste Tree *(Vitex agnus-castus)*

2.4 Coriander *(Coriandrum sativum)*

2.5 Crampbark *(Viburnum opulus)*

2.6 Dandelion *(Taraxacum officinale)*

2.7 Dandelion root rot

Plate 2

Loading set-up for spreading herb onto the screens.

well. The sides catch any herb spilled off the edge of the screen when loading. About ten empty screens can be stacked in a tray this size. It needs to be set at a height such that you can easily reach right across the screens without holding your arms uncomfortably high.

Located at one end or one side, there needs to be a platform or screen that can hold a sheet full of fresh herb. If this is just above the level of the top of the loading tray and overhanging the empty screens slightly, armfuls of herb can be easily grabbed and strewn over the screens.

PROCEDURES FOR AMBIENT AIR DRYING

Drying Leaf Herbs with Ambient Air

Ambient air drying of leaf herbs (includes leaf and flower, and aerial parts) is a reliable system in most parts of temperate Australia. West or north of the Great Divide, drying conditions will be good for much of the harvesting period, though spring or late autumn drying may present difficulties with prolonged humid conditions.

In coastal and mountain regions and in Tasmania, drying conditions tend to be a bit more fickle, but still it will be possible to dry leaf crops using ambient air to a large extent in many locations, provided a good heated back-up system is available.

Loading Screens

Cleaning Screens

Brush any dust, cobwebs or residual debris from the previous crop off the screens and stack them in the loading tray.

Amount of Leaf Herb per Screen

The next step is to spread the herb: how thickly depends on the particular herb, prevailing weather conditions and the screen's location in the drying shed. The following quantities are for standard-sized screens 1500 x 940 mm.

In Very Good Consistent Conditions: In the peak of summer when drying conditions are excellent, up to 3 kg fresh herb can be loaded per screen. However, it would be unwise to put this much on unless you are absolutely confident you will get hot dry conditions that will easily dry this thickness of herb.

In Reasonably Good Conditions: For most of the season when reasonable drying conditions prevail, load around 2 kg per screen.

In Marginal Conditions: You may want to reduce screen loadings to around 1.25 kg, or even to 1 kg, in early spring or late autumn. A thinner layer of herb will dry better in marginal conditions, when a thicker layer would be too slow to fully dry.

The other alternative in these conditions is to stay with the normal loading and use a heated back-up system for the second stage of drying after the herb has lost some of its moisture. This second option is more reliable in terms of quality.

Top Screens: The screen in the top position on the rack, which receives the radiant heat from the roof as well as the hot air at the top of the shed, can safely be loaded with up to 3 kg of easy-to-dry herbs during most of the season, but cut this back a bit in spring and late autumn.

These warmer top spots are useful for difficult-to-dry herbs or for finishing herbs in marginal conditions.

Optimal Loading of Screens

These are rough figures only and will vary from one species to another, from one location to another and from one shed to another. With experience you will get a feel for how much herb to put on the screens. Basically you need to put on as much as will effectively dry, but no more. This way you will make the most efficient use of your facilities.

Overloading will result in the herb drying too slowly and discolouring. In a thick layer of herb, the middle part does not have much contact with the air and has to wait until the layers above and below it dry out before it can begin to dry. If conditions are marginal, by the time this happens, the herb in the middle will be starting to go off.

Even Spreading

Correct loading of the screens isn't just a matter of throwing 2 kg of herb onto a screen and whipping it onto the rack. The herb needs to be spread evenly over the screen to make optimum use of the drying surface and ensure that it dries evenly. Any places where the herb is too thick on the screen will be slower to dry. This can cause problems if the rest of the herb is ready to process. If you don't notice this insufficiently dried herb in a batch that is going through the chaff-cutter, you could end up putting herb into storage that is too moist and having the whole batch go mouldy or be downgraded by a customer.

Leaf and aerial part herbs should be loaded onto the screen with a strewing or scattering motion. Grab an armful of herb from the harvesting sheet and shake it while holding it loosely and moving it across the screen so it falls in a thin layer, repeating this until the screen is generally covered. Finally, tidy up, taking herb from any thick spots and putting it into low or empty spots.

When loading herb onto the screens, watch out for compressed clumps of leafy stuff that look about the same height as everything else but are actually much denser and will be slower to dry. These can be avoided by loading with the shaking, strewing motion described. If you just drop large bundles onto the screen and then spread these out, you are likely to get some compressed areas of herb, which are much denser than others.

Judging Optimum Loading

Once you become familiar with the operation, it should not be necessary to weigh each screen, in fact I had never weighed a screen until I came to write this. I always judge them by feel: how thick the herb looks, how the weight of the loaded screen feels in my hands, and how good the drying conditions seem.

It is not difficult to cultivate this feel for the correct loading. It is the cheapest and quickest way of getting it right. Relying on scales would be too time consuming unless you devise an electronic system, which would cost a heap and not necessarily give you a better result. But certainly, check your weights now and again to make sure you aren't drifting to too light or too heavy in your loading.

When assessing a load visually, an appropriate amount of fresh herb for good drying conditions covers the screen so that the fabric just shows through the leaves in the odd place.

Contamination

While the herb is going onto the screen, keep an eye out for anything that shouldn't be there: weeds, other species of herbs, bird droppings, lumps of soil, snails, little frogs, caterpillars, too many dead and yellow leaves etc.

I usually keep a container handy to accommodate any little creatures and then let them go where they aren't going to cause problems.

Of course you cannot afford to spend a lot of time on this manicuring process. It is best to ensure that the quantity of extraneous material brought in with the harvest is kept to a minimum by good weed control and appropriate timing and technique of harvest. This should be just a final check in a system designed to produce high quality unadulterated and uncontaminated herbs.

If there is lot of extraneous material that takes too long to clean out of the herb, it may be better to throw the lot on the compost and try harder next time.

Avoid Tipping Sheets onto the Screens: When loading the last few leaves from a harvesting sheet, resist the temptation to just tip the sheet onto the screen, as any soil particles, insects, snails, lumps of compost or other debris picked up in the harvesting tend to settle to the bottom of the sheet. The last few leaves should be carefully picked up from the sheet to avoid including any undesirables.

Placement of Screens

As each screen is loaded, it is slipped onto the rack. It is usually best to put the top screen in position first and work down, because it is easier to slide a screen in from below than above an installed screen, especially high up on the racks.

Avoiding Mix-ups: Try to keep screens of individual herbs together in one bay or in adjacent bays. This reduces the risk of mixing different herbs or losing track of them. It may be necessary to vary from this policy a bit to ensure that all the screens of one herb are ready to process at about the same time. Some difficult-to-dry herbs may need to be placed in the warmest positions across the top of the drying racks.

Avoid placing herbs of very similar appearance next to each other on the racks, as they could accidentally be mixed up.

Fine Material: A few herbs tend to break up during drying and small particles sift through the shadecloth screens onto the screen below. Where this is the case, fine cloth screens should be used or the batch kept separate. Sometimes just the bottom screen of the batch needs to be fine cloth to protect herbs below from contamination.

Large Batches: Large batches of one herb are going to occupy so much space in the shed that there will inevitably be some variation in their rate of drying according to where they are placed. This can be overcome by loading less onto the lower screens, or by changing the screens about during the drying period. Usually I just wait until the whole batch is ready or, if necessary, process it in two or more stages.

Length of Time to Load Screens

It generally takes one person 2–3 minutes to load a standard-sized screen and place it on the rack. If two people work on opposite sides of the screens, this can virtually halve loading times.

Preliminary Chopping

Most leaf herbs will be a better quality if they are dried in the whole state, with the leaves attached to the stalks. If the stalks are not wanted, they can be sifted out in the processing of the dried herb. If the herb is required to be cut or milled, this is also usually best done once the herb has dried. However, there are a few leaf herbs that dry faster and turn out a better quality, if they are first chopped into short pieces: 13–50 mm (depending on the herb). These include Lemon Grass, Oats (green plant), Parsley, Rue, Chives, Mullein, Artichoke and no doubt a few others that I am not aware of.

The problem with some is a stem or midrib that is reluctant to dry. In the case of Lemon Grass and Oats, the stem is enclosed by leaves, which slow its drying. Others have a fleshy stem and/or leaves that are resistant to drying and don't produce a good quality unless the drying process is hastened by chopping them first.

Fortunately, most of these herbs don't seem to suffer greatly from bruising as long as they are chopped before they start to wilt. Their colour is not adversely affected by preliminary chopping: in fact it comes out better because they dry faster.

Pre-chopping is a possibility for a few easy-to-dry herbs whose stem is normally included in tea grade, such as Vervain. However, it is important to do a small test run first to ascertain whether they will be bruised by the operation.

Management

When good conditions prevail, managing an ambient air drying system is relatively passive. It is mostly a matter of opening doors and vents in the mornings and closing them up at night. If they are secure against vermin, vents can be left open overnight during prevailing warm dry weather unless remoistening is a problem.

If wind is excessive, the shed may need closing down somewhat to keep the herbs from blowing off the screens. During periods of high relative humidity, it is usually necessary to close down the ventilation to reduce re-moistening of nearly dry herbs. But just because it is raining does not necessarily mean the relative humidity is high: water can come out of the sky when the relative humidity is quite low.

This may sound like a contradiction, but if you observe the drying of your herbs in various types of weather, you will come to realise that this can occur. Steady rain is usually accompanied by high relative humidity, especially if it is fine drizzly rain, but showers don't necessarily mean a high relative humidity.

In Tasmania it is not uncommon to get a strong westerly or south-westerly wind that is quite dry and yet driven on that wind come frequent light showers. Sometimes at the end of such a day it can be rather surprising to find herbs dry enough to rub.

On the other hand, a sea breeze coming in the afternoon of a hot day brings a rapid increase in relative humidity. This can be very frustrating towards the end of a good drying day if the sea breeze arrives just as the herbs are nearly dry enough to start processing.

In the Huon, in southern Tasmania, the sea breeze blows through the valley with some intensity in the late afternoon. As my drying shed was made of wooden vertical board, it was not possible to shut the moisture out for long: it penetrated everything and brought processing to a standstill.

Ambient Air Drying in Less Favourable Conditions

If conditions are good, managing an ambient air drying system is very straightforward but during less favourable weather the drying facility needs to be managed more closely. With skill and timing it is possible to make this system work adequately in less than optimum conditions.

The Two Stages of Drying: First you need to appreciate that there are two stages to drying in this situation. The initial drying is when the bulk of the moisture in the plant is lost. This stage can proceed even in fairly humid conditions, though of course it will not proceed as quickly.

The greatest need for ventilation is when the drying shed is heavily loaded with freshly harvested herbs. During the first few days of this period a lot of moisture is released and if the air is not continuously changing, the relative humidity around the herbs rises too high and drying is greatly slowed.

While the first stage of drying can be carried out in fairly humid conditions with plenty of ventilation, above about

40% relative humidity it is not possible to complete the drying process as the plant will not release the last of its moisture to humid air.

The second stage of drying needs air with a relative humidity of less than 40%. Depending on prevailing conditions, there may be a number of things you can do to help achieve this.

Utilise Warm Upper Levels: The warmer upper regions of the shed can be used for finishing the drying process. If some heat from the sun is warming the drying shed, the upper part of the shed may be warm enough to finish off the drying of herbs that have been in the lower section.

Reducing Through Ventilation: If all or virtually all the herbs in the shed are nearly dry, reducing through ventilation (the amount of air entering and leaving the shed) can enable the sun on the walls and roof to build up more heat inside the shed. This higher temperature lowers the relative humidity, which will help finish the drying.

As the herbs approach a dry state, there is a much slower release of moisture and smaller volumes of through ventilation are required.

Reducing Remoistening: In marginal drying conditions, if you have a batch of herbs that is nearly dry, refrain from putting more freshly harvested herb into the shed as this will cause the humidity to rise and prevent the finishing of herbs that are almost dry.

By closing up the vents at night or in damp weather, you will reduce remoistening that has to be driven off before further drying can proceed.

Seizing the Moment: Sometimes during a period of difficult weather there will be the odd day when the humidity is lower, so that by late in the afternoon the herb is dry enough to process. Other tasks and engagements (even the weeding!) need to be put aside to clear this herb through while it is ready, because the opportunity may not arise again for a while, if at all.

The Need for an Adequate Back-up System

There will be occasions when the above strategies will not be sufficient to get your herb to an adequately dry state and you will have to resort to a heated back-up system to complete the drying.

Having a good back-up system with sufficient capacity can make drying in difficult conditions a lot more enjoyable. It takes away the element of risk in ambient air drying and ensures that the quality of your dried product can be maintained despite unfavourable weather.

When to Use the Back-up System

Some judgement needs to be exercised as to when and how to utilise the back-up system. If the weather is particularly adverse, it will need to be used for the whole drying process.

Often the back-up system can be used just to finish the last stage of drying when this is not possible in ambient air. This way the herb is dried for 5–10 days in ambient air and in heated air for a final period: usually 12–24 hours.

This is much more economical than drying the herbs the whole way with artificial heat and can be nearly as fast. With good ventilation, the initial drying of most leaf herbs in ambient air proceeds fairly quickly, even in marginal conditions, but slows down as the equilibrium point between moisture in the herb and in the air is approached.

Monitoring and Assessing

During the drying process, you will need to regularly monitor the progress of your herbs to assess when they are ready for processing.

There is some considerable variation among different species of herbs as to the rate they lose their moisture. To a large extent this depends on the mechanisms they have which reduce moisture loss, such as a waxy coating, furry hairs etc. Herbs that are difficult to dry need to be watched to ensure they are making adequate progress: if they are not, they may need moving to a better drying position.

When subjected to dehydration, most plants close off the movement of moisture from their stems to their leaves. This allows the leaves to first wilt and then die before the rest of the plant succumbs. It is a protective mechanism to reduce water loss and increase the plant's chances of survival.

This mechanism continues to function when the leaves and stems are cut from the plant. The first sign of progress in drying is for the leaves to start to wilt. This will happen in a few hours in good conditions.

While the herbs are drying on the screens, there is not much to do until they are dry. The important thing is to know when they are dry enough and to get them processed before they reabsorb moisture as it gets cooler in the evening or with a change of weather.

This is an area where a lot of growers have difficulty and it is all too common to receive samples of herb or even large commercial shipments that are inadequately dried: their moisture level being too high for safe storage.

Assessing Dryness of Leaf for Tea and Culinary Grade

Checking leaf for dryness is relatively easy. If it will crush and shatter in your hands, if it feels dry and crackly, then it is dry enough to process and store safely. But if it is at all limp or leathery, or feels soft and slightly pliable, it has too much moisture.

For processing to tea and culinary grades, if the leaf will easily rub through a wire screen, it is dry enough, but if it is still soft and resists breaking up when rubbed on the screen, it needs more drying.

A crop for tea or culinary grade can be processed as soon as the all the leaves are dry enough to shatter. For most herbs the part to check is the younger leaves at the ends of the shoots, as these are usually a bit slower to dry than the older leaves.

There is no need to wait for the stalks to dry completely before rubbing, in fact the processing will be much easier if the stems are still a bit moist and leathery. When they are in this state, the stems are less inclined to break up and fall through the rubbing screen, so the rubbed leaf takes less work to clean.

As far as moisture content goes, rubbing is generally a foolproof system for processing herbs: if they are dry enough to rub through the screen, they are dry enough to safely store; if they won't break up and pass through the rubbing screen easily, then they aren't dry enough. With rubbed herbs, stalks that still contain excessive moisture are not a problem because they are separated and discarded.

Assessing Aerial Parts for Dryness

When the stem is included in the finished product, as in dried aerial parts for manufacturing liquid medicines, assessment becomes trickier and there is a danger that excessively moist material can easily be fed through the chaff-cutter or hammer-mill and go into storage unnoticed.

Stems usually retain substantial amounts of moisture for some time after the leaves are thoroughly dry. The stems may feel very dry on the outside, but still be quite moist inside. The method of assessing them for dryness varies a bit, according to the species.

Fleshy Stems: An example of the most common type of herb stem is the stem of Peppermint. This is rather fleshy when fresh, leathery when partly dry, and when fully dry it becomes quite brittle and breaks with a snap.

Woody Stems: Another type of stem is woodier and breaks with a soft snap when still green, so it is not as easy to distinguish the difference when these are dry. An example of this type is Saint John's Wort. Its stems tend to have a thin fleshy bark that can be scratched off with the thumbnail if it is still moist, but bonds tightly when it is dry, so becomes very difficult to scratch off.

Pliable Stems: A third type is fairly fibrous and pliable and even when dry does not snap. An example is Oats. My test for this type is to bite a piece of the stem between my front teeth. If it is dry it will have a texture similar to paper, but if it is still a little moist, there will be a slight 'carrot-like' crunching between the teeth. The lower nodes are usually the slowest to dry.

Buds, Flower heads and Fruiting Bodies: Some aerial part herbs can retain significant amounts of hidden moisture in buds, flower heads or immature fruiting bodies. These need to be broken open to check for moisture inside.

Thoroughness in Assessing Moisture Content

Of course in any assessment of dryness, a good range of samples needs to be taken. Check the places where drying would have been the slowest, such as the lower screens that have other herbs all around them in the coolest part of the shed. Rather than just checking the front of the screen where the herb will have had a breeze blowing on it, reach in to the back and test some where the herb may be lying a bit thicker on the screen.

Look for the moistest part of the plant. This varies from one species to another, but it tends to be unopened flower heads or immature fruiting bodies, leafy tips or thick stem bases.

It is important to be aware of the phenomenon of re-moistening. You cannot count on dried herbs staying dry when they are exposed to ambient air, because when the relative humidity of the air rises, they will start to take up moisture again. So as soon as they are dry enough to process, you need to get started and, if necessary, keep going into the night to get them done before the damp night air moistens them up again.

If you miss that opportunity, they will probably be like a limp rag in the morning and you will generally have to wait till the next afternoon before they are again dry enough to process. If the weather changes though, it might be 3 or 4 days before they are ready.

This delay could result in a loss of quality, both for that crop and for other crops waiting for their place on the screens.

Drying Flower Herbs with Ambient Air

If drying conditions are consistently good, then ambient air drying of flowers can be considered, but flowers are somewhat more delicate than leaf herbs and can deteriorate badly if drying is too slow. If there is any doubt as to whether flowers can be dried quickly enough, a heated drying system should be used.

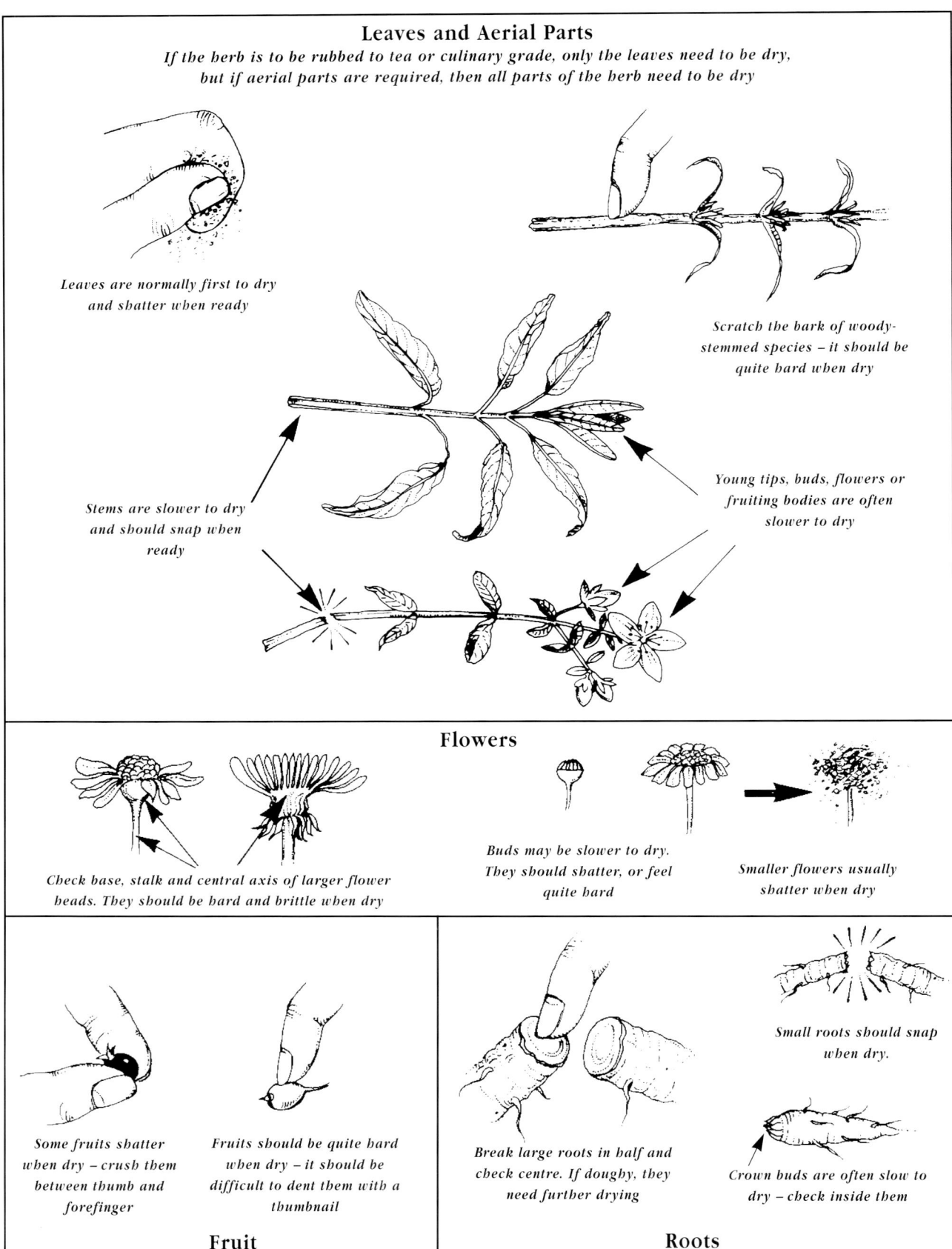

Assessing for dryness

Loading

The procedure for drying flowers is similar to that for leaf herbs, but the thickness and weight of herb on the screens needs to be a bit less. Most flowers need rapid drying to get good quality and colour retention. They need to be loaded more lightly on the screens and placed in positions where they will they dry quickly.

Management

If drying conditions are not adequate, steps need to be taken to ensure that flowers don't sit around and fade: get your back-up system going. Some flowers are more delicate than others, but none can be held for long without deteriorating. Good drying conditions for the first day or so are essential to 'fix' the flowers in their vibrant state. If initial drying is too slow, it allows the maturing process of the living flowers to continue for too long and they will lose their vibrancy as they proceed to seed formation.

Monitoring and Assessing

Flowers can be rather deceptive and need to be checked over closely to ascertain their dryness. It is often the small unopened or partly opened flowers that are the slowest to dry, and these may be overlooked when assessing the dryness of the crop, because they seem insignificant. With some species, the base of the calyx is fleshy and slow to give up its moisture, with others it is the immature fruiting bodies of the more advanced flowers that retain moisture.

Dry flowers should shatter easily when crushed. The base of the calyx and the stalk should be hard and brittle. Any softness or pliability is an indication of moisture. Buds should be quite hard or else should shatter easily.

If about 99% of the herb is very dry, then it will be capable of safely absorbing a little moisture from the 1% that is almost dry. Judging this can be tricky, though, and it is easy to get it wrong.

It is a good idea, at least until you are confident of what you are doing, to check the flowers a week or two after they have gone into storage to see if they feel damp at all. If there was too much residual moisture in some of the flowers, it may have been enough to take the overall moisture content over the safe limit, but this won't become apparent until this moisture has had a little while to move through the rest of the herb.

Drying Fruit Herbs with Ambient Air

There are a number of plants whose fruits are used as herbs. The most in demand are rosehips, Hawthorn berries and Chaste Tree berries. The first two of these are fleshy, while Chaste Tree Berries are relatively dry fruits. The fleshy fruits need good conditions for drying, but in warmer regions it is possible to dry some of them with ambient air provided you get them harvested as soon as they are mature enough and provided you are fortunate enough to have a warm, dry sunny autumn.

However, the fruit of the Sweet Briar (*Rosa rubiginosa*), the Rose commonly found growing wild in southern Australia, is especially difficult to dry because its skin is virtually impermeable: it needs consistent temperatures of 40–45°C. This fruit should be dried with artificial heat to ensure good quality because the hips will ferment inside their skins if the drying is too slow.

Loading

Fruits being fairly dense, a lot more weight will go onto the screens than with leaf or flower herbs. For ambient air drying, just spread them out one layer thick, otherwise progress will be too slow. Don't worry about leaf and other trash among these small fruits, as it is easier to remove them by rubbing and screening after everything is dry.

Be careful when handling the loaded screens and avoid tipping them, as the berries are liable to all go rolling off onto the floor. Fruit herbs really need the best spots for drying up next to the roof where the air is warmer and heat can radiate down onto them. This is where you will appreciate having a low roof, as going up a ladder with a screenful of fruits can be tricky.

Management

If persistent poor drying weather occurs, you will need to resort to your back-up system, as fleshy fruits should not be left sitting around in a moist state for too long.

Monitoring and Assessing

Fruits are quite hard when they are dry. As they approach dryness, they develop a sultana-like texture, but as their sugar content is not enough to preserve them in this state, drying needs to proceed until they are hard and dry. When they are adequately dry, they will not dent easily when pressed with the thumbnail, and Rose hips will shatter when squeezed hard between finger and thumb.

If the fruits are mixed with a proportion of leaf from harvesting, this needs to be dry enough to break up and rub through a screen, so it can be separated.

DRYING WITH HEATED AIR

A heated-air drying system can be a small system used as a back-up for an ambient air drying system, or it can be a system designed to handle the complete drying of herbs, possibly in large volumes, in situations where ambient air drying is not suitable.

Design

Heat Source

The first choice you will have to make is that of a heat source. The air for such a dryer could be heated by electricity using a heating element or heat pump; by burning a fuel such as wood, gas, oil or coal; or by using a solar collector.

Electric Heat

Electric heat is a feasible option on a small scale where there is not sufficient volume of herb to justify a larger fuel-fired dryer. It has the advantage of being easily linked to thermostat control and is clean (at the user's end, but in most places, fairly filthy at the generating end). Fire-safe electric heaters are also available.

The disadvantages are the running costs and the environmental unsoundness of using electricity as a source of heat, especially where it is generated by burning fossil fuels.

Wood Heat

If you are close to a supply of firewood, this is a good option, provided you are producing on a sufficient scale to be able to use the heat effectively: a wood-fired outfit has to be based on a stove large enough to burn for several hours or overnight with each stoking.

Wood-fired drying systems can be quite large and cost-effective.

In southern Tasmania there is a large apple drying plant which is wood-fired. It owes its continued existence to the conservative and community-orientated nature of its proprietors. Back in the 1950s and 1960s they resisted the trend to convert their old-fashioned wood-burning furnaces to oil-fired units, partly because it would have put a number of local families out of work. Today they find they are ahead, as their running costs are lower than if they were burning oil.

In regions where firewood is still plentiful and its harvesting is sustainable, it is a more environmentally sound source of heat as the carbon dioxide generated is part of the bio-cycle. However, smoke emission can be a problem if the stove is damped down too much.

Another consideration is the risk of fire: the installation of the stove must be done very carefully, as dried herbs are highly flammable. Wood-fired dryers have a habit of burning down.

Co-generation

Another possibility for small-scale drying is to combine household heating with herb drying. There are those of us who have installed a drying cabinet over the wood-stove, and those who have suspended the drying screens from the ceiling in the house.

The advantages of this combined heating system are the saving in capital and running costs and the interesting aromas, most of them pleasant, that permeate the house and everything in it.

The disadvantages are the dust and fine fragments of herb that fall through the screens, the steaming up of the house when a big load of fresh material goes in, the inconvenience of carrying herbs in and out of the house, the assault it makes on modern housekeeping values, and the potency of some of the vapours that come off some of the herbs.

Valerian is about the worst in its effects: it drives visitors away and makes the kids late for school, because they can't wake up properly in the mornings.

A heated dryer could be located so there is an option of venting its warm expelled air into a house, workshop or greenhouse, if required, or elsewhere, if not. This way the surplus heat could still be utilised if the system were fired up for just a partial load. A heat exchange system would be needed to solve the problems of humidity, condensation and odours.

Gas, Oil and Coal Heat

Fossil fuels are often used for drying conventionally grown herbs. Hops are typically dried with an oil-fired burner, usually with the burner's exhaust blasting through the drying herb as well. Combustion residues and the products of incomplete combustion are inevitably deposited onto the Hops: a rather unsavoury outcome. On a less intensively flavoured herb, they would perhaps be more noticeable. Such a system would not meet organic certification standards.

If you were considering oil or coal heat you would need to ensure that the operation did not fumigate the herb as it was drying. This could be done by incorporating some sort of heat exchange and venting the exhaust well away from the dryer's air intake.

LP gas is another possibility used by some growers. It is relatively (though not entirely) clean burning, but a heat exchange is still necessary. In its combustion it generates more water than carbon dioxide, so from an environmental point of view it is preferable to other fossil fuels.

Solar Heat

While the sun is a cheap source of heat, in southern Australia it can only be used in sunny weather in the warmer part of the year. In wet weather and winter, its output of heat is too low or too brief. Further north it can probably contribute more heat in winter, as the days are longer and the sun is warmer and out more often, but it still can not be relied upon as a sole source of heat.

For drying during the warmer part of the year, the sun can provide a significant amount of heat at low cost, but solar heating needs to be linked with another heat source that can cut in when the weather is unfavourable, or overnight if faster drying is required.

Regulating temperature can be a bit tricky with solar heat because of its variable nature.

Heat Banks

The use of heat banks is a possibility in a system where a surplus of heat is generated for a period of the day, such as with solar heat or in some cases wood-fired heaters.

While these are used in various forms in domestic heating, the throughput of air required for herb drying is proportionately much greater. It would take a very large heat bank to store enough heat to contribute significantly to the drying of herbs.

Heat Pumps

Heat pumps can be used in two ways: simply as a heat source or as a dehumidifier.

As a heat source, they operate like a refrigerator or air conditioner in reverse, effectively pumping in heat from the outside air. They are quite efficient at higher outside temperatures, so they are able to produce more heat per kilowatt than an electric heater, but their efficiency drops at temperatures below about 10°C.

A heat pump might be worth considering where it would be used when prevailing outside temperatures are not too low, though the capital cost is quite high.

For more detail on dehumidifying systems, see below under 'Other Drying Systems'.

Location of Heat Source

The heat source can be located in the same chamber as the herbs being dried or in a separate chamber with the warm air ducted to the drying chamber.

Some systems use an intermediate chamber, or plenum, where the air flowing from the heat source is allowed to mix and reach an even temperature before flowing on to the drying chamber.

If the heat source is in the drying chamber, the air inlet must be located so the cool air entering can flow directly to it. Baffles and/or fans may be needed to disperse and mix the warm air rising from it.

Temperature Control

It is important that air temperatures not exceed the recommended limits: not over 35°C for herbs with volatile constituents (generally aromatic) and not over 45°C for other herbs (generally non-aromatic).

Temperature can be regulated by increasing or decreasing the heat output, and/or by adjusting the air flow, depending on the system.

If it is possible to use thermostats to regulate temperature, it takes a lot of the worry out of drying. A thermostat can be used on the heat source to maintain a constant temperature and one can be used on a fan so that it cuts in at a certain temperature, increasing air exchange to keep temperatures from going too high.

Airflow

There are two aspects to airflow: air movement and air exchange. With some systems they amount to the same thing, as fresh heated air enters at one end of the system, passes through or over the herbs and out the other end.

In other systems, the air is exchanged as new air enters the chamber at one point and old air exits at another, but in the chamber itself the air could be still or moving around.

Air Exchange

Air exchange must be regulated to keep the humidity down. For instance, there is no point in maintaining a temperature of 45°C if the relative humidity goes up to 80%, because there is not enough air exchange. On the other hand, if air exchange is too great for the amount of heat being produced, temperatures will be below optimum and drying will be slower.

Regulating Air Exchange

At times it may be necessary to reduce or close down the air exchange, so both the inlet and outlet need to be adjustable and closable. In order to obtain sufficient air exchange, a fan may be required at the inlet and/or outlet.

The walls and doors of the drying chamber need to be reasonably airtight so warm dry air is not lost, and cool moist air doesn't get drawn into the middle of the system.

Air Movement

The movement of air enables it to pick up moisture from a surface and carry it away. Air movement must be even. If it is greater in some sections, these will dry more quickly than everything else. Some attention needs to be paid to achieving even air movement.

Vertical Airflow Systems

In a vertical airflow system, the heated air is made to pass up through the layer(s) of herbs and is then expelled. This results in sequential drying: in other words, the herbs that the air strikes first will be the first dried. In a cabinet dryer with a series of screens one above the other, the lower screens will be dry first. This can be used to advantage by sequential loading, or can be overcome by rotating the screens.

A sequentially loaded, vertical airflow system can be more energy efficient. It enables maximum utilisation of the drying capacity of the air as it flows through the driest herbs first and the moistest last.

Circulatory Airflow Systems

The other option is circulatory airflow, with the screens placed in a drying chamber with space around them where the air is kept continually circulating. Some air is vented from the chamber and replaced by fresh warm dry air.

With a circulatory airflow system, drying will be more even, but somewhat less energy efficient, as sequential loading is not possible and relatively dry air is expelled. For even drying of large batches, less labour is required with a circulatory dryer.

Increasing the Drying Rate

The rate of drying increases in almost direct proportion to airflow. From the point of view of quality, it is better to increase airflow rather than temperature if a greater throughput is required, as this does not result in such a loss of volatile constituents.

Containment of Flow

Vertical flow of heated air needs to be contained so that it is forced to flow through the herbs and does not take the path of least resistance and escape. There is no point in delivering warm air at the base of a bay of drying screens without walls around them, as the warm air will just flow away to the sides and be lost.

With circulatory airflow, the heated air is contained in a closed room or chamber and kept circulating over the herbs by means of fans.

Expelled Air

Air expelled from the chamber can be laden with moisture (and sometimes strong aromas). In cooler conditions, there can be a lot of condensation in the vicinity of the outlet vent, so be aware of this when locating your cabinet dryer. Provision should be made to vent the moist air outside where it can't drip back onto the herbs or anything else it might damage. If this moist air strikes an unlined iron roof, it can sometimes drip so much you would think it had been raining.

As for the aromas, while most of them are generally regarded as pleasant, there are a few, such as Valerian and Rue, that disagree strongly with some people.

Materials

Lining

It is important that the inside lining of the drying chamber is made of a material that will not deteriorate in warm moist conditions and will not contaminate the herbs. It should be impermeable so water vapour is not moving through the wall and condensing where it will cause problems. All in all, galvanised iron is about the most suitable economical material.

Insulation

While insulation may not be particularly important for a heated system operating in warm weather, it is vital for a dryer operating in cooler conditions. If significant heat loss occurs, there will be a drop in temperature around the walls and ceiling of the drying chamber, with reduced drying, remoistening of partially dried herbs, and a possibility of moisture condensation with water dripping back onto the herbs.

Building a Small Cabinet Dryer

What follows is a description of the system I have used for the past 15 years or so as a back-up for my ambient air drying set-up and for winter drying of roots and fruits. It can also be used for the complete drying process for leaves and flowers.

Size

A cabinet dryer, 950 mm wide x 1500 mm deep x 2400 mm high, has a capacity of 13 standard screens positioned 150 mm apart.

A cabinet dryer this size with a 1000W heat source and relying on convection alone for air circulation has proven just big enough as a back-up system for a 180–200 screen ambient air drying system, provided everything is organised to maintain steady operation. Often it has had twice the number of screens in it because they can be doubled and sometimes even tripled up on the lower shelves when the herbs are partially dry.

Throughput

Throughput with a 1000W heater for drying root herbs during winter is 1.5–2 kg dry weight per day. For finishing off leaf herbs as a back-up dryer for an ambient air drying system, the throughput is 5–10 kg per day, depending on how dry the herbs are when loaded.

This throughput could be increased considerably with a larger heat source and fan assisted ventilation.

Heat Source

Electric Heat

Initially I used a couple of 500W electric heaters of a type popular back in the 1950s: they are shaped like a long narrow box with a perforated top and sides. The actual element doesn't get hot enough to glow or start a fire. Their other advantage is their low profile so they don't occupy a lot of space in the bottom of the cabinet. They are no longer made, but sometimes can be found in second-hand shops. (Note: rewiring may be necessary as the rubber insulation on old flex may have perished.)

The modern equivalent to this type of heater is known as a panel heater and is normally attached to a wall but can be mounted on a stand. They are about 300 mm high and have an in-built thermostat.

An oil-filled column heater with a thermostat gives a good steady source of heat without risk of fire. Their main disadvantage is their high profile.

A fan-heater or a heater with a glowing element could be used, but they are more of a fire risk. If you do use one, it would be best to protect it with fine wire screen to keep flammable material away from the element and to contain any sparks.

Thermostat control of the heater has definite advantages, but the temperature can also be regulated by adjusting the vents.

A baffle above and/or a small fan next to the heater may be needed to mix the air so drying will be more even.

Solar Heat

A supplementary solar collector can save quite a bit of power or fuel if the sun is shining, though it can make operation of the dryer rather more complex as the two heating systems need to interface to maintain good drying conditions without temperatures going too high. For more detail on solar heat see 'Solar Heat Options' below.

LP Gas

LP gas is used by some growers who find it easy to regulate and competitive with electric heating costs. In those states where coal is being burned to generate electricity, LP gas is more ecologically sound as it produces less carbon dioxide when burned.

It is also a good option where you are not connected to the grid.

Wood Heat

A wood-fired heater is an option, but the average wood stove produces too much heat for a dryer this size. A small stove could be used, but it will require frequent stoking and more monitoring.

Alternatively a larger dryer could be built (see below under 'A Wood-fired Cabinet Dryer').

Construction of the Cabinet Dryer

If the materials used are light enough, the cabinet dryer will be easier to manoeuvre if you ever want to move it around.

Framing

I built the frame of my first cabinet dryer with green hardwood, which turned out to be a mistake as it was very heavy and the timber shrank enough to allow the screens to fall off the slats. Radiata pine is adequate for the purpose and fully dry timber will avoid shrinkage problems.

Lining

The interior lining of the floor, walls and ceiling of the cabinet dryer is best made from flat galvanised iron or a similar material that will resist abrasion and a few bumps, is impermeable to moisture and water vapour and is not going to contaminate the herbs. In my first cabinet dryer I used masonite for interior lining, but this allowed moisture to pass through and condense inside the walls, sometimes causing buckling of the outside cladding.

Insulation

Behind the lining there needs to be a layer of insulation all around. Unfortunately many commonly used insulation materials could be a problem if they ever found their way into the herbs through small gaps or cracks in the lining. Care does need to be taken to select materials that are not going to cause any possible harm.

Ordinary wool (untreated) and sawdust are two such options, but there are other possibilities. I would discourage the use of fibreglass, rock-wool, fluffed paper (which is treated with borax or sometimes other heavier chemicals). Even commercial wool batts are treated with something to stop moth infestation, so it is probably best to avoid them. Polyester insulation batts may be more suitable in that they are less likely to fragment and find their way through small gaps.

Reflective insulation ('Sisalation') can be used as an additional layer, but on its own is not really adequate for low outside temperatures.

I used some skirtings from my wool clip that weren't going to bring much of a price and this has proved quite adequate.

Design for a small cabinet dryer.

Exterior Cladding

The outside wall should be made of some material that will resist attempts of vermin to get in and make themselves comfortable.

Door

The door needs to be of similar insulated construction and to seal tightly. Any leaks can cause disruption of the convection – the loss of warm air and the intrusion of cold air – with resulting condensation.

Air Inlet

The air inlet needs to be at the bottom of the cabinet: either in the floor or at the base of a wall. This way fresh air can flow to the heat source and then rise up as it is warmed. The inlet needs to be adjustable so it can be closed right down at times or opened up for greater air flow.

The inlet capacity needs to be large enough for sufficient air to be drawn in to keep temperatures from going above optimum levels. By adjusting the inlet, the temperature inside the cabinet dryer can be regulated according to the flow of air. If the maximum opening of the inlet is about 0.09 m^2, this will be adequate for a cabinet dryer 950 x 1500 x 2400 mm with a heat source of 2000W.

Air Outlet

The air outlet at the top of the cabinet should be roughly the same size as the inlet and needs to be opened to approximately the same degree. If the outlet is opened too wide, it will allow cold air to flow down into the dryer, affecting its performance.

The expelled air should be vented in a place where condensation in cold weather won't cause problems.

Closing Down

The inlet and outlet need to be able to close tightly. If the herbs reach a state of dryness at a time when it is not convenient to process them, the vents can be closed down and the heat source turned off.

Vermin Control

The vents need wire screens over them which are strong enough to discourage rats from chewing through and fine enough to prevent mice gaining access.

Airflow

My first cabinet dryer relied on convection alone. To the second one I have added an exhaust fan at the outlet that can be used to increase airflow. It is linked to a thermostat just above the heat source, so it operates only when the temperature reaches a certain level.

As pulling air is not as efficient as pushing air, to achieve a greater airflow it would be better to install a fan at the air inlet. A centrifugal (squirrel cage) fan with backward-curved blades is more effective for developing pressure and maintaining air flow through a heavily loaded dryer.

Airflow for drying lightweight herbs, such as leaves, would have to be kept below the point at which they are lifted off the screens, but for heavier material, such as roots and fruits, greater increases in airflow are possible.

Reverse airflow from the top to the bottom of the dryer will also work, provided there is an adequate forced draught and/or extractor.

Operating the Cabinet Dryer

The method of operation for the cabinet dryer varies according to whether it is being used just for finishing almost dry herbs or for the complete drying of freshly harvested herbs.

Finishing Partially Dried Herbs

Often the dryer is loaded with leaf or flower crops that just need a few hours to get the moisture content down a further 5–10% so they can be processed. I have found the best system is to load about a dozen screens of almost dry herbs into the dryer first thing in the morning and leave them until evening. For finishing, this is usually enough to get them nice and dry. These herbs can then be processed and another load put in to dry overnight.

A thermometer placed on the lowest screen above the heat source can be used to monitor temperatures and the vents can be adjusted to keep within the optimum range. A maximum-minimum thermometer is good for keeping an eye on temperature extremes.

On a warm day, the airflow may need to be opened up a little more, or it may need to be reduced on a cold night, unless you have a thermostat to regulate the temperature automatically.

Until you become thoroughly familiar with the vagaries of your cabinet dryer, it is best to check it an hour or so after loading to see what temperature it is running at, and again a few hours later to see how evenly it is drying.

Often it is necessary to adjust the position of the heater as one end of the screens will be drying more quickly than the other. If this happens, it may also be necessary to compensate by reversing every second screen.

Drying Freshly Harvested Herbs

Putting freshly harvested herbs into the cabinet dryer requires something of a different approach. If too much fresh material is loaded at once, problems can occur.

First, because a lot of moisture comes off fresh herbs, especially from leaves and flowers, after the air has passed through several screens it will be getting close to saturation. If additional screens are loaded at this stage, the herbs on the upper screens will just be sitting in warm humid air until the lower screens have dried somewhat and the air coming through them is less humid.

Sequential Loading

Rather than subjecting your beautiful herbs to a steamy Turkish bath, it is better to keep them cool outside or still on the bushes until the first few screens have dried a little.

After half a day or so, when the herbs on the top screen show signs of drying progress, such as leaves or petals wilting or cut surfaces of roots drying, then another few screens can be loaded. By this stage the air coming through the first loaded screens will have some capacity to absorb moisture from the fresh material.

When placing screens of fresh material into a cabinet dryer which already has some partially dried herbs in it, it is usually best to keep the driest material at the bottom and the moistest at the top. This way the driest air can flow through the driest herb first and then through the moister herb above. This way remoistening will not occur.

As it approaches a fully dry state, the first-loaded herb requires very dry air to further reduce the moisture content. Because the release of this last moisture takes place at a slower rate, the air passing through it will still have plenty of capacity to take up moisture from less dry herbs on screens above. These moister herbs do not require very dry air for the first part of their drying.

As the dryer becomes full, the lower screens that are approaching the completion of their drying can be doubled or tripled up (either by putting 2 or 3 screens on each slat or by putting their contents onto one screen) provided this does not restrict airflow too much. Then all the other screens can be moved down, taking care to keep them in order of dryness. When the bottom ones are fully dry they can be taken out, the upper ones moved down and new screens loaded at the top.

Non-sequential Loading

Sometimes it may be necessary to depart from loading in sequential order. For instance, with aromatic herbs such as Elecampane or Valerian, temperatures at the bottom of the dryer may be too high. And if you particularly want all the screens of a batch to come dry at the same time, you might reverse the order and put the moister herb below, or rotate them during drying.

Trouble Shooting

Operating the cabinet dryer for the full drying process requires steady vigilance to prevent and rectify uneven drying. If you don't watch out you can easily end up with a dryer full of herb, all bone dry at one end of the screens, but still fresh and juicy at the other.

This sort of thing can be quite confusing until you realise why it is happening. There are several causes of uneven drying:

The Heat Source: Often the air rising from the heat source is not of an even temperature. Some heaters are hotter at one end than the other. If the air is not adequately mixed, there will be an updraught(s) of hotter air, usually directly above the heater.

This may need to be corrected with some sort of baffle. A small woollen blanket laid in the middle of a screen on the bottom shelf works quite well as it deflects some hot air and allows some to rise up through it. Alternatively, a small fan on a low setting, operating horizontally could be used to mix the air before it reaches the screens (too large a fan can cause uneven airflow).

Deflection: When the screens are loaded with heavy roots or fruits, they sag down in the middle. This sagging profile tends to deflect the rising air to the sides and ends of the screens.

In the lower screens, this deflection can help even out a central updraught, but halfway up the dryer there can be a distinct tendency for the middles to remain moist while the sides are drying well. This will get worse further up as the airflow is progressively deflected out to the sides.

To counteract this, pull some of the herb away from the middle and towards the ends and sides of affected screens so more air can flow up through the middle.

'Dry Chimney'

I use the term 'dry chimney' to describe a phenomenon that had me puzzled for a while, but it does have a logical explanation.

Should there be a tendency for drier spots to develop on the screens, a chimney effect can occur with these dry spots occurring one above the other. As the herbs in these spots dry out, they shrink and make more space for air to flow through. This accelerates the drying in these spots, with more shrinkage, more space and more air flowing through them and less through the rest of the screens.

As the herb in these patches approaches dryness, the reduced evaporation means the air flowing through them remains warm so it draws upward even more strongly. If you don't intervene in some way, you can have a situation where almost all the hot air generated by the heater is drawn up this dry chimney and out the top vent while the rest of the herbs sit on the screen, cold and damp.

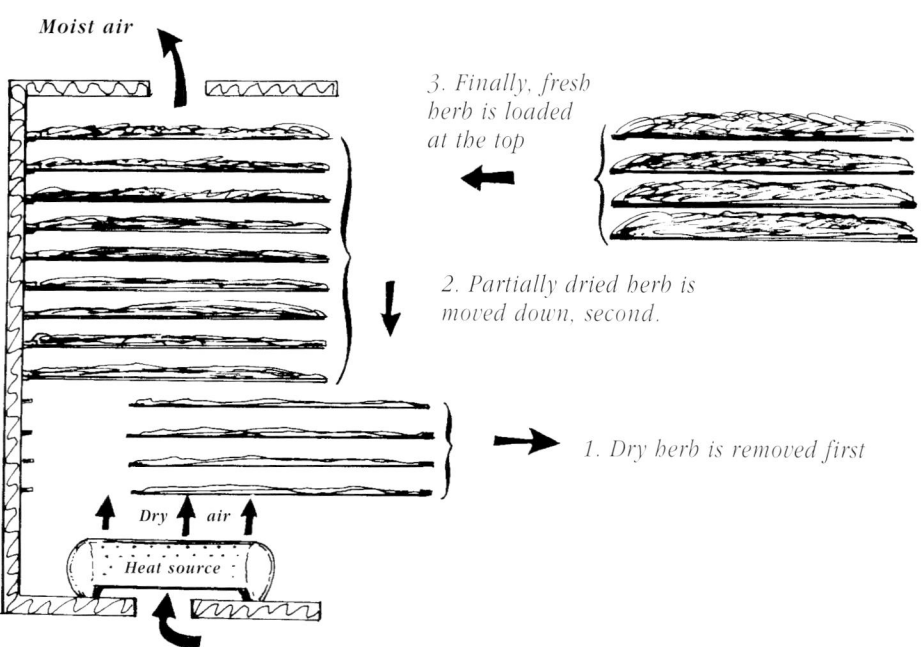

Sequential loading of the cabinet dryer: fully dried herb is removed from the lower part and partly dried herb is moved down to the lower positions. Fresh herb is loaded a few screens at a time in the upper part of the dryer.

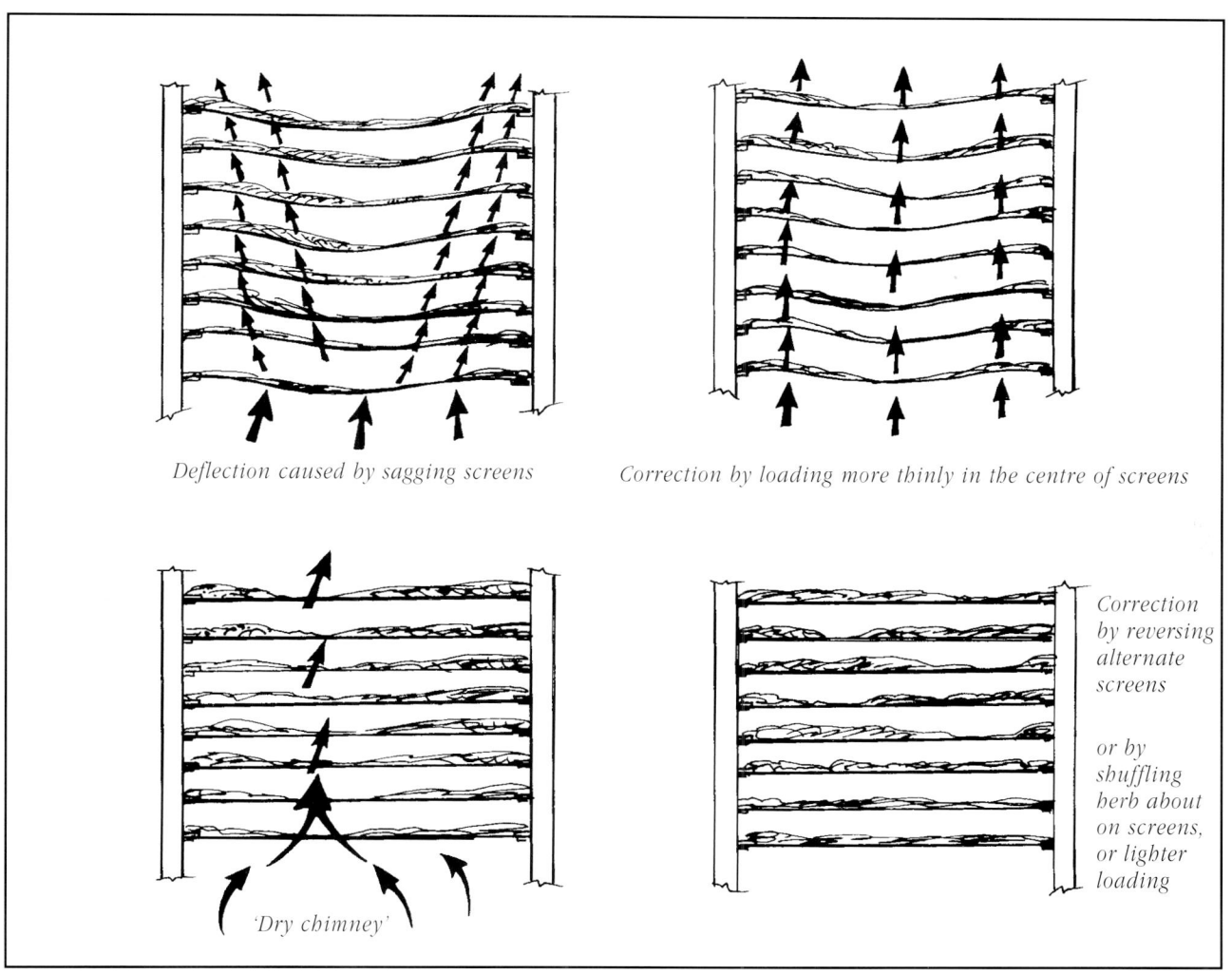

Trouble shooting in operating the cabinet dryer.

In an extreme case you can even get a cold down-draught at the other end of the dryer that brings moist and cooling air downwards to condense moisture on the herb. If this continues for long enough, the herb may even go mouldy.

Constant vigilance is the only solution to these problems. Every time you open the dryer, check several screens front and back for any unevenness of drying. If there is any evidence of this, shuffle the herbs around on the screens, or reverse alternate screens to obstruct any dry chimneys that have started to form.

If the dry chimneys always seem to occur at the same place in the dryer, look for an underlying cause, such as more heat at one end, a leaking door at the other, or a vent opened a bit wider at one end.

Dry chimney is also more likely to occur when the cabinet dryer is heavily loaded.

Drying Leaf Herbs in the Cabinet Dryer

Loading

If the herbs have been drying on screens in ambient air and just need a bit of artificial heat to finish them off, the screens can be put into the dryer as they are. Sometimes they can be doubled up: either by tipping one screenful onto another, or by putting two screens in one position.

When loading fresh leaf herbs to go into a convection dryer for the whole of their drying, be wary of the temptation to load too much more on, because it is important that air be able to flow up through the screens. Many leaf herbs will tend to mat down and close off the airflow if they are loaded too thickly. Those stalky enough to allow plenty of air to flow up through them can go on in a thicker layer.

Once leaf herbs that mat down are nearly dry, the screens can be safely doubled up because the leaves will no longer lie flat and close off the airflow.

Operating

Operating the dryer is mainly a matter of keeping an eye on the temperature and ensuring that the herbs are drying evenly. But don't assume you can just set it, check it once and then it will be okay until the herb is dry.

When a dryer is loaded with fresh herb, the temperatures it will achieve at first will be moderated by the high rate of evaporation of water. This takes up heat from the air. As the herbs get drier and evaporation slows, the temperature around them will rise and you may need to increase your airflow or decrease the heat output to keep within optimum temperatures.

Monitoring and Assessing

Monitoring and assessing the drying is basically the same as for ambient air drying, except that you have to check all over each screen because of the greater tendency for unevenness with convection drying.

Drying Flower Herbs in the Cabinet Dryer

Loading

Flowers can be loaded on the screens 50–100% thicker than for ambient air drying if there is a good flow of heated air, as flowers generally let air flow through better than leaves. Check drying progress to make sure you are not overloading, because it is important that flowers dry quickly.

Operating

As for leaves (see above).

Monitoring and Assessment

As for leaves (see above).

Drying Fruit Herbs in the Cabinet Dryer

Because fruits ripen in early to late autumn, they will often need to be dried with additional heat for at least part of the process. The hips of the common wild Sweet Briar Rose (*Rosa rubiginosa*) need to be dried at a sustained 40–45°C, as the skin of the hips is very reluctant to let go of any moisture.

Loading

Refer to section on drying fruit herbs under 'Ambient Air Drying'.

Sweet Briar hips, which need very good conditions for satisfactory drying, are best placed in the lowest racks in a cabinet dryer where they get first go at the hot air. Because the hips are so slow to release their moisture, they will have very little slowing effect on the screens above, even if the herbs there are almost dry.

Management and Assessment

Refer to section under 'Ambient Air Drying'. Of course you need to check all over the screens because of possible variations in drying rate.

Drying Root Herbs in the Cabinet Dryer

As root herbs are generally harvested when conditions are unsuitable for ambient air drying and because their fleshy nature makes them difficult to dry, they normally require artificial heat.

Chopping

Most root herbs will need some degree of chopping or splitting to enable them to dry quickly enough. The 'rule of thumb' for most roots is to chop or split any that are thicker than the average thumb (25 mm). Roots can be split in halves or quarters, but generally it is easier to chop them into short lengths: up to 40 mm long. This is usually quicker and it makes handling easier, too, as the short pieces are less bulky.

Chopping can be done by hand with a butcher's cleaver and a solid block of wood. If a large enough quantity is being loaded at one time, the roots can be fed through a chaff cutter. If they begin drying straight away this works well, but it is not good for roots to sit around in a chopped state for a few days waiting for a space in the dryer. They will deteriorate if drying is delayed because the chaff-cutter bruises and mashes them somewhat.

Putting small batches through the chaff-cutter does not save labour because of the time it takes to clean all the sticky mush off the machine afterwards.

If the chaff-cutter is used to chop up roots, they must first be washed very thoroughly as a small stone could carve a decent notch out of the blade.

Some species of root need special attention. Dandelion tends to get sections of root that slough and turn black. These parts need to be trimmed off and discarded. The same problem can occur in some other roots, especially if

Herbal Harvest

3.1 Dock, Yellow *(Rumex crispus)*

3.2 Echinacea, Broad-leaf *(Echinacea purpurea)*

3.3 Echinacea, Narrow-leaf *(Echinacea angustifolia)*

3.4 Elder *(Sambucus nigra)*

3.5 Fennel *(Foeniculum vulgare)*

3.6 Ginseng, American *(Panax quinquefolius)*

3.7 Hawthorn, berries and leaves *(Crataegus monogyna)*

Plate 3

Herbal Harvest

4.1 Hops *(Humulus lupulus)*

4.2 Horsetail, Field *(Equisetum arvense)*

4.3 Hyssop *(Hyssopus officinalis)*

4.4 Lemon Thyme *(Thymus x citriodorus)*

4.5 Lady's Mantle *(Alchemilla vulgaris)*

4.6 Marshmallow *(Althaea officinalis)*

Plate 4

they have been in the ground for 2 years or more before harvesting.

Most roots dry quite satisfactorily through their skins, but Marshmallow (*Althaea officinalis*) is an exception in that the skin of its root is virtually impermeable. In order for it to dry, even quite small roots need to be cut or broken in some way. Cutting it into 25 mm lengths works well, as does splitting or just peeling a strip off the thinner pieces with an apple peeler.

Loading

Once the roots have been suitably prepared for drying, they can be loaded onto the screens. The quantity to put on each screen is more determined by space than weight. The pieces of root should be spread evenly over the screen, one layer thick, and not crowding each other too much. If they sit flat against each other, this will prevent air from flowing over those surfaces and they won't dry much.

A screen loaded with fresh roots is often quite heavy, so take care in handling it to avoid spilling or having the weight break the screen frame.

Temperature

Aromatic roots should not be dried at more than 35°C, but non-aromatic roots can be dried at temperatures of up to 45°C.

Case Hardening

Case hardening is mentioned in some literature as being a problem in drying roots. With high initial temperatures, the surface of the root can become 'glazed' and this forms a barrier to the evaporation of moisture. An example of case hardening most of us have seen is the glazed 'skin' that forms on the cut surface of potato or pumpkin when baked in the oven, sealing in the moisture of the flesh underneath.

Case hardening does not seem to occur if the drying temperature is not over 45°, so I have never encountered it in drying roots myself. At these lower temperatures, the cut surface dries to a more porous texture and moisture is able to move through it more easily.

It is important that the initial drying of cut roots progresses at a reasonably good rate, otherwise mould will start to develop on the moist cut surfaces and bruised areas. If this occurs, the roots will have to be discarded.

During the drying of roots, a close eye needs to be kept on the dryer for the effects of overloading and uneven airflow described above, and measures will need to be taken to alleviate the problems which almost inevitably occur to some degree when drying roots.

Most root herbs will require chopping or splitting before going into the dryer.

Monitoring and Assessment

Roots require regular monitoring during drying to ensure the process is going evenly.

First, look out for the odd large piece that may have slipped past the blade and will need splitting to speed it up a bit.

When checking for dryness, roots need to be looked at carefully. It is important that you check all over each screen because roots tend to dry so variably. The size of the piece of root is a major factor in determining how fast it dries, so the small pieces will be dry long before the big ones.

Commonly there is one part of the screen where drying has been slower. Check the larger pieces from this section, bending them until they break. If they break with a clear snap they are probably dry enough, but to be sure try to dig into the centre of the broken face with your thumbnail. If it is quite hard, then the root is dry right through. If it is a little soft and doughy, it needs more drying (a few roots have a soft pithy centre, but this is not doughy when dry).

Roots with crown buds are often the slowest to dry out: sometimes the crown bud is hidden by a layer of little leaves that look quite dry on the outside, but conceal a moist heart. Splitting two or three of these down the middle will give you a picture of what is going on inside.

To confirm your impression of dryness, check a few more before putting them into storage. If all roots on the screen are very dry, except perhaps one or two that are nearly there, you can probably safely put them into the bag. Check them a week or two later, though, when the moisture has had a chance to even out to make sure your judgement was correct.

Drying Bark Herbs in the Cabinet Dryer

Pre-chopping

If bark is particularly thick – over 20 mm – it may need chopping prior to drying.

Most barks tend to curl up as they dry. In extreme cases this can result in a tight roll, restricting drying and even causing mould. Chopping prior to drying may be necessary to prevent this, especially with barks that are virtually impermeable through their outside surface.

Pre-chopping of bark can assist drying. Some thick tough barks are easier to chop in a fresh state.

Tannin Content

Quite a few barks have a high tannin content which reacts with iron to form a dark blue iron tannate. This substance is indelible and was used in the traditional blue-black ink. It could badly stain the bark and affect its appearance.

Barks containing high levels of tannin should be chopped when dry. If chopped fresh, a stainless steel blade should be used.

Equipment

If the bark is not too tough, a chaff-cutter could be used, otherwise use a cleaver, a hatchet or an axe.

Size of Pieces

Unless a fine chop is required, pieces can be 25–100 mm, depending on the thickness of the bark. Thick bark should not be in pieces that are too long.

Monitoring and Assessing

Watch the bark during drying to ensure excessive curling does not occur. In assessing bark, check the centre of thicker curled up pieces, which are slowest to dry. Most barks will snap when dry, but some softer spongy or fibrous types may be more deceptive. Compare the texture of an exposed edge with the inside of the centre.

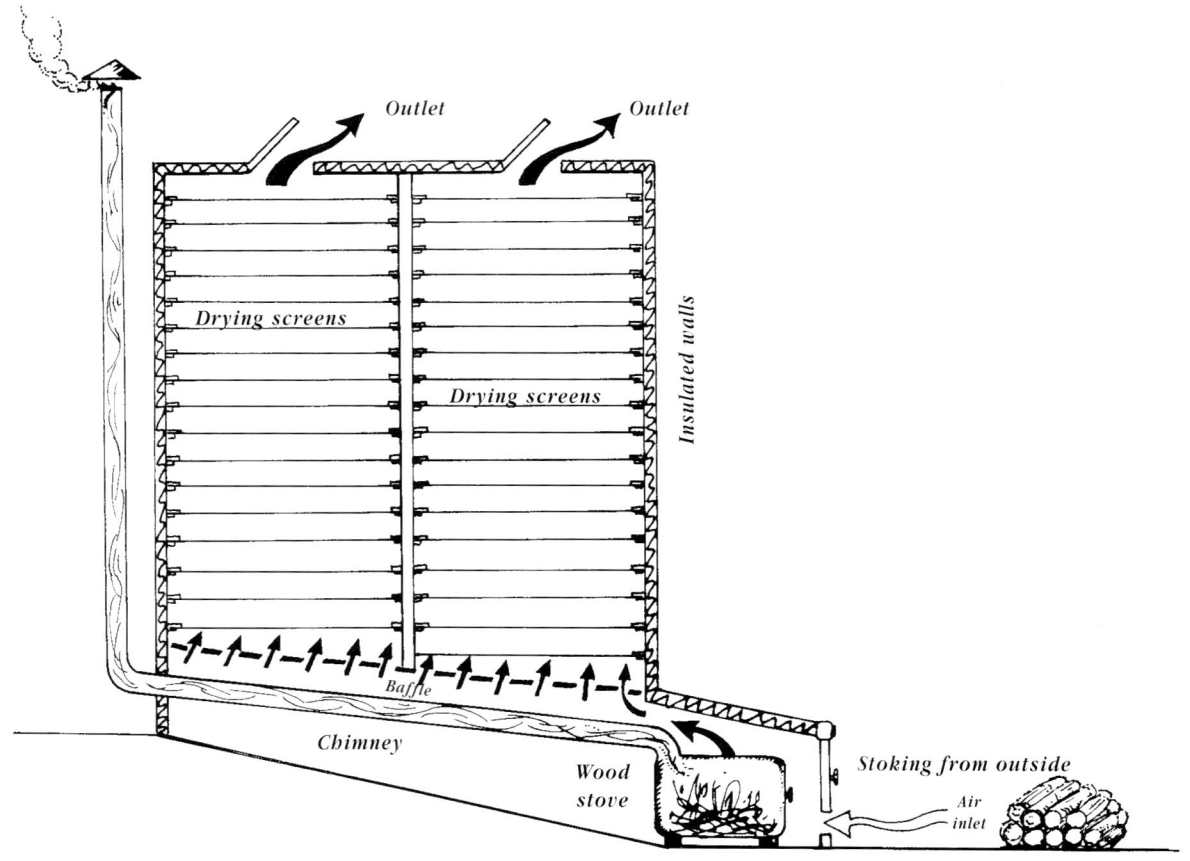

A design for a two-bay wood-fired dryer with external stoking stove below floor level.

A Wood-fired Cabinet Dryer

The nature of a wood-fired heat source requires a larger dryer and a somewhat different style of operation.

Because a wood-heater has to be of a reasonable size in order to burn for a number of hours between stokings, this means that the capacity of the dryer needs to be bigger to enable effective use of the heat produced.

For a medium-sized wood stove, a capacity of about 30 screen shelves is appropriate. There are two basic possibilities of design: the heater can be installed on the floor just below the screens, or it can be installed outside (preferably at a lower level) with the heat vented into the dryer.

I prefer the second option because it reduces fire risk and enables better mixing of the air before it reaches the herbs. It will also be more convenient for stoking and wood storage. In many situations, however, it may not be practical to site your stove below the floor level.

In this case, the stove could be set inside the dryer below the screens or out to one side, and the hot air directed into the dryer. This may need the assistance of a fan if convection is not sufficient. Baffles and/or a fan will probably be necessary to ensure good distribution of the heated air.

This dryer should consist of two bays, each holding about 15 screens at the normal spacing.

The size of the dryer's air inlet and outlet will depend on the size of the stove: it doesn't hurt to err on the large side, provided they can be adjusted and closed down tight. The stove can then be operated at full bore without overheating the herbs, and the dryer can be closed right down when the fire is out.

Convertible Cabinet Dryer

If a permanently installed cabinet dryer is going to occupy too much of your valuable ambient air drying space, you could consider the option that has worked well on one herb farm.

Their two-bay cabinet dryer is in a corner of their drying shed and is constructed so that the front and one side can be totally removed in good weather conditions, so those racks can be used for ambient air drying. When the cabinet dryer is fired up, the side and front panels are replaced, in probably less time than it would take to move all 30 screens into a cabinet dryer in another location.

Operation of the Wood-Fired Dryer

Careful monitoring of the temperature inside the dryer and regulation of airflow will be necessary to keep temperatures within the optimum range. The rate at which the wood is burned will depend on the amount of herb loaded in the dryer: for a small load, burn at a slower rate and reduce the air flow through the dryer. This will keep the temperatures higher and conserve fuel.

Because of the greater capacity of a wood-fired dryer and the time and labour involved in running it, you will probably operate it somewhat differently from the small electric cabinet dryer described above. It will be more efficient to fire up the stove when there is a large batch to be dried. You will usually want to get the herbs to all come dry at about the same time, so you are not stoking the stove for an extra week for just two or three screens that need finishing.

This will mean careful monitoring throughout the dryer, and rotating of screens, where necessary, during the drying process. Alternatively a circulation dryer could be used with wood heat.

While wood-heat is probably more ecologically sound than other sources of heat, the disadvantage of this system is that you have to keep a close eye on the stove and keep it stoked. You also have to cut the wood, which can be a bit much if you are already cutting a heap of wood to keep your house warm.

Solar Heat Options

A number of solar options exist that could be used with this or other heated drying systems.

The Solar Collector

Sunlight striking a dark surface is transformed into heat, which warms the air flowing over it. A black surface will absorb more light, producing more heat, and there are special paints for this purpose. There are several possible types of solar collector.

The Solar Panel

The solar panel is somewhat like the now-familiar solar hot water panel. The collector made with a black metal surface is insulated behind and covered with glass or plastic with a space through which the air flows. The inlet is at the base and the outlet at the top and as the air warms it rises and flows through the outlet and is directed into the drying chamber.

The Solar Tunnel

The solar tunnel consists of a series of arches covered with transparent plastic, and a floor of heat-absorbent black material. Air enters one end of the tunnel, is heated by the

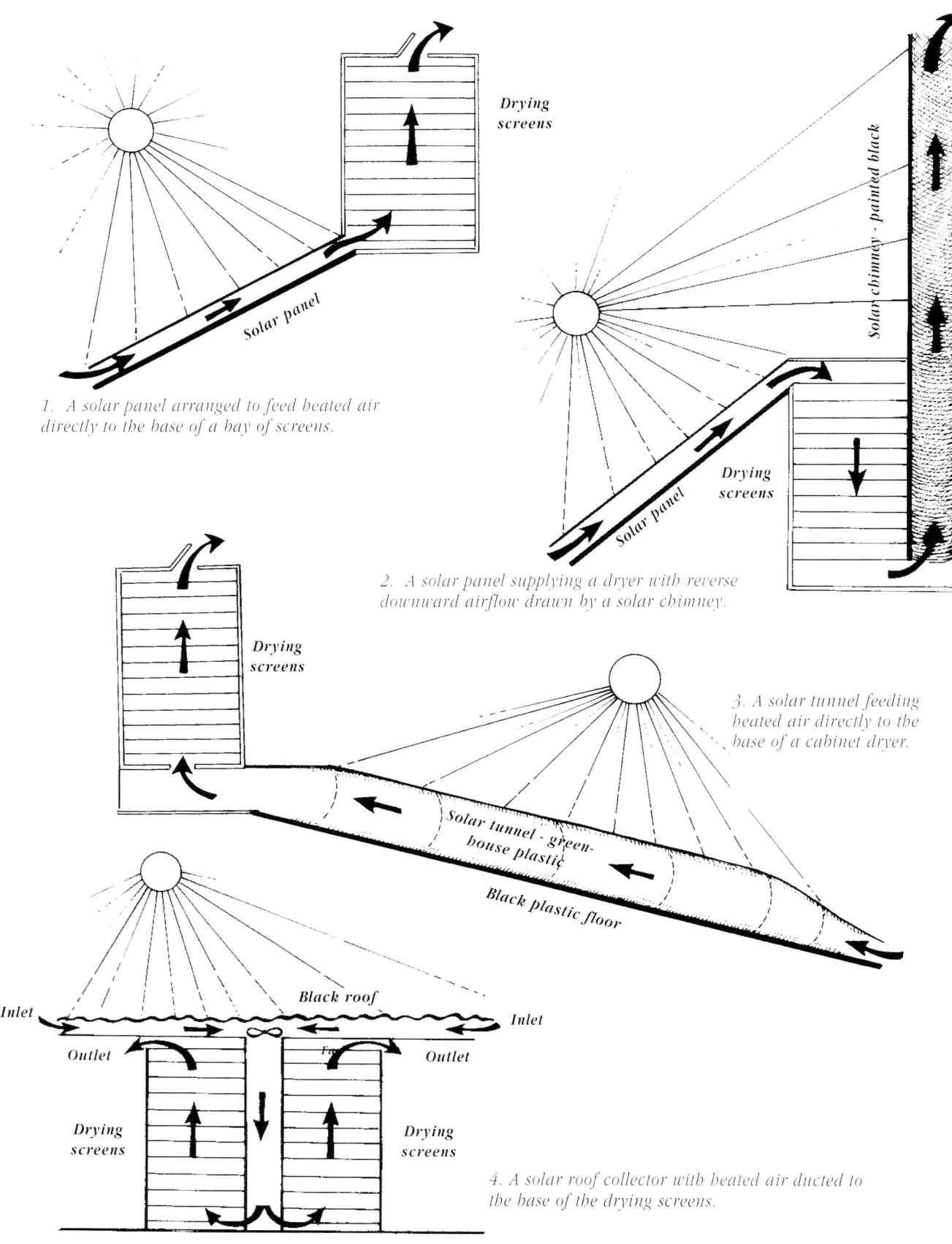

1. A solar panel arranged to feed heated air directly to the base of a bay of screens.

2. A solar panel supplying a dryer with reverse downward airflow drawn by a solar chimney.

3. A solar tunnel feeding heated air directly to the base of a cabinet dryer.

4. A solar roof collector with heated air ducted to the base of the drying screens.

A number of solar heat options can be used to reduce heating costs or improve drying efficiency.

floor, and is drawn off at the other end. This system is described in more detail below under 'Constructing a Solar Tunnel'.

Solar Roof Collector

The roof of the building, preferably painted black or a dark colour, is used as a solar collector. The roof is lined so that by means of a fan, air can be drawn in at the ends, heated in the roof cavity and ducted down to the base of the drying shed from where it rises through the screens. Alternatively, the hot air could be directed into the top of a reverse-flow cabinet dryer or circulation chamber.

The fan would need a thermostatic control so that it does not blow damp cold air through the dryer on overcast days or at night.

I don't know of anyone using this system for drying herbs, but it is being used for drying sultanas in Victoria and for drying grain overseas.

Airflow

Passive solar systems rely on convection for airflow, but in some situations this needs to be assisted with a fan or a solar chimney.

The Solar Chimney

A solar chimney can be used to augment airflow in sunny conditions in direct proportion to the output of the solar collector. A metal chimney with a diameter of 300–600 mm, is painted black. As it is heated by the sun, the air inside warms and rises out of the top, drawing air in at the base. It can be used in place of an exhaust fan to speed up air flow through a system.

One grower has used a solar chimney to operate a reverse flow drying cabinet. This chimney is used to draw away cool moist air from floor level while warm air delivered by the solar panel rises to the top of the drying chamber and is drawn downward through the screens.

Constructing a Solar Tunnel

My second cabinet dryer has been constructed to draw air from a solar tunnel made of UV-stable clear greenhouse plastic stretched over a series of metal arches, which span a black plastic floor.

The solar tunnel is on a slope, so the warmed air rises by convection and flows into the cabinet through the inlet in the floor.

The cabinet dryer works quite well as a solar–electric system, the solar heat being used when conditions are suitable and the electric heat coming on as a back-up, or at night if faster drying is required.

Size

Determining the optimum size for the solar tunnel is a rather complex matter as there are so many variables affecting its output, such as orientation, day length, cloud, wind, shade, air temperature, latitude and time of year.

The tunnel has to be oversized in order to obtain an adequate heat output in less favourable conditions. Measures need to be taken to damp it down on occasions when the heat output is too high.

A formula could probably be worked out for calculating the optimum size, but I just took a guess and made my tunnel 2 m wide and 18 m long. In southern Tasmania at 400 m above sea level with an eastern orientation and a bit of shade at times from trees and the house, this has turned out about right for a 950 x 950 x 2100 mm cabinet dryer. Nevertheless, the heat output is wildly variable and it takes a couple of layers of shadecloth over the tunnel to damp it down sufficiently in hot weather. If it is cool and cloudy, as it often is here, the output is insufficient for good drying.

In a sunny mainland location, a tunnel half this size, or even less, would probably be adequate in the warmer part of the year.

Slope

The steeper the slope of the tunnel, the greater the flow of air by convection will be. However, there will still be some convection on a level site as the cabinet will act as a chimney. This could be augmented by a fan or a solar chimney (see above). Avoid a reverse slope with the tunnel's inlet higher than its outlet, as it will be difficult to overcome reverse convection.

Floor

The floor needs to serve two functions: to absorb heat from the sun and to cut off moisture from the ground below.

Whether there should be insulation below the floor is a moot point. It will increase the output of the collector, but it will also make the peak output higher. Without insulation there would be some heat loss into the soil, but perhaps some heat bank effect when that heat is released later.

I put the black plastic down on the grass in my prototype. All plastic used in the solar tunnel should be UV-stable as breakdown products could be toxic.

Arches

I used a commercial cloche system with light channel section arches pushed into the ground.

Alternative methods of constructing the arches could easily be devised using pipe, heavy gauge wire, steel rod, or reinforcement mesh.

Plastic Covering

The clear UV-stable greenhouse film is laid over the arches and attached firmly at each end.

Wires that clip into holes at the base of the arches hold the plastic down in the channels. For added security it helps to drive in pegs on each side midway between the arches and tie a string over the tunnel to tension the plastic and keep it from flapping.

The plastic should be held down at the base or folded around the black plastic floor to make the tunnel reasonably airtight and also watertight, so water running off the tunnel can't find its way inside and defeat the whole purpose.

Double Skin Tunnel

One design for a solar tunnel incorporates a double skin to reduce heat losses. The inlet air is drawn through the space between the two skins, recovering some of the heat loss from the tunnel.

This system would probably need a fan to assist its airflow.

Orientation

If possible, the solar tunnel should be placed on a northern slope with full sun all day. However, this ideal arrangement may not be achievable, in which case the tunnel should be located and oriented in as good an arrangement as can be managed.

Damping

Because the solar tunnel has to be large enough to achieve an adequate output in less favourable conditions, its heat output is going to be too high in good conditions. Two lengths of 50% shadecloth can be kept handy to throw over the tunnel: a single layer for warm days and a double layer for hot days.

A more sophisticated system could make use of excess output by diverting it through a heat bank, possibly a chamber full of rocks, and storing the heat for use in the dryer in the cooler part of the day.

Circulation Drying with Heated Air

Circulation drying with heated air is another system that can be used with screen drying and produces a high quality dried product. It is based on a large chamber or room with adequate insulation. The screens are set up on racks, similarly to the arrangement already described for ambient air drying. Heated air is introduced or generated at floor level and the air kept circulating about the room by means of one or more fans. The inlet should be at floor level but the optimum positioning of the outlet would depend on where the humidity of the air is highest. Most designs seem to locate the outlet near the ceiling but the moistest air may be located in some other part of the chamber: it may be advantageous to remove cool moist air at floor level with an exhaust fan at the opposite end to the inlet.

Where a heated system is required for the complete drying process, circulation drying has some advantages over a cabinet dryer especially for larger volumes.

Advantages

Synchronous Drying of Batches: A batch of screens loaded together will all dry at pretty well the same time, which is more convenient for processing.

Even Drying: If the fans are well placed for good air circulation the herbs will dry very evenly.

Ease of Operation: Because the screens don't require moving or rotating during drying, operating involves less labour, especially when it is used for the complete drying process.

Disadvantages

Larger Size: The greater volume of space needed for this system means its capital cost will be greater. A circulation dryer needs to be approximately twice the size of a cabinet dryer of the same capacity.

Greater Energy Requirement: Because there will be greater heat losses from the larger drying chamber and because the exhausted air is still fairly dry, the running costs per kilogram of dried herb produced will be somewhat higher. These costs might be reduced by systems that reuse the exhaust air or cycle it through a dehumidifier.

Construction

The drying chamber should be airtight and well insulated if it is going to operate in cool outside temperatures. It should be lined with materials compatible with herb production and should be able to withstand bumps and scrapes (see 'Building a Cabinet Dryer', above).

One herb farm uses two second-hand insulated shipping containers with stainless steel lining. When fitted out with racks, each of these containers can hold 55 standard drying screens. Heat is supplied by an electric bar heater and the air circulated by three fans, two at floor level and one a metre or so off the floor. At this stage, ventilation is maintained simply by opening the doors slightly at one end, but a more sophisticated ventilation system is planned.

Operating at a temperature of 25–35°C, leaf herbs are dried in 4–5 days.

This system is in operation at 600 m above sea level in Tasmania. The calculated cost for electricity for drying a load of about 20 kg in summer is $2–3/kg at an average ambient temperature of around 15°C.

With a solar collector to supplement it, the running cost in summer could probably be halved.

Operating

Operating a circulation dryer with heated air is somewhat similar to ambient air drying when drying conditions are very good. Screen loadings should be approximately the same. As the air is flowing over the herb rather than up through it, avoid greatly increasing the thickness of the herb on the screens.

Roots and fruits can be dried quite effectively, provided temperatures and humidity are suitable. The temperature of the incoming air should be 30–35°C for aromatic herbs, and up to 45°C for herbs without volatile components.

Because the herbs are being dried in a confined space with the air circulated around them, if one batch of herb is almost dry when a fresh batch is brought in, they may be remoistened for a period as a lot of moisture will come off the fresh herb. For this reason it is better to have two chambers so loading can be alternated.

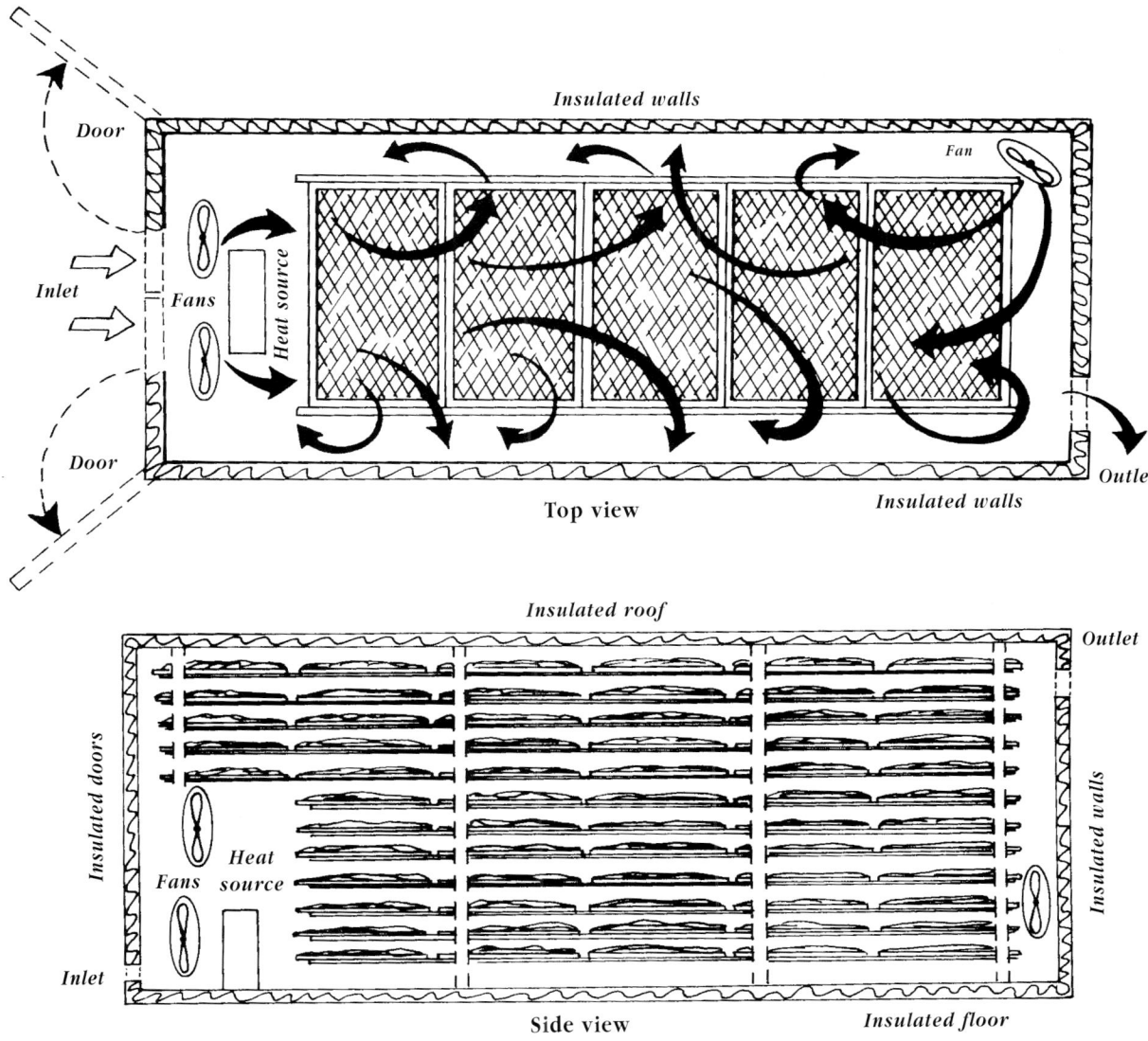

A circulation dryer set up inside an insulated shipping container.

As the initial part of drying will release a lot of moisture, ventilation should be increased during this period. If it is not possible to maintain good drying conditions with a full load of fresh herb, a partial load should be put in first, with the rest put in a day or so later, after the first load has dried a little.

For optimum drying, the relative humidity of the air circulating in the drying chamber should not be more than 50% for more than 24 hours and should become progressively drier as the herbs dry. As the release of moisture slows towards the end of the drying period, the ventilation of the chamber can be reduced, saving energy.

Two or more fans should be placed so the air is kept moving around in the chamber to maintain even temperatures and ensure rapid, even drying.

Care should be taken not to blow herbs off the screens or cause flapping or fluttering of leaves that are susceptible to bruising.

OTHER DRYING SYSTEMS

There are number of other drying systems sometimes used or advocated for drying herbs. Many of them cannot be relied upon to produce a quality product, but some may be a possibility in some situations or for certain herbs.

Outside Drying

Outside drying is a preferred method for products where low cost is the major consideration, such as in producing animal fodder. Large volumes are required at minimal cost and the consumers of the product don't have a lot of choice if they don't like its quality, they soon get hungry enough to eat it.

If outside drying is considered for herbs, the grower needs to be aware of the dangers of sun and rain. Even a heavy dew on partially dried herb can discolour it enough to make it virtually unsaleable, while any green plant will be severely faded if left exposed to sunlight for the whole drying process.

Field Curing

Field curing is the drying of herbs out in the field where they were grown. Even in its most refined form, field curing herbs has severe limitations, and these are covered in some detail in Chapter 11, 'Harvesting'. Field curing can't be recommended as a drying method for producing a quality dried herb.

Outside Floor Drying

Drying on a floor set up outside in the open is another method that is sometimes used, particularly in third-world countries. The herb might be spread out on a flat, clean dry surface, such as concrete or bitumen. In some places, an earth or sun-baked dung floor is used or maybe the herb is just spread out on a road to dry. Consequent contamination problems can occur with any of these situations. Little pieces of the surface can break off and get swept up with the herb, or petroleum vapours from the bitumen in hot sun could condense onto it: a nice thought!

Imported herbs sometimes have quite a range of unsavoury items in them that look as if they must have been swept up from the ground.

If the herb is not going to be left outside overnight, risking moisture damage, it will need to be raked up in the evening and spread out again in the morning, as it will normally take several days to completely dry the crop, even in hot weather. This handling of wilted material will cause bruising and discolouration. As the herb becomes drier, shattering of leaves and flowers will occur with resultant losses. And a gusty wind may blow the herbs away or drop dust and debris onto them.

The use of portable screens for outside sun drying of herbs would have the advantage of enabling one to bring them in at night or in inclement weather. This is a possibility for finishing shade-dried herbs where the last bit of moisture needs extra heat to drive it off. A few hours in the warm sun may do this without significantly fading the colour of the herb.

The disadvantage of all outside drying systems is the question of what you do when the sun doesn't work. It is all very well to plan on harvesting at the commencement of a spell of fine weather, but even with weather forecasts there are always going to be occasions when you will end up with a pile of partly dried herb on your hands in uncertain weather, wondering whether to put it out or bring it in.

My recommendation regarding outside drying is to avoid it if you want a high quality dried herb. There are just too many things that can go wrong too easily.

Drying in Bunches in Shade

On a small scale, herbs that have suitable stems can be tied in bunches and hung from strings or wires somewhere in the shade away from rain.

Be careful not to tie them in too big or too tight a bunch as the inner part will not dry well. If you are just growing a few herbs for yourself, this is probably the best system to use, but even on the smallest commercial scale it is soon going to prove too time consuming and labour intensive.

The only exceptions would be for drying branches of leaves of tree herbs which are too bulky to load onto screens, or if you have a special market for herbs dried in bunches.

Inside Floor Drying

Passive Floor Drying

In its crudest form, floor drying involves thoroughly sweeping the floor of a shed or putting down some sort of fabric, then simply spreading the herb out on the floor.

Passive floor drying will work in suitable weather if a thin layer of herb is spread out, but it is not a very efficient use of shed space. Being low down, the drying conditions are often not so good either.

Vermin could be a problem in some sheds.

I have used this method to dry a single large harvest of a wildcrafted herb on occasions when an empty shed was available.

Suspended Floor Drying with Forced Air

Suspended floor drying with forced air is a more sophisticated system often used on a commercial scale. The floor is raised and perforated, so the herbs can be spread in a relatively thick layer, usually 300–600 mm thick, and dried by forcing heated air up through them.

This system is used for some 'conventional' herb crops, such as Hops and Tobacco, and for large-scale herb production in many parts of the world.

Advantages

Efficiency of Handling: The labour involved in handling large volumes of herb is greatly reduced as many of the operations can be mechanised and where manual labour is required, the herb can be moved about with shovels or hayforks.

Economy of Scale: For a large-scale operation, the capital outlay per kilogram of production will be relatively low.

Disadvantages

Deterioration of the Herb: The drying proceeds from the bottom up, so the upper layers are subject to a prolonged period of high humidity and warmth, being 'steamed in their own juices'. This is likely to cause fermentation and deterioration, which will be apparent as discoloration. Many herbs will come out quite dark and there are likely to be metabolic changes in them, affecting active constituents.

The Need for Turning: In order to obtain even drying, the thick layer of herb on the floor may need to be stirred or turned over in a partially dried state. This will cause bruising, discoloration and shattering of leaf or flower herbs.

Large Volumes: For optimum utilisation of the drying facility, large volumes of each herb will be required. Small volumes will be inconvenient and more costly to dry.

Contamination: Because the herb is not spread out in thin layers manually, as it is for screen drying, weeds or other foreign material are likely to go into the dryer undetected.

Suitability for Producing a Quality Product

I have seen a lot of herbs dried using this system, though I have never used it myself. Quality is the major limitation as the herbs tend to suffer severely in the handling and drying, particularly those that are susceptible to bruising and discoloration.

Leaf herbs and flower herbs are most affected, roots not so much so. Essential oil levels are usually somewhat depleted. Even so the results can still be above general trade quality.

It may be possible to modify the system to overcome the problems and produce a higher quality, but this would incur higher labour costs and higher operating costs.

Heat Source

Fossil fuels are normally used in conventional dryers, though traditionally wood-fired furnaces were employed.

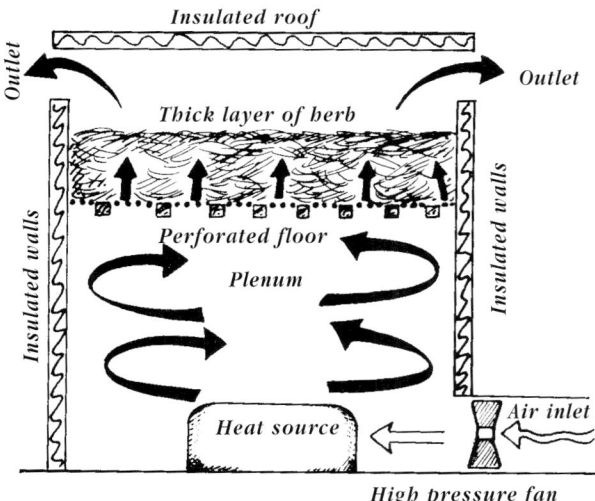

A suspended floor dryer: this system is often used for drying large volumes of herb, but the quality is usually adversely affected.

Electric heat is a possibility if it is economical. Solar heat can be used as a supplement, but another heat source will be required to maintain continuous operation.

If fuel is burned in the generation of heat, there needs to be a heat exchange. Direct flow of the burner's smoke or exhaust through the herbs, as occurs in some conventional dryers, will result in contamination with ash and products of incomplete combustion and is not acceptable for organic certification. The smoke or exhaust from the burner also needs to be vented well away from the dryer's air intake.

The heat source is located at ground level and the hot air rises up through the herb on the floor above.

Construction

The drying floor should be raised about 3 m above the heat source to provide a plenum or expansion chamber. This is important as it allows pressures to even out and the air to be mixed so temperatures are uniform. The expansion chamber should be airtight and insulated.

The drying floor can be constructed with wooden slats or a heavy screen, so that air can pass freely through it. It must be able to support a person's weight and the weight of the fresh herb: up to 100 kg/m^2. Hessian or shadecloth is laid over the floor.

There needs to be sufficient space above the floor to walk around. Walls and ceiling need to be insulated to reduce heat loss and prevent condensation and water dripping back onto the herb. Adequate provision needs to be made for the exit of air without obstructing flow or allowing condensation to run back into the dryer or cause problems anywhere else.

Airflow

Good airflow is vital for this system as the air must be forced through a thick layer of herb. The fan needs to be able to build up pressure against resistance in order to do this adequately.

To achieve this, a centrifugal fan (or squirrel cage fan) with backward-curved blades is most effective. Propeller-type fans are less satisfactory when operating against pressure.

Airflow must not be interrupted or spoilage of the herb may occur in such a thick dense layer.

Loading

Evenness of airflow is essential for even drying and this depends on uniform depth and density. Lines drawn around the wall can serve as a guide in loading. Stalky herbs are sometimes chopped to 150 mm lengths to enable even loading.

One grower recommends wilting the crop before chopping, to facilitate loading and speed up drying. However, this is likely to cause even more bruising and discoloration.

The loading procedure needs to be carried out quickly enough to cover the whole floor so the air can be forced to flow through the herb. Piecemeal loading will be less uniform and severe heating could occur in thick piles of herb with no air flowing through them.

One overseas organic grower loads aerial parts up to about 50 kg/m^2 fresh weight, this corresponds to a thickness of 300–600 mm. This mass of herb would be generating considerable heat of its own, which might contribute to the drying but would certainly be contributing to the deterioration of the herb.

Temperature

Temperature of the entering air should not exceed 35°C for aromatic herbs or 45°C for non-aromatic herbs. There may be a temptation to crank the system up to increase the throughput. Conventional practice dries Hops at 70°C or higher, but virtually all the essential oil is lost at that temperature.

Turning

It may be necessary to turn the herb in order to achieve more even drying and speed up the process.

Monitoring and Assessing

Thorough checking of the herb is essential, as pockets of moist herb could easily be hidden among the dried material.

Modifying this System for Improved Quality

In general, suspended floor drying is used for herbs where large volumes are going to be dried and where premium quality is not required. It would no doubt be possible to refine the system so that some of the problems were overcome.

The herb could be inspected on a moving belt before loading to remove weeds and other foreign material.

Lighter loading, probably not over 15–20 kg/m^2 would reduce deterioration, speed up drying, and might avoid the need for turning.

To avoid bruising and discoloration, the herb should be loaded in a fresh unwilted state. Chopping of leaf and aerial parts should be avoided with herbs susceptible to bruising.

These measures would ensure a higher quality dried herb, but the efficiency and throughput of the system would be lowered and the costs increased. Before adopting such a modified suspended floor system, it would need to be evaluated to ascertain whether it had any great advantage over a screen drying system.

Other Options

Variations on the suspended-floor drying theme could include the use of solar heat, convection rather than fan-driven air, the use of fan-driven ambient air, or the use of multiple floors or shelves, one above the other.

Convection systems were used for drying Hops in the days before electric fans, but the 'oast houses' were quite tall to get sufficient updraught.

Fan-driven ambient air could be used in hot dry conditions, but probably with thinner layers of herb spread on the floor and with some sort of a back-up system when conditions weren't suitable.

Recirculation of Used Air

The possibility of recirculating used air is sometimes raised as an energy saving measure. There is no point in recirculating warm air that has taken up as much moisture as it can from the drying herb, either on its own or when mixed with new air.

However, if the air still has some capacity to take up moisture from the herbs being dried, recirculation could be worth while. The practice would have to be carefully evaluated, taking into account heat losses in recirculating and loss of drying potential.

Where two dryers are operating, if one contains herbs that are almost dry, the used air from it could quite effectively be used for the initial stage of drying herbs in the other.

Another option might be a heat exchange where ingoing air was warmed by outgoing air without mixing them.

Drying with a Heat Pump Dehumidifier

Drying with a heat pump dehumidifier is another option for herb drying that is used overseas and is being used in Australia for food drying with good results.

The heat pump works like a refrigerator, cooling the air and condensing the moisture out of it on the cold side. It then warms the air up again on the hot side before it passes over the herbs again.

This salvages most of the latent heat of evaporation and saves having to heat new air to maintain drying.

Advantages

Independence of Ambient Conditions: Because the operation is a closed system, it can function effectively and maintain optimum humidity and temperature independent of outside conditions.

This would be particularly advantageous in sub-tropical regions where other systems can sometimes only lower humidity by raising temperatures above optimum levels.

Low Energy Requirement: This is a cheap drying system to run because the heat required to dry the herbs is absorbed in the condensation process and pumped back into the system. The cost of running the heat pump is usually much less than the cost of heating new air.

Quality: Because it is able to maintain good drying conditions, the outcome should be good quality, provided the heat pump dehumidifier is operated correctly.

Disadvantages

High Initial Cost: The system is not cheap. A heat pump dehumidifier with a throughput of around 10 kg/day, dry weight, might cost $10 000–15 000, depending on how much of the system you are able to fabricate yourself. A smaller system with a throughput of 2–3 kg/day dry weight might cost $3000–5000.

In order to make such a system a financially viable proposition, it would need to be in use at least 50% of the year. Even so it would take 7–10 years to pay for itself in energy savings, assuming it could save $1/kg.

Dependence on High-tech Equipment: Design, installation, maintenance and repairs will require the services of a specialist. This will be costly and could result in drying delays at critical times because this sort of equipment usually breaks down when it is in peak use.

For more information on this subject, a booklet that gives an excellent summary of information on heat pump operation, costs, and potential has been compiled by Professor Clifton Ellyet (see Bibliography).

Sun Drying under Cover

For most leaf, flower and fruit herbs, shade drying is considered necessary to avoid fading. Some roots, though, can be sun-dried if conditions are good enough.

If the roof is transparent, the roots can be in the sun while protected from rain and dew. In southern regions, this system is only worth considering towards the end of winter and early spring when the days are getting longer and the sun is becoming more intense. In northern regions, where there is more winter sunshine, it may be more useful.

There would need to be adequate provision for ventilation. I have experimented a little with such a system, but can't recommend it as reliable, because it depends very much on the weather being suitable, which it often isn't at the time of year when roots are needing to be dried.

Another major problem would be vermin, because the herbs would be difficult to adequately protect at a time of year when rats and mice are hungry.

A Greenhouse Covered with Black Plastic

A greenhouse covered with black plastic is sometimes advocated as a suitable set-up for drying herbs. An ordinary poly house is covered with black plastic to create a drying tunnel. Plastic is also laid over the ground as a moisture barrier and to give a clean surface.

While this may be a cheap option for drying in some situations, it would be risky.

Staying within the optimum temperature range would be difficult, as the heat generated inside in hot weather would be intense. Black plastic absorbs and radiates much more heat than galvanised iron.

Herbs drying in such a structure would also be vulnerable in the event the plastic covering was damaged. This could easily happen if something pokes or rubs against the walls inside, or something occurs outside, such as a stick falling from a tree and hitting the greenhouse.

Sooner or later the plastic will weaken with exposure to sunlight and will break up: this could happen suddenly in a strong wind or a storm, leaving the herbs totally exposed in a downpour.

Vermin would be difficult to exclude, as they can easily gnaw their way in. You might come out one morning to find that your valuable herbs are just a pile of rats' breakfast.

Freeze Drying

Freeze drying is sometimes touted as the bee's knees when it comes to drying herbs, but it does not seem to be used much, and the claim is perhaps questionable.

I once tasted a packet of Chives labelled 'freeze dried', which had about as much flavour as the paper of the label and were nothing like the quality of my own Chives dried in ambient air.

Freeze drying involves freezing the fresh material and sublimating the ice off by vacuum. Even at these low temperatures there can be a major loss of volatile components because in a vacuum they will vaporise more readily.

This is a high-tech system with a high capital cost and it would not be practical for the small grower.

Comparison of Drying Costs of Various Systems

The following are approximate figures for the energy cost of drying herbs using different systems, and are based on gas, oil or electricity at 7c/kW and arriving at a cost per kilogram dried weight.

What they don't take into account is the capital costs of the systems.

Drying System	Energy cost/kg
Ambient Air	0.00
Ambient Air/Electric or Gas Back-up	0.30–0.60
Cabinet Dryer – Non-sequentially loaded	1.50–2.00
Cabinet Dryer – Sequentially loaded	1.00–1.50
Circulation Dryer	2.00–2.50
Dehumidifier – Heat-pump type	around 0.50
Solar Heated Air	0.00
Solar Heated Air with Elect. or Gas back-up	0.30–0.60
Suspended Floor Dryer	0.50–2.00
Wood Heat – cutting own wood	mostly time

Sharing a Central Drying Facility

The establishment of a central drying facility to be shared by a group of growers is a possibility, but there are not many situations where this has worked successfully.

If it is done on a co-operative basis, the growers need to be able to co-ordinate with each other to ensure that the facilities are getting full use without leading to conflicting needs. If this works well there may be a mutual benefit, but if it doesn't, the growers may be better off on their own.

Alternatively, if the drying facility operates as a business, it could be a challenge to find enough reliable growers to supply high quality fresh herbs on a consistent regular basis to ensure the dryer is operating economically and profitably.

One service such a central drying system might provide would be to give growers a start on a trial basis, to see if they can grow the herbs before they invest money in their drying facility. If it were designed to be compartmentalised and self-operated, growers might be able to rent space for a period and do all the work (and carry all the risks).

Co-ordination

One challenge would be co-ordinating the harvesting of a group of growers so the herbs were all being harvested at optimum times and dried without any delays.

Operation

One person would probably have to take responsibility for the operation of the facility in order to ensure it ran smoothly and effectively.

Organic Certification

To avoid problems with certification, all produce going through the facility would need to be certified, or systems would need to be in place to ensure there was no contamination or substitution. Good records would need to be kept to enable inspectors to follow an audit trail.

Cost

A shared drying facility may not necessarily result in savings in running and capital costs.

Running costs would quite likely be significantly higher. The person employed to dry the herbs would have to be paid for their services. Unless the operation was co-ordinated very smoothly so the drying facility operated at a good percentage of its capacity, this cost could be disproportionately high.

Also, with a number of growers depending on the drying facility, it would have to be a very reliable set-up. This probably excludes ambient air drying, as it is too subject to weather fluctuations and consequent screen shuffling. One of the artificial drying systems would probably be more appropriate, but this would mean higher capital and running costs.

Consequently, by the time the growers had to pay someone to operate the dryer as well as meeting the higher running costs, they might not be any better off than if they had set up a cheaper drying system themselves at home and obtained the full price for their dried product.

Processing

Once the herb has been dried, the question of processing has to be considered. In a central drying facility it could be quite inconvenient to have growers travel long distances to process a few kilograms of herbs.

Alternatively, this could be entrusted to the individual operating the dryer, but again there would be an additional cost to the grower. If that person had at their disposal processing equipment that economised labour and ensured high quality, that may be a saving to the grower, but that equipment is not going to be cheap or easy to obtain.

Distance

If growers had to travel any significant distance to a central drying facility, this would add significantly to their costs. Quality could also be impaired with longer holding times and delays before batches of herb commenced drying.

Quality

A central drying system might ensure a more consistent quality of product, but it may not necessarily be better quality. This would depend on how well the operation of the dryer could be co-ordinated, the disposition of the growers, and the skill and reliability of the operator.

The activities of growers sharing a drying facility may affect each other, for instance white petals from one grower's Chamomile may blow all around onto other growers' leaf herbs, or one grower may load fresh herb and inadvertently remoisten another grower's herb that is just dry.

Problems that show up in drying and processing often have their origin in the grower's operation. If a grower is drying and processing their own product, they get to see and experience the consequences of how they went about growing and harvesting the herbs. But if someone else is doing the drying and processing, the grower may not appreciate the extent and cause of the problem.

Evaluation

The benefits of setting up a central drying system need to be weighed up very carefully against its pitfalls and shortcomings to ensure that such an undertaking would work to the growers' advantage.

13

Processing

Once they have reached a dry state, most herbs will need to be processed into a form suitable for the consumer, the wholesaler or the manufacturer. Even if the herb is to be placed in storage before going to its destination, immediate processing will generally be necessary to reduce its bulk. It can then be placed in sealed containers where it will be safe from moisture, insects and vermin.

While a few herbs are ready to use as they come from the drying screens, for the majority processing is important because it creates the finished product from the crude dried herb. The way processing is carried out can have a significant impact on quality. Hence it is essential to have suitable equipment and to use it skilfully with appropriate timing.

METHODS OF PROCESSING

For some reason the methods of processing dried herbs to a usable form seem to have been kept a well-concealed mystery. Information on this subject has generally not made it into most books and where it has, it is dealt with very superficially or could more properly be termed misinformation.

Unfortunately this has left many would-be growers stranded, trying to pluck the leaves one by one from their herbs and maybe finally deciding that working in an office was not so bad after all. Others, a bit more venturesome, have resorted to shoving their dried herbs through a hammermill, pulverising them into a disappointing product resembling the contents of a vacuum cleaner.

In this chapter we will focus on a number of techniques – using low-cost equipment – which can be used to prepare an attractive final product from the material that comes from the dryer.

How this is done depends on the use of the herb, and in what form the buyer prefers it.

Categories of Herbs According to Use

Culinary

Most herbs used in cooking need to be in small pieces or flakes so that diners are unlikely to disgrace themselves at table by choking on a stalk. The exception is coarse leaves such as Bay, which are better whole than in pieces, so they can be fished out easily when the dish is cooked. French Tarragon is normally in whole leaves as well, but its leaves are of a softer texture than most other culinary herbs.

A good quality culinary herb is usually in particles up to about 2.5 mm across and is virtually free of stem. As the stems of most culinary herbs contain little of the flavour, they are best excluded. The leaves can be broken up and the stems sifted out by rubbing the herb through a wire screen.

Infusion

For making infusions, or teas as they are commonly called, the herb needs to be in a form that can easily go into a teapot or a tea infuser.

Once again, for most purposes the stalks are superfluous. They usually don't contain much flavour or many of the active constituents, so it is better for the quality of the product that they be removed. Again, this is usually done by rubbing the herbs through a screen.

Leaf particle size is not as critical for making infusions as it is for culinary purposes, but for appearance, convenience of handling and volume, pieces of up to around 8 mm across are a good standard. This is somewhat larger than the average herb tea of mediocre quality you are likely to find in the trade, where leaf pieces are up to only about 3 mm.

Quality conscious growers find the larger sized pieces of leaf preferable because they look more attractive, they retain their flavour better (as there are fewer broken surfaces) and, being larger, they usually stay in the teapot rather than ending up floating around on top of the tea in the cup. They give the quality herb a distinctive signature that the customer can look for. For a manual rubbing operation, there is also less work involved.

The larger pieces also make for a bulkier product, so a 50 g packet will be about the same size as 100 g of the mediocre product they will be standing next to on the shelves. This helps overcome the difference there will be between the price per gram for the two products.

At first glance this may seem like deceptive packaging, but one must consider that there is often more than twice the flavour in the quality herb, and the same volume in a tea infuser will make a more delicious cup of tea. Consequently, presenting the quality herb in a bulkier form is really placing it on a more even footing with the denser, less flavoursome, inferior herb.

While the inferior herb may seem cheaper when compared simply on a price per weight basis, it may not be as cheap if it is compared on the basis of the amount required to make a decent flavoured cup of tea or to achieve the desired therapeutic effect; if, indeed, the consumer can get the inferior herb to make a cup of tea that tastes any good or does anything for them.

Decoction

A decoction is made by simmering the herb in water, usually for around 15 minutes. Decoctions are generally prepared from denser herbs, such as roots, which are reluctant to give up their constituents. For decoctions, particles of 2–8 mm are preferable. Dust can cause problems in handling and packaging so, if possible, the proportion of fines should be kept to a minimum. Particles larger than 8 mm take longer to extract and may not do so adequately.

These herbs are usually processed by hammermilling them or putting them through some sort of grinder. If they are being supplied in bulk to a wholesaler or manufacturer, they normally don't have to be milled by the grower, and can be sent off in larger pieces, as long as they are not too bulky for storage and shipping.

Manufacturing

Herbs for manufacturing into alcohol-based tinctures and extracts and other herbal products are usually supplied to manufacturers in a fairly coarse form. Leaf herbs are preferred as coarsely chopped aerial parts (including stalk), while roots can be as they come from the dryer: in large pieces.

As manufacturers normally do their own milling, the grower merely needs to reduce the bulk of the dried product to enable it to be stored and shipped conveniently. Coarse chopping into pieces 25–50 mm long is usually quite adequately done by a chaff-cutter. This enables the manufacturer to more easily assess the quality of the herb and check for correct identification and any contamination or adulteration.

Leaf herbs intended for manufacturing into liquid medicines usually include the stems. While in a few species the stem may contain significant amounts of active constituents, usually it is mostly woody tissue. The inclusion of the stalks makes a bulkier and cheaper raw material and this practice continues because the nature of the raw materials is not easily discerned by the end-user.

As herbal products become standardised according to their content of known active constituents, we are beginning to see a trend where manufacturers are becoming more concerned about the quality of their basic materials. Consequently, some are becoming more particular about the proportion of stem in the herbs they buy.

Determining End Use

Before you start processing you need to have an idea where the herb is going to end up. This is straightforward in the case of those herbs whose use is predominantly for one purpose, but many herbs have multiple uses. For example, Sage and Thyme are used as culinary herbs, as herb teas for their flavour, and as important medicinal herbs, either as teas or in tincture or extract form.

Once processing has been carried out, it usually isn't very easy to change the herb from one form to another. If the stalks have been sifted out, they have usually long been consigned to the compost by the time the order for aerial parts comes in. On the other hand, if the herb has been put through the chaff-cutter as aerial parts, it is not easy to sift the short pieces of stalk out to get it to a tea grade as the cut stalks tend to fall through the screen.

So the grower needs to have a fair idea of what form the market will require before starting on processing. (See colour plates 8.5, 8.6 and 8.7 for illustrations of the various forms of processed herb.)

Processing Facilities

Some equipment will be needed for processing herbs, depending on the species being grown and the form in which they are required.

Generally flowers and roots require very little processing by the grower after drying, but leaves need rubbing, chopping or stripping, and fruits and seeds usually need cleaning. Barks may need chopping.

Space

Provision needs to be made for an adequate work space for these operations. This is best located in the drying area, or next to it, so the herbs do not have to be carried far.

The processing area needs to be able to be closed up to protect it from wind and rain. If lighting is available, this will enable work to proceed after dark. Lights need to be placed or protected so they can not be broken by screens that are being manoeuvred around: broken glass is difficult to separate from many herbs.

If substantial quantities of leaf are to be processed during humid conditions, it is good if the processing area can be heated so the herbs are not being exposed to moist air for extended periods.

PROCESSING LEAF HERBS

The grower will need to have on hand the equipment required to process leaf herbs to their final stage for marketing. This usually means having your own equipment, though in some cases it may be possible to share with other nearby growers provided co-ordination is feasible.

Equipment for Preparing Tea and Culinary Grade Leaf Herbs

Some research and development could see major improvements in equipment used for preparing herbs for tea and culinary purposes.

There is a need among small- and intermediate-scale operations for mechanised equipment of an appropriate design and scale, which will separate leaf from stem, saving on labour, but still yielding a premium grade product.

The systems we have been using work reasonably well and can produce an attractive dried herb, but they are manual processes and are fairly time consuming.

The Rubbing Apparatus

Apparatus for rubbing herbs consists of a frame that holds the rubbing screen, under-screens and catching tray, with which the rubbing and screening operations are carried out. It is best to make the frame large enough for a batch of around 20 kg finished product to be put through in one go. The rubbing screen is about waist height, with the under-screens below it and the catching tray on the floor. The operation is entirely manual, and one person can do it alone or two people can work opposite each other.

The Rubbing Screen

I can no longer recall where I learned about the rubbing screen, but it is a traditional method for processing herbs. Most of my understanding of its design and use have come through trial and error, which is something of an ongoing process. There are further improvements continually being made to the system, some of which are outlined below. Nevertheless, the basic system works reasonably well for this scale of production and does not cost a lot to set up. On a larger scale, the operation could be mechanised.

The rubbing screen is essentially a woven wire screen with wooden sides high enough to contain the herb as it is being rubbed. To be efficient, it needs to be large enough for screenfuls of herb to be dumped into it, and so the operator(s) can use the full sweep of their arms when rubbing. This requires a size of at least 1100 x 1800 mm for the sides of the tray, but the width of the screen itself can be 910 mm if the sides are splayed out to 1100 mm to catch the herb as it falls from a drying screen.

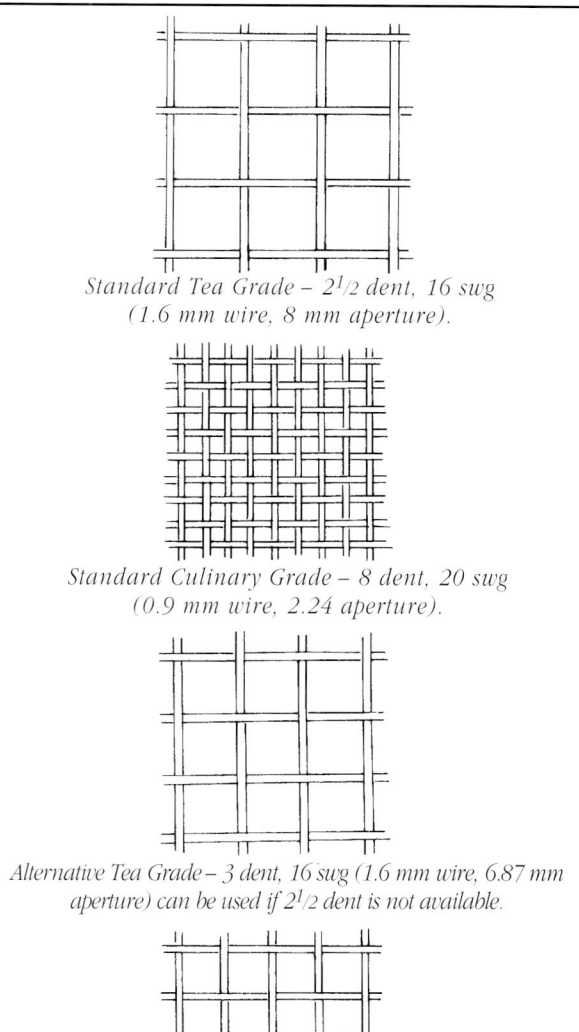

Standard Tea Grade – 2½ dent, 16 swg (1.6 mm wire, 8 mm aperture).

Standard Culinary Grade – 8 dent, 20 swg (0.9 mm wire, 2.24 aperture).

Alternative Tea Grade – 3 dent, 16 swg (1.6 mm wire, 6.87 mm aperture) can be used if 2½ dent is not available.

Intermediate Mesh Sizes – occasionally used for certain herbs – 4 dent 18 swg, (1.25 mm wire, 5 mm aperture) and 5 dent 18 swg (1.25 mm wire, 4 mm aperture).

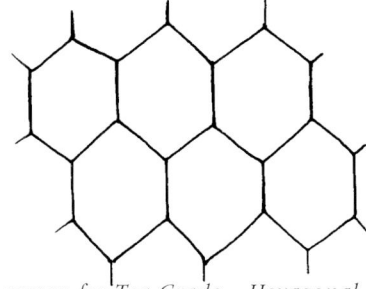

Under-screen for Tea Grade – Hexagonal mesh bird wire (known as 13 mm) can be used for under-screens when rubbing tea-grade herbs.

Mesh sizes for woven wire rubbing screens.

In fact the rubbing tray is easier to work with if the mesh screen itself is no more than 910 mm wide as the average person can then reach to the far side of the screen.

Mesh size of the screen will depend somewhat on the particle size desired. However, even for culinary use where a fine product is required, it is usually easier to first rub the herb through the coarser tea-grade screen first.

Woven Wire Mesh Sizes

Standard Tea Grade: For tea-grade herb, a screen with 1 mesh per cm, with the internal width of the apertures around 8 mm is about ideal. This is known as '2$^{1}/_{2}$ dent, 16 swg' or '8 mm aperture width, 1.6 mm diameter wire'.

Alternative Tea-grade Mesh Sizes: It may not always be possible to obtain a screen of exactly the above configuration. Sometimes only '2 dent, 16 swg' (aperture 11.1 mm across) and '3 dent, 16 swg' (7.1 mm aperture, 1.6 mm wire) are available Of these the '3 dent' is the preferable size. If the apertures are significantly larger than '2$^{1}/_{2}$ dent', in the case of most herbs too much stalk will pass through and the leaf will be coarser, making the herb too bulky to fit in standard-sized packets. It is better to go a little smaller than a little larger.

Culinary Grade: For culinary grade a much finer screen is needed, about '8 dent, 20 swg wire' (3.2 meshes per cm). This is 2.24 mm aperture, 0.9 mm wire. Avoid thinner wire as it will be too flimsy. Thicker wire is okay as long as the apertures are at least 2 mm.

If you are only processing small amounts of culinary herbs, you can get away with a smaller sized screen, 600–900 mm square, to rub them through after first putting them through a tea-grade screen.

Intermediate Mesh Sizes: Occasionally an intermediate mesh size is useful for certain herbs. I have 4 dent, 18 swg (5 mm aperture, 1.25 mm wire) and 5 dent, 18 swg (4 mm aperture, 1.25 mm wire) screens that I use now and then, but for most herbs the standard tea-grade and culinary-grade screens will be the only ones needed.

Suppliers of Woven Wire Mesh

Woven wire mesh can be a little difficult to locate as it usually isn't available in hardware stores. See Appendix 3 for a list of suppliers or look under 'Wire Products' in the Yellow Pages. Some suppliers expect you to buy 10 m minimum, but others will happily cut the mesh.

Galvanised woven wire screen is adequate, though stainless steel is available if you are willing to pay the price.

Under-Screens

A certain proportion of stem will find its way through the rubbing screen and the herb will have to be further processed to remove this. One way of doing this is to repeat the screening operation a number of times. To reduce the number of passes required, two under-screens can be placed under the rubbing screen so the rubbed herb falls through them and a proportion of the stems are caught.

The mesh size of these under-screens needs to be significantly larger than that of the rubbing screen used above them, as they will quickly choke up if the apertures are not large enough. Hexagonal mesh bird wire (13 mm across the shortest width) is a good size as an under-screen for tea grade. It has the advantage of being able to be tensioned to reduce sag.

The under-screens need to placed so that the herb falls about 300 mm between each screen. This tends to catch the stalks more effectively as the space allows them to become more horizontal as they fall. If the screens are placed too closely one above the other, the stalks will have no choice

Rubbing frame with rubbing screen, two under-screens, catching tray and sheet in place.

but to spear straight down through them all. The overall size of the under-screens also needs to be somewhat wider and longer than the rubbing screen, as the herb tends to drift sideways a bit as it falls.

For ease of cleaning, the under-screens should slide in and out on shelves. It is important that this sliding motion be quite smooth, so the screens can be gently slid out without disturbing the stalks on them: any jarring will dislodge stalks caught on the screen and will allow them to fall through onto the sifted herb below.

The Catching Tray

The catching tray can be an actual tray with sides about 450 mm high, or it can consist of detachable sides at the base of the rubbing frame. A sheet at least 2400 mm square can be attached to the sides of this tray with clothes pegs, so the rubbed herb falls into it as it passes through the screens. The sheet full of rubbed herb can be tied at the corners to make a bundle and lifted out for rescreening or putting into bags.

Alternatively, the catching tray can be mounted on castors so that it can be rolled clear of the rubbing apparatus.

The Winnower

Winnowing is an alternative to rescreening that some growers prefer. I did have a go at making a winnower myself, but I was not happy with its performance and went back to repeated screenings to get my herbs clean. The winnowing process was just too slow, even though the final product was quite good.

The principle of the winnower is to let the rubbed herb fall through the draught of a fan, so the lighter leaf is blown to the far end of a long narrow chamber and the heavier stem falls at the nearer end. A series of divisions or boxes catches the different fractions and keeps them separate.

The challenge is to create a winnower that will efficiently clean large volumes of leaf and separate the fractions precisely.

The Rubbing Winnower

The rubbing winnower is a combination of rubbing screen and winnower. It is being used by a few growers who find it very successful. It consists of a rubbing tray – as with the standard rubbing apparatus except that the working area of the wire mesh screen is at one end – the rest is covered with cardboard or plywood.

Below the rubbing screen are sheet-metal sides that taper down to a 450 mm wide, 2000 mm long, winnowing chamber the length of the rubbing tray. A normal 450 mm

The rubbing winnower: this device combines the operation of rubbing and winnowing.

fan is mounted below the screen at one end so the draught strikes the herb as it comes through and carries it towards the far end of the chamber, which is covered with shadecloth. The different weight particles fall into 3 or 4 boxes arranged along the length of the winnowing chamber.

I haven't had the opportunity to try it out yet, but it looks as if this could be a very useful system. By combining the operations of rubbing and winnowing, it should be quicker than rubbing first and then winnowing. With most herbs the end product should be better than with repeated rubbing and screening, as it will remove more stem.

There are some herbs which would probably still have to be cleaned by the screening method because they are not suited to being winnowed due to the weight of their leaves and the lightness of the stems.

The Chaff-cutter

The chaff-cutter is a useful machine for cutting up aerial parts of leaf herbs and for a few other tasks. For a herb growing operation producing a range of different herbs, a small machine designed to be turned by hand or driven by a belt from a small tractor will be adequate. A larger model would entail too much cleaning, so the time saved by its greater throughput will be lost in the extra time it takes to clean the machine before a different herb is put through.

If you can obtain a three-bladed chaff-cutter, this seems to cut more cleanly than the two-bladed type.

Cleaning

Any second-hand chaff-cutter will inevitably have been smothered in grease and oil and will need to be stripped down and thoroughly cleaned. It may well be coated in lead-based paint, so it should be sand blasted as well. This will ensure that there is no risk of contaminating the herbs that go through it.

Avoid just borrowing someone else's chaff-cutter, unless you know they have cleaned it and refrained from using petroleum lubricants. Contamination with pesticide residues from previous chemically grown crops can also occur with processing equipment that has not been thoroughly cleaned before the organic crop is put through it. This may seem a rather slim possibility, but it has occasionally occurred in the organic trade and certification guidelines require adequate precautions be taken to prevent it.

The problem is that the residues can be concentrated in the dust that is left behind when the chemically treated crop is processed. This contaminated dust then gets mixed in with the next crop to be processed.

Lubrication

A cleaner form of lubrication needs to be used as at least some of it eventually finds its way into the herbs. Most vegetable oils are inclined to go gummy, though olive oil seems to be more stable. Neat's-foot oil is non-hardening and can be used in place of mineral oil. For grease you can

The chaff-cutter is generally the best machine for chopping aerial parts.

The 'boat' or feeding tray holds herb as it is being fed into the chaff-cutter. It is also handy for a number of other operations around the drying shed.

substitute lard (unsalted): this is what people traditionally used to lubricate cart axles before the advent of petroleum products.

Alternatively, there are some commercial lubricants coming onto the market which are based on vegetable oils, though I don't know how genuinely non-toxic they are as they do contain some additives.

The 'Boat' or Chaff-cutter Feeding Tray

The feeding tray for the chaff-cutter can be around 2400 mm x 1200 mm, made of plywood, with sides about 225 mm high. One end comes to a point, with an opening that fits into the chute of the chaff-cutter. It looks rather like a flat-bottomed boat with a bit cut off its bow, hence it is generally referred to as 'the boat'.

It is large enough to allow the drying screens to be easily dumped into it without losing any herb over the sides.

It is quite a useful item around the drying and processing area, doubling as a tray for holding drying screens as they are loaded with fresh herb, for mixing different batches of herb, for holding processed herb while it is being loaded into bags, etc.

Power Source

Traditionally chaff-cutters were belt-driven by a horseworks, a steam engine or a tractor. A small chaff-cutter lends itself to being driven by an electric motor of around 1 hp, leaving your hands free to feed the herb through, but you may have to fit the motor yourself.

Hand-turned models can sometimes be found.

Catching the Cut Herb

Another modification you will probably need to make is to devise some means of catching and containing the herb as it falls from the blades. While it may be okay to scoop up chaff for animal fodder from the floor, it is unacceptable for a herbal product, due to the inevitable contamination. Before purchasing a chaff-cutter, take a close look at the action of the blades and figure out where the cut material will fall or fly and how you can manage to funnel it into a bag or a bin.

The Hammermill

Do You Really Need One?

Most growers will be able to manage quite well without a hammermill. It may be fast and efficient, but it is a bit like cracking a nut with the proverbial sledge hammer. It can do a lot of harm to your product, pounding it to dust so that it looks as if it could have been swept up off any hay shed floor.

I am always suspicious of any herb which has been processed through a hammermill. Because the material has been reduced to fragments that are difficult to distinguish, there can be a significant level of hidden contamination and an excessive proportion of stem. The shattering of the herb releases essential oils, making them vulnerable to loss by volatilising.

For good presentation of a high quality herb, hammer-milling should be avoided unless the nature of the herb particularly requires it.

A hammermill consists of a series of metal bars, which swing on a shaft that rotates at high speed over a metal screen in the shape of a half cylinder. These flying hammers repeatedly strike the particles of herb and fling them against the screen until they are small enough to pass through it.

The hammermill is useful for milling roots and for other herbs that are too hard to put through a chaff-cutter or those that are required in a finer form. However, the grower would need to know there was a market that required this form of herb before committing to acquiring a hammermill. Where milling is required, it is usually carried out by the manufacturer or wholesaler.

If, after all this, you are still convinced you need a hammermill, look for the following points:

Size: Most of the hammermills around would be simply too big for the scale of production we are looking at. Putting herb through a hammermill always results in a certain amount of loss as the speed and voracity of the machine pulverises the herb so that a portion of it is simply blown away as a fine airborne dust or left in the cracks and corners of the mill, to be laboriously cleaned out or simply mixed with the next product to go through the mill.

These losses will be too great with a large hammermill unless you have large volumes of one or just a few species of herb to mill. Unfortunately a small hammermill is not easy to find and can be relatively expensive.

Feeding Hopper: This needs to be large enough to take a reasonable volume of herb and should have a gate that can be opened to allow the herb to flow into the mill and then shut to contain the flying particles as they are belted around by the hammers.

Screens: You will need a number of different gauge screens for different products, ranging from about 2 mm to about 10 mm, and including a number of intermediate sizes. If you don't have a chaff-cutter, a 25 mm screen can be used for coarse chopping of aerial parts, though it will shatter the leaf more.

Catching: Some means of catching the milled herb is needed: some models have a bagging facility, enabling the milled product to blow straight into a bag. If you can afford it, this would be worth while because the herb tends to blow out of any gaps, and catching and containing it as it comes out the base of the hammermill is tricky.

Sealing: Check that the construction of the mill provides reasonably effective seals to keep the fines from being blown away by the force of the draught generated as it is operating.

Paint: As usual, these machines are painted to look good in the sales room without any consideration being given to where the paint will end up as it inevitably wears off during operation. Very likely you will need to dismantle the whole machine and have it sand blasted.

Dust Control and Noise: A hammermill is not exactly the sort of machine that can be set up on the kitchen table. Operating at high speed it is incredibly noisy and dusty. A separate milling room, preferably in a separate building, will be needed for it and some sort of dust extraction facility, plus a good respirator, earmuffs, goggles, and a fair measure of endurance.

If you can avoid having to use a hammermill, life will be more pleasant.

Protective Equipment Needed for Herb Processing

There are some aspects of herb processing which can be irritating or harmful. Even when operations are performed manually, you will need some protection from sharp stalks and dust.

Gloves

For rubbing herbs a pair of leather gloves is needed, preferably with long cuffs if you are processing irritating herbs like Nettle or Comfrey.

Even if your hands are thick skinned and well callused, they will not be able to withstand the painful sharp end of a dried herb stem. If the stems are still soft and pliable, you may be able to rub with bare hands, but with gloves on you can confidently apply more pressure and your hands become effectively larger so the rubbing action is more productive.

Dust Mask and Respirator

A disposable paper dust mask is adequate for most herbs, provided the dust is not too thick. For very dusty operations like hammermilling or for working with herbs with silica hairs, such as Nettle and Comfrey, a proper respirator is needed.

The fine silica hairs that break off these herbs rise in a thick cloud and are very irritating to the eyes, nose, throat and lungs. It is important to take precautions to protect yourself (or your employees) from the immediate and possible long-term harm that silica hairs can cause.

CAUTION: Before putting on your respirator, first heed a word of warning from someone who didn't look carefully inside the face mask first, but was fortunate enough to survive and tell the tale!

If you are in a red-back spider region, very likely a few will be lurking here and there in the shed and one may find the rubber face mask of the respirator a nice cosy place to camp. This possibility had never occurred to me.

One day I was about to put on the respirator when I noticed that the filters had a bit of dust caked on the outside of them, so I tapped them against the door-post and, just as I went to put it on, a rather large red-striped bum caught my eye. A fat red-back spider was sitting inside the face mask: it had evidently been nestled under the fold of the rubber rim and was dislodged by the tapping.

Goggles

Goggles are unnecessary for processing most herbs, but there is a need to protect the eyes from irritating herbs, such as Nettle and Comfrey, and from dust and flying particles from the hammermill. Plastic safety goggles are suitable. They are also handy when processing Peppermint in hot weather if there is not enough air movement and the essential oils coming off the herb make your eyes water too much.

Ear Muffs

Ear muffs are essential when operating noisy machinery like the hammermill and the chaff-cutter.

Overalls

For very dusty operations and for irritating herbs like Nettle and Comfrey, wearing overalls will mean you don't get your clothes covered in the silica hairs and spread them around everywhere.

Cloth Hat

A cloth hat helps keep the dust and bits of herbs out of your hair, especially if you like to rest the screens on your head as go up the ladder to load the racks.

Mechanised Rubbing Equipment

Technology is available to perform the rubbing operation, but it tends to be designed for large-scale operations processing several hundred kilograms of herb per hour. Unfortunately information about this sort of equipment is often surrounded by industrial secrecy, so it can be difficult to obtain.

Rotary Thresher

One design I have seen could possibly be scaled down, but so far I have not ever had the time and inclination to have a go at it.

This machine, which is used for processing culinary herbs by a large European herb-growing operation, consists of a large cylinder mounted horizontally. The lower half (that is the underside) is made of mesh screen of appropriate size. A shaft down the central axis has several beaters the length of the cylinder. These are attached so that as they rotate they sweep just inside the circumference of the cylinder and along its whole length.

The machine is mounted with a slight fall from one end to another. The herb is loaded in at the higher end and gradually works its way to the lower end while being swept around and around by the beaters. As the leaf breaks up, it falls through the screen while the stems continue down the cylinder and out the other end.

This machine is designed to be operated at different speeds to get optimum results with different herbs and conditions.

Technique

Rubbing

The rubbing process is mainly used for culinary and tea grade leaf herbs. The dried herb is stirred about over a wire mesh screen, separating the leaves from the stems and breaking them up into particles small enough to pass through the screen. Most of the stems are held back, though a few always manage to pass through the screen and need to be separated by further screening and/or winnowing.

Dryness of Leaf

For rubbing to work properly, the herb needs to be dry to the point that its leaves are brittle. They will then break up readily and pass through the screen. Ideally, the stems should still be a little moist and pliable, so they don't break up or spear down through the screen as easily.

However, it is not always possible to be there at the right moment and the stems may be quite dry by the time you get around to rubbing the herb. The stems tend to break up more easily when dry and more of these broken pieces pass through the screen. Being drier and stiffer, the longer pieces also tend to spear through the screen. This results in more stems going in with the leaf. Extra screening or winnowing is required to separate them out.

On the other hand, if you leave it too late in the day or if a change in the weather comes over, your herbs may re-absorb moisture and you will have to wait again until they are dry enough to rub.

Fortunately rubbing is a very safe method of processing herbs from the point of view of dryness. If they aren't dry enough, they will not break up easily and rub through the screen

If the leaves can only be made to break up and pass through the screen with vigorous rubbing against the wires, then they are a little too moist and should be dried further until they shatter easily.

Prevailing Air Humidity

If leaves are exposed to moist air during processing, they will gradually reabsorb moisture. If prevailing conditions are very humid and the herb is exposed for too long during processing, this could result in excessive remoistening.

In these circumstances it may be necessary to postpone processing or organise it so the exposure is minimised. For example one can just do a first screening and hold the partly processed herb until conditions are more suitable.

Protection

Unless the stems are quite soft, you will need to wear a pair of leather gloves and a dust mask, or for more noxious dusts, a respirator, goggles and protective clothing (see 'Protective Equipment' above).

Loading the Rubbing Tray

To begin, first check that your herb is dry enough and then dump the contents of as many as six screens into the rubbing tray If a larger amount is rubbed in each loading, it can result in more stem being pushed through, so it is better to do it in a series of smaller loadings.

Once this first loading has been rubbed through and the under-screens cleared, the contents of a further six screens can be put through and so on until the whole batch has been worked through the rubbing screen once.

The Rubbing Action

The style of motion in rubbing is quite important and it takes a while to develop a good technique. Most of us when first attempting it tend to force the herb through the screen with an action similar to that of grating cheese. While this may be effective in breaking up the herb and pushing it through the screen, it also forces a greater proportion of stem through the screen.

Instead, the aim should be to break up the leaf while minimising the amount of stems being broken or pushed through the screen. Usually the best way to achieve this is to begin with a sweeping, swirling action of the arms, reaching all through the piled up herb. This will knock a lot of the leaves from the stems by rubbing them against each other.

The motion needs to be varied somewhat according to the pliability of the stalks. If they are very dry and brittle, you have to be more gentle to minimise breaking them up, but if they are still moist and pliable, they can be worked quite vigorously without much problem.

The broken leaves can then be worked through the screen, and the remaining stems with some leaf still attached are gathered in bundles and kneaded in the manner of bread-making or hand-washing clothes. This action is aimed at breaking off the rest of the leaves, again more by rubbing the herb against itself than by rubbing against the screen.

Sometimes it works better to gather the remaining stem and leaf into a large sausage shape, which is rolled back and forth to shatter the leaves without breaking up the stems. The bare stems can then be discarded and the broken leaf rubbed through the screen.

Using the Under-screens

If under-screens are being used, you need to keep an eye on them while rubbing is proceeding. They are only of benefit while they are catching stems and allowing the leaf to fall through. As soon as the under-screens begin to be choked with stems, so that the leaf is being caught on them rather than falling through, the rubbing should be halted.

Those sections which are choked up need to be gently rubbed to work the leaf through, while endeavouring to retain the stems. Then carefully slide out each under-screen and shake the stems off onto the floor. However, if a close eye is kept on the under-screens, they can be shaken off before the leaf starts to build up on them. This way you avoid any rubbing on them, which inevitably makes a few more stems fall through.

How often the under-screens need clearing will depend on the nature of the herb, but it will need to be done at least once every loading of six screens. Always clear the under-screens before loading more herb into the rubbing tray.

There are a few herbs that will choke up the under-screens almost immediately, Raspberry Leaf and Coltsfoot, for example. These have fine furry hairs on the undersides of their leaves, causing them to cling together so they quickly block the holes in the under-screens. With these herbs you will need to remove the under-screens and resort to more passes through the rubbing screen alone, or else use a winnower to separate out the residual stems.

First Screening

Herb that has passed through the screens once is referred to as 'first screening'. It will inevitably have some stem mixed with it and if tea grade is required, further processing is necessary.

Some herbs intended for manufacturing into liquid medicines are required as leaf or leaf and flower, rather than as aerial parts including stem. For this purpose a first screening is usually quite adequate, because the actual proportion of stem in it is fairly small.

A few leafy herbs, such as Coltsfoot or Lady's Mantle, do not lend themselves to feeding into a chaff-cutter as they are too short for the feeding rollers to pull them through. For manufacturing-grade aerial parts, these herbs can be rubbed through the tea-grade screen and the stems can then be thrown in with the leaf, provided they are dry enough.

Depending on the amount of stem coming through the rubbing screen, under-screens may not need to be used for herb intended for manufacturing purposes.

With most herbs, a first screening will contain more active constituents than would aerial parts and would make a more potent extract or tincture because the stems, which are mostly cellulose, have been removed. However, manufacturers are currently going for the cheaper aerial parts of most herbs and the grower usually cannot get a premium price for first screening where aerial parts are used.

Rescreening Tea-grade Herbs

Most herbs will need to be passed through the rubbing screen and the two under-screens at least twice to get a product clean enough for tea grade. Often it will take three or four passes before they are clean enough.

When passing the herb leaf through the screen for the second and subsequent times, instead of rubbing with gloved hands, a simple wooden float (similar to that used for finishing concrete) about 125 mm wide and 600–750 mm long, can be used to sweep the herb back and forth over the screen. This is done until most of the leaf has been worked through and a bundle of stems with a few leaves intermingled remains. This bundle can then be carefully rubbed with the fingers to allow the leaves to pass through while retaining the stems, or most of them.

Of course this rescreening is not going to be 100% efficient and there will always be a proportion of stem that passes through. By careful screening and use of the under-screens this can be minimised and usually by the third or fourth screening, the proportion of stem will be sufficiently reduced to make it acceptable for tea grade.

Rescreening Culinary-grade Herbs

After first rubbing through a $2^{1}/_{2}$ or 3 dent screen, culinary herbs that are required in smaller sized particles will need to be rubbed through a finer screen. As the herb is being further broken down, this will be a similar operation to the first screening.

Repeated re-screening is necessary until the stems are virtually eliminated. Some of the methods outlined below under 'Eliminating Stem' may be able to be used to advantage.

How Much Stem Is Acceptable?

Tea Grade: Ideally there should be no stem in a premium quality tea grade of most herbs. However, a very small proportion of short fine stem is often difficult to totally eliminate and may be acceptable, provided it is a healthy colour.

At this stage there are no clear guidelines for stem content and herbs are assessed on a visual basis by individual

Processing

1. The dry herb ready for processing is first loaded into the rubbing tray.

2. The initial rubbing breaks the leaves off the stems and works them through the screen. Normally gloves are worn for this operation, but in this case the moist stem is soft enough to be rubbed with bare hands.

3. Clearing the under-screens of stems that have been caught by the mesh as the rubbed herb falls through.

4. First screening contains a proportion of stem that has passed through the screen.

5. The herb is then rescreened until it is sufficiently free of stem.

6. The fully screened herb is ready to put into bags.

wholesalers. The amount of stem acceptable will vary according to the herb and the buyer, but limits are around 0.5% to 3% by weight.

There should not be any stems longer than 40 mm left in the herb and very few longer than 25 mm. Coarse stems reduce the quality because they generally do not have much flavour or active constituents in them and they cause problems. The tea packager does not want stems poking through the walls of their neatly presented cellophane packets, and the consumer does not want stems sticking out of the jaws of their tea infuser, making it look like a reptile having difficulty swallowing its oversized prey.

Milling stem down to small particles and including it with the leaf is not acceptable in a quality tea grade, except with a few species where the stem is specifically included.

Culinary Grade: Culinary grade should be virtually free of any stem.

Manufacturing Grade: Where leaf is specified for manufacturing use, a small proportion of stem, up to perhaps 5%, may be acceptable, but this will depend on the individual manufacturer's specifications.

Eliminating Stem

There are a few refinements of the rubbing technique which can make stem removal a little easier to achieve. Most of them are based on the simple fact that the more times a piece of stem is pushed back and forth over the screen, the more likely it is to find its way through.

Consequently, the larger the volume of herb loaded, the greater the length of time the stem pieces are spending on the rubbing screen and the greater the proportion of stem that passes through.

In fact most of the stem comes through when the last part of each loading is being rubbed. This is when there are mostly stalks left on the screen with a bit of leaf among them. As one rubs this last portion to work the remaining leaf through the screen, inevitably a proportion of stem falls through as well. But when the first portion of the loading is being rubbed through the screen, very little stem comes through.

One can make use of this characteristic in a number of ways:

Surface Picking: After each loading of herb has been rubbed through, inspect the surface of the herb in the catching tray. The stems will be predominantly on the surface, having come through towards the end. Where there are just a few remaining stems, as in the final screening, these can be picked off by hand, or if there are a lot, the surface can be skimmed off for separate re-screening.

Splitting the Rubbing: When rubbing each loading, rub about half of it through and put the rest aside for separate screening later or for a different grade. Only a small proportion of the stems will pass through the screen in the first half, so the portion put aside will contain most of the stems.

For herbs that have both culinary and tea use, such as Spearmint and Thyme, the first portion can go for culinary use and the second portion can go for tea or manufacturing use, where a proportion of short fine stem will be acceptable. Of course both grades will probably need at least 3 screenings to get them to an acceptable quality.

Catching the Last Portion: Towards the later stage of rubbing each loading, place a drying screen in between the bottom under-screen and the catching tray. I usually use one with a fine cloth fabric as ordinary shadecloth will tend to let too much fine herb filter through. The screen can just rest on the surface of the herb and doesn't have to be the full dimension of the bottom tray as at this stage one can easily concentrate on a smaller area of the rubbing screen.

This screen then catches the last portion of the herb coming through the rubbing screen. This can be put it aside for re-screening separately (or it can be rescreened again straight away).

Sloping Screen: Another method employed by some growers for removing stems involves flowing the herb down a vibrating screen on a slope of around 45°. The herb will need to have been rubbed through the same sized screen at least once and probably twice, so the leaf particles are mostly broken down to a size that will fall through the screen easily.

As the herb flows over the sloping screen, the leaf particles generally fall through, while the stems will mostly flow on to the end of the screen. Because no pressure is being applied to them, the stems are less inclined to find their way through the screen, instead flowing on over it, aided by gravity and the vibration. If the screen can be arranged so the stems fall off the edge when they reach it, continuous operation will be possible.

There will probably be a line of division that can be drawn in the pile of finished herb under the screen, one side of which will be of acceptable cleanliness. The rest will require further passing over the sloping screen or perhaps an additional rubbing if there are oversized leaf particles still among the stems. When herb is rubbed through a screen, the particles produced include some that have been pushed through folded or on a diagonal, so they are still too large to easily fall through the screen without any pressure.

By following one of the methods described above, it is possible to reduce the volume of herb that needs to go through the repeated screenings necessary to remove the last few stalks. The preferred method will vary according to the herb and the characteristics of that particular batch.

Winnowing

Winnowing is another technique used by some growers for removing stem and it certainly can produce a very clear product.

The operation is either carried out with a winnower after the herb has been rubbed to a first screening, or during the first pass through the screen if a rubbing winnower is used.

Draught

The draught of air from the fan used for winnowing needs to be strong enough to carry the lighter leaf particles to the last third of the chamber, but not so strong as to carry the heavier stem pieces beyond the first third.

It is important that the herb be introduced to the draught only as fast as it can carry the particles away. If the herb falls too thickly or in clumps, the draught will not be able to separate the lighter and heavier particles. Rubbing the herb through a screen above the draught ensures that the particles are separate as they fall.

Separating the Fractions

Careful placement of dividers or boxes in the winnowing chamber helps separate the fractions. Normally there will be three fractions: pure leaf at the far end, a mixture of leaf and stem in middle and mostly stem in the first section.

The middle section is rewinnowed, sometimes repeatedly, until it has all fallen into the front or far sections. With some herbs or some batches of herb, there will always be a quantity of mixed leaf and stem remaining, while others separate out more completely.

Unfortunately not everything separates out as neatly as might be desired. Flower parts are sometimes on the heavy side and fall in with the stems. The coarser leaf of tea grade can be a challenge with herbs like Peppermint where some pieces of rolled up leaf tend to behave like stems in the draught.

Sometimes the mixed fractions can be rubbed through a finer screen to break up particles of rolled leaf or flowers.

Each herb behaves differently, according to the size, shape and density of its leaf and stem particles. This system works quite well for herbs like Spearmint, Raspberry leaf and Alfalfa, while others, such as Thyme and Sage with their thick leaves and light thin stems, do not separate as easily.

For a herb to be cleaned easily by winnowing, there needs to be a significant difference between the density of its leaves and stems.

Removing Contamination

Winnowing is also a valuable method for removing accidental contamination with material that is heavier than the herbs, such as soil particles or, heaven forbid, rat droppings. It is, however, an inadequate method of dealing with the problem of vermin infestation, because some herbs cannot be cleaned by winnowing and if the rats are defecating on your herbs, they will also be urinating on them.

Prevailing Air Humidity

Winnowing should only be carried out in conditions of where the humidity is low enough not to remoisten the herbs, so it is not a good job to put aside for a rainy day.

Other Processing Methods Required for Some Leaf Herbs

There are a few herbs in which the stem is included in tea grade, such as Oats, Vervain, and Rue. In some of these herbs, the stem performs the functions of a leaf and/or it contains active constituents, so it needs to be chopped or milled for a tea grade. With others the leaves are too fibrous to be broken by rubbing or stripping.

There are a number of options for dealing with these herbs. In some cases, chaff-cutting can be used to produce a tea grade. Sometimes this can be done before drying, as the chaff-cutter cuts them more cleanly and speeds up their drying (see 'Tea-grade Chopping' below, under 'Chaff-cutting').

With some herbs a more attractive product can be made by rubbing the dried herb first to separate off the leaf as larger sized fragments. The stems can then be chopped in a chaff-cutter or put through a hammermill and recombined with the leaves. This is preferable to putting the whole herb through the hammermill, which shatters much of the leaf to a fine powder, giving the herb a poor appearance.

A few herbs will not go through the mesh of the rubbing screen, notably Mugwort, or are required as whole leaf, such as Lemon Verbena and Bay. These need to be stripped rather than rubbed (see below under 'Stripping Leaves').

Producing a Quality Product While Still Maintaining a Degree of Freedom and Sanity

Rubbing can be a rewarding process, producing an attractive finished product when it is done with care. But it can become rather tedious and time consuming. Often your normal daily routine must be disrupted because herb needs to be rubbed while it is at the right stage.

If you are using a passive ambient air drying system, the herb will often be ready for processing only towards the end of the day. This certainly doesn't fit into the traditional 9 to 5 workday many of us are accustomed to.

Unless conditions are particularly favourable, if these ready-to-process herbs are left until the next day, they will reabsorb moisture overnight and still may not be ready to rub until late in the afternoon. If there is a change in the weather, conditions may not be right again for several days. Missing the opportunity to rub a batch of herb when it is ready can result in additional work and expense through having to resort to back-up drying facilities and/or loss of quality due to remoistening.

On the other hand, you have to be careful not to make 'getting the herbs rubbed' too high a priority in your life at the expense of meals, sleep, time with your family, leisure and the needs of the spirit.

If you are producing large volumes of quality herb and can't afford to have your drying and processing disrupted by the vagaries of climate, it is important to have a good back-up drying system that can be put into action without too much rigmarole. Then the volume and quality of your production will not be compromised and your sanity and enjoyment of the work not overly strained.

There are also a few techniques that are handy to enable one to gain a little more freedom in this regard, and these are outlined below.

Holding the First Screening

Usually leaf herb can be rubbed to first screening and held for an indefinite period until it is convenient to finish it off. This is safe, provided the moisture level is sufficiently below borderline to ensure the dry leaf can safely absorb any moisture in the few stems still in the rubbed herb. This is generally the case, but if moisture is borderline, that is if the herb is only just dry enough to rub through the screen, it is better not to risk the additional moisture of any juicy stems which have passed through with the leaf.

If you are putting a first screening aside to be finished off later, make sure it is adequately protected from re-absorbing moisture. Large volumes just being held overnight will usually be okay bundled in a sheet (provided air humidity is fairly low), but smaller volumes, or herb being held for longer periods, should go into drums or plastic bags.

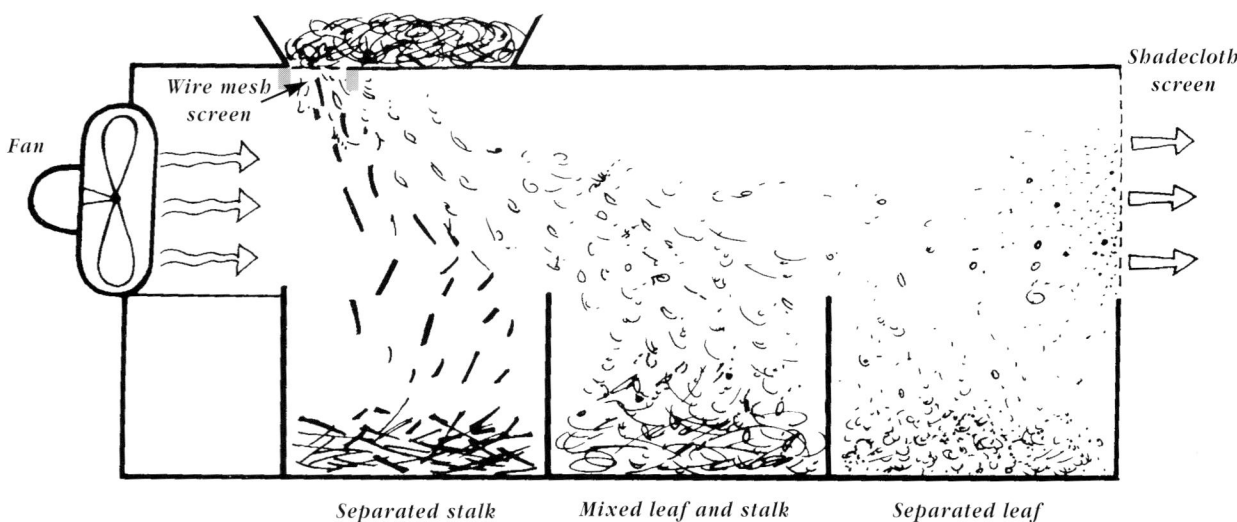

Winnowing: the draught from the fan separates the falling herb particles by blowing them varying distances according to their density. The heavier stem particles fall short, while the lighter leaf particles are blown the furthest. A mixture of leaf and stem tends to fall in between.

Closing the Shed up Tight

The remoistening of herb that is on screens can be slowed by having a drying shed that can be closed up tight. While the humidity will gradually increase or find its way in, a drying shed that will close out damp air can often give you time to have a meal or a break and then come back to finish the rubbing before the dampness shuts you down.

Stacking the Screens

Another little trick is to take the screens off the racks and stack them on top of each other with sheets or plastic over them to keep the dampness from penetrating.

Alternatively, load the screens into your cabinet dryer. If this is well constructed with tightly closed vents, it should be able to hold herb dry overnight, even if the dryer is not turned on.

Another option is to tip the screens off onto sheets that are bundled and covered until they can be rubbed.

Bear in mind, however, that if there is any moisture still in the stems, it will be gradually absorbed by the dry leaves. If held for more than a few hours like this, the leaves may become too damp to rub. How long they can be held depends on the herb's degree of dryness.

Chaff-cutting

The main use of the chaff-cutter is to produce a manufacturing grade of herb with the stems included. Chaff-cutting reduces the bulk of the herb to make it easier for storage and shipping. In this coarsely chopped form, the manufacturer can assess the herb's quality and confirm its identification. Chaff-cut herb is easily fed into a small hammermill to fine mill it for manufacturing purposes.

Moisture Content

Because the machine will just as happily chop up moist herb as it will dried herb, there is no natural safety factor to protect you from producing a batch of compost in your storage bins. To avoid this sort of disaster, all herbs need to be thoroughly checked for dryness before chopping.

Parts to watch closely are thicker stems, flower buds, immature fruiting bodies, and young growing tips. Be aware that those screens in less favourable positions may not have dried as fast as the rest and could be still too moist to safely process and store. Also with a very large batch of herb that has taken a good part of a day to load, the last screens could be about a day behind the first.

Herbs vary somewhat in their drying characteristics and in which parts are slowest to dry. Each has a different sort of feel when it is dry, and with a bit of experience you will become familiar with them. If in any doubt, check the bags a few days after chopping to see if they still feel crisp and dry: if part of the herb contained hidden moisture, this will gradually move into the rest of the herb, making it feel damp.

Concealed Moisture: What can often be misleading is a superficial very dry feeling of the exposed parts of the plant. Moisture may, however, be concealed in the stems or other parts that are more resistant to drying, and there could be enough of this moisture to bring the total moisture content of the herb over the critical level, with resulting problems.

Concealed moisture commonly causes problems for inexperienced herb growers. Often they are under the impression that their herb is fully dry, but when it reaches the manufacturer or wholesaler it is found to be in a dangerously damp state, perhaps even going mouldy. Consequently, the need for care in this area can't be stressed too highly. More detail on assessing herbs for adequate dryness is to be found in the Chapter 12.

Care of the Blades

Make sure the blades of the chaff-cutter are sharp and adjusted so they make good contact for their full length, producing a good shearing action and a clean cut. There is a series of adjustment bolts that press on the blade and hold it in the right position.

The blade should make good firm contact along its whole length as it slides over the striking plate, because this gives the cutting action. However, it should not be so firm as to unduly slow the rotation of the chaff-cutter.

Safety Precautions

A chaff-cutter is quite capable of cutting off a few fingers before it stalls to a stop, so always exercise due caution. Keep your hands well back from the feeding mechanism and use a piece of wood to push the last bit through. To help prevent accidents, it is best to install an extra switch at the front of the machine where you can easily reach it if you do get caught up in the works. Make sure the machine is also switched off at the wall after use, in case the front switch is accidentally bumped when cleaning the machine out after use, and in case children playing in the shed turn it on.

The Feeding Mechanism

The material to be cut is fed in by toothed rollers, the top ones being held down by a counterweight on the end of a lever. Make sure this is attached and functional as it is needed to hold the herb firmly for feeding it in at a steady rate. Up and down movement of the top roller is necessary to accommodate varying heights of material being chopped.

The feeding rate (the rotation of the rollers in relation to the rotation of the blade wheel) is adjustable on some machines by changing over some gears. This rate determines the length of the pieces cut.

Operating the Chaff-cutter

First dump the herbs off the screens into the boat. Check that the machine is clean before turning it on, then start it up and feed the herb in at a steady rate. The knack is to overlap the bunches so the feed mechanism is kept fairly full. If there are low spots in the herb that is being fed in, the rollers will drop down and it may be difficult to force a new bunch of herb in. You also need to keep an ear tuned to the motor to make sure you are not overloading it.

When you have finished cutting, turn the machine off at the wall, remove the guards and brush out all the dust and pieces of herb.

Tea-grade Chopping

A few herbs, for instance Oats and Lemon Grass, need to be chopped with the chaff-cutter to get a tea grade. Being grasses, they have leaves that are too tough to break up and pass through the screen. Others, such as Vervain, have only a small proportion of leaf and consist mostly of fine stems that serve the function of leaves.

Some (but not all) of these herbs will dry better and produce a better quality product if they are put through the chaff-cutter in the fresh green state before being put on the screens to dry. This is because they dry faster when their stems have been cut, and because the chaff-cutter cuts the fresh herb more cleanly, so there are fewer long raggedy pieces.

To minimise bruising, chaff cutting should only be done when the herbs are in a fresh unwilted state. It is important to first assess the effect of the cutting on the particular herb to ascertain whether chopping it fresh will cause bruising and discoloration during drying: chopping some herbs in the fresh state turns out to be a disaster.

There are also a few herbs, such as Parsley, which are rubbed when dry but which need preliminary chopping when fresh to promote their drying, as they are too slow to dry otherwise. Fortunately Parsley is not subject to bruising and darkening.

The temptation to process other herbs to tea grade with the chaff-cutter should be resisted, as it will result in an inferior product with a high proportion of stem.

Hammermilling

There is not a huge amount of skill involved in operating the hammermill, though it takes a bit of endurance. It is just a matter of pouring or pushing the herb into the mill at a steady rate without overloading it, and taking the usual precautions against accidents.

Protection

Wear appropriate equipment to protect your lungs, ears and eyes. In particular beware of particles flying back in your face as you feed the herb into the mill.

When I was wholesaling, I found hammermilling the least enjoyable task in herb processing, especially in hot weather. The work was carried out in a small enclosed room with the hammermill and the extraction fan both roaring and I was burdened with earmuffs, respirator, goggles, overalls etc., feeding the voracious machine and trying to get the job done as quickly as possible so I could get out of there and get all the protective gear off.

Moisture Content

What has been said under 'Chaff-cutting' regarding moisture content applies here as well. The hammermill will tear into anything regardless of moisture content, so make sure your herb is thoroughly dry before milling it.

Stripping Leaves

A few leaf herbs are sometimes better processed by stripping, in particular Bay leaves, Lemon Verbena and Mugwort (for tea grade). Mugwort has a peculiar woolly texture to its leaves, which felt together and simply refuse to go through the rubbing screen, no matter how dry they are. Lemon Verbena looks more attractive if the leaves are whole or in large pieces. Bay leaves need to be whole.

Stripping is a bit easier with some herbs if it can be done over a chicken-wire screen with a mesh of 40–50 mm so, as the leaves are stripped they can fall through out of the way.

Alternatively, stripping of leaves can be done in a large tray, such as the 'boat' described above for use with the chaff-cutter.

It is important that the leaves be dry enough to break easily from the stems. If you apply enough force you can strip damp leaves, but this will lead to problems in storage.

One method is to strip the leaves by running a gloved hand up and down the stems so the leaves are broken off. This tends to be rather slow because each stem has to be picked up and stripped individually.

Another method, which goes a bit faster, is to vigorously agitate a tray full of the herb so that the pieces are rubbed against each other, knocking the leaves off. Sometimes I take one of the larger stems and use it to stir and beat the other stems until they are mostly freed of their leaves. Care must be taken to minimise shattering of the stems and leaves.

As they are denuded, the stems can be picked out and discarded A few stubborn ones may need a bit more help to strip them off. Then the whole heap must be gone through, carefully to remove all the pieces of stem, large and small.

This system is only efficient for herbs with strong tough stems, otherwise picking through to get all the fine and broken pieces of stem is just too tedious. Even with tough stemmed herbs, picking out stems is usually slower than the normal rubbing process for other herbs.

Soil Contamination

Sources of Contamination

Soil contamination is a common problem with leaf crops. It can occur in various ways but soil splash and poor harvesting procedures are the most common causes.

Soil Splash

Soil splash occurs when soil is splashed up onto the leaves of the plants by rain or irrigation. It tends to be worse in row crops harvested close to the ground, especially in a crop's first season of production. Heavy rain or large droplet irrigation make it worse and some soils splash more than others.

Splashed soil then dries onto the leaf and is liable to be picked up in harvesting. Some species with rough or furry leaves retain more soil than others.

Harvesting and Drying Procedures

Lumps of soil are easily picked up when cutting close to the ground, or if cut herb is swept or raked up.

Walking on or close to a harvesting sheet spread out on the ground can deposit or flick soil onto it, which if not removed will end up in the finished product.

When spreading herb onto the screens, it should be lifted from the sheets rather than dumped from them, because heavier soil particles (and other unwanted debris) tend to settle to the bottom of the sheet, allowing them to be separated.

Acceptable Limits

In a quality product there should be no visible soil contamination. As soil particles are heavier than dried leaf, they will tend to settle out and show up as a thin layer at the bottom of the bag after it has been moved about a few times. If any soil contamination is evident, it may result in the product being downgraded or rejected.

Concentration of Soil Contamination

Because the soil particles have a tendency to settle out to the bottom of a container, a serious contamination problem can occur if, for example, small packets are filled from this container. The first packets will be fine, but towards the bottom of the barrel the last few packets will contain a significant proportion of soil. I have come across imported herb that has been as high as 30% fine sand (dry weight) as a result of this settling process.

Removing Soil Contamination

It is not going to be possible to prevent all soil contamination in the field and, even with careful management, some crops will inevitably carry a small amount of soil through to the harvested stage.

There are basically two options for removing this: either wash the harvested crop before drying, or separate the soil after drying and processing.

Washing

If the crop contains lumps of soil or is heavily soil splashed, washing before drying may be the only option. Immersion and agitation is usually easier than trying to hose off leaves and causes less damage.

A bathtub or a small tank full of water will be necessary. The herb should be dumped in and stirred around sufficiently to loosen and separate the soil particles, which will settle to the bottom. The herb should be allowed to drain for a while before being loaded onto drying screens.

A disadvantage of this method is that quite a lot of water may still be on the herb, especially with fine-leaved species and this could interfere with drying.

Sifting after Drying

If the soil is all in fine particles that are less than 1 mm across, it may be easier to separate it by sifting out the fines after drying and processing.

The herb can be sifted by passing it over a fine mesh screen, about 20 dent (20 meshes per 25 mm) with apertures of about 1 mm, which will allow the fine herb and soil particles to pass through. These can be discarded or it may be possible to separate the soil from them by agitation, gold-panning style, to make it settle, in which case the cleaned fines could be recombined with the herb.

With larger volumes of herb, it may be easier to pass it over fly-screen sized mesh first (preferably not previously walked on by flies). The portion that passes through could then be sifted through the fine screen described above.

Winnowing

Winnowing may be a possibility if the particles are heavy enough to separate satisfactorily.

PROCESSING FLOWER HERBS

Most flowers don't need processing beyond picking out the odd stalk, weed head and other trash from among them once they are dry.

Flowers that are in umbels are rubbed through a screen, usually 2½ or 3 dent. With Elder flower the fine umbel stalks that break up easily are included in the herb, but coarser stalks are separated.

PROCESSING ROOT HERBS

Milling

If the roots are in short enough pieces for bagging, storing and shipping, normally the grower does not have to do any processing. The wholesaler or manufacturer mills the roots to their own requirements.

If processing is required, most roots are simply run through a hammermill with the appropriate size mesh screen in place.

Rubbing

Rubbing of roots really only applies to preparing tea-grade Valerian, the roots of which are rubbed through a 2½ dent screen when dry. The aim is to break up the fine roots and sift out the crowns. These crowns can then be hammer-milled and recombined with the fine roots.

Removing Soil Contamination

A certain amount or residual soil that escaped washing will usually separate and fall from the roots in drying. Careful lifting of the roots from the drying screens can often avoid picking this soil up. Otherwise it should be separated and discarded, either by sifting with a screen or by shaking the herb so the heavier soil particles settle to the bottom.

PROCESSING FRUIT HERBS

Cleaning Fruits

Most fruits used as herbs are easier to harvest if you don't have to worry about the leaves too much. Herbs such as Rose hips and Hawthorn berries can be harvested with a comb or stripped off the bushes and the mixture of leaf and fruit spread on the screens to dry.

When the fruit has dried quite hard and the leaf shatters, it can be separated by rubbing the mixture all over a screen that has a mesh just small enough to contain the fruits and allow the particles of leaf and other debris to pass through. If there is a lot of larger trash and big stalks among the fruit, it may be advisable to first put the fruits through a coarser screen to break up the bunches and sift out the coarse trash. This will make subsequent rubbing over the finer screen easier.

The rubbing action used for this process is fairly aggressive, in that you are endeavouring to force the leaf and stalk through the screen. However, care still must be taken not to damage the fruits too much or break them up.

After screening, there is usually some slow hand-picking of residual stalks and trash to finish off. If seed cleaning equipment is available, this may be a quicker option, provided there is no possibility of contamination with residues from non-organic produce previously put through it, as there are cases of this occurring and organic products being rejected.

The fruits may retain their individual stems and pieces of calyx, but coarser stalks should be largely removed: how clean the end product needs to be depends on the final use. Usually if it is to be milled, a small proportion of stalk material, less than 1%, is acceptable. If it is for packaging in teas as whole fruit, the limit would be lower, probably in the range of 0.25–0.5%.

PROCESSING BARK HERBS

Bark herbs may need breaking up or chaff-cutting to reduce their bulk sufficiently to allow them to be packed into bags.

They will usually need to be milled before their final use, but the grower does not normally have to do this.

QUALITY ASSESSMENT

When assessing the quality of herbs there are a number of different aspects that need to be taken into account and in many of these there are no industry-wide standards. It is curious that this is an area that has been overlooked by much of the trade for a very long time.

This has resulted in the industry being dominated by poor to mediocre quality herbs, which has been to its general detriment. One of the factors contributing to the demise of herbal medicine early this century was the unreliability of herbal medicines due to their variable quality.

With the current surge of interest in the use of herbs, there is now an opportunity to establish guidelines for assessing quality. At present quality standards are varied and many aspects of quality are commonly overlooked. Nevertheless, standards for herbs are evolving and in some parts of the trade close attention is paid to quality.

Important Aspects of Quality

Correct Species and Identification

It is generally recognised that the accepted species – correctly identified – must be used. The species used in the trade are more or less standardised and most buyers do try to ascertain that herbs are correctly identified.

Some rely on an organoleptic assessment based on appearance, taste and aroma, comparing the herb with a known sample. Others perform thin layer chromatography or other analytical tests on the herb to confirm identification.

The grower is in a position to obtain positive botanical identification of the herb being produced. It is important that this be done because propagation material is not uncommonly incorrectly identified (see 'Identification' in the Chapter 5 on Propagation and Planting).

Freedom from Contamination

Various kinds of contamination are possible, and acceptable levels are relatively easy to quantify.

Foreign Plant Material

Foreign plant material includes weeds, other herbs and other parts of the harvested plant. Acceptable limits range from around 0.25% to 2%, depending on the level of quality and the buyer's standard.

Other Foreign Material

Officially the acceptable levels for soil, stones, dung, cigarette butts etc. is zero, but these items do show up not infrequently in poorer quality herbs.

Microbial Levels

Herbs intended for use as herbal medicines get a lot of attention with respect to their microbial levels. *Therapeutic Goods Act* regulations specify maximum acceptable microbial levels for herbal medicines.

Contamination with uncomposted animal manures, especially chicken manure, is a common cause of excessive levels of microbes in dried herbs. Soil, insect droppings and mould are other possible contaminants (see under *Therapeutic Goods Act 1991* in Chapter 14, 'Marketing and the Economics of Herb Growing').

Chemical Residues

For conventionally grown herbs the maximum residue levels (MRL) permissible are generally the same as apply to foods, as set by National Health & Medical Research Council. For certified organic herbs, the permissible levels are 2% to 5% of these levels.

Dryness

The generally accepted level of dryness required for safe storage of herbs is 10% moisture content, which allows a small margin of safety. It is surprising how much herb sold in the trade is not sufficiently dry, sometimes having a moisture content as high as 20%. This can be a result of inadequate drying or of poor packaging, which allows the herb to reabsorb moisture from the air.

Testing for Dryness

By Feel: The dryness of most herbs can be adequately assessed by their feel. The herb should feel crisp and the particles should fracture easily if rubbed in the hand, or else the herb should be too hard to easily dent or bend.

If the herb feels at all damp, or if the pieces have a limp or leathery feel about them and bend without breaking, there is probably too much moisture.

100°C Drying Test: A more precise test can be performed if very accurate scales are available. First chop 50 g or so of the herb into small particles (6 mm across), weigh it and dry it at 100°C, with adequate air exchange, for 30 minutes.

This can be done in a very slow oven or an electric frying pan on a low setting. A thermometer should be used to monitor the temperature. Take care not to lose any of the sample.

After 30 minutes reweigh the sample. Dry for a further 30 minutes and check the weight again. If there is no longer any loss of weight, the sample is down to 0% moisture. Otherwise continue drying and checking until there is no more weight loss.

The difference between the initial weight and the final 0% moisture weight is the amount of water in the herb being tested. This can be expressed as a proportion or percentage of the initial weight.

For example, if the sample weighed 50 g initially, and after an hour at 100°C its weight was a steady 42 g, it would have lost 8 g of water. This represents a moisture content of 8/50, that is 0.16 or 16%.

Overall Quality

The industry viewpoint becomes very diverse with respect to overall quality. While virtually everyone claims that the herbs they offer for sale are only of the highest quality, their assessment standards are very much coloured by the amount they are prepared to pay for their raw materials.

Where the quality of the herb used can be concealed, as with packaged or extracted products, the standards tend to be lower.

The 5-Star Assessment System

Two wholesalers of premium grade herbs in Australia use a 5-star rating system to assess the quality of the herbs they handle. To qualify as premium grade, herbs need to meet the following specifications:

1. Certified Organic or Biodynamic: This must be by a recognised certification body.

2. Good Colour and Vibrancy: The dried herb should look healthy, with no significant fading or discoloration, bearing in mind what is realistically achievable for that particular herb.

3. Good Flavour and Aroma: Flavour and aroma should be close to as good as can be obtained for each species in the dried form. Naturally there is going to be some seasonal and regional variation. There should not be any significant loss or lack of flavour due to the growing and drying conditions or the use of an inferior variety

4. No Significant Discoloured Leaf or Impurities and No Excess of Stem: While inevitably there will be a very small proportion of discoloured and faded leaf in most herbs, this should be an absolute minimum. The proportion of stem included should not exceed the buyer's standard for the form of that particular herb.

5. Correct Particle Size: The herb should be processed to a form appropriate for its intended use.

Premium Grade

If the herb meets the above standards, it is classed as premium grade. This is referred to as 5-star quality. If, for instance, it fails to meet the standard on one point, it will get a 4-star rating. Where this shortcoming is of a minor nature, the herb might sell for a slightly discounted premium price.

Second-grade Organic

Most of the organically grown herbs being imported into Australia rate only between 2-star and 4-star, because of discoloration, excess stem and/or lack of taste and aroma. Nevertheless, this is regarded as an acceptable standard by most manufacturers and some wholesalers.

Trade Quality

Most trade herbs would be lucky to rate 1 star. They usually are way below the standards that can be achieved. If they do make the grade on anything, it will be for correct particle size.

Levels of Active Constituents

Some manufacturers set standards for levels of known active constituents in some herbs. This is only done to a limited degree because in most herbs not all the active constituents are known.

Smaller wholesalers and manufacturers generally rely on organoleptic assessment, as used for the 5-star system. When 5-star quality herbs have been assessed for levels of active constituents, they generally rank very highly.

Therapeutic Effectiveness

Over the years there has been a lot of very positive feedback from herbalists as to the effectiveness of premium quality herbs. In some cases the action is seen to be stronger than for lower quality herbs, while in others the premium quality has been found to have therapeutic effects that the herbalist has been unable to obtain with the lower quality herb at any dose.

It would be very interesting to set up studies to see if this can be repeated in clinical trials.

The Need to Develop Overall Standards

At present, standards are fairly diverse and assessment of many aspects of quality is rather subjective and often influenced by monetary considerations.

There is a need to establish quality standards and to set up quality assurance programs. Work is already in progress in this direction and it is hoped that in a few years it will be possible to be a lot more definite on this subject. We need to be able put quality in more quantifiable terms so assessment can be carried out more objectively.

PACKING AND STORAGE

Once the herb has been processed into its final form, it needs to be protected so that it will retain its quality during storage and shipping. It also needs to be adequately labelled and treated (organically) for insect control during storage.

Storage Containers

To preserve their quality, dried herbs need to be kept dry and away from direct light, particularly sunlight and fluorescent light. Cool storage temperatures are preferable.

The herbs also need to be protected from vermin and insect infestation.

Quite a variety of materials and containers are used for storing herbs in the trade, but unfortunately it is quite common for unsuitable storage containers to be used, with resulting deterioration of the product.

Permeable Packaging Materials

Paper, hessian and unlined woven polypropylene bags, unlined cartons and fibre drums are often used for packing dried herbs, but they all have the problem of allowing some free movement of air and water vapour, so the contents might be gradually remoistened in damp weather. There is also the possibility of contamination or damage in shipment: you never know what else may be carted (and perhaps spilled) in the same truck, how the herbs will be handled, or whether they will be kept under cover.

Polythene Plastic Bags

For storage of herbs, it is hard to find a better container than a clear plastic bag. While many of us have objections to plastic, there is not much else that is strong enough and impermeable enough to do the job. It also has the advantage of being transparent, so you can see the herb inside. This helps avoid errors in labelling and other mistakes. It also facilitates assessment of quality.

Most growers pack bulk herbs in plastic bags stored in drums or in a dark, vermin-proof storage room. For storage and careful handling, plastic bags alone are usually adequate, either double bagging with 50 micron bags or a single 100 micron bag. Heavier herbs and those with hard sharp points, such as some roots, may need something a bit stronger, or a woven poly outer bag.

Size: A bag 1000 mm x 600 mm is a good size for packing 5–8 kg of leaf or flower herbs, and up to 20 kg of the denser root herbs. A plastic bag this size is difficult to obtain off the shelf: you usually have to order a large quantity to be made up. One herb wholesaler has bulk quantities made up and supplies smaller quantities to its growers. Smaller bags might be appropriate for small batches, but they are much slower to fill and create more handling.

For shipping, plastic bags need to be protected with something tough enough to endure being knocked around a bit.

Woven Polypropylene Outer Bags

Woven polypropylene is usually quite satisfactory for shipping herbs, as long as it has a good plastic liner, because it is not waterproof. A bag made of impermeable woven poly is available, but it can still leak at the sewn seams.

An ordinary stock feed bag is a good size: this is around 1000 x 600 mm.

The plastic bag used for a liner should not be smaller than the outer bag, because if it is compressed, the woven outer won't be supporting the smaller liner which could burst.

A thickness of 50 microns is usually adequate as a liner for a woven polypropylene bag, though this should be doubled for trips across Bass Strait or other long distances.

Used Bags

Used bags are not an option for most growers because of possible contamination from non-organic produce and the need for bags to be sound.

If you are using second-hand bags, make sure the old labels have been removed or crossed out.

Drums

Good second-hand drums for storage or shipping, with open end lids, a rubber seal and a hoop clamp can be obtained from food and fruit juice importers. Drums that have been used for food products are easy to thoroughly clean. Drums used for industrial chemicals may turn out to be difficult to clean and could pose contamination problems.

If you are producing large volumes of individual herbs, it is possible to obtain large polythene drum liners, so you can store the herb loose in 200 L drums. This is only practical if you are going to handle or ship your product in these drums.

Boxes

Cardboard boxes are another possibility, provided they have a plastic liner. They have the advantage of stacking closely for storage and shipping. Their disadvantages are that they do not resist damp conditions very well and need protection from vermin.

Cardboard boxes rely on tight packing of their contents for strength. If they are not completely filled or if the herb in them settles, they can collapse when a heavy weight is stacked on top of them.

Wool Bales

Use of wool bales is a possibility for storing and shipping large quantities of herb, as compressing the herb reduces bulk. Hops are commonly pressed into bales. The disadvantage of wool bales is that the herb is not adequately protected from becoming remoistened in humid conditions.

Putting a liner inside the bales would not work because of the pressing process. It may be possible to put a plastic liner over the packed bale of herb and then another woolpack over that to protect the liner from damage (it may be easier to put the three bags together before filling).

Ease of Inspection

Some buyers prefer their herbs to be supplied in containers which allow easy inspection so quality can be ascertained. Clear plastic has an advantage in that the herbs can be seen without having to open the bags.

Separation of Batches

The bags or containers used should be appropriate for the size of the batches produced.

While it is no problem for one batch to consist of several bags, if there are different batches or variations within one batch, these should be packed separately.

Length of Storage

One of the advantages of producing a storable product like dried herbs is that there is more flexibility in terms of marketing. If they can't be sold right away, they can be stored. If there is a surplus in a good year, it can be carried over to the next and the grower is not so much at the mercy of the market.

Normally growers can expect to have to store dried herbs for some period after harvest until needed by the buyers, be they wholesalers, retailers, manufacturers, herbalists or individuals. Most businesses operate on a limited trading stock, especially those that are going to pay the best prices, so the grower needs a capacity to store as much as a year's harvest.

In good storage conditions, most herbs will keep well for 2 years or so, but after that period a gradual deterioration in quality will become apparent. They fade a little, start to taste stale and their levels of active constituents diminish. If possible, I like to see herbs sold within a year, because they are likely to sit around on the wholesalers', retailers' and consumers' shelves for some time as well. If they are to be any good by the time they are used, they shouldn't wait too long before starting out in the distribution chain.

Old Stock

Rather than hold on to old stock that is not moving fast enough, it is better to clear it by offering it to some other buyer – even at trade prices – so you can keep your quality up for your regular customers. Some of them may have a use for cheaper old stock, too, but don't let it be used to lever the prices of your quality produce down to unrealistic levels.

Old stock that is never going to move is best consigned to the compost, where it can help in the production of new herbs.

Filling

Setting up the Bags

For filling bags it is useful to have some kind of holding frame so the bag is held open and stable while the herb is poured into it. If double bagging or using plastic liners in woven polypropylene bags, it is best to put these together before filling because it can be difficult and even disastrous to try to force an outer over a full bag.

Homogeneity

Make sure the product is more or less homogeneous before bagging it. This is not usually a problem with rubbed herbs, but can be with chaff-cut aerial parts, as the finer pieces of leaf and flower tend to settle and the coarser stem pieces come to the top. This is going to happen in the bag with handling anyway, but try to make the contents of each bag representative of the whole batch and not mostly stalk in some bags and mostly leaf in others.

Soil Contamination

When filling the bags, watch out for any soil contamination. This tends to settle to the bottom of the sheets or bins during processing and handling. If there is any amount visible in the bags after they are filled, the herb may need to be sifted to remove it – it tends to show up after they have been moved about a bit – see 'Soil Contamination' above.

Compressing

When loading bags, don't be afraid to push down hard on them to compress the herb and reduce its bulk. Leaves tend to be light and bulky, so packing them in tightly saves space in storage and shipping. I usually kneel on the top of the filled bag, top it up and compress it again to exclude as much air as is practical, and then tie it off.

Tying Bags

Bags should be tied off tightly with a piece of plastic bale twine. This should go twice around the neck of the bag, and then be cinched up firmly before tying a reef knot. The tie should be tight enough to prevent air flowing in and out of the bag, and to prevent moths from getting in. If a bag is only part full, it is better to simply tie a knot with the top of the bag, as this is quicker and it gets the loose end tucked out of the way.

Double bags of 50 micron polythene can be tied as one. The liner inside a woven polypropylene bag should be tied separately and the outer tied or stitched.

Labelling

When your production is small and of a limited range, it is easy to become a bit slack about labelling. Good labelling is important, however, not only to keep track of an increasing range and volume of production, but to convey essential information to your customers.

Good labels will help your stock management run more smoothly and will help avoid confusion and mix-ups, especially if you have other people working in your operation.

I prefer to use stickers rather than write on the bags. The stickers are easier to read and to replace, but most stickers will not adhere very well to woven polypropylene bags, so stick them on the liner inside and write brief details on the outer bag.

Each bag of herbs should carry vital relevant information about its contents, as indicated below.

1. Name of the Herb and Part

Use the name that is not going to be confused with anything else. Most of the English-speaking herb trade uses English names, but many herbalists like to use Latin names. Use of both together helps confirm identification for some customers, as both English and Latin names can vary and herbs often cross international frontiers. If you are supplying a particular wholesaler or manufacturer, it is best to use the names they use. Where different parts of a herb are in common use, such as Dandelion root and Dandelion leaf, the part should be indicated along with the name.

2. Form of the Herb

Tea grade, aerial parts or whatever can be abbreviated on the label, for instance:

AP	– Aerial Parts
CG	– Culinary Grade
FS	– First Screening
L & F	– Leaf & Flower
Lf	– Leaf
MG	– Manufacturing Grade
TG	– Tea Grade

3. Date of Harvest

The month and year at least should be noted on the label.

When filling bags, it helps to have some means of holding them open.

4. Batch Number

While adding a batch number to a label is not compulsory, it helps wholesalers and manufacturers who have to record a batch number to satisfy their paperwork requirements so any problems can be traced.

A batch number can be used that conveys quite a bit of useful information, while saving space on your label and time writing it.

Southern Light Herbs uses a system for batch numbers which consists of:

- three letters from the herb's English name (usually the first three letters);
- six digits, the first and second denoting the day of the month, the third and fourth denoting the month, and the fifth and sixth denoting the year;
- two or three letters, being the initials of the grower or growing enterprise.

Thus a Batch Number PEP040296BH signifies Peppermint harvested on the 4 February 1996 and grown by Beaufort Herbs.

Below is a list of the batch number codes used by Southern Light Herbs:

AGR	– Agrimony
ALF	– Alfalfa
ANE	– Anemone
ANL	– Angelica Leaf
ANR	– Angelica Root
BAS	– Basil
BAY	– Bay
BER	– Bergamot
BLE	– Blessed Thistle
BLF	– Blue Flag
BUR	– Burdock
CAL	– Calendula
CAT	– Catnip
CEL	– Celandine
CHM	– Chamomile, German
CHS	– Chasteberry
COL	– Coltsfoot
CML	– Comfrey Leaf
CMR	– Comfrey Root
DAL	– Dandelion Leaf
DAR	– Dandelion Root
DOC	– Dock, Yellow
EAN	– Echinacea angustifolia (Narrow-leaf)
EPU	– Echinacea purpurea (Broad-leaf)
ELD	– Elder
ELE	– Elecampane
EUC	– Eucalyptus
FEN	– Fennel
FEV	– Feverfew
FIG	– Figwort
GIP	– Gipsywort
GOA	– Goat's Rue
HWB	– Hawthorn Berries
HWL	– Hawthorn Leaf & Flower
HOP	– Hops
HBL	– Horehound, Black
HWH	– Horehound, White
HRA	– Horseradish
HTF	– Horsetail, Field
HYS	– Hyssop
LAD	– Lady's Mantle
LAV	– Lavender
LMG	– Lemon Grass
LMT	– Lemon Thyme
LMV	– Lemon Verbena
MAJ	– Marjoram
MSL	– Marshmallow Leaf
MSR	– Marshmallow Root
MEA	– Meadowsweet
MEL	– Melissa Balm
MOT	– Motherwort
MUG	– Mugwort
MUL	– Mullein
NTL	– Nettle
OAT	– Oats, Dried Green Plant
ORE	– Oregano
PRL	– Parsley Leaf
PRR	– Parsley Root
PAS	– Passionflower
PEN	– Pennyroyal
PEP	– Peppermint
PLA	– Plantain, Greater
RAS	– Raspberry
RCL	– Red Clover
RSB	– Rosehip, Sweet Briar
RSD	– Rosehip, Dog Rose
RSM	– Rosemary
RUE	– Rue
SAG	– Sage
SCU	– Scullcap
SPE	– Spearmint
STJ	– St John's Wort
TAR	– Tarragon, French
THY	– Thyme
VAL	– Valerian
VTH	– Variegated Thistle
VER	– Vervain
VIO	– Violet
WOB	– Wood Betony
WOR	– Wormwood
YAR	– Yarrow

5. Freeze Treatment

Indicate whether or not the herb has been through the freezer for insect control. If it has, this will ensure it won't start crawling and will reassure your wholesaler as well.

6. Grower's Name

Record your own name or the name of your enterprise.

7. Residential Address and Phone Number

Recording the address is a legal requirement, apart from which it is good promotion.

8. Certification Body, Certification Level and Grower Number

Your customers need to know the certification body, the certification level and your grower number. It adds credibility to your product if it is certified organic or bio-dynamic. Legislation in the pipeline will require that any produce claiming to be organic be certified by an approved certification scheme.

9. Weight

If you are putting the herb into the packaging in which it will be sold to the public or other end-users, the contents must be weighed on scales registered as legal for trade.

If you have an arrangement where the wholesaler or manufacturer does the final weighing, put an approximate weight and mark it as such if you don't have trade scales.

10. Quality

You may wish to put your own assessment of the quality of the herb on the label. This can be on a Q1 to Q5 scale as for the 5-star assessment system or you might choose to use some other scale.

Legal Requirements

If you are packaging herbs for sale to the public, you will need to include on your label items 1, 6, 7, 8 and 9 to meet legal requirements. If the product keeps for more than 2 years, you are not required to include an expiry date. Most dried herbs could be regarded as falling into that category, but you might prefer to add an expiry or a 'best before' date.

Herb	Part	Form
Quality Rating	Batch No./Harvest Date	
Weight kg	Freezer In Out	
Grower		
Address Ph.		
Certification Status Grower No.		

Example of a label layout for bulk herbs.

Recording

The grower needs to keep a record of the harvest and processing of each herb as it goes through the drying shed. For a small-scale one-person operation, it is really only necessary to record information that is not easy to keep in your memory, such as harvest dates and yields. However, if more than one person is involved in the operation, more detail needs to be recorded to enable good co-ordination and to avoid errors in processing, such as misidentification, accidental mixing of different herbs, and losing track of things in the drying shed.

This information can also prove very useful in later years in projecting yields, marketing, and general management of the herb-growing operation.

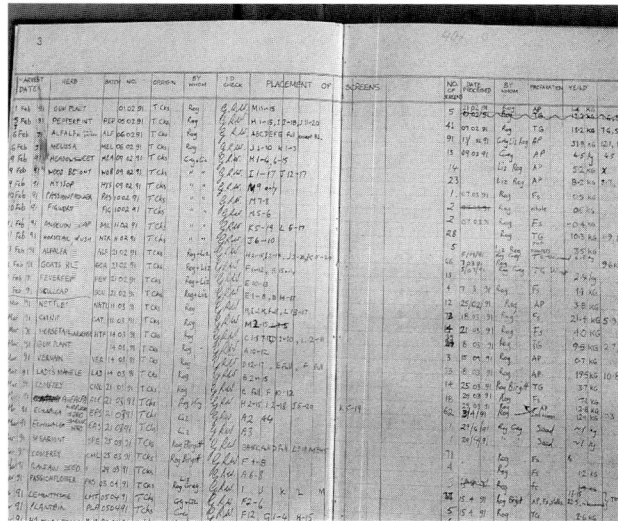

A harvest record book can help avoid errors and is useful for management and planning purposes.

A detailed harvest record book might contain such information as:
- Name and part of herb
- Date of harvest
- Origin of herb
- Batch number
- Person harvesting
- Person confirming identification
- Number and location of screens in drying
- Date of processing
- Person processing
- Form of herb
- Weight of processed herb
- Number of bags and their individual weights

Insect and Vermin Control in Storage

While your herbs are in storage awaiting the interest of your buyers, there may be a few others around that begin to take an interest in your products.

Vermin

Rats and sometimes mice find dried herbs quite attractive and their ravages can be devastating. Their activities are generally rather obvious, though, and can be dealt with and prevented with good construction methods, hygiene and diligence. Storage in 200 L drums is a good option for protecting processed herb from vermin.

Insects

Moths and weevils can be rather more insidious than rats and mice. They fly into your drying facility and lay their eggs on your herb where it is laid out on the screens. Because they look fairly insignificant in themselves and the eggs don't hatch for several weeks or more, the damage may not become evident for some weeks or months: perhaps not until you get a phone call from one of your customers complaining about grubs crawling around in their bags of herbs.

Upon inspection of your stock, to your horror you may find that many kilos of your precious hard-won produce has little grubs making webs all through it.

The culprit is a small brown-grey moth which flies around, usually in the warmer months, and is attracted to many herbs. It lays its eggs on them and these hatch and feed on the dried herb. The moth is usually introduced from food products in your house, where it typically attacks flour and other dried foods. Wholefood shops with many open containers commonly have a major problem with these larvae.

Many people refer to the larvae as weevils, but weevils are a type of beetle – some of which can be storage pests as well. I have come across real weevils in imported herbs occasionally, but have never had any problem with them in locally grown herbs.

False Alarms

A few herbs have a tendency to cling together as if they are full of moth larvae webs, but instead the clinging is caused by fine hairs on the pieces of leaf. Burdock root can also exhibit this tendency if the harvest includes some of the small furry leaves in the dormant crown. After Burdock root is milled, the fur from these leaves can cause clinging clumps of root particles that very much resemble the work of moth larvae and their webs.

Immune Herb Species

Some herbs never seem to be attacked by moths. Many very bitter herbs seem to be immune, as do Bay leaves.

Since having a few major outbreaks of moths, I have always freeze treated just about everything to ensure there are no problems.

Prevalence

The worst time of year appears to be spring and early summer. Flowers are most vulnerable, particularly Chamomile, but other herbs such as the Mints, Nettle and Alfalfa are common targets.

Storage in 200 L drums is an option for protecting processed herb from vermin.

You may find that for the first year or so you don't see the moths or their larvae, but eventually they will find their way to your drying shed They tend to breed and carry over in dust and residues that are inevitably left in nooks and crannies around the shed.

Freeze Treatment

A simple reliable remedy which will prevent moth larvae infestation is freeze treatment before the larvae hatch. After the herb is processed and bagged, simply put it in the deep freeze for 4 days. This period ensures that the cold penetrates a 1000 x 600 mm bag. A longer period may be required for larger bags, but 3 days is adequate for bags that are half this size.

After freezing, the bags are taken out, allowed to warm up, and when the condensation has evaporated off them, they are put into storage. Try to avoid opening the bag while it is still cold, especially in damp weather, because the condensation will remoisten the herb.

As long as the bags are then kept closed and there are no holes in them, the moths cannot reinfest them, as they cannot penetrate the plastic. The larvae, however, can chew through plastic, which is why you often see little holes in bags of infested herb. This is always the work of the larvae chewing their way out and not the adult moths chewing their way in.

This freezing treatment has proved very reliable, but you do need access to grid power (or a friend who has access to it) and you do need plenty of room in the freezer to cope with bags and bags of herbs. I found that 0.4 ha producing around 1000 kg of herb a year needed about 80% of the capacity of an average chest freezer through most of the harvesting season. At times quite a pile would be waiting to go through the freezer. The larvae usually don't become active for several weeks, so it is okay if there is a bit of a delay before freezing.

Carbon Dioxide Treatment

An alternative treatment to freezing developed by biodynamic grain growers is carbon dioxide fumigation. This is quite safe and can be done in a silo or even in a 200 L drum. A quantity of dry ice (frozen carbon dioxide) is placed in the container full of grain or herb and the lid is placed on, allowing a small escape at the top for the air that is displaced as the heavier carbon dioxide sublimates and occupies the container. This container is then left sealed for 10 days to ensure that any insects and their eggs are suffocated

It may be possible to use carbon dioxide to fumigate herbs in open bags inside a sealed room or in containers, but I don't know if this has been tried.

The Conventional Approach

The conventional approach to guarding against insect infestations is irradiation or fumigation with ethylene oxide or methyl bromide. The latter is not only harmful to all living organisms, but it affects the active constituents of some herbs and depletes ozone as well.

Cool Storage

Cool storage will prevent hatching and development of the larvae, but it does not actually kill the eggs.

Treatment of Larvae-infested Herb

If your herb already has larvae crawling around in it, you may be able to salvage it by removing the larvae and their webs by hand-picking and/or sifting and then treating the remaining herb in the freezer.

Shipping

Packaging

Herbs going via a commercial carrier need to be adequately packaged to ensure they will survive the trip. Bulk quantities sufficient to make up a sack can be shipped in a woven polypropylene bag with a sound plastic liner. Make sure that the liner is waterproof as the bag may be left out in the weather on a wharf or on the back of a truck somewhere. Double lining is probably good insurance. The bag should be well tied or stitched at the top – you don't need to invest in a fancy bag stitcher for a small-scale operation – lacing up the bags with a bag needle and twine is quite adequate, but take care not to puncture the liner.

Shipping in 200 L drums is an alternative if you have a source of good clean second-hand drums that are cheap enough or if you can get the empties back.

If drums are fitted with a polythene liner, your valuable herb is very safe from possible damage in transit. Four 200 L drums will stack onto a pallet.

Cardboard boxes might be another option for shipping.

Palletising is a good practice if shipping in volume: it makes freight cheaper and reduces the possibility of damage or loss in transit.

Freight Costs

It is best to ascertain in advance who is covering the freight costs. There can be enormous variation in charges from one company to another and even by the same company to different customers, so it is worth asking around. Your buyer may be able to get a better price from their end than you can from yours, or it may even pay to ship to an intermediate point and have the buyer pick the herb up from there.

Labelling for Shipping

Make sure each bag is clearly labelled with the name and address of the receiver and any other details required by the carrier.

Enclose a packing slip or invoice (and keep a copy) so your buyer can check that everything is there.

Accounts

A good accounting system is valuable, too: some herb buyers are very reliable payers, some are a bit slack but always pay, while a few manage their cash flow by delaying payments as long as possible.

14

Marketing and the Economics of Herb Growing

MARKETING

The reality of growing herbs commercially is that unless it is worth your while, you can't afford to keep doing it. Of course you may measure the value of your returns in more than simply monetary terms, but the financial aspect does come into the equation in most people's view. Growers will find they need to get involved in marketing to some degree.

An Overview of the Australian Herb Market

First it helps to have an overall picture of the distribution and marketing of herbs in Australia.

The grower needs to understand that the industry here is still in its early developmental stage. With changes constantly occuring, it is difficult to predict how markets will evolve. This chapter was written in 1996, in a few years time the situation could be quite different.

Trade Herbs

More than 90% of the herbs used in Australia are imported, mainly from places where they can be produced cheaply, such as Eastern Europe and third-world countries.

Some are grown and dried in large industrial operations with a heavy input of chemicals and little regard for quality. Others are grown where labour is cheap. In some countries children harvest wild herbs to help boost their families' meagre incomes.

These herbs may be traded through a number of other countries before they find their way to Australia. They usually pass through the hands of a small number of importing firms, which bring in large volumes to be distributed to manufacturers and wholesalers.

Herbs from these sources are referred to as trade herbs. They are generally cheap and available in large quantities. Many firms are quite happy with this arrangement, as it provides them with a steady supply of low-cost raw materials. However, there is a growing body of consumers, herbalists and retailers, plus a few discerning manufacturers and wholesalers, who are not satisfied with the quality of these trade herbs.

First, they are often of uncertain origin. While they may have a 'certificate' to ensure the herbs are what they are said to be, there are still major problems with contamination, adulteration and general poor quality.

What many people do not realise is that most trade herbs are grown with the usual barrage of chemical fertilisers, pesticides and herbicides used in conventional agriculture. They don't come from some pristine alpine meadow at all.

They are often grown in environments contaminated with industrial pollution, agricultural chemicals and Chernobyl fallout. Pesticides no longer permitted in Western countries can still be freely used in many Eastern European and third-world countries.

Analyses of herbs have sometimes revealed levels of chemical residues above the acceptable limits for foods, but authorities tend to ignore this because herbs are regarded as only constituting a small part of the average diet. When shipments of herbs have been found with excessive levels of radiation, the importer has been allowed to sell the herb after diluting it with sufficient uncontaminated herb to reduce the overall radiation level.

Having a certificate is no guarantee that the herbs have been harvested at the optimum time. Often they are obviously overmature, inadequately dried, dried at excessive temperatures, or stored in permeable containers in humid conditions to then reabsorb moisture.

Processing often leaves much to be desired. Excessive proportions of stem are commonly included in the herb, and it is often finely milled, no doubt to disguise the stems and other shortcomings.

It is normal practice for trade herbs to be subjected to fumigation and/or irradiation to control insects and reduce bacterial counts.

Trade herbs usually taste rather stale. There is no way of knowing how old they are, and they may have been in storage for years.

The result is a herb of poor to mediocre quality: lacking in colour, flavour and active constituents.

One might ask how businesses can survive marketing such poor quality products. There are a number of possible explanations.

Some may be genuinely unaware, having never seen a high quality herb.

Others are not too worried about quality because they manufacture and market products such as tablets, capsules, tea bags, tinctures, extracts, ointments etc. in which the quality (or lack thereof) of the starting materials is not readily discernible in the finished product.

Commonly, the people who make the decisions in this market chain don't actually get to see, smell and taste the herbs they are dealing in. All they do is handle the paperwork: price lists, orders, certificates, quarantine clearance, inventories, invoices etc. Being driven by money, their decisions are primarily based on price per kilogram (or per tonne), because that is the only aspect of the herbs they understand, so naturally they buy the cheapest.

Until about 15 years ago, trade quality was all that was available in the way of herbs in Australia, but then a few people began to introduce some better quality products onto the market.

Organically Grown Herbs

Since the early 1980s a few growers and enterprises have been producing or importing organically grown herbs for use as teas or culinary herbs, or for manufacturing liquid medicines and ointments. For the most part these are organically grown or wildcrafted on a relatively small scale in Australia, New Zealand, Western Europe or North America, but increasing volumes are coming from third-world countries.

Generally, established wholesalers and manufacturers have been reluctant to get involved in handling organically grown herbs unless these can be bought for the same price or cheaper than trade herbs, which is usually not the case. Consequently, we have seen the establishment of a number of enterprises specialising in products based on organically grown herbs Over the past 10 years, this has been the fastest growing sector of the industry.

Most of the organically grown herbs used in Australia are still imported and for the most part they are only second grade. While they are a notch above the general run of trade herbs, most of them are not the quality they could be. Out of a possible 5-star rating used to assess premium grade herbs, they usually rank 2 to 4 stars, being downgraded due to discoloration, excess stalk, or lack of taste and aroma.

Premium-grade Organically Grown Herbs

A number of Australian organic growers have found that there is a ready local market for their premium quality herbs. These stand out because of their brighter colour, superior flavour and aroma, and their greater effectiveness when used medicinally. They are mainly produced by small-scale operations where they are grown and dried under optimum conditions. Premium-grade herbs can command a higher price because of their higher quality and higher cost of production.

This category is a grade above the standard of other organically grown herbs available on the market. They are generally distributed by a few small wholesalers or, in some cases, by the growers themselves. The biggest demand in this area is for the more popular tea herbs with a smaller demand for culinary and medicinal herbs.

The Potential for Import Replacement

There is not much scope for competing with imported trade herbs on a price basis as our costs of production here are too high to be competitive on a world market. Two industries in which this has been attempted, Hops and Tobacco, are in dire straits because they have not been able to compete with overseas prices. (Some would say the Tobacco industry deserves to die out as it has been killing off too many of its consumers.)

Consequently Australian herb growers need to focus on obtaining prices that make their operations viable. This can be in the production of quality organically grown herbs.

Export Markets

Every now and again one hears talk of 'lucrative export markets' that are just waiting for Australian herb growers to exploit. So far very little has happened in this direction.

The fact is, prices for most trade and second-grade organic herbs overseas are lower than here and the cost of freight has to be subtracted from this. The question is whether Australian growers can produce these herbs cheaply enough to compete in overseas markets.

On the other hand, there may be a potential market for premium quality lines as consumer patterns in many countries must be similar to those in Australia. The penetration of overseas distribution networks presents a major hurdle. It is much harder to do this from a long distance, and unless your volume of production and/or profit margin is substantial, costs will be prohibitive.

At present, the best opportunities are still in supplying the local herb market.

The Nature of the Market

Most free markets are in a dynamic state. Variable forces are continually in action and the situation is always changing. On the 'level playing field' it usually works out that the big boys control the game.

Fortunately there are sectors of the herb trade relatively free of the less savoury goings on. If you can manage to stay clear of the big players and concentrate on the small manufacturers and wholesalers, whole food shops, herbalists etc., you generally find yourself dealing with people of some integrity.

In 10 years of wholesaling with a turnover of up to $80 000 a year, my bad debts for the period would not have exceeded $300 and this was during the 1980s when many small businesses went through difficult times.

When I started producing dried herbs in the late 1970s, none of the larger firms was very interested in sourcing their dried herbs in Australia and nobody was dealing in organically grown herbs. In 15 years this has changed significantly and now even some of the larger firms are making moves to secure local supplies.

Events in Eastern Europe have no doubt had an effect on the trade and there have been problems with supply from this region. Word is spreading among consumers about the need for quality. Locally produced herbs are in demand and are gaining a good reputation.

We now see herbal products advertised as 'being made from Australian grown ingredients wherever possible'. However, this usually means 'where it is possible to buy them as cheaply as imported herbs' or in other words, not very often at all!

Buyers of Herbs Grown in Australia

Today, Australian small- and intermediate-scale organic dried herb producers have several options for marketing and some decisions regarding marketing need to be made fairly early in the planning stage.

When I started out producing dried herbs, doing my own distribution was the only way of selling them. In those days there was no established marketing chain for high quality dried herbs. Now there are a number of other avenues open to herb growers.

Large Wholesalers and Manufacturers

To interest the big players you need to be able to produce large volumes of herbs on a consistent basis. As decision-making in these firms is money driven, they want their raw materials as cheap as possible. If you can produce the goods at a price that suits them, they can be a market for large volumes of herbs, but as soon as they find a cheaper source they are likely to drop your price or your product. Firm contracts are essential when dealing with these enterprises.

To produce the volume and achieve an economy of scale, the grower supplying these large firms will need to concentrate on large areas of one crop. This can mean a simpler specialised operation, but it can also mean the grower loses the benefits of diversity and is more vulnerable to crop failures and market forces.

In this sector there is a risk of getting sucked into the downward spiral that has caught most of modern agriculture as farmers struggle to maintain their incomes in the face of rising input costs and falling prices for their produce. Unable to apply any united leverage to obtain realistic prices, farmers endeavour to make ends meet by increasing production. This results in over-exploiting the land and a further depression of prices in an over-supply situation.

Smaller Wholesalers and Manufacturers

A number of small wholesalers and manufacturers are dedicated to supplying high quality herbal products and are willing to pay a higher price for organically grown dried herbs. Some will pay a further bonus for premium grade.

The volumes these businesses handle are smaller than the large manufacturers, so their suppliers need to grow a more diverse range of crops. This means that one grower can be producing several crops that differ in their requirements and timing of harvest. Workload and use of facilities can be spread over a larger part of the year, reducing costs, work pressure and risks.

Volumes required are still large enough to gain a reasonable economy of scale, but generally not so large that if a sale falls through the quantity will be impossible to sell elsewhere.

Growers will generally have to be willing to hold onto stocks or wait a while for payments, as these smaller enterprises do not have the facilities or finances to carry a lot of stock. In business relationships on this scale, human qualities can prevail a little more and people tend to look after each other.

Doing Your Own Distribution

Another option a number of growers have taken up is to do their own distribution. As the mark-ups in the herb trade are quite high, the increased return can be attractive. Many whole-food retailers like to deal directly with growers, as it helps make a more meaningful connection between the producer and the consumer.

You need to enjoy selling your produce and be able to establish a rapport with people. I did not find this part particularly difficult as high quality herbs sell themselves to a large extent.

The challenging side of doing your own distribution is that you have to handle two quite different operations at once. You may find your capabilities stretched to meet the demands of both.

To do your own distribution, you need to have a broad range of products to command a big enough niche in the marketplace. Maintaining year-round continuity of supply is important, which means ensuring that you have adequate quantities of all your range growing and being harvested.

You also need a place where you can safely store (and find when needed) almost a year's supply of everything, and a packing room which will satisfy food regulations. Your scales must be registered for trade and must be inspected every year.

Joint Marketing

A variation on the above theme is joint marketing, in which a group of growers market their products jointly, as a cooperative or some other form of association.

Value Adding

Rather than sell your herbs in bulk, you may be able to add value by packing them in 20–50 g retail packs. If you are doing your own distribution, this is recommended, as it identifies your herbs to the consumer and, if they are good products, gain the consumer's loyalty.

Bulk selling tends to be anonymous. Your product may end up being sold under somebody else's label or, worse still, the jar labelled as your organically grown herbs may get refilled with some other rubbish!

Mail Order Retailing

Mail order is another possibility. While I used to do a small number of mail orders, I never focused on building up this side of the business, though I know of one grower who did establish quite a large mail-order clientele.

Marketing Organic Herbs

In looking at the market for organic herbs, it helps if they are broken down into four categories according to their uses: 'Manufacturing' (for making herbal medicines), 'Tea' (for infusions and decoctions), 'Culinary' (for use in cooking) and 'Other Uses'.

Although quite a few herbs do fall into more than one category, the market for each of these categories has its own characteristics.

For detail on the form and processing requirements these categories see 'Categories of Herbs According to their Uses' in Chapter 13, 'Processing'.

Manufacturing Herbs

Quality

For most manufacturers, the quality of herb required is not as high as one might expect. This is because the finished product does not clearly display the quality of the starting materials.

Price

The prices offered for locally grown manufacturing herbs in Australia are largely influenced by prices for imported organically grown or trade herbs. The bottom line is the landed price for the imported product.

Currently the bonus for quality in this category is limited, but at least one manufacturer is adopting standards for levels of active constituents in its raw materials, and paying a little more for produce that meets these standards.

If you are growing for this market, you have to be willing to accept the prices offered by manufacturers. Growers need to look at their yields and costs of production closely and do their sums before committing themselves to growing any manufacturing herbs. At the prices offered, the viability of growing them here varies greatly from one herb to another (a comparison of yields, prices and gross returns is included in the Information Tables at the end of Section 2).

Aerial Parts and Leaf Crops: Overall, the prices for most species of which the aerial parts are harvested represent a reasonable proposition for the small- to intermediate-scale organic grower. Some leaf herbs may be worth producing if the manufacturer will accept first screening that still contains a few stalks.

Root Crops: Manufacturing prices for roots are a bit of a mixed bag: species with small roots tend to be rather low priced for a manual operation. These are harvested and washed mechanically overseas, but would need to be grown on a fairly large scale for this to be viable here. Species with larger roots that don't require so much labour in growing, harvesting and washing, generally offer a somewhat better return to the small- to intermediate-scale grower.

Seed Crops: Prices for seeds are typically much too low for most growers to consider unless they are running a fairly large operation and have access to a header or threshing machine and seed cleaning facilities.

***Flower Crops*:** Manufacturing prices for flowers tend to be inadequate because overseas they are harvested mechanically or with very cheap labour.

***Bark Crops*:** Bark prices may be a viable proposition if the bark is thick enough on the trees or branches and if you are able to peel it easily.

***Fresh Herb Crops*:** Because fresh herbs cannot be sourced overseas, manufacturers are obliged to pay local growers a price sufficient to ensure continued production of herbs required in the fresh state.

Consumption and Demand

Quite large volumes – in the order of tonnes – of some herbs are being used in manufacturing in Australia. Consumption varies according to the species: demand for some may be only a few kilograms per annum. The overall turnover of this category is growing significantly as herbal medicine gains recognition in wider circles.

***Volatility*:** My experience with the market for manufacturing herbs is that it can sometimes be rather volatile and difficult to predict.

Herbal medicine tends to have its flavour-of-the-month, which causes sudden peaks in demand. Conversely a particular herb can go out of favour and its sales dry up, leaving manufacturers overstocked.

Sometimes when a herb is difficult to obtain, a manufacturer will order it from several suppliers. If they all come up with the goods, this can result in an oversupply, which takes a while to clear.

***Usage of Premium-grade Organic Herbs in Manufacturing*:** Very little premium-grade organic herb is used in manufacturing at present. In the recent past one company found there was a good demand for this grade of quality in herbal medicines. Unfortunately the company was not in a position to expand to a scale large enough to remain viable under the regulations of the *Therapeutic Goods Act 1991*.

I think it is just a matter of time before some manufacturers grasp the opportunity to get a share of the market by producing herbal medicines using premium-grade organic starting materials, as there are many people who would purchase such products. Before this can happen, though, there needs to be a continuity of supply of premium-grade raw materials available.

Tea Herbs

A wide range of herbs are used as infusions and decoctions. Many of them are popular beverages, while others are more obscure. Virtually all of them have medicinal properties, but provided they are not sold with any therapeutic claims, many are able to be classed as foods rather than as therapeutic goods.

This does not prevent the people buying them from using them medicinally, but that is seen as their own decision. In fact most of the premium-grade herbs sold this way are going to be more effective than the poor quality 'therapeutic' trade herbs which have been packed according to the procedures required by the regulations.

Quality

The quality of tea herbs currently offered on the market varies considerably. While premium-grade tea herbs will attract very good prices, much of the organic herb being produced overseas and here falls short in quality and would rate as second grade.

Premium-quality herbs have been able to penetrate the market for tea herbs much more readily than the other categories. (For comparisons of quality, see colour plate 8, illustrations 8.1, 8.2, 8.3, 8.4.)

Price

***Second-grade Organic*:** Prices for second-grade organic tea-grade herbs are more or less determined by the prices of the imported organic or trade products, which are not particularly inspiring unless production is on a very large scale.

***Premium-grade Organic*:** Tea is the category where premium quality herbs can command a significantly higher price especially if the superior attributes of the product are obvious to the consumer at the point of sale. It is the end user who is making the choice about which quality they are going to buy and the consistent pattern has been that many consumers prefer the premium-grade herb tea, regardless of some substantial price differences.

Consumption and Demand

My experience with the demand for premium-grade organic tea herbs is that it is more stable and predictable than other categories, particularly in the main selling lines. Some of the minor species may fluctuate a bit, but on the whole, demand has been quite steady.

Volumes of turnover are quite good for the main selling lines, but vary from moderate to very small for many of the others. Some are only a financial proposition if the there is a market for the same herb in one of the other categories, or if a high enough price is obtainable.

The overall turnover of organic tea herbs is smaller than for organic manufacturing herbs, but as the market for premium-grade teas has always been undersupplied, there is definitely potential for further growth.

Culinary Herbs

Culinary herbs are used in cooking, ostensibly for their flavour, but many have digestive properties and also fall into the medicinal and tea herb categories.

The term 'culinary herbs' is sometimes used more broadly to include tea herbs. However, there are significant differences between the markets for the two, so I prefer to regard them as separate categories.

Quality

Depending on their origin, the quality of some imported organic culinary herbs is reasonably good, though not as good as the premium grade being produced here.

Price

One of the factors influencing the market for culinary herbs is the bulk price for most trade culinary herbs, which is little short of miserable.

Prices for Organic Culinary Herbs: Prices for organic culinary herbs are more or less on a par with those for tea herbs.

Yields of many culinary herbs tend to be a bit lower than the tea herbs, while the processing to a finer, cleaner form is more time consuming and tedious. This means that to make it worth your while, you need to be getting a good price for culinary herbs, but not too many businesses are prepared to pay this.

Consumption and Demand

While the turnover of premium-grade culinary herbs is pretty steady, it is small in comparison to that of tea herbs.

Even though large volumes of culinary herbs are used in Australia, there has not been a big demand in this category for premium-quality organically grown products and it has been difficult to penetrate the market.

My explanation for this is that most people who want quality in culinary herbs use them fresh. This hypothesis is supported by the fact that the best selling of the dried organic culinary herbs are Basil and Bay leaves, which are often not available fresh.

Herbs for Other Uses

There is also a market for herbs which have other uses, such as for potpourri, treating or feeding animals, making bio-dynamic preparations, insect repellents, dyes etc. Many of these herbs also have human medicinal uses, but cannot be sold as such unless packed by a licensed manufacturer and are not recognised as having food uses.

This category contains such an assortment of herbs and uses that it is not possible to make many generalisations about it. If the herb is not actually consumed, as in dyes or potpourris (which rely on added essential oils for their aroma anyway), the grower is not likely to get a premium for a higher quality organic product. For some other uses, though, quality organic herbs may bring a price bonus.

Price Stability

The Law of Supply and Demand

Price is inevitably influenced by the quantity of the product available on the market and the demand there is for it. This is called the 'Law of supply and demand' and is often spoken of as if it is immutable. In my own experience in marketing, this law does not seem to apply as clearly and simply as one might expect.

In an open market situation where a number of producers with similar products are competing, the law of supply and demand can lead to dramatic price fluctuations. The relationship between price and demand is not necessarily constant. In many instances a small shortfall or surplus can lead to a major rise or fall in market price.

A 10% shortfall can lead to a 50% increase in prices, if buyers have no alternative but to pay more for an essential item, until the price rises sufficiently high to force some buyers out of the market. If this increase in price tempts growers to increase their production, it does not take much to swing the pendulum the other way. A 10% surplus can cause a fall in price to 50% below where it started, when sellers compete to make a sale until the price falls low enough to increase sales or force producers out of the market. Potato prices and their fluctuations are a classic example of this effect in action.

How It Affects Herb Prices

Herb prices are somewhat cushioned from such dramatic swings by the fact that surpluses can be stored for several years, so an oversupply does not have such an impact.

In my experience, prices for trade and second-grade organic herbs have been more subject to market forces and fluctuations than premium-grade herbs. A large manufacturer or wholesaler usually has several sources of raw materials to turn to and will naturally buy from where they can get the lowest price.

When it comes to the market for premium quality herbs, the situation is somewhat different being based more on what is a reasonable price for the product concerned. Small wholesalers, retailers, herbalists and growers are more aware of their mutual dependence and try to ensure each other's viability.

Future Price Trends

It is difficult to give any firm indications of future price trends, because price is affected by a number of variables:

Second-grade Organic Herbs

The price of second-grade organic herbs has been fairly stable, but with some fluctuations. With increasing volumes of production in Australia, the trend among manufacturers and wholesalers has been to get the price down towards the bottom line or, in other words, the trade price. Because imported herbs of similar quality can be substituted for their product, the local grower of second-grade organic herbs does not have a great deal of leverage.

The question is whether growers for this market can manage to sustain production in the face of lower prices. It may be possible if there are sufficient economies of scale and if suitable techniques and equipment are adopted, but there does tend to be a large dropout rate among these growers.

Premium-grade Organic Herbs

In this sector prices have been more able to reflect the local cost of production. This is because there has always been an undersupply of this grade and there has never been any imported herb available of comparable quality.

As production has gradually increased, the availability and continuity of supply of premium-grade herbs has generated more markets as more consumers become aware of the enormous difference in quality and the value that premium grade represents.

If the current surge in interest in herb growing results in a big increase in the production of premium-grade herbs, some major changes could occur in this market. How much it can expand is a matter of speculation. The increased supply of locally grown second-grade organic herbs may also have an impact on the demand for premium-grade herbs.

My feeling is that if production increases gradually with a continuity of supply of a wide range of premium-grade herbs, there is considerable potential for expansion of this sector. If my own experience in marketing herbs is any indication, premium quality herbs could conceivably gain as much as 30% of the overall market.

I doubt that there will be any sustained increase in prices for this quality: more likely, prices may come down a bit, reflecting increases in productive capacity as growers develop more efficient techniques and equipment and achieve economies of scale. However, other factors may come into play. Demand could increase if the greater therapeutic efficacy – which many believe premium-grade herbs possess – gains wider recognition as a result of clinical trials and assays of active constituents. For example, the price of wild and woods-grown American Ginseng is way above the price of the artificially grown herb. There are people who are prepared to pay up to 20 times more for it because it is generally recognised to be of much better quality and more effective.

This sort of price differential is unlikely to be true of most common species of herbs, because they are easier to grow in quantity, but they might well continue to be around 2–3 times the conventionally grown or trade price.

The other area that is unexplored is the overseas market for premium-grade herbs. If there is a demand for them here, surely there would be a similar demand in other countries.

Projections

My own recommendation is that growers should try to keep their options open, as the markets are in a developing state and difficult to predict.

Some years ago I was of the opinion that large-scale operations would soon take over the production of the major tea herbs and small-scale growers would be obliged to turn to minor herbs for profitable production. However, this has not come about. Large-scale production of tea herbs such as Peppermint or Chamomile has been attempted, but has failed to come up with the quality (and often the volume) that small herb growers can supply. The demand for premium-grade tea herbs is greater than ever, while demand for premium-grade minor herbs has greatly diminished with the advent of the *Therapeutic Goods Act*, though there are signs that it may be starting to increase again.

Imports from overseas will continue to create a background to our local market, but so far no imports have matched the quality being attained by some local producers.

The Boom and Bust Syndrome

There is some concern that a boom and bust situation could occur in the herb industry. Growers need to be prepared to cope with some fluctuations in the market. If everyone who reads a book or does a course on herb growing goes out and produces a tonne of herbs, an oversupply situation would develop, which might take a while to sort out.

This sort of thing has happened in other fields of alternative agriculture and can be difficult to avoid if a lot of overenthusiastic promotion occurs. On the other hand, in spite of a wide prevalence of herb fever for nearly 15 years in Australia, the actual development of herb production has been relatively slow.

As long as development continues to be moderate and steady, the prospects for the industry look healthy. In this situation growers, wholesalers and manufacturers can make more accurate projections, and markets can expand gradually to absorb the increased production.

Growers' Options in Response to a Fall in Prices

In the event of a decline in prices, growers would have a number of options in dealing with the situation:

Get Bigger: Increase the area in production, streamline operations, adopt more efficient techniques and equipment, find more productive varieties etc.

Crop Flexibility: Move out of species that are in oversupply and change over to those that offer a better price.

If the production of major herbs were taken up by broad-acre operations, there would still be minor herbs whose market requirements are not great enough to interest large-scale growers. There should be opportunities for smaller growers to change over to minor herb crops that suit their scale of production and bring a better price, especially if current trends continue to generate an increased demand for a broad range of herbs.

Marketing: The mark-ups in the herb trade are quite high, so growers can increase their returns by wholesaling or retailing their own products. Other options are to form marketing co-operatives or enterprises, or to affiliate with other growers who are doing their own marketing.

Value Add: Produce a higher grade of quality and/or further process or package your products. This could involve putting herbs into retail packs and making up blends, bath-mixes, coffee substitutes, gift packs, tussie mussies and other herbal products.

Get Out: A final option is to forget about growing herbs and find something else you could be doing more profitably.

The reality is that unless growers can get good enough prices to make their herb growing operations viable, they are not going to stay in the industry. Many manufacturers and wholesalers tend to overlook this. While they can rely on overseas supplies, it doesn't greatly concern them if local growers drop out, but what if overseas sources ever fail?

The Therapeutic Goods Act 1991

The *Therapeutic Goods Act 1991* is rather controversial legislation, and has had a major impact on the herb industry. Effectively it has brought the production of herbal medicines under the regulations that deal with the production of pharmaceuticals.

This has been something of a mixed blessing. On the one hand it has given some credibility and legitimacy to the herb industry and has recognised many traditional medicinal uses of herbs; on the other hand it has imposed expensive regulatory controls on the production of herbal medicines.

While these regulations do not apply directly to the on-farm aspects of herb production, they have affected growers indirectly.

The Therapeutic Goods Act's Effects on Growers

Herbs for Therapeutic Use

The regulations are concerned with those herbs which have a therapeutic effect (that is, a physiological effect) on humans. Whether or not a herb product with therapeutic properties comes under the jurisdiction of the Act depends on whether that species of plant also has recognised human non-therapeutic uses.

If it has none, then manufacturing and packaging (beyond the farm production stage) for use 'in or on humans' comes under the Act.

If the species has recognised food or cosmetic uses as well as therapeutic uses, then it is the 'intention' of whoever puts the herb into its final packaged form for sale to the general public which determines whether it comes under Therapeutic Goods Regulations or not.

If that person or company 'intends' the herb to be used for therapeutic purposes, by making therapeutic claims on the label, then it comes under these regulations.

The processing and packaging (beyond the farm production stage) of herbal products that come under the Act needs to be done by a manufacturer whose premises and procedures have been inspected, approved and licensed by the Therapeutic Goods Administration. Specified procedures have to be followed for each batch of product, including many pages of paperwork, microbial counts, identification tests, high-tech analyses etc. The cost of meeting these requirements is beyond the capacity of small enterprises with a limited turnover.

Microbial Levels

The onus for meeting the Therapeutic Goods Regulations standards for levels of microbial contamination technically rests on manufacturers, but the regulations can present a few headaches for organic growers. The normal methods for controlling microbial levels are fumigation with ethylene oxide, methyl bromide, or irradiation. These are not acceptable treatments for an organic product, and can affect some of the active constituents.

Of course these treatments are standard procedure for imported trade herbs to get them past quarantine, so they normally easily comply with the maximum permissible microbial levels. An importer can bring in several tonnes of a particular herb, call it a batch, and get one test to cover the lot. They can then offer a certificate of compliance for every small quantity they sell.

This makes it much simpler for manufacturers to rely on imported herbs rather than source them from local organic growers who cannot supply certificates of compliance as the cost of testing is prohibitive for small batches. Only those manufacturers dedicated to using organically grown herbs are prepared to work with small growers and find alternative ways of keeping microbial levels down.

Procedures to Minimise Microbial Levels

Various farming practices can be adopted to reduce levels of microbial contamination.

Avoiding Raw Animal Manures: Raw animal manures are rich in micro-organisms and pathogens (chicken manure is a particular problem) and can cause excessive microbial contamination of the herb. Any animal manures to be used should go through a hot composting process before use.

Composting All Organic Fertilisers: The heat of the composting will reduce levels of pathogenic organisms, as long as the temperatures are high enough: around 60°C. To be fully effective, the heap will have to be turned (see Chapter 7, 'Compost').

Careful Harvesting: Harvesting techniques should avoid the inclusion of soil and any other material or items that might cause increased levels of micro-organisms.

Thorough Washing: This refers particularly to roots, but other herbs may need washing if they are contaminated with soil or with insect droppings.

Insect Control: This does not necessarily mean the use of insecticides as there are organic practices that can reduce or help avoid the build-up of insect numbers. Large numbers of aphids on the plants can raise microbial levels, as can droppings from grasshoppers or other insects.

Good Drying Procedures: Drying that proceeds too slowly or holds the herb in conditions of high temperature and humidity can cause a build-up of micro-organisms, as can storing inadequately dried herb.

Separation of Soil in Processing: Soil particles can be separated from the dried herb by sifting out fines during processing or by removing heavier particles that settle out.

'Appendix B'

In order to define which herbs have legitimate non-therapeutic human uses, a list of species known as Appendix B has been compiled and included in the Therapeutic Goods Regulations. This includes all herbs recognised as having human food or cosmetic uses.

A herb on this list packaged for sale to the public can be exempt from the regulations, provided it is intended by the packager to be for non-therapeutic use. The intended use is required to be stated on the label and can include culinary, beverage, food or cosmetic uses. What the consumer actually does with the herbs when they get them home does not seem to be a matter of concern, only the intention of the packager is considered.

If a non-therapeutic use involves human internal consumption, then the herb comes under ordinary health regulations for foods, but these are less demanding and meeting them is within the capacity of small enterprises.

Herbs not included on 'Appendix B' are regarded as having only therapeutic uses for humans. If they are intended for use 'in or on humans', they are required to comply with the Therapeutic Goods Regulations.

Alternatively, they may be sold to the general public if they are intended for some other use, such as feeding to animals, or making biodynamic preparations, insect repellents, potpourris, dyes etc.

'Appendix B' Herbs Covered in This Book

Note: This is not an exhaustive list of all the herbs included in 'Appendix B'

Agrimony	*Agrimonia eupatoria*
Alfalfa	*Medicago sativa*
Angelica	*Angelica archangelica*
Anise	*Pimpinella anisum*
Basil	*Ocimum basilicum*
Bay	*Laurus nobilis*

Bergamot	*Monarda didyma*	Passionflower	*Passiflora incarnata*
Blessed Thistle	*Cnicus benedictus*	Pennyroyal	*Mentha pulegium*
Borage	*Borago officinalis*	Peppermint	*Mentha x piperita*
Burdock	*Arctium lappa*	Plantain, Greater	*Plantago major*
Calendula	*Calendula officinalis*	Plantain, Narrow Lf	*Plantago lanceolata*
Caraway	*Carum carvi*	Raspberry	*Rubus idaeus*
Catnip	*Nepeta cataria*	Red Clover	*Trifolium pratense*
Celery	*Apium graveolens*	Rose, Dog	*Rosa canina*,
Chamomile, German	*Chamomilla recutita*	Rose, Sweet Briar	*Rosa rubiginosa*
Chickweed	*Stellaria media*	Rosemary	*Rosmarinus officinalis*
Chicory	*Cichorium intybus*	Rue	*Ruta graveolens*
Chives	*Allium schoenoprasum*	Sage	*Salvia officinalis*
Cleavers	*Galium aperine*	Sheep Sorrel	*Rumex acetosella*
Coriander	*Coriandrum sativum*	Shepherd's Purse	*Capsella bursa-pastoris*
Couch Grass, English	*Agropyron repens*	Tarragon, French	*Artemisia dracunculus*
Dandelion	*Taraxacum officinale*	Thyme	*Thymus vulgaris*
Dill	*Anethum graveolens*	Vervain	*Verbena officinalis*
Dock, Yellow	*Rumex crispus*	Violet, Sweet	*Viola odorata*
Elder, Black	*Sambucus nigra*	Witch hazel	*Hamamelis virginiana*
Elecampane	*Inula helenium*	Wood Betony	*Stachys officinalis*
Eucalyptus	*Eucalyptus globulus*	Yarrow	*Achillea millefolium*
Fennel	*Foeniculum vulgare*		
Garlic	*Allium sativum*		
Globe Artichoke	*Cynaria scolymus*		
Ground Ivy	*Glechoma hederacea*		
Hops	*Humulus lupulus*		
Horehound, White	*Marrubium vulgare*		
Horseradish	*Armoracia rusticana*		
Hyssop	*Hyssopus officinalis*		
Lavender, English	*Lavandula angustifolia*		
Lemon Grass	*Cymbopogon citratus*		
Lemon Thyme	*Thymus x citriodorus*		
Licorice	*Glycyrrhiza glabra*		
Linden	*Tilia* spp.		
Marjoram, Sweet	*Origanum marjorana*		
Marshmallow	*Althaea officinalis*		
Meadowsweet	*Filipendula ulmaria*		
Melissa Balm	*Melissa officinalis*		
Mugwort	*Artemisia vulgare*		
Mullein	*Verbascum thapsus*		
Nettle, Greater	*Urtica dioica*		
Nettle, Lesser	*Urtica urens*		
Oats	*Avena sativa*		
Oregano	*Origanum vulgare*		
Parsley	*Petroselinum crispum*		

Effects of the Therapeutic Goods Act on Markets for Local Herb Growers

Neither the therapeutic goods regulations nor food regulations apply to growing, harvesting, drying and initial on-farm processing of the dried product, such as rubbing, screening, chaff-cutting, milling, or bulk packaging. It is only when the product is further processed for sale to the general public that it comes under the regulations.

This means growers who do not do any final packaging are basically exempt from regulations. Nevertheless, they are still affected indirectly through the effects of regulations on small wholesalers and manufacturers who handle the bulk of local herb growers' products.

The marketing of herbs listed on 'Appendix B' has not been affected greatly because most small wholesalers were not putting any therapeutic claims on their packaging anyway, but the demand for the herbs regarded as solely therapeutic has fallen off somewhat in the retail tea market.

In the manufacturing category, the *Therapeutic Goods Act* has had greater impact. Because it places an emphasis on microbial levels and does not address many other important aspects of quality, it placed the one manufacturer specialising in premium quality herbal medicines at a disadvantage and was a factor in that manufacturer subsequently going out of business. Until other manufacturers take up this market niche, there will not be much call for premium quality manufacturing herbs.

While the *Therapeutic Goods Act* may have stimulated manufacturers to increase their turnovers to remain viable under the added cost burden, this has not translated into a significant benefit for growers. The increased demand for raw materials has not resulted in increased prices to local growers for most species, because manufacturers have sourced most of their herbs overseas.

The Need for a Grower Network

There are now a number of enterprises and organisations promoting the growing of herbs in Australia. Some are manufacturers in their own right, while others are wholesalers, bulk distributors or associations of growers.

At present, with most herbs in undersupply, there is potential for further growth in local production. However, we could soon see the situation where some lines are suddenly in oversupply unless growers are able to form an effective system of co-ordination and networking.

Most growers are still in a learning phase. While we may be trying to supply the same markets, we have a lot to gain by working together. At present there are a few associations of growers in various parts of the country, but mostly we operate substantially independently of each other.

If herb growers could establish a national network or umbrella body, it would be to the benefit of the rest of the industry. It could foster sharing of information, techniques, technology etc. and it might be able to foster regular production and help avoid dramatic price fluctuations, which do not really help anyone in the industry.

Problems and Questions to be Resolved

There are a few points which pose problems for information sharing and networking. One is the cost of acquiring and sharing information. Some growers have spent years researching and trialling to get to where they are now. How much is that information worth? How much should they expect to be paid for their time and effort in sharing it?

There are those who want to hang on to 'trade secrets'. How can we encourage them to see the mutual benefits in sharing information?

The other question is, how do we work out what information is worth sharing with other growers? Unfortunately there has been a tendency for people to get carried away in their enthusiasm for herb growing (or at the thought of the money to be made as a consultant to would-be growers) and present untested ideas and information obtained from questionable sources as if these were facts.

Financial Viability

It is difficult to give a watertight assessment of the capital costs and profitability of herb growing. Costs and returns are going to vary enormously as there is no one way to grow herbs, and individual growers' operations, resources and capabilities are all different.

The following examples are offered as a general guide, which can be used as a basis for estimating your own capital and operating costs, but naturally these will vary according to your particular growing situation, ingenuity, capabilities and availability of time and/or money.

Expenditure

Capital Outlay

Starting on 0.4 ha

The figures below are estimates for the establishment costs for a 0.4 ha herb-growing operation. No allowance has been made for the cost of the land. Many people on established farms will find they already have some major items, such as a suitable shed, a tractor or a rotary hoe, so for them the capital cost of establishing a herb-growing enterprise will be less.

A range of figures is listed for most items, the lower being a cheaper second-hand or home-made job, as opposed to a new(er) or contract-built one.

Capital Expenditure Items (on 0.4ha)	$
Heated Dryer (materials only)	200–3000
Chaff-cutter	200–500
Certification – Initial costs	400–600
Drying Screens (materials only)	1300–2000
Drying Shed	2000–10 000
Fencing	0–2000
Forage Harvester and Cart	600–1500
Hand Tools	200–800
Irrigation System	2000–4000
Mower	200–1000
Other Implements	1000–3000
Propagation Material	200–1000
Rotary Hoe	500–3000
Rubbing Screens and Frames	200–500
Scales	50–200
Tractor	2000–10 000
Sundries	500–1000
Total	*11 550–44 100*

These figures assume starting virtually from scratch and rapidly building up to full production in 2 years. If you want to start off with a big bang like this, do make sure you understand exactly what is involved in getting established and how everything needs to be done, as mistakes on this scale can be expensive in terms of both time and money.

Starting on a Smaller Scale

An alternative approach that minimises initial outlay (and risk) is to start off on a smaller scale, say 0.1 ha, and gradually expand as you learn what is required and when you can afford the capital outlay. That is how I started out, as did most of the other successful growers I know.

This way the initial capital outlay could be reduced to perhaps 40% of the lower of the above totals, provided you are prepared to work hard, make do with less than ideal facilities and equipment, and hire or borrow a tractor when needed rather than buy one.

There would be a need for additional capital expenditure later, but this could be taken on when you were in a better position financially and skill-wise.

Annual Costs

On the annual operating costs side, the items listed are those that are legitimate tax deductions, including items such as rates that you would incur anyway, even if your land were lying idle. If a trip to town involves some business as well as things of a private nature, then a portion of the cost of that journey can be claimed as a business expense. You would probably go to town anyway if you weren't growing herbs: you might even end up staying longer and spending more if you didn't have to get back to work on the herb farm!

The figures below are based on the typical expenses of a 0.4 ha operation run by one or two people on a self-employed basis. They are drawn from my own experience and that of a few other growers.

The labour input can be expected to be in the range of 60–80 hr/week during the peak periods of the year to perhaps 10 hr/week at quiet times. During the construction and establishment phase over the first 2 years, the labour input could be much higher, or additional labour could be employed.

Annual Operating Costs (on 0.4ha)	$
Books	50–200
Certification costs and levy	300–500
Compost Materials / Fertilisers	0–1000
Conferences, seminars, travel	100–350
Power or Fuel for drying	0–2000
Freight	200–800
Fuel – agricultural use	200–400
Insurance	400–800
Memberships and subscriptions	100–250
Motor Vehicle	500–1000
Packaging	100–400
Postage, Power, Phone	400–800
Protective equipment and clothes	50–200
Rates	300–500
Repairs	300–1500
Seed and other inputs	200–800
Tools	200–400
Sundries	500–1000
Total	**3900–12 900**

(On a smaller 0.1 ha operation, the annual costs might come to around 40% of this. For a larger operation, costs per 0.4 ha would be lower, but the need to hire labour could be an additional cost).

Comparisons of Gross and Net Returns

Example 1: Trade Prices

0.4 ha planted in several different aerial part and root herbs, harvested, dried, processed and sold at trade prices.

Assuming an average production over 4 years of 1200 kg/annum (200 kg roots, 1000 kg aerial parts), which is sold at an average of $9/kg for a

gross annual return of	$10 800
Less operating costs at an estimate of, say	$5 000
Net return	**$5 800**

Example 2: Manufacturing-grade Organic Prices

0.4 ha planted in several different aerial part and root herbs, harvested dried, processed and sold at manufacturing grade prices.

Assuming an average production over 4 years of 1200 kg/annum (200 kg roots, 1000 kg aerial parts), which is sold at an average of $14/kg for a

gross annual return of	$16 800
Less operating costs at an estimate of, say	$6 000
Net return	**$10 800**

Example 3: Premium-grade Organic Prices

0.4 ha planted in several different leaf and root herbs, harvested and sold at premium-grade organic prices, the leaf herbs being processed to tea grade. With the stalk removed, the quantity is reduced, but the value increases.

Assuming an average production over 4 years of 950 kg/annum (200 kg roots, 750 kg tea-grade leaf), which is sold at an average of $24/kg for a gross annual return of	$22 800
Less operating costs at an estimate of, say	$7 000
Net return	**$15 800**

There is somewhat more labour involved in the production of the premium tea grade, mainly due to the extra processing involved. Against this is the need for many manufacturing-grade aerial parts herbs to occupy the drying facility for a longer period to fully dry the stems.

This overall comparison shows premium grade as the most financially viable proposition at this scale of growing, while producing at trade prices offers a very marginal return, and manufacturing grade is somewhere in between.

However, growers need to take into consideration such factors as overall scale of production, yields, time, costs, labour requirements, skill involved, and available markets when assessing their own prospective returns.

Protective equipment – sundries that should be allowed for when developing your budget.

Comparing Prices for Aerial Parts and Leaf

In making a comparison between prices for aerial parts (leaf, flower and stem) and for leaf or leaf and flower (which have the stems removed) a number of factors need to be taken into consideration, including the proportion of stem in aerial parts, price conversion, different costs of processing, and different costs of drying. Let's consider these one at a time.

Proportion of Stem in Aerial Parts

The proportion of stem can vary greatly between species. It needs to be calculated from dry material, as the fresh weights can be rather different, many herbs having rather fleshy stems.

Herb	Stem
Horehound, White	33%
Hyssop	41%
Melissa Balm	37%
Motherwort (first year)	47%
Mugwort	46%
Nettle, Lesser	22%
Peppermint	33%
St John's Wort	31%
Yarrow	41%

Table showing typical percentages (dry weight) of stem in the aerial parts of a few species taken from the first harvest of the season: later harvests usually have less stem.

The majority of herbs normally comprise 30–40% stem for their first harvest of the season, but some are leafier, while others are stalkier.

The proportion of stem can vary greatly within the one species, depending on the location, the season, the stage of development, how high above ground the plants are cut, and whether it is the first, second, third or fourth harvest of the season.

The first harvest of overmature, tall, crowded plants, growing in harsh conditions and cut close to the ground will contain the highest proportion of stem, and in some species this can be well over 50% in poor quality samples.

With most species, the proportion of stem progressively decreases for the later harvests of the season. A first harvest of Peppermint, taken around the beginning of December, would typically contain around 33% stem, while a second harvest in January would be in the region of 25% stem, a third in March around 20% stem, while a fourth in May would have only around 15% stem.

How much this percentage will decrease during the season depends on the species and the growing conditions. With Alfalfa there is little difference between the early and late harvests, while with Motherwort and Yarrow there can be a big difference, with perhaps only 10% stem in a second or third harvest

Some herbs, such as Lady's Mantle and Coltsfoot, have very little stem to be removed in rubbing, while a few such as Meadowsweet and Vervain normally have the stem included in the tea-grade herb.

Price Conversion

To convert a price for a tea-grade herb to an equivalent price for manufacturing-grade herb, it is necessary to subtract the percentage of stem in the dried unprocessed herb from the price.

For example, on a yield basis, a price of $16/kg for tea-grade Alfalfa is equivalent to $10.72/kg for an aerial-parts manufacturing-grade containing 33% stem. For a herb such as Peppermint, a price of $24/kg for tea-grade leaf is equivalent to $16.08 for aerial parts containing 33% stem (a typical first harvest), but $19.20/kg for a third harvest of aerial parts containing 20% stem.

Different Costs of Processing

Another factor in the equation, is the time and labour involved in processing, which can be translated into a cost. If you figure your time at $15/hr ($10/hr for your labour, $5/hr for your overheads), then rubbing to tea grade typically costs around $3/kg (finished product), while chaff-cutting, which proceeds more rapidly and includes the stem in the finished product, amounts to around $1/kg chopped aerial parts, or $1.50/kg for the leaf component of the aerial parts if it is 33% stem.

Different Costs of Drying

The production of aerial parts rather than tea-grade leaf usually requires as much as 50% longer drying time. This can translate into a higher cost of drying: for fuel or power and/or the cost of a larger drying facility. If this is costed in as well, it would partially or wholly cancel out the reduced processing costs.

On the other hand, as the quality required by most manufacturers is not as high, a cheaper drying system – such as a suspended floor dryer – may be feasible. This would, however, preclude the marketing flexibility of also supplying the premium-grade market, as the quality produced by such a dryer is not good enough.

Production Costings

The costing of production is a difficult area to address as most herb growers, including myself, are not of an inclination to keep the intricate records necessary for time and motion studies. I don't have any firm figures for the growing side of the operation, but a few records of times taken to harvest, dry and process a few crops are interesting.

These figures should be taken as indicative only. Better yields, new technology and new techniques will no doubt bring improvements in many areas as the industry develops.

Tea-grade Leaf

5 kg Batch

In general, the harvesting, spreading on screens and rubbing had a throughput of around 2 kg/hr (finished product, dry weight) for plots harvested with the scythe and yielding 5 kg (dry weight) or more per harvest. This means they need to be grown on a scale of at least 50 m^2 (a typical herb would be yielding about 0.3 kg/m^2 or 15 kg per annum, which would be in three harvests of around 0.1 kg/m^2 or 5 kg each).

At a rate of $15/hr for your time, this amounts to around $7.50/kg for harvesting and processing these herbs.

2 kg Batch

On a smaller scale, the harvesting, drying and processing productivity decreases. Small batches of around 2 kg finished product had a throughput of about 1 kg/hr, while smaller or fiddly operations fell to 0.5 kg/hr.

At a rate of $15/hr for your time, this amounts to $15–30/kg for the harvesting, drying and processing alone, so small plots are of doubtful viability for most crops, considering you also have the cost of growing and your overheads to cover.

20 kg Batch

On a larger scale, the harvesting, drying and processing productivity rises to around 3 kg/hr (or $5/kg) in a straightforward operation for a herb processed to tea-grade at the optimum stage while the stems are still pliable, but higher productivity than this is unlikely as it is all manual work. If some of the operations could be mechanised, this would reduce the labour input, but for it to be worth while, the systems would have to manage to produce the same quality. So far nobody has done this successfully, though I still believe it is a possibility.

Tea-grade Leaf, av. 2 kg batch	±15.00/kg
Tea-grade Leaf, av. 5 kg batch	±7.50/kg
Tea-grade Leaf, av. 20 kg batch	±5.00/kg
Mfg-grade Aerial Parts, av. 2 kg batch	±10.00/kg
Mfg-grade Aerial Parts, av. 7 kg batch	±5.00/kg
Mfg-grade Aerial Parts, av. 40 kg batch	±3.00/kg
Marshmallow Root, 2 kg batch	±7.50/kg
Chicory Root, 10 kg batch	±5.00/kg
Dandelion Root, 11 kg batch	10–15.00/kg
Valerian Root, 10–15 kg batch	19–23.00/kg
Chamomile Flower, good harvesting	25–37.00/kg
Chamomile Flower, average harvesting	37–50.00/kg
Calendula Flower	15–30.00/kg
Red Clover Flower, combing second grade	15–30.00/kg
Red Clover, hand-picked premium grade	±45.00/kg

Table showing approximate cost of harvesting, drying and processing various herbs, based on a rate of $15/hr.

Manufacturing-grade Aerial Parts

The rate of throughput is higher for harvesting, drying and processing a manufacturing-grade of most aerial parts of herbs. Feeding the dried aerial parts through a chaff-cutter is quicker than the normal three passes through the rubbing screen required for tea grade. Apart from which the herb will typically contain 20–40% stem, which is separated from the tea-grade but included in the aerial parts manufacturing-grade.

7 kg Batch
Productivity should be in the range of 3–4 kg/hr for a plot of a typical herb yielding 6–8 kg dried aerial parts per individual harvest. At a rate of $15/hr, this amounts to a cost of harvesting, drying and processing of $3.75–5.00/kg.

2 kg Batch
Productivity falls off rather steeply for small batches, owing to the time required to set up and clean the chaff-cutter, so for a batch of 2 kg, it is around 1.5 kg/hr (a cost of $10/kg).

40 kg Batch
For a larger batch of say 40 kg, the cost of harvesting, drying and processing can approach 5 kg/hr ($3/kg).

Crops with a Low Proportion of Stem
Some herbs for manufacturing grade have only a small proportion of stem, which means there is less difference between productivity figures for manufacturing aerial parts and tea-grade leaf.

Root Crops
Productivity rates for the digging, washing, chopping and drying of Valerian, Dandelion and Marshmallow were calculated. The digging and washing were done by hand, the chopping was done with the chaff-cutter where possible, and the drying was done in an electrically heated cabinet dryer. No further processing was carried out once the roots were dry.

Large Roots
Large roots were the most efficient to handle. Marshmallow, had a productivity of 2 kg/hr for a relatively small batch, and Chicory was around 3 kg/hr on a larger scale (10 kg or more). These are representative of the large rooted herbs, such as Elecampane, Burdock, Chicory, Comfrey and Yellow dock. At a rate of $15/hr, this amounts to $5–7.50/kg for the harvesting and drying of these herbs.

At this cost of harvesting and drying, hand-digging and washing of these large roots could be a viable proposition, even at manufacturing-grade prices. However, crops with smaller, finer roots are more labour intensive.

Medium-Sized Roots
Dandelion ranged between from 1–1.5 kg/hr for a batch of 11 kg dried weight, depending on how much trimming of root rot was needed. Dandelion is typical of medium-sized roots, such as Echinacea, Parsley Root and Stoneroot. At a rate of $15/hr, this amounts to $10–15/kg for the harvesting and drying of these roots.

Fine Roots
Valerian is rather more labour intensive than other roots owing to the time it takes to separate and wash the fine matted roots to achieve adequate quality. Productivity for this root ranged from 0.65–0.8 kg/hr. At a rate of $15/hr this amounts to $18.75–$23.08/kg for the harvesting, washing and drying of this herb.

Overall, the cost of harvesting and drying would appear to make viability difficult for finer rooted herbs, such as Dandelion and Valerian, unless a price that realistically reflects the cost of production can be obtained, or unless some breakthrough in production methods can be achieved. At the price currently being offered for manufacturing-grade organic herbs, growers would be better advised to look to the premium-grade market or to more profitable crops.

Flower Crops

Productivity in harvesting and drying flowers tends to be fairly variable according to the nature of the herb's flowering and the method of picking. My main experience is in harvesting and drying Chamomile, with a productivity in the range of 0.2–0.6 kg/hr, dried flower, but usually around 0.3–0.4 kg/hr. At a rate of $15/hr, this means a normal cost for harvesting and drying of $37.50–50/kg.

The harvesting of flower crops such as Calendula and Red Clover can be a bit more productive. Hand-picking or combing and drying of these herbs showed a productivity of 0.50–1 kg/hr, representing a cost of harvesting and drying of $15–30/kg, but if premium quality is sought in Red Clover, hand-picking every 4 days means a productivity of around 0.33 kg/hr, or a cost of harvesting and drying of around $45/kg.

Processing for Flexibility in Marketing

In order to hedge their bets, growers may wish to keep their marketing options as open as possible. This can mean growing the one herb for several different end uses. While the herb tea market is at present the most stable and financially viable, this may not always be the case. Growers would be advised to spread their efforts over at least five or six different herbs and have some idea of alternative markets for them should these be needed.

Herbs such as Alfalfa, Nettle, Yarrow, Hyssop and Peppermint have a market as tea-grade leaf (or leaf and flower) and also as aerial parts. If you have a clear idea of your market's requirements, you can process the crop for these purposes as it comes from the screens.

However, you may find yourself in a situation where you don't know which form the market will require. If you chaff-cut the herb, it may be very difficult to convert it back to a tea-grade as the cut pieces of stem are difficult to separate by screening, or even by winnowing. The fragments of broken and shattered stem can make the finished product look unattractive.

On the other hand, if the herb is all rubbed through to tea-grade, and a manufacturing-grade is required, then you will be getting a meagre price for a smaller quantity of a value-added product with all the stem removed.

One option is to rub the herb to first screening, bag it, and chaff-cut the separated stem, packing it separately. These two portions can be held in storage until your market is known. Then if you have a request for tea-grade you can further process the first screening and discard the cut stem, but if your customer requires aerial parts, you can recombine the cut stem with the first screening.

Establishing and Maintaining a Market

In order to survive financially, you need to be able to hold your own in the market place. This doesn't mean you have to go out each day in a suit and tie, toting a mobile phone and reciting a well-rehearsed sales pitch. Your high quality dried herbs will do most of the selling for you, but if you have some basic idea of market dynamics, you can plan your strategies to enable the quality of your product to do its job.

Establishing your Market

The first hurdle is getting a leg into the market. This is much easier today than it was 15 years ago, because there is now a large body of consumers who appreciate high quality and want their herbs to be organically grown. There are also several wholesalers and manufacturers whose businesses are based on organically grown herbs.

If you are not interested in getting involved in wholesaling, there are people who will handle your products, provided those products are of high quality.

Establishing Credibility

You will probably find that you first have to establish credibility. Unfortunately, for quite a few years there has been a continual parade of people interested in growing herbs who have been approaching various manufacturers and wholesalers with glowing expectations of being able to supply large quantities of locally grown herbs. When I was managing Southern Light Herbs, I used to receive about two enquiries a week from prospective growers. This came to be rather exhausting and time consuming, particularly as virtually none of them ever came up with anything.

I'm afraid I became rather cynical after a while, and I know other wholesalers have done the same. Most people are reluctant to start talking serious business until they have a clear indication that you are likely to come up with the goods.

A Pilot Project

One good approach is to start some trial plots so you have an idea what is involved in dried herb production and can prepare some samples from your trial harvests. Meanwhile a bare fallow can be in progress in preparation for expanding your production.

You can then approach potential customers with your samples, telling them that you have been running trial plots and have propagation material available and ground ready to plant. This should provoke their interest enough for them to give you a clear indication of what crops and quantities they would like you to grow.

Try to be realistic as to how much you are capable of in your first season. It is far better to talk small and meet your targets than to talk big and fall short: remember your ongoing credibility is important.

Coming up with the Goods

The next hurdle is to make sure you come up with the goods and that they are of the high quality you have been promising. If you can do this, you have made the initial establishment of your market.

In many ways it is preferable to ease into the market rather than to come in with a big bang. Getting it right to begin with will help in establishing your reputation and it is easier to get things right on a small scale than on a big one.

Small quantities of high quality herbs will carve their way into the market, paving the way for larger volumes to follow. Large volumes of poor quality herb can damage your reputation and make it difficult for you to sell your next crop.

Sometimes it takes a bit of patience before things start to move. Often buyers are keen to have your product, but they have old stock they want to clear first.

Market Projections

At this point you should be able to make some reasonably good projections as to your future production and can put out further feelers for additional markets if necessary. This may just be a matter of communicating with your wholesaler or manufacturer. Be careful to be realistic as there is no point in cultivating a market that you are unable to supply: this can damage your reputation.

Maintaining Your Market

Now for the long haul: having established your markets you need to maintain them. Continuity of supply is important for most products. If you have a shortfall and are unable to supply the market for a while, it may take a while for things to pick up again when you get back into production, because your customers may have turned to other suppliers.

Communication

Communication can help overcome some of these problems. Try to establish a rapport with your customers, and help them understand the vagaries of primary production. If you have a good line of communication with them and can give them a clear indication of when your next crop will be available, they can plan around that, and try to be just running out of their replacement stock when your new harvest arrives. They will also have told their customers that the herb is grown on a real herb farm so it is subject to the vagaries of the seasons, that it is in demand, and when they can expect it to arrive newly harvested, full of vitality, not having been sitting in a warehouse somewhere for 5 years.

Don't rely on your customers remembering which herbs you are producing. Keep them up to date as to the herbs and quantities you have in stock, when you expect your next harvest and what volume is anticipated.

Reputation

Maintaining your good reputation is important. Of course, we all make mistakes, but usually what matters most is not the mistake but what we do about it when we have made one. It is good business to be honest about these things while taking measures to ensure they don't happen again.

Inevitably you will have some batches of herb that are not up to your usual quality. It is important to distinguish these, as they can damage your reputation badly. Sometimes they can be offered to your usual customers as second grade: they may be able to use them in blends, cheaper bulk lines or potpourri. Alternatively you may want to unload them elsewhere so they are not harming the market for your higher priced, premium quality herbs. They could be offered to other enterprises that normally deal in trade or second-grade herbs.

The same goes for old stock. If you have an excess of some herb and are holding stock in storage that is more than 1–2 years old, it is best to clear it at a reduced price (or perhaps to put it on the compost) and continue supplying your regular customers with fresher material.

Where to From Here

For many of us it is human nature to want to explore new areas, so having established our herb growing operation and found good markets, the restless mind seeks new avenues, new ventures.

Watch out for the expansion trap. One of the problems with a successful business is that it is hard to stop it getting bigger. Indeed, with a successful expanding business you can suddenly find yourself being dragged through life by the scruff of the neck, kicking and screaming. It can keep you poor feeding its appetite for new and bigger equipment, bigger facilities and an increasing range and quantity of stock, and if you don't get the sums right, it can leave you broke.

More businesses fail through over-expansion than through any other cause and this can apply to farms, too: even herb farms!

To remain sane and have some time for the things that are important to you, you need to realise that you do have a choice. You can say no, you can resist answering the call of surging adrenalin, you can decide where the limits are, put a lid on your expansion and let somebody else take up the opportunity.

Group Marketing

The production of high quality herbs lends itself to group marketing. In order to maintain quality (and sanity) there is a limit to the size of a herb growing operation. There can be advantages for groups of growers to work together under the umbrella of a company, a co-operative or simply a partnership, in order to be able to offer the market a complementary range of herbs and better continuity of supply.

Some firms prefer to work with such groups, because they find they are more reliable and it is simpler than dealing with a number of individual small growers. Nevertheless, a group will still have to go through the same process of establishing itself in the market.

Rather than becoming part of the wave of getting big or getting out, which is dominating many sectors of agriculture, the production of high quality dried herbs is more suited to the development of networks of small farms working in co-operation and co-ordinating their production.

Thus, growing herbs can be both a pleasant lifestyle and part of the rural regeneration.

Section 2

Individual Crops

SPECIES COVERED

Included in this section are individual entries for some 109 species of herb. This list is by no means exhaustive. For the most part the herbs described here are ones about which I have reliable information derived from my own experience or the experience of other Australian growers. I have tried to avoid relying on books and other literature for cultivation, harvesting and drying information: too often these are unreliable. Consequently there are some significant crops omitted or dealt with only briefly, while some of the species included do not offer great potential for large-volume production.

Nevertheless, the species covered do give a fair representation of the broad range of crop types and the techniques and equipment required to grow, harvest, dry and process them.

The 'Growth Type' System of Classification Used for the Herbs Grown as Crops

The system used here for classifying the individual herbs grown under cultivation places them in categories according to growth types.

This enables crops of similar growing and management requirements to be placed together. Detailed entries are provided for a number of major species in each group. These can be used as a guide for growing and managing other crops with similar habits and requirements.

This way, if growers encounter a new species whose cultivation and management requirements are unknown, they can get a fairly good idea where to start by looking at the other species of the same growth type. Wild-harvested species are placed in a category of their own.

This classification system is production orientated and cuts across botanical affinities and, of course, alphabetical order. The main drawback is that it can be difficult to find a species in the book if you don't know or can't remember which group it is in. To help the reader in this, an alphabetical plant index is provided at the very end of the book.

Major Herbs

Some species from each group are given detailed coverage. In general these are the ones for which a good market exists, but a few species are included because they are representative of a number of other herbs about which growers may need detail.

Minor Herbs

Minor herbs generally include herbs of smaller market potential, those whose detail is adequately covered under a similar species, and others which need further research and development. Many are quite significant for at least some growers, or could be so.

A brief description and general information is provided for each minor species and further detail is presented in chart form.

Wildcrafting and Weed Harvesting

A number of native species and weeds can be harvested wild in southern Australia. Detail is provided on the more significant species. A number of them are also duplicated as cultivated species, so the reader may need to refer to those sections for detail on some aspects: depending on whether the plant is more significant as a wild or cultivated herb.

EXPLANATORY NOTES ON THE INDIVIDUAL CROP ENTRIES

English Name of Herb

Generally I have adopted the name in most common use, but in some cases I have departed from this in order to distinguish from related species. In the case of a herb like Melissa Balm I have not used the common name, Lemon Balm, because in the dried herb the lemon flavour is hardly present.

Botanical Name

I have endeavoured to use the currently recognised botanical name. Following the botanical name is the abbreviated name of the author of the botanical name, for example, L stands for Linnaeus, the 18th century Swedish botanist who set up the binomial system of classification and gave names, which are still in use, to many species. Where a plant has been reclassified, the abbreviation for the first botanist to name it is given in brackets, followed by the name of the botanist responsible for the reclassification, for example: '(L) Link' indicates described by Linnaeus, then reclassified by Link.

Family

The family name gives an indication of other related species and helps with confirming identification. Scientific circles, in their wisdom, have in recent years adopted new names for several major families. Leguminosae is now Fabaceae, Compositae is now Asteraceae, Labiatae is now Lamiaceae, Gramineae is now Poaceae etc.

Part Used

The part or parts of the herb commonly used are listed as aerial parts, leaf, leaf and flower, root, flower, seed, or root. An item in brackets refers to a less common use.

Aerial Parts: All the above-ground parts are used in the final product, though the lower part of the stem and faded leaves would normally be omitted.

Leaf: The final product consists of leaf, or predominantly leaf.

Leaf & Flower: The final product consists of leaf with a significant proportion of flowers included.

Root: This means the root, usually the crown as well, and the rhizome, if present.

Flower: Flowers or predominantly flowers, but a certain proportion of stem and leaf may be acceptable.

Seed: Here the seeds – or what appear to be the seeds – are used. Many so-called seeds are technically fruits that contain the seed, but because they are dry and seed-like, they are commonly referred to as seeds.

Fruit: Where the fruit is fleshy, it is commonly referred to as a fruit.

Other Names

The listing of other names is not exhaustive, but tries to cover other English and botanical names that are likely to be encountered.

Usage and Tradition

Usage of herbs is a vast subject, so what is offered here does not claim to be extensive. Its intention is to bring the grower in touch with some of the currently recognised uses of the herb and some of its traditions.

Origin

Knowing a plant's geographical origin helps to give a general idea of its growing requirements.

Climate

A general description is given of the climatic range that suits the species.

Site

Specific details of the herb's preferred growing situation are given.

Exposure: This refers to the amount of light the plant enjoys and the amount of wind it will tolerate.

Drainage: This indicates how tolerant of waterlogging the species is. It is important to understand what is intended by the term drainage – refer to 'Site Assessment' in Chapter 3, 'Making a Start'.

Soil Type: This refers mainly to the texture of the topsoil.

Fertility: The level of fertility required for good productivity. A soil which is not fertile enough can usually be made adequate with sufficient application of compost.

Initial Weed Control: This refers to the preparation and the controlling of weeds on the site prior to planting. Where a bare fallow is recommended, if previous management of the site has greatly reduced weeds this is equivalent to a bare fallow.

Neighbouring Plants: An outline of the effects the species can have on neighbouring plants and how it can be affected by other plants.

Containment: Any measures that need to be undertaken to prevent the herb from spreading into the rest of the garden or elsewhere.

Identification

A description is given of the species, together with information on how to distinguish it from species it could be confused with.

Variety: The preferred variety/varieties from the point of view of production or market requirements.

Propagation

Method: An indication is given as to whether the herb grows from seed and/or by vegetative propagation, and the preferred method.

Timing: The optimum time of year for propagating or planting is given. There is often a lot of flexibility in this, but there are usually optimum times from the point of view of productivity the following season. Growers will have to make adjustments for their own particular location, as the seasons advance at different rates in different regions.

Planting: Method of planting.

Layout and Spacing:

- Layout involves the configuration of the plants in rows, beds or as a meadow-crop.
- Spacing refers to the distances between the plants,

measured from the centre of one plant to the next, and the distances between rows, again measured from the centres of the plants.

Special Requirements: Any particular growing or planting requirements that need attention.

Compost

The figures given for compost requirements are relative rather than precise, as every soil will differ in its fertility. The compost requirement listed is a guide to indicate how much compost the herb might require in an average situation. The grower will need to adjust these figures according to the soil on their property and the quality of their compost.

Compost varies greatly in its density and its nutrient content. A tonne of compost will vary from around 0.8 m to around 2.0 m according to its material make-up, moisture content and degree of decomposition. Wetter compost will be denser, while less mature compost will usually be fluffier and lighter. The rates refer to a compost with a moderate level of moisture: the grower will need to adjust the application rates if they are using a very wet or a very dry compost.

The rates are given in kilograms per square metre. Multiplying this figure by 10 gives tonnes per hectare. For metric conversion tables see page 556.

A measure of 5 kg/m² amounts to a solid compacted layer of 5–10 mm of compost spread over the entire area occupied by the crop, including the spaces between the plants and the rows. If the compost is just being put onto the soil between the rows and this is, say, half the total area occupied by the crop, then it would need to be twice as thick (10–20 mm) for the same rate of 5 kg/m² to be achieved.

Compost thickness on the ground can be hard to measure, as its fluffiness is so variable. I prefer to calculate how much the vehicle or trailer holds and then work out what area that quantity will have to be spread over to get the desired rate of application. After a while you develop a feel for how much should be going on to which crops.

Weed Control

The degree of weed control required and appropriate methods and strategies are noted.

Rotation

Rotation is referred to most commonly with respect to annual and biennial crops and those root crops that are dug and replanted on new ground each season.

Irrigation

A general guide is given for the crop's water requirement over the warmest, driest part of the season. It represents the total of rainfall plus irrigation. Naturally, the actual requirement of each site is going to vary according to the climate, the soil, and exposure to wind and sun and prevailing weather.

The water requirement of a herb is also affected by management practices such as mulching, the frequency and amount of irrigations, and the stage of development of the crop.

Pests and Diseases

Detail on likely or possible specific pest and disease problems is given. This coverage is not exhaustive as plant health is a dynamic situation and imbalances can manifest in many different ways in different situations.

Harvesting

Timing: A guide is given to the stage of development when harvesting should be carried out. Times of year mentioned are a guide only and will vary somewhat according to location and the nature of each season.

Tools: Appropriate tools and equipment for harvesting on a small to intermediate scale are noted.

Technique: Information here focuses on techniques and details specific to the crop. Technique is covered in general detail in Chapter 11, 'Harvesting'

Post-harvest Treatment: This refers to management practices necessary following harvest to prepare for the next harvest or crop.

Drying

Any specific problems or details of special techniques used in drying are noted.

Temperature: To minimise loss of quality in drying, the maximum recommended temperature should not be exceeded. Generally this is 35°C for aromatic crops with volatile constituents, and 45°C for non-aromatic crops. The optimum drying temperatures are in the range 5–10°C below these maximums.

Processing

Market requirements for the crop and any special techniques or problems are noted. For more detail on the various grades see Chapter 13, 'Processing'.

Storage: Usually this is just a reminder that freeze treatment to control moth larvae is necessary before sale or storage.

Yields: These figures are based on my own records and those of a few other Australian growers, usually over a number of seasons. Naturally, many perennial crops will not reach these levels in their first season, and yields are going to vary from one situation to another, depending on soil, management, variety, season, age of crop etc.

Yields are expressed in kg/m^2. To convert this to tonnes per hectare, multiply by 10. For metric conversion tables see page 556.

Marketing

A summary is given of the demand from the various markets. Bear in mind that at present the trade market has probably the largest turnover, the manufacturing market is next in overall size, while the premium-grade market is relatively smaller.

Price

A summary of the various prices offered or paid for different forms and qualities of the herb by various wholesalers and manufacturers 1993–96.

The premium-grade and manufacturing-grade prices are those actually offered to growers, or to growers' marketing bodies.

The trade price represents the prices importers are selling the herb into the Australian market. If you were looking at selling to these same importers or exporting on this market, the prices obtainable would be somewhat lower.

15

Spreading Herbs

Coltsfoot – *Tussilago farfara*
Couch Grass, English – *Elymus repens*
Gipsywort – *Lycopus virginicus*
Golden Seal – *Hydrastis canadensis*
Ground Ivy – *Glechoma hederacea*
Horseradish – *Armoracia rusticana*
Horsetail – *Equisetum arvense*
Licorice – *Glycyrrhiza glabra*

Passionflower – *Passiflora incarnata*
Pennyroyal – *Mentha pulegium*
Peppermint – *Mentha* x *piperita*
Raspberry – *Rubus idaeus*
Spearmint – *Mentha spicata*
St John's Wort – *Hypericum perforatum*
Tansy – *Tanacetum vulgare*
Yarrow – *Achillea millefolium*

The group of 'Spreading Herbs' is characterised by its venturesome nature. Spreading rampantly by means of rhizomes or extensive roots, the management of these herbs is characterised by the need to cope with the consequences of this manner of growth.

The two main problem areas are containment and weed control. Containment is necessary to prevent them invading other herbs and areas where they are not wanted. Weed control can be a particular challenge, because generally these plants do not advance as a solid front, but tend to wander off in various directions, leaving spaces between the new plants where weeds can easily establish. Consequently, maintaining a pure stand requires continual attention.

The general yardstick for inclusion in the 'Spreading Herbs' group is if the plant's rate of spreading is equal to or greater than the Mints, that is, 450 mm a year or more. If its rate of spreading is usually less than this and it forms a more solid stand, it will normally fall into the 'Expanding Clump Herbs' group.

Some species do not fall neatly into one or other group, their rate of spreading varying according to the conditions under which they are growing. While these two groups tend to grade into each other, there are general differences between them in terms of management, mainly in containment and weed control.

Containment

If a species spreads more than about 500 mm a season, particular attention to containing it may be needed. Some among this group have the potential to become serious problems in conditions which favour them, and they have been declared noxious weeds in parts of the country. Care must be taken not to contribute further to our already major noxious weed problems. Some of these weeds owe their origins in Australia to locations where they were being grown as medicinal herbs in the early days of the European invasion.

Containment Systems and Strategies

More than half the herbs in this group need special measures to keep them from invading other areas.

Growing in a Tub

An aggressive spreading plant can be contained by keeping it in a large tub on a concrete slab or similar surface, so any runners or roots emerging from the pot can't find their way into the soil. This system is limited to a very small scale, and is only suited to trials or situations such as herbal clinics where only a small quantity of the herb is needed.

Walls and Barriers

Constructing a physical barrier around the plot is one option. On anything but a small scale this is expensive and can be unreliable because runners may find a way under the barrier or through the tiniest crack.

Separated Containment Beds

In most parts of southern Australia where prevailing conditions in summer are dry, it is feasible to quite effectively contain an individual bed of most of these herbs, even some of the more aggressive ones. This can be done by giving each species its own watered and fertilised area, which is totally surrounded by a buffer zone of least 5 m (10 m for some species) of established grass, which is not watered during the growing season.

As long as they are not winter-active, spreading herbs will not be able to invade the unwatered grass buffer zone because there won't be enough moisture for them in summer and they won't be able to compete with the grass.

However, herbs which are winter-active, like St John's Wort and English Couch, will be able to invade the grass buffer zone in the winter and spring and will need some other system to contain them.

A containment bed will work best if it is circular with a sprinkler at the centre which reaches a little less than the circumference of the bed. Larger containment beds can be elongated – in the shape of two or more circles fused together – with a row of sprinklers in the centre. By sticking to these shapes you can avoid watering any of the grassed buffer zone.

Cultivated Buffer Zones

An alternative method, which works quite well for the less aggressive herbs, is to maintain a permanently cultivated buffer zone around the plot of the herb you wish to contain. The width of this cultivated zone will depend on the distance the herb is capable of sending out its rhizomes or extensive roots.

Why These Systems Work

Both of the above systems work because the runners don't extend indefinitely: they reach out a certain distance and then form another plant, which then extends onward. If runners are prevented from forming new plants outside the bed, by cultivation or by competition and lack of moisture, then the herb is kept from spreading.

Neighbouring Plants

With most spreading herbs, the main consideration in the choice of neighbouring plants is the need for maintaining good weed control and purity of the crops. Sometimes, too, neighbouring plants may harbour pests such as grasshoppers or snails, which can cause problems in some herbs in this group, particularly at emergence or at harvest (see Chapter 9, 'Pests and Diseases').

Avoid planting spreading herbs next to other species of the same group, particularly if they have similar foliage, as they are liable to spread into each other.

If two species of similar appearance, such as Peppermint and Spearmint, get mixed up together, you could have a problem separating them again as their leaves and growth habits are too similar.

Also avoid planting a spreading herb next to other growth types with similar shaped leaves.

One mistake I made in planting out the beds at 'Twin Creeks' was to put a patch of Vervain in between the Peppermint and Spearmint. As the Vervain was cultivated, pieces of rhizome from the edges of the Mint beds were carried into the Vervain where they would start to grow.

The problem was it was difficult to spot the Mint plants among the young Vervain leaves as they were too similar in appearance. It took regular diligent patrols to keep these rogue Peppermint and Spearmint plants from establishing themselves in the Vervain and possibly from there into the each other's beds.

The best species to have adjacent to spreading herbs are those which are easy to maintain free of weeds and which are not going to invade the spreading herb with runners or seeds. However, bear in mind that bushy perennials, particularly woody perennials, may provide a haven for snails.

Also avoid planting spreading herbs next to or down wind of crops such as Chamomile or Dandelion that are likely to shed a lot of seed.

Where possible, it is best to plant spreading herbs away from the edge of the garden. If this can't be avoided, keep a good wide buffer zone between the crop and the grass and weeds, which will be continually sending out invasive runners and seeds. Remember the outside rows of the spreading herb will expand, so allow an additional 0.5 m or so for this as well.

Propagation

Method

In general, vegetative propagation is preferred for spreading herbs. Their reproductive energy is channelled into producing prolific quantities of rhizomes, runners or roots, which develop vigorous new plants, while seed development tends to be fairly weak. Consequently many of the spreading herbs are difficult if not impossible to grow from seed.

Unless there is a specific reason for growing herbs in this group from seed, vegetative propagation will usually give the best and quickest results. A 1 m row will commonly provide enough propagation material to plant out up to 30 m of solid row after one season (or even much more if the runners are divided into short sections and spaced further apart).

This of course will vary according to the species and the vigour of the parent stock. Vigorous plants with lots of space around them will produce many strong runners or roots, but plants tightly confined in dense growth will produce fewer and smaller ones.

Plants vigorously invading fertile new ground will provide the best source of rhizomes: bigger ones, and lots of them. Look for the new rhizomes that have not shot any leaves yet, or ones that are only just beginning to do so. Older rhizomes that have already sent up shoots often have less vigour, but this depends somewhat on the species.

The fatter the rhizome, the more vigorous it will be. I also have the impression that underground rhizomes are more vigorous than green surface runners.

Some species in this group send up shoots from extensive roots. These shoots can be lifted from where they occur spontaneously. Alternatively, the roots themselves can be dug up and planted out in the new plot, or can be set in trays or nursery beds until they send up shoots.

Timing

Where practical, autumn planting is often preferred, provided the ground is relatively free of weeds and the crop can go in early enough for it to put on a bit of growth before winter. The plants will then be well established for the following season and will be much more productive than a spring planting. This is especially recommended in warmer inland regions where the hot weather comes on rapidly and young plants don't get much time to establish a good root system from a spring planting.

However, there is no point in putting spreading herbs into weedy ground in autumn, as the germination and growth of weeds among the crop while it is dormant will make weed control difficult. In these circumstances it is better to plant the crop very early in spring – around the end of August in Tasmania and Victoria (perhaps a little earlier further north), so it is emerging strongly soon after planting and can hold its own against the weeds. You will still have to maintain diligent control of the weeds, but conditions will be easier as spring progresses.

Planting

It is best to plant rhizomes or roots as soon as you can after digging them, but if necessary they can be held for a week or two, as long as they are kept moist and can breathe.

Each segment of a rhizome is capable of forming a new plant, but it is usually best to plant pieces 100–150 mm long. This length ensures there are at least several viable nodes in each piece. Usually each rhizome sends out several shoots, but in a long rhizome the growth is often more vigorous towards the growing tip. The shoots growing from the tip tend to inhibit the growth of other shoots. Cutting into shorter lengths encourages a more even shooting. It also helps make the rows straighter, as long curved pieces are eliminated.

To ensure a good stand, I like to plant rhizomes so their ends are touching in a continuous row. They are laid as straight as possible in a furrow and covered with 25–50 mm of soil. Depth of coverage depends on soil conditions: if moist, plant more shallowly at 25 mm, if dry, give them 50 mm of soil coverage.

Any leaves on the rhizomes are usually best covered with soil: these could transpire too much moisture and desiccate the rhizome in warm conditions, and they may be harbouring pests and diseases (such as Mint rust). By burying them you will break the cycle of any pest or disease (it may not be easily visible at this stage) and will ensure a clean start for the crop.

It is also worth going to some effort to make the rows as straight as possible. This will help in subsequent weed control operations, as it is much easier to follow a straight line than a crooked one. Rhizomes of spreading herbs often have a tendency to shoot sideways a bit, so you will find they come up as a strip 50–100 mm wide, but even so it is easier if these strips are as straight as possible, rather than wandering from side to side or converging.

This does not prevent you from laying the crop out in curved rows if your situation requires such, for example, you may be following a contour. A rotary hoe or a wheel hoe can follow a gently curved row almost as easily as a straight one, provided the curves are smooth and the spacing between the rows is maintained accurately.

Layout and Spacing

It is the nature of spreading herbs to expand vigorously, so, while you may initially lay the herb out in rows, usually by the end of the season it will have formed a solid stand or meadow-crop.

There are two schools of thought on lay out for spreading herbs, involving either close spacing or wide spacing.

Close Spacing

Some growers prefer to plant the rows close together – 300–500 mm apart – with a view to establishing a meadow-crop as quickly as possible. They feel that the sooner the spreading herb covers the ground to form a meadow, the easier weed control will be, because there will be no cultivation to stimulate weed germination, and early yields will be greater because the crop is denser.

If a mulch is being used for weed control this gets maximum yield for the quantity of mulch applied.

Wide Spacing

Other growers prefer to lay their spreading herbs out in rows 900–1200 mm apart. This makes it possible to cultivate between the rows with a rotary hoe or wheel hoe for much longer before the space is closed in by the expanding plants.

This cultivation does stimulate weed germination, but it also gets rid of some of the weed seed reservoir, so as the meadow becomes established it will be on cleaner ground. The additional space between the rows also makes it possible to work in large quantities of compost, which will help sustain good growth for a longer period as a meadow-crop.

Planting out rhizomes in a furrow using a string line to achieve a straight row.

My own preference is for the latter method. Unless the ground is very clean, planting in close rows is going to result in a lot of time-consuming, fiddly hand-weeding to maintain a clean crop. If the ground is reasonably clean I usually go for rows 900 mm apart, but if weed germination is still prolific, I prefer a wider spacing – up to 1200 mm – which will allow inter-row cultivation for a longer period.

Shape of Plots

Because of the spreading nature of these crops and the need to defend them from invasive weeds, the less perimeter the plot has, the easier it will be to maintain. In other words, a bed more approaching a square (or a circle) in shape will have less perimeter to defend than a long narrow one.

There is, of course, a lot of flexibility in this: it is just something to bear in mind. The minimum width of a plot is enough to allow room to swing a harvesting scythe: about 2.7 m.

Raised Beds

If natural drainage is not adequate for the crop, you may wish to use raised beds to obtain better growth. These will need to be very gently profiled so there is no shoulder at the edge to cause problems in harvesting. With most species harvested for their leaf, the last harvest is skimmed off very close to the ground, so any abrupt changes or irregularities in the profile of the soil surface are liable to get scooped up by the blade of the scythe, dumped in with the crop, and end up in the herb tea!

I prefer a broad raised bed with about 3 m between the drains: so three rows, about 900 mm apart, can be laid out in each bed This can then become a raised meadow-crop bed, with drains for access paths (see 'Flat Profile Beds v Raised Beds' in Chapter 5, 'Propagation and Planting').

Weed Control

Weed control in many of the spreading herbs is a real test of the grower's ingenuity and diligence. While a few species do manage to hold their own quite well and require minimum input, at least half of them are quite susceptible to weed invasion, especially during their establishment or while they are dormant in winter.

Weed control among the spreading plants can be very labour intensive, so strategies to reduce the labour input are essential.

Initial Weed Control

For many species in this group, a bare fallow prior to planting is essential to reduce weeds to a manageable level (see Chapter 4, 'Weed Management and Control').

Establishment Period

With vulnerable members of this group, weed control during the establishment phase is critical. From when they start to emerge, regular cultivation and hand-weeding should be maintained to ensure no weeds get the opportunity to establish a hold in what is to become a meadow-crop. Weed control at this stage is much easier than later on when the rows have closed up.

Ongoing Weed Control

Having got these crops off to a reasonable start, you will need to follow up with diligent weed control throughout the season. While the plants are growing strongly, there is a temptation to overlook the weeds, but the consequences can be disastrous later on if weeds manage to establish themselves among the expanding crop.

Regenerative Weeds

Some of the worst weeds for spreading herbs are those with similar growth habits, such as Couch and Sorrel. While you may have eliminated their runners with a bare fallow before planting the crop, there will inevitably be a reservoir of their seeds in the soil, and weeds will continue to germinate for some years. You need to be frequent and thorough in your weeding to prevent these weeds from ever becoming ensconced, for then they are very difficult to eliminate.

Winter-active Weeds

Most of these species of herb become dormant and die back to ground level in winter. This gives winter-active weeds a good opportunity to establish themselves, which can cause real problems.

Weeds that are active in winter can form a dense mat of growth over the beds, through which the herbs can have difficulty emerging in spring. Even if they are not so thick as to smother the crop, the weeds will be picked up when harvesting leaf crops, with a consequent loss of quality in the herb.

Winter Dormancy Plastic Mulch: Many winter-active weeds can be effectively controlled in meadow-crops on a small to intermediate scale with a black plastic 'dormancy mulch' during the middle of winter while the crop is dormant. This mulch will control all annuals and some perennials, such as clovers, which do not have a lot of reserves.

Plastic mulch is not effective against Sorrel or Couch or other weeds with enduring regenerative roots or rhizomes (for more detail on use of dormancy mulch, see Chapter 4, 'Weed Management and Control').

Post-harvest Weed Control

While a spreading herb is growing actively, it will hold its own against the weeds reasonably well. The main problem is in the weeds that have been able to get established beneath it. When the herb is harvested, these weeds have an opportunity to get a start, ahead of the herb's regrowth.

This early regrowth stage is a critical time for weed control. It is important to go over the beds shortly after harvest and, while the weeds are easily accessible, remove any that have gained a hold.

Weed Control in Spreading Herb Meadow-crops

Once a clean meadow-crop is established, it can be relatively easy to maintain free of weeds if suitable strategies are followed, particularly at critical times. In addition to those already mentioned, there are other techniques which may be useful (see Chapter 4, 'Weed Management and Control').

A Light Mulch: A mulch of light crumbly weed-free material, through which the herb is able to easily shoot up, can be used to suppress weed germination in the meadow-crop. This should be 25–40 mm thick for Mints, but can be thicker for species that shoot more vigorously.

Being crumbly, the mulch will still allow chipping with a hoe to get the odd weed which manages to come through and it is not inclined to harbour slugs. Of course, this thickness of mulch will do little to counter established weeds, so it must be put on when the crop is very clean.

This mulch will be more effective in fostering good growth if it contains ample levels of nitrogen and other nutrients. Many mulch materials are high in carbon – which can tie up nitrogen – and they need to be used with caution.

As the meadow-crop also needs surface composting, one option is to make sufficient suitable compost to serve both purposes.

One grower makes such a compost from sawdust and fish-waste and uses it with success on Mint crops.

Rejuvenation

While an established meadow-crop of a spreading herb can be easier to manage in terms of weed control, it will tend to decline in productivity: after two to four seasons (depending on the species and conditions) it may be producing only a fraction of its initial yields.

This is because the growth becomes very tight and root bound and applications of compost to the surface do not make it into the root zone. A meadow-crop of a spreading herb in this state needs to be rejuvenated by cultivation, to give the plants more room and so compost can be worked into the soil.

Technique

Rejuvenation is best done by tilling the plot back into narrow rows by 'tearing strips out of it' with the rotary hoe, so that you are left with narrow rows of herb with cultivated strips between them. These cultivated strips have to be somewhat wider than the cut of your rotary hoe, so you can lap the middle part of the cut.

By working the plot up in this way, you can give it a new lease of life. The next spring a good dressing of compost can be worked into the soil and, together with the breaking down of the dense mat of roots cut up and worked in by the rotary hoeing, this will nourish vigorous new growth.

Width of Rows

About 75 mm is the minimum width of herb that can be left intact without it being lifted out by the tilling, while 300 mm is about the maximum that can be left and still get a good rejuvenation of the plants. If rows are wider than this, the effect of the cultivation will not reach into the middle where the plants are still crowded.

If the bed is being reworked to rejuvenate it and the weed control situation is good, you can leave the rows about 300 mm wide: this will give you a bigger first harvest the next season than you would achieve from 75 mm rows.

Rotary Hoeing

A solid bed of spreading herb runners is a fairly tough engagement for the rotary hoe and it will take a number of passes to get the patch adequately torn up. You will need a capable rotary hoe and a bit of time to spend on the job.

It is important to give the herb a thorough tilling between the new rows: in a trial where we gave a patch just a light working over, this did not have much effect.

String Lines

As it is very difficult to create straight rows freehand, it is best to lay out string lines where you want the rows to be and work between them.

If you look at a spreading herb meadow-crop which has been established for a season or two, you usually notice growth is most stunted where the original row was planted, while the strongest growth is between the original rows. If the new rows of herb can be located over this stronger strip, the growth next season will be more vigorous.

Timing

This method works best if you are able to cultivate the soil at least in late autumn and early spring and preferably several times through winter as well. Heavier soils that are too wet during winter would have to be managed differently: perhaps by being rejuvenated earlier in the season. In any case, inter-row cultivation should follow through the spring to completely kill out the herb plants between the rows and maintain good weed control.

The best time to rejuvenate most crops is in autumn after the last harvest while the soil is relatively dry, as the repeated passes with the rotary hoe will do less damage. The new rows can then be hand-weeded. They can also be flamed to kill small germinating weeds, and they can be covered with a light mulch (about 25 mm thick) of sawdust or something similar to keep them free of weeds.

It may become obvious early in the growing season that a meadow-crop is too run-down to be worth persisting with. In that case, rather than wait for autumn, it is worth while rejuvenating the patch immediately after the first harvest, as it will give greatly increased subsequent harvests and the more vigorous growth will be less prone to pests and diseases.

Longevity of Crops

How long a spreading herb bed can be maintained with repeated rejuvenations I am not sure. My experience with this regime has been that after 5 years in the one place, Mint has still been as vigorous as it was initially, so I would expect with repeated rejuvenation it could be maintained somewhat longer.

While good farming practice would generally be to look at rotating the crop after a number of years, it would take the best part of a season to kill off the runners with a bare fallow. Consequently, the grower will probably want to extend the life of the crop as long as possible, provided the health of the plants can be maintained with applications of good compost and other organic or biodynamic practices

Rejuvenation as a Weed Control Strategy

Where a meadow-crop of one of these herbs has become unmanageable because the weeds are out of control and the area is too extensive to tackle on your hands and knees, the rejuvenation process can be an option for weed control.

Working the patch up again allows the resumption of inter-row cultivation and reduces the area which needs to be hand-weeded. Because this also invigorates the crop, yields are usually increased.

Post-harvest Treatment

After a leaf harvest, the remaining stubble of most herbs in this group should be mowed down very short to promote new growth from ground level. There are a number of benefits to be gained from this practice: more vigorous regrowth, cleaner subsequent harvests, and control of pests and diseases.

Timing is important as mowing needs to be carried out within 3 days of harvest to avoid setting back new growth.

Take care also not to throw debris and soil onto other crops (for more detail see 'Post-harvest Treatment' in Chapter 11, 'Harvesting').

Readers Please Note

As Peppermint is economically the most important herb in this group, I have covered it in some detail, much of which is relevant to the management and handling of other herbs in this group. Consequently the reader is encouraged to read the entry for Peppermint first.

Coltsfoot

Tussilago farfara L   ASTERACEAE

Parts Used: Leaf (Flower)

Coltsfoot is not as widely used as it was several years ago, as it is currently regarded by the Health Department as not safe for public use. Because it contains some pyrrilizidine alkaloids, it comes under a blanket ruling which schedules all such herbs. This is based on evidence of carcinogenic activity of these alkaloids at concentrations very much higher than occurs in normal therapeutic use.

There is some controversy surrounding this ruling because known carcinogens in other commonly consumed substances – such as tea, beer and chlorinated drinking water – are not considered a matter for such drastic action.

It is to be hoped that we can look forward to a time when decisions on these matters will be made by people who have a better understanding of herbal medicine, and then we can all enlarge our little plots of Coltsfoot, as it is a valuable herb for a variety of respiratory disorders.

Coltsfoot is also smoked, alone or in mixture with other herbs. By a quirk of the regulations, while Coltsfoot cannot be sold for human internal use – therapeutic or not – it can legally be sold for 'smoking pleasure': a typical anomaly, as smoking Coltsfoot is quite possibly the only harmful way of taking it. Meanwhile another herb commonly smoked is killing thousands of people in Australia each year and can be purchased at any corner shop.

It is not illegal to grow Coltsfoot, nor is it illegal to use it yourself. Selling it for 'smoking pleasure' or for non-human use is legal, and you can sell it to someone for their dog. So you can safely have it growing in your garden without worrying about receiving a visit from a car-load of heavy-shouldered gentlemen and without risk of a photo of your plants appearing on the front page of the local paper.

Origin

Native to Europe, north and west Asia, and north Africa. Naturalised in North America and elsewhere.

GROWING COLTSFOOT

In spite of the fears surrounding it, the Coltsfoot plant has a friendly feeling about it. Its soft fleshy green leaves with white woolly undersides are quite attractive and its delicate little flower gives the garden a bit of life in early spring. It is a native of temperate Europe and Asia where its yellow flowers are the first to emerge in spring: they sometimes even open and go to seed before the leaves appear. In the past the flowers and the leaves have been regarded as two distinct plants.

Given the right conditions, Coltsfoot is not difficult to grow, in fact the main problem is usually in stopping it.

Because of its invasive nature, you must make sure it is well contained. If it gets into other parts of your garden it can be difficult to eradicate. I speak from experience: it was thoroughly entrenched in our vegetable garden in Canada and though we battled it for five years, it always won.

It never became a problem elsewhere as stock ate it with relish. In 10 years of growing it in this country as a herb crop, I have been able to keep it contained within its own plot.

Climate

Coltsfoot needs a cool to mild climate and does not tolerate hot dry weather.

Coltsfoot *(Tussilago farfara)*.

Site

Exposure

Coltsfoot should be given one of your cooler moister sites, if possible, and sheltered from hot drying winds.

In Tasmania it can take full sun as long as it gets plenty of moisture and has some shelter from drying winds, but in north-east Victoria it never thrived until I set up shadecloth over it during the summer months (see 'Providing Shade' in Chapter 5, 'Propagation and Planting').

Drainage

Coltsfoot is quite happy growing in a swamp. Poor drainage is not a problem for it, but it will do well in a well-drained soil too.

Soil Type

Coltsfoot does well in a range of soils, though those with a reasonably good moisture holding capacity are preferred. If the soil is sandy, Coltsfoot will need very consistent watering. Weed control will be easier on the lighter soils.

Fertility

Coltsfoot needs a very fertile soil for good growth.

Initial Weed Control

A bare fallow is recommended before planting.

Neighbouring Plants

Coltsfoot shouldn't be planted too close to any other crops, in particular, keep it well away from other spreading herbs.

Containment

Coltsfoot can be a major problem if it gets loose in your garden. It is capable of sending out rhizomes up to 1800 mm long and going 450 mm deep, so the safest thing is to contain it in its own bed surrounded by an unwatered grassed buffer zone 5 m wide (as described in the general introduction to this section on Spreading Herbs). An alternative is to surround it with a 2 m cultivated buffer zone, provided you are confident you will be able to keep this zone continuously cultivated all through the growing season.

Dispersal by seed does not seem to be a problem. My Coltsfoot patch has gone to seed every year and the airborne seeds have been dispersed around the garden, both in Tasmania and Victoria, but in over 10 years I have never seen any sign of self-sown plants.

Coltsfoot is not likely to become a noxious weed outside the garden, because most Australian climates are too dry for it in summer. It is relished by livestock and, I imagine, also by native animals. It is dormant in winter, at least in Southern Australia.

Identification

Coltsfoot has long-stemmed single leaves, which grow from crowns and rhizomes. The leaves are round to heart shaped, with wavy edges, 80–180 mm wide, green with furry white underneath. The flowers appear early in spring, before the leaves emerge, and are usually gone by the time the leaves have developed. They are yellow, up to 40 mm wide, and look a bit like Dandelion flowers on single stems up to about 120 mm tall. The tiny seeds are borne on fluffy pappus, like Dandelion puff, but smaller.

The most important thing is to make sure you have Coltsfoot (*Tussilago farfara*) and not a species of Butterbur (*Petasites* spp.), which look similar and have a similar growth habit. *Petasites fragrans* is a garden escaper in Tasmania and Victoria. The main differences are as follows:

- The underside of the leaf of Coltsfoot is white, while in Butterbur it is pale green.
- The leaf of Coltsfoot is 'orbicular-cordate' while Butterbur is deeply cordate (see diagram illustrating leaf shapes in Appendix 1).
- The flower of Coltsfoot is up to 100 mm tall, solitary on the stem and yellow. Butterbur has flowers 100–400 mm tall, with 5–30 pink or purple flowers on each stem.

Propagation

Method

The best method of propagating Coltsfoot is by rhizomes, which it produces in abundance. Coltsfoot rhizomes are similar to those of Mint, except that they are finer, a bit more fragile and often deeper in the soil: I have found them 450 mm below the surface.

Coltsfoot produces seeds, but these are very fine, and I have had no success at all in germinating them. This does not necessarily mean Coltsfoot can't be raised from seed, but it is an indication that this is not a reliable method.

Timing

Optimum planting times for rhizomes are autumn and early spring.

Planting

Plant rhizomes end-to-end in furrows and cover with 25–50 mm soil.

Layout and Spacing

Coltsfoot should be set out as a continuous line of rhizomes in rows 900–1200 mm apart. These will normally spread to form a meadow-crop by the end of the first season.

For harvesting with the scythe, the rows should be in blocks at least three rows wide.

Coltsfoot should not be laid out in raised beds: the higher parts may get too dry and warm for it to thrive, and water-logging doesn't bother it at all.

Compost

Coltsfoot needs a heavy dressing of compost, 4–5 kg/m^2 per annum. As the leaves are harvested close to the ground, it is very important that the compost remains below the level of harvest, so it needs to be fine and crumbly, and to be spread early in spring just as the Coltsfoot leaves are starting to emerge, or a little earlier.

Weed Control

Good weed control is essential during the establishment period. In good growing conditions, weed control is less arduous once Coltsfoot forms a meadow-crop, but measures need to be taken to prevent weeds establishing among it. Black plastic mulch during winter dormancy is often needed to control winter-active weeds.

Rejuvenation

Growth declines after a few seasons in a meadow-crop, so rejuvenation is needed (as described in the introduction to this chapter). After rotary hoeing Coltsfoot, make sure the tines of your rotary hoe are clean and not carrying any pieces of rhizome to establish elsewhere in your garden.

Irrigation

Coltsfoot needs 25–50 mm of water per week through summer. Its soft leaves transpire freely and it needs consistent moist conditions for good growth.

Pests and Diseases

Leafminer

Coltsfoot sometimes gets a build-up of a little grub which lives inside the leaf and eats away the flesh, similar to the way that Beetroot and other fleshy leaved vegetables sometimes are attacked.

This is more likely to occur if the harvest is delayed. If it is a problem, the best thing is to mow the stubble down low after harvest to destroy any remaining older leaves that might be harbouring the grubs. This will greatly reduce the incidence of grubs in the new growth.

Grasshoppers

In hotter climates, if Coltsfoot is adequately shaded and watered, grasshoppers are not normally a problem for it, but they will attack it severely if it is under stress.

Snails

Being moist and shady, Coltsfoot is a favourite hangout for snails. If they are around, they can cause problems in harvesting (see management strategies under 'Snails' in Chapter 9, 'Pests and Diseases').

Harvesting

Timing

In Tasmania and Victoria you can take three harvests: the first early December, the second around late January, and the third around early April.

As the flowers are well gone by this time, harvest stage is judged by the condition of the leaves. The aim is to allow the leaves to develop fully, and then harvest before the oldest leaves start deteriorating. Because Coltsfoot tends to form a single layered canopy of leaves, the older leaves are in the midst of the crop, so it is difficult to avoid them.

Tools

Coltsfoot is best harvested with the catching scythe. Although it is quite low growing, with its soft fleshy stems it is easily cut and swept up by the catcher. Harvesting with a reaping hook is fiddly and the shorter leaves get left behind.

Technique

Some skill and accuracy with the scythe are required, but it is easier than it might seem. Coltsfoot has such soft and juicy stems it feels rather like cutting butter.

Post-harvest Treatment

Because the stems are so soft, there is no need to get the mower out to cut down the stubble, just going over it again very closely with the scythe is usually good enough.

Don't forget to put the shadecloth back on, if it is needed.

Drying

Coltsfoot is quite easy to dry and is not much subject to discolouration. It does have a tendency to get blown off the screens when dry, because it consists of large single leaves, which float off easily in the wind.

The stems are the slowest part to dry and need to be brittle if they are to be included in the finished product.

Temperature

Coltsfoot should be dried at temperatures up to 35°C.

Processing

Coltsfoot is used as a tea. For manufacturing, aerial parts or first screening are used (or they were until recent legislation took effect).

Tea Grade

Coltsfoot is one herb where most of the stem is pushed through the screen and included in the tea grade. This is because the stems are quite green and lack fibre. They break up easily into short lengths when rubbed on the screen, except for the odd coarser stem, which can be discarded.

Because of the woolly nature of Coltsfoot leaves, the broken particles cling together when rubbed, so under-screens cannot be used. Nevertheless, it is not difficult to get a good appearance with around three passes through the rubbing screen.

Aerial Parts

As it consists of mostly single leaves, Coltsfoot is very slow to feed through the chaff-cutter so it is easier just to rub it to a first screening for manufacturing use, and include the stems.

Storage

Freeze before storage to control moth larvae.

Yields

The best yields in north-east Victoria have been around 0.25 kg/m² per annum In Tasmania, where conditions suit it better, yields have been up to 0.4 kg/m² per annum. These figures are for both aerial parts and tea-grade leaf.

Marketing

There is only a small market for Coltsfoot at present owing to the restrictions on its use. It can be legally sold for 'smoking pleasure' or for non-human use.

Price

Coltsfoot used to bring around $25–30/kg for 5-star tea grade.

Manufacturing grade used to be $15–20/kg for organically grown Coltsfoot, but there is currently no requirement in this sector.

Trade price is around $10/kg.

Couch Grass, English
Elymus repens (L) Gould   POACEAE

Parts Used: Root (i.e. Rhizome)

Other Names: Twitch, Quack Grass, Quitch Grass, Dog Grass, Scutch, *Agropyron repens*, *Triticum repens*.

English Couch Grass is a frequent weed of cultivation in Tasmania, and does occur in other states. It is referred to as 'English' because in warmer regions the name Couch Grass usually refers to a quite different grass, *Cynodon dactylon*, which does not have the same properties. It is commonly grown in lawns, but also occurs as a weed of cultivation. Unfortunately there is some confusion about this and many people are under the impression that *Cynodon dactylon* is the Couch Grass of herbal medicine.

English Couch Grass is a significant herb, used as a herbal beverage (sometimes roasted), and as a diuretic.

GROWING ENGLISH COUCH GRASS

I don't know of anyone in Australia who has intentionally grown English Couch Grass, but over the years I have had a fair bit to do with trying to get rid of it.

This is a species which most people in their right minds would not invite into their garden. However, you may have an area where it is already well established and may wish to exploit it either on an ongoing basis or while in the process of getting rid of it. Realistically, for continuity of production and good quality, Couch Grass will need to be

English Couch Grass (*Elymus repens*).

managed as a crop. This is necessary in order to keep it free of weeds and maintain fertility for good production. Many of the management techniques for other spreading herbs should be appropriate for this species.

Identification

English Couch Grass grows to 1200 mm with green to bluish green leaves 60–300 mm long and 3–10 mm wide. The seed heads are in long spikes, with the spikelets 10–20 mm long arranged closely along the stem, and look similar to those of Ryegrass, except that the individual spikelets are rotated 90° in relation to the stem. Often English Couch Grass doesn't send up any seed heads. The plants spread by means of creeping wiry rhizomes.

Make sure you have got English Couch Grass, *Elymus repens*, and not 'lawn' Couch Grass or Bermuda Grass (*Cynodon dactylon*.) English Couch Grass can be distinguished by the following characteristics:

- It is winter-active in most parts of Australia, so if your plants have green leaves in winter you have the right one.
- 'Lawn' Couch Grass *(Cynodon dactylon)*, has stiff leaves. It is shorter in height (up to 600 mm), often forming a dense mat of leaves, usually less than 10 mm wide. The seed heads are palmate spikes: the seed-bearing branchlets radiate from the central stem like the stretched-out fingers of an open hand (hence the Latin name *dactylon*). Actually they more resemble the spokes of an umbrella, with the fabric removed. This species usually sends up a good show of seed heads.
- *Cynodon dactylon* goes dormant in autumn and does not emerge until the ground warms up in spring. The leaves are frost tender.

A number of species of *Agrostis*, variously known as Bent, Brown Top and Red Top, are temperate grasses with rhizomes. Their growth is similar to *Elymus repens* but shorter, with finer leaves and finer rhizomes and they form open feathery panicles. English Couch Grass has rhizomes 2–3 mm thick, while these other grasses usually have rhizomes only half this thickness.

Harvesting

Before commencing harvest mow the patch off very close and rake it clean, so none of the surface debris gets mixed up with the rhizomes.

The rhizomes will need to be picked over to remove any extraneous material and dead rhizomes. These latter could prove a problem in the first harvest of an already established patch: there could be a lot of them, as they are slow to decay. A regularly harvested patch would have a lower proportion of dead material in the harvest.

For more detail on Couch Grass see 'Information Charts'.

Gipsywort

Lycopus virginicus L   LAMIACEAE

Parts Used: Aerial Parts, Leaf

Other Names: Bugleweed, Gipsyweed, Water Horehound.

Gipsywort is used for certain thyroid and heart conditions, and as a sedative and antitussive.

GROWING GIPSYWORT

Gipsywort seems to prefer a cool temperate climate. At 'Twin Creeks' in north-east Victoria, it did very well in a site near the creek, but as it was crowding the Coltsfoot, I moved it to a more exposed site where it copped the full heat of summer. It did not grow as vigorously there and after several seasons it had virtually died out after being attacked severely by the Mint Flea Beetle.

Identification

Gipsywort (*Lycopus virginicus*) grows to about 1 m, with grey-green coarsely serrated ovate to oblong lanceolate leaves 60–100 mm long, which are opposite on the square stems. Flowers are small, whitish, sometimes marked with purple, in loose whorls in the axils of the leaves.

Lycopus europaeus is very similar, except that its lower leaves are pinnate, and its flowers are in denser whorls.

Native Gipsywort (*Lycopus australis*) is similar but grows taller and has narrower leaves with only a few teeth.

For more details see 'Information Charts'.

Gipsywort (*Lycopus virginicus*).

Golden Seal

Hydrastis canadensis L
RANUNCULACEAE

Parts Used: Root

Golden Seal is a highly valued medicinal herb used as a tonic, a respiratory herb and a digestive. It was adopted from the American Indians who used it as a dye and a medicine.

Origin

Shady deciduous forests in north-east North America.

GROWING GOLDEN SEAL

Golden Seal is probably responsible for more debilitating cases of herb fever than all other herbs put together. Just the sound of the name seems to be sufficient to arouse fantasies of fabulous fortunes to be made growing it.

The prospect of a return of $40 000–$100 000 per acre is enticing, but interested growers should do a few sums before plunging in off the deep end.

The first thing to be aware of is that Golden Seal takes at least 3–4 years to reach a harvestable stage. It is very specific in its growing requirements. Keeping the plant alive and healthy is not easy, as it is very delicate. In Australia, Golden Seal plants often fail to survive past their second year. At this stage the growing requirements and feasibility of Golden Seal as a crop here have not been worked out. Until sufficient trials have been carried out to establish its needs, growing it as a crop will be a very risky venture.

The cost of propagation material is very high. Golden Seal produces rhizomes that extend 200–600 mm and give rise to new plants. Rhizomes can cost up to $100 each in Australia.

Golden Seal (*Hydrastis canadensis*).

The plant does produce seed, but this usually proves to be non-viable. Golden Seal plants are notorious for not surviving in quarantine, so losses can be considerable.

Golden Seal needs 80% shade, either artificial or under trees: its native habitat is deciduous broadleaf forest, basically the same as Ginseng, so it will need additional shade in Eucalypt forest. It requires a cool to cold climate and a well-drained site.

The other big risk is that by the time the grower builds up sufficient propagation material and then grows that on to produce a harvest, the price may not be as attractive as it is now. Many new growers are reportedly planting it overseas, so if their production comes onto the market, the price may fall.

Even if Golden Seal does maintain its present price, and even if growers succeed in getting a good harvest, they would gross no more, and often less, than they would from many other herb crops that are easier and cheaper to grow. When the costs of producing Golden Seal in this country are deducted, net returns are likely to be much lower than for most other crops.

Notwithstanding all this, being such an important herb, growers who are interested in Golden Seal should be encouraged to give it a try, cautiously and on a very small scale until its needs and feasibility have been better assessed.

For more detail on this herb see 'Information Charts'.

Ground Ivy

Glechoma hederacea L LAMIACEAE

Parts Used: Leaf

Other Names: Gilly-over-the-Ground, Alehoof, *Nepeta glechoma*.

In Canada we used to occasionally make a somewhat bitter but not unpleasant tea of Ground Ivy, or Gilly-over-the-Ground as we called it. It is used internally to soothe coughs and as a mild tonic, and externally for bruises and sore eyes. It also has a long traditional use for clarifying and flavouring beer.

GROWING GROUND IVY

Ground Ivy is really a minor herb, and one I have not had much experience with. It used to grow as a weed in our vegetable garden in Quebec, together with Coltsfoot and Couch Grass, and we spent many hours endeavouring to get rid of it, with little success. In those days I did not

Ground Ivy (*Glechoma hederacea*).

know about bare fallowing, and in that climate it may not have worked anyway, as the growing season was short and humid with ample summer rain (in 5 years we only ever watered our garden once, in an unusually dry spell when it didn't rain for two weeks).

Identification

Ground Ivy is a low creeping herb that forms dense mats. It has thin wiry stems, which run over the ground, striking roots where they make contact. The leaves are dark green, hairy, round, up to 40 mm wide, with rounded teeth on the margins. The flowers are pink to mauve.

Harvesting

Because Ground Ivy grows in a low dense mat, harvesting with the scythe may not work too well: it may have to be cut with a reaping hook, which would be fairly slow.

For more details see 'Information Charts'.

Horseradish

Armoracia rusticana Gaert Mey et Scherb
BRASSICACEAE

Parts Used: Root

Other Name: *Cochlearia armoracia.*

Horseradish is quite well known in the form of Horseradish sauce, used as a condiment by those of us who enjoy the sensation of our insides being about to burn through. Horseradish also has medicinal uses, though, mainly for the respiratory and digestive systems.

It is one of those unfortunate plants which have been blessed with two botanical names, both in regular usage. Besides *Armoracia rusticana*, it is also known as *Cochlearia armoracia*.

Origin

Horseradish is a native of south-east Europe.

GROWING HORSERADISH

Horseradish is a fairly easy plant to grow. It establishes quickly and its attractive lush green foliage makes a strong claim over the ground it occupies.

Climate

Horseradish prefers a cool to mild climate. While it grows reasonably well with adequate irrigation in Victoria, it is not as vigorous as it is in Tasmania.

Site

Exposure

Horseradish will grow in most sites in cool to mild climates, but needs some moderation of temperatures in hotter regions.

Drainage

It will tolerate waterlogged conditions, but harvesting can be difficult if the soil is too wet and sticky.

Soil Type

It will grow in a variety of soils, though it does not seem to like red volcanic soil.

Horseradish *(Armoracia rusticana)*.

Fertility

It needs a high level of fertility.

Initial Weed Control

A bare fallow is recommended before planting.

Neighbouring Plants

It can be very invasive, so keep it away from plants where this could be a problem.

Containment

When Horseradish is growing vigorously, it sends out extensive roots up to about 2.4 m long, which give rise to new plants along their length. In Victoria these were not much of a problem, but in Tasmania they were heading all over the garden. Horseradish, therefore, is best contained in its own bed.

Being winter dormant and moisture loving, it is easily contained by an unwatered grass buffer zone.

Bear in mind that Horseradish will take a fair bit of effort to get rid of from where you plant it, so make sure you put it in the right place.

Identification

Horseradish can grow up to 1500 mm tall, with large basal leaves 300–1000 mm long, lanceolate with toothed margins and long petioles. The leaves on the flowering stems are shorter and are often divided almost to the point of being pinnate. The flowering stems are up to 1 m high, with clusters of small white flowers. The roots are stout, fleshy and tapering to about 600 mm long, creamy white with typical Horseradish aroma and flavour. It also forms numerous thick extensive side roots.

Propagation

Method

Propagation is normally by root division – any piece of root will grow, though it is best to plant pieces big enough to give rise to strong plants. Pieces 75–100 mm long and 12–25 mm thick are a good size. Horseradish is also supposed to grow from seed, but I can't recall mine ever having set seed. Once a bed of Horseradish has been planted, it is usually not necessary to plant it again after harvest as there are inevitably plenty of pieces of root left in the soil which will generate a new stand.

Timing

Optimum times for initial planting of Horseradish are in autumn or early spring.

Planting

Plant root pieces in furrows and cover with 50–75 mm soil. It may be of advantage to poke deep holes and set the pieces in vertically.

Layout and Spacing

The root pieces should be placed about 300 mm apart in rows about 900 mm apart.

Compost

A heavy dressing of compost, 4–5 kg/m^2, should be applied in spring. If this is done early, it can be worked in before the new shoots start to emerge.

Weed Control

As Horseradish is a very vigorous, strong growing plant, which shades its competitors, ongoing weed control is not very demanding, but don't let weeds get established amongst it as they could cause problems, reducing yields and making harvest difficult.

Irrigation

Horseradish needs 25–50 mm of water per week through summer.

Pests and Diseases

Leaf Blight

In Victoria the Horseradish I grew suffered from a rust fungus which destroyed most of the leaves in late summer. I think this was a result of stress, growing in hot dry conditions, because it seldom occurs on Horseradish in Tasmania. When I increased the frequency of irrigation in Victoria, the rust was significantly reduced.

Snails and Slugs

Snails and slugs are reported to be a problem if they are able to attack the newly emerging shoots in spring. As the snails in 'Twin Creeks' never made it down to the Horseradish bed, I didn't have this problem, but I have seen a lot of slug damage during a wet summer in Tasmania.

Cabbage White Butterfly

One book refers to cabbage white butterfly as a major pest of Horseradish, but I have never seen any evidence of it, in spite of cabbages being grown nearby.

Harvesting

Timing

Some authorities say to harvest Horseradish after 2–3 years and this may be better in places where the season is shorter or growth slower. I have always harvested mine every year, though in Victoria yields were not very good. However, I was afraid that if left it for a second year, the larger plants would compete too much with each other for available moisture and would be even more stressed.

Horseradish can be dug in autumn as soon as the leaves have died down, or during winter. If the soil is subject to excess moisture, it is best to harvest it as early as possible, while conditions are suitable.

Tools and Technique

You can dig Horseradish with a garden fork. Some of the taproots go down some way, and they may need a spade if you are going to chase them, but most of the roots are near the surface.

On a larger scale I imagine Horseradish could be harvested with a potato digger, though this may need modification.

As some root needs to be left behind to establish the next crop, there is no need to be totally thorough in harvesting.

Post-harvest Treatment

Level off the ground while it is still friable so the bed is ready for next season.

Drying

Large pieces of root – those thicker than your thumb – need to be split or chopped before spreading on the screens to dry. Artificial heat will be needed, but keep temperatures down to avoid loss of the volatile constituents.

While Horseradish is commonly used fresh, if carefully dried it will retain much of the characteristic hotness.

Temperature

Horseradish should be dried at a temperature of not more than 35°C.

Processing

There is not a great deal of demand for tea-grade or culinary-grade dried Horseradish in my experience: most of what I grew went for manufacturing liquid medicines. For this purpose it was sent to the manufacturer in the coarsely chopped form which came from the dryer.

Storage

Freeze treatment to control moth larvae is necessary before sale or storage.

Yields

My best yield in north-east Victoria was only around 0.2 kg/m² per annum. In a cooler climate where Horseradish grows more vigorously, yields would be higher.

Marketing

As the market is limited, the grower should have a firm undertaking from a purchaser before attempting this crop.

The fresh root is used for making sauce and sometimes in manufacturing herbal medicines.

Price

Premium quality organically grown Horseradish is currently around $30/kg for 5-star tea grade.

Manufacturing-grade, organically grown Horseradish is currently around $10–15/kg.

Trade price is around $8/kg.

Horsetail, Field

Equisetum arvense L   EQUISETACEAE

Parts Used: Aerial Parts

Other Names: Shave Grass, Bottlebrush.

Field Horsetail makes a pleasant mild-flavoured herb tea, which is quite popular. It is an important herb with a fairly wide usage medicinally as a diuretic and for genito-urinary disorders. It is also used in biodynamics as a fungus control known as Preparation 508.

Horsetails are a relic of an ancient vegetation form which once dominated the countryside. Our coal seams consist, for a large part, of the remains of members of the Horsetail family, but there were no survivors on this continent. Sheoak foliage does bear an interesting resemblance, but it is totally unrelated, botanically.

Origin

Field Horsetail is native to Europe.

Field Horsetail (*Equisetum arvense*).

GROWING HORSETAIL

Field Horsetail is a delicate, feathery looking plant, somewhat reminiscent of a horse's tail, upside-down. The foliage is in segments, not unlike our native Sheoak. Given the right conditions, it can spread very rapidly. Fortunately most parts of southern Australia are too hot and dry for it in summer, and it goes dormant in winter.

In spite of its reputation as a terrible weed, Field Horsetail can be a difficult plant to establish in many regions, but once you figure out what its growing requirements are, it can be very productive.

This species has been declared a noxious weed in some states, though there is no evidence of it becoming a problem other than in summer rainfall areas. In regions with a dry summer, it will survive only in moist irrigated sites and only then if it gets some protection from full sun.

Being dormant in winter, it is unable to take advantage of moist conditions at that time of year or to compete with winter-active species.

Climate

Field Horsetail prefers a mild moist climate.

Site

Exposure

Field Horsetail doesn't like dry winds at all: it can suffer severe burning in a windy site.

In areas with hot summers, it needs a cool site. It may also need shade until it is well established.

Drainage

Field Horsetail will tolerate poor drainage and waterlogging.

Soil Type

Good moisture retention is important, but as very thorough cultivation will be necessary for good weed control, friability is important, too.

Fertility

Field Horsetail needs a moderate level of fertility.

Initial Weed Control

Weed control must be very thorough. Field Horsetail does not compete well with weeds.

Neighbouring Plants

Because of its incredible invasiveness and its ability to grow in the shade of other plants, Field Horsetail mustn't be grown near any other herbs.

Containment

Given the right conditions, Field Horsetail can spread up to 9 m in a season, so it needs a wide buffer zone to contain it. However, in most of southern Australia where summers are on the dry side, it will not be able to invade an unwatered well-grassed buffer zone 10 m wide, (as described at the beginning of this chapter). It goes quite dormant in winter, so there is normally only a brief period in spring and autumn when prevailing conditions are moist enough for it for it to grow without irrigation.

Any growth that manages to get a start among the non-irrigated grass soon burns off in summer.

This containment system worked very well for more than 5 years in north-east Victoria, but I have been unable to persuade Plant Quarantine to let me try it in Tasmania. Field Horsetail is regarded as a serious weed in northern Europe, which scares the pants off local authorities, so it is hard to convince anyone that our conditions here don't suit it.

While it does form spores, these apparently only rarely ever succeed in getting established, and in 5 years I never saw any evidence of this happening.

In summer rainfall areas, Field Horsetail has a much greater potential of becoming a noxious weed and it has become naturalised near Sydney. Growing it in these regions is not recommended as it could be very difficult to get rid of.

Under cultivation it should be grown in a bed of its own. I made the mistake of putting it into a plot shared with a few other invasive herbs, expecting to be able to keep them separate with 2 m cultivated strips between them. While the Coltsfoot and Horseradish stayed within their bounds, the Horsetail spread through the whole plot and was a bit of a nuisance. But it never managed to go more than a 0.5 m into the surrounding grassed area which was left unwatered in summer.

Heavy Metal Uptake

Field Horsetail has the ability to take up significant quantities of heavy metals, such as lead, cadmium, zinc and copper, so it should not be grown in soils where high levels of these are present, such as in old orchards or places where heavy dressings of superphosphate have been applied in the past.

Its high silica content indicates it is able to pick up silica and convert it into soluble compounds, and it even has the ability to accumulate gold, though probably not in payable quantities.

Identification

Field Horsetail is a non-flowering plant producing two types of stem:

- Vegetative (sterile) stems are numerous, green, grooved, jointed, 1.5–5 mm thick, with a whorl of branchlets at every joint, up to 500 mm tall, appearing late spring and dying down to the ground in autumn. See colour plate 4.2.
- Fertile stems are few, emerging in early spring only, grey-brown, up to 300 mm tall and 8 mm thick, unbranched, jointed, with a spore producing cone at the tip. These die off after shedding spores.

The leaves on both types of stems are merely teeth which form a sheath around the joints.

The plants spread by means of black wiry rhizomes and they also form small tubers.

Make sure you have the true Field Horsetail (*Equisetum arvense*). It is not easy to obtain in Australia. Most herb nurseries only have *E. hyemale*, but they commonly call it *E. arvense* because they have never seen the real thing.

Equisetum arvense is usually no more than 500 mm tall. The stems are segmented, with a whorl of lateral branchlets coming from each segment of the main stem. The plant is a soft green colour. It sends up short, grey-brown, unbranched 'flowering' shoots in the spring, but these die after shedding their spores.

Equisetum hyemale grows taller: up to about 1000 mm. Its stems are thicker, only rarely branching and then just a single branch here and there. This species is a darker green colour and I don't recall ever seeing it form spore-cones. When you see the two species, there is really no mistaking them for each other.

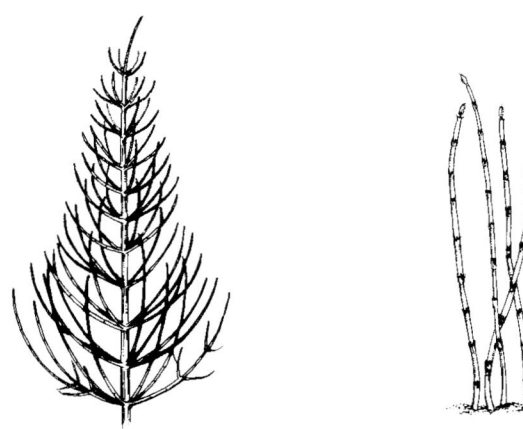

Field Horsetail *(Equisetum arvense)*, left; and Rush Horsetail *(E. hyemale)*, right.

Equisetum hyemale, Scouring Rush or Rush Horsetail, is not a preferred species, though it is very useful for scouring pots and pans. Unfortunately this species is often mistaken for *E. arvense*. Even more unfortunate is that it is very hard to get rid of, as it tolerates hot sun and is winter active into the bargain. *E. hyemale* has a much greater potential as a noxious weed in our climate and has apparently established over large areas in parts of New Zealand.

Propagation

Method

Propagation of Field Horsetail is by rhizomes. These are thin and black and can appear deceptively lifeless.

Theoretically Field Horsetail can also reproduce from spores, but this is rare.

Timing

Optimum planting times are in autumn or early spring.

Planting

Plant rhizomes end-to-end in furrows and cover with 25–50 mm soil.

Layout and Spacing

Plant in continuous rows 900–1200 mm apart. To ensure moister conditions, lay out a level culture rather than raising the beds at all. As the plants are harvested very close to the ground, the soil surface needs to be fairly smooth.

Special Requirements

In Victoria, and also in an exposed site in Tasmania, I found it impossible to establish Field Horsetail without shade. The hot sun of summer in Victoria and the drying winds in Tasmania would burn off the tender shoots.

Under 50% shade provided by a shadecloth tunnel, it soon built up to a dense stand, but after a while I noticed it was coming up thickly beside the tunnel and growing quite happily in the full sun.

Subsequently I found that once the stand is thick enough, provided it receives copious regular watering, Field Horsetail can grow in full sun. In this situation the densely growing plants are able to protect each other and the shade can be taken off.

However, in most locations shade would need to be provided at least for the first season or so, until the stand becomes dense enough. Shade could be in the form of tunnels or a cover as described under 'Providing Shade' in Chapter 5 on Propagation and Planting. Whether the shade can be dispensed with later will depend on the climate, site and watering regime.

Compost

Field Horsetail needs compost at the rate of 2–3 kg/m² per annum.

Having seen it growing in barren ditches in Canada, I was initially under the impression its prehistoric metabolism and powers of extraction enabled it to survive without the nutrient supply that most modern plants require.

Certainly good yields were sustained for some years without additional compost, but it was situated on some fertile alluvial soil. Eventually the growth started to decline and we began giving it moderate dressings of compost to which it responded.

Compost needs to be applied in early spring and should be in a friable form so it can be spread evenly over the surface in a thin layer that won't be picked up in harvesting.

Weed Control

Weed control is the big hurdle for successful growth of this crop. The problems encountered with weed control in other spreading herbs are even greater with Field Horsetail, because of its rapid spreading and because it produces very little shade. Even in a dense vigorous stand, it doesn't suppress weeds much at all.

Consequently success in weed control depends on your skills in management, and long hours on your hands and knees picking weeds out of the crop.

First there needs to be a very good initial weed control: inter-row cultivation is possible for a short while, but the plants soon fill up the spaces, and you are left with hand-weeding as your only option during the growing season. If this is done thoroughly immediately after each harvest, and followed up a couple of weeks later before the fronds become too tall, the crop will reach harvest without a weed problem.

Once it reaches a bit of height, it is best to avoid walking through Field Horsetail, as this knocks the fronds over, making harvest a bit difficult.

A black plastic winter dormancy mulch works well for eliminating many troublesome winter-active weeds, but it will not stop regenerative perennials like the various Couch Grasses and Sorrel.

If your crop is becoming too weedy, you may need to look at a partial fallow or rejuvenation to reduce the weed problem.

Partial Bare Fallow

One thing that did work successfully for controlling weeds was a partial bare fallow. The Field Horsetail had gradually become infested in places with 'lawn' Couch Grass and Sorrel – too much to ever clean up by hand – so we decided to bare fallow the affected area over summer.

After a couple of months of repeated cultivations, the Couch Grass and Sorrel, were obviously cleaned up, so we left the plot to rest, planning to replant it with Field Horsetail in autumn. But in a while a thick growth of Horsetail appeared, free of Couch and Sorrel, and we just had a few germinating weed seeds to deal with. Evidently the Field Horsetail's reserves were greater than those of the weeds. Mind you, the weed growth had never been allowed to proliferate, so weed roots and runners were quite small and thin, without a lot of reserves.

Rejuvenation

Field Horsetail does not need rejuvenating as frequently as Mint to maintain good growth, but it may need to be rejuvenated to eliminate a perennial weed problem.

Irrigation

Field Horsetail needs 25–50 mm of water per week through summer.

Careful attention must be paid to maintaining adequate moisture levels because if the plant is stressed at all, the foliage will be burned. Don't water during the heat of the day, and try to avoid watering immediately before harvest, as the plants tend to bend over with the weight of water and take a day or so to straighten up again. Large droplets or watering in windy conditions can flatten the crop.

Pests and Diseases

I have never seen any sign of anything eating Field Horsetail: evidently the creatures with the teeth and digestion to cope with this plant died out many millions of years ago, in Australia at least.

Harvesting

Timing

There are usually three harvests for Field Horsetail in Victoria: the first in November, the second in January, the third in March, with a possible small fourth harvest in May.

Tools

Field Horsetail is easily harvested with the catching scythe. The reaping hook can be used, but is much slower.

Technique

Harvesting is straightforward. Horsetail cuts very easily, though the silica in it tends to take the edge off the blade fairly quickly.

Post-harvest Treatment

After harvest, Horsetail needs to be mowed close to the ground and then thoroughly weeded and followed up by further weeding during early regrowth.

Drying

Horsetail is about the fastest drying herb I know of. In hot weather it is sometimes dry enough to rub after only a couple of days on the screens. This rapid loss of moisture would explain why it suffers so easily from hot sun or drying wind.

Processing

Horsetail is widely used in tea form and also to manufacture liquid medicines.

CAUTION: In handling dried Field Horsetail, it is important to take precautions against breathing the dust, as it contains significant amounts of silica. If you are handling any volume, a proper respirator should be used.

Tea Grade

The dried herb is easily rubbed through a 2½ or 3 dent screen, usually two passes are sufficient to produce a good tea grade.

Manufacturing Grade

Because Field Horsetail is so easy to rub, for manufacturing grade I used to do a first screening, with one pass through the rubbing screen.

Alternatively it can be put through the chaff-cutter.

Storage

To control moth larvae, freeze treatment should be done before storage.

Yields

At 'Twin Creeks' in north-east Victoria, a well established patch of Field Horsetail was yielding 0.35–0.5 kg/m^2 per annum.

Marketing

Field Horsetail is used in moderate volumes for a variety of purposes.

Price

Price to growers for premium quality organically grown Field Horsetail is currently around $25/kg for 5-star tea grade.

Manufacturing-grade, organically grown Field Horsetail is currently $12–16/kg.

Trade price is around $9/kg.

Licorice

Glycyrrhiza glabra L FABACEAE

Parts Used: Root, Rhizome

Other Name: Liquorice.

Licorice is a major herb with wide usage medicinally, mainly as a demulcent, expectorant and to flavour other less palatable herbal combinations. It should be noted, though, that excessive use can lead to potassium depletion. It is also used as a herb tea – alone or in combination with other herbs – and to flavour Tobacco and confectionery. The licorice sticks we buy today are mostly aniseed and other things, but traditionally they were made from the dried juice of the Licorice plant.

Origin

Licorice is native from southern Europe to Pakistan.

GROWING LICORICE

Licorice is not difficult to grow, if you have the conditions which meet its requirements. It usually takes 2–3 years to produce a harvestable crop, though you can dig it sooner if you need some. Growth the first season is quite slow while the Licorice establishes itself, and you may feel a bit disappointed because the plants haven't done much. However, during the second year they will start to put on some growth and you may find new plants emerging in unexpected places as it sends out its long rhizomes.

Harvesting is the big challenge. The roots grow in all directions – including straight down – for as much as a metre, and are very long and fibrous.

Licorice *(Glycyrrhiza glabra)*.

Climate

Licorice is generally grown commercially in climates with a hot summer – Spain, Persia, India, southern Russia – but it has been grown commercially in warmer parts of Britain. It thrives in northern Victoria and should be a productive crop over most of the southern mainland.

It will grow in Tasmania, though less vigorously. Nevertheless, it is worth exploring as a crop in cooler regions because British-grown Licorice (grown in Yorkshire, latitude 53°N) was noted for its delicate flavour.

Site

Exposure

Locate Licorice in as warm a site as possible. Excessive wind may burn the leaves to some extent.

Drainage

Licorice needs good drainage.

Soil Type

A very deep alluvial soil is preferable, as its roots like to go a long way down. Harvesting is difficult on heavier soils and where there are stones. It does not do well on clay soils.

Fertility

Licorice needs a very high level of fertility for good growth.

Initial Weed Control

A bare fallow prior to planting will help reduce weed problems.

Neighbouring Plants

In favourable conditions, Licorice sends out long invasive rhizomes, so it is best to keep it on its own, or at least down one end of your beds or next to crops where you can cope with the incursion of a few underground rhizomes giving rise to Licorice plants here and there.

While the rhizomes may extend up to 8 m, the plant doesn't totally take over like some invasive species.

Containment

If containment is desired, an unwatered mowed grass buffer zone of about 10 m will be necessary to contain Licorice.

Left to its own devices, it will wander around and would probably survive in most climates in southern Australia without irrigation. It is mentioned as a persistent weed of cultivation in its native habitat.

Identification

Licorice (*Glycyrrhiza glabra*), grows up to 1500 mm high. The leaves are pinnate with 4–7 pairs of ovate leaflets, 25–50 mm long. Inflorescence consists of 20–30 lilac-blue flowers 10 mm long in loose racemes 10–150 mm long, in the axils of the leaves. These form separate pods 10–25 mm long.

The plant forms several long horizontal stolons or rhizomes, which can extend up to 8 m in length. The roots and rhizomes are yellow-brown inside, with a characteristic strong sweet flavour.

Several species and/or varieties of Licorice are grown commercially, but *Glycyrrhiza glabra* is the only species recognised in herbal medicine. There are also several other species of *Glycyrrhiza*, including an Australian species. Getting the right propagation material is where you can run into problems.

In one herb identification fiasco we bought a quantity of tissue-cultured supposed Licorice plants from a prominent nursery in New South Wales and put them in a large trial plot. These grew magnificently – over 3 m high – but at the end of the season the roots totally lacked the characteristic sweet taste. In our ignorance we concluded that because the crop has to grow for 3 years, it must take that long for any sweetness to develop.

Finally at the end of 3 years, when the roots still tasted bitter, with a rough bean-like flavour and no sweetness at all, I realised something was wrong.

A specimen was sent off to the National Herbarium, which referred it to Kew in England. The closest they could arrive at was that it was definitely not *Glycyrrhiza glabra*, nor any other cultivated species, but probably *Glycyrrhiza fetida*: its leaves certainly had a fetid aroma when crushed, rather reminiscent of an oil refinery.

It took another year to get rid of the stinking things, and the seedlings kept coming up for some time afterwards.

One of the points which had kept us on the wrong track was an incorrect illustration in a reference book. Instead of *Glycyrrhiza glabra*, it shows some other species with a short, dense clustered inflorescence, similar to that of our trial plants. *G. glabra* has a longer, more open inflorescence, almost the length of the pinnate leaves, which is, incidentally, what the book's text describes, but I hadn't read it closely enough.

This same false stinking Licorice (probably *Glycyrrhiza fetida*) has shown up in some herb display gardens, labelled as *Glycyrrhiza glabra*. Quite possibly it has been distributed widely throughout Australia.

It can be distinguished by its taller growth, 2–3 m, its leaves which have a strong smell when crushed (*G. glabra* leaf is virtually odourless), its flowers and pods are in dense spiny clusters (*G. glabra* has separate individual flowers and pods), its lack of rhizomes, the grey-brown colour of the inside of the root and the lack of sweet flavour in the root (*G. glabra* root has a yellow-brown colour and characteristic sweet flavour, even in younger plants).

Propagation

Method

Propagation is normally by planting the rhizomes. These extend horizontally from the plant and look much like roots. They can be identified by the whitish buds at the tips.

Licorice can also be grown from seed, but this can be difficult to obtain. In view of the prolific amount of seed produced by the suspected *Glycyrrhiza fetida*, identification of plants grown from seed should be confirmed.

Timing

Early spring planting is probably the most convenient as this is when rhizomes will be available from a previous, winter-harvested, crop.

Planting and Layout

Sections of rhizome about 150 mm long can be planted 300 mm apart in rows, with 900–1200 mm between the rows. Make sure the sections have their bark undamaged as it tends to peel off easily.

Licorice is commonly grown in rows hilled up quite high to facilitate the harvesting of the deep roots.

As its development is limited in its first season, an annual crop of another species can be grown in the inter-row spaces, provided it is compatible with the Licorice. In traditional Licorice regions overseas, vegetable crops are grown between the rows in the first season.

By the second and third seasons, the Licorice should occupy all the space between the original rows.

Rotation

Good practice would suggest rotating the crops rather than continuing to grow Licorice in the same place. There will inevitably be a fair bit of Licorice coming up afterwards from pieces left behind in the soil, so subsequent crops would need to be able to cope with this.

Compost

Licorice needs a heavy dressing of compost at the rate of 5 kg/m^2 per annum, or possibly more.

Weed Control

During the first season, good weed control should be maintained while it is still possible, with inter-row cultivation, followed up with hand-hoeing. Licorice does go dormant in winter, but emerges strongly in spring, so winter weed growth is not as much of a problem as it is with Mint and other spreading herbs.

In Licorice's second and third seasons, weed control will be limited to hand-hoeing among the plants, as inter-row cultivation will no longer be possible. If the weed control has been thorough prior to and during the establishment period, this later hand-hoeing should be manageable.

The crop is not likely to be smothered by weeds, but they will reduce yields and possibly interfere with harvest.

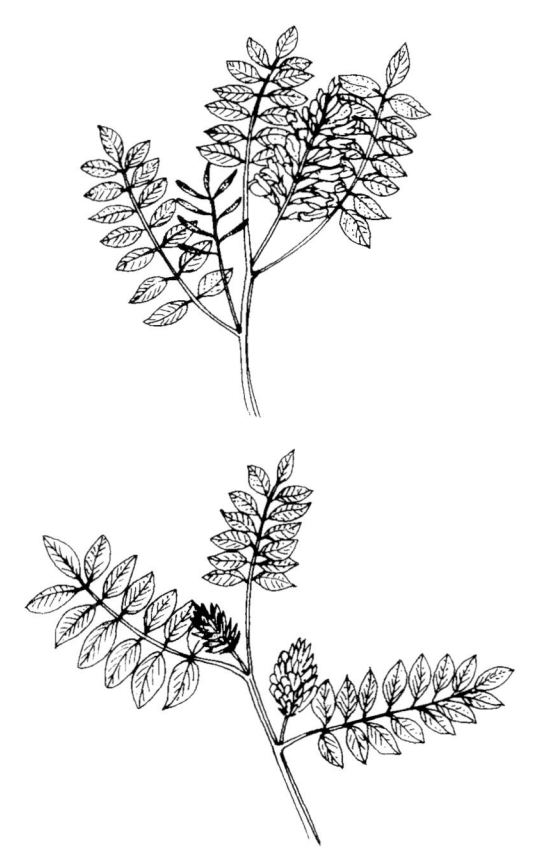

Comparison of the flower and pods of preferred species of Licorice, *Glycyrrhiza glabra* (top), with an unacceptable species, believed to be *G. fetida* (bottom), which has been widely distributed in Australia.

Irrigation

Licorice is very deep rooted and is adapted to hot climates, so while irrigation is necessary for good growth, it doesn't need to be as closely monitored as for some herbs. About 25 mm of water per week should be sufficient for good growth, but monitor soil moisture and growth to ascertain what your crop's needs are.

Pests and Diseases

So far none of the Licorice in various trial plots has suffered significantly from any pests or diseases.

Harvesting

Timing

In Portugal and Spain Licorice is harvested after 3–4 years. The operation can be carried out in autumn, winter or very early spring, as Licorice does not emerge until the ground has warmed up a bit.

Tools

For digging by hand, a good spade or shovel is needed as the roots are too long, too deep and too fibrous to dig with a fork.

Technique

Traditionally Licorice was harvested by first digging a trench 600–900 mm deep beside the first row, and then excavating across the Licorice patch, advancing the trench as the harvest progressed. Sounds like heavy work, but if the soil is light, deep and free of rocks, it should be easier. With this system, once the first trench is dug, the rest of the soil is basically just moved sideways. Care should be taken to keep the topsoil on top.

One grower described harvesting his trial plot as the hardest job he had ever done: the roots were extending in all directions as well as down to some depth.

If it is any consolation, the roots don't seem to go as deep in heavier soil, but the going being harder, they will still be as hard to dig anyway, and the crop does seem to do better in deep light soils.

The remains of the stalks must be cut off where they join the roots: either before or after digging, whichever suits your system best.

Mechanical Harvesting

If Licorice is to be grown on any significant scale, mechanical harvesting needs to be adopted, as the labour of hand-digging is going to be too daunting for most Australian growers.

One large-scale grower uses a blade plough 600 mm deep to lift the roots, which are then picked up with a modified rock picker.

Care would need to be taken in mechanical harvesting not to damage the roots excessively, as the bark is easily stripped off. Damaged roots will be harder to wash effectively.

Mechanical harvesting would need to somehow avoid inverting or mixing the soil profile and burying valuable topsoil.

Saving Propagation Material

During harvest, good rhizomes for propagation should be selected and set aside so they are not damaged during the handling and washing process.

Post-harvest Treatment

After harvesting, the plot will look like a World War I battlefield, so a fair bit of soil will probably need to be moved back into place.

Drying

Before washing, propagation material for the next crop should be selected and put aside. The rest can then be washed thoroughly.

The roots should be put through the chaff-cutter. Chopping would be difficult to do after they have dried, as they are so fibrous and tough.

Drying will require some artificial heat.

Temperature

Licorice should be dried at temperatures approaching, but not exceeding 35°C, as it contains some volatile oils.

Processing

Licorice is used as a tea and in liquid medicines, however, growers would not normally have to do the final processing to tea grade, unless they are doing their own packaging for distribution, in which case they would need a hammer-mill.

Storage

Freezing should be carried out before storage to control moth larvae.

Yields

Mrs Grieve (*A Modern Herbal*) refers to average yields in Europe, (presumably in the Mediterranean) of 4–5 tons/acre after 4 years. This amounts to a yield of 0.25–0.3 kg/m^2 per annum.

Yields in northern Victoria have been comparable.

Marketing

There is a big market for Licorice, but whether it is a viable proposition in Australia is another question. Small-scale growers would need to get a return of $20–30/kg to make it worth while, depending on yields and harvesting methods.

Price

Price to growers for premium quality, organically grown Licorice is around $25/kg for 5-star grade.

Price to growers for manufacturing-grade, organically grown Licorice is currently around $12–18/kg.

Trade price is around $9/kg.

Passionflower

Passiflora incarnata L   PASSIFLORACEAE

Parts Used: Aerial Parts

Other Name: Maypop.

The 'passion' of Passionflower refers to an association of the form of its flower with Christian symbology.

In favourable conditions, it sends out extensive roots – sometimes 10 m long – from which arise vigorous new plants, but the crown which gave rise to these off-shoots often dies out over winter.

Passionflower is a close relative of the cultivated Passionfruit and, like it, is a perennial climbing vine. It differs from Passionfruit in that it dies back to ground level every winter and emerges again in spring, though often in a different place.

Its main use is for its sedative action, especially for insomnia. The fruit is edible – with a taste very similar to Passionfruit – and is quite popular in the area to which it is native.

Origin

Passionflower is native to the south-east of the United States, as far north as Pennsylvania and Illinois.

GROWING PASSIONFLOWER

Because of its unusual growth habits, management of this species is somewhat different from that of most other herbs. Provision needs to be made to enable it to climb, and you need to be able to cope with its walkabout nature. On the other hand, weed control is much less demanding than for many other herbs, as the Passionflower emerges through and climbs away from most weeds.

Climate

Passionflower prefers hot summers, but winters in parts of its natural habitat are very cold, so no part of Australia should be too cold for it in winter. While it thrives in northern Victoria, it has not done well in cool windy sites in southern Tasmania.

Site

Exposure

Full sun is preferred, with protection from cold winds.

Drainage

Passionflower needs good drainage.

Soil Type

Passionflower does well on medium to heavy loams: I haven't tried it in lighter soils.

Fertility

Passionflower needs a reasonably fertile soil.

Initial Weed Control

Weed control is not as critical as it is for many other herbs, because Passionflower grows vigorously and tall, so it is not bothered much by the presence of weeds.

Neighbouring Plants

Neighbouring crops need to be able to cope with emerging Passionflower plants, which tend to come up all over the place.

Passionflower (*Passiflora incarnata*).

Identification

Passionflower is a climbing vine, up to 5 m long. The dark green leaves, 70–120 mm long, have 3 (occasionally 5) lobes, which are pointed. The exquisite flowers are white to purple, 40–80 mm wide, forming ovoid edible fruit about 50 mm in diameter and light green turning yellow when ripe. (See colour plate 5.4)

The plant sends out extensive roots up to 10 m long, which give rise to new crowns along their length.

Distinguishing Passionflower from Other Plants

The main plants to distinguish this species from are the related Passionfruits.

One distinctive feature is that the Passionflower dies back to ground level in autumn, while the Passionfruits are evergreen. The leaves of Passionflower have a characteristic sweet, tobacco-like aroma due to the alkaloids present. The lobes of the Passionflower are a little narrower and diverge more than those of the Passionfruits.

Propagation

Method

My original stock was grown from seed, but some growers have had difficulty getting seed to germinate.

If you have access to some established Passionflower, the easiest thing is to dig up rooted shoots as they appear in late spring, and transplant them. Make sure you get down deep enough to get a piece of the horizontal root, with the shoot attached.

Another option might be to dig up the root system and take root cuttings, though I haven't tried this method.

Timing

Seed can be started in a greenhouse in early spring, or in the open when conditions have warmed up.

Shoots can be transplanted in late spring. It is hard to know where to find them until they emerge. This is around the end of October in north-east Victoria, but will be later in cooler areas.

Planting

Shoots can then be set in prepared ground, 300–600 mm apart, in rows 1.5–2 m apart.

Layout

While Passionflower vines will happily ramble over the ground, harvesting is easier and growth is better if some sort of trellis is erected for the vines to climb on. I just stretched lengths of cyclone wire – 300 mm above the ground at the bottom, and reaching to about 1200 mm high – held up by a few steel posts, so the vines could hang on and then reach out into the air from there.

Because it is such a tall plant, it can interfere with sprinkler irrigation by creating sprinkler shadows on the far side of it and depriving neighbouring plants of water. For this reason, and because of its invasive nature and its tendency to become a haven for weeds, Passionflower is best located at the edge of the garden or in its own separate plot.

Containment

Passionflower may not be easy to contain in conditions that favour it, as its extensive roots travel up to 10 m, and it has no trouble pushing up through thick grass. At 'Twin Creeks' in north-east Victoria, I initially set it along the edge of the garden: it subsequently sent up new shoots some distance into the garden and also out into the grassed area for some 10 m, but did not seem to venture too far where there was no moisture in summer for it.

It is best to have Passionflower in its own separate patch or, failing that, adjacent to crops maintained in rows so, without having to walk all over plants, it is possible to knock off Passionflower shoots that come up in the wrong place.

In climates with sufficient summer rainfall, Passionflower might take off into the bush, like another Lantana: growers should exercise some caution.

Compost

Passionflower needs compost at the rate of 3–4 kg/m^2 per annum.

In the first year, compost can be worked into the soil beside the plants as they are put in. In subsequent years, the compost can be spread on the surface after the vines have appeared in late spring, or earlier if you wish: just try to anticipate where they will come up. The vigorous Passionflower shoots will have no trouble pushing up through compost.

Weed Control

In favourable conditions where it is a vigorous plant and growing over the top of everything else, Passionflower is not greatly troubled by weeds. However, a reasonable degree of weed control is necessary to reduce competition for moisture and nutrients, and to avoid problems when harvesting.

The Passionflower vine's tendrils curl tightly around everything they meet, so if there are tall weeds growing through the vine, they will be gathered up in the harvest.

Also if the Passionflower is next to a total weed control zone, you won't want a lot of weeds going to seed among it and blowing into your other herb beds.

Mulching

This is one herb mulching works really well for. Because it grows well above the level of the mulch, there is no danger of it being picked up in harvesting, and weed seeds in the mulch are not a problem for Passionflower. Grass mowings from around the perimeter of the gardens are usually handy at the time the Passionflower needs mulching.

I found there were two ways I could go about mulching it: early mulching or late mulching.

Early Mulching

The grass growing on the Passionflower patch can be mowed in spring a few weeks before the vines are due to emerge from their dormancy. The whole patch can then be covered in a heavy mulch, and the Passionflower will emerge up through it.

Mulching this way gives a good suppression of weeds for most of the summer, but it has the disadvantage of slowing down the warming of the soil due to the insulation of the mulch, so the shoots are slower to emerge. In a hot summer climate this is not a big problem, and the Passionflower does not seem to be greatly affected by getting a start a week or two later. However, in a cooler climate the soil would take a lot longer to warm up and growth could be severely set back with this method.

Late Mulching

Alternatively, the weeds over the Passionflower can be kept mowed until just before the Passionflower is due to emerge, then left for a few weeks, until the plants are all 300 mm or so high. Then the mulch can be put around them in a heavy enough application to suppress the weeds.

This method allows the Passionflower to get away earlier, but the mulching involves more work and is not as effective close up to the plants: you will probably have to pull a few weeds here and there.

With both methods there is still a need to maintain weed control during the season by going over the patch periodically, putting a bit more mulch on where necessary, and cutting any weeds down which are among the vines, but weeding is not a big job. Mowing and mulching, with a bit of hand-weeding, will give adequate control over the growing season.

Cooler Climates

In cooler climates mulching the Passionflower may not be the best approach. Mulch definitely should not be put on until the soil has thoroughly warmed up, but even so, the lower soil temperatures under the mulch over summer could affect the growth of the Passionflower. It may do better with the greater warmth of bare soil.

Irrigation

Passionflower is quite deep rooted, but still benefits from regular irrigation. About 25 mm of water per week over summer should be sufficient.

Pests and Diseases

Passionflower does not seem to be prone to any pests or diseases. Its leaves have a sedative action, so no doubt if anything starts to eat them, it just goes to sleep.

Harvesting

Timing

Passionflower is normally harvested towards the end of the growing season, after the fruit have formed. It is harvested at this late stage because its main active constituents are alkaloids and these develop their highest concentration as the plant approaches maturity. However, one should not wait too long, as the leaves will lose their colour, and the plant will be difficult to dry with ambient air if it is too late in the season. It is best to harvest it at the end of March or the beginning of April.

Tools and Technique

The easiest way to harvest Passionflower is to cut the stems with a reaping hook, just above the yellow leaves, and pull the vines away from the trellis. This progresses reasonably quickly as each stroke with the reaping hook yields a good armful of vine.

Post-harvest Treatment

The remains of the vines can be left to die off. Later on they should be pulled off the trellises, so there is no old material left to get mixed in with the next season's crop.

Drying

Most parts of the Passionflower vine dry easily, except for the fruit, which are extremely slow to dry. Consequently, it is best to pull any fruit off as you harvest and spread the herb on the screens. If you wish to include the fruit in the dried herb, you can cut them up so they dry faster. Otherwise they can be excluded: they don't seem to contain active constituents.

If any whole fruit is left in with the leaves and stems, it is either going to take ages to get them dry, or the moisture in the fruit could cause the herb to go mouldy later on in storage.

Temperature

Passionflower can be dried at temperatures of up to 35°C.

Processing

Passionflower is used to some extent as a tea. It is also used a lot as liquid medicine, for which aerial parts are needed.

Tea Grade

Tea grade should be rubbed through a $2^{1/2}$ or 3 dent screen until a tea grade free of excessive stem is obtained. Sometimes the stem is finely chopped or milled and included with the leaf to make a tea-grade aerial parts.

Manufacturing Grade

Manufacturing grade Passionflower can be prepared by putting the dried vines through the chaff-cutter. The main thing is to watch out for any fruit on the vines that haven't dried adequately.

Storage

Freeze treatment should be carried out before putting the bags into storage.

Yields

We used to get 35–50 kg aerial parts per annum from about 50 m of trellised vines. As these were mostly along the edge of the garden in a single line, it is a bit hard to extrapolate this into a yield figure, but it would be in the region of 0.35–0.5 kg/m^2 per annum.

Marketing

There is reasonably large demand for Passionflower as aerial parts for manufacturing and a small demand for tea grade.

Price

Price to growers for premium quality, organically grown Passionflower is currently around $24/kg for 5-star tea grade.

Price to growers for manufacturing-grade, organically grown, is currently around $16-20/kg.

Trade price is $10–12/kg.

Pennyroyal

Mentha pulegium L   LAMIACEAE

Parts Used: Leaf & Flower, Aerial Parts
Other Name: Pudding Grass.

Pennyroyal is a relatively low growing herb, which can be found growing wild in parts of southern Australia. It is usually in pastures subject to waterlogging, where it thrives, owing to its enjoyment of wet situations and its unpalatability to stock.

Pennyroyal is very useful for keeping mosquitoes off when doing a job like transplanting in the evening when clouds of them descend to take advantage of a sitting target. A bunch of fresh Pennyroyal rubbed firmly over the skin is an effective insect repellent. The herb is used in some commercial repellents.

Traditionally it was used as a condiment in meat puddings, and a few people enjoy it as a herb tea.

Pennyroyal has a long history of medicinal use, including for some types of indigestion and as an emmenagogue, so pregnant women should go easy on the herb and should definitely avoid the oil.

GROWING PENNYROYAL

I have had mixed success growing Pennyroyal. It thrived in one garden in Tasmania, but at 'Twin Creeks' it gradually died out. This was surprising as it was growing wild in the district, but it may have been something to do with the soil.

The Pennyroyal did well where it was subject to waterlogging in winter. Wherever it is growing in the wild, it will be found to be subject to winter waterlogging, so evidently this situation suits its nature. The patch which died out was well drained.

While Pennyroyal is often found growing wild, it is seldom in a pure enough stand. It usually has a small amount of grass scattered through it: just enough to look terrible in the finished product. For this reason it may be necessary to bring Pennyroyal into cultivation.

Identification

Pennyroyal has much-branched stems, prostrate or erect, growing to 300 mm tall, and forms over-ground runners. The leaves are dark green, aromatic, more or less hairy, and oval shaped, 8–20 mm long, and strongly aromatic. The flowers are reddish purple to lilac blue, and in dense whorls in the axils of the leaves. In winter the plant dies down to a mat of leafy runners.

Variety

There is some variation in the height of growth and even a low-growing variety used for lawns. For dried herb production, the taller variety which grows to about 300 mm is needed, otherwise harvesting is too difficult.

Weed Control

Avoid using a black plastic dormancy mulch for weed control. It can kill off Pennyroyal because the plant doesn't really go dormant, but reverts to low leafy growth in winter.

Processing

Aerial parts are used for manufacturing. In preparing a tea-grade leaf and flower, repeated screening may be necessary on account of the fine stalks which readily pass through the screen.

Marketing

As this is a low-volume herb, market-wise, growers need to be sure they have a market for it before committing themselves to any significant production.

For more detail see 'Information Charts'.

Pennyroyal (*Mentha pulegium*).

Peppermint
Mentha x *piperita* L   LAMIACEAE

Parts Used: Leaf, Aerial Parts

Peppermint is the most popular of the tea herbs. Nearly everyone enjoys its refreshing flavour, and it also has important medicinal uses, mainly for indigestion and respiratory problems. It is used to a degree as a culinary herb – it usually features as a flavouring for sweets and desserts – though not as much as Spearmint.

While other Mints have ancient traditions, Peppermint is relatively new on the scene, but its flavour has made it a favourite herb, being very agreeable to most palates. Its essential oil content makes it a valuable herb medicinally, both in its own right and to mask other less pleasantly flavoured herbs.

Origin

Peppermint first appeared in England about 300 years ago as a naturally occurring hybrid of Water Mint (*Mentha aquatica*) and Spearmint (*Mentha spicata*).

GROWING PEPPERMINT

Even though it grows well in quite a range of conditions, Peppermint is a challenging crop. To achieve consistently good yields, you need to have a good understanding of its growing requirements, its pests and diseases, and, above all, you need to know how to control weeds growing with it. There are effective organic techniques of management which can be used, though, and it can be a productive and rewarding crop to grow.

It is also a crop that can be confidently expanded as there is a big demand for high quality dried Peppermint. This gives the grower the opportunity to benefit from economies of scale which are not possible with many other herb crops.

Peppermint is an attractive plant with its dark green foliage tinged with purple. A patch of Peppermint in its flush of growth just prior to harvest is a powerful sight. Its wonderfully strong aroma make handling it quite an experience, too.

Climate

While Peppermint will grow in a wide range of climates, it is better suited to regions where there is a contrast of temperatures. It seems to grow more vigorously in these climates and to develop more essential oil.

In Australia the best regions for growing Peppermint, from the point of view of producing a fine quality for tea use, seem to be Tasmania, Victoria and at least the tablelands and Dividing Range of New South Wales. These are the areas in which I know Peppermint will develop a good strong flavour. I would expect western New South Wales and most of South Australia would be suitable, too. The critical factor seems to be contrasts of temperature during the growing season.

However, much of this zone would not be suitable for commercial Peppermint oil production, as the market for this has very exacting requirements.

Tasmania, with its erratic climate, stimulates a high oil content in Peppermint, and has acquired a reputation of being the only place in Australia where decent Peppermint can be grown.

When I moved to north-east Victoria, however, I was very pleasantly surprised to find the Peppermint I grew there was perhaps even stronger than my Tasmanian harvests, and it yielded well. The leaf seemed to be thicker, and the oil content in late summer was quite overpowering when it was being rubbed. The actual flavour was probably a bit coarser than Tasmanian Peppermint, which seems to me to be a bit more subtle in its fragrance.

While the weather of north-east Victoria is less changeable, the daily contrasts of temperature stimulate good oil development in the Peppermint. This is particularly so in the valleys of the foothills of the Dividing Range, where summer day temperatures in the valleys typically reach 40°C, but at night cool air flows down from the surrounding hills and temperatures drop off markedly.

This region also has quite cold weather in winter, with good frosts, and this is probably important for the full development of plants like Peppermint, which emanate from temperate climates.

Each region's Peppermint, and even each farm's crop, seems to have its own character, which changes through the season, no doubt in response to subtle differences in climate and soil.

In regions whose climate lacks contrasts of temperatures, such as along the east coast of New South Wales and Queensland, there will be lower levels of essential oil in Peppermint and similar herbs. The oil levels seem to decrease as one goes north.

The flavour of one sample of Peppermint tea I tasted from the south coast of New South Wales was quite good, but others from the north coast of New South Wales and southern coastal Queensland have had distinctively less flavour. These were grown in a warm subtropical climate where there is not much contrast of temperature on a daily or seasonal basis.

From a therapeutic point of view, being of lower oil content, such Peppermint would probably not be as effective. However, growers in this area may still find they have a market for Peppermint tea. While it may be a little inferior to that grown in the preferred regions, if skilfully dried it will be of much better quality than most imported Peppermint.

Site

Exposure

While Peppermint will grow in a range of situations, including quite shady positions, it needs as much sun as possible for best growth. It quite thrives in hot conditions, and will tolerate windy sites as well.

Drainage

Good drainage is preferable for best growth. Peppermint will tolerate some excess of moisture, but even though it is a Mint, it does not thrive in situations subject to severe waterlogging – growth will be adversely affected and it will even die out in places.

Soil Type

Peppermint will do well in a wide range of soils. The ideal is a soil light enough to allow cultivation for weed control in moist conditions, but still with sufficient ability to retain moisture for good growth in summer, as Peppermint has a high water requirement. The best soils would be volcanic red soils and alluvial loams.

Heavy loams and clay soils can deteriorate badly with the cultivation necessary to maintain weed control and stimulate growth. They will also be unworkable for too much of the time when they are wet, which means weeds will get away at critical times.

Fertility

Peppermint needs a very fertile soil. Of course this is going to require heavy dressings of compost, but the soil needs to be in pretty good heart to start with. The pH of the soil does not appear to be particularly critical: Peppermint does well in the 5 to 7 range.

Initial Weed Control

Good weed control is the linchpin for successful production. Peppermint's spreading nature means that if the weeds are not under control in the ground it is spreading into, they will be continually coming up among the crop, where it is difficult to deal with them.

With Peppermint it is vital that all the regenerative roots of any perennial weeds have been eliminated before planting: this usually means a summer bare fallow before planting (for more detail on this technique see Chapter 4, 'Weed Management and Control').

If possible, avoid sites where weeds have become established in the past leaving a big reservoir of seeds in the soil, as these will continue to germinate for a long period and you will have to deal with them coming up in the Peppermint. If you do have to use such a site, try to reduce the reservoir of weed seed by repeated cultivations alternating with conditions favourable for the germination of these weeds. It might be preferable to precede the planting of the Peppermint with a crop that allows cultivation and clean weed control, such as an annual row crop or one of the short-term root crops, like Marshmallow or Burdock.

Neighbouring Plants

There is a traditional belief that Nettle growing in the proximity of Mints will increase their oil content. I can't say I have seen anything in my experience to verify this. I did have Spearmint growing next to Nettle at 'Twin Creeks' and while it did very well there, I couldn't detect any significant difference between it and another patch of Spearmint growing well away from the Nettle.

On the other hand, I can be fairly definite about which plants not to put next to Mint. This is more from the point of view of maintaining good weed control and purity of the crop than because of some deleterious effect the other plants will have on it: though sometimes neighbouring plants may harbour pests, such as grasshoppers or snails, which can cause problems.

First, avoid planting Peppermint next to other spreading herbs as it they are liable to spread into each other. In particular, keep Peppermint well away from other Mints and Pennyroyal. If these manage to invade each other and get mixed up, you could have a problem sorting them out, as their leaves and growth habits are so similar.

Identification

Peppermint grows 300–900 mm tall, the stems are more or less purple and square, the tops branching somewhat. The leaves are green to purple-green, lanceolate to ovate-lanceolate, pointed, with toothed edges, 40–80 mm long, 10–25 mm wide. Flowers are usually mauve, occasionally white, in a conical spike 30–75 mm long, and are borne at the top of the main stem (see colour plate 5.5).

The plants form rhizomes, and die back to the surface of the ground in winter.

Variety

There are different strains of true Peppermint, falling into two categories called Black (with purple stems) or White (with lighter coloured stems). However, the colour of the stems of the Black Peppermint varies according to growing conditions and at times they can be quite green.

The main variety grown for essential oil production in Tasmania is known as Black Mitcham. It has proved to be an excellent variety as a dried herb for use as a tea. There are other varieties of Peppermint used for oil production, but I have not had the opportunity to try them out. They may well be quite suitable from the point of view of flavour, as these varieties are selected for their high levels of essential oils in proportions that appeal to the palate. Growers should assess the acceptability to their market before commencing with such varieties.

Black Mitcham Peppermint (*Mentha* x *piperita*).

Varieties Grown from Seed

Being a hybrid, Peppermint will either be sterile or, if it does form seeds, they will not grow true to type: anything could come up.

Unfortunately, seed of a number of spurious and inferior varieties is often sold as Peppermint by seed companies. These are not worth trying: at best, they will lack the intensity and quality of flavour of Black Mitcham, and often they just taste horrible.

One of my early learning experiences in herb growing was to try growing Peppermint and Spearmint from seed I bought from a seed catalogue. They both came up as a woolly murky-green Mint, the plants indistinguishable from each other, and really not fit for human consumption.

Nursery Stock Varieties

Many herb nurseries also offer plants of what they call Peppermint, but these are usually just a relatively strong flavoured form of Spearmint: again not worth bothering with when the Black Mitcham is so superior. Why waste all the work of growing a crop by starting off with an inferior variety which is going to leave you with a product difficult to sell?

Propagation

Method

Peppermint is grown from rhizomes. These need to be young, healthy, and white if they are underground, or pink and green on the surface. Older rhizomes that have already sent up shoots will be weaker or not viable at all.

Timing

Mint can be successfully planted out from March to November in southern Australia, but planting in early spring or mid-autumn is preferred.

If weed control over winter is not a problem, autumn planting will give the Peppermint a better chance to establish and will produce better crops the following season. Otherwise, plant very early in spring.

Planting

Plant rhizomes end-to-end in furrows and cover with 25–50 mm soil.

Make sure any leaves on the rhizomes are covered. These could transpire too much moisture and desiccate the rhizome in warm conditions, and they may be harbouring the odd rust colony. By burying them you will break the development of any rust (it may not be easily visible at this stage) and ensure a clean start for your Peppermint.

Layout and Spacing

Peppermint can be laid out in continuous rows 900–1200 mm apart. For harvesting with the scythe, these should be in blocks at least three rows wide.

Peppermint in its first season, growing in rows. Being on cleaner ground, this crop has been planted in rows 900 mm apart.

By the second season the rows have expanded to form a solid meadow-crop.

Usually by the end of the first season, the Peppermint will have spread to form a solid meadow-crop which can be maintained for two seasons or so.

An alternative approach is to plant the rows closer together: 300–600 mm apart. Provided weeds are not a problem, this can give higher initial yields and a meadow-crop is established sooner. However, weed control can be more labour intensive with closer spacing, as there is soon no space to pass with the larger implements.

My own preference is for the wider spacing, as it allows for more inter-row cultivation. This means the Peppermint is expanding into cleaner ground and there are fewer weeds in the subsequent meadow-crop.

Compost

Being a 'heavy feeder', Peppermint needs generous applications of compost to sustain good growth. Depending on your soil's inherent fertility and the quality of your compost, it will need 5 kg/m^2 or more per annum, or 50 t/ha. This is normally all applied in early spring.

For an autumn planting, give the rows a light side dressing on planting and then a full application the following spring.

Peppermint needs your best most mature compost if it is going to make optimum growth. The compost can be worked into the soil the first season, as there is still plenty of space between the rows, but usually by the second spring the crop will have spread to a solid stand or meadow-crop. Getting heavy applications of compost onto a meadow-crop can be a bit tricky, as you need to avoid smothering the plants and to avoid pieces of compost being retained on the leaves where it could find its way into the harvest (for detail on technique see under 'Spreading Compost' in Chapter 7, 'Compost').

If a good application of compost is made in spring, this should sustain growth through the rest of the season. With crops like Peppermint there can be problems in putting on mid-season applications of compost if there is no inter-row space left to work it in. Surface applications later in the season are subject to excessive drying out, and there is more likelihood compost will get picked up with the harvest, as the later harvests are cut quite close to the ground.

If necessary a mid-season application of compost can be made if very fine crumbly material is available. This should be put on as a light dressing immediately after harvest and watered in. Care must be taken to ensure none of it will be picked up in the next harvest.

Weed Control

Weed control is where we come to the crux of the matter, as success in growing Peppermint depends very much on this being effective. In this regard, Peppermint is one of the most challenging of herb crops to grow.

From their initial establishment, the Peppermint plants are characterised by an open spreading growth, with spaces between which allow weeds to develop where they are difficult to get at. When the first shoots emerge, they never come up in the straight line they are planted in, instead they will be up to 75 mm away on either side.

Meanwhile the weeds are inevitably germinating, too. While cultivation between the rows is easy, this broad strip of young Peppermint shoots mixed with weeds needs to be cleaned up or the future of the crop is in jeopardy.

On a small scale you can simply get down on your hands and knees and remove weeds with your fingernails or by scratching among the young shoots with a weeding claw. On a larger scale this is too labour intensive, so the grower needs to find ways to reduce the magnitude of this task.

Initial Bare Fallow

A good bare fallow prior to planting is really an essential prerequisite for growing Peppermint, as it can greatly reduce the numbers of weeds you have to deal with.

Later Planting

Plant a little later, allowing time for additional shallow cultivation(s) to reduce germinating weeds. This is something of a trade-off, with a reduction in yield in the first season being accepted in order to reduce the labour of weed control.

A Light Mulch

A light mulch, applied before both the Peppermint and the weeds emerge, will still allow the young Peppermint shoots to come through while suppressing the germination of many weeds.

Materials

The mulch material needs to be of a friable nature, like sawdust, mature compost or peat. One grower has had good success with 'rotted log': the crumbly remains of fallen forest giants obtained from the nearby bush.

Caution needs to be taken with high carbon, low nitrogen mulches, as they can cause nitrogen starvation if they are worked into the soil.

It is important this mulch material be weed and seed free, as you don't want to be introducing your next weed problem as you try to solve this one.

Some care needs to be exercised with sawdust – on its own or in compost – as it can find its way into the harvested crop through the action of the wind, splashing caused by rain or irrigation, or being picked up during harvesting. It can't be removed from the processed dried herb by winnowing, as it is too light.

Thickness

The mulch should be no more than 25 mm thick over newly planted Peppermint to avoid smothering the emerging shoots, though it could be thicker elsewhere between the rows. It needs to be crumbly so the weeds that do emerge through it can be easily cultivated.

A thicker layer – around 50 mm – can be used over established Peppermint as it will be able to push up further.

Effectiveness

This technique can be very effective in reducing the levels of weeds germinating from seed, but it is of limited effectiveness against the larger seeded grasses, and of no avail against established weeds or the roots and rhizomes of regenerative weeds such as Couch Grass and Sorrel.

It is quite effective when used over a newly planted crop following a thorough bare fallow prior to planting.

Narrow Strip Mulch

If the availability of mulching material is limited, you can just spread a narrow strip of mulch over the newly planted rows.

Provided sufficient good quality compost is used, with only a small quantity of a carbonaceous material like sawdust, and most of it remains on the surface of the soil, nitrogen starvation will be avoided.

Permanent Light Mulch

One option is to use a light mulch on the Peppermint on a permanent basis. A special compost should be made for the purpose, using materials that will provide a good level of nutrients while remaining crumbly. One herb grower makes this from sawdust and fishwaste, with good results.

Large Scale

While complete sheet mulching may not be practical for larger scale operations, a narrow strip of light mulch over the Peppermint rows may be feasible. The application of this could conceivably be mechanised on a broad-acre scale.

Flame Weeding

Flaming can be carried out on newly germinated weeds just before the Peppermint emerges: this will work on broadleaf weeds, but not too well on grasses as their growing point is protected below ground. Grasses are often a serious weed problem in Peppermint.

Tickle Weeding

Tickle weeding, or similar techniques where light springy tines or brushes are worked over the emerging Peppermint, can be effective when the weeds have just germinated and are still soft and fragile.

I have had success with a metal spring rake, of the kind used for raking lawn clippings. The tines are scratched over the strip with sufficient vigour to break or flick out the tiny weeds, while the young Peppermint shoots, being a little more robust, remain in place and suffer only a little damage, from which they are able to recover.

Tickle weeding could also be done with a stiff broom or, on a larger scale, with a tractor-mounted tickle-weeder or a brush-weeder. Timing of the operation is critical so that the weeds are struck at the vulnerable cotyledon stage. If left too late, they will have toughened enough to survive the operation, while the Peppermint will be more advanced and may suffer more damage.

The Use of Grazing Animals to Control Weeds

There are reports of Peppermint growers using grazing animals to control weeds. One essential oil grower told me how a few sheep were sometimes run in the Peppermint to keep the weeds down, and a large organic grower in the USA has been using Chinese weed-eating geese in Peppermint grown for tea.

I have tried both sheep and geese (ordinary geese) for weeding Peppermint, but they did too much damage to the crop. Some sheep ate the Peppermint rather than the weeds, and while the geese didn't actually eat the Peppermint, they gave it a good pruning.

The other problem was their droppings, which were sticking to the leaves of the Peppermint. Needless to say I didn't pursue this avenue of weed control any further.

Critical Times for Weed Control

Once it has reached a certain height, Peppermint will hold its own reasonably well during the growing season. There are critical periods in its cycle when weeds are more able to take a hold, so if attention is paid to these, weed problems can be reduced

After Harvest

When the Peppermint is harvested, this provides light for weeds that have been growing under it. The early regrowth period can be critical, as this is an opportunity for the weeds to get a good start in the crop. It is also an opportunity for the grower to get at these weeds while they are accessible.

Over Winter

The main test of your weed control will come when the Peppermint goes dormant.

This is when any winter-active weeds that have become established in the bed, or those that germinate in the moist conditions of late autumn, will be able to grow on unabated, without competition from the crop.

It is remarkable how rapidly what was a well-weeded patch of Peppermint can degenerate to a mass of weeds. If you don't do anything about it, the crop can be beyond salvaging by spring.

As the Peppermint by this stage will probably have developed into a meadow-crop, hand-weeding can be a formidable task if there are more than just a few weeds amongst it.

Black Plastic Dormancy Mulch: Many winter-active weeds can be controlled with a black plastic dormancy mulch (see under 'Mulches' in Chapter 4, 'Weed Management and Control').

Rejuvenation as a Weed Control Technique: The rejuvenation procedure described below can be effective in greatly reducing the amount of weeding required to maintain weed control over winter. Because the crop is reduced to narrow rows, hand-weeding these is a more manageable proposition.

In the past I have used this system of winter weed control successfully for 5 years on a plot of about 800 m² of Peppermint.

Rejuvenation

In a meadow-crop Peppermint steadily declines and after two or three seasons will probably be producing only a fraction of its initial yields as it becomes very tight and root bound.

Peppermint in this situation will respond vigorously to rejuvenation by cultivation in a manner that leaves narrow strips of Peppermint to re-establish the crop (as described in the introduction to this chapter).

Peppermint can be kept growing vigorously this way for at least 5–6 years, and possibly a good deal longer, though it would be good practice to rotate it with some other crop after a while.

Irrigation

Mints are a great lovers of moisture and their root systems are relatively shallow, so good growth depends on an adequate supply of moisture from regular watering.

During peak growth in hot midsummer weather, Peppermint can require 50 mm of water per week in the period leading up to harvest when there is a large bulk of leafy growth. In the earlier stages of development, the requirement would be closer to 25 mm per week.

The water needs to be applied every 4–6 days, depending on the soil type, stage of growth and prevailing weather conditions. As with most crops, it is important to ensure there is always adequate moisture available, otherwise growth checks will occur.

Irrigate before the soil gets too dry, rather than wait until the crop is starting to wilt, as photosynthesis stops somewhat before the wilting stage.

Peppermint is one plant you can safely water in the middle of the day without any risk of burning the leaves. I have even turned the water onto it – with no ill effects – when it was wilting in the hot sun. This treatment would have badly scorched some species.

As a basic check as to whether your Peppermint is getting enough water, compare the growth of those sections of the bed that receive additional moisture, such as where the sprinklers overlap or where a low spot exists. If the growth in the wetter sections is obviously better than the rest, this is an indication that the patch could do with more water to achieve its optimum growth.

Pests and Diseases

Peppermint is subject to more pest and disease problems than many other herbs. This may have something to do with it being a hybrid selected for its particular qualities, rather than being a species that has evolved in the wild.

In an organic regime, many of the pest and disease problems that afflict conventionally grown crops don't develop, but there are still a number which may need some attention in order to get maximum yields of premium quality.

Mint Rust

Symptoms

Mint rust is a fungus that forms rust-coloured colonies on the undersides of the leaves of the Peppermint. These break open to release clouds of airborne spores, which then infect other leaves. While there are many types of rust, this one is mainly specific to Peppermint.

It attacks the lower leaves first and progressively moves up the plant (see colour plate 6.5). As the pustules spread, the leaves die and drop off, so the culmination of the disease process is a ragged plant with the odd dead leaf hanging on the stems and a few young green leaves at the tip. The rate of development of the disease accelerates as it progresses. So while the initial spread may seem fairly slow, once it reaches a certain stage it can rapidly destroy a promising crop.

Occurrence

Rust is a problem in most regions of Australia where conditions suit the development of high quality Peppermint. It does not seem to be occur in humid sub-tropical areas, like coastal Queensland, and may be less virulent at higher altitudes.

Nature of the Disease

Mint rust is something of an enigma. In contrast to most fungus diseases, it seems to be at its most virulent in dry conditions rather than in the humid situations that normally favour fungal development.

I suspect the relationship between rust and Mint is something of a symbiotic one: that perhaps the function of the rust is to limit the rampant spread of the Mint. Never killing off its host, the rust loses its virulence in late autumn and allows the Mint to recover.

Mint rust has a complex annual life cycle, which goes through several stages. It persists through winter in greatly reduced incidence, which allows the Mint to get a good start in its growth. Then the rust progressively invades the Mint until it peaks in late summer.

The youngest leaves seem to be immune to the rust and it normally does not infect the stems of Peppermint.

There may be a link between the high levels of essential oil in the Mints and their susceptibility to rust. Rust fails to develop in climates that do not foster high levels of essential oil. The disease attacks the Mints at the time of the year when their oil content is highest, and varieties of Mint that have been bred for their resistance to rust are disappointingly low in their essential oil content.

Rust seems to run counter to the nature of other plant diseases, in that it attacks apparently healthy plants. At least that is my impression to date. Perhaps there is some factor stressing the Peppermint we have not yet come to understand.

There do seem to be cases where one bed of Peppermint will remain consistently free of rust, while another nearby will be heavily infected.

Some friends who grow Peppermint under a light mulch find older established beds suffer much less rust than new beds

Mint Rust Strategies and Controls

Fortunately there are a few things that growers can do to limit or prevent rust development:

Foster Vigorous Growth: Vigorously growing plants seem to be less susceptible and they reach the harvesting stage earlier, giving the rust less time to develop. This is done by giving the Peppermint optimum conditions for growth, in particular, plenty of compost and irrigation when needed.

Harvest Early: Harvesting the crop earlier can limit the development of the disease.

Harvesting every 4 weeks to prevent the rust from ever getting to serious levels is a possibility, but yields would be lowered.

On the other hand, delayed harvests encourage rust development as they give it more time to build up.

Mow the Peppermint Stubble: The stubble should be mowed as close to the ground as possible after each harvest. An ordinary suburban lawn mower set as low as it can go without damaging the blades will do a good job. This will reduce the number of old leaves left on the plants. It is these old leaves that carry the rust over to infect the new growth.

A number of growers – all at altitudes of 600–800 m above sea level – find just mowing the Peppermint stubble down very short after the first and second harvests is enough to give adequate control.

In general, though, most growers find that Peppermint needs more drastic intervention to keep the rust under control.

The Biodynamic Approach

One grower I know has had reasonable success with weekly applications of 501 (horn silica) and 508 (Equisetum tea spray). These are traditional biodynamic treatments for fungus. The 501 needs to be stirred for 1 hour before spraying it on.

Although I'm a biodynamic grower myself, I've never tried it. I've always been somewhat daunted by the prospect of having to spend so much time stirring and spraying the Peppermint once a week throughout the whole season, and I have been concerned as to what I would do if it didn't work for me. Flaming once a season seems easier and I have always had good results with it.

Flaming

For Essential Oil Production: Flaming was the traditional rust control used by conventional broad-acre Peppermint growers until about 15 years ago when, unfortunately, an effective fungicide was developed, which would stick to the underside of the leaves. (A nice thought: this toxic substance is still going to be sticking to the leaves when they are harvested!)

Essential oil growers formerly used to plough their Peppermint under in autumn and then flame it in September when the first growth was 50–100 mm high. This timing and double-barrel approach was necessary because Peppermint intended for oil production has to be kept rust free until harvested in February, when its oil content is highest.

For Dried Herb Production: When I tried this method, I found it really knocked the stuffing out of the Peppermint, which was slow to recover and patches died out.

I speculated that because Peppermint for tea use is harvested much earlier than it is for oil, it should be possible to cut it before the rust did too much damage, and then flame it after this first harvest in early December. This turned out to work very well and the Peppermint recovered much more vigorously after this mid-season flaming and remained free of rust until the end of the season.

This system has consistently given good results over a 15-year period.

Types of Flamers: There are several types of flamers available. The principle is basically the same as for flame-weeding and the same equipment can be used. Similar equipment is also used for sterilising greenhouses, chicken sheds etc. Flamers or flame guns burn either LP gas or kerosene. They shoot out a hot flame about 400–600 mm long, somewhat like a miniature jet engine. (Flame guns should not to be confused with flame throwers, which project a fiery ball for several metres, and which are used in trench warfare and for starting fires.)

Single-burner, hand-held flamers are obtainable. These get quite hot, so the operator has to wear woollen trousers while using it.

A 75 mm model LP gas flamer is probably the best option. Fitting a piece of sheet iron (about the shape of a large square-mouthed shovel) over the burner will spread the flame and hold it close to the ground. This improves its efficiency considerably.

Timing: In the period leading up to the first harvest, a close eye needs to be kept out for the development of rust in the crop. If this is starting to spread, the harvest may need to be taken earlier.

It is best to flame the Peppermint stubble within 3 days of the first harvest, so you don't knock back any regrowth.

For flaming you need windless conditions so the flame is not dissipated, and it is easier to work in the cool of the morning or in the evening.

Preparation: The stubble needs to be mown off as low as possible and left for half a day or so to dry off, so the moisture of the leaves isn't impeding the flaming. It also proceeds faster if there is no moisture from dew or rain.

Technique: It is important to understand a couple of things that will make flaming easier.

First, you are not trying to sterilise the soil to destroy any lingering rust spores which might reinfect the Peppermint, you are just killing all the green leaves on the Peppermint to break the rust's cycle. Consequently there is no need to flame the bare soil in the vicinity of the plants: any spores present will have lost their viability by the time the new growth of Peppermint emerges.

Secondly, you don't have to burn the leaves to a cinder to kill them: they just need to be singed enough to cook them. The leaves will change colour slightly when they have been done enough: they will still be green, but it will be a sort of 'cooked' green. It took me a little while to recognise the difference.

Gas flamer on Peppermint to control rust.

The flame can be passed over the Peppermint at a slow walking pace. In damp conditions or when there is denser growth, the rate of travel will need to be slower to ensure sufficient heat build-up. This is also the case if a breeze is dissipating the heat.

It is important to get a complete leaf kill, so go over the patch thoroughly. Also make sure you hit any other Peppermint that may be around the place in trial plots, nursery pots, or as escaped plants that have established themselves in adjoining grass or in other herb crops. If a plant is in a situation where you can't flame it, take all the leaves off by some other method, otherwise these plants are likely to reinfect your Peppermint crop.

Cross Infection: While there has been some disagreement among authorities as to whether rust on Peppermint can infect Spearmint and vice versa, the consensus seems to be that they each have their own specific strains of rust, which can't infect each other.

However, Scotch Spearmint is apparently susceptible to both strains of rust.

Monitoring: It will usually be the next day before the singed leaves turn dark and you can spot any patches that were missed. You can then nip off any leaves that have survived or give them another burst with the flamer.

Large Plots: Large plots of Peppermint that have to be taken in a staggered harvest can be a problem. It is not practical to delay flaming for more than a few days after harvesting, as it will set the Peppermint back considerably if it starts to regenerate only to be knocked down again. Try to keep staggered harvests as close as possible and flame the later harvested patches as soon as possible to reduce the possibility of them reinfecting the earlier harvested ones.

Flaming larger areas can be carried out using tractor mounted LP gas flamers, where several flamers are mounted together in a row, so the tractor can flame a 2 or 3 m strip in one pass.

For flaming to be successful on large areas, close attention needs to be paid to thoroughness, as a small patch missed can reinfect the whole crop.

Fire Precautions and Permits: Woollen trousers, woollen socks, leather boots and preferably a woollen shirt (or at least a cotton one) should be worn. Water should be kept handy in case of an accident or a fire.

Because the flaming of the Peppermint usually needs to be done at the beginning of the fire season, a permit may need to be arranged, and fire prevention precautions need to be taken.

This is also a good reason for flaming after the first cut, even if there is no evidence of rust. If your district is prone to rust, then the chances are it will be present in your patch, and, though it may not yet be apparent, it will build up and be a problem later on in the season. By then, however, fire conditions may be such that it is unsafe to flame.

After the Flaming: After flaming, go over the plot and douse any smouldering material. Following up with an irrigation will stimulate regrowth.

Alternatives to Flaming

For small plots and for those who don't like the idea of using a flamer, the same effect can be achieved using other methods that destroy all the remaining leaf on the Peppermint stubble.

Rust is only carried on live leaves, so if the plant is stripped of its leaves by some method, the rust's life cycle will be broken and the new growth will remain clean, usually for at least 3 months, until a spore floats in from somewhere and the rust slowly builds up again.

Mowing and Nipping: Sometimes when Peppermint is growing densely, the lower leaves will have fallen from the stalks by harvest time. When this is the case, after mowing the stubble off very low there will be very few green leaves left. These can then simply be removed by hand or chipped off with a hoe and if this is done thoroughly, the rust will have nowhere to survive.

Burning Straw or Leaves: Traditionally straw was spread on the beds and burnt. I have done a similar burn using dry gum leaves. Judging the thickness of fuel necessary to get a good surface kill of the Peppermint without cooking the rhizomes just below the surface is a bit tricky.

Using a Weed-trimmer: Another alternative that works quite well is to go over the Peppermint stubble with a nylon cord weed-trimming attachment on a brush-cutter, and whizz off all the remaining leaves. If possible, trim the cord before it becomes too ragged, so you don't leave pieces of nylon among your herbs.

The disadvantage of this method is that the spinning cord tends to cut into the surface of the soil and expose more rhizomes: it is hard to judge how low to go. The trimmer also throws a lot of debris and soil around.

Harnessing Grasshoppers: Every cloud has a silver lining and even a cloud of grasshoppers can be put to good use, as I discovered on one occasion.

A patch of Spearmint had no sign of rust at the first harvest in December, so I didn't bother to flame it. However, this proved to be a poor decision because by the time of the second harvest the patch was badly stricken with rust. The problem was that it was then late January – the height of summer in north-east Victoria – and everything around was tinder dry. Flaming would have been very risky and there was no chance of getting a fire permit.

I mowed the stubble as close to the ground as I could, but there was still a good coverage of green leaves remaining. While pondering how to remove them, I noticed that the grasshoppers – prolific in the garden at that time of year – were tucking into these old Spearmint leaves. So I decided to wait and see how many of them they would clean up.

A few days later there was barely a leaf remaining, and those I was able to nip off by hand in a few minutes.

When the new growth came up it was free of rust and remained so until the end of the season, allowing me to take a good third harvest.

Grasshoppers

Overall the impact of grasshoppers on Peppermint is usually not too devastating, provided the Peppermint is growing vigorously. However, when its vigour has declined after several years as a meadow-crop, it can suffer more severely.

Flea Beetle

The flea beetle showed up at 'Twin Creeks' in north-east Victoria a few years after the Peppermint was established. Among the herbs, it only seems to attack the Peppermint, Spearmint and Gipsywort (*Lycopus* sp.). It is apparently a native insect that has shown a liking for these plants, which are possibly related to its natural diet.

It feeds on the root systems of the Mint over winter, and emerges as an adult to feed on the leaves. It is a bronze-coloured beetle 2–3 mm long, which jumps when it realises you are looking at it.

These diminutive beetles do manage to make their mark, however, and their numbers can build up to the point where they are doing significant damage. This appears as small holes in the leaves. These look like black spots from a distance. While numbers remain small and growth is vigorous, the damage is merely cosmetic, and not a problem in a crop that is going to be dried and processed. It would be a much more serious problem for someone supplying the fresh herb market.

As the vigour of the Peppermint declines during its period as a meadow-crop, the numbers of beetles seem to increase and the damage is more severe. Patches of the crop may even die out due to the ravages of the larvae underground and the lack of vigour in the Peppermint, and a large proportion of leaf will be lost to the adults feeding on it.

The best way to manage the problem seems to be to rejuvenate the Peppermint as described earlier. The re-invigorated Peppermint will leave the beetles behind and damage will no longer be as significant.

Alternatively a new planting of Peppermint will also get the jump on the beetle for some time, perhaps for two or three seasons, until the growth in this new patch becomes too tight and crowded.

So far the flea beetle has not appeared in Tasmania. To prevent it getting there, Tasmanian growers should avoid bringing in propagation material of Peppermint, Spearmint or Gipsywort from Victoria, as the larvae can be carried in the rhizomes.

Aphids

At times in the spring, big numbers of aphids can build up on Peppermint. In the days when I was using chicken manure, I used to get heavy infestations of aphids on my Peppermint, but since adopting the use of a balanced compost with lower nitrogen levels, I hardly ever see them and then only in small numbers (see 'Aphids' in Chapter 9, 'Pests and Diseases').

Snails

Snails can be a problem in Peppermint, not so much from the point of view of damage to the crop, but because they get picked up in the harvest and can find their way into the finished product (see 'Snails' in Chapter Chapter 9, 'Pests and Diseases').

Slugs

In moist climates, slugs can build up in Peppermint. Actual damage is usually not significant, but in damp conditions they can be up in the leaves and from there they are picked up in the harvest. Once dried, the slugs can't be separated by screening, which can be a problem.

Harvesting

Timing

Peppermint will normally give three harvests in a season and maybe a fourth in warmer regions.

The first harvest is usually late November to early December, though spring-planted Peppermint in its first year of growth will be somewhat later than this. The second harvest will be ready 6–8 weeks after the first, so it comes mid-January to mid-February. The third harvest is somewhat slower to develop, as growth slows with the shortening days, and comes 9–12 weeks later, usually late March to early May.

Peppermint is an exception to the general rule of harvesting leaf herbs at the early flowering stage. Because it is slow to develop flowers, it is best to harvest Peppermint somewhat before even the buds appear, as the leaf will be of better colour, though the flavour may not be as strong in the first harvest. Keep an eye on the lower part of the plants and try to time your harvest before there is too much fading and discoloration of the lower leaves.

Rust Development

Another factor to take into consideration in timing of harvest is rust development. Usually it is possible to get the first harvest off before the rust affects it too much, but in some years rust will show up early and necessitate an earlier harvest to save the Peppermint before it is too damaged.

The same can occur with the later harvests, but if rust control measures were thoroughly carried out on the Peppermint after the first harvest, the rust usually doesn't show up again until these harvests are completed. However, this isn't quite 100% reliable, so you still need to be on the lookout and take a pre-emptive harvest if necessary.

A small amount of rust is acceptable in the dried Peppermint, as long as it doesn't affect the appearance. So the occasional spot on the leaves is okay and won't affect flavour.

If rust is present in a large planting, which is being harvested over an extended period or in staggered harvests, this period of harvest needs to be kept as short as possible in order to synchronise rust control measures. There is no point in getting rid of the rust in one part of the crop if it is going to come in from an adjacent area which can't be treated until it is harvested.

Height of Growth

The first growth of Peppermint will be relatively tall: it can be 600 mm or more.

The second growth is shorter, the stems are thinner, and the internodes are shorter, so don't keep waiting expecting it to reach the height of the first growth. You will be pleasantly surprised at the density and leafiness of this second growth, and yield will be higher than you might expect at first glance.

The third (and fourth) growth is shorter again and has an even greater proportion of leaf. While the plants may not look substantial, the high proportion of leaf makes it worth while. If it is getting late in the growing season, this last harvest needs to be taken before the leaf starts to deteriorate as the Peppermint goes into dormancy and while ambient air drying conditions are still capable of drying your crop. Even so, some assistance is likely to be needed to complete the drying process.

Tools

The harvesting scythe is my implement of choice for Peppermint, which is usually grown in a patch of a reasonable size. Harvesting can be done with a reaping hook or sickle, but this is much slower and, for the low growing last harvest, may not be very practical.

For the initial harvest of newly planted Peppermint, you may find the scythe tends to pull some plants out by the roots. If this is happening, it is best to resort to the reaping hook. By the time the second harvest is ready, the roots will have established a more secure hold on the ground. The stems will be less tough, too, so using the scythe should not be a problem.

Mechanical harvesting on a large scale is a possibility (see Chapter 11, 'Harvesting').

Technique

Harvesting Peppermint is usually fairly straightforward. Try to cut the plants just above the layer of yellow or discoloured leaves: the height of these will vary through the patch.

Keep a close eye out for rust, which may occur as pockets of badly affected plants in an otherwise sparse infection. A few spots of rust on a small proportion of leaves will not be a problem, but leaves severely damaged by rust should be avoided, as they will affect the appearance and quality of the finished product.

If you are harvesting large volumes of Peppermint, you will need to take care that the harvested material does not build up too much heat before it is placed on to the drying screens.

Harvesting in the morning while it is still cool will delay the onset of the composting process, which accelerates as the temperature increases. Peppermint harvested in the heat of the day will start to compost very quickly.

Large harvests should be taken in manageable portions so you aren't holding the harvested material more than 2–3 hours. Put the harvested Peppermint in a cool place, where there is good air movement, and separate the bundles to reduce the mass of herb you are holding together, so some heat can dissipate. If excessive heating does develop, it can be slowed or stopped by throwing the bundles open, so they can cool.

Post-harvest Treatment

Within 3 days of harvest, the remaining stubble should be mowed down very short to promote new growth from the rhizomes at or below ground level.

This stimulates more vigorous regrowth, assists in rust control and makes subsequent harvests easier and cleaner.

After the last harvest of the season, it may be better to leave the stubble standing: this will help suppress weed development to some extent, and being so short, will not cause a problem for the next season's crop.

For more detail see 'Post-harvest Treatment' in Chapter 11, 'Harvesting'.

Drying

Peppermint is relatively easy to dry, but it does wilt fairly quickly. If it is handled in this wilted state, it will bruise and turn dark. This can be avoided by loading it onto the screens promptly and by not attempting to harvest too large a volume at a time. Don't disturb the drying herb: this is not necessary with screen drying.

Drying is best if it is completed within 7–10 days. Try to avoid slow drying (longer than 2 weeks) or heavy re-moistening, as these can result in darkening.

A good quality Peppermint when dried should be an attractive deep green colour, perhaps with some purple coming through.

Most of the Peppermint on the market is a dark murky brown-green colour without much flavour, the result of poor drying. Handling in large quantities, composting, deep-layer floor drying, field curing, remoistening, drying at high temperatures and too slow drying can result in the leaf bruising and going dark and/or significant losses of essential oils and flavour.

Temperature

Drying temperatures should not exceed 35°C, but if the temperature at the top of an ambient air drying shed gets a little above that for a few hours in the heat of the day, this does not seem to matter too much. A good range of temperature is 25–35°C.

Processing

Peppermint is primarily used as a tea, for which a tea grade is required. It is also used in the manufacturing of liquid medicines, for which purpose good quality aerial parts should be acceptable, though most manufacturers are currently using second- and third-grade leaf.

I have seldom sold any Peppermint to manufacturers as the price offered was always way below the return I could get selling a premium grade for tea use.

While Peppermint does have some limited use as a culinary herb, I have never had any request for a finer screened culinary grade.

Tea Grade

For tea use, the dried herb should be rubbed and screened to a tea grade free of stem. The optimum sized rubbing screen is $2^{1}/_{2}$ dent, though 3 dent is acceptable.

If the Peppermint stems are allowed to become totally dry, they will tend to break up and go through the rubbing screen, creating added work in separating them. If you can catch the Peppermint at the point where the leaves have dried to the brittle stage, but the stems are still moist and pliable, you can greatly reduce the amount of work required to produce a good tea grade.

Rubbing Peppermint can be a rather cleansing experience, especially for the mucus membranes of the nose and the eyes. Rubbing the second harvest during the hot weather that prevails at that time of year can release a lot of essential oil vapour and start your nose and eyes streaming: you may have to resort to a respirator and goggles to complete the task.

Winnowing

Peppermint is relatively easy to winnow, though some of the leaves are inclined to curl up, which makes them a bit denser, so they tend to fall in with the stems. If there is a high proportion of these they can be salvaged by rubbing the separated stalk and leaf through a finer screen to break up the curled leaves, followed by rewinnowing.

Market Preferences

Particle Size: Most of the trade Peppermint you will see, and even the imported organically grown herb, is processed to small particles, which disguises discoloured leaf and impurities and results in a further loss of essential oils.

Our experience has been that people specifically request the larger leaf Peppermint once they become familiar with it: they have come to associate it with quality and are prepared to pay a significantly higher price for it.

Variations in Flavour: Because Peppermint develops varying levels of essential oil, and varying ratios of individual essential oils according to the time of year, you will notice some quite distinctive variations in flavour through the season between the different harvests from the same plot.

The first harvest tends to be rather milder than the second which is normally the strongest. The oil content is highest between January and March, after which it declines. The third harvest, taken in autumn, is usually intermediate in strength between the first two.

While Peppermint oil producers have to be very particular as to the timing of their harvest to ensure they meet the market requirements for their essential oil levels and ratios, the market for herb teas is not nearly as demanding: witness the sort of rubbish being sold in most places as 'Peppermint tea'.

Imported Peppermint in tea bags is usually boosted with essential oil to give it enough flavour: sometimes it is possible make a cup of tea that is just as strong by tipping out the herb and using the paper tea bag alone!

Consequently a good quality of even the milder harvests of locally grown Peppermint have been so superior to the average trade Peppermint (and still a notch above imported organic Peppermint) that we have not felt it necessary to blend the different harvests of the season to bring up the strength of the first harvest or to create a more standardised product. Our impression has been that some variation in flavour is acceptable as long as the flavour is strong enough and good.

On the other hand there has been a noticeable market preference for the stronger Peppermint from southern and inland regions over the milder Peppermint from north-coast New South Wales.

Aerial Parts

Peppermint for manufacturing use should be chaff-cut in the normal way. Check the thickest stems for dryness, and also the tips near the growing point. These are the slowest parts of Peppermint to dry.

Storage

Peppermint needs freezing or carbon dioxide fumigation to control moth larvae in the herb before putting it into storage.

Yields

Peppermint yields 0.25–0.45 kg/m² (2.5–4.5 t/ha) tea grade or 0.3–0.6 kg/m² (3–6 t/ha) aerial parts per annum.

Marketing

Because Peppermint is really the flagship of the herb tea range, and because the quality of most of the Peppermint on the market is so poor, the demand for premium-grade, organically grown Peppermint in Australia has never been satisfied to date.

There is also a large market for manufacturing-grade, which is currently not being supplied from local production.

Price

Price to growers for premium quality organically grown Peppermint is currently $24–26/kg for 5-star tea grade.

Price to growers for manufacturing-grade, organically grown Peppermint leaf is currently $10–15/kg. Some buyers are offering a lower price for aerial parts.

Trade price is around $6–9/kg.

Raspberry

Rubus idaeus L ROSACEAE

Parts Used: Leaf

Raspberry leaf is a well-known herb tea, the drinking of which is widely regarded as an essential part of preparation for childbirth. However, some people quite enjoy it as a tea for its flavour alone. The fruit of this plant is well known and widely enjoyed. It, too, is sometimes an ingredient in herbal medicines.

Curiously, Raspberry is often overlooked in books on herbs, yet it is always in the top six best-selling tea herbs.

Origin

Raspberry is a native of Europe and Asia.

GROWING RASPBERRY

Raspberry is easy to grow, given the right conditions. However, to obtain good yields of high quality leaf, the crop needs to be managed somewhat differently from Raspberry being grown for the fruit.

Raspberry (*Rubus idaeus*).

Most varieties of Raspberry produce canes that live for two seasons. The first season the canes produce just leaves (though some varieties may produce some fruit in the autumn on these first-year canes). In autumn they drop their leaves and then the following summer, as second-year canes, they produce leaves and fruit. At the same time the plants produce a new crop of first-year canes.

For maximum leaf production, only the first year canes are worth considering, as they produce larger quantities of better quality leaf. The leaf on the second-year canes is smaller and its colour is not as good, it is stalkier and the fruit are a problem in drying. The second-year cane has more abrasive prickles, too, and is rather unpleasant to handle.

A patch of Raspberry grown for fruit can produce some quality leaf as a sideline. Each year Raspberry fruit growers cut out a lot of first year canes and suckers, which could be dried for Raspberry Leaf. The yield would not be anything like as great, but in a big patch of Raspberries being grown for fruit, it could be a viable proposition.

Climate

Raspberry prefers a cool to mild climate, but it will grow in regions with a hot summer, provided winters are cold enough for its dormancy, and it is amply irrigated and not exposed to hot drying winds.

Site

Exposure
Raspberry prefers a sheltered site with full sun, though it will tolerate some shade.

Drainage
Good drainage is essential, as Raspberry will not tolerate waterlogging.

Soil Type
Light to medium loams are preferred.

Fertility
Raspberry needs a fairly fertile soil.

Initial Weed Control
Weed control is not as critical for Raspberry as it is for many other herbs, but good initial weed control will make ongoing weed control much easier.

Neighbouring Plants
As Raspberry is a large tall plant which invades neighbouring crops, it is not a good neighbour for many smaller herbs and is best kept separate.

Containment
Raspberry expands at a good rate and suckers from its roots are thrown up some distance away from the plant. Containment is best achieved by surrounding the Raspberry bed with a strip, about 2 m wide, of mowed grass and clover.

Identification

Raspberry can grow up to 2 m tall, though most varieties are about 1.5 m, usually with bristly stems. The leaves are green and glabrous above and light grey tomentose underneath, with 3 or 5 ovate leaflets up to 100 mm long (see colour plate 6.1). Flowers are white, about 15 mm across, in drooping panicles of 1–6, followed by fleshy cone-shaped red to yellow fruit.

Raspberry should be distinguished from other species of the same genus, which include cultivated berry fruits and several species of Native Raspberry.

Variety
There are a number of cultivated varieties of Raspberry. Lloyd George has proved to be a good variety, as it is leafy, not too tall, and its regrowth after harvest is vigorous. Leaf yield from Red Antwerp is lower, its height is a problem and regrowth less vigorous. Chilcotin is also a good leafy variety that seems to have a vigorous regrowth. No doubt there are others suitable for leaf production.

The qualities needed are leafiness, moderate height, and good regrowth after harvest. Varieties which fruit on the first-year canes may not be suitable, though Lloyd George's habit of producing a light crop of fruit in autumn is not a big problem.

Propagation

Method
Propagation is by division of the rootstock. The rootstock tends to divide itself as it sends up new canes.

If you dig up a clump of canes, you will find they separate out into crowns, with one or a few canes and roots arising from each. These canes should be cut back to about 150 mm, but their roots should be kept as long as possible.

Timing
The canes should be dug and transplanted while the plants are still dormant during winter or very early spring.

Planting
Set the plants in holes about 150 mm deep and cover them, so the crowns are just below the surface and the stumps of

the canes are sticking out of the ground. The stumps are mostly there to mark where the plants are until the new canes shoot up from the crown. If shoots emerge from buds above ground on the old cane, it is best to cut them off as they will be fruit bearing.

Layout and Spacing

There is some diversity among Raspberry leaf growers as to the preferred layout:

Close Spacing: Some growers lay their Raspberry out on 450–600 mm centres in rows 1200 mm apart and allow it to form a solid stand.

Wide Spacing: My own preference is for planting on 450–600 mm centres in rows 3–4 m apart. This wide spacing gives the rows enough room to expand to 2 m width while leaving enough room to mow a strip of White Clover growing between them.

The advantage of this system is that it encourages dense growth in the rows, which shades out the weeds and makes it hard for them to invade the Raspberry. If the crop is planted in closer rows, it soon fills up the spaces and becomes a solid stand. If this results in more plants than the ground can adequately sustain, the growth may become more open, allowing weed invasion, which is hard to control in a meadow crop.

The White Clover fixes nitrogen, which benefits the Raspberry, and also prevents erosion. It responds well to regular mowing, which keeps it short, reduces its water consumption and keeps weeds among it from dominating it or from going to seed.

If a grower had enough space and irrigation water, it would be possible to consider an even wider spacing of the Raspberry rows and grow White or even Red Clover strips several metres wide between the rows. The mowings could be thrown around the base of the canes as a mulch or as a sheet compost. This way it should be possible to grow enough to sustain the Raspberry with very little extra compost.

Compost

Raspberry needs good dressings of compost to get optimum growth, 3–4 kg/m^2 per annum. This needs to be applied in early spring. If the Raspberry is maintained with inter-row cultivation, the compost can be worked in beside the plants.

Alternatively compost can be placed on the surface of the ground beside the canes or over the beds as a mulch before the new canes start to emerge.

Weed Control

While Raspberry grows strongly and shades out many of its competitors, if regular weed control isn't carried out, Raspberry will gradually be overtaken by tall weeds, particularly grasses.

With Raspberry there is a temptation to disregard weeds if they are below the level of harvest. Some weeds such as White Clover and Sorrel will not cause problems, but if taller growing weeds such as grasses, docks etc. are ignored early in their development, it will take a lot of work to remove them later on.

Mowing

Maintaining Raspberry in widely spaced rows with frequent mowing between them helps weed control considerably, especially if the mower can reach under the overhanging canopy to some extent, as the growth in the rows is stronger and denser than if the Raspberry spreads out to form a solid stand. It also reduces the area you have to keep weeded.

Spring Weeding

The main problem weeds are mostly grasses that get established while the Raspberry plants are dormant in winter. So in early spring it is important to go over the patch and give it a thorough weeding. If the Raspberry can go into winter with the stubble of the last harvest carrying some leaf on it, this cover will discourage the weeds a bit. These old canes can then be mowed off in early spring before the buds shoot, so the plants put all their energy into the new canes.

Mulch

If you have enough clean mulching material, Raspberry is a herb which could be maintained with a mulch, as it is harvested well above ground level, and its emergence in spring is strong enough to push through a good thick layer of mulch.

Geese

Another option for Raspberries is to run geese among them. Geese will not eat Raspberry leaves, and the canes are robust enough to withstand their activities, and tall enough that the droppings are well below the level of harvest.

Geese eat many of the weeds that are a problem in Raspberry, particularly grasses and clovers, but they do tend to be selective in what they eat, and I found that after a while weeds such as Plantain and Cocksfoot started to dominate, and had to be dealt with by other means.

They say the Chinese train their geese, while young, to eat the weeds they want them to control and to leave the growing crops, but I don't know anyone with any experience in this.

Maintenance

As the Raspberry rows expand, after a number of years they will need to be cut back to maintain the space between the rows. Alternatively, they can be allowed to form a solid stand, if you are able to maintain good growth and weed control

Irrigation

Raspberry needs plenty of moisture for good growth – 25–50 mm of water per week – particularly in hot weather.

Pests and Diseases

When Raspberry is grown in optimum conditions, it usually does not suffer many problems. However, it is not always possible to make sure optimum conditions prevail.

Grasshoppers

Grasshoppers love to eat Raspberry leaves and in a bad year they will greatly reduce yields. Poultry such as ducks or chickens are probably the best control, and they won't harm Raspberry leaves.

Alternatively, one can focus on improving the conditions for the Raspberry, so it can better resist or tolerate the grasshoppers. Additional moisture and compost and some protection from heat stress will foster stronger growth.

Raspberry Rust

Raspberry rust is somewhat similar to Mint rust, but nowhere near as serious. Late in the growing season, rust-coloured spots appear on the lower leaves and gradually multiply and move up the plant. If the second harvest is left too late, it may be affected. Normally all that is necessary is to monitor the development of any rust and harvest before it becomes a problem: the next season's growth will be free of rust until late in the season.

Harvesting

Timing

First harvest is December to early January. There is some flexibility in timing, but the growth reaches a point where the gain in new leaf at the top is balanced by the loss of leaves lower down the stem. Raspberry is a crop that can be flexible in its harvest to fit in with the needs of other crops and the available drying space. However, the first harvest should not be excessively delayed as this will cause the second harvest to come too late, with consequent problems with rust and possibly with drying.

Raspberry's first harvest usually slots in nicely after the Mints. By the time their first cut is dry, Raspberry is usually at its optimum stage.

Tools

While I have harvested Raspberry leaf with a catching scythe, it is quite hard work. The stems are fairly robust and the canes have to be cut a fair way up, so the scythe has to be swung at an awkwardly high angle, with a heavy load of leafy canes in the catcher.

For this reason, unless you are very keen on using the scythe, I recommend harvesting at least the first harvest with the reaping hook. Being large leafy bushes, harvest progresses reasonably quickly, and as it is cut fairly high on the stems you don't have to bend much, so it is not too hard on your back.

You may need to wear a glove on your grasping hand for harvesting, as the canes of Raspberry can be rather prickly.

Technique

Partial Harvest

When I first grew Raspberry, I treated it as a woody perennial and left some green leaf on the stems at the first harvest. The canes will then send out new shoots from the axils of these upper leaves.

One problem with this method is that it is hard to judge the vitality of these lower green leaves, and they can burn off when exposed to full sun. Regrowth is poor when this happens.

For the second harvest, late in the season, the leaves can be harvested down to the point where they are too faded.

Total Harvest

Some growers have adopted the practice of taking all the green leaf off the first growth and then mowing the stubble in the same manner as other spreading herbs are treated. The Raspberry seems to regenerate quite vigorously from ground level and the quality and quantity of the second harvest is better, so this appears to be a better system.

It may be possible to get three cuts this way if the first cut is taken earlier, but this would depend on how long a recovery period the Raspberry needs.

Leaf Picking

An alternative method of harvest, used by one or two growers, is to pick the individual leaves, tea-picker style.

This probably is not very efficient time-wise, though it does eliminate the need to remove the stems from the dried leaf. For a berry fruit grower who is not processing any other dried herbs, this may be a viable option. The leaves would need to be picked above the main leaf stalk, dried and then somehow crushed or worked through a screen, to achieve a suitable grade.

Post-harvest Treatment

Post-harvest treatment depends on the harvesting system used.

Partial Harvest

With this system, the canes are left with at least one set of fully green leaves after the first harvest. After the second harvest, the stubble can be left standing until early spring as this tends to reduce weed invasion. These old canes will need to be mowed down close to the ground before the new shoots emerge in early spring.

Total Harvest

After harvesting this way, the stubble of the first harvest, and possibly the second, is mowed to ground level. The stubble of the last harvest, whether second or third, can be left and mowed in early spring, as above.

Drying

Raspberry leaves lose their moisture very quickly, so there is normally no problem drying this herb. It is subject to discoloration if handled in a wilted state or excessively re-moistened, so a little care needs to be taken in this regard.

Temperature

Raspberry leaf can be dried at temperatures of up to 35°C.

Processing

Raspberry leaf is widely used as a tea, and also for manufacturing liquid medicines.

Tea Grade

For tea grade rub the leaf through a $2^1/_2$ or 3 dent screen. Because the fine down on the underside of the leaves makes them cling together, it is not possible to use underscreens while rubbing Raspberry leaf, so you have to be content with repeated screenings or winnowing to remove stalks.

The leaf needs to be quite dry to rub easily, but the stalks can be a bit of a problem to eliminate as they are quite short. They fall through the screen easily, and it is difficult to get 100% separation with a winnower.

This problem can be reduced if you can time your rubbing so the leaf is dry, but the stalks of the leaves are still a little leathery. This way the stems don't break up as much, but this is a fine line of judgement and is not always convenient to put into practice.

If you are also producing a manufacturing grade, the last part of your tea-grade screenings, with a small proportion of stalk, can be combined with the manufacturing-grade first screening, as it will be roughly equivalent.

Splitting the Rubbing

Another method that can be used with Raspberry leaf to help separate the stems in preparing a tea grade is to 'split the rubbing' when rescreening.

The first part of each loading will pass through the $2^1/_2$-dent rubbing screen with very little stem. After about $3/_4$ of the loading has been rubbed through, put the last $1/_4$ or so aside. These set-aside portions can be subsequently rubbed through a finer 4-dent screen which will be more effective in removing the short pieces of stem.

Manufacturing Grade

This is one herb where the trade recognises the value of removing the stems from the dried herb, as the aerial parts of Raspberry contain more wood than leaf. For manufacturing use, the first screening is quite adequate, as a small amount of stem in the herb is acceptable.

Storage

While I have never known Raspberry leaf to be attacked by moth larvae, freezing it before storage is advisable as a precaution.

Yields

Yields of 0.2–0.3 kg/m² tea-grade leaf per annum have been obtained in Victoria and Tasmania.

Marketing

As Raspberry leaf is one of the more popular herb teas, it is in good demand and there is also a moderate demand for manufacturing grade.

Price

Price to growers for premium quality organically grown Raspberry leaf is currently $24–26/kg for 5-star tea grade.

Price to growers for manufacturing-grade, organically grown Raspberry leaf is currently around $8–16/kg

Trade price is around $7–8/kg.

Spearmint

Mentha spicata L LAMIACEAE

Parts Used: Leaf

Other Names: Garden Mint, Pea Mint, Mint.

Spearmint is somewhat less salubrious than Peppermint: it is more the workhorse of the herb stable. Widely used in cooking, especially in Arabian cuisine and in the well-known mint sauce, it is the plant people are usually referring to when they speak of 'Mint'. (But in this book the word 'Mint' refers to both Peppermint and Spearmint.) Spearmint is a popular tea in its own right, and makes a flavoursome blend with other herbs. A good quality Spearmint has a somewhat softer sweeter flavour than Peppermint, but it penetrates other flavours more.

While Spearmint's medicinal properties are similar to Peppermint, they are regarded as being milder, so it is not much called for in this regard.

Spearmint is the common Mint you will find growing in gardens and in the wild. However, many of these varieties are inferior, with low levels of essential oils and poor flavour.

Spearmint *(Mentha spicata)*.

Origin

Spearmint is native to southern Europe.

GROWING SPEARMINT

The growing and management of Spearmint is almost identical to that of Peppermint, so readers can use the entry for Peppermint as guide, while noting the points mentioned below for where the information varies.

Spearmint is more robust than Peppermint, tolerating a wider range of conditions and often producing better yields. In particular it is one of the few major herbs that will do well in sites that suffer severe waterlogging.

It is a somewhat more delicate plant when it comes to drying and needs good drying conditions to produce a good colour.

Site

As Spearmint will tolerate severe waterlogging, it can be put in sites unsuitable for most other herbs. However, it will do better where the drainage is good.

Identification

Spearmint grows 300–900 mm tall, with square stems, and tops somewhat branching, leaves lanceolate, light green to green, pointed with toothed edges (see colour plate 7.1). Flowers are pale lilac, in a spike at the top of the main stem.

The plants form rhizomes and die back to the surface of the ground in winter.

Variety

The situation with propagation material for Spearmint is not much better than for Peppermint. Many of the varieties of Spearmint commonly available (and grown commercially) are not very satisfactory: some of them having very poor flavour.

Some years ago I was given some propagation material of a variety of Spearmint selected for oil production. It has turned out to be an excellent variety for dried herb production as a tea and as a culinary herb, with a very strong, clear sweet flavour.

Pests and Diseases

Mint Rust

Rust normally attacks the leaves, but occasionally it will infect the stems as well. These develop swollen twisted sections, with blotches and pustules on them. This can be treated by travelling a little slower with the flame, so enough heat builds up to kill these infected stems as well.

Sometimes a spring outbreak of stem rust occurs, which can badly affect growth. An early harvest followed by flaming will control it.

Drying

Spearmint is a bit more challenging to dry than Peppermint. The leaf dries easily enough, but needs optimum conditions through the whole drying process. It bruises easily and will discolour if it is too slow drying or if it is subject to re-moistening at all. This will darken it to a sort of army green. Given good drying conditions, Spearmint dries to an attractive green colour.

Spearmint should be processed when the leaf is dry and the stems are still flexible.

Processing

Spearmint is used as a tea and as a culinary herb.

Tea Grade

As Spearmint stems are very straight and smooth when dry, they tend to spear down through the screen very easily, necessitating further screening and/or winnowing to remove them. If the herb can be rubbed when the stems are still moist enough to be flexible, this will save time and effort in producing a stem-free finished product.

Rubbing Spearmint is usually a milder, less overpowering experience, than rubbing Peppermint.

Culinary Grade

Spearmint for culinary use should be rubbed through the standard culinary screen after it has gone through the tea-grade process.

The aim is to produce a finer leaf and remove the fine stems present in the tea grade.

With a herb like Spearmint, which is processed to both tea grade and culinary grade, the grower can take a short cut and save time and effort in processing the culinary grade.

Tea-grade first screenings can be used to rub down to culinary grade. But with each screenful there is no need to try to push all the herb through the fine screen: just rub about half of it through, then tip the rest aside. This first half which has gone through will contain hardly any stem, while the discarded half, with its greater proportion of stalk, can go back into the batch being further screened to tea grade.

This way the culinary grade can be produced in about two passes through the fine screen: otherwise it would take perhaps four passes to adequately screen the lot.

Yields

Spearmint could be expected to yield 0.25–0.5 kg/m^2 per annum tea or culinary grade.

Marketing

Spearmint does not sell in the volume that Peppermint does, but there is a steady market for it. In the premium-grade market, much more is used as a tea than for culinary purposes.

Price

Price to growers for premium quality organically grown Spearmint is currently around $24/kg for 5-star tea or culinary grade. Trade price is around $3–9/kg.

St John's Wort

Hypericum perforatum L   HYPERICACEAE

Parts Used: **Leaf & Flower, Aerial Parts**

Other Names: *Hypericum perforatum* var. *angustifolium*.

St John's Wort is covered in detail under 'Wildcrafting and Weed Harvesting'. There are vast areas in the grip of this species, and so far no satisfactory method of controlling its spread. It does quite well under cultivation, but growers should think twice before considering bringing it into their district, because of the risk of it going wild. The market for this herb can be more than adequately supplied by wild-crafting in infested areas, so there is no need to consider growing it, except perhaps where none is growing locally and the fresh herb is required for making St John's Wort oil: an extract in olive oil used for healing wounds or sunburn.

GROWING ST JOHN'S WORT

St John's Wort is an easy herb to grow, but it is not so easy to ensure it will stay within the confines of the garden.

Being a widely declared noxious weed, approval would need to be obtained from Plant Quarantine for growing it under controlled conditions from where it could not escape.

Containment

As St John's Wort is winter-active and will invade an unwatered grassed buffer zone, it needs to be surrounded by a continuously cultivated buffer zone of at least 1.8 m width.

Physical containment with a barrier may also be feasible, as the rhizomes spread on or close to the surface.

Precautions need to be taken not to let seed ever form: if the plant is harvested every time it goes to flower this is not a problem.

If there is a surplus above your requirements, it should be cut and disposed of. The traditional practice of burning it to keep away evil spirits no doubt helped to prevent it from spreading.

Pests and Diseases

A number of natural pests and diseases of St John's Wort have been introduced as biological controls. So far none have proved very successful, though some appear promising.

For more detail see 'Information Charts' and Chapter 21.

Tansy
Tanacetum vulgare L ASTERACEAE

Parts Used: **Leaf & Flower, Aerial Parts**

While Tansy does not enjoy wide usage, it is well known and there is a potential for increased demand in the future as it is used as a drench for livestock as an alternative to the chemical vermifuges which are in widespread misuse. It is also, to a limited extent, used by humans as a vermifuge, but there is some concern about its toxicity.

David Hoffman, in *The New Holistic Herbal*, says Tansy is quite safe to use for short periods, but not for prolonged periods or during pregnancy.

Tansy was traditionally used in pancakes in Lent, the bitterness seen to symbolise Christ's suffering.

Bunches of Tansy flowers are popularly held to be an insect repellent, but my observation in the garden is that flies just love to sit on them.

GROWING TANSY

This is one herb which is very easy to grow and does not demand much attention. It holds its own against weeds, being winter-active. It is somewhat invasive, but not to the degree of say Coltsfoot or Couch Grass. It just advances steadily on a solid front and smothers any plants growing next to it.

Containment

While Tansy only expands at about 300–600 mm a year, its advance is solid and relentless, so it would be wise to surround it with a cultivated buffer zone at least 1 m wide.

Identification

Tansy grows densely and somewhat straggly, 600–1200 mm tall, with aromatic pinnate leaves to 120 mm long, the leaflets deeply toothed. Inflorescence is terminal umbels of golden yellow composite flowers, which lack petals and last a long while on the plant.

The plant tends to stay green over winter, and spreads by means of rhizomes.

Variety

There is a frizzy-leaved variety which growers should avoid as it is even more straggly and difficult to harvest.

Drying

Tansy needs good drying conditions, as it is slow to dry.

For more detail see 'Information Charts'.

Tansy (*Tanacetum vulgare*).

Yarrow

Achillea millefolium L ASTERACEAE

Parts Used: **Leaf & Flower, Aerial Parts**

Other Name: Milfoil.

Yarrow has had a prominent place in herbal medicine since antiquity: its botanical name comes from the tradition that Achilles used it to treat the wounds of his warriors. It is also employed in treating fevers, and has a number of other diverse medicinal uses.

As a fresh herb, Yarrow can be included in salads and, as it is relished by livestock, it is an ingredient of herbal pasture mixtures. It can be used in brewing in place of Hops and can be drunk simply as a herbal beverage for its flavour. It is also used as a tobacco substitute, and in cosmetics.

In Chinese tradition the sticks used in consulting the 'I Ching' are made from Yarrow stalks.

Yarrow plays a special role in biodynamics. Its flowers are made into the biodynamic compost preparation 502 and in extolling its virtues, Rudolf Steiner recommended it should be growing on every farm.

Origin

Yarrow is native to Europe, but naturalised in temperate regions all over the world.

GROWING YARROW

Yarrow is quite easy to grow, but it does need some attention at specific times. It is a crop which establishes vigorously, giving a fairly good yield the first season and reaching its peak in the second.

Climate

Yarrow does well in a wide range of climates.

Site

Exposure

Yarrow prefers full sun, and will thrive in exposed and windy sites, too.

Drainage

It prefers good drainage, but has some tolerance for wetter sites.

Soil Type

It does well in medium to light loams.

Fertility

It needs at least a moderate level of fertility and is more productive in fertile soils.

Initial Weed Control

A bare fallow is recommended before planting.

Neighbouring Plants

Keep Yarrow away from other spreading herbs, and small plants with a high moisture requirement.

Containment

Yarrow's expansion takes place at about the same rate as Peppermint, so it can be contained by a cultivated strip about 1 m wide around its bed.

Identification

Yarrow is an aromatic herb, with grooved stems up to 600 mm tall. Leaves are 20–100 mm long, bipinnate, very finely divided, slightly hairy. Flower heads are normally white or light pink, composite, 4–6 mm across and in dense terminal umbels (see colour plate 7.6).

The plants form rhizomes and normally die back to the ground in winter.

This is another herb that has related species often confused with it. In particular, *Achillea tanacetifolia*, which is sometimes cultivated in gardens, and may be found growing wild, often along roadsides in Tasmania.

**Leaf of Yarrow, *A. millefolium*, left;
and leaf of *A. tanacetifolia*, right.**

A. tanacetifolia looks very similar, but it is a somewhat larger coarser plant, growing to 1000 mm, with red to light pink flowers, leaves not so finely divided, and broader leaflets. It has the same aroma as Yarrow. It is best distinguished by the broadness of the leaf segments and the winged rhachis or central stem of the pinnate leaf.

Achillea millefolium flowers are normally white, but pink and crimson forms do exist.

To be on the safe side, obtain propagation material from plants with white flowers (but beware, there is a very light pink form of *A. tanacetifolia*).

Variety

There is some variation within *A. millefolium* as well. Forms of this species exist with flowers of varying shades from pink to a deep crimson. Some varieties seem to have much tougher stalks than others. One strain produces a greater number of flowering stalks, particularly on the regrowth after harvest and for a longer period as a meadow-crop.

The Yarrow which grew wild where I lived in Quebec gave a distinct numbing sensation to the tongue when a fresh leaf was chewed, but this does not happen with the Yarrow here. I don't know if this is a genetic or a climatic difference.

Preference should be for a white-flowered variety which maintains a good level of flowering.

Propagation

Method

Propagation by rhizomes is easiest and these are produced in good quantities.

Propagation by seed is also feasible, but as these are small, seedlings will take longer to develop.

Timing

Optimum planting times are in autumn or early spring.

Planting

Plant rhizomes end-to-end in furrows and cover with 25–50 mm soil.

Layout and Spacing

Yarrow should be set out in continuous rows 900–1200 mm apart. For harvesting with the scythe, these should be in blocks at least three rows wide.

Compost

Yarrow needs compost at the rate of 3–4 kg/m².

Weed Control

Once established, Yarrow tends to hold its own a bit more strongly than many other spreading herbs, so it is less prone to weed invasion. It is somewhat vulnerable to White Clover, which can spread quite thickly through it if allowed to do so, but grasses and other winter-active weeds don't seem to be such a big problem, as it forms a fairly tight mat.

With some diligent attention at critical times, I have always been able to maintain good weed control in Yarrow without resorting to a black plastic dormancy mulch.

Rejuvenation

While it is invading new ground, Yarrow will flower prolifically, but after one or two seasons it will have expanded to become a meadow-crop. This suppresses flowering and you can end up with a dense feathery lawn with a few feeble flower stalks emerging.

This suppression effect seems to vary a bit with the variety and the site, but it can greatly reduce Yarrow's productivity.

The solution is to give the Yarrow some new ground to invade by rotary hoeing strips out of it, as described for rejuvenation in the introduction to this chapter. This usually has to be done every year or two in autumn. Repeated cultivation is required, to thoroughly kill the Yarrow between the new rows. As the remaining plants reinvade these cultivated strips, they will flower vigorously again.

Irrigation

Yarrow needs 25–40 mm of water per week through summer.

Pests and Diseases

Apart from the ravages of grasshoppers, I haven't had any problems with pests and diseases in Yarrow.

Harvesting

Timing

Yarrow should be harvested in the early flowering stage. The first harvest is usually in December, a little later than Mint. It should be possible to get three harvests in a season, though often there is little or no flower in the third harvest.

Don't leave the second or third harvest too late, hoping for the growth to get taller, as the thick sward of feathery leaves tends to shade the under layers, so while it looks green on top, it can be quite yellow or even black underneath.

Tools
Harvest with the catching scythe.

Technique
The second and third harvests can have a lot of discoloured leaf hidden in the under layer, so some care needs to be taken to avoid harvesting too much of this with the crop.

If you are harvesting flowers for making 502, you will probably want to take them from the first growth. If a large enough plot is grown, you can take a portion of the flowers for making 502 and then harvest the rest as a dried herb (see Chapter 10, 'Biodynamic Aspects').

Post-harvest Treatment
Mow the Yarrow stubble down close to the ground after the first and second harvests.

Drying
Yarrow dries very easily, but it can discolour if subject to excessive remoistening during drying.

Temperature
The drying temperature should not exceed 35°C.

Processing
The rubbed leaf and flower of Yarrow is used as a tea, and aerial parts for manufacturing liquid medicines.

Tea Grade
Follow normal procedures.

Blending of Different Harvests
As there is quite a difference between the proportion of flowers in the first harvest and the later harvests, you may wish to mix them together to get a more standardised product.

Aerial Parts
Chaff-cut as per usual.

Storage
Yarrow needs freeze treatment before storage.

Yields
Yarrow can yield as much as 0.35 kg/m^2 tea grade or around 0.5 kg/m^2 of aerial parts per annum, but this depends on having a vigorous crop.

Marketing
Yarrow enjoys moderate popularity as a herb tea, and a moderate level of usage for manufacturing.

Price
Price to growers for premium quality, organically grown yarrow is currently around $24/kg for 5-star tea grade.

Price to growers for manufacturing-grade, organically grown Yarrow aerial parts is currently around $10–15/kg.

Trade price is around $7–8/kg.

Yarrow (*Achillea millefolium*).

16

Expanding Clump Herbs

Bergamot – *Monarda didyma*
Chamomile, Roman – *Chamaemelum nobile*
Comfrey, English – *Symphytum officinale*
Dandelion – *Taraxacum officinale*
Hops – *Humulus lupulus*
Melissa Balm – *Melissa officinalis*
Mugwort – *Artemisia vulgare*

Nettle, Greater – *Urtica dioica*
Oregano – *Origanum vulgare*
Scullcap – *Scutellaria lateriflora*
Tarragon, French – *Artemisia dracunculus*
Valerian – *Valeriana officinalis*
Violet, Sweet – *Viola odorata*

The species in this group of herbs are in many ways similar to the spreading herbs in their habits and requirements.

Expanding clump herbs also spread by means of rhizomes or roots, but these are shorter than those of the spreading herbs. Consequently their expansion is at a slower rate and their growth is denser.

This results in two fundamental differences:

- They are plants that expand rather than invade, so they normally don't require containment.
- They are not as vulnerable to weed invasion, because their growth is denser.

Apart from these two points, the similarity between the two groups is great enough to allow us to continue to use the detailed introduction of the previous chapter as a guide to their management, with attention to the points outlined below, where any variations are covered.

Nevertheless, the natural order of things seldom falls neatly into our attempts to classify it. There are a number of species intermediate between these two groups and also a number intermediate between this group and perennial crown herbs.

Site

Containment

This is not a problem for most of this group, as their rate of expansion is on a smaller scale.

Neighbouring Plants

Expanding clump herbs are far more flexible in the available choices of neighbouring plants.

Propagation

The timing and methods of propagation are basically the same as for the spreading herbs.

Though growing from seed is an option for more of them, if sufficient vegetative propagation material is available, it is usually the preferred method and in some cases the only method.

Layout and Spacing

Most are planted in furrows with continuous rows of rhizomes; some are established in clumps about 300 mm apart. Rows need to spaced somewhat closer to get optimum density, usually 600–900 mm apart, but there are some exceptions.

Weed Control

Weed control is not as challenging with this group, because they form fairly dense clumps, which expand as a solid front and offer less opportunity for weeds to get established. In general they can be maintained free of weeds with regular inter-row cultivation and hand-weeding, without it becoming too demanding.

Rejuvenation

Most of this group will not require rejuvenation as they are slower to crowd themselves out and the plants are more tolerant of being squeezed up together. Consequently, when they do fill up their space to form a meadow-crop this can usually be kept going productively for several seasons.

Bergamot

Monarda didyma L LAMIACEAE

Parts Used: Leaf & Flower, Flowers

Other Names: Bee Balm, Oswego Tea.

Bergamot enjoyed a big boost in popularity when American patriots threw the heavily taxed imported Oriental Tea overboard at the Boston Tea Party in 1773 and people adopted Bergamot, under the name 'Oswego Tea', as the beverage of the new republic. Since then, Coffee and Coca-cola have displaced it as the national American beverage and Bergamot has sunk into relative obscurity there.

In Australia this herb is not well known, but whenever people try it, Bergamot is always well liked.

It is a popular belief that Bergamot is added to Oriental Tea to make Earl Grey. This is not the case, but there is a connection. The herb Bergamot gets its name because its aromatic qualities are reminiscent of a species of Citrus called Bergamot (*Citrus bergamia*), the oil of which is used to flavour Earl Grey tea.

Bergamot is not much used as a medicinal herb, being weak in its actions. It is more used as a herbal beverage, usually the leaves and flowers together, but in some European circles, the flower petals alone are used.

Origin

Bergamot is a native of north-east North America.

Bergamot (*Monarda didyma*).

GROWING BERGAMOT

Bergamot adds a delightful touch to the garden, with its blue-green foliage and its magnificent bright red flowers. However, like many other species from this part of North America, it does not take too kindly to the rawness of the climate in many parts of southern Australia. Given the right situation, it will thrive, but in many places it will just linger and maybe die.

Climate

Bergamot needs a cool to mild climate. It does not seem to be suited to regions where summers are hot.

Site

Exposure

Shade or semi-shade would be preferable for hotter regions. It will do well in full sun in Tasmania, provided it gets some protection from winds. In a windy site it will suffer badly from wind-burn and breakage: its stems snap off easily at the base.

Drainage

Bergamot will not tolerate waterlogging at all.

Soil Type

Bergamot seems to do well in lighter soils, provided it has sufficient moisture.

Fertility

It needs a fertile soil.

Initial Weed Control

A bare fallow is recommended before planting.

Neighbouring Plants

Bergamot is not particular about its neighbours nor does it cause problems to other plants. It might benefit from the protection of taller plants, such as Rosemary or Elder, provided they did not deprive it of moisture.

Identification

Bergamot grows 400–1000 mm tall, with erect quadrangular stems. Leaves are opposite, green to blue-green, sometimes tinged with red, up to 150 mm long, lightly toothed. Flowers varying in colour from pale pink to purple to scarlet red, resembling honeysuckle, in large whorls, terminal or in the top two or three axils (see colour plate 1.4). The flowers appear in summer, but if the plant is not growing vigorously it often won't flower.

Variety

There is quite a range of flower colours available, but for the dried herb the bright red is often preferred, as the red petals look very attractive in the finished product and this is the only variety harvested for the petals.

Nevertheless, flavour is important and can be variable: some varieties are poor flavoured. One variety I am growing at present has mauve flowers and a delightful flavour, which makes up for its lack of colour.

Propagation

Method

Bergamot produces small pink rhizomes, like a miniature version of Mint rhizomes. Being small and delicate, they should be planted as soon as possible after digging.

It can also be grown from seed, but you may not get the colour you want.

Timing

Plant rhizomes in autumn or early spring, as for Mint. Plant seed in early spring.

Planting

The rhizomes being so small, it is often easier to plant little clumps 25–40 mm across rather than to separate the rhizomes individually. I usually make a furrow and set these little clumps in it. If you have plenty of propagation material, you could plant larger clumps or, if your supply is limited, smaller clumps, or even individual rhizomes.

Layout and Spacing

Bergamot should be set out on 150–300 mm centres in rows about 900 mm apart. For harvesting with the scythe, these should be in blocks at least three rows wide.

Broad low-profile raised beds, as described in the introduction to 'Spreading Herbs', are recommended if you have any doubts at all about your drainage, because Bergamot is very touchy about excess water.

Compost

Bergamot needs a fertile growing situation, but if it is too fertile it is likely to become top-heavy and fall over. Consequently your application of compost needs to be in accordance with the nature of your soil, usually in the range 2–4 kg/m² per annum.

Weed Control

Ongoing weed control is important, but if the Bergamot is doing well, it will not be too hard to maintain it free of weeds as it forms fairly dense clumps. However, in conditions that don't suit it, Bergamot is likely to be overrun by weeds.

The important thing is to go over the Bergamot plot thoroughly in early spring and after each harvest, and not let any perennial weeds get a hold in it.

Irrigation

Bergamot needs 25–50 mm of water per week over summer. While the species requires plenty of moisture, it can easily be harmed by overwatering, especially in heavier soils, as it won't tolerate wet feet.

I found this out the hard way, growing Bergamot on a medium loam with a clay subsoil. An erratic irrigation timer kept the water going for many hours longer than intended, which the Bergamot did not appreciate at all.

Pests and Diseases

When it is doing well, Bergamot does not seem to suffer badly from pests and diseases, though there are a few that can cause problems.

Grasshoppers

Grasshoppers will attack Bergamot if it is growing under stress. The first thing they like to chew off is flowers, so your Bergamot may lack the red petals which normally highlight it. It is not a good crop for grasshopper areas.

Snails

Bergamot is the sort of crop snails like to get amongst. For managing and avoiding problems with them, refer to 'Snails' in Chapter 9, 'Pests and Diseases'.

White Flies

While white flies may sometimes be found on Bergamot, they don't seem to do any noticeable damage.

Harvesting

Timing

Ideally Bergamot should be harvested when the plant is in full flower. First harvest is November–December, with a second in February–March, and possibly a third later on.

Don't delay a harvest excessively while waiting for the flowers to appear. Sometimes the growing conditions are not conducive to flowering and the leaves will start to deteriorate if you wait too long.

If the first growth of the season is very tall and leafy, it may have to be cut early, otherwise it will fall over, making harvest difficult. If this is happening to your Bergamot, it can possibly be avoided by moderating your compost application.

Tools
The catching scythe is the best tool for harvesting, except for the initial harvest of a newly planted crop.

Technique
The usual procedure.

In Switzerland and other parts of Europe, just the petals of the flowers of the Bergamot are harvested, to make a high class, luxury herb tea, which is quite expensive and enjoyed by the elite. The petals are picked off the plants by hand, so the tea contains no green material.

Post-harvest Treatment
Mow stubble off but not so low as to damage rhizomes near the surface.

Drying
Bergamot dries easily, though it is subject to bruising and darkening if handled in the wilted state or if it is excessively remoistened.

Temperature
Dry Bergamot at temperatures up to 35°C.

Processing
Bergamot is used as a tea on its own or blended with Oriental tea to make a mock Earl Grey. As there can be great variation in the amount of red petals showing in the finished product, you may wish to make a more standardised product by mixing different batches.

Tea Grade
Rub the dried herb through the tea-grade screen to remove stems, using normal procedures.

Storage
Bergamot needs to be freeze treated before sale or storage to prevent moth larvae infestation.

Yields
In a suitable situation, Bergamot is fairly productive, with yields of up to 0.4 kg/m^2 tea grade per annum.

Marketing
The market for Bergamot is limited, but this is partly because it has not been available. My experience has been that when there is a continuity of supply, the market for this herb will steadily grow as people become familiar with it.

Price
Price to growers for premium quality, organically grown Bergamot is currently around $24/kg for 5-star tea grade.

Chamomile, Roman
Chamaemelum nobile (L) All. ASTERACEAE

Parts Used: Flowers

Other Names: English, Garden, Lawn or Common Chamomile, *Anthemis nobilis*.

Roman Chamomile is well known for its use as a herb lawn, particularly for Corgies to run around on.

Even though it is mentioned in most herbals, it is really not much used as a herb tea because it is so bitter and unpalatable. Virtually all the Chamomile available in the trade is the German Chamomile (*Chamomilla recutita*), which is a more pleasant flavoured herb, and also much more productive.

Unfortunately there is some confusion in many people's minds as to which Chamomile is which. This situation is exacerbated by herb books, which often fail to distinguish between the species adequately and sometimes have them hopelessly confused.

The two species have similar medicinal uses, and references to Chamomile go back to ancient Greece and Egypt, though there is some uncertainty as to which Chamomile the ancients used. The Chamomiles are mainly used for their relaxing, carminative and anti-inflammatory actions. Of the two species, Roman Chamomile has the reputation of being slightly stronger.

The difficulty for the beginner in unravelling the confusion between these two species is further compounded by the unfortunate array of botanical names which have been in recent use.

Roman Chamomile used to be called *Anthemis nobilis* but has now become *Chamaemaelum nobile*.

German Chamomile was once *Matricaria chamomilla*, then it became *Matricaria recutita* for a while, but is now known as *Chamomilla recutita*.

For detail on German Chamomile see Chapter 20, 'Annuals, Biennials and Short-lived Perennials'.

GROWING ROMAN CHAMOMILE
Roman Chamomile is not at all difficult to grow, but as a dried herb there is not much market for it. It is only worth considering for clinic gardens or if there is a specific requirement for this particular species. Otherwise German Chamomile is preferred, is easier to harvest, yields better and has a big market.

Identification

Roman Chamomile is a perennial plant with a creeping habit, growing up to 300 mm tall. Leaves are finely divided and aromatic. Flowers are composite, the heads borne singly on the ends of stalks, 15–30 mm long, yellow-centred with white petals, sometimes double.

For detail on distinguishing this species from German Chamomile (*Chamomilla recutita*), see the entry for the latter in Chapter 20, 'Annuals, Biennials and Short-lived Perennials'.

Variety

Make sure you have a flowering variety, not the non-flowering variety known as 'Treneague'.

Harvesting

Timing

Harvest Roman Chamomile when the flowers are open but not too advanced. Roman Chamomile normally flowers once a year, in late spring.

Tools

Roman Chamomile can be harvested with the same flower comb as German Chamomile.

Technique

The technique is basically the same as for German Chamomile, but the flowers are much more reluctant to separate from the stalks, so progress is slow and is frustrated by long pieces of stalk breaking away from the plants.

For more detail see 'Information Charts'.

Roman Chamomile (*Chamaemelum nobile*).

Comfrey, English

Symphytum officinale L   BORAGINACEAE

Parts Used: Leaf, Aerial Parts, Root
Other Names: Comfrey, Common Comfrey, Knitbone.

English Comfrey is not exclusively 'English'. This name is commonly used to distinguish it from other Comfreys, of which there are several.

Comfrey used to be the flagship of herbal medicine, but it has been effectively blown out of the water by regulations restricting its use. These have come about as the result of a panic triggered by the misinterpretation of a few papers on the toxicology of pyrrilizidine alkaloids and some misunderstanding of Comfrey's usage in herbal medicine. While pyrrilizidine alkaloids do occur in Comfrey, most herbalists are confident normal usage presents no significant risk to human health.

Levels of alkaloid vary in the different parts of the plant: the root and very young leaves have higher levels, while dried Comfrey leaves harvested from plants at a mature stage are very low in alkaloids.

Comfrey is traditionally used internally and externally to facilitate healing, as a demulcent and an expectorant.

It is still legal to grow Comfrey, and its preparations can be sold for external use on humans, but not for internal use. Non-therapeutic use for humans is not recognised, but you can sell it to someone to give to their dog or their cow.

GROWING ENGLISH COMFREY

English Comfrey is one of the easiest of plants to grow. It is very vigorous and will tolerate a wide range of conditions. One caution, though: it is very hard to get rid of, so make sure it goes into the right place and make sure you have a market for it, as there is no longer much demand for it.

Containment

While English Comfrey's expansion is not particularly rapid, once it has established a hold, it is hard to drive back. Putting it on the edge of the garden means you only have one side of it to contain. It is best to keep a strip of about 900 mm cultivated along the sides of it.

Identification

English Comfrey grows 300–1200 mm tall. The leaves and stems are covered with stiff hairs. Lower leaves are up to 250 mm long, lanceolate, with slightly wavy margins, the upper leaves are narrower. The flowers are purplish, pinkish, or yellowish-white, on terminal cymes (see colour plate 2.2).

The main problem is to distinguish English Comfrey from the more common and very similar Russian Comfrey (*Symphytum peregrinum*, syn. *S.* x *uplandicum*), a natural hybrid which appeared some time last century.

English Comfrey has a number of distinguishing features, but the best thing is to look at the flowering stems. Where the leaves are attached to the stem, a wing of leaf-like material continues from the stalk of the leaf and down the stem, past the leaf below it. Each leaf has two of these wings coming from it, so on the lower part of the stem there are a number of wings beside each other.

The stems of the Russian Comfrey do not have wings developed to this degree. They will be wingless or, if there are wings, they will be narrower and extend no more than half way down to the leaf below.

Alkanet (*Alkanna tinctoria*) and Common Alkanet (*Anchusa officinalis*) have leaves similar to Comfrey. Their flowers are rather different though, resembling Forget-me-not.

Another plant to avoid confusing with Comfrey is, of course, Foxglove (*Digitalis purpurea*). Foxglove is deceptively similar when in the rosette stage, but as soon as it flowers there is no confusion. As Foxglove is a biennial and grown from seed, it is unlikely to become accidentally mixed in with your Comfrey when you are planting it.

Processing Leaf

CAUTION: The silica hairs on Comfrey leaf can be quite irritating when dry, especially to the eyes and respiratory tract. Adequate protective equipment should be worn.

For more detail see 'Information Charts'.

Comfrey, English *(Symphytum officinale)*

Dandelion

Taraxacum officinale Weber   ASTERACEAE

Parts Used: Leaf, Root, Flower

Dandelion does not need much of an introduction, as it is one of those plants which have followed Europeans everywhere throughout the temperate zones.

It is a valued medicinal herb: the leaf as a diuretic and the root for its action on the liver. It is also eaten as a salad herb, especially in early spring in cold climates where it is the first green thing available. In warmer climates it tends to be less palatable, being more bitter.

The root is roasted to make Dandelion 'coffee', a popular beverage.

The flowers are gathered and dried to make the bio-dynamic compost preparation 506.

Origin

Dandelion is native to Europe and Asia, and has naturalised elsewhere.

GROWING DANDELION

There is something very congenial about Dandelion: the way it seems to enjoy human company, and almost every part of the plant is useful. A big healthy Dandelion plant in flower almost seems to glow like an image of the sun.

But for every shining light there is a dark side. Dandelion is so well adapted to proliferating and surviving in our footsteps, it is apt to start coming up in all the wrong places and become a weed. The main problems are the seed, which blows around and gets started everywhere it can, and the pieces of root, which break off and regenerate when you try to dig it out.

Climate

While Dandelion prefers cool to mild climates, it will do reasonably well in hotter situations, provided it gets plenty of moisture.

Site

Exposure

Dandelion grows well in shade but also in full sun. Wind is not a problem for it, except for very hot dry winds.

Drainage

Dandelion will tolerate poor drainage, but the quality of the roots may be affected.

Soil Type

Light to medium soil is preferred for ease of weed control and digging, and it is easier to wash roots that have grown in light soil.

Fertility

Dandelion needs a fertile situation.

Initial Weed Control

A bare fallow is recommended before planting.

Neighbouring Plants

Being low growing, Dandelion is easily overshadowed, and should not be placed alongside plants that will compete with it for moisture.

The main problem it presents for neighbouring plants is the seeds it tends to send floating on the breeze when it is in their direction. The wild strain of Dandelion is the worst for this, as it flowers all throughout the growing season. For this reason Dandelion should ideally be grown some distance away by itself.

If this is not possible, Dandelion should be grown on the down-wind side of the garden, in a corner with neighbouring plants that are managed as a row crop and where it is easy to maintain weed control. Every now and then you will have to go over the garden with a tweaker or a large screwdriver and grub out Dandelions that have come up where they shouldn't.

The cultivated 'broadleaf' strain of Dandelion is not as bad as the wild one, as it only flowers for a period in spring and then it just concentrates on making leaves and roots, so it doesn't send out as much seed.

Identification

Dandelion is a low growing perennial, 150–300 mm tall, with a basal rosette of leaves and a thick branching rootstock. The leaves are variable, the tip is usually broader than the base, and the leaf may be entire or irregularly cut into deep triangular lobes shaped like a lion's tooth with the teeth pointing backwards towards the base of the leaf (hence the name Dandelion, from the French *'dent de lion'*). The midrib of the leaf is distinctly hollow.

The flower-heads are yellow, 30–50 mm across and single on hollow stalks. These ripen and turn into the characteristic fluffy 'clocks' we all used to try to tell the time with, by counting how many puffs it took to blow all the seeds off (see colour plate 2.6).

There are a few weeds similar to Dandelion, and they are often mistaken for it.

The true Dandelion can be distinguished by the following characteristics:

- The flower stalks of true Dandelion are single and hollow. Other plants may have hollow stalks, but these will be branched or, if they are single, the stalks will normally be solid.

- The leaf of the true Dandelion has a distinctly hollow midrib, while the other species have leaves with solid midribs.

Variety

Dandelion is a somewhat variable plant and there are varieties selected for particular attributes.

I believe there may be a variety selected for root production, but I have not been able to obtain it.

The choice seems to be between a 'broadleaf' or 'thick leaf' strain available from seed suppliers, or the wild strain you will find growing in lawns, parks and pastures.

From the point of view of leaf production, the 'broadleaf' Dandelion is superior and its roots are larger and easier to dig. However, it is more susceptible to a rot which sloughs large sections of root, reduces yields and takes time to trim

Dandelion (*Taraxacum officinale*).

away. Having fewer flowers is an attribute if you want better quality leaf and fewer Dandelions coming up all over your garden, but it is not as good if you are after a good supply of flowers for making the biodynamic preparation.

The wild strain of Dandelion usually suffers less from root rot, but its roots are smaller and more branched. This makes for lower yields and more work digging and washing, but as the roots are smaller, there is less splitting or chopping necessary.

When the wild strain is harvested for leaf, its quality is affected by the proportion of flower stems that end up in the finished product.

If one of your purposes for growing Dandelion is to obtain flowers for the biodynamic compost preparation 506, then the wild strain is preferred, as it flowers more prolifically.

Probably the best thing would be to try some of each variety, grown separately, and compare your results.

Propagation

Method

Dandelion can be grown from seed, from pieces of root, or from transplanted crowns.

From Seed: If you are planting a significant area of Dandelion, then seed will be the preferred option as it is the easiest way to obtain large numbers of plants. Seed can be started in trays in autumn or early spring. Sometimes germination can be difficult. This may due to dormancy or some suppression factor, such as temperatures being too high. Usually I have had good success with seed sown in autumn.

Dandelion plants in cultivation often give rise to large numbers of seedlings in their vicinity during autumn and winter, so you may be able to find enough seedlings without much effort. Seedlings give a good success rate if planted out when they are 25–50 mm across. They can be transplanted at earlier and later stages, but they will have a lower success rate.

From Root: Any piece of Dandelion root is capable of producing a new plant, so another option is to dig up plants, break them into pieces and plant them where they are to grow, or start them in trays or pots and transplant them when they have started to shoot.

One problem with this method is that sometimes a proportion of the roots will rot rather than grow. It seems they tend to get infected with rot when the root is damaged or broken. If you are transplanting freshly shooted roots, you need to be careful to avoid damaging the new shoots, as they are very delicate.

From Transplanted Crowns: This is probably the quickest way to get started, but also the least reliable. Often it is possible to get a plentiful supply of crowns by digging them from somebody's lawn or a park, or from self-started plants around your Dandelion bed.

Success rates are sometimes poor with this method because root rot seems to get in when the crowns are damaged in digging them up. Smaller crowns transplant more easily, and digging with a garden fork will damage the roots less than getting them out with a tweaker or a screwdriver. This may seem a bit rough on a front lawn, but the soil does not need to be turned right over, just loosened with the fork so the crown can be slipped out with its roots intact.

Timing

Autumn or spring is the time for planting, depending on the weed situation in your site. Autumn planting will give the plants a better start for the next season, but if you anticipate a lot of weed growth over winter, then early spring planting is better.

Planting

Root pieces can be laid out in furrows and covered with about 50 mm of soil.

Layout and Spacing

Dandelion's optimum spacing is in rows 600 mm apart with 225–300 mm between plants. This allows enough room for inter-row cultivation for most of the season. If your weeds are well under control, you may be able to place the rows closer together.

After the crop has been harvested for the roots, a large number of plants will come up from the broken pieces left behind, and if the weed situation is not too bad, the Dandelion can be allowed to develop as a meadow-crop. However, if weeds are too abundant, it will be necessary to cut the patch back into rows to make cultivation for weed control easier.

Compost

While Dandelion enjoys a fertile soil, some caution needs to be exercised to avoid over-stimulating it. If it grows too vigorously, the roots tend to crack open and develop root rot. Consequently the compost should be balanced, not too nitrogenous, and should be applied in moderation. Dandelion will probably need a rate of 1–3 kg/m^2 per annum, depending on your soil's state of fertility and the quality of your compost.

Compost should be applied in early spring and worked in. For previously harvested patches being managed as a meadow-crop, the compost should be spread and worked in very early, before the plants start to develop. If soil conditions are too wet to allow this, then the compost can be spread on the surface later and watered in, but it will not be as effective and may cause problems when you come to harvest the leaf of the Dandelion.

Weed Control

In the early stages of Dandelion's development, the plants are rather vulnerable to domination by weeds. Diligent weed control at this stage is necessary. If you can be thorough about weed control in your Dandelion, you will gradually deplete the weed seed reservoir in your soil and the job will be much less demanding.

For the first few years it will probably be necessary to maintain the Dandelion in rows to make weeding easier, using the wheel hoe. But as the weed situation improves, you should be able to let the Dandelion's development become more random and occupy most of the space as a meadow-crop. Weeding is then maintained by hand-hoeing among the plants.

Management

As it would take a bit of effort to eradicate Dandelion from the ground it has been growing in, it can be maintained in the same bed each year, while it continues to grow well. At 'Twin Creeks', after growing in the same place for 5 years, the Dandelion still seemed to be as productive as ever.

Irrigation

Dandelion needs plenty of water for good growth in regions with a hot summer. This means 40–50 mm of water per week during the hotter part of the season. In cooler regions somewhat less will be needed.

Pests and Diseases

Being a fairly soft and palatable plant, Dandelion is subject to a few pests and diseases, but with its remarkable regenerative powers, it usually manages to recover.

Dandelion Root Rot

There is probably a technical name for the affliction known as Dandelion root rot. Essentially it is a rotting away of portions of the root (see colour plate 2.7). It does not seem to affect any other herbs, but has consistently occurred in Dandelion grown in a range of different climates and soils.

This rot infects the root where it is damaged. The root sometimes splits as it is growing and the rot gets started in these splits. It seems to be worse if the plant is growing very fast in fertile soil. The outside skin of the root can't keep up with the expansion and it splits, often in the main taproot, just below the crown.

How far the infection progresses seems to vary. Usually the Dandelion will eventually overcome it and form a new skin under the sloughed off part, but sometimes it goes right across the root. If this happens just below the crown, the top will die and after a while the remnants of the root system will send up new plants, as the rot usually does not extend very far along the root.

The 'broadleaf' strain of Dandelion, with its faster growth and thicker taproot, is more susceptible to root rot, but it can affect the wild strain as well. The reduction of yields is usually not great, perhaps 20% at the most, but it means you have to trim the affected parts from the roots before drying.

By reducing the amount of compost applied to the Dandelion, I found it was possible to reduce the incidence of root rot. Irrigation management may also be a factor: it is possible one cause of roots splitting is the rapid uptake of water by plants that are a bit dehydrated.

Leaf Spot

In hot or humid conditions, the leaves of Dandelion sometimes develop brown spots, which reduce their quality. It is no doubt the result of the plant being stressed. The main consequence is the leaves are not suitable for harvesting. If this condition is starting to develop or is likely to, it is best to harvest the leaves before they deteriorate.

The regrowth is often less affected.

Grasshoppers

Grasshoppers can sometimes be a problem with Dandelion in regions with a hot summer. They tend to attack it if it is under stress. By keeping plenty of water up to it in hot weather, the effect of grasshopper attacks can be reduced.

I also noticed the grasshoppers seemed to show a preference for certain individual plants: they would strip these bare but not touch other Dandelions nearby. My Dandelion patch was established with plants from a number of different sources, so there must have been a fair bit of genetic variation in it. After several seasons of stripping the leaves off their favourite Dandelions, the hoppers seemed to lose interest in the patch. My assumption was that they had eliminated the ones they liked and the rest were less palatable to them.

Harvesting Flowers

If you have a nice stand of Dandelions you may be interested in harvesting the flowers for making the biodynamic compost preparation 506. This is a fascinating operation and brings the grower in close touch with the cycles of the Dandelion.

This operation is covered in detail in Chapter 10, 'Biodynamic Aspects'.

Harvesting Leaf

Timing

Dandelion leaf can be harvested when there is enough leaf to make it worth while and the plants are big enough to spare some leaf without affecting their growth too much.

If the roots have been harvested the previous winter, it may be January (though sometimes earlier) before the plants have developed enough leaf to take a harvest. It is usually possible to get a second and maybe a third harvest later in the season, but this varies.

If Dandelion were grown for the leaf alone, then harvesting could begin much earlier and there might be 4–6 cuts a season.

Tools

It is usually possible to harvest the leaf with a scythe. However, if the rosettes are too low, it may be necessary to use a reaping hook.

Technique

As you don't want to set the development of the roots back too much, in harvesting the leaf of Dandelion the aim is to ttake just the surplus and leave a good cover of green leaves. This is not difficult with the scythe, as it only picks up the higher growing leaves and enough green cover is left.

Careful cutting is important to avoid getting too many old or scruffy leaves in the harvest. With the wild strain, it is difficult to avoid getting a lot of flower stalks as well, but these can be sifted out to some extent after the herb is dry. This is where the 'broadleaf' strain is better because it is usually free of flowers by the time the leaf is harvested.

Post-harvest Treatment

You may wish to cut down any scruffy leaves or stalks which might get picked up in your next harvest, though make sure you leave a good cover of green leaf.

Drying Leaf

Dandelion leaves dry easily, provided the screens are not overloaded. If placed too thickly on the screens, Dandelion tends to mat down and be slow to dry. Because the leaves are very light when they are dry, they should be placed somewhere in the shed where the wind won't catch them and blow them around. The puff from the flowers can be a bit of a nuisance at times, getting in your nose and eyes.

Dandelion remoistens easily, so you may find it is too damp to rub when other herbs are dry enough. It discolours a bit if it is too slow to dry or if it is excessively re-moistened.

Temperature

Dandelion leaves can be dried at temperatures of up to 45°C.

Processing Leaf

If flower stalks are present in the leaf, then rubbing should be done in a manner to minimise breaking them up so they don't go through the screen. Inevitably a proportion will go through. A certain amount is acceptable, but if there is too much the leaf will be downgraded. The old brown stalks are the main problem. as they break up easily and look terrible in the finished product.

The midribs of the leaves should be included in the herb, but sometimes they are reluctant to break up. Unless conditions are fairly dry, they tend to remain leathery, especially at the base of the leaf. If necessary, the unrubbable portion can be put back onto a screen in a warm spot and rubbed through later, when conditions make this possible.

Dandelion leaf is used as a tea and also in manufacturing liquid medicines.

Tea Grade

Tea grade is rubbed through a $2^{1}/_{2}$ to 3 dent screen, The 'broadleaf' strain normally only needs re-screening once, but the wild strain will need re-screening a number of times to reduce the proportion of pieces of flower stalk. Using under-screens will help here.

Manufacturing Grade

Manufacturing grade is usually just a first screening, but if there is an excessive proportion of flower stalk, it sometimes might need to be re-screened. It is easier to rub Dandelion leaves than to try to push them through the chaff-cutter, and if flower stalks are present, the screen will separate them out to a degree.

Storage

Freeze treatment to control moth larvae is necessary before sale or storage.

Yields

Yields of Dandelion leaf at 'Twin Creeks' in north-east Victoria were not very impressive. The leaf harvest was a side-line of the root production, so the plants took half the season to reach a reasonable size. The growing conditions in the hot summer were not the best for producing Dandelion leaf, besides which the plants were kept a bit lean to reduce root rot.

The best season we had yielded about 0.1 kg/m² per annum, but most seasons were well below that.

However, I would expect a patch of Dandelion grown specifically for the leaf in milder growing conditions and given heavier dressings of compost, would give yields two or three times what we were obtaining.

Harvesting Root

Timing

While most of the literature talks of digging Dandelion at the end of its second year, it must be remembered these are European sources and our growing season is much longer. I have always dug it at the end of the first year, provided the Dandelion had a good early start. My yields have usually been comparable with those obtained in Britain after 2 years growth.

If Dandelion is left to grow for 2 years here, it is liable to deteriorate as the roots become over-mature and die back in places. This affects their quality.

Dandelion is normally dug after the leaves have died down in autumn, or during winter.

There is another school of thought which prefers the root to be dug in spring after the plant has come out of dormancy. Roots dug at this stage are lower yielding, some of their content having gone into making the new leaves, but this results in higher concentrations of some of their active constituents. The root is more bitter when harvested at this stage. But unless you have a particular request for spring-harvested Dandelion, it should be harvested in autumn or winter.

The soil needs to be friable enough to be shaken easily from the clumps of root. If your soil is on the heavy side, it is best to dig the Dandelion as soon as it dies down, before the rains make the soil unworkable.

Tools

A garden fork is required for digging, and a knife or a small cleaver and board are needed when trimming off the leaves.

Technique

The Dandelion plants are loosened with the garden fork, lifted and the soil shaken off them. They then need their leaves trimmed off.

The system I use is to take a wheelbarrow along and throw the dug and shaken roots into it. One person then works from the barrow, trimming the leaves off the roots and throwing them into an open bag.

The leaves need to be trimmed off above the crown, as the plant does not go completely dormant in our climate. Cut them off at the very base of the leaf, at the top of the crown.

Trimming the leaves is quite time consuming, especially on the wild strain, as it forms a multitude of small crowns The 'broadleaf' strain is better in this respect: one large crown which only takes one chop.

If you are getting bogged down with leaf trimming, it might be possible to use a mower set very low or a weed trimmer to trim the leaves off before digging. However, this could make it difficult to find the crowns if you are digging a meadow-crop.

Mechanical Harvesting

As the market for Dandelion root is quite substantial, there is potential to grow it on a broad-acre scale. For this to be viable, it would need some system of mechanical harvesting, as hand-digging barely pays wages.

It may be possible to harvest Dandelion with a carrot or potato harvester, but these may need modification, as Dandelion roots are not as regularly shaped as these vegetables are.

If conditions are suitable, it might be possible to lift the roots with a plough-type attachment sometimes used for potatoes. This is like a two-sided plough with several rods on each side extending from the mouldboards. It lifts the soil and breaks it open, and any solid material rides along the rods and is left on or near the surface. You can then go along and gather up the roots. It works quite well for potatoes.

The soil would have to be light and friable for it to work and the Dandelion would have to be in rows. It may have to be in hills as well. There would be some loss, particularly of the smaller roots but these would left in the ground to make good propagation material for the next year's crop.

Post-harvest Treatment

After harvest, level off the surface with a rake while the soil is still loose.

Drying Dandelion Root

The roots of Dandelion need a thorough washing before being dried. Soil tends to get trapped in the fine roots, where the roots fork, and in the multiple crowns, so they need a good blasting with a jet of water directed from various angles. The denser clusters of roots will need to be broken apart to facilitate washing.

Washing on a large scale could probably be mechanised using machinery developed for washing vegetables.

After washing, the roots need to be checked over for dark sections of root rot, which should be trimmed off.

Any roots greater than the thickness of your thumb should be split or chopped to help them dry. If you are sure there are no small stones among the roots, you can put them through the chaff-cutter before drying. However, I usually prefer to chop them with a small cleaver on a solid board as it is very easy for small stones and grit to be trapped among the roots of Dandelion and this could damage the blades of the chaff-cutter.

It is best to let the roots air dry for a while after washing, to dry off surface water.

Chopped roots should not sit around for more than a day or two before going into the dryer, because the cut and damaged surfaces will start to rot or go mouldy if they aren't dried quickly.

The herb is dry when the biggest roots are hard and snap clean when broken. The interior of the roots of the large pieces should be checked for remaining moisture. When they are dry, they will be too hard to dent with the thumbnail.

Temperature

Dandelion roots can be dried at temperatures up to 45°C.

Processing Root

As some soil and grit tends to be trapped in fine roots and crevices, it is best to tumble the roots around on a fine screen – shadecloth is ideal – to remove the dust and the finest roots. If the roots are lifted off the screen rather than poured off, any little stones and soil particles will be left behind.

Tea Grade

Quite a bit of Dandelion root is used as a decoction, made by simmering in water for 10–15 minutes. For this the dried root needs to be milled or ground to a broken state, with pieces up to 6 mm across. This milling is usually done by the wholesaler.

Roasting Root

For roasting Dandelion root, it needs to be thoroughly dried first. The roasting process is rather tricky as the variations in size and colour of the particles cause them to absorb heat at different rates. Consequently the smaller and the darker particles will tend to burn before the larger and the lighter-coloured particles are sufficintly roasted.

Manufacturing Grade

Dandelion root for use in manufacturing liquid medicines can be sent in large pieces as it comes from the dryer.

Storage

Dandelion root needs to be freeze treated to control moth larvae before sale or storage.

Yields

In north-east Victoria, our yields of dried Dandelion root were fairly consistently around 0.25 kg/m^2 per annum, which is equivalent to 2.5 tonnes/ha – comparable to the figures quoted by Mrs Grieve (*A Modern Herbal*) of '4 or 5 tons of fresh roots to the acre in the second year' (equivalent to 2400–3000 kg/ha, dried).

Marketing

There is a reasonable demand for premium-grade, organically grown Dandelion leaf, and particularly for premium-grade, organically grown Dandelion root, for beverage use.

There is a market for manufacturing-grade Dandelion, but the price offered is not adequate for this scale of production. It would have to be a large-scale mechanised operation to be worth while.

Price

Prices to growers for premium quality, organically grown Dandelion leaf and Dandelion root are both currently around $30/kg for 5-star tea grade.

Price to growers for manufacturing-grade, organically grown Dandelion are, at time of writing:

    Root – around $10–17/kg
    Leaf – around $10–15/kg

Trade prices are currently:
    Root – around $12–13/kg
    Leaf – around $9/kg

Hops

Humulus lupulus L CANNABINACEAE

Parts Used: Strobiles (i.e. Cones or fruiting bodies)

Other Name: Hop Bine.

While Hops are well known for their use in beer-making, what is not so well known is that this use has only become universal in the past few hundred years when Hops replaced a number of other bitter herbs traditionally used in brewing.

Hops also have medicinal uses, mainly for their sedative properties, which makes the Hop a valuable herb for our often over-stimulated and stressful lifestyle, but it is contra-indicated for depression.

The Romans used the young shoots of Hops as a vegetable, and in north-east Victoria local Italian families still do likewise.

Origin

Hops are native to northern temperate zones. They are occasionally found growing wild in Australia, usually as a consequence of earlier cultivation.

GROWING HOPS

Hops are grown on a large scale in parts of Australia, mainly in north-east Victoria and Tasmania, and a lot of specialised machinery and large capital investment is involved. There is also widespread use of chemicals in weed and pest control, and the crop is normally dried at temperatures around 70°C using diesel burners. With these production methods, the quality of the dried Hops produced must be questionable.

Hops (*Humulus lupulus*).

There is some organic cultivation of Hops in Australia: one small grower in north-east Victoria who was squeezed out in the 'rationalisation' of the Hop industry in the past 20 years has been able to use his picking machinery and drying facility to grow Hops organically.

So while premium quality Hops can be grown organically on a small scale, without the picking machinery it is not going to be an economically viable proposition.

Another problem is that flowers of Hops contain a powdery yellow resin, which falls from them and sticks all over everything. This resin is very bitter and very difficult to clean off equipment such as drying screens, so it is likely to contaminate other herbs.

Nevertheless, a few Hops could have a place in a clinic garden or in a display garden. They are climbers with large green leaves and a stand of Hops fully grown is a magnificent sight, as they can reach up to 6 m.

Climate

Hops will grow in a wide range of climates.

Site

Exposure

Hops like full sun, but need shelter from strong winds.

Drainage

Hops need good drainage. They are normally grown on deep alluvial soils where waterlogging never occurs.

Soil Type

Light to medium loam is preferred, though weed control will be easier on the lighter soils .

Fertility

Hops need a fertile humus-rich soil.

Initial Weed Control

Weed control is not too critical, though an initial bare fallow to get rid of perennial weeds will help.

Neighbouring Plants

Hops climb over everything, so they are not good neighbours for most herbs.

Other Requirements

Hops need some means of support that will allow you to harvest the vines. On a large commercial scale, tall poles are set up throughout the hop-garden and heavy wires are permanently strung overhead. Above each plant several twines are attached to these wires and these are pushed about 125 mm into the soil, using a dibbler or knife with a 'V' cut in the end, so the vines can climb up them. An alternative sometimes used is to tie the string to the base of the plant.

Alternatively, a teepee-like frame or some other contrivance using long thin poles could be erected, or the hops could be allowed to trail up a nearby tree or fence.

Identification

Hops may become 6 m tall, but die back to ground level each year. Leaves are long stalked, heart shaped and usually 3-lobed – but can be 5-lobed or 7-lobed – and coarsely toothed (see colour plate 4.1). The plants are dioecious (male and female flowers are borne on separate plants): the male flowers are in loose bunches 75–125 mm long, the female flowers developing into leafy cone-like strobiles about 20 mm across when mature.

Variety

There are many different varieties, including a number recently developed with a much higher content of the bitter principle, alpha-acid. However, what is not known is whether the active constituents required for medicinal use are present in their usual proportions in these new varieties.

The traditional varieties have more fragrance and less bitter principle. If possible, growers should stick to these older varieties, though these may be hard to obtain.

Californian Cluster is recommended for north-east Victoria

Propagation

Method

Hops are usually propagated from suckers or cuttings. For the cuttings, pieces of stem about 150 mm long with fresh buds on them are taken from under the surface of the soil, where they emerge from the crown.

Hops will also grow from seed, but seedlings would not be true to type and will produce at least 70% of useless male plants.

There is some disagreement as to how many male plants are needed: some growers maintain they are not necessary at all, while others keep one or two male plants in about a hectare. The rest should be female plants, as only the female flowers can be used.

Timing

Planting can be carried out in autumn or spring, with autumn being preferred as the plants get off to a better start.

Planting

Start the cuttings in nursery beds until they have formed roots. Plant the rooted cuttings or suckers in holes in the soil deep enough to cover them vertically.

Layout and Spacing

Layout and spacing depend on how the Hops are being cultivated and how much space is required to accommodate machinery. The plants can be placed 900–1200 mm apart, in rows 1.2–3 m apart.

Hops may be difficult to eradicate, as they will regrow from seed or small pieces of stem left in the ground.

Compost

Compost needs to be applied at a rate of 40–70 t/ha or 4–7 kg/m^2, in early spring.

With large-scale operations there is a lot of good compost material available from the Hops themselves, as only a very small proportion of the plant is used.

Weed Control

Traditionally Hop-fields were always cultivated for weed control. However, a shallow-rooted cover crop of White Clover, kept mowed at frequent intervals, will have more benefit, provided there is sufficient irrigation water for the White Clover not to be competing with the Hops for moisture.

Some growers find they can safely graze sheep in their Hop-fields once the vines are well up the strings: the sheep keep the grass and Clover down and trim off all the lower leaves of the Hops, without damaging the stems.

Management

The Hop bines or stems need to be given something to climb up. They twist upwards in a clockwise direction. Four to twelve bines can be sustained by each plant, depending on spacing.

Conventionally this is done using polypropylene strings attached to the system of overhead wires described earlier. However, the polypropylene poses a major disposal problem, and it means much of the vine cannot be composted and recycled. If a fibre of vegetable origin could be used in the twine, this problem could be solved.

Any extra stems emerging from the crown should be cut back so all the plant's energy goes into the main ones which are trailed up.

On a small scale, long thin poles can be provided for the Hops to climb up: this was the system used 100 years ago.

Irrigation

Regular irrigation is required, as Hops are heavy users of water, requiring 25–50 mm per week.

Pests and Diseases

Two-Spotted Mite or Red Spider

In warm dry conditions, the leaves of Hops can become infested with mites. Some call this mite the red spider, but officially it is the two-spotted mite. Mite infestation is no doubt a symptom of stress the plants are experiencing.

Conventionally, two-spotted mite used to be controlled using heavy chemicals. With the increasing understanding of the use of predatory mites to control mites and the realisation that healthy plants resist diseases and pests, it should be possible to manage the problem within organic and bio-dynamic guidelines.

One grower in north-east Victoria used to have trouble with mites until he stopped using superphosphate and changed over to organic management.

Harvesting

Timing

The strobiles, or cones, are harvested when they are fully developed. They become crisp and the bracts can be pulled out individually without the cone splitting. This is usually late summer to mid-autumn. If left too long, the yellow resin, which contains some of the active principles, will start to fall out.

Tools and Technique

Traditionally Hops were picked by hand into large baskets. More recently they were picked into large canvas trays suspended over a frame the bines could be draped over, so the pickers could just break off the cones and let them fall onto the canvas.

These days, on industrial hop farms, the picking is all done by machine. The bines are cut down and hauled into the machine, which flicks the cones off with wire loops and then puts them through a cleaning process to remove any pieces of leaf.

Hand-harvesting is sometimes recommended for 1–2 year old plants, as it can be done without cutting the bines, so it does not set the young plants back as much.

Drying

Large-scale commercial Hops operations use suspended drying floors, where the hops are piled 300–600 mm thick and heated air at 70°C or more blown through them.

This system leaves a lot to be desired from the point of view of producing Hops for medicinal use. First, the Hops are subjected to too high a temperature. While the heat and the humidity in the middle of the thick layer of Hops may not affect the bitter principle, it could very likely affect *other* active constituents. Humid heat is very destructive to many biologically active substances.

Virtually all the volatile oils are lost when Hops are dried at this temperature. For beer production this does not seem to be a concern, because these will be lost anyway when the beer is boiled, and the brewer is mainly concerned about the bitter principles. In drying Hops for brewing, the bitter principles are actually better retained if the herb is subjected to an initial period of 80°C, but this need only be brief in order to destroy a certain enzyme.

The other no-no is the direct blasting of the exhaust fumes of the diesel burners into the drying herb. This causes contamination with residues such as sulphur compounds and products of incomplete combustion, and is unacceptable for organic certification.

Hops can be ambient-air dried like most other herbs, and the result is an especially fragrant finished product. Some artificial heat may be required in cooler climates as Hops are often not harvested until April. To check for dryness, the flowers should be dry right to the centre and feel crisp. If in doubt, check a few days after putting the flowers into storage, to see if any dampness has developed. Hops left in a moist state in storage can develop some off-flavours due to fermentation.

The bitter yellow resins in Hops stick onto tools and equipment, building up in a thick layer until they flake off. Precautions need to be taken to ensure this resin doesn't contaminate other herbs and give them off-flavours.

Temperature

The drying temperature should not exceed 35°C if the herb is intended for medicinal use as Hops contain volatile oils.

Processing

All the processing having been done before drying, the Hops are ready when dry. However if you have any volume of harvest, you may need to press the Hops, as they are very bulky. Traditionally Hops are pressed into wool bales, which can weigh 70 kg or more and are very difficult to handle. A smaller pack could be devised by modifying the press to make a bale about half that size. Alternatively it may be possible to press the Hops directly into 200 L drums.

Hops are used in the same form for brewing, for tea, and for manufacturing into liquid medicines.

Storage

Hops do not seem to be susceptible to moth infestation, but must be adequately sealed to prevent them from becoming remoistened in storage.

Yields

Hops are not particularly high yielding: Mrs Grieve (*A Modern Herbal*) quotes an average of 450 kg/0.4 ha per annum for the period 1898–1907 in Britain. This equates to 0.11 kg/m^2. Hops give a similar yield today in Tasmania.

Marketing

The market for premium-grade organic Hops is limited, but increasing.

As some conventional Hop growers are doing organic trials, it may be a risky market to dive into, because conventionally grown Hops have been in overproduction for some years now.

Price

Premium-grade organic Hops are currently bringing $17/kg, but the price paid by manufacturers for organic Hops has been around $8–14/kg.

Trade Price is around $8–10/kg.

Prices to conventional growers have been in the range of $3–5/kg.

Melissa Balm

Melissa officinalis L LAMIACEAE

Parts Used: Leaf, Aerial Parts

Other Names: Lemon Balm, Balm.

Melissa Balm was anciently cultivated only as a bee plant, but the Arabs recognised its medicinal properties as a carminative, antispasmodic, diaphoretic and sedative, and it has been in use ever since.

This herb is probably better known by the name of Lemon Balm, because of its lemon scent and flavour in the fresh state. When dried, though, it retains only a vague reminiscence of this lemon aroma, so Melissa Balm is the name I prefer to use. It is often referred to simply as 'Balm' in the herb trade.

There are those who have doubts as to the value of dried Melissa Balm, because some of its properties are attributed to the aromatic oils, which are lost in drying.

GROWING MELISSA BALM

While Melissa Balm will grow reasonably well in many places, it often suffers severely from two-spotted mite, which causes severe fading and burning off of the leaves: an indication the plant is under stress.

Neighbouring Plants

Because Melissa Balm is susceptible to two-spotted mite, it should not be placed near other species that might also suffer from this pest, such as Hops, Greater Celandine or Meadowsweet.

Identification

Melissa Balm has slightly hairy square stems, branching near the top, growing 300–800 mm tall. The leaves are opposite, ovate, light green, lightly toothed, with a characteristic lemon scent when fresh. Flowers are whitish, occasionally pink or yellow, 7–15 mm long, in small clusters in the axils of the upper leaves (see colour plate 5.1).

This is one herb that is not too likely to be confused with anything else.

Pests and Diseases

Two-Spotted Mite or Red Spider

Often Melissa Balm does not look very healthy. It may start off in spring with vigorous green growth, but as everything else blossoms forth, it begins to turn yellow in places.

Melissa Balm (*Melissa officinalis*).

The fading spreads through the plants and parts will even turn brown and die off. This is the result of the activity of tiny mites, which live under the leaves, sucking the sap.

The cause of the problem is really one of the plant being under a degree of stress. Lemon Balm needs humidity and shelter from wind. If it is growing in a situation where these needs are met, it usually will not suffer from mites.

I had two patches of Melissa Balm at 'Twin Creeks', one was in a hot exposed site and it always was badly hit by mites. The other patch was in a cooler, more sheltered plot, closer to the creek, with trees around, though it received almost full sun. This second plot was hardly ever affected by mites, nor was the Melissa Balm that grew wild alongside the creeks in the district.

An infestation of mites can usually be controlled by cutting the plants and mowing the stubble very close to the ground.

Drying

Melissa Balm dries very easily, but unfortunately the lemon aroma of the leaves is virtually lost in drying, even at quite low temperatures.

Temperature

Try to keep drying temperatures below 30°C.

For more detail see 'Information Charts'.

Mugwort

Artemisia vulgaris L   ASTERACEAE

Parts Used: **Leaf & Flower, Aerial Parts, (Root)**

Other Names: Felon Herb, Wild Wormwood, Artemisia, St John's Plant.

Mugwort has a long tradition of use in magic and ceremonial practices and is regarded as a powerful herb in many cultures. John the Baptist was said to have worn a girdle of it in the wilderness for protection from evil spirits.

Most of the usage today in our culture stems from Eastern influences. It features strongly in Chinese medicine and is an ingredient of macrobiotic cuisine.

In Western tradition, it is mainly used for digestive disorders and as a tonic, but the occasional request comes for it as an ingredient in an ancient witchcraft formula for flying ointment.

It was traditionally used in brewing before the days of Hops (hence the name Mugwort), and as a culinary herb in stuffing for poultry or fatty fish.

GROWING MUGWORT

Mugwort is a vigorous herb, which is easy to grow and holds its own well, perhaps too well, against other plants.

Neighbouring Plants

Because Mugwort grows so tall, it can create sun and sprinkler shadows. It draws a lot of moisture from some distance around owing to its extensive root system, so it can have an adverse effect on adjacent plants, I suspect it is also allelopathic: that is, it exudes substances that have a detrimental effect on other species.

Containment

Containment is not really a great problem with Mugwort, in most situations, as it only expands at the rate of 50–150 mm a year. However, in cool moist climates, it may become a weed, so try not to let it go to seed in these situations.

Identification

Mugwort has angular reddish purple stems and grows up to 1800 mm tall. The leaves are 50–100 mm long, dark green above, whitish and downy underneath, pinnate or bi-pinnate with toothed leaflets. The numerous small flowers are brownish yellow to red and are arranged in panicles towards the top of the plant.

Mugwort *(Artemisia vulgaris)*.

While there are no other plants commonly confused with Mugwort, there are a large number of species of Artemisia, some of which are likely to be similar.

Drying

Mugwort is quite reluctant to dry and quickly re-absorbs moisture if the humidity starts to rise.

It will discolour if drying conditions are not good enough. The drying of aerial parts can be hastened by coarsely chaff-cutting the herb prior to drying.

Processing

Mugwort is required as leaf and flower for tea and culinary use, and as aerial parts for manufacturing liquid medicines. Other uses, such as moxibustion, require special preparation of the herb. Mrs Grieve alludes to this briefly in *A Modern Herbal*.

Tea Grade

Mugwort is in a class of its own here too, as it is the only broad-leaved herb I know of which cannot be rubbed through a 2 1/2 dent screen, no matter how dry it is. The down on the leaves felts them together into fluffy balls and very little of the herb will go through the screen.

The only option is to strip the leaves rather than rub them: for this to work easily, the leaves need to be very dry and brittle (see 'Stripping Leaves' in Chapter 13, 'Processing').

For more detail see 'Information Charts'.

Nettle, Greater

Urtica dioica L URTICACEAE

Parts Used: Leaf, Aerial Parts, (Root)

Other Names: European Perennial Nettle, Stinging Nettle, Common Nettle, Nettles.

Most people are painfully familiar with Nettle. It features in much folklore, and childhood experiences of running into it are not easily forgotten. But most Australians will not be familiar with the Greater Nettle, which is usually regarded as *the* Nettle of herbal medicine. In this country we are more likely to come across the annual Lesser Nettle (*Urtica urens*), which is often found in chook pens, stockyards and house gardens. Some may have even encountered the Native Nettle (*Urtica incisa*) growing in the bush in high rainfall areas, usually in moist gullies, which it protects with its sometimes vicious sting.

While Lesser Nettle is sometimes accepted as being of equivalent value medicinally, it is difficult to bring into cultivation. Although it grows as a prolific weed in places, it is very hard to get it to do what you want it to. It is dealt with more fully under 'Wildcrafting and Weed Harvesting'.

The Greater Nettle (*Urtica dioica*), is much easier to grow and does very well under the right conditions. It has a long tradition of use medicinally, as an astringent, a diuretic and a tonic. The young leaves are also used as a vegetable and in soups. Nettle can be used to make beer and the fibres were used to make fabric until the early part of this century.

The root of Greater Nettle is also used to some extent, mainly in anthroposophic medicine.

Origin

Greater Nettle is a native of Europe and Asia, and is naturalised in North America.

GROWING GREATER NETTLE

This Nettle is a vigorous perennial that strongly dominates the area given to it and is very productive. It goes dormant in winter but comes away vigorously in spring.

Its botanical name refers to its dioecious nature: it has separate male and female plants. This does not affect its growth or requirements and it doesn't seem to matter if one has male plants, female plants or both, there is no apparent difference between them in productivity or qualities.

Climate

Greater Nettle grows more vigorously in a cool to mild climate, but it will tolerate hot summers provided it gets plenty of moisture.

Site

Exposure

It will grow well in exposed sites and in full sun.

Drainage

It needs good drainage: it will not thrive in situations subject to any waterlogging.

Soil Type

Greater Nettle will grow on a range of soils. Light to medium loams are preferred, but on one site on volcanic red soil it has not been much of a success.

Fertility

It needs a very fertile situation.

Initial Weed Control

All perennial weeds need to be eliminated with a bare fallow prior to planting. While Greater Nettle is able to hold its own fairly well against weeds, it still needs a good start.

Neighbouring Plants

Greater Nettle is a reasonably well-behaved neighbour for most other plants and is not easily overwhelmed by other plants either.

Location

Because of the hazard it poses to the unwary, Greater Nettle should be put in a safe corner where unsuspecting children and others won't too easily wander into it, to their dismay.

In my experience it has never spread beyond the bed it was planted in, but if you had male and female in the same bed, it might be a different story if Greater Nettle were allowed to go to seed. Regular harvesting of the leaf prevents this.

In most regions it will not persist as a weed beyond fertilised and irrigated sites as it requires high fertility and summer moisture to survive and is winter dormant. In this regard it differs from the annual Lesser Nettle, which is winter-active and has been able to establish itself as a weed in many regions, though never beyond very fertile sites.

Identification

Greater Nettle grows 800–1800 mm tall, and the leaves and stems are covered in fine stinging hairs. The leaves are opposite, toothed, ovate, up to 140 mm long (see colour plate 5.6). The flowers are minute, appearing in long pendulous racemes in the axils of the upper leaves. Its roots are a distinctive yellow colour, as are the larger of its underground rhizomes. The plant goes dormant from June to August in Victoria and Tasmania.

It is fairly easy to distinguish Greater Nettle from Lesser Nettle, which is shorter, usually only up to 600 mm, with darker green leaves, which are more oval-shaped and with larger teeth on the margins. The flowers are smaller and the plant forms flowers in the axils all the way to the bottom of the stem: indeed, it starts to flower almost as soon as it has germinated. The roots are a whitish colour and it does not form rhizomes.

Lesser Nettle is winter-active in southern Australia, normally germinating in the autumn and growing through the winter and spring. It usually dies off by late spring or early summer.

The Native Nettle (*Urtica incisa*), is less easily distinguished from Greater Nettle. Native Nettle is a quite variable perennial plant, forming rhizomes, but these are white or pink and don't turn yellow as they get older, nor do the roots. Another distinguishing character is that Native Nettle lacks the covering of fine hairs on the stems; instead it just carries a few larger stingers.

Variety

While there is some variation in Greater Nettle, there does not seem to be any particular preference as to variety.

Propagation

Method

Propagation is normally by rhizomes. All the Greater Nettle rhizomes seem to be viable, even the older ones, which grow quite thick. But make sure you have rhizomes and not roots, as they tend to grow together. The rhizomes can be distinguished by their jointed nodes. It may be necessary to hose the soil off so you can distinguish them.

Greater Nettle can also be raised from seed.

Timing

Planting out can be done in spring or autumn. If weeds are not a major problem, autumn planting will allow better development and higher yields the following season. However, if weeds are likely to be a problem over winter, a very early spring planting will be easier to maintain free of weeds.

Planting

Rhizomes can be broken into sections about 150 mm long and planted end-to-end in furrows, covered with around 50 mm of soil.

Greater Nettle *(Urtica dioica)*.

A stand of Greater Nettle in its first season, ready for harvest.

By its second or third season Greater Nettle will have expanded to form a solid meadow-crop.

Layout and Spacing

Normally Greater Nettle is planted in rows 900–1200 mm apart. This allows for inter-row cultivation and compost application for a season or two until the plants expand into a meadow-crop.

If drainage is not quite adequate for Nettle, it may benefit from being grown in low-profile raised beds (as described in Chapter 5, 'Propagation and Planting'). However, these would probably not be sufficient for salvaging a badly waterlogged situation, as Greater Nettle does need very good drainage.

Compost

Greater Nettle is a very heavy feeder and will not thrive unless it receives a heavy dressing of compost each year. It needs 40–60 t/ha (4–6 kg/m^2). That means a bed 20 x 20 m will need 1.5–2.5 t of good quality compost each spring. The compost should be worked in, where possible. As Greater Nettle gradually spreads and becomes a meadow-crop, it may be necessary to spread the compost over it in spring. For greater detail on compost spreading in this situation, see under 'Meadow-crops' in Chapter 7, 'Compost'.

If growth declines during the season, a light dressing of suitably crumbly compost, applied after harvesting, will help keep it from yellowing.

Weed Control

Greater Nettle is fairly straightforward when it comes to weed control and is easier to manage than many other herbs. It still does need weeding, but it is easier to stay in control than it is with spreading herbs like the Mints. Because it goes dormant over winter, it usually needs a big clean-up in spring, before the growth gets too tall. Then it will hold its own fairly well, provided you maintain inter-row cultivation and go over it thoroughly after each harvest. A pair of washing-up gloves is handy for this, unless you enjoy getting stung (some people do!). Dock leaf juice or the juice or tea of the Nettle itself can be used as an antidote.

Winter Grass and other winter-active weeds often invade Greater Nettle while it is dormant. While the early growth of Greater Nettle is never at risk of being smothered by weeds, there can be problems if the first harvest has Winter Grass amongst it. It may be necessary to harvest the Nettle a bit higher than usual to avoid the Winter Grass.

Another option for control of winter weeds is a black plastic dormancy mulch placed over the beds for 4-8 weeks or so in midwinter. Nettle seems to tolerate this fairly well, but mow off any soft shoots it sends up under the black plastic.

Rejuvenation

Greater Nettle will maintain good yields as a meadow-crop for a number of years, provided it has adequate nutrition. If growth starts to decline it can be reinvigorated by cultivating strips through the crop to bring it into rows for a season (see under 'Rejuvenation' at the beginning of Chapter 15, 'Spreading Herbs').

How long Greater Nettle can be maintained in the same bed is uncertain, but it can last for at least 5 years.

Irrigation

Greater Nettle needs regular irrigation in dry weather, as it has a relatively shallow root system and a large water requirement, particularly when in heavy leaf. This is particularly important in hotter regions, where it will not thrive unless it has plenty of moisture available at all times. Normally it needs 25–50 mm of water per week over summer.

Pests and Diseases

Nettle Butterfly

Nettle is frequently the host to a dark brown caterpillar, which matures to a red, white and brown butterfly. Normally these are not a significant problem. They form a beautiful iridescent chrysalis, which is often found attached to the screens after drying Nettle.

Grasshoppers

Grasshoppers can take their toll of Greater Nettle, particularly if it is feeling any moisture stress. However, if it is well watered and growing vigorously, they usually aren't too much of a problem.

Harvesting Leaf and Aerial Parts

Timing

It is usually possible to get four harvests of Greater Nettle during the growing season. First cut is around early November as the herb comes into flower or shortly before. The second growth will flower a bit but less prolifically, so you have to watch the leaf development to decide when to cut. It is ready to harvest 6–8 weeks after the first. The third and fourth harvests usually don't flower before the leaf starts to deteriorate and they are usually about 8 weeks apart.

As with most species, the regrowth gets progressively shorter and leafier as the season progresses.

Tools

Normally Greater Nettle is harvested with the catching scythe. It is best to wear gloves, so you can safely pick up any fallen pieces.

Technique

Greater Nettle is quite easy to cut, but watch out for faded yellow or white leaves on the lower parts of the stem and try to avoid them: their height tends to vary a fair bit.

Post-harvest Treatment

After harvest, mow the stubble off as close to the ground as possible. This stimulates more vigorous regrowth and makes subsequent harvesting easier and cleaner.

Drying Leaf and Aerial Parts

Gloves are essential for spreading Nettle on the screens.

Drying Greater Nettle is rather tricky and it can be difficult to produce a good colour. Although it dries easily, if re-moistened at all or bruised by handling in a wilted state, it goes very dark, almost black.

Dried in good conditions, Greater Nettle should come out a dull green colour, but in marginal drying conditions it will take some heat supplement to achieve this quality. (Note: This does not apply to the Lesser Nettle, which dries to a brighter green and is less subject to darkening.)

Temperature

Nettle can be dried at temperatures up to 45°C.

Processing Leaf and Aerial Parts

CAUTION: Before doing anything with Nettle once it has dried, you need to equip yourself with some protective gear. While the dried Nettle has lost most of its sting, the fine hairs present a perhaps greater hazard.

Once the stinging hairs are dry, they break off and tend to float around. As they contain silica and are rather sharp, they can be very annoying to the skin, and very irritating to the eyes and respiratory tract. What their long-term effect on the lungs and the eyes is, I don't know. Personally, I don't care to find out, so I always wear a respirator and goggles when processing Nettle.

A pair of long gloves is a good idea, too, plus a hat and a boiler-suit over your clothes, so when you finish you don't carry Nettle dust with you into the house. If you are handling large quantities of Nettle, it would be worth getting the sort of protective equipment that asbestos workers wear.

All this protective gear makes processing Nettle rather hot work, so try to do it on a cooler day or towards evening.

Nettle is always a steady selling herb tea, being fifth of sixth in order of popularity. It is also used a great deal as a liquid medicine, for which the aerial parts are used.

Tea Grade

Rub through the standard $2^1/_2$ or 3 dent screen. On the plus side, Nettle is usually a little easier than most herbs to get free of stalk, owing to the fibre in it, which reduces shattering.

The hairs which break off tend to congregate in the dried herb and affect the appearance. They can be separated by passing the herb over a fine screen, as for removing soil contamination (see 'Soil Contamination' in Chapter 13, 'Processing').

Aerial Parts

Chaff-cut aerial parts. Check that the stems are completely dry: the young tips are usually the slowest part to dry. Very thick stalks are slow too and these should snap when dry.

CAUTION: When you have loaded your Nettle into the bags, you may feel that at last you can take off the sweaty mask and goggles while you tie the bags up. However, this is a particularly dangerous operation. As you compress the bags to squeeze the air out of them to tie the neck, it is very easy to get a puff of dried Nettle stingers right in your face. If they get in your eyes you could be in trouble.

Storage

Nettle needs to be freeze treated for moth larvae control before sale or storage.

Yields

Yields have been in the range $0.25–0.4$ kg/m^2 per annum of tea-grade leaf. This would equate to around $0.35–0.5$ kg/m^2 or $3.5–5$ t/ha per annum of aerial parts.

Harvesting Root

For Nettle root, the roots and rhizomes of the Greater Nettle are harvested.

Timing

Greater Nettle root should be harvested at the end of the growing season or during winter while the plant is dormant. As the soil needs to be relatively dry and friable, it is best to do the digging before conditions get too wet and difficult.

Tools

A vine hoe, which is like a forked hoe, is the best implement to use to dig the roots up.

Technique

Before commencing digging, mow the tops off as low as possible and rake up any debris, so it doesn't get picked up with the roots.

As Greater Nettle roots and rhizomes are quite fibrous and tough and trap a lot of dirt, it is necessary to tear them apart as much as possible as they are dug, somewhat in the manner of digging Valerian.

Washing is something of a challenge, owing to the fine roots, which cling to bits of debris and soil.

Post-harvest Treatment

As harvesting Greater Nettle roots and rhizomes effectively removes the source of next season's plants, the bed will have to be replanted.

Of course, the harvest could be taken from part of the bed where there is surplus growth, with strips of Nettle left intact as new rows for next season.

Drying Root

Nettle root dries very easily and there is no need to chop it prior to drying.

Temperature

Nettle root can be dried at temperatures up to 45°C.

Processing Root

When dry, the Nettle root can be worked around over a $2^1/_2$ or 3 dent screen to separate off the very fine roots, which should break away and fall through the screen, leaving the coarser tougher roots and rhizomes behind. Any root big enough to be yellow can be included, so your rubbing should be soft enough to leave these all unbroken. If you find these are breaking up and going through, then try rubbing the roots over a shadecloth screen.

If a tea-grade Nettle root is required, you may be able to chaff-cut it, depending on how good your blades are, or it may need to be hammer-milled. Before putting it through the chaff-cutter, make sure there are no small stones trapped among the fine roots.

Marketing

While there is a steady market for Nettle leaf and aerial parts, growers need to co-ordinate with each other to avoid an oversupply situation.

The market for Nettle root is extremely limited in this country, so you would need a firm undertaking from a buyer before commencing production.

Price

Price to growers for premium quality, organically grown Greater Nettle is currently around $24/kg for 5-star tea-grade leaf.

Price to growers for second-grade, organically grown Greater Nettle is currently $10–16/kg for leaf and $9–13/kg for aerial parts.

Trade price is around $9/kg for aerial parts.

Oregano

Origanum vulgare L LAMIACEAE

Parts Used: **Leaf & Flower**

Other Name: Wild Marjoram.

Oregano is a traditional Mediterranean herb and a popular ingredient in Italian and Greek cuisine. A good Oregano is hot and spicy and imparts a distinctive flavour to a dish.

It also has some medicinal uses, mainly for respiratory and gastrointestinal disorders, but these are not widely recognised.

Origin

Oregano is native to Europe, Middle East and the Himalayas.

GROWING OREGANO

Once established, Oregano is easy to maintain, but getting it started can sometimes be tricky. Other problems can be sorting through the confusion between Marjoram and Oregano and finding a good strong-flavoured variety.

Climate

Oregano does well in a range of climates. Some authorities state that it needs a hot summer to develop its characteristic pungency, but I have not found this to be the case.

While climate may be a factor, in my experience the variation in flavour is predominantly genetic. Both strongly pungent and weak flavourless Oregano can be produced in the same location, regardless of whether it is a hot or cool climate. The critical factor is the variety of Oregano grown.

Site

Exposure

Oregano enjoys full sun and will tolerate exposure to wind.

Drainage

Oregano needs good drainage: it will not tolerate wet feet.

Soil Type

A light to medium loam is preferable.

Fertility

A moderate level of fertility is required.

Initial Weed Control

A bare fallow is recommended before planting.

Neighbouring Plants

Generally there are no problems with Oregano. It is not tall or aggressive enough to be a problem to neighbouring plants and its own growth is dense enough to resist invasion. It can also tolerate moisture deprivation, so greedy neighbours don't distress it as much as they might some other species.

Identification

Oregano is an erect plant, up to 750 mm tall, usually bushy, growing from creeping rootstocks. The leaves are opposite, broadly ovate, around 25 mm long without any distinct teeth, often with a reddish tinge and flavour varying from bland to sharp. Flowers are rose-purple to pink to white, 6–8 mm long, borne in spikes or clusters at the top of the plant.

It is important to be able to distinguish Oregano from Sweet Marjoram (*Origanum majorana*) and Pot Marjoram (*Origanum onites*). These latter two differ from Oregano in that their growth-type is woody perennial, while Oregano grows from a creeping rootstock.

Oregano (*Origanum vulgare*), white-flowered variety.

The flowers of Oregano are more open than those of the Marjoram, which form in tight clustered spikes or knots.

Sweet Marjoram leaves have a delightful sweet aroma, Pot Marjoram is a duller version of this. A good-flavoured Oregano is more pungent.

Variety

Make sure you have a suitable variety: a good hot variety will really burn your tongue just chewing on a leaf, a milder variety will have some flavour, while a bland one can be almost flavourless. The flavour can really only be ascertained if you can taste the plant. If possible, get a super hot variety, as this is the quality the market appreciates. A mild-flavoured variety may be acceptable to some markets but is likely to be difficult to sell if the hot Oregano is available.

It appears Oregano varieties will grow true to seed and some reputable seed suppliers do offer a hot Greek Oregano. However, if you just purchase ordinary Oregano seed you may end up with a variety with no flavour.

There is also a golden variety of Oregano, which would not be suitable for commercial production.

Propagation

Method

Oregano forms something like rhizomes, but these are tightly packed together and as the plant is somewhat intolerant of being pulled apart, it is best to plant it out in small clumps.

Timing

Propagate Oregano in spring or autumn.

Planting

Plant the clumps in furrows, cutting the tops right back to ground level, but just firm them in and don't cover them. There is often a high failure rate, as Oregano is curiously touchy about being transplanted. It seems to do better if you can get it into the ground as soon as possible after digging it up. If it is held out of the ground for a day or two, it often seems to fail.

Layout and Spacing

Oregano can be set out in rows about 900 mm apart with the plants on 150–300 mm centres. If you have good weed control, there may be an advantage in setting the rows 600 mm apart so the Oregano will grow into a solid meadow-crop sooner. This will be easier for harvesting, though production will also decline sooner when the plants begin to crowd each other.

Compost

Oregano does not require such heavy applications of compost as some other herbs: an application of 2–3 kg/m^2 should be adequate.

Weed Control

Oregano grows quite densely with a solid base of leaves, which do not allow much space for weeds to establish. Nevertheless, it will require regular inter-row cultivation and going over the beds hoeing or pulling out those weeds that manage to get a toehold.

Oregano cannot be covered with black plastic for a winter mulch for weed control, as it never goes fully dormant: such a mulch is likely to kill it.

Rejuvenation

As it takes a few years to fill out the spaces between the rows, crowding does not become a problem for 4 or 5 years. However, once Oregano runs out of space to expand, it will gradually decline.

It will eventually need to be replanted in a new patch, or rejuvenated as described in the introduction to Chapter 15, 'Spreading Herbs'. As I have never grown it for so a long period in one place, I am not sure which method would be preferable.

Irrigation

Oregano needs about 25 mm of water a week: it will tolerate drier conditions than many herbs.

Pests and Diseases

Apart from the tendency to die off in transplanting, Oregano is pretty well disease free.

Harvesting

Timing

Harvest in the early flowering stage: there are usually two harvests or possibly three in the season. The first harvest is late December–January; the second, March–April.

Tools

Once the clumps have become large enough, the catching scythe can be used effectively on Oregano, but until that stage it may be better to harvest with a reaping hook.

Technique

Although its stems do appear rather woody and it is closely related to Marjoram, which is a woody perennial, Oregano's regeneration is quite different. When its top growth is cut down, Oregano regenerates strongly from ground level. Consequently all the green leaves can be harvested, but cut it above any lower yellow leaves.

Post-harvest Treatment

After harvest, mow the stubble off, but not too low: leave about 25 mm of the low surface growth.

Drying

Oregano dries quite easily, but wilted material will bruise and darken if handled.

Temperature

Dry at temperatures up to 35°C.

Processing

Almost all Oregano goes for culinary use. If small amounts are required for other purposes, culinary grade will suffice.

Culinary Grade

Oregano needs to be rubbed through a standard culinary-grade screen a number of times to obtain a fine-leafed product free of stalk. Alternatively Oregano can be winnowed after screening to get it clean.

A fine white dust comes off the leaves of Oregano. This is best separated and discarded.

Storage

Freeze treatment is recommended, even if you think moth larvae wouldn't appreciate such a spicy herb.

Yields

Up to around 0.4 kg/m^2 of rubbed leaf per annum can be expected from a fully established bed.

Marketing

Good quality Oregano with a pungent flavour is one of the more popular of the culinary herbs as most people do not have it in their own gardens. However the market for it is limited at this stage, as it is for most organic culinary herbs.

Price

Price to growers for premium quality, organically grown Oregano is currently around $24/kg for 5-star tea grade.

Trade price is around $6/kg.

Scullcap

Scutellaria lateriflora L LAMIACEAE

Parts Used: Aerial Parts

Other Names: Side-flowering Scullcap, Hoodwort, Mad-dog Scullcap, Quaker Bonnet. Also spelt as Skullcap, sometimes incorrectly referred to as *Scutellaria laterifolia*.

Scullcap has sedative properties and is used in the treatment of many conditions where tension is a factor.

This valuable medicinal herb has been the subject of some controversy in recent years. It would appear that much of the Scullcap sold in the general herb trade has not been the correct species: other species of *Scutellaria* have sometimes been substituted for it, or even a species of *Teucrium*.

Origin

Scullcap is a native of North America.

GROWING SCULLCAP

Scullcap is a delightful little herb with delicate blue flowers and a tender gentle nature.

Given the right conditions, Scullcap is not difficult to grow. Its growth habit is fairly open and spreading – so much so that it might even be classed as a spreading herb.

Scullcap (*Scutellaria lateriflora*).

Climate

Scullcap will grow in a range of climates.

Site

Exposure

Scullcap grows well in full sun but will not stand up to much wind, as its brittle stems can be easily snapped off at ground level.

Drainage

Scullcap is very sensitive to poor drainage and will die if subjected to any waterlogging.

Soil Type

Light to medium loam is preferred.

Fertility

A reasonably fertile situation is adequate.

Initial Weed Control

A bare fallow before planting is recommended, as weeds can be a problem during Scullcap's establishment.

Neighbouring Plants

Avoid neighbouring plants that might overshadow Scullcap.

Identification

Scullcap grows to 300–900 mm, with square stems, petioles (leaf stalks) 5–20 mm long, leaves opposite, 30–80 mm long, deltoid-ovate to lanceolate, toothed. The flowers are blue, 5–8 mm long, in slender one-sided racemes in the leaf axils.

The initial flowers appear singly in the axils, but as development progresses, one-sided racemes of flowers are formed.

The plant forms numerous small rhizomes and dies back to ground level in winter.

There is a lot of confusion about Scullcap and growers need to make sure they have the species that is going to be acceptable to their buyers. Before commencing production on a significant scale, samples should be sent off to ascertain their suitability.

Propagation

Method

Scullcap can be grown from seed or by vegetative propagation: either from crown divisions or from rhizomes.

Timing

Plant seed in early spring. Crowns are preferably divided in autumn, or else in early spring. Rhizomes can be planted in autumn or spring.

Planting

Seed should be sown in flats or nursery beds and transplanted out when large enough. Scullcap has quite an extensive root system, so it needs some care in transplanting.

Crown divisions should be taken with as much root as possible on them, but with the leaves cut well back. Set the root in at its normal depth, not too shallow.

The small rhizomes can be separated and planted out in shallow furrows.

Scullcap divisions and rhizomes need to be put in the ground promptly: they won't hold out of the ground for any length of time.

Layout and Spacing

Scullcap should be set out on 300 mm centres or in continuous rows 900 mm apart.

For harvesting with the scythe, these should be in blocks at least three rows wide.

Compost

Scullcap needs a moderate application of compost at the rate of around 2–3 kg/m^2, worked in beside the plants each spring. Avoid excessive fertilisation as the stems are brittle and may not support a heavy growth of leaf.

Weed Control

Scullcap should be kept free of weeds with regular inter-row cultivation and hoeing around the plants. Winter-active weeds can be a problem for this crop as it goes fully dormant and dies back to ground level in winter, giving weeds the opportunity to invade it.

Irrigation

Scullcap needs regular irrigation of around 25–40 mm of water per week through summer, but care needs to be taken to avoid overwatering.

Pests and Diseases

Grasshoppers

At times growers in northern Victoria have had trouble with grasshoppers in Scullcap, but this can be reduced with more consistent watering so the plants aren't stressed. Ducks can help, too.

Harvesting

Timing

Harvest when the Scullcap is in flower or earlier if the leaf is starting to deteriorate. The first cut should be ready in November–December, with subsequent cuts ready at 6–8 week intervals.

Tools

Normally it is possible to harvest with the catching scythe, but if this is breaking the plants off, Skullcap should be harvested with a reaping hook.

Technique

Scullcap seems to make a better recovery if it is cut a little above the base of the plant so some green leaf remains. All the flowers should be picked up in the harvest.

Care needs to be taken in the cutting of Scullcap that the plants are not snapped off at the base, because the stems are quite brittle.

Post-harvest Treatment

Mow stubble off low to encourage new leaf growth from ground level.

Drying

Scullcap dries quite quickly with no problems.

Temperature

It should be dried at a temperature of up to 35°C.

Processing

Scullcap is used to some extent as a tea and it is also manufactured into liquid extract.

Tea Grade

For tea grade, the herb is rubbed through a 2½ or 3 dent screen to remove stems.

Manufacturing Grade

The dried Scullcap can be put through the chaff-cutter. Alternatively it could be rubbed to a first screening and then the stem mixed back in.

Storage

Freeze treatment to control moth larvae is necessary before sale or storage.

Yields

Once established, Scullcap can yield around 0.4 kg/m^2 aerial parts per annum. Yields of tea-grade leaf are somewhat lower, around 0.3 kg/m^2.

Marketing

Being a valued medicinal herb, there is some demand for good quality genuine Scullcap. However, there have been recent increases in production, so growers should ascertain they have a firm market before considering this as a crop produced on any significant scale.

Price

Price to growers for premium quality, organically grown Scullcap is currently around $24/kg for 5-star tea-grade leaf.

Price to growers for manufacturing-grade, organically grown Scullcap aerial parts is currently $12–20/kg.

Tarragon, French

Artemisia dracunculus L   ASTERACEAE

Parts Used: **Leaf**

Other Name: Tarragon.

French Tarragon is used today solely as a culinary herb, although there is mention of its having been used for toothache in the past.

It is primarily used in fish and chicken dishes, salad dressings, and for making Tarragon vinegar.

French Tarragon has an exquisite piquant flavour, some of which is inevitably lost in drying. For this reason many chefs prefer the fresh herb if they can obtain it.

Unfortunately French Tarragon has a rogue brother, Russian Tarragon, which is often passed off as the real thing. Growers sometimes think they have French Tarragon when, in fact, it is the Russian impostor.

Origin

French Tarragon is a native of southern Europe.

GROWING FRENCH TARRAGON

Given the right conditions and adequate attention, French Tarragon is not difficult to grow and can be quite productive. It has a gentle nature, somewhat different from many of the Artemisias, which are often quite aggressive.

Climate

While the literature often stresses that French Tarragon is easily killed by heavy frosts, this is in reference to European and North American climates where the winters are much colder than ours. There are few places in Australia where it would get too cold for French Tarragon.

It does not seem to thrive in hot summers, though, and tends to lack flavour when grown in these conditions. The best growth and quality seem to be obtained in Tasmania and the cooler parts of the mainland.

Site

Exposure

In hotter climates French Tarragon needs a cooler site. In mild and cool climates it thrives in full sun and tolerates wind.

Drainage

French Tarragon needs good drainage as it will die if subjected to waterlogging.

Soil Type

Light to medium loam is preferred.

Fertility

French Tarragon needs a fertile soil

Initial Weed Control

A bare fallow is recommended prior to planting, as French Tarragon is fairly vulnerable to weed invasion.

Neighbouring Plants

Avoid planting French Tarragon next to spreading herbs, or next to tall plants that might overshadow it.

Containment

Containment is not necessary. Although French Tarragon gradually expands, this is a relatively slow process and it is not a dominating plant.

Identification

French Tarragon grows up to 900 mm tall, with slender branched stems, and leaves that are dark green to blue-green, lanceolate, and 30–50 mm long. Flowers are grey-green or white, woolly, and in clusters at the end of the branchlets.

The critical thing is to distinguish French Tarragon from the very similar Russian Tarragon. The Russian species is generally a little taller and coarser, more vigorous, and the foliage is slightly greener. It also sets seed, which French Tarragon never does.

French Tarragon (*Artemisia dracunculus*).

The easiest way to tell the difference is by tasting a fresh leaf: fresh French Tarragon has a sharp aniseed component to its flavour, while Russian Tarragon is only vaguely aromatic and quite bland.

Russian Tarragon is not worth growing. Many seed catalogues list Tarragon among their herb seeds and some even call it French Tarragon, however, as French Tarragon cannot be grown from seed, this will always be Russian Tarragon.

Propagation

Method

Propagation is by small rhizomes.

Timing

If weed control is not a problem, an autumn planting is preferred, otherwise plant in early spring.

Planting

Small clumps of rhizomes can be set in furrows and covered with 25 mm or so of soil. These clumps can be as small as 25 mm across if you need to stretch your propagation material, or up to about 100 mm across if you have plenty.

Layout and Spacing

The clumps can go in on 150–300 mm centres, in rows about 900 mm apart. As it can be harvested with the scythe, French Tarragon should be set in beds that are three rows or more wide.

If you have any doubts about the drainage of your site, you should lay it out in broad low-profile raised beds (see Chapter 5, 'Propagation and Planting').

Compost

French Tarragon needs compost at the rate of 3–4 kg/m² per annum.

Weed Control

Its canopy of leaf being relatively open, French Tarragon is more vulnerable to weed invasion than most of the other herbs in this group, particularly over winter.

A black plastic dormancy mulch should work well for it, as it goes completely dormant in winter.

Rejuvenation

French Tarragon does get a bit crowded after a number of years and yields go down. I imagine it could be rejuvenated in the manner described for spreading herbs, but growers usually replant it in a new bed using fairly large clumps. The Tarragon in the old bed is not very hard to kill out with cultivation.

Irrigation

French Tarragon needs about 25–40 mm of water per week during its peak growth in summer, depending on site and prevailing conditions.

Pests and Diseases

There don't seem to be any particular pest and disease problems for this herb.

Harvesting

Timing

French Tarragon should be harvested before the plant gets too advanced. This is somewhat before the flowering stage.

French Tarragon first sends up a stem, which carries single leaves up its length. The optimum stage for harvest is when the plant starts to send out new shoots or branchlets from the axils of the leaves.

It will continue to grow if it is not cut at this stage, but leaf increase is slowed as its stems become woody, flowering commences, and older leaves start to deteriorate. Consequently overall production and quality is better if it is harvested earlier and started on its next cycle of regrowth.

Normally you can get three harvests a season, the first being in late November to early December, the second around the end of January, and a third around late March.

Tools

Harvest with a catching scythe, except perhaps the initial harvest after establishing a new bed, which may need to be cut with a reaping hook to avoid pulling the plants out by the roots.

Technique

Follow usual harvesting procedures.

Post-harvest Treatment

After harvest, mow the stubble off close to the ground to encourage more vigorous regrowth.

Drying

French Tarragon is sometimes difficult to dry as the leaves on a plant vary greatly in their rates of drying. The young soft tips tend to discolour if dried too slowly and the leaves turn brown if remoistened even fairly slightly.

The secret is to dry French Tarragon as quickly as possible without interruptions to the process and within the recommended temperature limits.

In marginal drying conditions, you will need some artificial heat to achieve the best quality, or you may be able to do it by putting your Tarragon up in the highest, warmest spots in your drying shed.

Temperature

French Tarragon should be dried at temperatures not exceeding 35°C. This limit should be adhered to as strictly as possible, owing to the volatile nature of some of the essential oils in it. The aniseed component, particularly, is easily lost and even the best drying conditions will not conserve all of it.

Processing

All French Tarragon is sold for culinary use. It is one of the easiest of the culinary herbs to process.

Culinary Grade

Generally French Tarragon is preferred as mostly whole leaf, which is prepared by rubbing it through a normal tea grade screen (2½ dent). It is relatively easy to rub and the stalk is easy to sift out, so it can be cleaned in two or three screenings.

If the herb can be rubbed as soon as the leaf is dry, while the stems are still leathery, it will be even easier.

Storage

French Tarragon needs freeze treatment to control moth larvae before sale or storage.

Yields

Yields have been in the range of 0.15–0.3 kg/m^2 per annum.

Marketing

French Tarragon is one herb where the trade price is something approaching that obtainable for premium organic herb. Climate and other factors, such as lack of propagation material, may restrict the production of French Tarragon to a few countries where the price of labour is relatively high.

This would appear to make it an attractive proposition, as it is a relatively easy crop for small-scale herb producers to grow and process.

However, we have had very little enquiry from other sectors of the trade, and the market for premium organic French Tarragon is fairly well supplied at this stage. Growers would need a firm commitment from buyers before proceeding to plant significant areas of this herb.

Price

Price to growers for premium quality, organically grown French Tarragon is currently around $24/kg for 5-star culinary grade.

Trade price is around $22/kg.

Valerian

Valeriana officinalis L   VALERIANACEAE

Parts Used: **Root & Rhizome**

Other Names: Garden Heliotrope, Setewale.

Valerian is a potent herb that usually evokes a strong response in some form or other, often quite emotive. For many it is very effective as a sedative, especially for insomnia and it has a long tradition of medicinal and even culinary use. It is also used as one of the biodynamic compost preparations, which is prepared from the flowers.

But there are quite a few people who have an adverse reaction to it. Some find its odour quite repulsive, even nauseating, and just the thought of Valerian being freely available to the public has been known to send shivers down the spines of Health Department officials.

My own experience with Valerian has usually been quite positive. Working with it during the day, I find it has a mellowing influence, though I am no longer able to use it for insomnia: it keeps me awake at night instead. I seem to have developed a tolerance to its sedative action, which apparently happens with some people.

The idea of using Valerian in cooking may not sound very appetising, but in the past it was commonly regarded as a vegetable and an essential culinary herb. To most of us today, though, Valerian smells and tastes too much like old socks to be able to contribute much to the enjoyment of a meal.

Personally I find the smell of freshly dried Valerian quite sweet and attractive, though I must admit not everyone agrees with me.

Origin

Valerian is a native of Europe and western Asia.

GROWING VALERIAN

Valerian is not hard to grow, provided it has the right conditions. The harvesting and washing part of the operation is the limiting factor. The structure of its root system is different from most other herbs in that the rootstock or crown is relatively small, but it has a lot of fine rootlets growing out from it and usually a number of horizontal rhizomes or runners as well. These fine rootlets tend to get tangled together in dense mats, which trap a lot of soil.

Occasionally growers have been known to develop an allergic reaction to Valerian.

Climate

Valerian prefers a cool climate. There is a variety that can be grown in regions with a hot summer, provided the plants get a very early start and plenty of moisture. However, yields in these conditions will never be as good as in cooler climates.

Site

Exposure

Valerian will grow in full sun or semi-shade. In hot conditions it needs some protection from hot drying winds and the flowering tops are vulnerable to being blown over in exposed sites. It is best to locate Valerian in a sheltered part of the garden where it is protected from hot winds, even in cooler climates, because the occasional spell of hot weather can knock the stuffing out of it. It commonly wilts on a hot day, even when the soil is saturated.

Drainage

Valerian will tolerate waterlogging, but these sites can be a problem when it comes to harvesting. In wet conditions it is very difficult to dig, and very difficult to shake the soil from the rootlets.

Soil Type

For ease of harvesting and washing, Valerian needs to be grown on lighter soils. These still need to have reasonably good moisture holding capacity, as the plant is shallow rooted and has a high moisture requirement. It is also important that the soil be virtually free of small stones and other debris, such as charcoal, as these will inevitably end up in the dried herb through being entangled in the fine matted rootlets.

Fertility

Valerian likes a fertile situation, but if the plants grow too well, the roots will be so dense that separating them for washing and drying becomes quite difficult and time consuming.

Initial Weed Control

A bare fallow is recommended before planting.

Neighbouring Plants

Valerian needs to be placed away from aggressive plants that would compete with it for moisture.

Valerian may have a tendency to self-sow in neighbouring crops if conditions are cool and moist enough, but this is uncommon and does not usually present a major problem.

Identification

Valerian grows up to 1800 mm tall, with grooved hollow flowering stems. The leaves are pinnate, up to 300 mm long, with lanceolate leaflets, toothed or entire. The small white or pink flowers are borne in terminal inflorescence on the tall stems (see colour plate 7.2).

The plant has aromatic fine roots, small crowns and, in some varieties, long white rhizomes.

Do not confuse with Red Valerian (*Centranthus ruber*), which is a quite dissimilar plant from the same family and grows as a weed in some regions.

There are a number of other species of Valerian, but as far as I know, they are not available in Australia.

Variety

There is considerable variability in the species. Apparently there are diploid, tetraploid and octoploid varieties. These refer to the number of chromosomes the plants possess: it seems at some stage, long before the advent of genetic engineering, a Valerian plant developed which had double the normal number of chromosomes. As these are always in pairs, the normal complement is referred to as diploid and a double complement, as tetraploid. Possibly this tetraploid variety then had another little genetic accident and produced a plant with double again, or four times the original number of chromosomes, which is referred to as octoploid.

The octoploid varieties have been found to contain the highest levels of active constituents, while the tetraploids have somewhat lower levels and the diploids lower again. Consequently, the octoploid strains are preferred for medicinal use.

One octoploid variety with 56 chromosomes has proved to be very vigorous in a wide range of conditions and produces large numbers of rhizomes. It is able to withstand hot dry conditions better than other varieties which have been tried, and it is less subject to devastation by grasshoppers.

It is possible varieties that produce few rhizomes may be easier to harvest and wash, as their root systems might be less tangled. This is an area that could be worth exploring.

Valerian (*Valeriana officinalis*).

Propagation

Method

Rhizomes

Some varieties produce abundant strong horizontal rhizomes or runners, which will give rise to new crowns the following season. These rhizomes are formed in late autumn and are good for spring planting.

Varieties which produce few or no rhizomes will have to be propagated by division or from seed.

Young Crowns

Alternatively, young crowns can be planted in early autumn. These can be obtained by digging up well-developed clumps and tearing them apart to separate the crowns. Only young crowns that have not yet flowered should be used: old crowns will die off. The young crowns will establish themselves and send out rhizomes in late autumn, so by the next spring strong clumps of Valerian are developing.

Seed

Valerian can also be grown from seed, but initial development will be slower and crops grown from seed in spring are unlikely to be ready to harvest until the end of their second season.

There can be problems with supply of seed and with germination. Unless the source of the seed is definitely known, it may turn out to be an inferior variety.

In soil mixes for seed raising and potting, avoid using expanding mica, perlite or other materials which might become entangled in the fine roots of Valerian as these can cause contamination problems.

Timing

Early Autumn

Early autumn is the preferred time for planting out Valerian in regions with a hot summer. Valerian will do best if young crowns can be planted early enough in autumn to give them time to get well established before winter. Then in spring they will make rapid growth and develop their root systems and leaf canopies before the heat of summer.

The secret to Valerian being able to endure summer in the hotter regions is to get it established early, so there is a good development of roots and a good cover of leaf to keep the soil cool in hot weather. In a solid clump, the leaves tend to protect each other, while a weaker plant with sparser leaves burns more easily in the heat.

Early autumn planting is also suitable for cooler regions.

Spring

Rhizomes can be planted in very early spring. This timing works well in Tasmania and other cooler regions. Best results will be gained from large strong rhizomes.

Don't plant crowns in spring as they will just go straight to flower, with little leaf and root development.

Seed should be sown in spring.

Note: There appears to be some confusion in the terminology used to refer to the different parts of the Valerian root system.

Crown: This term is used here to refer to the central solid rootstock. This gives rise to the leaves and flower-stem above ground, and the rootlets and rhizomes below ground. Much of the literature refers to this central crown of Valerian as a 'rhizome' or an 'erect rhizome', which can be rather confusing.

Rhizome: Technically a rhizome is a horizontal underground stem, with nodes and bud(s). This is the terminology adopted here. In Valerian this is a long thin white runner, up to about 6 mm thick, and up to 150 mm long. It gives rise to a new crown from the bud at the end.

Planting

Set the crowns in with the growing point just covered (remove most of or all the leaves, of course). For rhizomes, plant them end-to-end in a continuous line in furrows and cover with 25–50 mm of soil.

Layout and Spacing

Crowns should go in on 150–300 mm centres, in rows around 900 mm apart. Single rows are okay, or blocks, it doesn't matter.

Rotation

Valerian should be rotated with other crops as there can be a build-up of pests like cock-chafers if it is grown continuously in the same place. Rotation should be with crops that can cope with a few Valerian plants coming up from rhizomes that have escaped the harvest.

Compost

Valerian likes fertile conditions, so give it a dressing of 2–4 kg/m^2 applied in early spring. In favourable conditions it may be advisable to moderate the use of compost. If too vigorous a growth of Valerian develops, the densely matted root systems can be very difficult to pull apart and wash clean.

Weed Control

Valerian needs a bit of attention to get it through the early part of its development, but once it has formed good leafy crowns, these will suppress weeds fairly well. It is important to keep it very clean, as the roots of any weeds will become entwined with the Valerian rootlets, making harvest more difficult.

The form of the Valerian leaf often makes it difficult to spot the weeds growing amongst it, as the leaves of many weeds just don't stand out. I am often appalled at the number of weeds I have missed when I go back over the row, so it is usually best to go over it twice.

Irrigation

Reliable regular irrigation is essential for Valerian, as its large soft leaves transpire a lot of moisture and its root system is quite shallow. It needs as much as 50 mm of water per week.

Pests and Diseases

In a cool to mild climate, Valerian doesn't suffer many problems, but in the climate of most parts of southern Australia, which is rather stressful for it, it seems to be subject to more pests.

Grasshoppers

If the Valerian is under too much stress, it seems to attract the grasshoppers, which will devour it. The answer is to grow a variety that will tolerate the heat better, keep it amply watered and give it plenty of compost. A few chooks or ducks could help too.

Cockchafers

Cockchafers are beetle larvae that live underground and chew on various roots for their sustenance. They are fat grey-white grubs with brown heads, and are usually curled up when you find them.

While Valerian root's sedative action is no doubt to protect the plant from creatures eating it by making them go to sleep, it doesn't appear to bother cockchafers, which seem to be curled up asleep most of the time anyway.

I found they tended to be more of a problem if the Valerian was left for 2 years before harvesting, or if it was replanted in the same place. In that situation large numbers of cockchafers can reduce yields.

If cockchafers are a problem, then it is best to harvest the Valerian every season and rotate it to a new place each year.

Management

Valerian forms a crown, which initially produces a rosette of leaves. Some crowns will flower in their first year, while others take 2 years to flower. Those that have reached this stage send up a tall flowering stalk in early summer, taking a fair bit of the plant's energy. The crowns that form these flowers die after flowering, the growth being continued by new crowns forming from rhizomes produced by the old crown. The rootlets to be harvested are produced by young crowns that have not yet flowered.

If the flowers are not required for making biodynamic preparation 507, they can be cut down as they appear, which will encourage more leafy growth. If you want to harvest some of the flowers, the stalks can be cut down after the flower harvest is finished. They usually need to be cut off two or three times.

Harvesting Flowers

The biodynamic preparation 507 is made from the fresh flowers. There are two or three different methods of preparing this, but they all start off with picking the flowers in the early opening stage (see 'The Compost Preparations' in Chapter 10, 'Biodynamic Aspects').

Harvesting Root

Timing

Valerian can be dug in late autumn or early winter. It is one of the last plants to die down in autumn and one of the first to shoot in spring.

In mainland Australian conditions, it is usually best to dig it in the first season. While growth will thicken up if left for the second year, this is often not an advantage. As there are more plants drawing moisture from the same area of ground in hot weather, the resulting moisture stress can mean the plants go backwards and yields may be no greater after the second year.

In milder conditions, such as in Tasmania, where the denser growth of the second season can be supported, the root systems can become so densely matted that harvesting and separating the rootlets becomes very difficult.

Tools

The most efficient tool for harvesting Valerian is the vine hoe: a fork with four prongs bent at right angles. Failing this, a garden fork can be used.

A large screen around 1500 x 900 mm with a 25-50 mm chicken-wire mesh is needed for separating the loose soil from the rootlets. This needs to be propped up with something.

Technique

Clearing the Surface

Before you start digging, clear the surface of any debris that might get mixed up with the Valerian roots. This includes the remains of the tops of the Valerian itself, which may need to be cut or pulled off.

Rake these tops and any loose debris well aside, but unless weeds will come out cleanly, roots and all, it is best to leave them in situ with their crowns or leaf bases attached. During the digging, shaking and washing processes, it is easier to separate out whole root systems of weeds than little pieces of torn off root.

Digging

Making a go of growing Valerian requires the development of a reasonably efficient harvesting system and optimum use of tools and techniques. The main problem in harvesting is getting the soil separated from the matted rootlets. In a good crop rootlets are densely intertwined and it requires some dedication and skill to get them apart.

This task can be made much easier by digging the Valerian in such a way as to tear it apart so it comes away in small clumps. This can save a lot of frustrating work later on when trying to pull large clumps apart.

The vine hoe is very handy for this operation. By starting at one end of a row and working along with a deep chipping action, you can tear away at the mat of Valerian roots, taking 50–75 mm thick pieces off at each stroke. By flicking these pieces up and back a little as you dig them, they can be left on the surface as you work your way forward. Often they will stick a little to the vine hoe, which can allow you to give them a good shake to loosen up some of the soil.

Digging with a blade, such as a spade or a blade hoe, is not recommended as it will cut the rootlets rather than pull them apart and these short pieces will be difficult to retain in the washing process.

The soil must be dry enough for this operation: if it is too moist, the soil won't shake out of the roots very well. For this reason lighter soils are to be preferred for growing Valerian.

Sometimes, if the prongs dig into a tough part of the clump, they just stick into it rather than break it off. This can be overcome by pushing the handle sideways to lever the clump apart. You may need to place one foot on the ground just in front of the fork to stop the rest of the clump from lifting out.

When you get to the end of the row, you can turn around and, with the vine hoe, work your way back along the same row, raking and pulling up the clumps, shaking a bit more soil out of them and leaving them sitting in little piles on the surface.

Drying out the Clumps

If your soil is dry enough you can proceed directly to the next stage, but if it is too wet and is reluctant to shake out, you may need to leave the pieces of clumped roots in windrows until they have dried out a bit more. This can be speeded up by covering them in wet weather and exposing them when it is fine.

Don't let the soil get too dry as it could become very hard, especially with heavier soils.

Shaking

The next step is to shake the roots over the chicken-wire screen. There will be a lot of soil still among them which is best left in the garden, rather than being washed away into the grass somewhere or, heaven forbid, down a creek.

The screen should be set up where you are digging. Rest it on or against something so there is plenty of space under it. Place a tool handle or a thin round pole across the screen – this is to thump the pieces of clumped root on to shake more dirt out of them.

Before shaking and thumping each piece, you will probably want to put aside any good rhizomes needed for propagation material. It is best to select them at this stage, as they will get bruised and damaged going through the shaking and washing process.

This is one of those jobs where you have to strike a balance and trade the loss of some of your soil for a saving in time. So just shake out as much soil as you reasonably can. The soil that goes through the screen should be resifted to catch some of the roots that find their way through.

An alternative method, which might save some time, would be to arrange some sort of trap to retain the soil and organic debris that washes out of the roots. This would enable you to bypass the shaking process and go straight from digging to washing.

Washing

Washing Valerian root is a lengthy process and needs to be done on a warmish day, if possible, as handling wet roots can be miserable work when its cold.

First Washing

The first stage of washing is to lay all the clumps out on the washing screens and give them a good blast with the hose – from all directions and on both sides – to wash out as much of the remaining soil as you can.

Separating the Crowns

The next step is to pull all the clumps apart, to free up any soil still trapped in the matted tangle. This involves pulling and teasing apart all the crowns to separate them from each other so the soil can be washed out. It is best leave the rootlets attached to the crowns as this will prevent the rootlets from going through the washing screen. A good jet of water will wash them clean once they are no longer tangled around each other. Extra large crowns may need to be split by tearing them in half, leaving all the rootlets attached.

There is a temptation to cut the matted roots apart with a knife or a chopper, but this makes washing much more difficult. Cutting the clumps creates a lot of short unattached pieces of rootlet which will wash through the screen and are very difficult to separate from the soil particles and debris.

Instead, if the crowns are pulled apart, most of the rootlets remain attached to them. These little clusters of rootlets will then be held by the screen, allowing the jet of water to force the soil and debris out of them.

Second Washing

Once the crowns are all separated from each other, they should be laid out on the washing screens and hosed down again thoroughly on both sides, until they are quite clean. They should be allowed to drain for a while before being put into the dryer.

Large-scale Harvesting

If production of large volumes is going to be viable at the prices offered by manufacturers, some means of streamlining harvesting and washing operations needs to be developed.

Unless higher prices can be obtained, the methods outlined above are too labour intensive. Harvesting, washing and drying a 20 kg batch of Valerian, whole crowns, rhizomes and roots, dried weight, typically involves 25–30 hours labour, so at $15–18/kg, returns would barely cover the cost of labour in harvesting.

Possible avenues to explore would be a means of mechanising the digging, washing and teasing apart of the root systems.

Some varieties may prove to be more easily separated and washed than the variety I have been growing, which does put out a lot of rhizomes, forming many crowns with intertwined root systems.

Plant spacing and fertility also influence the nature and degree of tangling of the roots.

Post-harvest Treatment

Rake over the patch to level it off while the soil is still friable.

Drying

After spreading the roots on the screens, you may want to allow the surface moisture to dry off for a day or so before putting them in the dryer. Valerian normally does not need chopping or splitting before drying, though this would speed up the drying of the crowns. These are the slowest part to dry and they can be deceptive, as they can feel quite hard on the outside and still be moist inside.

The bud of the crown is the last part to dry, so if you push into this part with your thumbnail you can judge if there is any moisture left in it. If you have limited drying space, you could rub the rootlets and rhizomes off the crowns and put the crowns back in until they are dry. Because of Valerian's penetrating aroma, I prefer to dry it by itself or else at the top of the cabinet dryer, so the air coming off the Valerian, laden with aroma, goes out the exhaust vent rather than through other herbs.

When dry, Valerian should be lifted carefully from the screens, as there is usually some grit, soil, etc. which remains trapped in the rootlets despite your most diligent efforts at washing. As the roots dry and shrink, they let go of some of this and it falls onto the screen.

Tea-grade Valerian root.

Temperature

Valerian should be dried at temperatures of not more than 35°C, as it contains volatile constituents.

Processing

Quite a lot of Valerian is used as a tea and it is also used for manufacturing liquid medicines. There is one school of thought that maintains Valerian should be allowed to age for 1 or 2 years before use, but I don't know how much evidence there is to support this.

Tea Grade

For tea grade, Valerian should be rubbed firmly through a standard ($2^1/_2$ dent) screen which will break up the rootlets and rhizomes and separate the crowns. This is best done as the herb comes from the dryer, because if there is any residual moisture in the crowns it could remoisten the rootlets somewhat and make them hard to rub.

Usually two passes through the screen are sufficient to produce a good tea grade (under-screens are not necessary).

The crowns can be hammer-milled down to particle sizes up to 6 mm, after which they are recombined with the rootlets and rhizomes. If this is beyond your facilities, exclude the crowns from your tea grade: you may be able to sell the them separately to a manufacturer, or your wholesaler may be able to mill and recombine them.

Manufacturing Grade

For manufacturing grade, the Valerian is used just as it comes from the dryer and comprises whole crowns, rootlets and rhizomes.

Storage

If the crowns are included in the Valerian, after it has been in bags for about 2 weeks, its moisture content should be checked in case of undetected residual moisture in the crowns.

Valerian needs to be freeze treated for moth control before sale or storage. Some care needs to be taken that Valerian is securely stored, as it is very attractive to vermin: it is quite effective as a bait in rat traps.

As Valerian emits a rather strong, penetrating odour, which passes through plastic bags, it is best to store it on its own, so all your other herbs don't end up smelling like Valerian. It is also recommended that it be stored somewhere outside your dwelling, as the aroma coming off it may have effects you might not have intended.

I had some experience of this at 'Twin Creeks' when I was living in a caravan there. I had an order for tea-grade Valerian, which needed packaging, so I left a large plastic bag of it just inside the door of the caravan to remind myself to do it in the morning.

I slept very badly that night and, perhaps as a consequence, the following day was one of those when nothing goes as planned. I never got around to packaging the Valerian, which stayed in the caravan.

The next night I found myself again unable to get off to sleep, until it dawned on me I was experiencing the same effects as if I had taken Valerian. Upon realising this, I took the bag out of the caravan, opened the windows, went back to bed and was then able to sleep soundly.

Yields

Yields of Valerian (dried rootlets, rhizomes and crowns) in north-east Victoria, were around 0.25 kg/m^2 per annum, while in more favourable conditions in Tasmania yields of up to 0.4 kg/m^2 per annum can be expected.

Marketing

The market for premium grade, organically grown tea-grade Valerian is rather limited at present.

Significant volumes of Valerian are used for manufacturing, but growers need to ensure the price offered is sufficient to justify the labour and costs involved in producing it.

Price

Price to growers for premium quality, organically grown Valerian is currently around $37/kg for 5-star tea grade.

Price to growers for manufacturing-grade, organically grown Valerian is currently around $15–18/kg.

Trade price is around $13-16/kg.

Violet, Sweet
Viola odorata L   VIOLACEAE

Parts Used: **Leaf**

The aroma of Sweet Violet flowers is one of the most pervasive of plant perfumes. For me it still recalls images of English countryside from a period I spent there as a young child. Sweet Violet is a plant that seems out of place in many parts of Australia. We do have our own native Violets, which are much hardier and have no perfume.

Violet is used medicinally, mainly for respiratory conditions. The flowers are also used in confectionery, to decorate cakes and desserts and in making a French liqueur, 'Ratafia de Violettes'.

GROWING SWEET VIOLET

Sweet Violet is not suited to the climate of many parts of southern Australia, as it does not thrive in our hot sun.

While it needs shade in summer, it likes sun in the cooler part of the year. In north-east Victoria, I found the best way was to grow it under 50% shadecloth, which was put over the plot in mid-spring and taken off in mid-autumn (see 'Provision of Shade' in Chapter 5, 'Propagation and Planting').

Identification

Sweet Violet is a perennial with long stolons, or surface runners. It grows 100–200 mm tall with the leaves arising from a small crown. The leaves are more or less heart shaped, up to 100 mm long, with small teeth. Flowers are up to 20 mm across, violet, white or pink, and very sweet smelling.

Variety

There is some variation in flower colour. I have not heard of any preferred variety for herbal use, though it is probably best to stick to varieties that are closer to the wild form.

Pests and Diseases

If Violet is not thriving, a likely cause is too much exposure. Additional shade may be required.

Drying

Sweet Violet leaf is supposed to be best used fresh, but the trade mostly uses the dried herb. This is partly because of the inconvenience of sourcing fresh Violet leaves, and because the fresh preparation can't be made as concentrated as the dried.

To preserve its qualities, it is important to dry Sweet Violet as quickly as possible under good conditions. The leaves are easily blown off the screen when they are dry so it needs to be in a safe spot.

Dry at temperatures of up to 35°C.

For more detail see 'Information Charts'.

Sweet Violet (*Viola odorata*).

17

Perennial Crown Herbs

Agrimony – *Agrimonia eupatoria*
Alfalfa – *Medicago sativa*
Artichoke, Globe – *Cynara scolymus*
Celandine, Greater – *Chelidonium majus*
Chives – *Allium schoenoprasum*
Echinacea, Narrow-leaf – *Echinacea angustifolia*
Elecampane – *Inula helenium*
Fennel – *Foeniculum vulgare*
Figwort – *Scrophularia nodosa*
Garlic – *Allium sativum*
Ginseng – *Panax* spp.
Goat's Rue – *Galega officinalis*
Horehound, Black – *Ballota nigra*

Lady's Mantle – *Alchemilla vulgaris*
Lemon Grass – *Cymbopogon citratus*
Marshmallow – *Althaea officinalis*
Meadowsweet – *Filipendula ulmaria*
Motherwort – *Leonurus cardiaca*
Pasque Flower – *Anemone pulsatilla*
Plantain, Greater – *Plantago major*
Poke – *Phytolacca decandra*
Red Clover – *Trifolium pratense*
Stoneroot – *Collinsonia canadensis*
Vervain – *Verbena officinalis*
Wood Betony – *Stachys officinalis*
Wormwood – *Artemisia absinthium*

Herbs falling into the 'Perennial Crown Herbs' group grow from a central crown at the head of the root system. The crown typically has a number of buds or shoots and gives rise to several or many stems, which normally die down to ground level at the end of the growing season.

After their aerial parts are harvested, the focus of regeneration of perennial crown herbs is from the crown at ground level.

The crown usually enlarges progressively as the plant grows, developing more buds and shoots. Many of these species divide spontaneously, with portions of the crown separating to form independent plants.

This gives rise to a clump of crowns, which gradually increases in size in a manner somewhat similar to the expanding clump herbs, except that the rate of expansion is slower, so perennial crown herbs can be maintained as row crops for a good many years. Consequently, weed control is quite easy for most species in this group, as most of the weeding can be done by inter-row cultivation and weeds generally find it hard to get established in the solid growth of the clumps.

This makes production of most perennial crown herbs relatively straightforward.

The boundary between this group and the previous one, 'Expanding Clump Herbs', is not clear-cut and some species could arguably be placed in either group, depending on growing conditions etc.

Propagation

Method

With a few exceptions, this group of herbs can be grown both from seed and from crown divisions. The exceptions are those that are selected cultivars, such as Lemon Grass and Garlic, and a few non-dividing species, such as Red Clover and Alfalfa.

In general, crown divisions will give faster developing plants, but quantities may be limited, so it can take a while to build up stock this way.

Many of this group can be grown quite easily from seed, so this is often the preferred method of propagation, particularly if large areas are being planted.

Layout and Spacing

Most species are best maintained in rows, though at least two can be directly sown as a meadow-crop.

In laying out rows, bear in mind that expansion will be slow, so if weed control is not a problem, rows can be spaced closer. The optimum spacing usually ranges from 600–900 mm, just enough to allow continued inter-row cultivation with a wheel hoe.

Management

Weed Control

Weed control is fairly straightforward with most of these species as the strong growth from the crowns means that they can hold their own well. Usually it is just a matter of regular inter-row cultivation and a bit of weeding around the individual plants.

Most herbs in this group are not likely to be lost to winter-active weeds as there is usually enough room to cultivate around the plants in spring and they come away strongly enough to get above most weeds.

Many perennial crown herbs are useful in rotations when a crop is needed which can cope with a heavy germination of weeds or self-sown herbs.

Some members of this group are best grown separately from other species because they are able to tolerate a certain level of weed infestation, which can spread to neighbouring crops where it can be a real problem.

Root Crops

A number of these herbs are grown for their roots, and as the plants only regenerate from the crown, plots will need to be replanted after harvest. This can be done with pieces of crown, which gives faster development than seed.

Longevity

Most perennial crown herbs are quite long-lived. Apart from those dug for their roots, they are generally long-term crops that take several years to reach peak production, but will maintain it for a good period.

Post-harvest Treatment

Virtually all of this group regenerate most strongly from ground level. After harvesting their aerial parts, mowing off the stubble will generally stimulate more vigorous regrowth and give a cleaner subsequent harvest.

This needs to be carried out within 3 days of harvest to avoid setting back any new growth (for more detail see 'Mowing Stubble' under 'Post-harvest Treatment' in Chapter 11, 'Harvesting').

Agrimony

Agrimonia eupatoria L ROSACEAE

Parts Used: **Leaf & Flower, Aerial Parts**

Other Name: Sticklewort.

Agrimony has a long history of medicinal use, mainly as an astringent, a diuretic and a vulnerary. It doesn't have any tradition of culinary use, though the tea is sometimes drunk for its pleasant flavour.

GROWING AGRIMONY

Agrimony is quite a robust plant that is fairly easy to grow and its bright yellow flowers add a little colour to the garden in late spring.

It prefers a cool to mild climate, though it will survive in regions with a hot dry summer. In that situation it makes most of its growth in spring and autumn and just hangs in during the heat.

Agrimony (*Agrimonia eupatorium*).

Neighbouring Plants

At 'Twin Creeks' Agrimony seemed to be a chronic grasshopper haven, which was a bit of a problem because they would feed on it until it was stripped and would then move onto vulnerable plants nearby.

Containment

As the fruit has small bristles that enable them to cling to clothes, animals coats etc., you may find the odd Agrimony coming up in unexpected places. For this reason I would keep sheep away from it, though in most regions it is not likely to become a weed.

Identification

Agrimony grows to 900 mm tall, with downy reddish stems. The leaves are pinnate, up to 200 mm long, and mildly aromatic. The flowers are bright yellow, 5–8 mm across, numerous, and borne on tall spikes. The fruit resembles a miniature acorn with a bristly cap.

Pests and Diseases

Grasshoppers

Hot summers are rather stressful for Agrimony. This makes it vulnerable to grasshoppers, which will often strip the leaves off the plants.

At 'Twin Creeks' it was usually possible to take the first harvest before the grasshoppers moved in. As the first is the biggest Agrimony harvest, I wasn't too concerned about losing the second harvest: the herb was usually in oversupply anyway.

Supportive measures such as increased irrigation and some protection from the heat will probably help. Our Agrimony always recovered strongly once the grasshoppers disappeared in autumn.

Alfalfa

Medicago sativa L FABACEAE

Parts Used: Leaf & Flower, Aerial Parts

Other Names: Lucerne, Purple Medic.

Alfalfa, or Lucerne as it is known in Australia outside the herb trade, is probably the most widely grown herb in Australia. Yet most manufacturers have, at least until recently, relied on overseas sources for the herb.

The reasons are basically because manufacturers have not been very interested in sourcing their materials locally and because most of the Alfalfa grown in Australia is intended for animal fodder.

You can't just pick up a bale of Alfalfa from a paddock somewhere as it is liable to have grass and other weeds in it, plus trash, soil and dung picked up by the baler. Alfalfa hay can be tricky to dry, so there could be mould in the bales if it was too moist, or loss of leaf if it was too dry. Often it is too stalky as a result of leaf loss, late harvest or use of a stalky variety, and inevitably there is some bleaching as a result of field curing, and possibly rain damage as well. In some areas Alfalfa is sprayed to control weeds and pests.

This is not to say that the quality of imported Alfalfa is any better. Most of it is shocking, having many of the defects described above. In fact people are so used to the chopped stale brown stalks commonly sold as Alfalfa, that they often don't recognise the bright green leaf of premium quality Alfalfa as being the same herb.

Alfalfa is used medicinally for its nutritive and diuretic properties. It is a common herbal beverage and the sprouted seeds are popular in salad.

Origin

Alfalfa is native to the Mediterranean region and western Asia.

GROWING ALFALFA

Alfalfa is quite different from most other herbs in its growth and management. For one thing it does not transplant very well and must be grown from seed. Once established, its great strength is its exceptionally vigorous regrowth which bursts forth and leaves most weeds far behind.

Alfalfa has to be direct seeded. Once established, it can be maintained by regular harvesting. This keeps weeds as a low under-storey most of the year, as the Alfalfa outgrows them, keeping them shaded and short of moisture. Being very deep rooted and fixing its own nitrogen, Alfalfa is able to draw on moisture and nutrients not available to other plants, and this gives it a competitive edge.

Climate

Alfalfa is suited to a wide range of climates.

Site

Exposure

Full sun is required. While wind is tolerated, it can be problem if it blows a heavy crop of Alfalfa over just prior to harvest.

Drainage

Good drainage is essential as Alfalfa is very deep rooted and will not tolerate waterlogging. It can endure brief flooding on alluvial soils.

Soil Type

Alfalfa prefers lighter soils with a pH of 6.0 or higher: it does not do well in acid soils, which may need the application of ground limestone, dolomite or wood ashes to make them suitable. This should be done far enough in advance for it to have an effect before sowing.

Fertility

Moderate fertility is needed, but if soil is too fertile, growth may be top heavy and fall over.

Initial Weed Control

A good bare fallow is recommended prior to seeding as this is the only opportunity the grower has to eliminate or reduce weeds to ensure the crop can get established.

Neighbouring Plants

It is best to establish Alfalfa in its own separate area. This is not because of problems the Alfalfa itself poses to other plants, but because its management technique allows weeds to get established among it as an under-storey and these will tend to spread, by seed or runner, into neighbouring crops. I found this out the hard way when I sowed a long strip of Alfalfa down the middle of my herb garden at Huonville in Tasmania. The Alfalfa did very well but there was a continual battle on both sides of it, as I tried to contain the Couch Grass and Sorrel and stop them spreading from the Alfalfa crop to the adjacent beds.

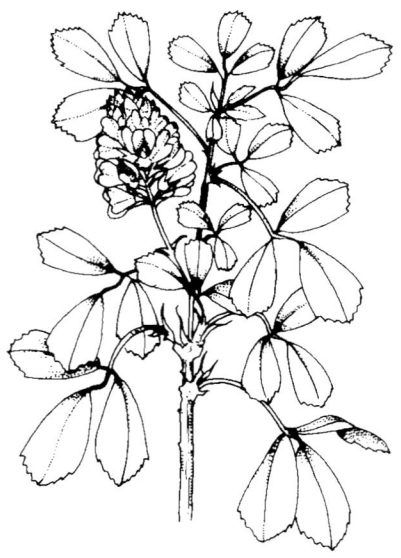

Alfalfa *(Medicago sativa)*.

Identification

Alfalfa grows 300–1000 mm tall, with a much branched stem. Leaves are trifoliate, leaflets toothed, obovate to oblong, up to 30 mm. Flowers are 15–30 mm, violet-blue in axillary racemes (see colour plate 1.1). Pods are pubescent, spiralled.

Alfalfa has a thick deep taproot. Most varieties show a degree of growth through winter, and some are winter-active.

Alfalfa does bear a resemblance to some clovers, but its leaflets are longer and the flower head of Alfalfa is more open with larger individual flowers.

A number of other species of *Medicago* occur as weeds in different parts of Australia. The most similar is Sickle Medic (*Medicago falcata*), sometimes known as Yellow Lucerne, which can be distinguished by its yellow flowers and its sickle-shaped seed-pods. Most of the others are annuals and they all seem to have yellow flowers.

Variety

There is a multitude of varieties of Alfalfa available in Australia, which have been bred for particular climates or growth requirements.

Some have been bred for resistance to disease or to various insects that attack it (probably when it is under the stress imposed by modern farming methods). There has also been a selection for leafiness in some varieties and stalkiness in others. Some varieties are winter-active, others are dormant in winter.

From the point of view of herb production, the most important thing is to choose a leafy variety as the leaf is the most important part of the herb and the main ingredient of the tea grade.

Resistance to pests and diseases may be a factor to consider if you are in a region that suffers from these problems. In my own experience, these have not been a problem, but it was not in a major Alfalfa growing area.

From the point of view of dried herb production, winter-activity is of no benefit, as the crop is difficult to keep clear of weeds at this time and difficult to dry. But if an otherwise suitable variety is winter-active, this is not a problem, though I believe the winter-dormant varieties may be more productive in summer.

Over the years I have had good results with Hunter River and Si-River, which superseded it. They are both good leafy varieties.

When enquiring about seed at your local stock and station agent, don't forget to ask for 'Lucerne' or they may not know what you are talking about.

Propagation

Method
Alfalfa needs to be direct seeded.

Timing
The optimum time to sow Alfalfa is just before the onset of the prevailing dry conditions of summer. This is around October in northern Victoria, or late December in Tasmania. This timing allows a number of passes to kill spring germinating weeds and then the Alfalfa can go in just before conditions become less favourable for the germination and growth of the shallower rooted weeds. The Alfalfa quickly develops a deeper root system, which gives it an advantage in these conditions and it can get ahead of the weeds. If it is sown too early, weed growth can smother it when it is in the seedling stage.

Planting
Before planting, the seed should be inoculated with nitrogen-fixing bacteria if it is going into soil which has not grown Alfalfa or the related Medics before, as these species have their own particular strain of bacteria, which probably won't be present in the soil.

The seed can be broadcast, or sown in drills (rows about 200 mm apart). If broadcasting you may need to sow it at a higher rate. Sow at around 12 kg/ha or $1.2 g/m^2$. Cover the seed by harrowing it in lightly and then, if possible, roll it to get it flat and firm up the seed-bed, even if the roller is just a 200 L drum full of water. This will help germination and get the surface reasonably level, which will make harvesting and stubble mowing much easier later on.

Layout and Spacing
For ease of management, the length and width of the bed should be determined by the dimensions of the range of your irrigation sprinklers, so the Alfalfa is growing where the water is available and water isn't being wasted by pouring it onto the grass.

Rotation
A patch of Alfalfa is generally good for at least 5 years, after that it may start to get sparse. The bed may need to be rotated to another crop and a bed sown down on new ground.

Compost
The rate of compost application depends on the fertility of the soil and the growth of the Alfalfa. The growth of the crop should be monitored to assess how much compost it needs.

As it is deep rooted and can fix its own nitrogen, it may not require much compost in many situations, but on poorer soils it will need at least moderate applications of compost to maintain it.

Excessive compost can favour weed growth and cause the Alfalfa to grow too lushly and fall over. On the other hand it is important to put at least a small amount of compost onto biodynamically grown Alfalfa so it can receive the influence of the compost preparations.

If wood-ash is available, it is a good organic fertiliser for Alfalfa, providing a balanced range of minerals.

Alfalfa often looks very poor for a while when first established. The leaves may have a yellowish look and the plants may be spindly. I don't know whether this is due to some lag in the proliferation of its nitrogen fixing bacteria, but I have noticed it on each of the three occasions I have established a new planting.

On the first occasion I gave it a good dressing of something I would not use now: chicken manure. This was on the advice of an Agriculture Department consultant who only understood the problem in nitrogen, phosphorus and potassium terms, and saw chicken manure as the organic equivalent to a bag of fertiliser.

By the time I was establishing the second bed of Alfalfa, I was wise to the problems with chicken manure and instead just put on a moderate dressing of wood-ash.

The third patch of Alfalfa was in a section to be used as a source of compost material, and I didn't give it anything at all.

The interesting thing was they all responded in about the same period of time, including the untreated patch. In each case, by the second season the growth of the Alfalfa had filled out and was a healthy dark green colour.

Weed Control

Mowing
Weed control in Alfalfa can be maintained almost exclusively with the mower. The first growth in spring will be thick with weeds carried over from winter, so this is best mowed off close to the ground in September or October. The subsequent regrowth will see the Alfalfa coming away ahead of most weeds, but if moisture is still plentiful there may still be too many tall weeds among the crop for ease of harvesting. If this is the case, it should be again mowed back close when in flower. By this stage the Alfalfa will be still coming away as strongly as ever, but the growth of grass and weeds will be slowed and they will remain as an under-storey.

Recovery Period

It is important to understand that an adequate recovery period after mowing or grazing is vital if Alfalfa is to replenish its reserves after regenerating. The plant puts everything into its regrowth and it can be killed if it does not get 4–8 weeks recovery period (or longer in winter, according to the growing conditions) after harvest. The general rule is to wait until it flowers or until small leaf shoots are starting to emerge at the base of the stems.

After harvest, it is important that the stubble be mowed off immediately, or at least within 2 days, so it is done before any regrowth commences.

Winter Weed Growth

During winter there will be a growth of weeds and grass which will outgrow the Alfalfa and by spring the crop may look rather like a lost cause. In fact it is best to allow this weedy growth to continue until the Alfalfa has had sufficient time to recover before mowing it.

No matter how weedy the Alfalfa gets over winter, by the third harvest (or often even the second one) the crop will be very clean.

Weedy growth or harvests not required can be taken off for composting or can be slashed and mulched down with repeated passes with the slasher. If growth is thick, you may need to kick it around a bit afterwards to spread the mulch evenly so it isn't smothering any of the Alfalfa.

Irrigation

Although Alfalfa is deep rooted, in dry conditions it will eventually deplete deeper reserves of moisture and growth will slow. It will then respond vigorously to irrigation but it usually only needs about half of what the rest of the herb garden is getting. Irrigation of 12–25 mm a week will usually be adequate during summer.

If growth is excessive and the crop is tending to fall over before harvesting, limiting irrigation will help prevent rampant growth.

Pests and Diseases

While Alfalfa has an impressive list of pests and diseases, I have not had any real trouble with any of them. I rather think most of them are the consequence of plant stress due to one cause or another and in a well-managed organic or biodynamic system they are less likely to occur and if they do, they usually aren't very serious.

Harvesting

Timing

Alfalfa should be harvested when it is just coming into flower. At 'Twin Creeks', there was normally a first cut in late September–early October, which was mulched or composted as it was full of weeds. The second cut was usually good quality and came around mid-November. The next four cuts came at 4–5 week intervals and there was usually enough to cut again in mid-May, but by this time it was getting hard to dry the harvest. That made a total of seven cuts a season, of which five were good quality Alfalfa, coming at a time when drying conditions were good.

Tools

Provided it hasn't fallen over, which can make harvesting a bit tricky, Alfalfa is easy to cut with the harvesting scythe.

Technique

The levels of weed growth and faded leaves need to be watched carefully so you can cut just above them as you swing into the Alfalfa.

If the crop has fallen over, it can be more challenging to harvest. To harvest it with the scythe you need to orient yourself so you are facing in the same direction the crop is leaning, so the blade is moving at right angles across the leaning stems and hooking under them to cut and pick them up. It works best if you only try to pick up a little each stroke, rather than bury the blade into the crop.

Sometimes the crop has fallen in a number of different directions and you have to approach each section from a different angle according to the way it has fallen.

This problem may be avoided by reducing irrigation so the growth is less rampant, or by harvesting a bit earlier. Also try to avoid irrigating tall growth when it is windy, as the tops fall over more easily if they are heavy with water.

Post-harvest Treatment

Immediately after harvest, the stubble should be mowed down to the ground. This will cut the weeds back hard so they get less of a start on the Alfalfa. It is important to do this within 2 days of harvest as if the Alfalfa has started its regrowth it will be severely weakened by a second mowing off at this stage.

Drying

Alfalfa is reasonably easy to dry in good conditions, though it is a little more retentive of its moisture and remoistens more readily than some other herbs.

It is one herb that is hardly subject to bruising when wilted, and remoistening does not affect its colour much.

On the other hand, it fades very quickly in sunlight or fluorescent light, so these need to avoided during drying.

While it is drying, Alfalfa is particularly attractive to rats, which no doubt recognise its nutritive value. Precautions need to be taken to prevent them getting at it while it is on the screens.

The dried herb normally has an odd mousy smell about it.

Temperature

Alfalfa can be dried at temperatures of up to 45°C.

Processing

Alfalfa leaf (including flowers) is used as a tea, and the aerial parts, including stalk, are used in manufacturing liquid medicines.

Tea Grade

Alfalfa needs to be rubbed through a 2½ or 3 dent screen (with under-screens in place) for tea grade. The ideal time is when the leaves are brittle dry and the stems are still moist enough to be pliable, because then they will be less inclined to pass through the screen. This can greatly reduce the number of rescreenings necessary, because when dry the stalks are very smooth and straight and will spear down through the screens very easily.

Aerial Parts

For aerial parts, the Alfalfa needs to be put through the chaff-cutter. Make sure the herb is thoroughly dry before doing this: the leaves should shatter when crushed in the hand and the stems should snap when bent. The upper part of the stem is usually the slowest part to dry.

Storage

Alfalfa should to be freeze treated for moth control before sale or storage. It needs to be stored away from light and vermin.

Yields

Yields of 0.6 kg/m^2 tea grade, or 0.9 kg/m^2 aerial parts, per annum can be expected from an established bed of Alfalfa yielding four to five usable harvests in a season. It is usually the third season before Alfalfa reaches its full potential, but then yields can be sustained for a number of years.

Marketing

Alfalfa is used in fairly large volumes, but because this crop can be produced on a large scale relatively easily, the market is subject to oversupply. Growers need to ascertain that they have a market for it at a reasonable price.

Price

Price to growers for premium quality, organically grown Alfalfa leaf is currently around $16/kg for 5-star tea grade.

Price to growers for manufacturing-grade, organically grown Alfalfa aerial parts is currently $3–8/kg.

Trade price is $3–7/kg.

Artichoke, Globe
Cynara scolymus L ASTERACEAE

***Parts Used:* Leaf**

Globe Artichoke, which is commonly regarded as a vegetable, may have some potential as a dried herb crop. The leaves are used in herbal medicine for liver conditions, which are not uncommon in our society with the amount of alcohol consumed.

Globe Artichoke (*Cynara scolymus*).

GROWING GLOBE ARTICHOKE

Globe Artichoke is relatively easy to grow and should give good yields, especially in cooler regions. It holds its own well against weeds once it is established, but don't make the mistake of allowing your geese access to it as I did once. They consumed the whole Artichoke plants and even drilled deep holes into the ground with their beaks as they worked down into the roots.

The leaf should be harvested when the buds or globes begin to form, so it would probably not be possible to combine globe and leaf production.

The midrib of the leaf is very fleshy, so it will need to be chopped before drying.

Identification

Globe Artichoke grows 1000–1800 mm tall. Leaves are large, up to about 800 mm long and 300 mm across, deeply divided and almost pinnate, grey-green above and whitish beneath. The flower heads are large, more or less globe-shaped with a fleshy receptacle and bracts and purple florets.

The plant is usually winter-active in Australia.

This species should not be confused with the Jerusalem Artichoke (*Helianthus tuberosus*) which is quite different in appearance, being a relative of the Sunflower.

Cardoon or Artichoke Thistle (*Cynara cardunculus*) is closely related to Globe Artichoke and grows wild in parts of Victoria and some other regions. It has spiny leaves and spinier flower-heads and is a bit more silvery in its appearance.

For more detail on this herb see 'Information Charts'.

Celandine, Greater

Chelidonium majus L   PAPAVERACEAE

Parts Used: Aerial Parts

Greater Celandine has a long tradition of medicinal use, though today it is a lesser herb in terms of volume. Its use is chiefly for liver problems, but it must be used with caution as it is toxic in excessive doses.

GROWING GREATER CELANDINE

Greater Celandine has a rather soft delicate appearance, which belies its inner caustic nature. When the stem is broken an orange latex exudes, and some care needs to be taken when handling Celandine as this juice can cause skin irritation. The harvested plant, fresh or dried, also gives off an acrid vapour, which can tear into the respiratory tract if accidentally inhaled.

Identification

Greater Celandine grows 300–900 mm tall, with branched slightly hairy stems. Leaves are pinnate, with 5–7 leaflets, ovate or oblong, with rounded teeth, light green above, whitish-green below. Flowers are pale yellow, 4-petalled, approximately 20 mm across, forming green capsules 30–50 mm long.

The only other species that might cause some confusion would be Lesser Celandine (*Ranunculus ficaria*), which is quite dissimilar, not even being of the same family. This is low growing, no more than 250 mm tall, with yellow flowers with 8–12 petals.

Pests and Diseases

If Greater Celandine plants are under stress from growing in a situation too warm for them, they will tend to be attacked by two-spotted mite (red spider). This will show up as yellow areas on the leaves, with minute many-legged creatures living on the undersides. These creatures seem oblivious to the toxic nature of the juice.

The best approach is to reduce the stress on the plant by providing shade or additional moisture.

Processing

Aerial parts are used for manufacturing liquid medicines.

Put the herb through the chaff-cutter, but before doing so, make sure the fleshy stems are completely dry.

For more detail on this herb see 'Information Charts'.

Greater Celandine (*Chelidonium majus*).

Chives

Allium schoenoprasum L LILIACEAE

Parts Used: Leaf

Somehow Chives are always referred to in the plural. Though the Oxford dictionary lists the herb as Chive, one never hears the word used in the singular.

Chives are commonly thought of as purely a culinary herb, but they do have medicinal properties as well, being an appetite stimulant and digestive.

Origin

Chives are native to cooler parts of Europe and north Asia.

GROWING CHIVES

Chives are quite easy to grow, but as a dried herb their production is a bit different from most other herbs.

As a dried herb crop per se, they have limited potential, but they could be a useful sideline for a fresh herb grower during periods of surplus growth.

Climate

While Chives do prefer a cool climate, they will grow through fairly hot summer conditions provided they have plenty of moisture.

Site

Exposure

Chives will grow in full sun or semi-shade, and will tolerate wind.

Drainage

They need a well-drained site.

Soil Type

They will grow in a range of soil types.

Fertility

A very fertile soil is needed.

Initial Weed Control

An initial bare fallow is recommended to make subsequent weed control easier.

Neighbouring Plants

Chives are easily affected by large neighbours shading them out or stealing all the available moisture, so they need to be located next to plants which are not too agressive.

Identification

Chives are a perennial, growing in clumps with small bulbs, which produce hollow grass-like leaves, dark green, 200–300 mm long and 2–5 mm thick, with a delicate onion-like aroma and taste. Flowers are pink to purple, compact spherical, at the top of hollow stems.

Chives can be confused with some types of onions that also grow in clumps. These onions are coarser in appearance and flavour, and tend to be a bluer green.

Garlic Chives differ in that they have flat leaves.

Variety

Chives tend to vary a bit in flavour and appearance. Make sure you start off with a variety with a good flavour, but not too short in its growth. The size may be difficult to gauge as Chives do vary greatly depending on their growing conditions and how crowded the clumps are.

Propagation

Method

Chives can be propagated from seed or by dividing the clumps.

Timing

They should be planted in autumn or early spring.

Planting

Individual crowns can be planted, but small clumps will reach harvestable dimensions sooner.

Layout and Spacing

Plant Chives on 150–300 mm centres in rows 450–600 mm apart.

Chives (*Allium schoenoprasum*).

Compost

Chives need heavy dressings of compost for optimum growth: 4–5 kg/m² worked in beside the plants in spring.

Weed Control

As Chives do not compete with weeds very strongly, they need regular attention, with inter-row cultivation and careful weeding around and in the clumps. Grass can be a particular problem as the leaves of some grasses, particularly Rye Grass, are difficult to see in the clumps and they may end up contaminating the crop if not noticed.

Rejuvenation

The growth of Chives will decline after two or three seasons when the clumps get too big and crowded. They can be rejuvenated by digging them up, dividing them into small clumps and replanting.

Irrigation

Chives have a shallow root system and need regular watering, 25–50 mm a week, during summer.

Pests and Diseases

Apart from a bit of slug damage during early spring, they are fairly free of pest and disease problems.

Harvesting

Timing

Chives should ideally be harvested before the flowers emerge, as the flower stalks are rather woody and should not be included. Once the plant flowers, it goes into decline, so to maintain vigorous growth, the plants need to be cut regularly to prevent flowering.

If your Chives do manage to elude your diligence and start to flower, you should harvest right away and remove the flowers as the plants are harvested (see below under 'Technique'). The following harvest should be taken as early as possible as there will be flower buds already formed at ground level which will continue to grow. Early harvesting will help revert the growth to leaf production

Tools

Use a reaping hook or hand shears to harvest Chives.

Technique

Cut the leaves about 25 mm above the ground. Any leaves that have fallen over, and ones that are discoloured or have soil adhering to them, should be left behind.

As Chives must be chopped into short sections before drying, they need to be harvested in a manner that will keep them all lined up together to facilitate chopping. This can best be done by carefully grasping each bunch as it is cut and laying them down, all in the same direction, in a basket or a box.

If there are long flowers among the Chives, they can be removed by holding the bunches fairly low down and pulling the flowers out by the heads. Any shorter flowers and flower buds can be separated by then grasping the bunch close to the top and shaking it, so the shorter flower buds fall away from the loose end.

Post-harvest Treatment

After harvest, go over the plants and trim off any leaves or flowers still attached to the plants

Drying

If you just spread whole leaves of Chives out to dry, the result will be disastrous, as they will turn yellow and brown before they finally dry. This is why it is commonly stated in books that Chives cannot be dried. This problem can be overcome by chopping the Chives into short lengths before they are spread on the screens to dry in good conditions.

The optimum length for drying is 6 mm. Some care needs to be exercised in the chopping operation to ensure there are no pieces more than 20 mm long, as these will not dry fast enough and will end up looking very faded, and will downgrade the rest of your product.

A good chaff-cutter can be used to chop the green Chives before drying, or else chop them by hand with a big sharp knife and a board. If the harvesting and handling of the Chives has been done carefully with all the leaves aligned, the chopping operation will be much easier and the problem of longer pieces will be avoided, as these are usually caused by leaves approaching the blade sideways.

Temperature

While Chives should be dried at temperatures not higher than 35°C, it is important they be dried fairly quickly to prevent deterioration. Ambient air drying is possible through much of the season, provided the Chives are placed in the best positions high on the drying racks. Otherwise use artificial heat.

Make sure the Chives are properly dry before taking them off the screens; they should feel quite crisp. As they will readily reabsorb moisture, check them carefully.

Processing

If the Chives were cut evenly, then once they are dry no further processing is required for a culinary grade. However, if any long pieces are present, these can be removed by rubbing the dried herb through a 2½ or 3 dent screen. This will improve the final product.

Storage

Chives should be freeze treated before sale or storage to control moth larvae. Another thing to be aware of is that they fade very easily if exposed to light in storage or at point of sale. They must be the fastest fading herb in the range, so if care is not taken they will soon look like a bag of shredded white paper. Sunlight and fluorescent lighting are the worst offenders.

Yields

As I have only produced this herb on a trial basis, and that was many years ago, no figures are available for yields.

Marketing

There is a rather limited market for dried Chives as a culinary herb.

Price

Price to growers for premium quality, organically grown Chives is currently around $24/kg for 5-star culinary grade. Trade price is around $12–14/kg.

Echinacea, Narrow-leaf

Echinacea angustifolia (DC) Heller

ASTERACEAE

Parts Used: Root, Seed

Other Names: Coneflower, Purple Coneflower, Black Sampson, *Brauneria angustifolia* (*E. angustifolia* is sometimes mistakenly called *E. augustifolia*).

From being a rather obscure plant once used by American Indians for snakebite, Echinacea has rocketed into prominence in recent times by virtue of its stimulating effect on the immune system. As such, it offers an alternative to the often harmful administration of antibiotics.

Echinacea is pronounced 'eck-in-ay-see-ah', with the accent on the 'ay' in the middle of the word.

There are a number of species of Echinacea but Narrow-leaf Echinacea (*E. angustifolia*) is the one that features in most literature on herbal medicine and is preferred by many herbalists as it is considered to be the most potent.

Another widely used species is Broad-leaf Echinacea (*E. purpurea*). Although it does not feature widely in the literature, Broad-leaf Echinacea is well regarded among herbalists and, because it can be easily cultivated on a large scale, it is cheaper to produce. Being somewhat different in its growth habit and management, it is covered separately under 'Annual, Biennial and Short-lived Perennial Herbs'.

Narrow-leaf Echinacea (*E. angustifolia*) is mainly gathered from the wild in the USA, but unfortunately increased use is depleting wild stocks of this species, which has not been brought into cultivation to any great extent. Consequently, the price is increasing significantly and there are common occurrences of adulteration and substitution with the roots of other plants.

Origin

Narrow-leaf Echinacea is native to mid-western and south-western USA.

GROWING NARROW-LEAF ECHINACEA

There is something very inspiring about a bed of Echinacea. A strong energy seems to emanate from the plants and the flowers are quite striking.

While the high prices offered for Narrow-leaf Echinacea make production of it sound lucrative, after growing it for

Narrow-leaf Echinacea (*E. angustifolia*).

several years, I still had not found the key to making it as profitable as many of the other herbs I was producing. The main problems were its slow growth, its low yields, the susceptibility of its roots to rotting, and the labour involved in digging it.

The form of this species, with its deep roots and leathery leaves, is an adaptation to its natural habitat of semi-arid prairie and open woodland. It just doesn't seem to respond much to the increase in moisture and nutrients provided under cultivation.

This seems to be the case overseas as well: in the USA efforts to bring it into cultivation do not seem to have resulted in much being produced.

Quite a few trial plots are being established around the Australia, so perhaps we will have some answers to these problems in a few years.

It may take a period of selective breeding to develop varieties of this species of *Echinacea* which will be sufficiently productive. This selection would have to be made bearing in mind the need to retain the same therapeutic qualities in more vigorous plants. As the actions of the herb are not fully understood, this will need to be done in some way other than just by simply measuring the 'active constituents'.

Climate

Narrow-leaf Echinacea is very much at home in hot dry summers. It will grow in climates with a cool summer, but more slowly.

In its native climate, the winters are very cold, with the ground freezing to some depth but summers are quite hot and dry.

Site

Exposure
Full sun and exposure seem to be preferred.

Drainage
This species needs extremely good drainage: it cannot tolerate periods of excess water, which are likely to kill off the roots.

Soil Type
Lighter soils seem to be preferred. As Narrow-leaf Echinacea roots go down very deep, the soil needs to be deep with no clay horizon. Excessive stone can make digging very difficult. In its native environment it grows in soils with a relatively high pH.

Fertility
Narrow-leaf Echinacea needs a moderate level of fertility.

Initial Weed Control
Because of its slow open growth, good weed control is important. To facilitate this, a thorough initial bare fallow is important.

Neighbouring Plants
Narrow-leaf Echinacea will not hold its own very strongly against aggressive and invasive plants.

If snails are likely to be a problem, care must be taken to locate Narrow-leaf Echinacea in a part of the garden where there won't be any cover for snails over winter, as the young shoots in spring are very vulnerable to attack by snails.

Identification

Narrow-leaf Echinacea (*Echinacea angustifolia*) grows to about 500 mm tall, with leathery leaves covered with short sparse hairs, 70–200 mm long, more or less lanceolate, but usually with blunt tips (the leaves are similar in shape and texture to Blackwood leaves). The flower-head is single on the end of a solid stalk. Petals are around 30 mm long, purple, pink or rarely white, and held more or less at right angles to the stem (see colour plate 3.3). The centre of the flower is a raised rounded cone, usually purple and quite tough and bristly, even spiny (hence the name Echinacea, which is from the Greek for hedgehog).

E. angustifolia usually has several main roots, which go straight down to a depth of 600 mm or more, with very few fibrous roots. The roots snap easily, revealing a dark grey centre.

Echinacea pallida is a similar species, but perhaps a little taller – up to about 800 mm – with longer and narrower leaves, which are a little darker green. Its flower is usually white or pink, with long drooping petals, which give it an almost melancholic appearance. Its root system is similar to *E. angustifolia*.

E. pallida is regarded by some authorities as being an acceptable alternate species, but others regard it as inferior. It is lacking in iso-butylamide, one of the main active constituents present in other Echinacea species, so it does not give the same tingling sensation to the tongue.

The most common species you will have to distinguish it from is *Echinacea purpurea*. Quite often seed sold as *E. angustifolia* turns out to be *E. purpurea*, which can easily be distinguished by the nature of the leaves. *E. purpurea* has softer, broader leaves which taper down to a distinct point.

It has one central vein with other veins branching from it, while the other two species have several parallel veins running the length of the leaf.

E. purpurea is also taller – up to 1000 mm – with more abundant flowers and has a shallow fibrous root system.

Distinguishing the three *Echinacea* species is not very difficult, but unfortunately they tend to hybridise, so if you are growing the plants near each other, you may find a few coming up that appear intermediate in their characteristics.

There may be some potential in this direction for breeding more vigorous growth into *E. angustifolia* by crossing it with *E. purpurea* and selecting from the progeny.

Assessing Potency

A crude taste test for the potency of Echinacea is to chew on a piece of the plant. One active constituent, iso-butylamide, has a strange effect on the tongue, giving it a tingling buzzing sensation. A potent dose can get to the point of being almost intolerable and bringing tears to your eyes.

This test can be used to roughly assess the potency of Echinacea, be it part of the fresh plant, the dried product or the liquid medicine.

Variety

More work needs to be done in finding or breeding a more vigorous variety, which is higher yielding and better suited to cultivation.

Propagation

Method

Propagation can be by seed, by root cuttings or by crown division. Pieces of root will send up new shoots, which form new crowns, and crowns can be divided.

In my experience, vegetative propagation is often erratic, as the cuttings and divisions are very susceptible to rotting.

Growing *Echinacea angustifolia* from seed can be quite successful, provided the seeds are given a period of cold stratification first, to break their dormancy. This can be done by holding the seeds in a moist cold situation for about a month (see below under 'Planting').

Timing

Root cuttings and crown division can be carried out in autumn, winter or spring. Material cannot be held for any length of time as it is liable to rot as a consequence. One area of experimentation which might be worth while would be to do trials involving planting root cuttings and crown divisions to establish whether there are times of the year or certain conditions that give a better success rate.

Seed needs to be started so germination occurs in early spring. As the early growth is quite slow, the seedlings need plenty of time to develop to a reasonable size by the end of the season.

Planting

Set the root cuttings in the ground with about 50 mm of soil over them. Crowns should be set in place with the crown at the surface.

Seed can be stratified by mixing it with moist peat-moss or sand and putting it in a plastic bag in the fridge where it is just above freezing for about a month.

Alternatively, seed can be planted in a seed flat or box in autumn with a normal covering of soil and left outside in the open, in a place where it will be kept moist and subject to cold, preferably frosts.

The seedlings can be transplanted into the field when they have reached sufficient size, around 25–50 mm. Don't let them get too big before transplanting, as they quickly develop their deep root systems.

Layout and Spacing

The plants can be set on 150–300 mm centres, in rows around 600 mm apart. Single rows are okay but there needs to be enough room to get in to dig up the roots at harvest, as they go down to 600 mm or more.

This is one species that would probably benefit greatly from being grown in high raised beds, as it is drought tolerant and needs very good drainage. This would be to supplement existing good drainage, not to substitute for it. Raised beds would also facilitate harvesting but they could interfere with weed control.

Compost

Composting of Echinacea is another area where there is room for more knowledge. I just used to give it a moderate dressing of compost, about 2 kg/m^2 each spring. It did not seem to respond to heavier applications.

Weed Control

Weed control is important in the cultivated situation, as such a slow-growing species can very easily be overtaken by weeds and smothered. Its foliage is relatively sparse, so it doesn't compete strongly with weeds. Regular inter-row cultivation is necessary, followed up by hoeing around the plants.

Irrigation

As my Echinacea beds have always been in the midst of all the other herbs, they have received the same amount of water as everything else. Consequently, I have never had the opportunity to determine what its actual requirements are. It may well have been receiving a harmful excess of water, contributing to its root rot problem.

It would be interesting to set up a trial where it was grown without irrigation to see how it responded.

Certainly autumn irrigation should be restricted, if possible. It is best for Narrow-leaf Echinacea not to go into winter with the soil saturated, as this can contribute to the roots rotting in wet winters.

Pests and Diseases

Being plants of dry climates, the problems likely to be encountered are those associated with excessive moisture. Grasshoppers don't seem to be interested in Echinacea, apart from chewing a few petals.

Root Rot

After all the effort of looking after the plants for 2 or 3 years, it is frustrating to find a large proportion of them have rotted in the ground. The rot usually goes right down the roots, often wiping out the whole root system, though sometimes only part of the root system is affected.

Root rot seemed to strike worst in sections of the garden that suffered from periods of excess moisture, though this situation never lasted for very long as the drainage was reasonably good. There would be periods of heavy rain over several days when a temporary excess would occur, so evidently Narrow-leaf Echinacea is very sensitive to any waterlogging.

Snails

Snails can be a problem in spring when the leaves are just emerging from the crowns. The snails will attack them every night, setting the Echinacea back severely and sometimes eventually killing it.

The best solutions are to keep the snails away from the Echinacea by good quarantine, duck patrols, and/or by not providing cover nearby for them to hide in during the day when birds are around. This can be managed by careful choice of neighbouring plants.

Alternative Management Strategies

As essentially a plant of dry habitats, it may be possible to grow Narrow-leaf Echinacea in a semi-wild situation and reduce management costs.

With its deep roots, it should be more tolerant of dry conditions than many of our weeds and pasture plants. It may be possible to grow it with no irrigation or with only occasional watering over summer. And it might be possible to control the weeds by grazing with animals.

This would depend on how palatable Narrow-leaf Echinacea is to stock, which would have to be determined by means of trials and observation. If it worked, it would reduce the cost of maintaining the Echinacea over the several years required to grow it large enough to harvest.

Harvesting Seed

There is sometimes a good market for Echinacea seed, which is the most potent part of the plant. For more detail see under 'Broad-leaf Echinacea'.

Harvesting Roots

Timing

The roots can be dug when they have reached sufficient size, usually after at least 2 or 3 years. Dig in late autumn, after the leaves have died down and before the ground gets too wet.

Tools

A good spade is necessary, preferably a long narrow one like those used for digging deep drains. A narrow pointy trowel is very helpful, too, and secateurs or reaping hook are needed for trimming off the dead tops.

Technique

The aim is to follow the roots down as far as possible, digging them out with the spade and then using the trowel to extract the last part. They break off very easily, so some is usually left behind: it is a matter of judging how deep it is worth going, as the roots taper off as you go deeper.

Digging is a slow job because of the depth the roots go, and being quite sparse there is a lot of digging to be done for not very much root.

The root of Narrow-leaf Echinacea is normally grey or grey-brown. During harvesting you need be on the lookout for sections of root that are rotting. These can range from being just a bit off-colour to being very soft and slimy. Smell is usually the best indication.

The thin and brittle nature of the roots means some care needs to be taken in digging them out. It is easy to lose pieces of root that break off in the soil.

Mechanical Harvesting

Mechanised harvesting of Narrow-leaf Echinacea could be difficult. The depth of the roots would make it beyond the scope of conventional machinery, and the thin fragile nature of the roots could be a challenge for a mechanical harvesting system: a large proportion of the roots would be broken into small pieces and easily lost.

It might be feasible to simply take a partial harvest, skimming off the top 300 mm or so of roots and leaving the crop to regenerate from the deeper roots left behind.

Post-harvest Treatment

Often Narrow-leaf Echinacea will regrow from pieces of root left in the ground during harvesting, so if the bed is raked level and maintained free of weeds, you may get a second crop coming up in the same place.

Drying

Echinacea is relatively easy to dry. Any pieces thicker than your thumb should be split or coarsely chopped to facilitate drying. Avoid fine chopping of the root before drying as this can result in a loss of active constituents in Echinacea.

Temperature

Echinacea should be dried at temperatures of not more than 35°C as it contains volatile oils.

Processing

Echinacea root is used to some extent as a tea, but mostly in manufacturing liquid medicines.

Tea Grade

Tea grade should be milled into pieces up to 6 mm across, but usually this does not have to be done by the grower.

Manufacturing Grade

For manufacturing grade, the roots can be sent off in the form that they come out of the dryer.

Storage

Freeze treatment is necessary to control moth larvae before sale or storage.

Yields

At 'Twin Creeks' our best yields after 2 years growth were 0.2 kg/m^2 (or 0.1 kg/m^2 per annum, equivalent to 1000 kg/ha per annum). Often the yields were much lower than this.

Marketing

There is a big demand for Echinacea Root from *E. angustifolia* and the price has been increasing in recent years. This high price is causing a move to use Broad-leaf Echinacea as a cheaper alternative, which could affect demand.

Price

Price to growers for manufacturing-grade, organically-grown Narrow-leaf Echinacea Root is currently $60–80/kg.

Elecampane
Inula helenium L
ASTERACEAE

Parts Used: Root
Other Names: Scabwort, Elf Dock.

Elecampane has a long tradition of use as a respiratory herb and also as a culinary, though in recent times its use in sweets has ceased. It is still used in alcoholic beverages in some countries.

GROWING ELECAMPANE

Elecampane is a striking plant: with its very large leaves and bright yellow flowers, it has some resemblance to the Sunflower, though the flower is smaller.

This is an easy plant to grow and harvest. Its substantial root sends up a strong crown of leaves and a thick flowering stalk.

Elecampane (*Inula helenium*).

Being a root that is harvested each year, it needs to be replanted, preferably in a new site each season. Elecampane is a good crop to follow herbs that leave a weed legacy for subsequent crops, as it is easy to keep cultivated.

It will grow in a wide range of climates, though it becomes a little stressed in very hot dry conditions and needs some protection from wind.

Its management and harvesting is very similar to Marshmallow.

Identification

Elecampane grows up to 2 m tall. Leaves are light green, alternate, elliptical, up to 450 mm long and 150 mm wide, lightly hairy, and toothed. The thick hairy stems bear flower-heads up to 70 mm across, single or grouped, with numerous long yellow petals.

The plant dies down to the ground in autumn. The root is quite large, the thick branches extending up to 200 mm, mostly in lateral directions.

For more detail on this herb see 'Information Charts'.

Fennel

Foeniculum vulgare Mill.   APIACEAE

Parts Used: Seed

Other Name: *Foeniculum officinale.*

Fennel is used as a culinary herb for flavouring in foods cordials and beverages and is also used medicinally, mainly as a digestive and respiratory herb.

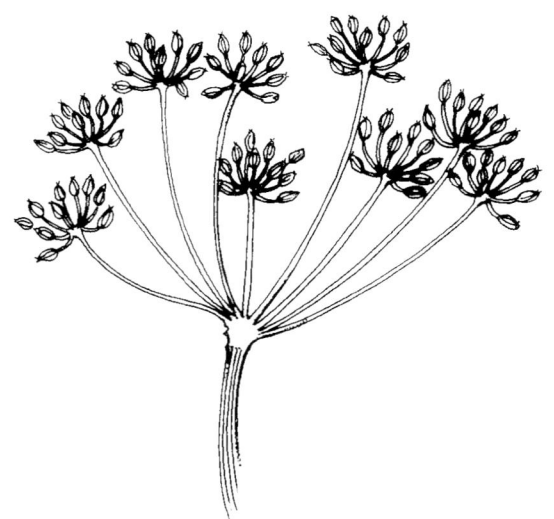

Fennel (Foeniculum vulgare).

It is grown as an essential oil crop in Tasmania and grows wild in much of southern Australia. Fennel is a declared noxious weed in some states.

GROWING FENNEL

While seed can be harvested from wild plants, it is not always easy to find suitable locations. It could be considered by growers as a crop if wildcrafted supplies prove inadequate, or if a certified organic or biodynamic supply is required.

The crop can be grown from seed and a good flavoured variety with a high essential oil content should be obtained. Seed harvesting and cleaning equipment would be needed if any significant scale of production was envisaged.

Fennel needs a long enough growing season for seed to ripen while weather conditions are still suitable for harvesting. The plant usually doesn't produce seed until its second year. Fennel is a fairly long-lived perennial and judging by the wild stands, weed control, compost and irrigation requirements would be minimal in areas of moderate rainfall and fertility.

Containment

To prevent Fennel spreading, all seed should be harvested and the crop should be surrounded by a mowed or grazed buffer zone.

For more detail on this herb see 'Information Charts' and the entry for Fennel in Chapter 21, 'Wildcrafting and Weed Harvesting'.

Figwort

Scrophularia nodosa L
SCROPHULARIACEAE

Parts Used: Leaf & Flower, Aerial Parts

Other Name: Knotted Figwort.

Figwort is what we might refer to as a minor herb, but finds occasional use in the treatment of skin problems and as a diuretic. It was also traditionally used for haemorrhoids.

GROWING FIGWORT

Figwort is an easy herb to grow. It is an interesting plant, with its knobbly rootstock and delicate subtle flowers enjoyed by bees. Possums are said to be repelled by its rather rank-smelling foliage.

It seems to prefer a climate with a mild summer, but it will grow in hotter situations with a bit of protection from

Figwort *(Scrophularia nodosa)*.

extreme heat. Weed control, harvesting and drying are all very straightforward.

Identification

Figwort has square stems, growing 400–1200 mm tall, with strong smelling leaves which are opposite, ovate, up to 80 mm long, and dark green. Flowers are greenish-brown, up to 10 mm long, and in panicles at the top of the stem. Figwort has characteristic underground tuberous swellings at the base of the crown and dies down to ground level in autumn.

For more detail on this herb see 'Information Charts'.

Garlic

Allium sativum L   LILIACEAE

Parts Used: Corm (i.e. bulb)

Garlic has been widely used since ancient times, both for culinary and medicinal purposes. For the latter it is valued for treating infections and for respiratory conditions.

Origin

Garlic is native to Asia.

GROWING GARLIC

Garlic is one of the few herbs grown as a winter crop, normally being planted in autumn and dug in late spring or early summer.

Growing Garlic is not entirely compatible with dried herb production. I did produce them together for a number of seasons, but the Garlic needs its own separate space and racks while it is curing inside, as it sheds a lot of soil, skins and debris, which float around and get into other things.

Climate

Garlic will grow in a range of climates, but the quality will be better in regions that dry off in mid-to-late spring, as cleaner looking Garlic will be obtained and curing is easier.

Site

Garlic likes full sun and is not bothered by wind at all.

Drainage

Drainage needs to be very good, as Garlic will not tolerate wet feet.

Soil Type

Light to medium loams are preferred: heavier soils are not suitable. Red soils can stain the outside skin of the corms.

Fertility

A reasonable level of fertility is needed, but high levels of nitrogen should be avoided. A pH of at least 6.0 is preferred.

Garlic *(Allium sativum)*.

Initial Weed Control

Initial bare fallow is recommended if weeds are likely to be a problem.

Neighbouring Plants

Because the soil needs to dry out as the crop approaches maturity, Garlic should to be placed in a situation where this is going to be possible without interfering with the irrigation of neighbouring crops at a critical time.

Identification

Garlic is a perennial, growing from a bulb that sends up several leaves, which are 10–25 mm wide, up to 300 mm long, blue-green, and strongly aromatic. Flowers are borne at the top of a round unbranched central stem, in a dense spherical umbel, pink to green, sometimes displaced by small bulbils. Some varieties do not flower.

The bulb or corm is divided into a number of individual cloves. The plants normally die down in summer.

Variety

There is a great deal of variation in Garlic, so it is important to select a strong-flavoured variety suited to your climate, time of planting and market requirement.

Varieties with large cloves are usually preferred but be careful about Russian or Elephant Garlic, which has very large cloves, but only a mild flavour (there may be a limited specific market requirement for this variety, though).

Propagation

Method

Garlic is normally propagated by crown division. The corms are broken apart into individual cloves, which are then planted.

Timing

For most varieties mid-autumn is the optimum time. Planting too early can result in the corms developing side shoots, which don't look very attractive and don't keep as well. Late planting usually results in lower yields.

At least one variety can be planted in spring in cooler climates, which enables a later harvest that can still be in saleable condition for 2–3 months after other varieties. Most varieties, however, will yield very poorly if planted in spring.

Planting

If possible, start off with stock from a local organic grower, but if you have to buy it elsewhere, make sure it is locally grown. Imported Garlic is often irradiated or fumigated and will not sprout.

Plant the cloves individually. Only the larger cloves should be planted as these yield larger corms when mature. It is false economy to save your smaller corms for seed as these will have smaller cloves.

The cloves should be planted right way up, with about 25 mm of soil over them. If they are planted too shallowly, they can lift themselves out of the ground when the roots start to shoot.

In warmer regions, a period of chilling below 10°C is recommended prior to planting to encourage better development.

Layout and Spacing

Garlic should be set out on 100–150 mm centres in rows 300–450 mm apart.

Compost

Depending on soil fertility, Garlic needs only a moderate dressing of compost at the rate of 2–3 kg/m^2. If it is following a crop where a high rate of compost was used, it may not need any.

Good corm development depends on availability of potassium and phosphate, and these could be augmented with commercial organic mineral fertilisers or wood ash, among other things.

Too much nitrogen causes side-shoots to develop.

Weed Control

Good weed control is critical for success with Garlic as it does not compete well with weeds and yields will be badly affected by any competition. Regular inter-row cultivation and hoeing around the plants is needed right through the growth of the crop.

Managing to kill weeds in winter can be difficult. One technique I found helps is to rake them into little piles between the rows. This way the top ones will smother the bottom ones and the piles can be turned over after a few weeks to kill the top ones.

Rotation

Garlic should be grown in rotation with other crops. It needs to follow a crop that does not leave a legacy of weeds and preferably one where a heavy application of compost has been used, as this will give the Garlic a better balance of nutrients without an excess of nitrogen.

If weed control has been well maintained, Garlic leaves the ground fairly clean for the next crop.

Irrigation

Often Garlic can be grown without irrigation, but it may need a little water when the corms are filling out in spring if conditions are particularly dry.

Once the corms have filled out and the outer leaves start to dry off, the soil should be allowed to dry out as this prevents staining and helps prevent side-shoots developing.

Pests and Diseases

Garlic is relatively free of pests and diseases, but a few problems can arise which may need attention.

Plants Dying off in Spring

Plants dying off in spring is usually the result of waterlogging over winter. The plants may appear to be all right for a period, but then they start to yellow and die off prematurely as the roots and corm rot.

Glassy Cloves

One problem that had me puzzled for a while was that sometimes the insides of the cloves would turn glassy and flavourless, and later dry out or go mouldy. The damage tended to be on one side of the corms and finally I figured out that it was sunburn caused by leaving the Garlic to cure in the open in hot weather.

The problem can be alleviated by curing the corms under shadecloth or under their own leaves (see below).

Harvesting

Timing

Garlic should be pulled when the tops and the outer layers of skin on the corms start to die off. In dry conditions there is some flexibility, but if the soil is still moist, the plants should be pulled a little on the early side to reduce staining of the skins.

Tools

In light soils it may be possible to pull the Garlic by hand, but usually a vine hoe or a fork is needed to loosen the roots first. On a larger scale, a tractor-drawn implement could be used to loosen them or run a blade under them.

Technique

Once loosened, the Garlic can be pulled up by the leaves. Several can be held in a bunch and the soil shaken out of the roots by tapping them against your boot. They can then be laid on the ground to cure for a while in the sun and the wind.

Curing

In order for Garlic to keep and present well, it needs to be cured. This involves drying all the outer skins of the corms, and needs to progress reasonably quickly to prevent staining and mould.

The initial week or so of curing is usually best carried out in the field. If conditions are hot, then the corms need to be protected from sunburn, either with shadecloth or by laying each bunch down so its leaves cover the corms of the one before.

Alternatively, Garlic can be fully cured on racks inside, but this requires spreading it quite thinly to ensure it doesn't go mouldy.

Normally the Garlic is brought inside after a few days to a week when the leaves and outer layers of skin have dried off somewhat. It is then dried on racks or screens: chicken-wire screens are probably the most suitable but it can be dried on shade cloth screens if they are strong enough. Good air circulation is important.

In Tasmania it usually takes about a month to fully cure Garlic, by which time it is ready to trim and store.

Trimming

The leaves and roots need to be clipped, unless the Garlic is to be plaited, in which case only the roots should be clipped.

Various knives and clippers can be used for the job but my preference is for a pair of dagging shears. Whatever you use must be able to keep a reasonably good cutting edge and will probably require frequent sharpening.

The roots can be trimmed in the field and this job is easier if it is done when the Garlic is freshly pulled, before the roots dry and get tough. This also reduces the amount of soil you bring into your drying facility.

The tops are best left for a few weeks to allow the sap to move down into to the corms, but they don't have to be totally dry before trimming them off.

Cut the stems off about 10 mm or so above the corm. If the stems are still moist inside, the corms should be further cured until all the inner layers of skin are dry: check a few by breaking them open.

Cleaning

Once the corms are fully dry, they are easily cleaned by tumbling them around on a chicken-wire screen when conditions are dry and the outer skins will flake away. It is important to have the Garlic looking bright and clean: then it will bring a better price and sell well.

Storage

Whole Garlic corms can be stored safely for some period under the right conditions. It needs to be dry with sufficient air circulation. While low humidity prevails during the summer and early autumn, Garlic will keep well, but as the humidity rises problems occur.

With the rise in humidity, the cloves will begin to sprout. Once the skin of the cloves is opened by the emerging roots, the clove is vulnerable to desiccation and attack by mould.

In order to prevent this, Garlic should be stored in a location where the temperature is around 20°C, the humidity kept low, and there is good air circulation around the corms.

I found that on an open storage shelf up near the ceiling in a warm part of the house I was able to safely keep the Garlic for 2–3 months longer than if it was stored outside. It is important for Garlic to be moved into dry storage before prevailing outside humidity causes the root buds to start swelling.

This storage system enabled me to supply retailers for 6 months or more each year.

There are sophisticated cool storage regimes used for Garlic, but these may be beyond the scope of the small grower.

Yields

My yields in Tasmania were around 1 kg/m² fresh Garlic, or 10 t/ha.

Producing Dried Garlic

There is a significant market for good quality dried Garlic, but this is an area I have never explored. It would be a specialised operation involving skinning and fine chopping of the Garlic before drying under controlled conditions.

With standard shadecloth screens, the sticky, pungent nature of chopped Garlic could give rise to contamination problems for other crops dried on them afterwards.

Temperature

Garlic should be dried at a temperature of not more than 35°C.

Processing

Dried Garlic is marketed in flake, granule or powder form, predominantly for culinary use.

Marketing

Fresh Garlic is usually marketed through fruit and vegetable outlets, though many health food stores will carry it, too. There is a small demand for fresh Garlic for manufacturing into liquid medicines. Plaited Garlic makes an attractive presentation and brings a higher price.

Price

Fresh organic Garlic is normally around $8/kg, but varies in price according to time of year, available supply and the person to whom you are selling.

Organically grown dried Garlic granules should bring around $20–25/kg.

Ginseng
Panax spp.
Panax quinquefolius L: **American Ginseng**
Panax pseudoginseng Wallich: **Korean Ginseng**
ARALIACEAE

Parts Used: Root

Other Names: *Panax ginseng, P. schinseng* (both for *P. pseudoginseng*).

Ginseng has been attracting a lot of attention among would-be growers because of the high monetary value placed upon it. It is used for a number of stress-related conditions, as it has adaptogenic properties, which help increase physical and mental endurance and efficiency.

American Ginseng is shown on colour plate 3.6.

The two species are very similar, with similar properties.

Origin

American Ginseng is a native of north-eastern North America. Korean Ginseng is a native of Korea and Manchuria.

GROWING GINSENG

Curiously the growing of Ginseng has developed its own culture somewhat separate from other herbs, so Ginseng growers tend not to identify themselves as herb growers.

While there has been enormous interest in growing this herb in many areas around the world and some have been successful, there has also been a high failure rate, especially among first-time growers.

The main problems confronting production are Ginseng's need for a very precise growing environment that imitates its native habitat. In the wild, it is a forest floor plant, growing under a dense canopy of deciduous broadleaf trees, which provide around 80% dappled shade.

Ginseng is a very delicate plant and does not recover easily from setbacks. If it loses its leaves during the season, it will not regenerate until the next spring and only then if it has enough reserves.

Three different systems have been developed for its production:

Artificial Shade: This is an intensive system, which is expensive to set up and relies on shadecloth over the crop. Growing the Ginseng plants in close spacings in this artificial environment results in a heavy dependence on chemical inputs, particularly pesticides and fungicides.

This high input system is only viable on a scale of at least 0.8 ha, while a crop is produced in 3–4 years and yields are high. The price obtained for the dried root is lower and the overheads are high.

Wild Simulated: This approach to growing Ginseng is based on simulating its natural environment. The crop is grown under the shade of trees. Some cultivation, weed control and crop protection is undertaken, but overheads are lower than for artificial shading.

This system lends itself to organic methods. The slower growth produces a crop after 5–7 years, with much lower yields, but of a quality much sought after, so prices are much higher.

Korean Ginseng (*Panax pseudoginseng*).

Woods Grown: This involves planting Ginseng in its natural habitat (or what appears to be its natural habitat) in the forest. The plants are left to grow on their own and are harvested 6–12 years later. Where this is successful, it produces very low yields but of a very high quality, which bring the highest price. As there is little input in maintaining the crop, the returns can be quite good, if the system works.

Identification

American Ginseng (*Panax quinquefolius*) grows to 600 mm tall, the single stem bearing a whorl of 3- or 5-palmate leaves. Leaflets are obovate, 80–130 mm long. Flowers are pink, in a single terminal umbel, forming a cluster of red berries.

Korean Ginseng (*Panax pseudoginseng*) grows to 750 mm tall, the single stem with a whorl of 3- or 5-palmate leaves. Leaflets are lanceolate, finely toothed, and up to 320mm long. Flowers are greenish, in several terminal umbels, forming bright red berries.

Growing in Australia

While these systems work in North America and other northern temperate regions, Australian growing conditions are different. Some trials have been successful in producing a good quality wild-simulated Ginseng in Victoria, growing them under Mountain Ash (*Eucalyptus regnans*).

It was found additional shade was needed as eucalypt leaves do not cast the same shade as broadleaf deciduous trees. How successful this system will be in different parts of Australia is not yet clear, but it is possible Ginseng could be grown as a crop in the cooler highland parts of the mainland and in much of Tasmania.

Ginseng is very particular in its soil requirements, especially with regard to drainage, which must be very good. Many of its needs are poorly understood, so success is a bit of a hit and miss affair, even in its native habitat.

What pests and diseases will affect it here is not yet evident. If possums take a liking to it, it may be difficult to economically exclude them from a plantation under trees, particularly in Tasmania where their pressure can be very high.

As there is a significant cost in getting established with this crop and a big risk of failure, growers should evaluate the crop carefully, get good information on its production, and start out on a trial basis to assess the crop's potential in their situation.

For more detail on these herbs see 'Information Charts' or contact Gembrook Organic Ginseng, PO Box 44, Gembrook VIC 3783, telephone/fax (03) 5968 1321.

Goat's Rue
Galega officinalis L FABACEAE

Parts Used: Leaf & Flower, Aerial Parts

Other Name: French Lilac.

Goat's Rue has a reputation of promoting milk flow in animals and humans, and this probably has something to do with the origin of its name. It is also used as a vegetable rennet (the fresh juice) and has the ability to reduce blood sugar levels.

It is sometimes grown as an attractive ornamental.

GROWING GOAT'S RUE

Once established, this herb produces bountifully for a minimal input on the part of the grower. It adds a softness to the garden with its light green pinnate foliage and its gentle pink flowers.

While it will grow in a range of climates, it prefers a mild climate and withstands exposure to winds.

Its growth is fairly prolific, so it could overshadow smaller plants growing beside it. It is a good plant to put in next to other herbs that are a bit aggressive.

Goat's Rue needs good drainage and will grow in a range of soil types. It needs a moderate level of fertility for good growth.

Goat's Rue (*Galega officinalis*).

Identification

Goat's Rue is a bushy plant growing up to 1500 mm with hollow stems. The leaves are pinnate, with 11–17 leaflets, oblong or oblong-ovate, 10–40 mm long. The flowers are light pink (some varieties white, purple or lilac), 10 mm long and in terminal racemes.

The roots have a characteristic milky whiteness to them, and the plants die back almost to ground level in winter, but usually there is some low green growth retained.

Variety

There are cultivars selected for their flower colours.

Weed Control

This is fairly straightforward for Goat's Rue as it forms a solid clump and grows over most weeds, though some inter-row cultivation and hoeing around the plants will be necessary.

Pests and Diseases

At 'Twin Creeks' in north-east Victoria, Goat's Rue sometimes became variegated and lost its vigour. As this affected some plants but not others, I think a virus may have been responsible, perhaps as a result of heat stress.

Harvesting

Two to three harvests can be taken in a season, usually mid-December or so for the first, and then one around February and possibly a third in April or May.

Marketing

As the market for Goat's Rue is only very small, growers should ensure they have a firm commitment from purchasers before producing any quantity.

For more detail on this herb see 'Information Charts'.

Horehound, Black
Ballota nigra L LAMIACEAE

Parts Used: **Leaf & Flower, Aerial Parts**

Other Name: Stinking Horehound.

Black Horehound would have to be one of the most unpleasant-flavoured herbs I know. Yet, incredibly, it has a reputation as an antiemetic, particularly for morning sickness during pregnancy.

It is hard to imagine someone feeling nauseous being able to drink a cup of Black Horehound tea. Perhaps for this reason it is mostly used in blends or as a liquid extract.

GROWING BLACK HOREHOUND

While Black Horehound is hardly a plant to get wildly excited about, it has its qualities and is not hard to grow. One's impression of the plant is transformed when the little pink flowers emerge, as they seem to show that, despite its rough exterior, Black Horehound has a gentle inner nature.

While Black Horehound can be grown in a range of conditions, it reaches its full potential in climates with a mild summer, as in Tasmania.

Black Horehound is somewhat similar in appearance to White Horehound, to which it is only distantly related and it is not ever likely to become a noxious weed as that species has.

Identification

Black Horehound grows 400–1000 m tall, with hairy angular branching stems. Leaves are heart shaped, crenulate, 20-60 mm long, strong smelling and covered with a fine down.

The pink flowers are borne in whorls in the axils of the upper leaves and appear fairly late in the plant's development.

Black Horehound can be distinguished from White Horehound (*Marrubium vulgare*), by its darker appearance, its pink flowers, its very repulsive aroma and taste, and its lack of burrs.

White Horehound is lighter in general appearance and taller, more erect in its growth and has white flowers. While its leaves taste extremely bitter, it has a more 'noble' flavour and aroma. It forms burrs, which catch in the wool of sheep.

Variety

I was initially under the impression there were two different varieties of Black Horehound, a large vigorous growing one and a diminutive one. However, when I took divisions from my rampantly growing Black Horehound in Tasmania and planted them in north-east Victoria, they turned into the diminutive form, so the variation is evidently climatic.

Propagation

Propagate by crown division or seed.

Black Horehound forms very solid clumps that sometimes need the assistance of an axe to break them apart. While this breaks some of the crowns, as long as the pieces have some sections of crown that have both leaf and root attached, they will grow.

Set the pieces in the ground as deep as practical with the crown just below the surface.

For more detail on this herb see 'Information Charts'.

Black Horehound (*Ballota nigra*).

Lady's Mantle

Alchemilla vulgaris L ROSACEAE

Parts Used: Leaf & Flower, Aerial Parts

Other Names: Lion's Foot, *Alchemilla mollis*.

Lady's Mantle is used medicinally, mainly to help relieve the discomfort of menstruation and menopause. As a herbal beverage it has a pleasant mild flavour which can be enjoyed by male and female alike although it is currently not very widespread.

Alchemilla means 'little magical one', which is a rather apt description of this plant.

Diminutive though it is, it has subtle powers, like the ability of its leaves to collect and retain drops of water. Even in hot weather, these drops are very slow to evaporate and will last for days. This water was traditionally held to have special properties. People used to collect it to wash their faces with, and alchemists used it as an ingredient in their formulae for turning lead into gold.

GROWING LADY'S MANTLE

This little plant always holds a fascination for me, with its soft green foliage, unimposing yellow-green flowers, and the drops of water held in the centre of its leaves.

Some herb growers have found it difficult to grow Lady's Mantle, but where it is not subjected to heat stress and has a soil that suits it, it can be fairly productive.

It prefers a climate with mild summers. In regions with a hot summer, semi-shade is preferred, but it will grow well in full sun in cooler regions.

Identification

Lady's Mantle is a low-growing herb, up to 300 mm tall, with light green, round, ribbed leaves 30–80 mm across and with 7–11 lobes. The flowers are yellow-green, 3–5 mm across, in terminal panicles, which appear among the leaves (see colour plate 4.5).

The plant dies back to ground level in winter.

For more detail on this herb see 'Information Charts'.

Lady's Mantle (*Alchemilla vulgaris*).

Lemon Grass

Cymbopogon citratus (DC) Stapf
POACEAE

Parts Used: Leaf

Other Name: *Andropogon* sp.

Lemon grass is a very popular herb tea, ranking only behind Peppermint and Chamomile in usage. It is also very popular in Asian cuisine, but does not have any great claim to medicinal qualities.

Origin

Lemon Grass is a native of south-east Asia.

GROWING LEMON GRASS

Although the leaves of this tropical plant are frost sensitive, it can be successfully grown in many parts of southern Australia where the summers are hot and winter frosts are not too hard. Because the leaves give quite a bit of protection to the crown, it can withstand a reasonable frost and still recover.

Climate

Lemon grass needs a hot summer for good growth. While it may survive in cooler climates, it can never be a

Herbal Harvest

5.1   **Melissa Balm** *(Melissa officinalis)*

5.3   **Mountain Pepper** *(Tasmannia lanceolata)*

5.6   **Nettle, Greater** *(Urtica dioica)*

5.2   **Pasque Flower** *(Anemone pulsatilla)*

5.4   **Passionflower** *(Passiflora incarnata)*

5.5   **Peppermint** *(Mentha x piperita)*

Plate 5

Herbal Harvest

6.1 Raspberry *(Rubus idaeus)*

6.2 Red Clover *(Trifolium pratense)*

6.3 Rose, Dog *(Rosa canina)*

6.4 Rue *(Ruta graveolens)*

6.5 Rust on peppermint

6.6 Rose, Sweet Briar *(Rosa rubiginosa)*

6.7 Sage *(Salvia officinalis)*

Plate 6

commercial proposition in places like Tasmania where there is only about one hot day a year, as good growth depends on continuous hot weather.

Very heavy frosts can kill it, but it will survive lighter frosts.

Greenhouse production is a possibility in cooler regions, but is unlikely to be a financial proposition.

Site

Exposure
Full sun is essential for Lemon Grass, and it will tolerate some wind.

Drainage
Lemon Grass needs good drainage.

Soil Type
It will grow on a wide range of soils.

Fertility
It needs a fertile soil.

Initial Weed Control
A bare fallow will help make ongoing weed control easier, but as the plants grow quite strongly in solid clumps, weed control is not so critical.

Identification
Lemon Grass is a tall grass, up to 1000 mm. The foliage is blue-green and has a characteristic lemon aroma when crushed. The plant rarely flowers in southern Australia.

The best test for distinguishing it is to smell and taste the leaves.

Variety
There is some variability in essential oil content, so you need to start off with a good variety.

Propagation

Method
Crown division.

Timing
Plant Lemon Grass in spring after danger of frost is past. Early plantings could be protected with cloches.

Planting
Set the divisions in the ground with the crown just below the surface. Excessive leaf should be cut well back.

Layout and Spacing
Lemon grass should be set out on 300–400 mm centres in rows about 900 mm apart.

Compost
Lemon grass needs a good dressing of compost at the rate of 4–5 kg/m^2 (40–50 t/ha), worked in beside the plants in early spring.

Weed Control
Inter-row cultivation followed by hand-hoeing around the plants. Mulching is another possibility as the plants are cut well above ground level.

Irrigation
Lemon grass needs 25–50 mm of water per week through summer.

Pests and Diseases
Rabbits can be a problem, especially with young plants, so if they are around, the Lemon Grass will need to be protected.

Young Lemon Grass (*Cymbopogon citratus*).

Frost Protection

Provided the crown is well insulated by the leaves, Lemon Grass can withstand a fairly good frost. It will burn the leaves severely, but if they are thick enough, the crown will be protected and the plant will recover.

The critical thing is to ensure the plants get a good start in their first season, and then they will be big enough to survive the winter.

It is also important they go into winter with some surplus growth on them.

Avoid heavy mulching around the Lemon Grass in winter, as this keeps the soil's warmth away from the frost sensitive leaves: bare soil helps keep frost from settling, as it radiates some warmth to the plants.

If you want to give your plants added protection, you could consider covering them right over with something.

Harvesting

Timing

As Lemon Grass does not appear to flower, at least not in southern Australia, the plants are harvested when there is sufficient growth: around 400–500 mm. Depending on the weather, subsequent harvests can be taken after 4–6 weeks regrowth.

In northern Victoria, four harvests are possible in a good year, though in milder seasons only three are possible.

In areas subject to frost, the surplus growth left on plants to protect them through winter will discolour and will need to be trimmed off in spring once the danger of frost is past and before new growth commences.

Tools

The reaping hook is the best harvesting tool, as Lemon Grass is too tough to cut with the scythe and is harvested above ground level. Some growers cut it with a brush cutter.

Technique

Cut the Lemon Grass at the base of the new growth. Avoid including brown leaf, as this affects the quality and appearance of the final product.

As the leaves need to be chopped before drying, they should be kept lined up together during harvesting and handling. Placing them in boxes as they are cut can make this easier.

Post-harvest Treatment

Lemon Grass plants may need periodic trimming back to the base if they get too tall.

Drying

Tea Grade

If it is intended for tea use, Lemon Grass is best cut into pieces around 1 cm long with the chaff-cutter before drying. This ensures faster, more even, drying. Lemon Grass cuts much more cleanly in the fresh state and is not subject to bruising.

Culinary Grade

Lemon grass for culinary use should be cut into pieces 100–150 mm long. This is the way it is traditionally used in Asian cuisine, as it allows the Lemon Grass pieces to be retrieved prior to serving. Lemon Grass does not soften in cooking and short pieces can cause some embarrassment when they lodge in the gullets of dinner guests.

Assessment

When checking for dryness, look closely at the inside of the thicker pieces of stem. These can feel very dry on the outside, but still be quite moist inside.

Temperature

Lemon Grass should be dried at a temperature of up to 35°C.

Processing

No further processing is needed for Lemon Grass if it was chopped prior to drying, except perhaps to remove any long pieces.

Storage

Freeze treatment is necessary to control moth larvae before sale or storage.

Yields

In Northern Victoria, yields have been around 0.35 kg/m² tea grade per annum, but they would be higher in warmer regions with a longer growing season.

Marketing

There is a fair bit of Lemon Grass already in production in Australia, including one quite large certified organic operation in Queensland, so growers need to ascertain that there is a market for their product before going in at the deep end.

Price

Price to growers for premium quality, organically grown Lemon Grass is currently around $16/kg for 5-star tea grade.

Price to growers for second-grade, organically grown Lemon Grass is currently around $12/kg.

Trade price is around $8/kg.

Marshmallow

Althaea officinalis L MALVACEAE

Parts Used: Root, Leaf & Flower

Other Names: Sweet Weed, Althaea, Guimauve.

Although the marshmallows we commonly toast on the fire today have nothing to do with this herb, it was originally their main ingredient.

Marshmallow is a bountiful plant and has a long tradition of medicinal and culinary use. It was considered a delicious vegetable by the Romans and is still eaten today. It contains a lot of mucilage, which gives it demulcent properties, so it is used in treating inflammation of the digestive tract and also for respiratory conditions.

Origin

Marshmallow is native to Europe and Asia.

GROWING MARSHMALLOW

If you have the right conditions, Marshmallow is an easy herb to grow, with few problems.

Its grey-green foliage contrasts with light pink flowers, and its domination of weeds and ease of harvest give the grower a break from some of the more intensively managed herbs.

Climate

Marshmallow prefers a climate with a mild summer.

Site

Exposure

Marshmallow likes full sun or semi-shade, and will grow happily in a windy situation.

Drainage

It grows well in sites subject to waterlogging, but also does well with good drainage.

Soil Type

It grows in a wide range of soils.

Fertility

A fertile situation is needed.

Initial Weed Control

Marshmallow does not really need an initial bare fallow as the plant sends up such strong growth that it gets above most weeds. It is a good crop to follow something that leaves a lot of seed in the soil, such as Chamomile.

Neighbouring Plants

Although Marshmallow is quite tall, it does not impose on neighbouring plants much. It is a good crop to put beside invasive herbs.

If allowed to form seed in cooler climates, you may find it coming up by itself in the vicinity.

Rotation

It is handy to have a species like Marshmallow in production, because it can be used to follow crops that leave a legacy of weeds or self-sown plants.

Identification

Marshmallow is an erect perennial growing to 1250 mm. The stem and leaves are softly hairy. The leaves are grey-green, with 3–5 lobes or are undivided, broadly ovate, with toothed margins and tapering to a point. The flowers are white to pink, 5 petalled, 30–40 mm in diameter and clustered in the leaf axils (see colour plate 4.6).

The plant dies down to the ground in autumn. The root is quite large, with smooth pale yellow skin. The thick branches extend up to 400 mm, mostly in lateral directions.

There is sometimes a confusion between this species and the Mallows, *Malva* spp. and *Lavatera* spp., some of which are unfortunately popularly also known as 'Marshmallow', though they never grow in marshes.

Marshmallow (*Althaea officinalis*).

The leaves of the Mallows have a more rounded outline than true Marshmallow (*Althaea officinalis*). Most of the Mallows have petals notched at the tips. The roots of these species are white skinned and lack the size and development of true Marshmallow, which is not found wild in Australia. So if it is growing wild, you can be almost certain it is *Malva* or *Lavatera*.

Propagation

Method

Marshmallow can be grown from seed, but it is small and the crop will be slow to reach a harvestable size, probably taking 2 years to do so.

If sufficient stock is available, crown divisions are easier, have a very good success rate, and will produce a harvest at the end of the first season.

Timing

Plant seed in spring. The crowns can be divided in late autumn or early spring.

Planting

Seed

Seed is fairly small and should be sown in a nursery bed or in seed trays.

Divisions

Crowns need to be cut up with a knife, with at least one active bud on each piece. Sections of root without buds will not grow.

During harvest it is best to save crowns from a number of the more vigorous plants. The thick roots can be harvested off these crowns for drying, leaving the centre of the crown, with its active buds, to be divided up for replanting. Each piece of this centre should have a crown bud which will generate leaves, and some outside skin which will generate roots.

Layout and Spacing

Plants should be set on 450–600 mm centres in rows 900 mm apart.

Compost

Marshmallow needs a good dressing of compost each spring, 3–4 kg/m², worked in beside the plants.

Weed Control

Being large, strong, widely-spaced plants, Marshmallow is easy to maintain free of weeds, as inter-row cultivation and hand-hoeing are quite straightforward.

Irrigation

Marshmallow needs around 25–50 mm of water per week during summer.

Pests and Diseases

In grasshopper regions, Marshmallow can be attacked badly. This can reduce yields, especially of leaf and flower. If the plants are in a cooler situation, they seem to suffer less. A few ducks can help with grasshopper control.

Marshmallow growing in a hot situation may sometimes develop deep black cracks around the crown. This cracking can become severe enough to kill the plant. I believe this is due to heat stress, as affected plants in my garden were growing on the northern side of a line of young tress, and I have never seen the cracking in cooler situations.

Harvesting Leaf & Flower

Timing

Harvest the leaf and flower when the plant is in flower. This is usually mid-to-late summer. It should be possible to take two harvests each season.

Tools

The reaping hook is used to harvest Marshmallow tops.

Technique

As the main crop is the root, just the upper portion of the aerial parts is harvested, so growth is not set back too much.

Drying Leaf & Flower

Marshmallow leaves and flowers dry easily.

Temperature

They should be dried at temperatures up to 35°C.

Processing Leaf & Flower

Rub the dried Marshmallow through a screen of 2½ or 3 dent, with under-screens in place. Marshmallow rubs easily, with few problems.

Storage

Freeze treatment is necessary to control moth larvae before sale or storage.

Yields

Marshmallow yields around 0.1–0.2 kg/m² dried leaf and flower per annum. Higher yields could be obtained if total harvests were taken and the roots were not being harvested.

Harvesting Root

Timing
Dig Marshmallow root in late autumn or winter. As it is easy to dig and get a large volume in a short time, Marshmallow is a good crop to keep in mind when there is a need to get some root ready in a hurry to put into an empty dryer.

Tools
Harvesting roots will require a garden fork, and a reaping hook or secateurs for trimming tops.

Technique
Being such a large root, Marshmallow may take a bit of levering to heave it out of the ground. Two people working from opposite sides with forks can make the job easier. I usually prefer to trim the tops off afterwards, as they provide a handle to grab when pulling the roots up.

Post-harvest Treatment
Following harvest, the ground needs to be raked level, while it is still loose, in preparation for the planting of the following crop.

Pieces of root left in the ground are not a problem as they will not grow unless they have crown buds attached. They can take a long time to die, though.

One year, while digging a crop of Burdock, I kept finding thin pieces of yellow root. They were rather soft and pliable, but still obviously part of a living plant and definitely not Burdock. As the Burdock was weed free, I was rather mystified until I recalled a crop of Marshmallow had been harvested from that bed the previous autumn. The broken pieces of Marshmallow root, unable to grow because they had no crown buds, had lasted a whole year in the ground without rotting.

Drying Root

Marshmallow roots need to be washed before drying. Being such a large simple root system, this is quite easy, though they will usually require some breaking apart to free the soil.

Propagation material should be selected from a number of good yielding crowns, bearing in mind the benefit of retaining some genetic diversity among the material you are selecting.

Chopping or Splitting
Marshmallow is the only herb root I know of which will not dry through its skin: it forms a total seal. In order for it to dry, every piece of the root needs to be either chopped into short sections no more than 40 mm long, or else split or scraped. Even very thin pieces need some treatment, otherwise they simply will not dry.

I find the easiest thing is to put the whole lot through the chaff-cutter. As the root system is quite open in its form, there is not much likelihood of little stones being hidden among the roots and damaging the blades.

The roots can then be loaded into the dryer. Beware of overloading the screens, as a thick layer of Marshmallow roots can be rather heavy.

Assessment
Marshmallow roots need to be checked very thoroughly for total dryness, as the impermeable skin can prevent or greatly slow the drying of some pieces.

Temperature
Dry Marshmallow root at temperatures from 35–45°C.

Processing Root

Marshmallow root is used to some extent as a tea, and is also used to manufacture liquid medicines. Usually further processing does not need to be done by the grower.

Storage
Freeze treatment is necessary to control moth larvae before sale or storage.

Yields
Marshmallow yields up to around 0.4 kg/m^2 dried root per annum.

Marketing

There is a small demand for Marshmallow leaf and a moderate demand for Marshmallow root, both mostly for manufacturing.

Price
Price to growers for premium quality, organically grown Marshmallow is currently around $24/kg for 5-star tea grade, both for leaf and flower, and for root.

Price to growers for manufacturing-grade, organically grown Marshmallow is currently around $10–15/kg for leaf and flower, and $12-18/kg for root.

Trade price is around $9/kg for leaf and flower, and $12-13/kg for root.

Meadowsweet

Filipendula ulmaria (L) Maxim

ROSACEAE

Parts Used: Aerial Parts

Other Names: Queen of the Meadow, *Spiraea ulmaria*.

Meadowsweet is highly regarded in herbal medicine as a digestive remedy and has traditionally been used in flavouring wine and mead. There is no doubt a connection here: heavy partakers of wine and mead must have found their digestive tracts needed the assistance of Meadowsweet.

Aspirin has a connection with this herb. Salicylic acid was first isolated from Meadowsweet and later aspirin was synthesised from this. The word 'aspirin' is derived from *Spiraea*, which was the botanical name in use at that time.

Origin

Meadowsweet is native to Europe and Asia.

GROWING MEADOWSWEET

Meadowsweet is a soft, gentle plant, which ennobles the garden with its foamy cream blossoms. Coming from a cool northern temperate climate, it may need some protection in the harshness of our hotter regions.

Climate

Meadowsweet prefers a climate with cool to mild summers.

Site

Exposure

Meadowsweet likes full sun in cooler climates, and semi-shade in hotter regions. It is a little sensitive to wind.

Drainage

My experience has only been in growing Meadowsweet in well-drained soils. Some of the literature describes it as growing in swamps and marshes.

Soil Type

Meadowsweet does well on heavier soils, but can be grown on a range of soils.

Fertility

It needs a fertile soil.

Initial Weed Control

Bare fallow is recommended prior to planting as this will make subsequent weed control easier.

Neighbouring Plants

As it is susceptible to two-spotted mite (red spider), don't locate Meadowsweet next to plants such as Melissa Balm, which also suffer from this pest. And place it away from aggressive plants that might compete with it for moisture.

Identification

Meadowsweet grows 600–1200 mm tall, with reddish stems. Leaves are pinnate, with 2–5 pairs of leaflets, plus a few little tags of leaf along the midrib. The leaves on the flowering stems are whitish underneath, but the lower leaves are green underneath. Flowers are faintly aromatic, creamy-white, 2–5 mm, 5-petalled, with numerous long stamens in dense irregular umbel-like clusters at the top of the stems. The plant dies down to ground level in winter, and has little pink rootstocks which smell of wintergreen, as do the stems, but not the leaves.

Variety

The wild form of Meadowsweet with single flowers is probably preferred, but the double-flowered form seems to have the same properties.

Propagation

Method

Crown division is preferable, but it can be grown from seed.

Timing

Divide in autumn or early spring. Plant seed in spring.

Planting

Set the crowns in the ground just below the surface.

Meadowsweet (*Filipendula ulmaria*).

Layout and Spacing

Meadowsweet should be set out on 150–300 mm centres in rows about 900 mm apart.

For harvesting with the scythe, these should be in blocks at least three rows wide.

Compost

Apply compost at a rate of 3–4 kg/m^2.

Weed Control

Weed control consists of regular inter-row cultivation and hoeing around the plants.

Irrigation

Meadowsweet needs 25–50 mm of water per week through summer.

Pests and Diseases

Two-Spotted Mite (Red Spider)

At 'Twin Creeks' in north-east Victoria, I had two beds of Meadowsweet, one in full sun and one which was shaded in the mid-afternoon. The patch with full sun grew less vigorously and at times suffered badly from mites.

The partly shaded patch was always more vigorous. In hot conditions, though, the end which received the most sun would suffer from mites, but never the most shady end.

If the stubble is mowed off very short after harvest, this will break the mites' hold on the plant and the new growth will be free of them.

Grasshoppers

Grasshoppers also took a fancy to the Meadowsweet growing in full sun, but left those in the partly shaded plot.

Harvesting

Timing

Take the first harvest when Meadowsweet comes into flower. There will normally be two more harvests in the season, but the regrowth rarely flowers, so this should be harvested when the leaf has reached a good stage of development, and before it starts to deteriorate.

If you notice any leaf burn caused by mites (red spider), then harvest right away, before too much is lost.

Tools

The catching scythe is the most efficient harvesting implement.

Technique

Usually Meadowsweet can be cut quite close to the ground, as there is not much deterioration of the lower leaves.

Post-harvest Treatment

Mow the stubble off as cleanly as possible, as this helps break the life cycle of any mites present.

Drying

Meadowsweet dries rapidly without any great problems.

Temperature

It should be dried at a temperature of not more than 35°C.

Processing

A fair proportion of Meadowsweet is used as a tea and it is also made into liquid medicine.

Meadowsweet is one herb in which the stalks should be included in the final product, because they contain one of the active constituents, methyl salicylate, which does not appear to be in the leaves.

There is a considerable difference between the first harvest of the season and the later harvests. The first cut contains a lot of flowers and the undersides of the leaves are predominantly white. The later cuts contain virtually no flowers and the undersides of the leaves are predominantly green.

In fact, to look at the finished products side by side, they appear to be two different herbs. For this reason it may be best to combine them to obtain a more standardised product, provided they are of similar quality.

Tea Grade

The herb should be rubbed through a 2^1/$_2$ or 3 dent screen with under-screens in place, repeating until the stalks have been separated. If possible, these stalks should be hammer-milled or otherwise chopped into pieces that are a maximum 12 mm long, after which they are recombined with the rubbed leaf.

Manufacturing Grade

Chaff-cut the aerial parts for manufacturing grade. There is usually no problem with residual moisture as the stems dry quickly.

Storage

Freeze treatment is necessary to control moth larvae before sale or storage.

Yields

At 'Twin Creeks' we obtained yields of 0.25–0.35 kg/m^2 tea grade aerial parts, but in better growing conditions yields should be higher.

Marketing

There is a moderate demand for Meadowsweet, but growers should ascertain that they can sell it before commencing production.

Price

Price to growers for premium quality, organically grown Meadowsweet is currently around $24/kg for 5-star tea grade.

Price to growers for manufacturing-grade, organically grown Meadowsweet is currently around $10–18/kg.

Trade price is around $8/kg.

Motherwort
Leonurus cardiaca L   LAMIACEAE

Parts Used: **Leaf & Flower, Aerial Parts**

Motherwort is a herb used to some extent for menstrual and uterine conditions, but more as a heart herb and sedative. In fact one herbalist suggests its name is actually derived from its value in helping the distressed and overwrought mother who is having difficulty coping with a household of unruly children.

GROWING MOTHERWORT

Motherwort can become quite a matriarchal tower of strength in the garden with its tall stems adorned with small pink flowers and its blue-green, drooping, deeply cut foliage.

It tends to grow prolifically in its first year and then steady down to a more moderate level of production.

For maximum yields it may be worth growing Motherwort from seed each year.

Containment

Motherwort is inclined to reseed itself around the garden, but the seedlings are not hard to kill. Nevertheless, some precautions should be taken to prevent it from dropping seed where it might cause problems.

I made the mistake of giving a plant to my mother-in-law, thinking that, as a mother of five children, it would be an appropriate herb for her garden. What I forgot was Motherwort's own capacity for giving rise to offspring and unfortunately they came up all over her beautiful flowerbeds. It took a couple of years to get rid of them all.

Motherwort (*Leonurus cardiaca*).

Identification

Motherwort is an erect plant 900–1500 mm tall with stout square stems. The lower leaves are deeply 5-lobed, the upper are 3-lobed, all with toothed margins, strong smelling, and up to 100 mm across. The pink to purple flowers are in axillary whorls up the stem, and become quite prickly as they mature.

Motherwort dies back to a low crown of leaves in winter.

Harvest

Post-harvest Treatment

Following harvest, go over the plants and trim growth down to about 75–100 mm above ground level. Don't mow right down to ground level as this can severely weaken the plants. Motherwort needs some green leaf to help it recover after harvest.

For more detail on this herb see 'Information Charts'.

Pasque Flower

Anemone pulsatilla L RANUNCULACEAE

Parts Used: Aerial Parts

Other Names: Pulsatilla, Anemone, Wind Flower, *Pulsatilla vulgaris*.

Pasque Flower is a diminutive plant which is sometimes grown as an ornamental for its magnificent flowers with petals of deep purple and bright yellow centres. It is a valuable medicinal plant, used as a sedative and analgesic.

GROWING PASQUE FLOWER

This delicate plant adds its special qualities to the herb beds, with delightful flowers and fluffy seedheads.

It can be tricky to get Pasque Flower started, though, and it needs regular attention to keep it free of weeds.

Pasque Flower prefers a climate with a mild summer, but it can be grown in hotter regions where it will put on most of its growth in the cooler parts of the season.

Pasque Flower prefers full sun and it is traditionally held that it requires wind to make the flowers open. There may be some truth in this, as it flowers quite early in the season and the flowers won't open unless they are dry. So if it is windy, the chances are the flowers will be dry enough to open.

There are some reports of skin irritation being caused by the fresh juice. The plant is toxic in the fresh state.

Identification

Pasque Flower is a soft hairy plant growing 50–400 mm tall, with bipinnate or tripinnate leaves. The flowers are 30–50 mm across, hairy, they have 6 petals and are dark blue to violet with yellow centres (see colour plate 5.2). White- and red-flowered forms exist. After flowering a fluffy pappus forms.

Pasque Flower usually continues to grow slowly through winter.

Variety

If possible, obtain the purple variety as it is closer to the wild form.

For more detail on this herb see 'Information Charts'.

Plantain, Greater

Plantago major L PLANTAGINACEAE

Parts Used: Leaf

Other Names: Broad-leaf Plantain, Rat-tail Plantain, Waybread, White Man's Foot.

This little plant has followed European invasion all over the temperate zones of the Earth. The American Indians used to refer to it as 'White Man's Foot'.

Greater Plantain is quite common in the milder parts of southern Australia, especially in damp places and under deciduous trees, but in the wild the plant is usually small and mixed with other species.

It is used as a respiratory herb and has antiseptic and healing properties. It is very useful to have growing around the farm as a handy treatment for insect bites and stings: just crush a leaf and rub the juice onto the site.

The young leaves were traditionally used as a pot herb.

GROWING GREATER PLANTAIN

Once its competition is removed and it is given a dressing of compost and regular watering, it is amazing how big this herb becomes, justifying the name 'Greater Plantain'.

Greater Plantain prefers a climate with mild summers, but with plenty of water and a little shade it will grow in hotter situations.

Containment

As it drops a lot of seed, a buffer zone of cultivated soil about 900 mm wide should be maintained around it.

Pasque Flower *(Anemone pulsatilla)*.

Identification

Greater Plantain has a short perennial rootstock, which sends up a basal rosette of leaves, which are oval to broadly ovate, 50–300 mm long, dark green, with 7 prominent parallel veins. The small inconspicuous flowers are borne along the length of the stalk, which somewhat resembles a rat's tail, 50–400 mm tall.

The plant dies down in winter, but retains some green leaves.

The main identification problem is in distinguishing it from the related Narrow-leaf Plantain (*Plantago lanceolata*). This usually has much narrower leaves, but sometimes develops broader leaves in early spring, which can make it look very much like Greater Plantain.

Once the flower-stalks emerge, there is no longer any confusion, as Narrow-leaf Plantain has tall flower-stalks, with a short thick flowering spike at the end. As boys, we used to 'shoot' these flower heads at each other by folding the stem, catching the head in the fold, and pulling until it flew off.

Greater Plantain's flowers extend most of the way down the stem, so there is a long thin flower head which cannot be 'shot' in the same way.

Narrow-leaf Plantain has similar uses and is regarded by some as an alternative species (see the entry for 'Narrow-leaf Plantain' in Chapter 21, 'Wildcrafting and Weed Harvesting').

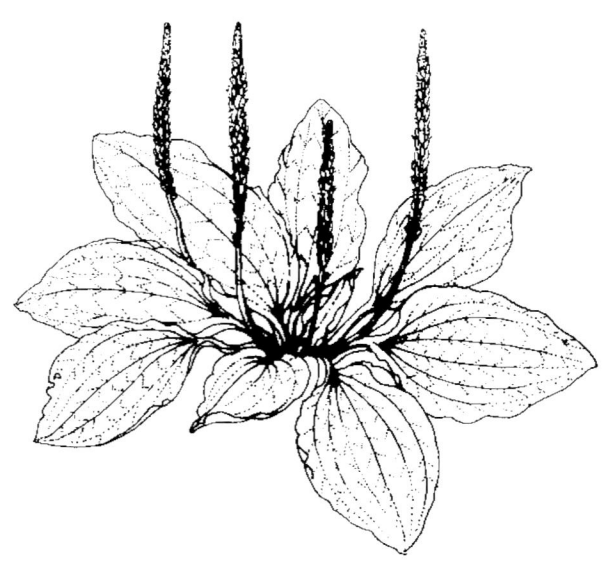

Greater Plantain (*Plantago major*).

Drying

The Greater Plantain flower-stalk is not part of the finished product, but stalks can be separated from the dried herb by screening.

Greater Plantain's leaves are quite thick and fleshy, so they need good drying conditions to dry quickly without fading. If they are too slow drying or are subject to remoistening, they will discolour badly.

The fleshy leaf-stalks are the slowest part of the plant to dry and can be rather deceptive.

Processing

This species is mostly used in manufacturing liquid medicine.

Before processing, check that the leaf-stalks are completely dry. These stalks can pass through the screen even when they are a bit moist and can cause moisture problems in storage.

Manufacturing Grade

As Plantain leaf is required without too much flower-stalk, a first screening is usually adequate. Chaff-cutting is not an option because too much flower-stalk would be included.

Storage

This herb can be a bit deceptive as to its dryness so it is important to check its moisture a week or two after processing.

For more detail on this herb see 'Information Charts'.

Poke

Phytolacca decandra L
PHYTOLACCACEAE

Parts Used: Root

Other Names: Poke Weed, Pocan, Red Ink Plant, *Phytolacca americana*.

Poke is used in small doses in treating respiratory and other infections and for its action on the lymphatics.

While parts of this plant are quite toxic, there is a widespread tradition in North America of using the young leaves as a food, after special treatment to remove the toxins.

Poke (Phytolacca decandra).

GROWING POKE

Poke is quite a striking plant with its large leaves, purple stems and drooping clusters of purple-black berries. It looks almost 'other-worldly' and to some people it has a rather sinister air about it.

It is easy enough to grow, but herb growers should be very cautious with it.

If possible, supplies should be obtained from wild plants. However, Poke is not to be found everywhere in Australia, and much of it is another related species, *Phytolacca octandra*, which is regarded as being inferior. So it may be necessary to grow *Phytolacca decandra* in order to have a reliable supply.

The problem with Poke root is it can be quite toxic in larger doses, so special precautions and procedures must be adopted to avoid it accidentally getting mixed with other herbs.

Containment

Poke is a potential noxious weed and at one stage was declared such in Victoria. It can spread quite rapidly and some distance by means of birds eating its berries.

Preventing this requires some diligence on the part of the grower, as in the latter part of the season it forms fruits continuously.

The birds are quite fond of these and after eating them, disperse the seeds wherever they perch, so Poke can quickly spread all over the rest of the garden and the nearby countryside.

The only ways of controlling this are to keep the birds off or to pick all the berries every few days before they form viable seeds.

I was growing Poke for a while in north-east Victoria, where it did very well. I tried to keep the berries from forming by regularly picking them off, but it was difficult to keep this up, owing to the continuous flowering of the plant and the pressure and distraction of other work.

When I started to find young Poke plants coming up in various parts of the garden, I became rather anxious about the possibility of them being overlooked and accidentally included with other herbs in harvesting, so I stopped growing it.

One possible alternative would be to grow it in a region where Poke is growing wild already: in parts of the north coast of New South Wales it is a common weed.

One saving grace is that it regrows only from the crown and broken pieces of root left in the ground will not grow if they don't have any active buds attached. This means it can be killed by hoeing, provided the plant is cut below the crown.

For more detail on this herb see 'Information Charts' and the entry for Poke in Chapter 21, 'Wildcrafting and Weed Harvesting'.

Red Clover

Trifolium pratense L FABACEAE

Parts Used: Flower, Leaf and Flower

Other Name: Cow Grass (for the variety Hamua).

Although Red Clover appears to have been known to the ancients, its use has only become widespread in relatively recent times. Today it is popular as a herbal beverage.

It makes a pleasant cup of tea, which may be enjoyed for its flavour alone or medicinally as an alterative, as an expectorant and as an antispasmodic.

It is widely grown as a pasture plant in southern Australia.

Origin

Red Clover is native to Europe, but has naturalised widely in temperate regions of the world.

GROWING RED CLOVER

Red Clover is a very useful soil builder because it fixes nitrogen, has a deep root system, and produces a mass of leafy growth. In fact, it is one of the few herb crops that generates a credit in terms of fertility.

Most herb crops consume more than they produce and need to draw on the rest of the farm for compost to sustain their production but, in terms of nitrogen and organic matter, Red Clover returns more to the soil than it consumes, as the harvested flower represents only a small proportion of its growth.

This crop could be grown on a large scale, on an area of 2–3 times the rest of the herb-growing operation, as a source of material for making compost to sustain the cultivated crops. Where inputs of fertiliser are needed to build up poorer soils, this could be done by using organic mineral fertilisers on the Red Clover to produce a surplus of compostable organic material.

Red Clover is easy enough to grow and weed control is minimal. The tricky part is in getting a good crop of flowers from it. Flowering seems to vary according to soil, climate and variety: growers are still in the process of working out the best combinations for good flower production.

While I usually have managed to get reasonable harvests of flowers in Tasmania, when I attempted to grow Red Clover in north-east Victoria I encountered a number of problems with it and never did harvest any flowers.

Climate

Red Clover will grow in a wide range of climates, but seems to flower more heavily in a cooler climate.

Site

Exposure

Red Clover prefers full sun and will tolerate exposure to wind.

Drainage

Good drainage is needed.

Soil Type

It will grow on a wide range of soils.

Fertility

A moderate level of fertility is preferred. If the soil is too high in nitrogen, it will tend to produce more leaf and less flower.

Initial Weed Control

Bare fallow is recommended prior to sowing, to reduce weeds which compete with the crop and can make harvesting difficult.

Neighbouring Plants

It is best to grow Red Clover separately, or perhaps next to other crops managed in a similar way, as an under-storey of weeds develops, which can spread into neighbouring crops and cause problems. Red Clover needs to be located where it can be mowed off at intervals.

Identification

Red Clover grows up to 600 mm tall. The stems are softly hairy, the leaves are trifoliate, with ovate to obovate leaflets up to 50 mm long. Flowers are rose-purple, in dense spherical heads up to 30 mm across (see also colour plate 6.2).

The plant continues to grow through winter. It is a perennial, though sometimes short-lived.

Red Clover can be confused with several other clovers:

- Crimson Clover (*Trifolium incarnatum*) is an annual, with crimson flowers and oblong (about twice as long as broad) flower-heads, while Red Clover is a perennial, and its flowers have a touch of purple in their colour and are spherical, or close to it.
- Strawberry Clover is a shorter plant with smaller pinkish flower-heads which retain their colour as they mature to resemble a strawberry, while Red Clover flowers turn brown and shrivel as they mature.
- Alsike Clover is similar to Red Clover in its growth, but its flower is more like a large White Clover flower with a pink tinge to it.

Red Clover (*Trifolium pratense*).

Variety

Red Clover is widely grown as a fodder crop. The current trend is to new low-oestrogen varieties. Livestock feeding on traditional varieties of Red Clover can develop fertility problems as a result of a plant oestrogen they contain.

New low-oestrogen varieties are being promoted and old high-oestrogen varieties are being phased out.

There is some diversity of opinion among herbalists as to whether oestrogen is a desirable component of Red Clover for medicinal use. Some manufacturers and other buyers prefer old high-oestrogen varieties.

Montgomery Red is a good traditional high-oestrogen variety, which flowers heavily, but it is unobtainable these days. Hamua (or 'Cow Grass', as it is often known) is still available and has a high oestrogen level, but its flower yields have been disappointing. Most other varieties currently available are low oestrogen. Redquin does seem to produce an abundance of flowers, but it is low in oestrogen.

If you have trouble finding the desired variety, it may be possible to collect seed from roadside or pasture plants growing in a place where a new crop of Red Clover has not been sown in the past 15 years or so. Chances are these will be Montgomery Red, which was the main variety of Red Clover used for many years.

Propagation

Method

Red Clover is propagated by direct seeding.

Timing

It is sown in early spring or autumn, when conditions are such that the surface of the ground can be kept moist enough to allow germination. The best time is just before a good rain, which will help cover the seed and keep it moist long enough for germination to take place.

Planting

Red Clover seed is quite small and needs to be close to the surface for good germination. It can be sown by itself, or under-sown with a cover crop, usually a grain crop such as Oats or Wheat.

Sowing with a Cover Crop

Because Red Clover's initial growth is slow, this can allow a faster growing crop to be taken while the Clover grows as an under-storey and then takes over after the cover crop is harvested or mowed off. It is important that the cover crop be sown lightly so as not to smother the Red Clover. There needs to be sufficient nutrition and moisture for both crops.

A cover crop can help prevent erosion as it quickly establishes a good hold on the soil.

Sowing Red Clover Alone

For optimum growth of Red Clover, it is best to sow it in a pure stand as the competition of the cover crop will slow its development and delay flowering.

On a small scale, it is easiest to sow Red Clover seed by broadcasting it by hand or with a seed spinner or fiddle. It is important to get an even distribution. It should be sown at the rate of 10 kg/ha which is 1 g/m^2. If there has been no Clover of any species growing in the patch where the Red Clover is going in, the seed should be inoculated with nitrogen-fixing bacteria.

The seed-bed should be free of weeds, and finely tilled but firm. Rolling is very helpful in achieving a firm seed-bed and levelling out the lumps and bumps so mowing will be easier. Failing this, a smudge made of planks or an old gate can be dragged over the ground, before seeding.

The seed should be lightly covered with soil. This can be done by dragging a leafy branch or a chain over the ground. A good fall of rain or a generous irrigation can have a similar effect of covering the tiny seeds with soil and also provides moisture for germination.

Layout and Spacing

Red Clover should be planted in beds with dimensions large enough to mow easily, probably with a tractor. Irrigation is easier if these beds correspond to areas covered by your irrigation system.

Compost

Some caution needs to be exercised with regard to compost application. Bear in mind that Red Clover will flower more heavily in soils that are not excessively fertile: if there is too much nitrogen it will produce mostly leaf and not much flower.

If the soil is reasonably fertile, it may well need very little compost. In poorer soils, a moderate application at the rate of 2–3 kg/m^2 (20–30 t/ha) should be adequate.

As Red Clover fixes its own nitrogen and creates a surplus of organic matter, organic mineral fertilisers with a balanced make-up but no nitrogen, such as wood ashes or some commercially available products, could be used. These may be better for encouraging flower development than nitrogen-rich materials like compost.

Weed Control

Early in the establishment of a new stand of Red Clover, it will help to mow it off when it is 200–300 mm tall, depending on the growth of the crop and the weeds in it. This will cut back competing grasses and weeds and will give the Red Clover an advantage because of its strong regrowth.

Once established, Red Clover will compete strongly with other plants, but in the first growth of the season there can be a lot of grass and other weeds amongst it. If this weed growth is too thick to allow easy harvest of Red Clover flowers, it should be mowed off.

Provided there is enough moisture, this mowing will stimulate a strong regrowth of Red Clover and it should leave the weeds behind long enough for several harvests of flowers.

If weed growth becomes a problem later in the season, mow it off again, provided it has had enough time to build up its reserves. Be aware that mowing too frequently will kill it.

Weeds such as Plantain and Flatweed, which send up flower heads at about the same height as Red Clover flowers, can be a problem in harvesting.

Irrigation

Red Clover needs around 25 mm of water per week through summer.

Pests and Diseases

Grasshoppers

In Tasmania there are not many pest or disease problems with Red Clover, but in some regions the grasshoppers will eat the flowers.

If you are planning to grow Red Clover where grasshoppers are a problem, then you will need to control them or else get around them by having your crop flower before they emerge or after they depart.

Red Clover can produce a good quantity of blossom on regrowth late into autumn.

Management

Red Clover needs to be managed carefully to maintain its vigour and to maximise flower harvest.

In hayfields and pastures, it typically only lasts for one or two seasons, but in a situation where it is grown for its flowers, mowed only once or twice a year and never grazed, Red Clover has consistently lasted much longer.

In a pasture situation, it tends to die out early because it won't stand up to repeated grazing: it needs a good recovery period of 6–8 weeks.

The best production of flowers tends to be on the regrowth. Often the regrowth after a hay crop will carry quite a lot of flowers on it. Growers will need to experiment a little to get the right combination of factors that promote good flowering. These tend to be lower levels of nitrogen and possibly moisture: sometimes when plants are stressed a little they flower more heavily.

If you have a large enough area in Red Clover, it can be a useful forage or compost crop. Parts of the crop could be mowed at different times on a staggered basis, so there are sections at different stages of regrowth and flowering. Trials of different management techniques can be run to see how Red Clover is best managed for maximum flower production at the optimum times of the year.

If drying depends on ambient air, the flowers need to be available for harvest when good drying conditions prevail. Later autumn harvests will be difficult to dry this way, resulting in loss of quality.

Harvesting

There are at least three methods used for harvesting Red Clover and these give products of different qualities.

Hand-picking can harvest a high proportion of flowers at an early stage of maturity and with a low proportion of leaf, producing 5-star quality Red Clover whole flowers.

Combing produces a high proportion of flowers at various stages of maturity, together with a low proportion of leaf. The best this method can produce is 4-star quality Red Clover whole flowers.

Mowing the Red Clover flowering tops produces a high proportion of leaf and a variable proportion of flowers at various stages of maturity: this is Red Clover leaf & flower.

Timing

Red Clover usually flowers from late spring through to late autumn. Each flower head is a cluster of florets, which open progressively. By the time the upper florets have opened, the lower ones will be turning brown. For best quality, flowers should be picked before they have fully opened.

The optimum time is when the flower heads are between a quarter open and fully open, but very little of the Red Clover available on the market even approaches this level of quality: the flowers usually look very brown.

As Red Clover flowers need to be dried very quickly, if you are relying on ambient air drying, picking will need to be done when conditions are very good. They should be harvested in the morning of a warm sunny day, so their critical first period of drying can progress rapidly.

If artificial heat is available, there is more flexibility regarding timing, but avoid picking later in the day in hot conditions. Combing will be easier in the morning and picking in warm conditions can result in deterioration of the flowers.

Because it flowers sequentially, Red Clover continuously produces new flowers, while the older flowers fade, turn brown and darken. Harvests should be frequent in order to minimise or avoid faded brown flowers.

Hand-picking Flowers

For premium quality flowers, hand-picking should be done every 3–4 days (longer in cool weather), in order to catch the flowers before they have quite fully opened. Flowers which have fully opened, even though they may look okay when picked, get bruised in picking and will partly turn brown before they are dry. With this method, all the flowers can be picked at the optimum stage of between a quarter open and fully open.

Combing Flowers

If combing is carried out at intervals of about 10 days (shorter in warm weather, longer in cool weather), there will be sufficient flowers to make combing worth while, though a small proportion will be starting to turn brown.

With this method the flowers will range from unopened buds to slightly over-mature flowers starting to fade.

Mowing Leaf & Flower

In order to get a good proportion of flowers in the mown product, some of them will have to be allowed to go brown, but the harvest should be taken before too many of the older flowers darken and while the leaf is still a dark green colour.

With this method the flowers will range from small buds to rather over-mature blooms.

A period of 6–8 weeks between mowings will be needed to allow the Red Clover to recover.

Tools

Hand-picking Flowers

Hand-picking requires a nimble pair of hands and a small basket or bucket attached to the waist in front, so both hands are free. A larger basket or sheet is needed to empty the flowers into at intervals.

Combing Flowers

The same type of comb used for Chamomile can be used for Red Clover (see 'Harvesting Flowers' in Chapter 11, 'Harvesting'). A basket should be carried along into which the flower comb is periodically emptied.

Unless a steady breeze is blowing, a fan will be needed to winnow excess leaf from the harvested flowers.

Mechanical harvesting of the flowers may be a possibility, as the combing action is similar to that used for Chamomile.

Mowing Leaf & Flower

A harvesting scythe is appropriate for small- to intermediate-scale production. On a large scale, it might be possible to use some sort of mechanical harvester which can be set high enough to harvest and pick up the upper part of the plant.

Technique

Hand-picking

To be efficient, the fingers need to move quickly, nipping the flowers off between finger and thumb or between index and middle finger. The flowers can have stalks of up to about 25 mm and one or two sets of leaflets attached (that is 3–6 leaflets). Any over-mature flowers showing brown discoloration should be nipped off and dropped.

The advantage of this technique is that it can produce the absolute best quality Red Clover flowers. If a good price is

Hand-picking Red Clover flowers. This can produce the absolute best quality and is worth while doing if a good price can be obtained.

obtainable, this can justify the time it takes to harvest by hand-picking, which is around 2–3 hours per kilogram of dried weight.

It is also very flexible as far as weeds are concerned, because it is still possible to pick when the patch is quite weedy or in a mixed stand of grasses and Red Clover.

Where possible, work uphill so you don't have to reach down so far: this is easier on the back.

In order to reduce trampling and keep the Red Clover standing upright, it is best to follow the same paths through the crop each time you walk through it to pick it.

Combing

The comb catches the flower heads and plucks them off. The leaflets attached to the flower heads can be included in the finished herb but other leaf and stalk should be kept to a minimum.

Combing style will necessarily vary according to the way the Red Clover is growing. When it is more or less erect with a predominance of flowers, these are relatively easy to comb. It may even be possible just to walk through the clover, holding the comb at a set angle and quickly filling it, but usually it is more a matter of diving in and out of the bushes with the comb to pick up the flowers without getting too much extra leaf.

Combing can be quicker than hand-picking if there is a good stand of flowers and 1 kg or more (dried weight) of flowers can be combed in an hour. The quality, however, is never as good as with hand-picking, because inevitably there are unopened buds and over-mature flowers included in the harvest.

Harvesting more frequently to avoid the flowers becoming over-mature will usually make combing about as slow as hand-picking, but with the disadvantage that it gathers the unopened buds, and so sets back the development of new flowers.

As the Red Clover grows taller, it starts to lean over, so it is necessary to stroke the comb in the direction of this lean. As the plants become more and more recumbent, often in different directions, the flowers become sparser, grass and weed seed heads stick up through it, and combing becomes more and more difficult and less productive. It will eventually reach a point where it is best to mow it all off and allow the crop to regrow.

Winnowing

Red Clover flowers which have been combed will inevitably contain some leaf picked up in harvest. The leaflets attached to the flower heads can remain, but loose leaves should be separated by winnowing. This need only be a crude set-up, using the breeze or a fan and does not take long.

A sheet can be put down on the ground and the flowers poured slowly across the draught. This is best done with a shaking motion, so they fall singly rather than in clumps. The draught should carry the lighter leaves to the far end or off the edge of the sheet, while the flowers, being heavier, should fall short. There will be an area of overlap which needs to go through a second time.

Winnowing is not necessary for hand-picked flowers, as any excess leaf gathered is minimal.

Note: If held in volume for any length of time, Red Clover flowers will generate heat quite rapidly, which will cause deterioration of the product. If they are likely to heat before they can be loaded onto the drying screens, they should be spread out in a cool place to ensure dissipation of heat.

Mowing Leaf & Flower

Mowing involves cutting the flowering tops off at a height where there is a good proportion of flowers and where the leaf is in good condition.

If the scythe is being used, the Red Clover will need to be standing up fairly straight, otherwise mowing could be difficult.

Post-harvest Treatment

The crop needs to be assessed to determine whether it will continue to produce more flowers satisfactorily or whether it needs mowing off.

If a patch of Red Clover is big enough, this mowing and regrowth could be staggered for different sections of the patch.

If the crop is being harvested for leaf & flower, the stubble should be mowed off after each harvest to ensure more vigorous regrowth. This will also help ensure that plants are standing up straight at harvest.

Drying

Red Clover needs to dry rapidly, particularly for the first day or two, in order to set good colour. If this initial drying is too slow, the flowers will continue maturing and turn brown. If very good drying conditions don't prevail at the time of harvest, Red Clover should be dried with artificial heat, or else the harvest should be postponed.

When checking for dryness, the flowers should fall apart when rubbed between the fingers and thumb.

Herbal Harvest

7.1 Spearmint *(Mentha spicata)*

7.2 Valerian *(Valeriana officinalis)*

7.3 St John's Wort *(Hypericum perforatum)*

7.4 Wood betony *(Stachys officinalis)*

7.5 Thyme *(Thymus vulgaris)*

7.6 Yarrow *(Achillea millefolium)*

Plate 7

8.1   Red Clover: trade quality, left; premium grade, right.

8.2   German Chamomile: premium grade, left; trade quality, right.

8.3   Peppermint: premium grade, left; trade quality, right.

8.4   Spearmint: premium grade, left; trade quality, right.

8.5   For culinary purposes, most herbs are used in small particles but some are used whole. A larger particle size is preferred for infusing to make teas.

8.6   Decoction: denser herbs that need to be simmered in water to extract their constituents need to be in particles of suitable size.

8.7   Manufacturing Grade: herbs that will be further milled or otherwise processed by the manufacturer can be shipped in a fairly coarse condition.

Plate 8

Temperature

Red Clover should be dried at a temperatures of 25–35°C.

Processing

Red Clover is widely used as a tea and it is also made into liquid extract.

Tea Grade

The market prefers whole flowers for tea-grade Red Clover, but where these are not available, rubbed leaf and flower is sometimes substituted.

Whole Flowers

Red Clover flowers do not need any further processing, except to pick out any superfluous leaf and stalk and any brown flowers. If there is variation in quality between batches, they should be separated into two or three categories.

To achieve 5-star quality, there should be virtually no brown discoloration on the flowers. If there is some browning evident, the herb will fall into a lower category and bring a lower price.

Rubbed Leaf & Flower

Rubbing may be an option for tea-grade Red Clover, but whole flowers are preferred. Rub the dried herb through the standard tea-grade screen.

Manufacturing Grade

Because manufacturers will probably not want to pay a premium price, hand-picking for this market is unlikely to be a viable proposition. Harvesting with a comb may be a possibility but at present manufacturers are using either cheap imported flowers (of mediocre trade quality) or Red Clover rubbed leaf and flower (flowering tops), which is mostly leaf. This latter is rubbed in the same manner as for tea grade, though a first screening may be acceptable.

Storage

Freeze treatment to control moth larvae is absolutely necessary before sale or storage, as flowers are very attractive to the moths and massive infestations can occur, which will render it unsalvageable.

Yields

Red Clover yields have been around 0.05–0.15 kg/m² tea-grade dried whole flowers per annum, but I feel this could be increased as optimum management techniques are developed.

Marketing

There is a moderately large market for organic whole flower Red Clover, as there is currently very little available and it is a widely used herb.

Price

Price to growers for premium quality, organically grown Red Clover whole flowers is currently around $50/kg for 5-star tea grade, and $30–35/kg for 4-star.

Price to growers for manufacturing-grade, organically grown Red Clover rubbed leaf and flower is currently around $10–15/kg.

Trade price is around $9–10/kg for whole flowers.

Stoneroot

Collinsonia canadensis L   LAMIACEAE

Parts Used: Root

Other Names: Horse-Balm, Richweed, Knob Root.

Stoneroot is a medicinal herb used as a diuretic and in the treatment of urinary calculi and haemorrhoids. Its name may have been derived from its use in treating bladder stones, or from its incredibly hard roots, which are like rocks when they are dry.

GROWING STONEROOT

Stoneroot is quite attractive with its soft green foliage and yellow flowers, but it is a rather delicate plant adapted to the sheltered environment of shady forests of north-east North America where there is a very regular rainfall throughout the growing season. While it is a bit more adaptable than some plants coming from this environment, it does need some protection from the harshness of hotter climates in order to succeed. It is easily damaged by wind.

It needs a reasonably fertile situation but lush growth is more vulnerable to wind damage.

Identification

Stoneroot grows 400–1200 mm tall, with erect unbranched stems, bearing opposite, ovate, toothed leaves, 100–200 mm long. Flowers are yellow in a loose branching pyramidal head, and lemon scented with a suggestion of ginger.

The plant dies down to ground level in winter and has a very hard, knobby tuberous crown.

Drying

Before washing, propagation material should be saved from a number of the best crowns.

The roots need to be washed before drying. To do this effectively they need to be broken apart: they are extremely hard, but will cleave into segments.

After washing, any piece of root thicker than your thumb will need to be chopped to facilitate drying.

This is one root you wouldn't dare put through the chaff-cutter: it will probably need an axe to chop it and some set-up to catch all the flying pieces.

Tea Grade

It is to be hoped that the grower will not be asked to prepare tea-grade Stoneroot. When it has dried, Stoneroot becomes rock-hard and is almost indestructible, so it sounds like a load of blue metal going around inside the mill. It takes ages for the hammermill to eventually break it down to size.

For more detail on this herb see 'Information Charts'.

Vervain

Verbena officinalis L VERBENACEAE

Parts Used: Aerial Parts

Other Names: *Herba sacra*, *Herba veneris*, Herb of Grace.

Vervain has a long history of association with sorcery, magic and other spiritual traditions. The name Verbena was used by the Romans for altar plants. They also called it *Herba sacra* and *Herba veneris*. Naturally Vervain became woven into Christian tradition as well and there is a legend it was found growing on the Mount of Calvary. Another name given to it is Herb of Grace.

Today it is used medicinally primarily as a nervine.

Origin

Vervain is native to the Mediterranean region.

GROWING VERVAIN

Vervain is a rather sparsely leafed, scruffy looking plant and is pretty easy going. To look at it, it conveys the impression of an ungenerous nature, but it is in fact quite a productive herb.

Climate

Vervain will grow well in a wide range of climates, though in regions with a cold winter it can suffer from frost damage as it is winter active.

Site

Exposure

Full sun is best for Vervain, and it will tolerate a windy site.

Drainage

Good drainage is required.

Soil Type

Light to medium loam is preferred.

Fertility

Moderate fertility is required.

Initial Weed Control

Bare fallow is recommended prior to planting, as Vervain is somewhat susceptible to weed invasion because its canopy is fairly sparse.

Neighbouring Plants

Provided you grow the tall variety of Vervain, it isn't much trouble for neighbouring plants. The shorter wild form is a bit more inclined to spread itself around the place by means of its seeds.

Containment

Seedlings that come up in the wrong place are a nuisance, but are not hard to kill with cultivation. Nevertheless, it is best to avoid letting Vervain go to seed.

Identification

Vervain grows 350–900 mm tall, with ribbed angular stems. It is loosely branched and only sparsely leafy. The opposite leaves are up to 60 mm long. They vary from ovate and coarsely toothed to deeply lobed or pinnatifid, up to 60 mm long. Flowers are small, lilac-coloured, and borne in long terminal racemes.

The plants tend to remain green over winter and will grow actively in milder climates.

Variety

There is a shorter variety of Vervain growing wild in parts of Australia. It has broader leaves and goes to flower very quickly. It is less productive than the taller variety and because it goes to flower so quickly, it inevitably drops a lot of seed on the ground, which germinates easily and can get out of control.

The preferred variety is taller, sparser leaved and slower to flower and form seed. Because of these attributes it is easier to manage in cultivation and is more productive.

Propagation

Method
Crowns can be divided for propagation or Vervain can be grown from seed. Crown division is good if you have vigorous plants that separate easily. Often there are a lot of self-sown seedlings around an established bed, and these can be used as propagation material.

Timing
Plant seed in early spring or autumn in warmer climates. Transplant in early spring or autumn. In colder climates, spring planting is preferred, as a good frost can heave the crowns out of the ground and expose them.

Planting
Sow seed in a nursery bed or in trays, unless the ground is fairly free of weeds, in which case it could be direct sown.

Transplanted divisions or self-sown plants need to be taken with as much root as possible and the leaves need to be cut right back. They should be set with the crowns well covered, just below the surface.

Vervain (Verbena officinalis).

Layout and Spacing
Vervain should be set out on 150–300 mm centres in rows about 900 mm apart.

For harvesting with the scythe, beds should be in blocks at least three rows wide.

Compost
Vervain needs a moderate application of compost at the rate of 2–3 kg/m^2, worked in next to the plants in spring.

Weed Control
Good weed control needs to be maintained with inter-row cultivation, hoeing around the plants and hand-pulling or clawing out any weeds that have nestled themselves in among the crowns.

Rejuvenation
A bed of Vervain will be quite productive for several years. Self-sown seedlings replace the older plants to continue production, but after a while the plot will start to decline.

In this situation, the beds need to be replanted, preferably in a new place.

Irrigation
Vervain needs 25–40 mm per week through summer.

Harvesting

Timing
Harvest when Vervain is in the early flowering stage. Sometimes there is a harvestable crop in spring (around October) if the winter has been mild, but if this harvest is not good quality, it is best to mow it off and take the next one.

There are generally three good harvests a season and sometimes a fourth.

Tools
Harvest Vervain with the catching scythe.

Technique
The technique used is quite straightforward. As the leaf is sparse, the light penetrating it usually keeps the leaves green down to a low level.

Post-harvest Treatment
After harvest, mow the plants off to 25–50 mm above the ground. Cutting too low may damage the crowns.

Drying

Vervain is a little slow to dry, so it needs good conditions. If dried in marginal conditions, or if it is subject to excessive remoistening, it can turn a dull brown. Normally it is a grey-green when it is good quality.

If it is intended for tea-grade, chopping before drying may be an option (see 'Tea Grade' below).

Temperature

Vervain should be dried at a temperature of not more than 35°C

Processing

Because there is really not much leaf on Vervain, the green stems largely perform the role of leaves and should be included in the finished product.

Some Vervain is used as a tea, but most of it is used as a fluid extract.

Tea Grade

For tea-grade the dried Vervain can be put through the chaff-cutter and then screened to remove the longer pieces, which are then put through the chaff-cutter again, until it is all in pieces not over 25 mm. But this is quite tedious.

Another option is to rub it first, to separate the leaf in large-sized particles, and then hammermill the stems and mix them back in.

Perhaps a better method is to put the herb through the chaff-cutter before drying. This definitely chops it shorter and cleaner and if the chopped Vervain is dried quickly it won't be bruised or discoloured provided it was not wilted when chopped. The result is more attractive than hammer-milling, which tends to smash the stems and expose the white insides.

Aerial Parts

Aerial parts for manufacturing should be put through the chaff-cutter when dry. Just check that the thicker stems are totally dry.

Storage

Freeze treatment to control moth larvae is necessary before sale or storage.

Yields

Vervain yields 0.5–0.7 kg/m² tea grade or aerial parts per annum.

Marketing

Demand is very small for tea-grade, but moderate for manufacturing use.

Price

Price to growers for premium quality, organically grown Vervain is currently around $20/kg for 5-star tea grade.

Price to growers for manufacturing-grade organically grown Vervain is currently around $12-14/kg.

Trade price is around $10–11/kg.

Wood Betony

Stachys officinalis (L) Trevisan

LAMIACEAE

Parts Used: Leaf & Flower, Aerial Parts

Other Names: Betony, *S. betonica*, *Betonica officinalis*.

Wood Betony has a long history of medicinal usage and association with magic. It was used by the ancient Egyptians, Greeks and Romans, and also by the Anglo-Saxons. It was held to be a powerful herb with many virtues.

Today it is used as a nervine tonic and a sedative.

GROWING WOOD BETONY

Wood Betony is a pretty sight when in flower and is sometimes grown as an ornamental. In spite of its name, which comes from its natural habitat as a woodland plant, it seems to adapt to growing in the open fairly well.

It prefers a climate with a mild summer, but will withstand hotter conditions, provided it has plenty of moisture.

Wood Betony *(Stachys officinalis)*.

Identification

Wood Betony forms a dense rosette of leaves and sends up a square flower-stalk, 300–600 mm tall. The leaves are dark green, oblong to lanceolate, toothed, up to 150 mm long, with a thick central vein. The flowers are pink or purple, in a dense terminal spike at the top of the stalk (see colour plate 7.4).

The plant dies down in winter, but remains green.

Harvesting

Harvesting is fastest with the catching scythe, but until the crowns are large enough and more or less contiguous, it may be difficult to cut them cleanly: the tufts of leaf tend to bend away from the blade if they have empty space behind them. If this is happening, it may be better to harvest with the reaping hook until the plants are larger.

For more detail on this herb see 'Information Charts'.

Wormwood

Artemesia absinthium L ASTERACEAE

Parts Used: Leaf & Flower, Aerial Parts

Other Names: Absinthe, Green Ginger.

Wormwood has a long tradition of use for indigestion and intestinal parasites, dating back to ancient Greece and probably long before.

It also has featured as an ingredient in wines and aperitifs, though in some countries its use is prohibited in alcoholic beverages because of its harmful effects if taken in excess.

It is an extremely bitter herb and probably for this reason it is not much used today: most people don't like the taste.

GROWING WORMWOOD

Wormwood is an adaptable plant which holds forth strongly in the garden. It has a straggly nature, being unable to grow straight, its stems prefer to lean and fall in an irregular fashion.

Although its stems are quite woody, it falls into this group of perennial crown herbs, because it dies back to the ground each year and reshoots from the base.

Neighbouring Plants

Wormwood is not a very good neighbour to many other herbs, tending to smother them or have an adverse effect on some. The important thing is to make sure it has plenty of room to flop around.

Identification

Wormwood grows to 1000 mm tall, with woody stems, covered with fine silky hairs. The leaves are about 80 mm long, aromatic, the lower leaves are much-lobed or divided, more or less bipinnate or tripinnate, while the upper leaves become progressively less divided (the segments are narrow and blunt), grey-green to almost blue, also covered with fine silky hairs. The flower heads are 3–4 mm across, grey-green with many minute, dull yellow florets.

The plant dies down to ground level in winter, but there is usually some leaf around the crown.

There is often confusion between this species and other *Artemisia* species, particularly *Artemisia arborescens* (Tree Wormwood). This latter species is often found growing around farms, particularly in poultry yards, and many people are under the impression it is Wormwood *A. absinthium*.

A. arborescens can be distinguished by its taller, straighter, growth of up to 2 m, its silvery-white foliage (*A. absinthium* has a green or blue tinge to it) and its milder aroma and flavour: it is not as bitter. *A. arborescens* retains its foliage and branches through winter, while *A. absinthium* dies back to the ground.

Wormwood.

There are some hundreds of species of *Artemisia* throughout the world, so there are possibly others with which it may be confused.

Rejuvenation

After a number of years, Wormwood will lose vigour and some plants may die. It is probably best to start a new bed in a new location, as replanting the gaps gives poor results, in my experience.

Drying

Wormwood is very slow to dry and needs good conditions. Place it in the best position available. Late harvests will probably need artificial heat to finish them.

The stems are the slowest part to dry and need careful checking if you are doing aerial parts.

Temperature

Wormwood should be dried at 25–35°C.

Marketing

Growers should make sure they have a definite commitment from buyers before going into production of this herb, as the market has been rather volatile.

For more detail on this herb see 'Information Charts'.

18

Woody Perennials

Feverfew – *Tanacetum parthenium*
Gum Plant – *Grindelia robusta*
Hyssop – *Hyssopus officinalis*
Lavender, English – *Lavandula angustifolia*
Lemon Thyme – *Thymus x citriodorus*

Marjoram, Sweet – *Origanum majorana*
Rosemary – *Rosmarinus officinalis*
Rue – *Ruta graveolens*
Sage – *Salvia officinalis*
Thyme – *Thymus vulgaris*

Woody perennial herbs are characterised by the woody nature of their stems. The focus of their regrowth is high up on the aerial parts of the plants. They typically retain their leaves all winter or, if they shed them, they regrow from buds high on the stems and not from ground level.

Their growth pattern is typically thick and many of them will form into a hedge, though this may be in miniature. Once established they generally don't take a lot of maintenance in terms of compost or weed control.

Quite a few of this group come from a Mediterranean climate, where the best growing conditions are in autumn and spring, while summers are dry. They consequently tend to have a lower water requirement than other groups, and some tolerance to hot dry conditions. This generally makes them well adapted to conditions in southern Australia.

There are a few herbs more or less intermediate between this group and perennial crown herbs, and there is an arbitrary line drawn between woody perennials and shrubs. If plants can grow to more than 2 m in height, they are included in the 'Trees and Shrubs' group.

Propagation

Propagation of this group is by seed, division, cuttings or layering. As seed is often small, vegetative propagation is preferred for many of them: cuttings and layering tend to give good success rates.

Layout and Spacing

The larger species lend themselves to management as a hedge with access space on either side. The shorter species are often successfully managed as a solid bed 3–5 m wide, as this effectively excludes weeds and facilitates harvesting.

Weed Control

While most species in this group require some attention during their establishment period, once they form a solid hedge or bed, their dense growth usually excludes weeds quite effectively. Because most of them are winter-active, weeds don't get much opportunity to make inroads during winter.

Harvesting

When harvesting woody perennials, it is important to leave enough leaf on the stems to enable them to regrow. Harvesting should be like a moderate pruning all over the bushes, leaving a $1/4$ to $1/3$ of the healthy green leaf. If species in this group are cut too low, they will be severely set back or may even die. This is because their life force or focus of regeneration is high up on the plant.

Rejuvenation

Many of this group are not long-lived, so they may need replanting after 4–5 years, sometimes less. This is best done in a new site.

Feverfew
Tanacetum parthenium (L) Shultz
ASTERACEAE

Parts Used: **Leaf (Leaf & Flower), Aerial Parts**

Other Names: Featherfew, *Chrysanthemum parthenium*, *Pyrethrum parthenium*.

Feverfew is a common plant in herb gardens, but for many years it had slipped into obscurity in the field of herbal medicine. It has recently come back into the light as a very effective treatment for migraine and as being of assistance in the treatment of arthritis.

Even though its effectiveness has been proven in clinical trials, it still has not been much embraced by herbalists, and it is not used in any great volume. This may be because the Feverfew generally available in the trade is of very poor quality. The clinical trials that established its effectiveness for migraine were done with fresh Feverfew leaf.

In spite of its bitterness, sometimes small amounts of Feverfew are used in cooking to 'cut' the grease.

Origin

Feverfew is a native of south-east Europe.

Feverfew (*Tanacetum parthenium*).

GROWING FEVERFEW

Feverfew is hardly a typical woody perennial and is somewhat between this group and the previous one in its characteristics.

It starts off in the form of a perennial crown herb, but as it gets older it becomes woodier and regrows from higher up on the plant.

It is quite easy to grow, in fact it is rather astounding how quickly it takes off in its first year.

Occasionally people have developed a strong allergic reaction to Feverfew: one English grower was obliged to discontinue production.

Climate

Feverfew will grow in a wide range of climates.

Site

Exposure

Full sun is preferred by Feverfew, and it will grow in an exposed windy site.

Drainage

Good drainage is required.

Soil Type

Feverfew grows well in a wide range of soils.

Fertility

It needs a moderate level of fertility.

Initial Weed Control

A bare fallow prior to planting will help with weed control, but as Feverfew is not severely troubled by weeds, initial weed control is not as critical as it is with some herbs.

Neighbouring Plants

Feverfew holds its own fairly well and does not encroach on other plants. Like most other woody perennials, it provides shelter for snails, so it should not be put in next to vulnerable plants where snails are a problem.

Identification

Feverfew grows to about 900 mm tall, with much branched stems, initially soft but becoming woody. Its leaves are strongly scented, light green, feathery, pinnate or bi-pinnate, and up to 75 mm long. The many flower heads are 10–20 mm across with yellow centres and white petals. The plant retains its leaves through winter. While it is usually a perennial, it sometimes behaves as a biennial.

Feverfew is often mistakenly called Pyrethrum, which differs in that the flower head is larger and solitary on a long stalk, and the segmenting of the leaves is more open and narrow.

Variety

There is a golden-leaved variety and a double-flowered variety, both of which should be avoided if you are growing for medicinal purposes.

There is some variation in the proportion of leaf to flower the plants produce, so if you are selecting stock, try to choose leafier plants.

Propagation

Method

While Feverfew can be grown from cuttings or divisions, plants grown from seed are more vigorous.

Timing

Seed can be started in spring or autumn, cuttings and divisions in early spring.

Planting

Set the cuttings or divisions well into the ground with no more leaf attached than they can support. Seed is best started in trays or in a nursery bed, as it is so small.

Layout and Spacing

Plant out Feverfew in rows 900 mm apart with 300 mm between plants. For harvesting with the scythe, the rows should be laid out in blocks of three or more.

Compost

Give Feverfew a moderate dressing of compost of 2–3 kg/m^2 in spring, worked in beside the plants. If production declines, Feverfew may respond to more compost.

Weed Control

Regular inter-row cultivation should be kept up as long as possible, with hoeing around the plants.

Rejuvenation

Feverfew takes off with a burst in its first season, but it declines in the second year and levels off to moderate yields, which can be sustained for a number of years. If maximum production is required, it should be replanted from seed each year.

Irrigation

Feverfew needs irrigation at around 25 mm per week over summer.

Pests and Diseases

Being quite bitter, Feverfew is not a favourite food for any pests, but it does become a real snail haven, which can cause problems for nearby plants.

Harvesting

Timing

Feverfew should be harvested just before flowering. Usually four cuts a season are possible.

Tools

The catching scythe is best for harvesting.

Technique

When the Feverfew is young, harvesting is easy and the plant can be cut quite low down, just leaving a small amount of green leaf. If the final product is going to be aerial parts, avoid including an excessive proportion of stem, as only the leaf is active.

As the plants get older and woodier, they become more ragged and irregular in their growth, and harvesting with the scythe becomes a bit more challenging. Older plants need to be harvested quite high up, so just skim off the leafy growth leaving a little green on the bushes.

Post-harvest Treatment

Post-harvest treatment is just a matter of trimming so the bushes are more or less even, making the next harvest easier.

Drying

Feverfew is reluctant to dry and re-moistens easily. If it is excessively remoistened or too slow to dry, it will lose its bright colour. If good drying conditions aren't forthcoming, it may need artificial heat, especially at the end of the season.

Temperature

Feverfew should be dried at temperatures of up to 35°C.

Processing

Only very small volumes of Feverfew are used as a tea, while moderate quantities are made into fluid extract.

Tea Grade

Rub tea grade through a 2$^1/_2$ or 3 dent screen, with under-screens in place, repeating until free of stalk.

Manufacturing Grade

This is one herb where it has definitely been established

that the active constituents are in the leaf, but still the Feverfew offered in the trade is mostly stalk. It is hardly surprising that the properties of Feverfew were forgotten for many years and only rediscovered by people using leaves from their own plants.

If aerial parts are required, they should be leafy without too great a proportion of stem. They can be chopped with a chaff-cutter. Make sure the stalks are completely dry, because they are rather slow to become so.

If the proportion of stem in the aerial parts is too high, this can be overcome by doing a first screening for a manufacturing grade. This will give a much leafier product than chaff-cutting.

Storage
Feverfew should be freeze treated before sale or storage to control moth larvae.

Yields
A bed of Feverfew, which had been established for a number of years at 'Twin Creeks' in north-east Victoria, was yielding 0.3–0.4 kg/m^2 tea-grade leaf per annum, but first year growth can yield up to 0.8 kg/m^2.

Marketing
Feverfew is a herb you need to be careful with. Demand is small to moderate, but can be volatile.

Price
Price to growers for premium quality, organically grown Feverfew is currently around $24/kg for 5-star tea grade.

Price to growers for manufacturing-grade, organically grown Feverfew is currently around $15–20/kg.

Current trade price is around $22/kg.

Gum Plant

Grindelia robusta Nutt. ASTERACEAE

Parts Used: Aerial Parts

Other Names: Tarweed, Gumweed, Shore Grindelia, *Grindelia camporum* (alternative species).

The name Gum Plant is applied to a number of species of *Grindelia* that have similar properties.

It is primarily a respiratory herb, specific in asthma, but also with some use externally for skin irritation, burns etc.

GROWING GUM PLANT
Gum Plant has proven to be very adaptable. While it comes from a climate with hot summers and is reputed to thrive on poor gravelly soil, it will also do very well in cooler regions, in more fertile situations, and on heavier soils.

This is another atypical woody perennial in that it only becomes woody as it gets older.

Identification
Gum Plant grows to about 500 mm tall, with light green, leathery lanceolate leaves, up to 150 mm long. The flowers are yellow, about 30 mm across. The flower buds are covered with a sticky aromatic resin.

The plant stays green through winter.

A number of other species of *Grindelia* are alternative species for the same purposes.

Drying
Gum Plant is very slow to dry. The main problem is the buds covered with aromatic resin, which keeps the moisture sealed in. Putting Gum Plant through the chaff-cutter before drying will not solve the problem, as this would not cut all the buds.

Gum Plant needs to be placed in the warmest part of the drying shed. Allow about three times the normal drying period for other herbs and then carefully check the inside of the buds for dryness.

Two or three weeks after processing and bagging, check Gum Plant again to see if any moisture has come from the buds, making the herb feel damp.

For more detail on this herb see 'Information Charts'.

Gum Plant (*Grindelia robusta*).

Hyssop
Hyssopus officinalis L LAMIACEAE

Parts Used: **Leaf & Flower, Aerial Parts**

Hyssop is another herb with ancient associations, its name being mentioned by Greek herbalists and in the Old Testament. It is has a range of medicinal uses, including use as an expectorant, an antispasmodic, a diaphoretic and a sedative. At times Hyssop has enjoyed some usage as a culinary herb and is still occasionally used in soups and stews.

It makes a very pleasant-flavoured herb tea, quite enjoyable for its flavour and slightly bitter aromatic fragrance.

Origin

Hyssop is a native of central and southern Europe and temperate western Asia, where it grows in sunny situations on dry calcareous rocky soils.

GROWING HYSSOP

Hyssop is a fairly typical woody perennial and adds a special touch to the garden with its deep blue flowers and low hedge-like growth. With a little attention to its requirements, it is not difficult to grow and is a pleasant herb to harvest and dry.

Climate

Hyssop will grow in a wide range of climates.

Site

Exposure

Full sun is needed for Hyssop, and a windy site is okay.

Drainage

Very good drainage is essential, as Hyssop will die if there is any waterlogging.

Soil Type

Hyssop needs a light to medium loam with an adequate level of calcium.

Fertility

Moderate fertility is adequate.

Initial Weed Control

Bare fallow prior to planting will make subsequent weed control easier.

Identification

Hyssop is a small woody-stemmed shrub, growing to about 800 mm. The leaves are dark green, linear, 20–30 mm long and opposite. The flowers are normally dark blue (violet, red, pink and white forms exist) and 7–15 mm long in one-sided whorls in the upper leaf axils (see colour plate 4.3). The plant retains some of its leaves in winter.

Variety

The blue-flowered form is the traditional variety and is preferred for dried herb production.

Propagation

Hyssop can be grown from seed, cuttings or by division. Seed and cuttings are best started in spring, while division can be carried out in autumn or in early spring.

Planting

Hyssop seed is fairly small and the surface needs to be kept moist for germination. Cuttings should taken from semi-hardened stems.

Division is quite easy with Hyssop, provided the parent plant is not too old. A whole clump should be dug up and then carefully pulled apart, endeavouring to obtain pieces of stem with both roots and leaf attached.

Layout and Spacing

Hyssop should be set out on 150–300 mm centres in rows about 1800 mm apart. For harvesting with the scythe, beds should be located so there is room to approach each row from both sides. Alternatively, a single row can be set at the

Hyssop *(Hyssopus officinalis).*

edge of the garden or alongside a low-growing herb, such as Dandelion, which does not mind being walked upon. In either case, allow enough room for the Hyssop to expand to about 1200 mm wide.

Compost

Hyssop needs a moderate dressing of compost at the rate of 2–3 kg/m², worked into the soil beside it in spring.

Weed Control

Weed control consists of regular inter-row cultivation followed by hoeing around the plants. Weed control is more demanding in the establishment phase, but once the Hyssop has developed a continuous canopy, it holds its own fairly well.

Rejuvenation

Hyssop is not particularly long-lived and will tend to die off after 4–5 years. To maintain continuity of production, a new planting should be started every 3 years or so. Then as the old planting fades out, a new one will be coming into full production.

Irrigation

Hyssop needs about 25 mm of water per week through summer.

Pests and Diseases

Snails

Snails love to hide in the base of an established Hyssop hedge and if there are enough of them, they can do considerable damage to it by feeding on the bark and cambium of the stems and on the young leaves. If this is allowed to go on unhindered, it will kill the plants.

If no other controls are available, carefully picking the snails all out each day over a number of days will reduce their numbers to endurable levels.

Grasshoppers

Grasshoppers are generally not too much of a problem with Hyssop, but they do like to eat the flowers. A second harvest won't have much blue in it if they are around.

Harvesting

Timing

Harvest Hyssop early in the flowering stage. This is November–December for the first harvest and January–February for a second. Sometimes a third harvest is possible late in the season.

Tools

Choice of harvesting tools will depend on the form of the Hyssop plants and your ability with the scythe. I imagine most people will prefer to use the reaping hook, but on a good solid stand of Hyssop, the catching scythe can do a reasonable job and is much quicker. It does take some skill though, because the blade needs to be held in an awkwardly high position and there is not much tolerance in the depth of herb to be cut off.

Technique

Hyssop should be cut fairly high on the plants, leaving a short length, say 25 mm of leaf-bearing stem to enable good regeneration. Generally the point to cut them is a little below the lowest flowers. How far below will depend on the length of the new growth and the desired shape of the bushes.

Harvesting young plants should be more conservative, to allow them to fill out. The aim is to establish a continuous hedge as soon as possible.

Older plants should be cut closer, so as not to allow them to get too leggy and break apart. Once they have gained full size, they should be cut back to pretty much the same level each harvest. There just need to be a few pairs of green leaves left on the branchlets.

Post-harvest Treatment

Post-harvest treatment is just a matter of going over the plants and evening up any ragged parts so there is a smooth profile to the bushes.

Drying

Hyssop is quite easy to dry.

Temperature

Hyssop should be dried at a temperature of not more than 35°C.

Processing

Hyssop is used to some extent as a tea, and aerial parts are used in making a fluid extract. Its culinary use is too limited to warrant a separate culinary grade.

Tea Grade

Tea grade is processed by rubbing the dried herb through a $2^{1}/_{2}$ or 3 dent screen, with under-screens in place, and repeating the process until it is free of stalk. It should come through with a fairly dark green leaf contrasting with the dark blue of the flowers.

Aerial Parts

Aerial parts can be fed through the chaff-cutter. Check first that the stems are all thoroughly dry.

Storage

Freeze treatment to control moth larvae is necessary before sale or storage.

Yields

In north-east Victoria we obtained yields of up to 0.4 kg/m^2 tea grade or 0.65 kg/m^2 aerial parts per annum in the second and third years, but in the fourth year yields dropped as the Hyssop went into decline.

Marketing

In spite of its wonderful qualities, Hyssop is only a minor herb with a small but diverse demand.

Price

Price to growers for premium quality, organically grown Hyssop is currently around $24/kg for 5-star tea-grade.

Price to growers for manufacturing-grade, organically grown Hyssop is currently around $10–16/kg for aerial parts.

Current trade price is around $11–12/kg.

Lavender, English

Lavandula angustifolia Mill. LAMIACEAE

Parts Used: Flower

Other Names: *Lavandula officinalis, L. vera.*

English Lavender is widely recognised as a herb used in perfumery, cosmetics and to be kept in little sachets to sweeten up undies drawers. The dried herb also has medicinal uses as a carminative, antispasmodic and antidepressant.

GROWING ENGLISH LAVENDER

English Lavender will grow well in most climates, except for humid subtropical areas where it will not flower successfully. It is grown conventionally on a large commercial scale on a few farms in Victoria and Tasmania for the oil and potpourri trade, but there is a small market for organic English Lavender.

Identification

English Lavender is a woody perennial growing to about 800 mm tall. Leaves are opposite, narrow-lanceolate or oblong-linear, 20–60 mm long, up to 6 mm wide, grey to green, covered in fine hairs. Flowers are purple, but can be white, grey-blue, pink or dark purple, tubular, 6–15 mm long in loose spikes to 80 mm long. The plant retains its leaves all winter.

English Lavender can be distinguished from other lavenders by its long, narrow, smooth-edged leaves. Other species have leaves which are toothed or frilled or else shorter.

Variety

There is a considerable variation in colour and oil characteristics, so a variety should be selected which is suitable for the intended market. For the dried herb, the dark purple variety 'Vera' looks more attractive.

Propagation

Propagation is by means of ripened soft-stem cuttings in spring or autumn, or by layering. Growth from seed is very slow.

Harvesting

Much Lavender is dried and sold in bunches, which brings a better price.

To produce rubbed flowers, the Lavender is cut with a reaping hook, dried on the stems, separated by rubbing through a 2½ or 3 dent screen, and cleaned by further screening or winnowing.

On a larger scale, the flowers are harvested mechanically.

There is one main harvest December–January, with a possible small second one March–April.

For more detail on this herb see 'Information Charts'. Seminars on Lavender production are run by Yuulong Lavender, telephone (03) 5368 9453, fax (03) 5368 9175.

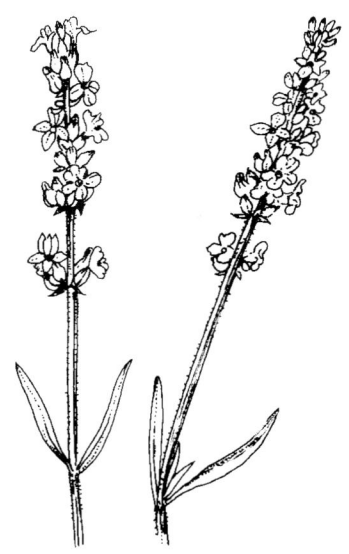

English Lavender *(Lavandula augustifolia).*

Lemon Thyme

Thymus* x *citriodorus (Pers.) Schreb. ex Scheigg. & Korte   LAMIACEAE

Parts Used: **Leaf and Flower**

Lemon Thyme is a species of Thyme with a delightful lemon aroma and flavour and makes a delicious full-flavoured herb tea.

It is primarily a beverage herb and does not have any reputation for medicinal use.

Origin

Horticultural: it arose as a hybrid between *Thymus vulgaris* and *T. pulegioides*.

GROWING LEMON THYME

Lemon Thyme is a fairly easygoing plant, which creates a delightful carpet of blossom when in flower.

Climate

Lemon Thyme will grow in a wide range of climates.

Site

Exposure

Full sun is required, and Lemon Thyme will grow in a windy situation.

Drainage

Good drainage is needed.

Soil Type

Lemon Thyme prefers a light to medium loam.

Fertility

It needs a fertile soil for sustained production.

Initial Weed Control

Initial bare fallow before planting is recommended, as weed control during the establishment phase is fairly demanding.

Neighbouring Plants

As Lemon Thyme is a low-growing species, it needs to be away from plants that will overshadow it.

Identification

Lemon Thyme is a small woody-stemmed shrub growing up to 300 mm tall. Leaves vary in colour from dark to light green, sometimes variegated, opposite, ovate to lanceolate, up to 10 mm long. The flowers are pale blue, in small oblong inflorescences at the tops of the stems (see colour plate 4.4).

The plant retains its leaves all winter.

The main distinguishing characteristic of Lemon Thyme is the lemon aroma of the leaves, though during winter the lemon component of the plant's aroma declines. Lemon Thyme's leaf is broader than that of Common Thyme.

One authority mentions inferior lemon-scented varieties of Common Thyme (*Thymus vulgaris*), but I am not familiar with these.

Variety

For a dried herb product, the dark green variety is preferred for its better appearance when dried.

Propagation

Method

Propagation is by means of division or layering. Being a hybrid, it cannot be grown from seed.

Timing

Division can be done spring or autumn, layering should be started in autumn.

Planting

Quite often Lemon Thyme plants will spontaneously layer themselves where the straggly edges rest on the ground. This can be encouraged by throwing some soil onto these prostrate stems in autumn to foster the development of roots.

Layering gives good propagation material because the stems are fairly fine and numerous and not too woody. The pieces can be separated easily without much damage once they are cut off from the main bush.

Alternatively, plants can be dug up and divided by pulling them apart. Relatively young plants with a multitude of stems are best for this.

Layout and Spacing

Lemon Thyme should be set out on 150 mm centres in rows about 600 mm apart. After a season or so, these will have filled out all the spaces so the crop is growing in a solid mat. This solid growth makes weed control and harvesting much easier.

For harvesting with the scythe, the rows should be in blocks of at least four across.

Lemon Thyme can be laid out in rows, which will expand to form a solid block after two seasons or so.

Compost

Lemon Thyme normally needs a moderate dressing of compost at the rate of 2–3 kg/m².

While there is still space between the rows, it is easy to get compost on, but once the bushes fill out and meet, putting compost on becomes a problem because the soil between the plants is virtually inaccessible (see 'Problem Crops' in Chapter 7, 'Compost').

Weed Control

Weed control needs to be attended to very thoroughly in the first season or so, with regular inter-row cultivation, hoeing and hand-weeding around the plants. If good weed control is maintained during this establishment stage, then when the Lemon Thyme has filled out all the gaps, there will be a solid carpet of growth, with very little opportunity for weeds to get established.

Rejuvenation

In my experience, Lemon Thyme goes into a slow decline after about five seasons. I suspect this is mainly because it is difficult to get adequate compost onto it. If no means of overcoming this is available, it may be necessary to replant the Lemon Thyme in a new bed.

Irrigation

Lemon Thyme needs about 25 mm of water per week through summer.

Harvesting

Timing

Harvest when the Lemon Thyme is fully flowering, which is usually November for the first harvest of the season, with another cut 2 months later, around January, and a third about 3 months later, in April.

As the regrowth tends to flower irregularly, with some shoots starting well before others, you will need to strike a balance in determining when to harvest.

Tools

When I first started growing herbs I always found harvesting the Thymes tedious and slow. Initially I used shears and reaping hook, then I experimented with 12 volt hedge trimmers, but found them cumbersome and hard on my back to operate at such a low level.

Then I discovered it was possible to cut these low-growing woody perennials with the catching scythe. Now I only use the other harvesting tools when these species are in the establishment stage and there are still gaps between the plants.

Technique

It is important to cut the Lemon Thyme at the right height, neither too low nor too high. There need to be a few green leaves remaining on each sprig for regrowth: generally you can take up to ³⁄₄ of the green leaf. Where the plant is cut back too hard, it will be very slow to recover and may die off.

On the other hand, if not enough is taken off each harvest, the plants will become very long and floppy, which makes subsequent harvesting difficult.

I had been afraid to use the catching scythe on the Thymes because I was afraid of scalping the bushes or leaving long tufts. I had always imagined the tops of the plants as being too springy to cut easily, but when I tried scything them I found it wasn't so difficult after all, provided there was a good density of growth and the blade was nice and sharp. Cutting early in the day makes it easier, as the plants are stiffer then because their water content is higher.

One thing to be wary of when cutting Lemon Thyme in full flower is that there will be a lot of bees avidly working the

Harvesting of Lemon Thyme can be done by skimming it with the catching scythe, leaving sufficient leaf on the stems for regeneration.

flowers. I have never been stung while harvesting, but I have cut a few unfortunate bees' wings off. I found this can generally be avoided by sweeping the scythe backwards over the plant before cutting it. As the back of the blade strokes the tops, the bees are disturbed and fly on a little way.

As the harvest proceeds, the bees become concentrated, thicker and thicker, in the last remaining corner of the Lemon Thyme, which can be fairly seething with buzzing wings.

Scything the edges of the patch can be a bit tricky, particularly in getting a clean cut without bringing up soil and debris.

Post-harvest Treatment

Usually it is necessary to go over the bed with a reaping hook or shears after harvest and tidy up the edges and any lanky tufts.

Drying

Lemon Thyme dries easily.

Temperature

The temperature should not be more than 35°C.

Processing

Lemon Thyme is purely a tea herb.

Tea Grade

First rub Lemon Thyme once through a $2^1/2$ or 3 dent screen, with under-screens in place. For best results, a 4 dent screen (with apertures 5 mm across) can be used to further screen the Lemon Thyme until it is sufficiently free of stalk. (Screening and winnowing is another option.) Some short stalks remaining in the finished product are inevitable, as they are very fine and difficult to separate. If they are less than 25 mm long, they won't cause problems.

Rubbing can be made much easier if you can catch the Lemon Thyme at a stage when the leaf is brittle dry but the stalk is still a little moist. This slightly pliable stalk will be less inclined to spear down through the screens and so will be easier to separate.

Storage

Freeze treatment to control moth larvae is necessary before sale or storage.

Yields

At 'Twin Creeks' in north-east Victoria, we obtained yields of $0.2–0.3$ kg/m^2 tea grade per annum. This level was reached in the second season, continued about 5 years, then started to decline.

In cooler regions yields should be similar.

Marketing

Lemon Thyme is not widely available in the herb trade, but virtually everyone who tastes it enjoys its rich lemon flavour. With continuity of supply it has the potential to develop into a major herb tea, but this may take a while because it is still largely unknown.

Price

Price to growers for premium quality, organically grown Lemon Thyme is currently around $24/kg for 5-star tea grade.

Price to growers for second-grade, organically grown Lemon Thyme is currently around $20/kg.

Marjoram, Sweet

Origanum majorana L LAMIACEAE

Parts Used: Leaf & Flower

Other Names: Knotted Marjoram, *Majorana hortensis*.

Sweet Marjoram was well known to the Greeks and Romans and has a long tradition of culinary use. It also has a history of medicinal use and is still occasionally used as a digestive and an expectorant.

GROWING SWEET MARJORAM

Sweet Marjoram is a somewhat delicate plant and in some situations it is hard to get it to survive the winter. This can be overcome with some care in selecting the right position for it: good drainage is essential.

Its flavour is so delightfully sweet and fragrant that it is worth going to some trouble to succeed with it.

It will grow in a range of climates, though very heavy frosts will kill it. This is more of a problem in Europe and North America than in Australia where there are not many places too cold for it.

Don't make the mistake of mulching Sweet Marjoram heavily to protect it from frost as is sometimes advised: this will only increase frost damage in woody perennials. The mulch prevents warmth from the soil reaching the leaves, so the frost hits them harder.

Identification

Sweet Marjoram is a small woody shrub 300–400 mm with branched stems covered with short fine hairs. The leaves are sweetly aromatic, elliptic, opposite, grey-green, 10–30 mm long. The flowers are small and insignificant, in tightly clustered spherical spikes or 'knots'.

The plants retain some green leaf all winter.

There is a lot of confusion between Sweet Marjoram, Pot Marjoram and Oregano, and many people don't draw much of a distinction between them. From the point of view of high quality herb production, there is a vast difference in flavour, so it is important to have them sorted out.

Sweet Marjoram lives up to its name, with its lovely sweet fragrance; Pot Marjoram is a dull-flavoured herb with just a hint of the fragrance of Sweet Marjoram; a good Oregano will be spicy and burn your tongue.

Sweet Marjoram is different from the other two as it retains its miniature woody shrub form throughout its growth. Pot Marjoram and Oregano are expanding clump herbs, which send up an erect stalk carrying leaves and flowers, but die down to a low leafy growth over winter.

Variety

A good sweet-flavoured variety is needed.

Weed Control

As Sweet Marjoram usually doesn't seem to form into a solid bed like the Thymes, weed control is more demanding. They are not very competitive plants, especially when young, so it easy for weeds to get a hold among them. Regular inter-row cultivation followed by hoeing and hand-weeding around the plants is needed.

Management

Beware of soil splash with this herb, because it is low growing with fine hairs on the leaves and there is usually bare soil around the plants. If there is any amount on the leaves, it can be a serious contamination problem, especially in a culinary herb, as people don't appreciate gritty omelettes.

If soil splash is a continuing problem in your beds, then susceptible herbs like Marjoram might need a light layer of non-blowing mulch, such as chopped herb stalks.

Rejuvenation

Sweet Marjoram is not a long-lived plant in the best of conditions, and needs to be replaced every 3–4 years or so. It is probably best to start it afresh in a new bed, rather than keep filling gaps in an old bed of declining plants.

For more detail on this herb see 'Information Charts'.

Sweet Marjoram (*Origanum majorana*).

Rosemary

Rosmarinus officinalis L LAMIACEAE

Parts Used: Leaf, Leaf & Stem

Rosemary has been recognised as a medicinal herb since earliest times and today is mainly used as a circulatory and nervine tonic. Like many popular culinary herbs, it also has digestive properties.

Users of Rosemary should be aware that very large doses can be toxic. Curiously, this fact has not drawn the attention of regulatory authorities, and being such a widely used culinary herb, it would be a rather awkward one to legislate on.

Origin

Rosemary is native to the Mediterranean coast.

GROWING ROSEMARY

Rosemary is one of those herbs you have to plant and then wait for a while until it gets big enough to harvest. It can be grown as a hedge or allowed to develop into large bushes.

Climate

Rosemary will grow in a wide range of climates. In Australia it is only slightly affected by the hardest frosts.

Site

Exposure

Rosemary enjoys full sun, but it will grow in semi-shade. It will tolerate a windy site and can be used to provide some shelter for other herbs.

Drainage

Good drainage is essential, as Rosemary won't tolerate waterlogging.

Soil Type

A light sandy loam is preferred.

Fertility

Only a moderate level of fertility is required.

Initial Weed Control

A bare fallow prior to planting is preferable but not essential as the plants are so widely spaced inter-row cultivation is easily maintained.

Neighbouring Plants

Remember Rosemary keeps getting larger and larger, so room must be left to allow for this. If space is valuable, in the first year or so annual crops could be grown between the rows.

Rosemary tends to be a snail haven, so it should not be placed next to vulnerable species. Being tall, it can cast a sprinkler shadow on neighbouring plants.

It is fairly drought resistant, so it can safely be placed alongside the aggressive thirsty herbs that are sometimes a problem to find neighbours for.

Identification

Rosemary is a woody-branched shrub, which will grow up to 1800 mm tall. The aromatic leaves are opposite, linear, green above and white below, up to 35 mm long.

The flowers are small, normally pale blue (but deep blue, white and even pink forms exist) and in short racemes in the leaf axils. The plant retains its leaves all winter.

Variety

There are a number of cultivars, some selected for flower colour and some for growth form. Avoid the prostrate and dwarf varieties, as they are not as productive. The best thing is to take cuttings from plants with desirable growth characteristics.

Propagation

Method

Growing from cuttings is the easiest form of propagation. Rosemary can also be grown from seed, but germination can be slow.

Timing

Cuttings can be started in autumn or in spring. Seed can be started in spring.

Planting

Take 150–225 mm cuttings with a 'heel' and with the leaves stripped from the lower two-thirds. Set them to that depth in sandy soil, which is kept moist. This is easiest to maintain in the open garden if it is done in mid-to-late autumn. The cuttings will then be rooted and may be ready to plant out next spring, though they may be best left until the following autumn. It will be obvious by then which are good vigorous plants, and autumn is a good time to transplant them. A 50–90% strike rate can be expected with cuttings started in the open like this.

The seed is rather small and will need to be started in a tray (as described under 'Growing Herbs from Seed' in Chapter 5, 'Propagation and Planting').

Layout and Spacing

Rosemary should be set out on 600 mm centres in rows about 1800 mm apart. If a dense hedge is required, closer spacing would achieve this sooner.

Compost

Rosemary needs only a modest dressing of compost each spring at the rate of around 2 kg/m^2. In fertile situations it will need very little, if any.

Weed Control

Weed control is easily maintained with regular inter-row cultivation and hoeing around the plants. Once Rosemary reaches some size, weeds are not a problem and the area around the bushes could be maintained by mowing, provided this did not create a weed problem for adjacent crops: otherwise it should be kept cultivated.

Irrigation

Rosemary does not need much water, perhaps 13 mm per week through summer for better growth. In many locations it could be grown without any irrigation as it is quite drought tolerant. Avoid overwatering, however, as this may harm it.

Pests and Diseases

Not too many creatures are likely to take a fancy to chewing on Rosemary.

Sometimes the odd bush will die and it is hard to figure out why. As Rosemary is very sensitive to excess water, waterlogging or overwatering could be the cause of this. Excess fertiliser may also be a cause.

Harvesting

Timing

Harvest Rosemary once a year in late summer to early autumn, after the new growth has hardened off and before flowering. If the new growth is cut too early, it will turn black as it dries. If Rosemary is harvested while flowering, the flowers dry to trashy looking stuff which mixes in with the leaves and affects the appearance of the dried product.

Tools

Harvesting Rosemary is a job for the reaping hook.

Technique

Cut the stems just above the hard wood and above any brown or faded leaves, making sure there is enough leaf retained on the bush.

Harvesting is really just pruning and should be done with an eye for the desired form of the bushes.

Sometimes larger branches need to be taken off but rather than dry and rub these as whole branches, the good material should be cut from them for drying and the rest should be discarded, as the older parts of the branch can retain a lot of trashy dead leaf and debris, which will affect the quality of the finished product.

Post-harvest Treatment

Following harvest, prune off any yellow or otherwise undesirable foliage remaining on the bushes.

Drying

Rosemary needs to dry reasonably quickly to retain a good green colour. If it is dried too slowly, or excessively re-moistened during drying, it will turn a dull brown colour.

Temperature

Rosemary should be dried at a temperature of not more than 35°C.

Rosemary (*Rosmarinus officinalis*).

Processing

While some Rosemary does get used as a tea, most of it will be used as a culinary herb. The culinary grade is mostly whole leaf, so this is quite satisfactory for making tea. There is also some requirement for Rosemary for manufacturing fluid extract.

Culinary Grade

For culinary grade, rub through a 2½ or 3 dent screen, with under-screens in place and repeat this, or winnow until a stem-free product is achieved.

Manufacturing Grade

The requirements for manufacturing grade should be determined before processing. We always used to supply a first screening for manufacturing purposes, which removed 80–90% of the stem, but I note some manufacturers are specifying leaf and stem. Considering the price they are offering, the herb would need to be bulked out with stem to make it worth while for the grower.

Leaf and stem would need to go through the chaff-cutter.

Storage

Freeze treatment to control moth larvae is necessary before sale or storage.

Yields

Yields of up to 0.2 kg/m² culinary grade per annum can be expected after 2–3 years. This may increase somewhat over the following 5–6 years.

Marketing

There is only a very limited market at present for organic, culinary-grade dried Rosemary, and a small requirement for manufacturing use, so growers would need to confirm they had a definite market before committing themselves to significant volumes.

Price

Price to growers for premium quality, organically grown Rosemary is currently around $24/kg for 5-star tea grade.

Price to growers for manufacturing-grade, organically grown Rosemary is currently around $10–13/kg.

Current trade price is around $5/kg.

Rue

Ruta graveolens L   RUTACEAE

Parts Used: Aerial Parts

Other Names: Herb of Grace, Herby Grass.

Rue has long been an important element in herbal medicine. It was a major ingredient in Mithridates' poison antidotes and was also used as an antimagical herb. Its use today is mainly as an antispasmodic and for sprains and ligament injuries.

It also is used as a culinary herb, but some discretion is necessary on account of its strong flavour. It can be used in salads and is an ingredient of the Italian grape spirit, 'Grappa con Ruta'.

Rue is also known as Herb of Grace. When I first came to Tasmania I was staying in a small rural town where I noticed one of the houses had a little sign up which read 'HERB OF GRACE'. Curious as to what this designated, I was told the fellow who lived there used to work for Grace Removals and his name was Herb. Consequently he had become known as 'Herb of Grace'.

GROWING RUE

Rue is quite a hardy and adaptable plant that is easy to grow, though some strains are short-lived and it does need good drainage.

It is about the strongest smelling herb in the garden, with an aroma which is not to everyone's liking. Just brushing against it can be rather overpowering.

Identification

Rue is a many-branched, woody and fleshy stemmed shrub growing to about 1000 mm. The alternate leaves are intensely aromatic, bluish green, bipinnate or tripinnate, with segments up to 15 mm long. Flowers are yellow, in terminal inflorescences (see colour plate 6.4).

Some strains are semi-deciduous, while others retain their leaves all winter.

Variety

There seems to be some considerable variation in the form and longevity of Rue. I started out with three different forms and found the bushier, leafier strain which retains its leaves all winter was longer-lived and more productive than other strains with a more leggy form. These latter lost most of their leaves in winter and only lived for 2 or 3 years.

Drying

Rue would to be one of the most difficult herbs to dry. Its stems are rather fleshy and if it is just spread on a screen to dry in the normal way, even in the heat of summer it will not be dry a month later.

The only way to get a good result is to chop it before spreading it on the screens. This is best done with the chaff-cutter, which cuts it nicely. This way it will dry through the cut surfaces, though still rather slowly. Fleshy herbs like this do not seem to bruise: Rue comes out an attractive green colour when dried this way.

It is important to check the herb thoroughly in case there are any longer or thicker pieces and fruiting bodies that have not yet dried.

Temperature

Rue should be dried at a temperature of not more than 35°C.

For more detail on this herb see 'Information Charts'.

Sage
Salvia officinalis L LAMIACEAE

Parts Used: Leaf & Flower, Aerial Parts

Other Names: Red Sage, Broad-leaf Sage, Narrow-leaf Sage.

Sage is well known as a culinary herb and also enjoys some usage medicinally for its antiseptic, anti-inflammatory, astringent and digestive properties. It has a tradition going back into ancient times.

Origin

Sage is a native of southern Europe, particularly the northern shores of the Mediterranean, where it grows on limestone soils in full sun.

GROWING SAGE

Sage is quite easy to grow, given the right conditions, but the preferred varieties do not seem to be available in Australia. It is something of a fiddly plant to harvest and process for culinary grade using small-scale techniques, though tea grade is less demanding.

Climate

Sage will grow in a wide range of climates, though it may suffer a little from the heaviest of frosts.

Site

Exposure
Full sun is preferred and Sage will tolerate some wind.

Drainage
Good drainage is essential as Sage will not thrive if there is excess water.

Soil Type
Sage will grow in a wide range of soils.

Fertility
A moderately fertile soil is required.

Initial Weed Control
A bare fallow prior to planting is important, as this will reduce weed problems later on.

Neighbouring Plants
Sage is a wonderful haven for snails, so place it away from vulnerable plants.

Identification

Sage is a small woody-stemmed shrub, up to 700 mm tall. The young stems are square, white and woolly. The aromatic grey-green leaves are opposite and ovate to lanceolate, up to 80 mm long with rough surfaces. Flowers are bluish-purple, up to 30 mm long and in terminal spikes (see colour plate 6.7).

Sage loses most of its leaves in winter, but does retain some.

Sage *(Salvia officinalis)*.

Variety

There are a number of varieties of Sage. Some have been selected for ornamental qualities, such as variegated or different coloured leaves, while others are preferred for medicinal usage or for their productivity.

The commonly available variety is Narrow-leaf Sage, which can be grown from seed. However it tends to go to flower rapidly, which reduces yields of leaf.

English books mention preferred varieties that don't seem to be available in Australia, perhaps because they can't be grown from seed. They are Broad-leaf Sage, which is reluctant to flower and produces greater amounts of good quality leaf, and Red Sage, which has reddish stems and leaves and is the traditional medicinal variety.

I did find a reddish coloured variety of Sage here, but in the end I discarded it as being merely ornamental. Its essential oil content was quite low so it did not have the flavour of the ordinary Sage.

Propagation

Method

Narrow-leaf Sage can be grown from seed, from cuttings, by division or by layering.

Other varieties will not grow true from seed, so they must be propagated vegetatively from cuttings, by division or layering.

Sage does not layer much spontaneously, as its growth is too upright, but it can be encouraged to do so by pegging stems down or by heaping soil up around the plant.

Timing

Seed can be sown in spring. Cuttings can be started in spring or autumn, as can divisions. Layering is best started in autumn.

Planting

Sage seed is relatively large as herb seeds go, so it is not hard to get started. In some places it is treated as an annual and sown from seed each year.

Cuttings can be handled as described for Rosemary.

Division is sometimes possible if the plants are not too leggy and they have a number of stems arising from the base.

Layering is a good option, you just need to plan in advance, so your stems have struck roots when you need them for transplanting.

Layout and Spacing

Sage should be set out on 150–300 mm centres in rows about 600 mm.

For harvesting with the scythe, these should be in blocks at least three rows wide. These blocks will usually fill out so the plants are more or less contiguous, but Sage won't form as dense a cover as the Thymes.

Compost

Sage needs compost at the rate of around 2–3 kg/m², worked into the soil beside the plants.

Weed Control

Regular inter-row cultivation with hoeing around the plants will keep Sage free of weeds. Once established, it holds its own fairly well.

Rejuvenation

Sage is often short-lived and the beds may need to be replaced in a new spot every 3–5 years.

Irrigation

Sage needs about 25 mm of water per week through summer for good growth.

Harvesting

Timing

Narrow-leaf Sage must be harvested as soon as it starts to go flower, as it stops producing leaves at this point and quality deteriorates as flowering progresses, mainly because of the additional stem produced and the coarse fruiting bodies, which don't look very attractive in the dried herb.

Harvesting will encourage new leaf growth. First harvest is quite early in the season – around November – and there should be at least three cuts a year.

Tools

Young plants should be harvested with the reaping hook, but this is tediously slow as only a small amount can be cut with each stroke. When the plants are big enough, they can be cut with the catching scythe.

Technique

As with other woody perennials, Sage needs to be harvested with care to leave enough leaf on the stems to enable it to regrow. Try to avoid the old discoloured leaves: these tend to hang on to the stems and get picked up in the harvest.

Post-harvest Treatment

Any dead or deteriorating stems should be pruned out after harvest.

Drying

Sage is a bit slow to dry and will turn brown if drying is too prolonged or if it is subjected to excessive re-moistening during drying.

Temperature

Sage should be dried at a temperature of not more than 35°C.

Processing

Sage is used as a culinary herb and as a tea herb. For these it needs to be rubbed through appropriate sized screens. It is also used for manufacturing fluid extract, for which aerial parts are used.

Tea Grade

There is enough demand for Sage for tea use to justify doing a tea grade. For this it is rubbed through a 2½ or 3 dent screen, with under-screens in place, and then re-screened or winnowed until it is free of coarse stem.

Culinary Grade

For culinary use it is best to do an initial first screening through a 2½ or 3 dent screen and then rub it through an 8 dent screen, usually two or three times until it is free of all fine stem and trash.

Aerial Parts

For this simply feed the dried leaf, flower and stem through the chaff-cutter. Check carefully that they are thoroughly dry as Sage is a bit slow to release its moisture. The upper parts of the stems and the young leaves are the slowest parts to dry.

Storage

Freeze treatment to control moth larvae is necessary before sale or storage.

Yields

By its second year, Sage should be yielding approximately 0.2–0.35 kg/m² tea or culinary grade or 0.3–0.5 kg/m² aerial parts per annum, but this will probably decrease as the plants age.

Marketing

There is a moderate demand for Sage and being a herb with multiple uses, there is some flexibility in marketing.

Price

Price to growers for premium quality, organically grown Sage leaf is currently around $24/kg for 5-star tea or culinary grade.

Price to growers for manufacturing-grade, organically grown Sage is currently around $12/kg for aerial parts and $12–16/kg for leaf.

Current trade price is around $5–9/kg.

Thyme

Thymus vulgaris L LAMIACEAE

Parts Used: **Leaf & Flower, Aerial Parts**

Other Names: Common Thyme, Garden Thyme.

Thyme is very well known as a culinary herb and also has important uses medicinally, mainly for respiratory and digestive complaints.

Origin

Thyme is a native of the western Mediterranean and southern Italy.

GROWING THYME

Thyme is a delightful plant to grow and is easy to get on with. It has a special quality about it, especially when it has formed a solid carpet, pink with flowers and humming with bees.

Its requirements and management are very similar to Lemon Thyme.

Climate

Thyme grows in a wide range of climates.

Site

As for Lemon Thyme.

Identification

Thyme is a small shrub with fine woody stems, often gnarled and twisted. It grows 100–300 mm tall. The grey-green aromatic leaves are opposite, linear to elliptic and up to 15 mm long. The flowers are white to lilac, small, and in many-flowered terminal inflorescences (see colour plate 7.5).

Thyme retains its leaves through winter and in most parts of southern Australia is winter-active.

Thyme *(Thymus vulgaris).*

Wild Thyme (*T. serpyllum*) differs in that it normally has a creeping habit, with broader bright green leaves. It lacks the flavour of common Thyme.

There is a multitude of varieties and cultivars of Thyme, with different growth characteristics, leaf colours and flavours.

Variety

The preferred variety is the Common or Garden Thyme, but make sure you have a good upright growing strain, as Thyme is short enough at the best of times. Growers should ascertain whether the variety they are intending to grow is suitable for their prospective buyers in terms of flavour and essential oils.

Propagation

Method

Thyme can be grown from seed, or can be propagated by division and layering.

Timing

Seed should be started in early spring, division can be carried out in early spring or autumn, and layering is best done over winter, starting the process in autumn.

Planting

The seed of Thyme is very small, so it needs to be started in a seed tray (as described for fine seeds under 'Growing Herbs from Seed' in Chapter 5, 'Propagation and Planting).

Division and layering

As for Lemon Thyme.

Sometimes in spring in cooler climates quite a few self-sown seedlings can be found beside mature Thyme plants and these are quite good for planting out.

Layout and Spacing

As for Lemon Thyme.

Compost

As for Lemon Thyme.

Weed Control

As for Lemon Thyme.

Rejuvenation

As for Lemon Thyme.

Irrigation

Thyme needs around 25 mm per week through summer for good growth.

Harvesting

Timing

Thyme should be harvested in the early flowering stage in spring, usually October to early November. It normally does not flower again during the season, but in good conditions there can be one or two further harvests of leaf.

It virtually stops growing in hot summer conditions, but in milder regions it will make reasonable growth in summer and three harvests per season are possible.

Tools

As for Lemon Thyme.

Technique

As for Lemon Thyme.

Post-harvest Treatment

As for Lemon Thyme.

Drying

As for Lemon Thyme.

Processing

Thyme is used as a culinary herb, as a tea herb, and for manufacturing fluid extract.

Tea Grade

For tea grade, the herb is rubbed through a 2$\frac{1}{2}$ or 3 dent screen with under-screens in place. This is best done while the stems are still a bit flexible, as then they will be less inclined to break up and go through the screen. The Thyme should be rescreened through a finer screen or winnowed until it is free of all long pieces of stem, but some short pieces are acceptable in a tea grade as long as there are not too many of them and they are less than 25 mm long.

If you are also preparing a culinary-grade Thyme, you will find it easier to take your first screening and use it to process your culinary grade first and then prepare your tea grade from the screened out portion (see below).

Culinary Grade

Thyme for culinary use needs to be free of stems, which are a bit hard to swallow. For this the herb is best screened first through a 2$\frac{1}{2}$ or 3 dent screen, as above, and then further rubbed through an 8 dent screen to remove all the stem.

This is most easily done by rubbing each loading until about half of it has gone through. Very little of the stem will go through with this first half: most of it would go through with the second half, if you continued rubbing.

Instead, just tip this latter half to one side each time. You will end up with two piles: the rubbed pile will be mostly leaf with very little stem and the remainder will be partly leaf and partly stem.

The first pile will take only a little further screening to achieve a very clean culinary grade. The second pile would take forever to make a decent culinary grade, but with a few screenings enough stem can be removed from it to make a good tea grade or manufacturing grade.

The alternative of trying to make culinary grade out of the whole batch is very time consuming and tedious, as the herb will have to be rescreened many times. Winnowing Thyme is difficult because the leaf is small and relatively dense, while the stem is fine and relatively light, so they behave similarly in the winnower and don't separate easily.

Manufacturing Grade

There are several possibilities for processing manufacturing-grade Thyme. The dried herb could be put through a chaff-cutter for aerial parts. This will contain a high proportion of stem, but if your manufacturer is insisting on a rock-bottom price, this may be one way of meeting that price, with a bulkier but poorer quality product. However, this would not meet the specifications of some manufacturers.

Alternatively, a first screening could be done, which would give a better quality product.

A third possibility is to offer the screened-off portion from culinary-grade screening. This could be rescreened once, if necessary, to remove some of the stem. If the culinary grade brings a good price, you can afford to sell the leftovers at a price the manufacturers are willing to pay.

One advantage of screening the manufacturing grade, rather than chopping it, is that in the event of fluctuations in demand, it can be converted into a tea grade or even a culinary grade with further screening.

Storage

Freeze treatment to control moth larvae is necessary before sale or storage.

Yields

At 'Twin Creeks' in north-east Victoria, yields varied from 0.2 to 0.35 kg/m^2 tea and culinary grade per annum.

Marketing

There is a moderate demand for Thyme and some flexibility in marketing, as it is used in several different aspects of the herb trade.

Price

Price to growers for premium quality, organically grown Thyme is currently around $24/kg for 5-star tea or culinary grade.

Price to growers for manufacturing-grade, organically grown Thyme is currently around $12–15/kg.

Current trade price is around $4–5/kg.

19

Trees and Shrubs

Balm of Gilead Poplar – *Populus candicans*
Bay – *Laurus nobilis*
Cascara Sagrada – *Rhamnus purshiana*
Chaste Tree – *Vitex agnus-castus*
Crampbark – *Viburnum opulus*
Elder, Black – *Sambucus nigra*
Ginkgo – *Ginkgo biloba*

Hawthorn – *Crataegus monogyna, C. laevigata*
Horse Chestnut – *Aesculus hippocastanum*
Lemon Verbena – *Aloysia triphylla*
Linden – *Tilia* spp.
Rose, Dog – *Rosa canina*
Willow – *Salix* spp.
Witch Hazel – *Hamamelis virginiana*

A number of trees are used as herbs, though technically this is a botanical contradiction, as a herb to a botanist is a non-woody plant. The term 'herb' is used here to refer to those plants with medicinal, culinary or beverage uses.

As a group, trees and shrubs are a little more diverse than other groups, varying from mere bushes to enormous trees 80 m tall.

Planting trees and shrubs is a long-term project, which doesn't mean it should be put off for another time when the future may seem a bit clearer or when you can better afford to expend the energy on planting them. Rather it means if you don't get them in now, it will be that much longer you or your children will have to wait until they have grown.

It also means you will have to give a bit of thought to the long-term effects of planting them. A location needs to be suitable not only for the growth of the trees or shrubs, but also in terms of the influence they will have on the area around them.

Trees cast shade, reduce wind, take up moisture, drop leaves and branches, shelter birds and animals, and create a variety of microclimates around themselves, including frost pockets and frost shelters.

Trees can take up a lot of space, which may not be easy to visualise when you plant a tiny seedling. They also vary enormously from species to species in rate of growth and the period between planting and harvesting.

A well-planned tree and shrub planting program should be an integral part of any herb garden. Trees can be used to effectively improve the microclimates of your growing area by creating shelter from wind, by creating a greater diversity of growing situations, with cooler sites on their south sides, warmer sites to the north of them and drier sites close to them.

Trees can create problems though: for one thing, they drop leaves. While these make an important contribution to the soil when they break down, before they reach the ground they can lodge among leaf-crop herbs nearby. If the foliage of the herbs is of a nature which tends to catch and hold any debris, the fallen leaves may be gathered in with the harvest.

Another major problem with trees and shrubs, particularly the larger ones, is their roots. Most have extensive surface root systems, which travel quite a distance in search of water and nutrients. When they find your composted and irrigated garden beds, they quickly spread through them to take up as much as they can, to the detriment of your crop.

If subsoil conditions are suitable, you can limit this invasion of roots by regular deep ripping or trenching alongside your beds. Most of these extensive roots are in the top 600 mm of the soil.

Propagation and Planting

Propagation, establishment and management of trees and shrubs is somewhat different from most herbs.

Propagation methods vary according to the species, but they are usually grown from seed or from cuttings. Generally, these need to be started in pots or in a nursery bed prior to planting out as 1- or 2-year-old stock.

Soil Preparation

Tree and shrub planting ordinarily does not require a lot of prior soil preparation for weed control, though often grass and weed competition can be detrimental to the early growth until the tree or shrub begins to dominate the site.

Deep ripping prior to planting helps loosen the soil for better root development and also increases moisture penetration and retention.

Weed Control

Grass and weeds do need to be kept down while young trees and shrubs are being established.

Mowing and Mulching

In some situations, weed control can be effectively maintained by mowing up close to the trees or shrubs, then raking the mowings up to them as a mulch. If this is done in spring before the soil has dried out too much, it will conserve moisture as well as suppress competition.

Where trees and shrubs are planted direct into grass, a sheet of cardboard with a hole or slit in it for the trunk, placed on top of the ground around the newly planted tree or shrub can help make the mulch more effective by offering a little more resistance to grass and weeds trying to push up through it.

It is important when mulching trees and shrubs to leave a little space – say 50 mm – around the trunk down to the surface of the soil. This will allow some air circulation so the base of the trunk is not kept continually moist. If the mulch is laid right up to the trunk, this can cause collar rot and kill the tree.

For this system to work effectively, trees and shrubs need to be planted with access for mowing in mind. While random planting can look more natural, mowing can be frustrating if spaces are too small to pass through or are oddly shaped, making for tricky manoeuvres and a greater risk of accidentally knocking trees or shrubs over. On the other hand, dead straight rows can look dully artificial.

An attractive compromise can be curving rows. These might follow the natural curves of the site, such as contours, creeks or soil type boundaries. Even where a straight line such as a fence or property boundary is being followed, planting along a wavy line will help break up the formality. A curved or wavy row is relatively easy to follow when mowing, as long as the curves are gentle.

Cultivation

Cultivation can be used for weed control around trees and shrubs. Where they are planted among other herbs, it is important not to let them become a harbour for weeds, which will then invade or drop seed among these neighbouring crops.

Where the trees and shrubs are planted separate from other crops, cultivation is often unnecessary. Once established they can usually suppress the grass and weeds immediately around them, and the rest can more efficiently be managed by mowing. This is also better for the soil and provides mulching material to go around the trees and shrubs.

Irrigation

Being deeper rooted, trees and shrubs generally need less irrigation than other herbs, though in dry summer regions many species will benefit from extra water, particularly those from milder summer rainfall climates.

Some thought needs to be given to irrigation when planting to ensure water can be provided to those species needing it, and to ensure that as trees and shrubs grow they don't interfere with irrigation systems and create sprinkler shadows over lower growing crops.

For the most part, it is best to plant trees and shrubs to the outside of your row and meadow crops, as they cause too many problems if they are placed among them. The best layout for mixed plantings is to place trees and shrubs around the boundaries of cropped areas. Another workable arrangement is to crop the spaces between the rows of trees or shrubs until they have filled out.

Either way, sprinklers can be set up in the middle of the low-growing crops so they cover all the central area and throw to the base of the trees or shrubs. Row and meadow crops benefit from the shelter provided by trees and shrubs, as long as they are not too close to them.

Irrigation management for trees and shrubs planted on their own is a bit different from the management for herbs. They are generally better off with less frequent but heavier waterings, which soak right down into the subsoil. This encourages deeper rooting and benefits them more than the shallower rooted grass around them.

For ease of management and conservation of available water, drip irrigation has its advantages but it needs to be managed carefully to avoid overwatering or underwatering. Personally I prefer to use mini-sprinklers or micro-sprays rather than drippers, or else to use several drippers for each tree so the water is applied over a larger area.

Harvesting

The harvesting of trees and shrubs for herbal use introduces another factor in their management. For this they need to be placed and managed for ease of harvest and so that optimum yields and quality can be obtained.

Some species may need wider spacing so they can reach their full development, while others may need to be shaped or maintained as hedgerows.

If the trees or shrubs are serving a dual purpose as shelter or ornament as well as being available for harvesting, then some sort of a compromise may be necessary. This should not be too much of a problem for the harvesting of leaves, flowers or fruit. Bark harvesting is more destructive as it requires the lopping or felling of the tree.

It is fairly common to read instructions on harvesting bark which recommend taking it off one side of the trunk, on the premise that if you don't ringbark the tree, it will continue to live. Although the tree may indeed live for a while, the removal of such a large piece of bark will create a permanent injury, which is likely to ultimately lead to the tree's death. Fungi will invade the wound and either eventually spread throughout the tree, or will weaken it at that spot so eventually the wind blows the tree over.

An alternative is to lop branches off and peel the bark from these, or else cut down the whole tree or shrub and harvest the bark.

A replacement planting program should be established to ensure continuity of production, though some species will regenerate by coppicing or sending up new shoots from the stump.

A tree which is to be felled for harvest should be planted clear of anything it may damage as it falls.

Pollarding is another option for harvesting the bark, and perhaps the leaf, of some species which are capable of coppicing. This is the cutting of the trunk/s of the tree about 2 m above the ground, so they can coppice and regenerate above the reach of livestock. The trees can be repeatedly cut every 5 or 10 years. This method is particularly suitable for harvesting species where thinner younger bark is preferred.

Harvesting leaf usually involves a degree of pruning, so the regenerative capacity of the species has to be borne in mind. Fruit and flower harvesting are less destructive, but trees may need to be kept pruned to an accessible size.

Planning

One of the most difficult things with slow-growing tree crops is judging what the market will require in 10 or 20 years time when the crops are in full production, and getting an idea as to whether they are going to be a worthwhile proposition.

With some trees and shrubs it is possible to find full-grown specimens in your district, which can give you an idea of how they grow, their size, their impact on the area around them, how you feel about them, and the effect they will create in the scenery around you. It is possible, too, that you may be able to harvest from them, which will give you an idea of their feasibility and viability as a crop.

Balm of Gilead Poplar
Populus candicans Ait.   SALICACEAE

Parts Used: Winter Buds

Other Name: *P. gileadensis*.

Balm of Gilead is a name applied to a number of quite unrelated plants, which is the cause of much confusion at times. The original Balm of Gilead was a tree of the Middle East (*Commiphora opobalsamum*), to which miraculous powers were attributed. This species is now almost extinct, but its name has been passed on to other plants, which also include the Balsam Fir (*Abies balsamea*) and a rather smelly shrub with trifoliate leaves, *Cedronella canariensis*.

Balm of Gilead Poplar (*Populus candicans*) is regarded as a hybrid between Balsam Poplar (*P. balsamifera*) and Cottonwood (*P. deltoides*).

Balm of Gilead Poplar buds are sticky with a strong fragrance. They are used in commercial preparations for respiratory and skin conditions.

GROWING BALM OF GILEAD

The Balm of Gilead Poplar grows quite quickly in suitable conditions. Its preference is for a cool to cold climate, but it will grow well in warmer regions if it has adequate water. It can withstand very heavy frost and grows well in a waterlogged site.

Containment

Balm of Gilead Poplar tends to send up a few suckers, which come up several metres away from the tree and need cutting back periodically.

Identification

Balm of Gilead Poplar is a deciduous tree up to 30 m tall, with a spreading form and a broad irregular head when mature. The bark is smooth and light brown on younger trees, perpendicularly furrowed and darker when older. The leaves are broad, heart-shaped, alternate, 100–150 mm long, dark above, lighter below, with hairs on the leaf stem and on the underside of the leaves. Flowers are drooping scaly catkins up to 150 mm long. The winter leaf buds are covered in an orange sticky highly aromatic resin.

There are a number of other balsam poplars sometimes grown here, notably Balsam Poplar (*P. balsamifera*) and a Chinese Balsam Poplar (*P. yunnanensis*). They lack the hairs on the leaf stem and underside of the leaves, which characterise Balm of Gilead Poplar.

Balm of Gilead Poplar (*Populus candicans*).

Make sure you don't end up with the shrub *Cedronella canariensis*, often offered by nurseries as Balm of Gilead. It is hard to understand how this species could have gained the name, as it doesn't produce any resin and the aroma of its leaves smells rather more like boot polish than balsam.

Variety

Balm of Gilead Poplar is the preferred species, but other balsam poplars are sometimes substituted for it.

Pests and Diseases

Balm of Gilead Poplar is reported to be prone to canker in warmer climates, but it has done well in north-east Victoria with no sign of this affliction.

Quite commonly the young leaves develop black spots and look very deformed as they mature. This typically occurs in late spring or early summer. Usually the first few leaves are normal and then the later ones start becoming deformed.

I first thought this was some kind of disease, but one day I noticed there were a lot of bees hanging around the trees. On closer observation they were seen to be scurrying all over the young leaves, scraping away at them diligently with their tongues to get the traces of balsam.

It was then I realised the black spots and subsequent deformities were caused by this industrious activity: the bees were scraping so hard they were damaging the young leaves. Apparently they use this resin to make propolis. The trees grow so vigorously that this bit of leaf damage hardly seems to affect them.

Harvesting

Timing

The buds are best harvested in late winter, when they have reached full size, but before they begin to open.

Technique

I found picking the buds off the trees by hand tediously slow. There was no way I was going to pick enough to justify the time spent harvesting. Not only was it slow, but the number of buds accessible from the ground was limited. The biggest buds were out of reach on the top branches.

I don't know how it is harvested overseas, but it is not an expensive herb.

An alternative method might be to prune the tops out of the trees, pollard style and then pick the buds off these.

Drying

Balm of Gilead Buds are very slow to dry, owing to their coating of resin. When thoroughly dry they are fairly hard, with no sign of moistness inside.

For more detail on this herb see 'Information Charts'.

Bay

Laurus nobilis L   LAURACEAE

Parts Used: **Leaf**

Other Names: Laurel, Sweet Bay, Sweet Laurel

The Bay has been held in great respect since ancient times, figuring in the spiritual traditions of the Greeks and Romans, and no doubt of other cultures of its native region.

Today it is used as a culinary herb, though in the past it has been used medicinally. It is also used as an insect repellent, mainly to keep moths out of dried foods.

Origin

Bay is a native of Asia Minor and southern Europe.

GROWING BAY

Bay is an attractive tree, but slow growing, especially in its early years.

Climate

Bay has proved quite adaptable to a range of climates and there are few parts of southern Australia where it would not grow. It will withstand dry summers and a fair amount of frost.

Site

Exposure

Bay prefers full sun, but it is tolerant of partial shade. It will tolerate some wind, but does prefer a sheltered site.

Drainage

Bay needs good drainage.

Soil Type

Bay will grow well in a range of soils, as long as they are not too heavy.

Fertility

Bay does better in a fertile soil.

Tree Size and Form

Bay grows up to 20 m tall, with a relatively compact rounded form.

Neighbouring Plants

Bay will grow in the vicinity of larger trees without being overshadowed, but for best development it should be allowed plenty of room.

Identification

Bay is an evergreen tree with smooth grey bark. Leaves are dark green, leathery, shiny on the upper side, lanceolate to ovate-lanceolate, 30–120 mm long. Flowers are small, creamy yellow and borne in clusters in the leaf axils. Fruit is a dark purple berry, 15 mm across.

The species grown in Australia may be *L. canariensis*, which is very similar and closely related to *L. nobilis*.

The Californian Bay Laurel (*Umbellularia californica*) is a more distantly related species, somewhat similar in appearance and is reportedly sometimes offered by nurseries in place of Sweet Bay. The leaf is perhaps a little lighter in colour and the berry is green when ripe. The aroma of the leaves is more rank. A crude differentiating test is to crush a handful of fresh leaves, hold them to your nose and breathe in cautiously. If it is the California Bay Laurel, your nose will immediately start to pour with mucus and you will probably start to sneeze. If you get a bit too much, a dreadful sensation sets in: your head feels as if the top half has been split off as a sharp pain strikes across about the level of the ears.

This test works with the fresh leaves, but I haven't tried it with dried (once was enough!). California Bay Laurel leaves are used to some extent as a substitute for Bay leaves in USA, but the flavour is not as noble: in fact, I found it somewhat repulsive and I doubt if they would be acceptable to the trade here. Hugh Johnson, in his *Encyclopaedia of Trees*, says 'I would no more put it in my soup than I would petrol'.

Variety

There is some variation in the size of the leaves but this is possibly climatic, because in Tasmania the Bay trees have larger leaves than on the mainland, and one tree growing beside the remains of a cottage at Cockle Creek near the southernmost tip of Tasmania has the largest leaves of all.

Medium-sized leaves, 80–100 mm long are preferred. If the leaves are too large, they tend to curl more in drying and this, together with the length, can make packaging difficult.

Bay (*Laurus nobilis*) foliage.

Propagation

Method
Propagation is by means of seed or cuttings.

Timing
Seed should be started in spring, and cuttings in spring or autumn.

Planting
Plants should be fairly advanced, at least 600 mm high, before planting out, so they don't get lost among the vegetation, as their initial growth is slow.

Layout and Spacing
When planting out there is a choice between wide spacing (6 m centres) to allow the trees room for their full development, or much closer spacing (3–6 m x 1–2 m) so they can be harvested as soon as they have filled out the inter-spaces. I am uncertain as to which will give better results in the long term, but the closer spacing will provide much greater yields on an area basis in the short term.

Harvesting from large trees is limited by what you can get access to. The upper part of large crowns is difficult to reach.

Compost
Moderate dressings of compost in the vicinity of the young trees should help them establish a bit faster.

Weed Control
Being quite slow to get started, young Bay trees tend to suffer from grass competition, so it is important to keep grass mowed and a mulch maintained around the base of the trees, at least until they reach a good size.

Irrigation
Bay trees will probably benefit from some irrigation in dry conditions and will make faster growth, but they can survive without it. Be careful not to over water as this may cause root damage.

Pests and Diseases

Grasshoppers
In spite of Bay's reputation as an insecticide and insect repellent, grasshoppers can significantly damage the young trees until they are tall enough to be out of reach. If grasshoppers are likely to be a problem, use taller stock when planting out, or protect the trees in some way.

Harvesting
Until Bay trees reach a harvestable size, there may be some opportunity to harvest from established trees elsewhere. Quite large trees are sometimes to be found in gardens or around old homesteads and ruins and if they are not close to spraying activity, roadsides or other sources of contamination, they can be harvested. A good-sized tree can yield a substantial amount of leaf.

Timing
Harvest mid summer to mid-autumn, after the leaves have hardened off and any fruit has fallen. If harvested too early in the season, the young leaves will turn black.

It is usually possible to harvest once a year.

Tools
Secateurs and/or loppers are required for the harvest, and a ladder for large trees.

Technique
Harvesting is basically a pruning operation, done with an eye to obtaining good quality leaf and not cutting the foliage back so hard it won't recover. Bay trees can stand a lot of pruning and recover surprisingly quickly, considering how slow they are to grow when young. How much you cut off depends on the situation, whether you are taking a light pruning from a tree which has ornamental value as well, or whether you are harvesting from a plantation established specifically for leaf production.

Larger branches that have been cut should be trimmed down so you are just taking leafy branchlets up to 900 mm long.

Avoid foliage with debris or old leaves amongst it: often this accumulates in parts of the tree.

Post-harvest Treatment
Go over the trees or bushes and trim off any old growth which is sticking out or was unharvestable, so the profile is evened up. If you are harvesting from somebody else's tree, it is important you leave it looking attractive and to clean up any mess you have made.

Drying
Bay leaves are quite easy to dry. In good conditions under shade, they come out looking so much brighter green than the sad looking little leaves you see in the supermarket.

The leaves are easier to remove when dry, so they are best dried on the branchlets. Smaller branchlets can be placed on screens, larger ones can be hung from wires.

When dry, the leaves should shatter when crushed and come off the branchlets easily.

Temperature

Bay leaves should be dried at a temperature of not more than 35°C.

Processing

Bay leaves are generally used whole as a culinary herb, as this allows them to be removed after cooking. Broken leaves are less popular.

Culinary Grade

To remove leaves, the branchlets should be put into the 'boat' or the rubbing tray and stirred around, shaking them against each other to knock the leaves off. Care should be taken not to shatter too many leaves with rough handling.

Storage

Freeze treatment to control moth larvae is not necessary as this herb repels them. Try not to crush the leaves in storage.

Yields

Modest pruning of a large tree can yield 10–20 kg of dried Bay leaf per annum.

Marketing

Bay leaf is one of the more popular of the culinary herbs, particularly in the organic market, as most people don't have a tree in their backyard. It is widely used in cooking, even by people whose main culinary talent is the liberal use of pepper and salt.

Price

Price to growers for premium quality organically grown Bay leaf is currently around $30/kg for 5-star culinary grade, but this price reflects the time and effort involved in harvesting from trees with ornamental value, which can only be lightly trimmed.

I would expect the price of Bay leaves to eventually come down to somewhere on par with other organic culinary herbs if plantation grown rather than harvested from occasional trees.

Current trade price is around $4/kg: quite a difference!

Cascara Sagrada
Rhamnus purshiana DC
RHAMNACEAE

Parts Used: **Bark (Aged at least 6 months)**

Other Names: Cascara Buckthorn, California Buckthorn.

This herb's common name, Cascara Sagrada or 'Sacred Bark', seems peculiar, but it may be a euphemism, as perhaps a previous name commonly used described the herb's action a bit too succinctly.

Cascara Sagrada is used as a mild purgative and a bitter tonic.

GROWING CASCARA SAGRADA

My experience with this species is limited to running a trial with it in north-east Victoria, which gave mixed results, but we didn't ever harvest any bark as the trees were still too small when we left.

It never thrived out in the open where it didn't get any summer irrigation, but it did do quite well on the edge of the garden where it received some extra moisture.

Identification

Cascara Sagrada is a deciduous tree up to 10 m tall. It has reddish brown bark when young. The leaves are elliptic to ovate, 50–150 mm long. The flowers are small, greenish-yellow, and the fruit is a black berry about 7 mm across.

Harvesting

Technique

If the tree is carefully felled to leave a clean, undamaged butt, Cascara Sagrada may be able to regenerate.

Drying

The bark is fairly thin and apparently rolls into quills, a bit like cinnamon. Care would need to be taken to ensure the inner part of these dries fast enough to prevent mould developing.

The dried bark needs to be aged at least 6 months, preferably a year, before use.

For more detail on this herb see 'Information Charts'.

Chaste Tree

Vitex agnus-castus L
VERBENACEAE

Parts Used: Fruit

Other Names: Chasteberry, Monk's Pepper, Agnus Castus.

Chaste Tree berries are used in herbal medicine as a tonic and normaliser for the reproductive system. They are widely used to help normalise hormone levels in women and have the reputation of reducing sex drive in men. Medieval monks used the herb as pepper to help keep their minds on their prayers.

GROWING CHASTE TREE

Chaste Tree is quite an attractive small tree with light blue flowers and graceful palmate leaves, which might raise the suspicions of some of your neighbours, because at first glance a vigorous young tree can bear a striking resemblance to Marijuana!

Chaste Tree appears to be fairly specific in its growing requirements, preferring a warm but not too hot climate. Generally Tasmania is too cool for it to make good growth. Melbourne's climate seems to suit it really well but in the relentless dry heat of summer in north-east Victoria, its growth was less vigorous.

Identification

Chaste Tree is a deciduous tree up to 6 m with a light grey bark and an open spreading form. The aromatic leaves are opposite, compound, palmate, the 5–7 leaflets up to

Chaste Tree (*Vitex agnus-castus*) foliage and fruit.

100 mm long, linear-lanceolate, finely toothed, green above and grey and closely felted below.

Flowers are small, light blue to lilac, in whorls on racemes up to 150 mm long forming at the ends of the branchlets (see colour plate 2.3). The fruit is a small hard berry, which is aromatic.

Apart from one notorious weed, there is not really anything Chaste Tree might be confused with. Chaste Tree can be distinguished easily by its woody stems and fine toothed foliage. The flower and fruit are also quite different.

Variety

There does seem to be some variation in form and productivity, so there may be some potential for improving yields by propagating from the more productive trees.

Harvesting

Harvesting should wait until the leaves have fallen around mid-May, and then it should be carried out promptly but in dry conditions.

Tools

For harvesting the berries you need a finely woven basket, which can be worn in front of you, similar to that used for Hawthorn Berries and Elder Flowers. If the bushes are low, a broad container can be placed on the ground and slipped under the branches.

Technique

The berries are stripped by hand into the basket. This operation is rather slow and there would need to be a heavy crop for it to be a viable proposition.

Drying

Drying is quite easy, as there is very little moisture in the berries. Take care they are not accessible to vermin.

Processing

Processing is just a matter of sifting and perhaps winnowing any debris out of the dried berries.

Yields

Our yields at 'Twin Creeks' were pretty discouraging. After 5 years growth, a 40 m row of bushes yielded only 4.7 kg dried berries, which took about $3^{1}/_{2}$ hours to pick. In a milder climate yields might be better.

For more detail on this herb see 'Information Charts'.

Crampbark

Viburnum opulus L CAPRIFOLIACEAE

Parts Used: Bark

Other Names: Guelder Rose, Snowball Tree, Cranberry Tree, Highbush Cranberry, Pimbina, *V. trilobum* (for American subsp.).

Crampbark is a valuable medicinal herb used as an antispasmodic, a sedative and an astringent. The fruit can be eaten cooked and is somewhat similar in flavour to the true Cranberry. The berry of the European Crampbark is not safe to eat raw, but the American subspecies, *V. opulus* var. *americanum* (= *V. trilobum*), can be safely eaten off the bush, though you probably wouldn't eat many, as it is terribly sour. In Canada it is the last of the wild fruit to be eaten by birds, the red berries hanging on the trees long after snow has started to fall.

The variety most commonly found in cultivation is the Snowball Tree (*V. opulus* var. *sterile*), which is grown as an ornamental for its globular clusters of white flowers. It is sterile and does not form berries, but its bark does seem to have the same therapeutic properties as the wild form.

Origin

Native to Europe, north Asia and northern North America.

GROWING CRAMPBARK

Crampbark is reasonably adaptable and once established it is fairly hardy. It makes an attractive bush or small tree and prefers a cool to mild summer, but can be grown in hotter conditions with adequate moisture and a suitable microclimate.

In cooler regions it prefers full sun and will tolerate cold winds. In warmer regions it will need some protection from hot drying winds and will perhaps need partial shade.

Identification

Crampbark is a deciduous multi-stemmed shrub up to 4 m tall with grey-brown bark. Leaves are opposite, maple-like, 3–5 lobed, coarsely toothed. The flowers are white, in flat-domed clusters 70–100 mm across (globular in Snowball Tree). Fruit is a soft juicy berry, red, purple or sometimes yellow (see colour plate 2.5).

Variety

The wild form is preferred, but may be difficult to obtain here. There are some cultivated varieties with berries, which are close to it, and these would be preferred to the non-fruiting Snowball Tree (*V. opulus* var. *sterile*).

Crampbark Stems.

Crampbark or Snowball Tree (*Viburnum opulus* var. *sterile*).

Harvesting

Technique

As Crampbark is continually sending up new stems from the base, older stems can be harvested when they are large enough. The size at which they are harvested will depend on their growth and the need. Small stems are fiddly to peel and don't yield much bark. They need to be at least 50 mm at the base to be worth doing.

For more detail on this herb see 'Information Charts'.

Elder, Black

Sambucus nigra L CAPRIFOLIACEAE

Parts Used: Flowers, (Fruit, Leaf)

Other Names: Elder, Common Elder.

More details of this herb, including identification and harvesting, are covered in Chapter 21, 'Wildcrafting and Weed Harvesting'.

GROWING ELDER

While it may be possible to wildcraft Black Elder in some areas, there never seems to be enough of it and timing of harvest is much easier if it is growing at hand. It is an attractive tree with special qualities and much to contribute to the garden, provided it is treated with respect. It is said to have a positive influence on composting processes, so it is good to locate it next to the compost-making area. It will also benefit from the compost as it likes a fertile site.

Elder prefers a climate with a cool to mild summer, but it will grow in hotter areas if it gets adequate water and some protection during summer.

Neighbouring Plants

The foliage of Elder casts a dense shade and it seems to discourage the growth of many other plants close to it.

While the tree will grow in shade, it bears more flowers in a site where it gets at least some sun.

Containment

In some districts Black Elder will spread widely with birds eating the fruit, and seedlings will even come up in undisturbed bush. Keeping the flowers picked clean will prevent fruit forming, as it flowers for only a few weeks in spring.

Variety

There is a variety or species of Elder sometimes grown in cultivation which flowers continuously throughout the season. Its flowers are odourless, however, and are not suitable for herbal use.

Propagation

Elder can be grown from seed, but it is very easily started from hardwood cuttings set in open ground when the plant is dormant.

As Elder is very late to lose its leaves and very early to come into leaf, the optimum period for planting and transplanting can be very short. In mild winters the buds may burst before the old leaves fall.

For more detail on this herb see 'Information Charts' and Chapter 21, 'Wildcrafting and Weed Harvesting'.

Ginkgo

Ginkgo biloba L GINKGOACEAE

Parts Used: Leaf

Other Names: Maidenhair Tree, *Salisburia adiantifolia*.

The Ginkgo is regarded as having preceded conifers in evolutionary development and is the sole surviving species of its order. This unique tree bridges time, making a connection between our era and the distant prehistoric past.

Ginkgo fruit has been used in Chinese herbal medicine for centuries. Western herbal use is rather different and only a recent development. Clinical trials have shown Ginkgo leaf to be quite effective in assisting blood circulation to the brain, particularly for the aged.

Origin

Ginkgo is a native of China, but it probably no longer grows in the wild. It is a sacred tree planted around Buddhist temples.

GROWING GINKGO

Ginkgo is relatively slow growing, but is worth while planting because the falling autumn leaves are preferred, so harvesting can commence on a small scale from an early age.

Ginkgo is quite adaptable with regard to climate: there would be few, if any, parts of Australia too cold for it and it is apparently fairly tolerant of dry conditions as well.

Neighbouring Plants

If the intention is to catch the leaves as they fall, Ginkgo should be located where this is going to be possible and away from other trees whose falling leaves might contaminate the harvest.

Identification

Ginkgo is a tree growing up to 30 m, with a spreading open form. The leaves are deciduous and there are separate male and female plants. The bark is grey-brown, furrowed. The leaves are fan-shaped, 50–100 mm across, with parallel veins and typically with a cleft in the centre, though this is variable: some individuals have leaves with two clefts, some have none (this variation in leaf shape can also occur on the one tree). Flowers emerge as the leaves appear in spring: catkins on the male plants, yellow-green bulbous flowers on the female trees. Fruit is a yellow drupe, ill-smelling, 20–30 mm long.

Variety

There are a number of cultivars, but the wild form would be preferable.

Harvesting Leaf

Fortunately the yellow autumn leaves contain the highest level of active principles, so harvesting does not harm such a special and sacred tree.

Problems

The main problem is waiting until the trees are big enough to yield a substantial harvest or finding a tree where you can set up apparatus for gathering the leaves as they fall. Ginkgo is not a commonly planted tree in Australia.

Timing

Harvesting takes place as the leaves fall in autumn.

Tools

Sheets, shadecloth or plastic bird-netting and some means of suspending them under or around the tree are needed to catch the falling leaves.

Alternatively, the leaves could be hand stripped onto a catching sheet, or some mechanical harvesting system may be feasible, using vacuum to strip the leaves off the tree.

Technique

Catching Falling Leaves

A leaf-catching set up will have to be designed for the particular site, with some means of regularly picking up the leaves, probably every day. It is not good for them to be subjected to rain and dew after they have fallen. The tree, of course, would have to be in a sheltered location and away from other species, especially deciduous trees. As the leaves still contain moisture when they fall, they need to be dried in the normal drying facility.

Sweeping or vacuuming up fallen leaves from the ground is another option, but this could be risky, considering the other sorts of things likely to be found among the grass.

Even a well-manicured lawn will produce some contamination, which would need to be removed somehow. Gathering would also have to be done daily if quality were to be ensured, and this would be rather labour intensive.

Stripping Leaves

Another option could be to strip the leaves off the tree after they turn yellow. Trials would have to be conducted to determine whether stripping has a detrimental effect on wsubsequent growth, and whether some leaf should be left on the tree.

For more detail on this herb see 'Information Charts'.

Ginkgo (*Ginkgo biloba*) foliage.

Hawthorn

Crataegus monogyna Jacq.
Crataegus laevigata DC
ROSACEAE

Parts Used: Leaf & Flower, Fruit

Other Names: May, Whitethorn, *C. oxyacanthoides* (= *C. laevigata*).

The usage, identification, harvesting, drying, processing and marketing of Hawthorn are covered in detail in Chapter 21, 'Wildcrafting and Weed Harvesting'.

GROWING HAWTHORN

If you are considering planting hedges for fencing or shelter, Hawthorn is a possibility where it is suited, as it can provide a harvest of berries or leaf and flower.

It prefers cooler climates, but will do quite well in regions with a hot dry summer. It is very tolerant of poor drainage, and adaptable to most soils.

Neighbouring Plants

When planting Hawthorn, consider the need for access for harvesting and give it a bit of space from other plants. It is well-suited to planting along pasture fence-lines, as it makes a solid impenetrable hedge.

As Hawthorn is a host for black spot and pear slug, it should probably be located away from Apples, Pears, Cherries and Plums.

Containment

Hawthorn can be liable to spread, thanks to birds eating the berries and scattering the seeds about the countryside in their droppings.

In some districts it may be unwise or even illegal to grow Hawthorn because it can be a noxious weed, coming up all over the place. In other regions it is less invasive and authorities are not too worried about it. It is really a matter of looking around and seeing how it has behaved in your region. The Adelaide Hills are one of the worst areas for Hawthorn getting away, as are parts of central and eastern Victoria, and the Southern Tablelands in New South Wales.

Variety

There are a number of Hawthorn cultivars and hybrids grown as ornamentals, but only the wild form should be planted for herbal use.

Propagation

Hawthorn can be grown from seed but it is reluctant to germinate, owing to dormancy. The dormancy can be overcome by stratifying the berries in moist sand for 12 months, or by freezing them for several weeks before sowing. Sometimes a good crop of seedlings can be found beneath wild trees.

Hawthorn can also be propagated from cuttings, although these may be slow to strike, or from suckers by dividing these off from the base of the tree.

Special Requirements

As the bushes need to be picked from a ladder when they get taller, plant them on reasonably level sites.

Pests and Diseases

Pear Slug

Pear slugs are slimy little creatures, which tend to show up late in the season and can be in such numbers that they strip all the leaves off the bushes. It doesn't seem to greatly affect the trees' productivity though, probably because the defoliation doesn't occur until fairly late in the season. They don't seem to touch the fruit and are usually gone by harvest.

Black Spot

The fungus which causes black spot on apples can make Hawthorn berries rather spotty. It is not harmful, just cosmetic, so affected berries are fine for use in milled form or for manufacturing extracts, but they won't be as attractive in whole form.

For more detail on this herb see 'Information Charts' and Chapter 21, 'Wildcrafting and Weed Harvesting'.

Hawthorn

Horse Chestnut

Aesculus hippocastanum L
HIPPOCASTANACEAE

Parts Used: **Seed**

Other Names: Conkers, *Hippocastanum vulgare*.

The Horse Chestnut is yet another herb whose name connotes some association with the horse. Perhaps it comes from the idea that their strong flavour makes them only suitable for horses, though most horses would not readily eat the nut of this tree, or the Horseradish either.

Horse Chestnut is used medicinally for its action on the vessels of the circulatory system.

Horse Chestnuts are not normally regarded as edible: the outside skin of the fresh seed is toxic.

GROWING HORSE CHESTNUT

Horse Chestnut is an attractive and majestic tree, which casts a dense shade. It has very much a European look, if you want that sort of effect in part of your garden. It produces a magnificent display of blossom in spring and is quite fast growing in the cooler and moister parts of southern Australia, but suffers in hot dry conditions.

Horse Chestnut (*Aesculus hippocastanum*) foliage and fruit.

Identification

Horse Chestnut is a deciduous tree growing to 30 m. The bark is smooth at first, becoming scaly as it gets older. Leaves are divided palmately, with 5–7 leaflets, obovate, 100–200 mm long. The flowers are in showy erect conical inflorescences 300 mm long at the ends of the branchlets, white, and sometimes tinted pink or yellow. Fruit is a capsule with fairly soft spines, with 1–3 large brown seeds up to 40 mm across.

The Horse Chestnut bears a superficial resemblance to the true Chestnut, but is easily recognised by its palmate leaves, and its seed capsules, which are not as prickly. While Horse Chestnut capsules do have raised points on their outer surface, they can be comfortably gathered with bare hands. The true Chestnut has simple leaves and its capsule is painfully prickly, even to the most calloused hands.

Variety

There are some cultivated forms of Horse Chestnut, including a pink-flowered one which is actually a hybrid of Horse Chestnut with another species of *Aesculus* and should be avoided.

Harvesting

Timing

The capsules or the seeds fall to the ground when ripe.

Tools

A rake might be helpful

Technique

Harvesting is basically just a matter of picking the capsules up off the ground and husking the seeds out of them. Be careful you don't accidentally pick up other less savoury things that may also have been dropped on the ground around the trees.

Drying

Unless they are to be used fresh, the seeds will need to be split or chopped to facilitate drying.

Temperature

Horse Chestnut should be dried at a temperature of not more than 45°C.

For more detail on this herb see 'Information Charts'.

Lemon Verbena

Aloysia triphylla Britt.   VERBENACEAE

Parts Used: Leaf, Leaf & Flower

Other Names: *Lippia citriodora, Verbena triphylla.*

The fragrant lemon flavour of Lemon Verbena makes a delicious tea, which is quite popular. Some authorities attribute medicinal qualities to it, mainly for nausea and indigestion, but it is little used for these purposes.

Origin

Lemon Verbena is a native of Chile, Argentina and Uruguay.

GROWING LEMON VERBENA

Lemon Verbena has a character of its own. While the foliage has a soft green colour, left to their own devices, the bushes become very straggly which does not fit in with a formal European-style garden setting.

The delightful fragrance of this small tree makes it an enjoyable species to grow, but propagation can be tricky.

Lemon Verbena is unusual in that its deciduous cycle seems to be out of rhythm with the seasons. The leaves remain green on the bushes until very late in the season, often well into winter, but the new leaves don't emerge until it is almost summer. In tropical and sub-tropical conditions, it remains evergreen.

Climate

Lemon Verbena will grow in a range of climates, including hotter situations if it has enough moisture. It will stand some frost and is hardy in most parts of southern Australia.

Site

Exposure

Full sun is preferred by Lemon Verbena, but it does best where it has some protection from winds.

Drainage

Good drainage is essential as it will not tolerate excessive moisture at all.

Soil Type

A range of soils are suitable.

Fertility

A reasonably fertile soil is preferred.

Tree Size and Form

Left to its own devices, in good conditions Lemon Verbena will grow to 5 m with a very open straggly form.

Neighbouring Plants

It doesn't cast a dense shade and if harvested regularly, it will be kept fairly small.

Identification

Lemon Verbena is a shrub or small tree, deciduous in temperate regions, with an open straggly form and brown bark.

The leaves are aromatic, lanceolate and normally in whorls of three, 50–100 mm long with short stalks. Flowers are around 5 mm long, white to pale purple, in axillary and terminal panicles.

Variety

A lime-scented variety exists, which should be avoided for dried herb production.

Propagation

Method

Lemon Verbena can be grown from cuttings, but its success rate is somewhat erratic. It seems to do better in the hands of someone with a green thumb.

Timing

There is a lack of consensus on when to propagate cuttings, with different people recommending spring, summer, autumn or winter. The best thing would be to just keep trying different times and methods until you have some success.

Planting

As Lemon Verbena only sends a few fibreless roots out when it strikes, the tops need to be cut back heavily when planting out, which should be done when it is dormant.

Layout and Spacing

Lemon Verbena should be set out on 2–3 m centres in rows about 4 m apart. As it is pruned back fairly hard when harvesting, the bushes don't reach a large size.

Compost

A moderate dressing of compost, 2–3 kg/m^2, around the young trees will help their growth.

Weed Control

Mow and mulch around the trees to reduce competition.

Irrigation

Lemon Verbena needs around 25 mm of water per week through summer, but avoid overwatering as it is sensitive to excessive moisture.

Pests and Diseases

Lemon Verbena bushes may suddenly die in early summer, but this can usually be attributed to excessive moisture or inadequate drainage during the previous winter.

Harvesting

Timing

Harvest when the bushes start to flower. This is January–February in Tasmania, perhaps a bit earlier in warmer regions. A second and possibly a third harvest can be taken later in the season.

Tools

A reaping hook or secateurs are used for harvesting.

Technique

The bushes should be cut back quite hard to stimulate straighter and more vigorous regrowth.

I always used to harvest and dry it with the leaves on the branchlets.

Another method described by one grower is to cut it back very hard so its regrowth produces tall straight stems. These he cut at the base and pulled through his hand to strip the leaves off them while they were still in the fresh state. The stripped leaves were then dried whole.

I have never tried this method myself, being afraid it would bruise the leaves, but it might be worth a trial. If bruising could be avoided it would produce whole leaves which look very attractive when dried and packaged. It would be important to strip the leaves before any wilting occurred.

Post-harvest Treatment

The bushes may need additional pruning to maintain a suitable form.

Drying

Lemon Verbena dries very quickly, but if it is subject to re-moistening it can discolour somewhat.

Temperature

It should be dried at a temperature of not more than 35°C.

Processing

Lemon Verbena is used as tea and in potpourri. The herb looks more attractive if it is in whole leaf form or at least in large pieces, but it can be rubbed similarly to other leaf herbs.

Tea Grade

To produce tea grade, when the leaves are dry, the branchlets can be stirred around against each other to gently knock the leaves off.

Alternatively they can be rubbed through a $2^{1}/_{2}$ or 3 dent screen.

Storage

Freeze treatment to control moth larvae is necessary before sale or storage.

Yields

Yields after about 3 years growth in Tasmania were around 0.2 kg/m^2 tea grade per annum.

Marketing

There is a limited demand for organically grown Lemon Verbena at present, but this is partly because of its lack of availability. If greater quantities were consistently available, I'm sure the demand would increase.

Price

Price to growers for premium quality, organically grown Lemon Verbena is currently around $34/kg for 5-star tea grade but is as low as $10/kg for second grade.

Current trade price is around $11/kg.

Lemon Verbena (*Aloysia triphylla*) foliage and flowers.

Linden

Tilia cordata Mill.   **Small-leaved Linden**
T. x europaea L   **Hybrid Linden**
T. platyphyllos Scop.   **Large-leaved Linden**
TILIACEAE

Parts Used: **Flowers**

Other Names: Lime, *T. x vulgaris* (= *T. x europaea*).

The noble Linden tree has held a special significance for many peoples throughout history. In Europe it is widely used as a herbal beverage, but it has important medicinal properties as well, being highly regarded as a nervine and a diaphoretic.

Origin

The Linden is a native of Europe.

GROWING LINDEN

Lindens are stately trees which seem to convey a higher presence, a feeling of great strength. A friend tells me that in Switzerland Lindens are left to grow on hilltops as a symbol of independence. In a similar vein, perhaps, they were planted around Parliament House in Hobart.

They grow at a moderate rate and are long lived.

Climate

Lindens grow best in a climate with a mild to warm summer, they don't thrive in hot dry conditions, but they can withstand plenty of cold.

Site

Exposure

Lindens need full sun and can tolerate a fair amount of wind, but not hot dry winds.

Drainage

Good drainage is essential.

Soil Type

Lindens grow well in a wide range of soil types.

Fertility

They prefer a fertile soil.

Tree Size and Form

Lindens grow to around 40 m in good conditions, with a graceful symmetrical form. The crown is usually quite dense and taller than its breadth.

Neighbouring Plants

For the purpose of harvesting, free access needs to be available around the tree and, remember, it does grow to be quite a large tree, casting a dense shade.

Identification

Lindens are deciduous trees. The bark is grey, smooth when young, becoming rougher with age. The leaves are broadly heart shaped, dark green above, light green below, 60–100 mm long with toothed margins. The flowers are yellowish white and hang (or sometimes stand erect) in clusters of 5–10 from the axils of the leaves. The stem of the cluster has a light green wing or spathe attached to it. The fruit is round and nut-like.

Small-leaved Linden (*T. cordata*) has leaves and flowers at the smaller end of the range described, while Large-leaved Linden (*T. platyphyllos*) has the larger leaves and flowers. *T. x europaea* is a hybrid between these two species.

T. americana (American Linden or Basswood), should be avoided as its flowers can cause nausea. As far as I know, it is seldom planted in Australia.

T. petiolaris and *T. tomentosa* are silver-leaved species, with the undersides of the leaves and the flower parts tomentose, or densely covered in short fine hairs. These species should be avoided: their flowers are toxic to bees so they may have ill-effects on humans as well.

Large-leaved Linden *(Tilia platyphyllos)*.

Variety

In my limited experience harvesting Linden flowers, I found the Large-leaved Linden (*T. platyphyllos*) somewhat better than the other species, as its flowers are bigger, enabling faster picking.

Propagation

Method

Lindens can be grown from fresh seed, or else by cuttings or layering. Sometimes suckers can be taken from around standing trees.

Timing

Plant the seed as soon as it is ripe in late summer. Cuttings are said to be more likely to succeed in autumn. If a tree is cut down, it will send up a lot of suckers and these will form roots at their bases if sandy soil is thrown around them.

Planting

Lindens transplant easily when dormant, but don't attempt to move them when they are in leaf.

Layout and Spacing

Lindens should be set out on 10–20 m centres to allow them room for full development and to ensure they retain foliage down to the base to make harvesting easier.

Although it is almost sacrilege to suggest disfiguring such a noble tree, it may be worth experimenting with pruning the tops out of the trees to encourage shorter broader growth that is more accessible for harvesting, as most of the flowers of large Lindens are quite out of reach.

Compost

A moderate dressing of around 2–3 kg/m^2 of compost around the young trees will help their growth.

Weed Control

Mow and mulch around the trees to reduce weed competition.

Irrigation

A Linden needs 13–25 mm of water per week through summer in drier regions, though if the soil is deep enough to provide sufficient reserves of moisture, it may be able to endure the summer without additional watering.

Pests and Diseases

Some European books mention problems with aphids, but I have not observed this in Australia.

Harvesting

While waiting for your trees to grow large enough, it may be possible to find good harvestable trees away from sources of contamination. In southern Australia it is sometimes planted in parks and gardens, particularly in cooler regions. It does very well in Tasmania and there are some beautiful trees around Parliament House and in St David's Park in Hobart, but unfortunately these locations are too polluted for harvesting.

Timing

The trees flower around Christmas in Tasmania, but probably earlier in warmer regions. It is important to harvest them in the early flowering stage, as the quality deteriorates quickly as they mature.

Tools

A tall stepladder and a basket that can be worn in front of you, as described for harvesting Elder flowers and Hawthorn berries, are needed for the harvest.

Technique

The clusters are picked with the winged stems (spathes) attached, but not the true leaves. It is best to pick cleanly, as the leaves are tedious to get rid of later.

The main problem is the slow rate of picking, about 0.3 kg/hr dried weight. Most of the flowers are way out of reach, even from a ladder. Because the flowers are mingled with the leaves, it is hard to see how the harvesting could be speeded up. It would take some very sophisticated technology to mechanise the operation.

Drying

Linden flowers are easy to dry, but they should be dried quickly to preserve their quality.

Temperature

They should be dried at a temperature of not more than 35°C.

Processing

Once dry, no processing is necessary, although some buyers may require it to be cut.

Tea Grade

The whole clusters are usually acceptable for tea grade. It may be possible to chop Linden flowers with the chaff cutter.

Manufacturing Grade

The requirements for manufacturing grade are the same as for tea grade, but I rather doubt much Australian-grown Linden flower will ever get used for manufacturing: the cost of harvesting will always be too high.

Storage

Freeze treatment to control moth larvae is necessary before sale or storage.

Yields

On the one occasion I harvested Linden flowers it took about 6 hours of picking to get enough to produce 2 kg dried flowers.

Linden flowers have traditionally come from Eastern Europe where people get paid next to nothing for this sort of work.

Marketing

There is a moderate demand for Linden flowers and currently no local product available.

Price

Price offered to growers for premium quality, organically grown Linden flower is currently around $40/kg for 5-star tea grade.

Price offered to growers for manufacturing-grade, organically grown Linden flower is currently around $14–18/kg.

Current trade price is around $16/kg.

Rose, Dog

Rosa canina L   ROSACEAE

Parts Used: Fruit ('Hips')

The Dog Rose is dealt with in more detail under Wildcrafting and Weed Harvesting, but as it is the preferred species of rosehip and its wild occurrence in Australia is limited, it is worth considering cultivating it.

GROWING DOG ROSE

Dog Rose is a hardy adaptable species, traditionally used as rootstock for ornamental roses because of its vigour.

Climate

It thrives in a wide range of climates.

Site

Exposure

Dog Rose likes full sun and it will tolerate a windy site.

Drainage

Good drainage is preferred.

Soil Type

Dog Rose will grow in a range of soils.

Fertility

A fertile site is preferred.

Size and Form

Dog Rose grows to 5 m tall, but is usually 2–3 m with a breadth of several metres.

Neighbouring Plants

A bit of space around the bushes is necessary for access when harvesting.

Containment

Dog Rose can go wild in some situations and is a declared noxious weed in parts of South Australia. However, over most of southern Australia it has not proved particularly invasive and it would not be difficult to keep under control. It usually occurs as scattered bushes in pasture.

Harvesting the hips will further reduce the risk of it spreading, as there won't be any for the birds to eat and drop around the place.

Distinguishing This Species

Dog Rose should be distinguished from Sweet Briar, the species most commonly found growing wild in southern Australia, which has become a troublesome weed in many regions.

Dog Rose has an array of large thorns on its main stems, but no smaller thorns among them. The hips and their stalks are free of prickles. When fully ripe, the hips of the Dog Rose lose their sepals or flower bracts, so they have clean ends with small scars where the flowers were attached. Sweet Briar has smaller thorns and prickles among the large thorns on its stems. The hips have prickles on their stalks and sometimes on the hips as well. The flower sepals remain attached to the hip when ripe.

While Sweet Briar hips can be used for some purposes, it is not the preferred medicinal species and harvesting can be unpleasant because of the prickles. Its hips are also much harder to dry, as their skin is virtually waterproof.

For these reasons the Dog Rose is the preferred species for planting.

Propagation

Method
Dog Rose can be easily started from cuttings. Growing from seed is also possible, but germination is sometimes difficult: maybe they need to pass through a bird's digestive tract first.

Timing
Plant seed and start cuttings in autumn.

Planting
One-year-old rooted cuttings can be set out when dormant.

Layout and Spacing
Dog Rose should be set out on 1–2 m centres in rows around 4 m apart.

Compost
A moderate dressing of compost around the bushes will help their growth.

Weed Control
Mow and mulch around the bushes to reduce competition.

Irrigation
Dog Rose seems very tolerant of dry conditions but additional moisture may improve yields.

For more detail on this herb see Chapter 21, 'Wildcrafting and Weed Harvesting' and information charts.

Willow
Salix spp.
SALICACEAE

Salix alba L: White Willow

Parts Used: Bark

Willow Bark is used as an anti-inflammatory, analgesic and antipyretic. A number of species are used for this purpose and more research is needed to determine which ones are the most suitable, as there is a significant variation in active constituents.

GROWING WILLOWS

Willows would be some of the easiest of all trees to grow. In the right places they are quite attractive, but their soft green clashes with native vegetation.

Willows will grow in a wide range of climates, including those with a hot summer, provided they have adequate moisture.

Tree Size and Form
Some Willows can grow to 25 m tall with an upright spreading form.

Neighbouring Plants
Make sure they have enough room, as their rapid vigorous growth tends to crowd out other trees nearby.

Containment
As any piece of branch or twig can strike roots, they can take over creeks and river banks, crowding out native vegetation and restricting water flow. Willows can choke a stream, causing floodwaters to divert and start erosion along a new course.

Planting Willows may be beneficial in some degraded sites along streams, but where native vegetation can be retained or re-established, Willows should be kept out of the picture.

Another problem is caused by their soft foliage which breaks down more rapidly than the leaves of native plants, causing changes in associated flora and fauna in the river systems and the loss of native fish.

However, if they are planted away from creeks and rivers, there is not much likelihood of these problems occurring.

Identification
White Willow (*Salix alba*) is a deciduous tree up to 20 m tall, with a spreading crown and often branching close to the ground. The twigs are shiny yellow to red-brown, the older bark is rough and grey. The leaves are narrowly elliptic to oblanceolate, finely toothed, 30-80 mm long and 6-18 mm wide, the upper surface green and the under surface silky, light blue-grey. Flowers are slender, male and female catkins 30-90 mm long, fruits are small and conical.

Because this species is naturally quite variable and freely hybridises with other Willows, positive identification may be difficult. Hybrids between White Willow and Crack Willow (Salix fragilis) are common and widespread in Australia: they may be distinguishable by the longer leaves and the twigs which are brittle at the base.

Species

Some authorities specify White Willow (*Salix alba*), while others accept other species, including Crack Willow (*Salix fragilis*), Black Willow (*Salix nigra*), Bay-leaved Willow (*Salix pentandra*) and Purple Willow (*Salix purpurea*). Prospective purchasers should be consulted as to which species they prefer and an analysis should be carried out to determine whether the species and growing conditions are going to produce an adequate level of active constituents in the bark.

There is some doubt as to the suitability of a hybrid Willow (Salix alba x fragilis) which grows wild and is common throughout much of southern Australia. An analysis of one sample showed a very low level of salicin, one of the active constituents in Willow Bark.

Weeping Willow *(S. babylonica)* and Pussy Willow *(S. cinerea)* are not regarded as acceptable species for medicinal use.

Propagation

Method
Propagation is easiest by cuttings, but may be possible from seed.

Timing
Cuttings should be set in the ground in autumn or winter, before the leaves develop.

Planting
Because they strike so readily, Willow cuttings can reliably be set in the ground where they are to remain permanently. Quite thick and long pieces can be planted.

Pests and Diseases

A leaf rust is commonly seen on Willows but as they are so vigorous and the rust develops late in the season it doesn't seem to affect them much.

For more detail on Willow see Information Charts.

Witch Hazel

Hamamelis virginiana L
HAMAMELIDACEAE

Parts Used: Bark, Leaf

Witch Hazel is a herb traditionally used by native North Americans and was adopted by European settlers. It is used in herbal medicine and in cosmetics as an astringent.

GROWING WITCH HAZEL

My experience with this species is quite limited. While we did manage to keep it alive in north-east Victoria, it never thrived and we never harvested anything.

In the right conditions it is quite an attractive small tree. It prefers a cool climate. In hotter climates it should have partial shade, as its leaves are inclined to burn in the hot sun.

Identification

Witch Hazel is a deciduous shrub or small tree, 1.5–4 m tall. The trunks are single or in groups of up to four and the bark is brown and smooth. Leaves are elliptic to obovate, 50–100 mm long, the edges coarsely toothed and undulating. The surface of the leaf is corrugated, and somewhat downy when young. Flowers are strap-like, in clusters, bright yellow outside, brownish-yellow inside, about 20 mm long and appear in late autumn.

Variety

Make sure you have the American species, *Hamamelis virginiana*, as there are a number of other species in cultivation.

For more detail on this herb see 'Information Charts'.

Witch Hazel *(Hamamelis virginiana)*.

20

Annuals, Biennials and Short-lived Perennials

Angelica – *Angelica archangelica*
Aniseed – *Pimpinella anisum*
Basil – *Ocimum basilicum*
Blessed Thistle – *Cnicus benedictus*
Burdock – *Arctium lappa*
Calendula – *Calendula officinalis*
Caraway – *Carum carvi*
Catnip – *Nepeta cataria*
Celery – *Apium graveolens*
Chamomile, German – *Chamomilla recutita*
Chicory – *Cichorium intybus*
Coriander – *Coriandrum sativum*
Dill – *Anethum graveolens*
Echinacea, Broad-leaf – *Echinacea purpurea*
Mullein – *Verbascum thapsus*
Oats – *Avena sativa*
Parsley – *Petroselinum crispum*

The species in the 'Annuals, Biennials and Short-lived Perennials' group are almost all short-lived plants which need to be started from seed every year or so.

Annuals are typically started from seed in spring, though autumn sowing is preferable for some species. Many of them can be direct seeded. Most reach maturity, flower, form seed and die in their first season, though one or two species will sometimes live a little longer than a year.

Biennials normally don't flower until their second season. They then form seed and die. Some can be started in early autumn and will flower the following season, others won't flower until their second full season.

Short-lived Perennials are a bit more variable, but most of them will live for 2 years or so before they die, though some of them will linger on a bit longer. To maintain vigorous growth, it is best to start fresh plants from seed about every second year.

Also included in this group are perennials, grown as a root crop, which need to be started from seed.

Propagation

Propagation from seed is the only option for most herbs in this group and the best option for the remainder.

Species with larger seeds can be direct sown, but for those with smaller seeds it is usually best to start them in trays or nursery beds as this enables the provision of better conditions for germination and early growth. The seedlings can be planted out when they are sufficiently advanced to make weed control easier.

Rotation

The species in this group are best grown in rotation with other species. They can be rotated with other annual, biennial or short-lived perennial crops and/or with some of the perennial root herbs that are dug and replanted every year. Some vegetable or fodder crops could also be part of a rotation.

If possible, a species should be rotated with another crop where a different part of the plant is harvested, for example, root crop followed by leaf crop followed by seed crop followed by flower crop, or where that is not practical, with a crop belonging to a different family. This helps provide the diversity the soil and plants need.

A few herbs in this group can present problems in planning a rotation, because they reseed themselves very thickly and their seedlings come up in subsequent crops. With these species the choice of following crops needs to be made according to their ability to cope with this self-seeding.

Management

Most of this group are into full production in their first year and often they don't require a long initial preparation of the soil for effective weed control. Many of them are well worth considering by growers who are getting started, as they can provide a return sooner than most other herbs.

However, herbs that are grown for their seed are a whole different ball game. With seed crops the mechanised large-scale grower can generally produce quality as good as or better than the smaller grower, and at a much lower price.

It is only worth while for a small grower to produce seed crops if there is a particular requirement for small quantities, or where a good price can be obtained or a specific need exists.

Weed Control

Managing this group of herbs is somewhat different from most of the perennial herbs. For many of them 100% weed control is not as critical, as they are fast growing, and in the case of flower, seed and root crops, the presence of a few weeds can usually be tolerated.

Where these crops are laid out in rows, there is usually a good opportunity for weed control early in the crop's development.

For direct seeding of smaller seeds, a bare fallow and stale seed bed is recommended (see 'Bare Fallow' in Chapter 4, 'Weed Management and Control').

Harvesting

Leaf Crops

Leaf crops can be ready to harvest 8 weeks or so after sowing. For detail on harvesting herbs in this group, see under 'Annuals, Biennials and Short-lived Perennials' in Chapter 6, 'Herb Growth Types'.

Flower Crops

Annuals grown for flowers are into full production 8 weeks or so from planting. Usually a number of harvests can be taken as they flower for a prolonged period.

Root Crops

Annuals: If the root is required from an annual herb, it will need to be harvested prior to or during flowering, before the plant starts to go into decline.

Biennials: Biennials are normally harvested for root at the end of their first full season. If they reach the flowering stage, the root is usually woody and of little value.

Short-lived Perennials: The roots of these are usually in their prime at the end of the first full season, but sometimes they can be allowed to go for a second season.

Seed Crops

Seed is harvested at the end of the plant's life, except for short-lived perennials, which may form seed two or three times.

Angelica

Angelica archangelica L   APIACEAE

Parts Used: **Root (Leaf, Seed)**

Other Name: Angelica officinalis.

Angelica has a long tradition of use in northern European countries. The root is used medicinally, mainly as a respiratory and digestive herb (the leaves are occasionally used also). The seeds and the stems are used in confectionery. The seeds are also used in various liqueurs and have some culinary uses.

Then there is that green stuff found in fruit mixes and cakes that is supposedly candied Angelica stems. It once was Angelica, but today it is more likely to be something else that is flavoured, plasticised and dyed green.

Origin

Angelica is native to northern Europe and Asia.

GROWING ANGELICA

Angelica is a most impressive plant in full growth, but it is fairly particular in its requirements.

It is a biennial, though in some conditions it won't flower until its third year or even later, after which it dies. In Australia it usually flowers in its second full season, but if established late the first season, it may not flower until its third season.

While there is possibly some use for other parts of the plant, it is primarily the root which is required in the herb trade.

Climate

Angelica prefers a cool climate with a mild summer. In hotter conditions it will need some protection.

Site

Exposure

Angelica enjoys full sun in climates like Tasmania, but partial shade is preferred in hotter regions.

Drainage

Good drainage is essential.

Soil Type

Angelica prefers a light to medium loam.

Fertility

A high level of fertility is required.

Initial Weed Control

Weed control is not particularly critical if seedlings are raised and transplanted into place.

Neighbouring Plants

Angelica grows quite tall and broad, so it can smother neighbouring plants if it isn't given enough room.

Identification

Angelica is a biennial or short-lived perennial herb 1–2.5 m tall. The leaf stems are ribbed and hollow. The leaves are 2–3 times pinnate, with soft, green toothed leaflets. The flowers are greenish-white and in large spherical umbels on a tall flower-stalk (see colour plate 1.2).

Some green leaf is retained through winter, unless it is very cold.

Angelica should not be confused with Glossy Angelica (*Angelica pachycarpa*), which is commonly grown as an ornamental and sometimes referred to as 'Angelica'. It has much darker glossy green leaves and a smaller growth habit.

Propagation

Method

Seed

Angelica needs to be grown from seed. It is generally held that the seed needs to be fresh and loses viability after a few weeks, but I have heard of instances where relatively old seed has been successful.

Seed stored dry in the fridge or freezer will retain reasonable viability for a year or so.

Division

Older plants produce side-shoots at the base which look as if they could be broken off and planted out. Some books describe early spring division of these second-year plants as an alternative propagation method.

I once tried dividing these shoots off and planting them, but they never took root. Mrs Grieve, in *A Modern Herbal*, says division gives inferior results.

Timing

The best results are usually with fresh seeds, which should be planted within a few weeks of forming in midsummer to early autumn. To ensure these are available, it is necessary to have a few Angelica plants going to seed each year.

There is also a dormancy factor involved as often seed sown in summer or autumn will not germinate until the following spring.

Alternatively, seed could be kept in the fridge or freezer for spring planting.

Planting

The seed is best started in a nursery bed or a seed-tray. It should only be covered very lightly with soil, but must be kept moist until it germinates.

Germination can be erratic. It usually takes several weeks, but may not occur until the following spring.

Alternatively, if a plant is allowed to mature and drop seed on the ground, self-sown seedlings may be available. These seem to germinate better if there is a light covering of fine mulch or old compost on the surface, so the seeds can fall amongst it and germinate when conditions suit them.

Layout and Spacing

Set Angelica out on 450–600 mm centres, with the rows 900–1200 mm apart.

Compost

Angelica needs a heavy dressing of compost at the rate of around 5 kg/m².

Weed Control

Weeds can be kept under control with regular inter-row cultivation and hoeing around the plants. Being large and dominating, Angelica is easy to maintain.

Rotation

Angelica which is being transplanted is a good crop to follow Chamomile or other crops which drop a lot of seed, as it won't be affected by the heavy growth of seedlings.

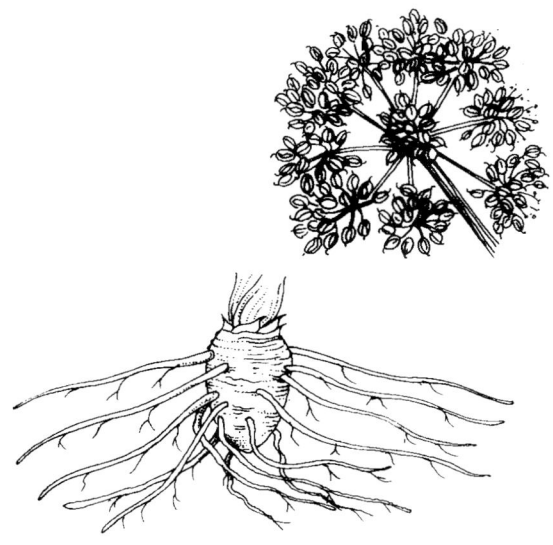

Angelica (*Angelica archangelica*).

Irrigation

Angelica has a high moisture requirement, needing 25–50 mm per week through summer.

Pests and Diseases

Grasshoppers

Grasshoppers can be a problem in some regions, as they devour Angelica.

If they are present in significant numbers, some poultry should be enlisted to keep them down.

Harvesting

Timing

Angelica should be harvested in late autumn or winter. Normally they will be a little over a year old. Plants that have flowered are no good for harvesting.

Tools

A garden fork is needed for harvesting.

Technique

Angelica has a fairly large root system, but as this is not deep, it is fairly easy to dig.

Post-harvest Treatment

The ground should be raked over after digging, so it is ready for the next crop. Pieces of Angelica root left in the ground are no problem as they won't strike.

Drying

After washing, any pieces thicker than your thumb should be split or chopped to facilitate drying.

When dry, the centres of the larger pieces should be checked to ensure they have completely dried.

Temperature

Angelica should be dried at a temperature of around 35°C.

Processing

Angelica is used as a decoction, for which a tea grade is required, and it is also used in manufacturing herbal medicines.

Tea Grade

For tea grade the dried roots should be milled down to pieces around 6 mm across, but the grower is not normally required to do this.

Manufacturing Grade

The pieces of root can be sent off as they come from the dryer.

Storage

Freeze treatment to control moth larvae is necessary before sale or storage.

Yields

Angelica yields around 0.2–0.3 kg/m^2 dried root per annum.

Marketing

There is a moderate demand for Angelica.

Price

Price to growers for premium quality, organically grown Angelica root is $25–35/kg for 5-star tea-grade.

Price to growers for manufacturing-grade, organically grown Angelica Root is currently around $15–17/kg.

Current trade price is around $15/kg.

Aniseed

Pimpinella anisum L   APIACEAE

Parts Used: **Seed**

Other Name: Anise.

Aniseed has a long history of use as a culinary and medicinal herb and was well known to the ancient Egyptians and Greeks.

It is used medicinally mainly as a digestive herb and an expectorant.

Aniseed (*Pimpinella anisum*).

GROWING ANISEED

Being a seed herb, Aniseed will require specialised equipment for harvesting and for cleaning any volume. I have only grown it once in a small trial in Tasmania, which was not particularly successful, owing to its short stature and its tendency to fall over when mature.

Aniseed requires a certain amount of warmth to produce ripe seeds, but most regions, including the warmer parts of Tasmania, appear to be suitable for growing it.

Identification

Aniseed is an annual herb growing to about 400 mm tall. The lower leaves are simple, coarsely toothed, more or less kidney shaped, the upper leaves are more feather-like, blue-green, aromatic. The flowers are white, in open compound umbels and the seeds ovate, brownish and ribbed, about 5 mm long.

Variety

There is apparently some variation in the size and volatile oil content of the seed.

Rotation

Aniseed needs to follow crops that leave the ground fairly clean, as weeds can be a problem in the crop. It is also inclined to drop a bit of seed, but this shouldn't be a worry in ensuing crops as it is a smallish plant that is easily killed by cultivation.

Irrigation

Aniseed may need additional moisture in dry conditions, but avoid aggressive sprinklers as these could flatten it.

Harvesting

Timing

The seed is inclined to fall from the head very easily when ripe, so on a small scale it is probably best to harvest Aniseed a little early, when the tips of the fruits or seeds have turned greyish-green, and let it cure for a week or two before threshing the seed out with a flail.

Traditionally, Aniseed was often pulled up by the roots, rather than mown, but this would only be advisable where the soil does not contain any fine stones that might end up among the seeds.

On a large scale, harvesting with a header, seed maturity would have to be watched closely so the Aniseed could be harvested before too much seed was lost onto the ground.

For more detail on this herb see 'Information Charts'.

Basil

Ocimum basilicum L LAMIACEAE

Parts Used: Leaf

Other Names: Sweet Basil.

Basil is a popular culinary herb, but it is regarded by some as having medicinal properties as a digestive herb (as are most culinary herbs), and as a weak sedative and stimulant of milk flow.

Origin

Basil is native to southern Asia and the Middle East.

GROWING BASIL

Apparently, in tropical conditions, Basil is a woody perennial which forms a bush and continues for several years, but in most of southern Australia it is an annual and needs to be sown from seed every year.

Basil is a good crop to consider if you are just getting started and have a hot summer and good soil. Because it is quick to establish, it can be in full production early in your first season, but remember it is frost tender.

Basil *(Ocimum basilicum)*.

Climate

Basil needs warm growing conditions and is sensitive to frost and to cold winds, which will burn the leaves almost as badly as a frost.

In cooler climates it can be grown successfully if planted late in a warm position or under cloches, although yields will be lower and quality may be poorer.

Site

Exposure

Basil needs a warm position with full sun. It thrives in hot conditions, provided it has plenty of moisture, and it needs good protection from cold winds.

Drainage

Basil won't tolerate wet feet, but being a summer crop it can often be planted in a position that is a bit wet in winter but dries out in summer, making it unsuitable for perennials.

Soil Type

Basil will grow in a wide range of soil types.

Fertility

Basil needs a high level of fertility.

Initial Weed Control

If direct seeding is planned, a period of bare fallow prior to planting is recommended to reduce weeds.

Neighbouring Plants

Avoid placing Basil where it will be shaded by other plants, as it usually needs all the sun it can get. Basil may benefit from being planted on the leeward side of taller plants to give it some protection from wind.

Identification

In temperate Australia, Basil is an annual herb, 30–60 cm tall, with branching four-sided stems. Leaves are more or less ovate, 5–10 cm long, shiny, aromatic. Flowers are white in whorls in terminal racemes.

Variety

Make sure you have Sweet Basil (Large Sweet), not Bush Basil or Opal Basil, which are dwarf varieties with inferior flavours.

Propagation

Method

Basil is grown from seed.

Timing

The seed should be planted when the ground and the weather have warmed up and all danger of frost is past. For an earlier start, Basil can be established under cloches, which are removed as the weather gets warmer.

Planting

Basil tends to be slow to germinate, but this can be overcome by soaking the seed in warm water for 24 hours. During this time it will go through an amazing transformation, turning into a slimy mess, somewhat resembling frog's spawn.

This has to be poured into the furrow. I used a tin can shaped to make a beak for pouring. To overcome the problem of the seeds clinging together, add extra water to the mixture and give it a swirl every now and then as you pour it out along the furrow, adjusting your rate of pouring so the seeds are more or less optimally spaced.

Transplanted seedlings can be used, but care must be taken as they don't enjoy being transplanted. Small seedlings transplant more easily than large ones.

Layout and Spacing

Basil should be thinned to (or set out on) 100–150 mm centres in rows 300–450 mm apart.

For harvesting with the scythe, these should be in blocks at least four rows wide.

Compost

Basil needs a good dressing of compost at the rate of around 5 kg/m^2. If the leaves are not a deep green colour, it needs more compost or growth will be poor.

Weed Control

Being an annual row crop, a moderate level of weeds can be coped with, as long as these are kept under control. Good weed control prior to planting is important if direct seeding, but not as critical for transplanted seedlings, though the better the preparation, the easier the crop will be to manage. As the plants are short and closely spaced, thick weed growth can be difficult to keep under control.

Thorough weeding while the plants are small is essential for good growth, and diligent weed control during the season is important, to eliminate competition and contamination.

Rotation

As only a moderate level of weeds can be coped with in Basil, it is not a good crop to follow weedy crops with or those that self-sow heavily.

If weeds are kept down in the Basil, the ground should be reasonably clean for subsequent crops.

Irrigation

Basil needs plenty of moisture for good growth: 25–50 mm of water per week through summer.

Pests and Diseases

In good growing conditions, Basil doesn't seem to suffer much from pests and diseases. The main problem I had growing it in marginal conditions in Tasmania was windburn blackening portions of the leaves.

Harvesting

Timing

Once Basil starts to flower, the growth of new leaves is diminished. Careful harvesting at this stage will stimulate the growth of new leaves, but if the plant is not harvested, its quality will deteriorate as it goes to seed.

In warmer locations at least three harvests should be possible over the growing season.

Tools

Basil can be harvested with the catching scythe if you are accurate enough, otherwise use a sickle or reaping hook.

Technique

Basil needs to be cut off about a third of the way up the stem, taking about two-thirds of the leaf and removing all the flowers. Being effectively an annual, it should not be cut back too hard or it will not regenerate well.

For the last harvest, the plants can be cut lower without worrying about regeneration.

Post-harvest Treatment

After harvesting, go over the plants and remove any remaining flowers. This will stimulate more leaf growth.

Drying

Good colour is difficult to obtain if Basil is dried too slowly: even with good drying conditions it comes out looking rather dull. Handling the herb in a wilted state will cause bruising and darkening.

Temperature

As Basil is very aromatic, keep drying temperatures below 35°C.

Processing

Basil is primarily a culinary herb, and it is all sold in that form.

Culinary Grade

Rub through an 8 dent screen to produce culinary grade: it is okay if the flowers go in, but try to get the herb free of stalk. Rubbing when the leaf is brittle but while the stalk is still a little pliable will make this task easier.

Storage

Freeze treatment to control moth larvae is necessary before sale or storage.

Yields

It is a long time since I have grown Basil and I don't have any accurate records of yields, but it should make about 0.2–0.3 kg/m^2 dried culinary grade.

Marketing

Basil is the most popular of the dried culinary herbs and at present (1996) there isn't any significant Australian supply of certified organic or biodynamic dried Basil: the bulk of it is imported from the USA.

Price

Price to growers for premium quality, organically grown Basil is currently around $24/kg for 5-star culinary grade.

Current trade price is around $3–4/kg.

Blessed Thistle

Cnicus benedictus L ASTERACEAE

Parts Used: Leaf & Flower, Aerial Parts

Other Names: Holy Thistle, *Carduus benedictus*, *Carbenia benedicta*.

Blessed Thistle is a herb with a tradition going back at least to medieval times when it was regarded as something of a panacea, with a reputation of even being able to cure the plague.

Today it is mainly used medicinally as a bitter tonic and an astringent. It is also features in liqueurs, in particular Benedictine, and various parts of the plant can be eaten as a vegetable.

GROWING BLESSED THISTLE

Blessed Thistle is an annual that prefers cool to mild growing conditions. In cooler regions, it can be grown as a summer crop, but in warmer regions it does better as a winter crop, starting from seed in autumn and flowering fairly early in spring.

It is quite frost hardy but it does not thrive in hot dry conditions, tending to suffer from two-spotted mite.

Containment

Blessed Thistle does tend to self-seed somewhat but it does not appear to be a potential noxious weed. It only seems to get established in cultivated ground and the seed is quite heavy, so it doesn't travel far.

Identification

Blessed Thistle is an annual herb, with multiple branching stems. Leaves are light green, lanceolate, with large irregular teeth ending in rather soft spines. Flowers are yellow, 30–40 mm across, at the ends of the branchlets, covered in sharp spiny bracts so they are partially concealed.

Blessed Thistle is fairly easy to distinguish from other Thistles with its soft, yellow-green leaves and its yellow spiny flowers.

Rotation

Blessed Thistle could be used in various rotations. As a winter crop it could be grown where ground is available around March, provided the plot is sufficiently free of weeds. This planting should be harvested by late October, enabling a summer crop of, say, Basil to go in.

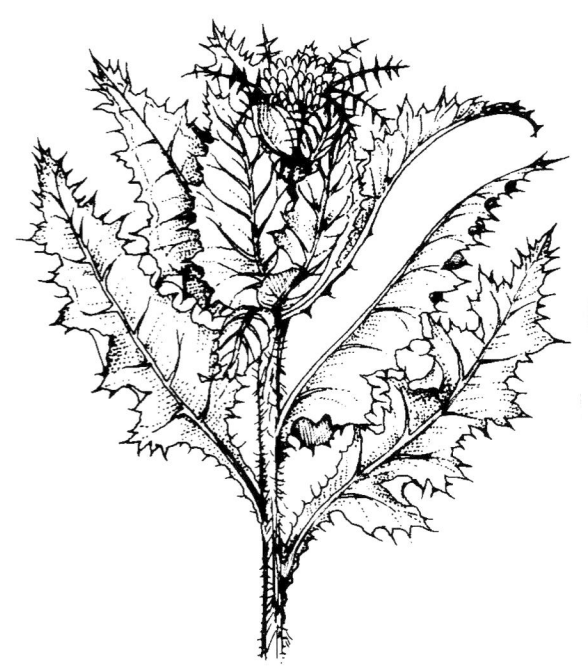

Blessed Thistle *(Cnicus benedictus)*.

Pests and Diseases

Two-spotted Mite

In warm dry conditions, Blessed Thistle is inclined to suffer from two-spotted mite (red spider). This is exacerbated if the plants are overcrowded. Basically this is because the plants are stressed: if they are adequately spaced and grown in the cool conditions they prefer, the problem does not arise.

Drying

There can be problems with drying Blessed Thistle as the flower heads take forever and a day to dry, while the leaves dry relatively quickly. This makes it very easy to misjudge the dryness and put the herb into storage without realising there is still a lot of moisture in the flower heads. This moisture is gradually released, causing the herb to go mouldy in storage.

With ambient air drying, allow Blessed Thistle to dry for about 4 weeks. To check the herb for complete dryness, split a few of the flower heads and look inside them. After processing and storage, the herb should be checked again for dryness 2–3 weeks later.

For more detail on this herb see 'Information Charts'.

Burdock

Arctium lappa L   ASTERACEAE

Parts Used: Root, (Seed, Leaf)

Other Names: Greater Burdock, Gobo, Lappa.

Burdock is much esteemed medicinally, as an alterative, as a diuretic and in treating skin conditions. It is also used as a vegetable in Japanese cuisine, where it is known as *gobo*. The root is sliced into thin cross-sections and soaked in several changes of water to remove tannins before cooking.

Origin

Burdock is a native of Europe.

GROWING BURDOCK

In good conditions Burdock grows into a mighty plant, with a tap root nearly as thick as your arm and more than a metre deep. So before taking it on, you need to consider what you are letting yourself in for, as harvesting can be something of a battle in heavy or stony soil.

Burdock is a biennial and the root is harvested after the first season.

Climate
Burdock will grow in a wide range of climates.

Site

Exposure
Burdock likes full sun.

Drainage
It needs good drainage.

Soil Type
From the point of view of harvesting, a deep light soil is best, free of stones and with a subsoil that is easy to dig.

Fertility
Burdock needs a very fertile soil for good growth.

Initial Weed Control
If Burdock is direct sown, the ground needs to be relatively free of weeds to enable it to establish itself.

Neighbouring Plants
Burdock develops huge leaves, which will overshadow any smaller herbs planted too close to it.

Burdock *(Arctium lappa).*

Containment
This species is regarded by some as a potential noxious weed, but as growing it for the root means it is dug before it goes to seed, and as broken pieces of Burdock root left in the soil will not strike, it is unlikely to cause problems if it is managed properly.

The main area of concern is if it is grown for seed, as the burr is the king of all burrs, clinging to anything that touches it. The most important thing would be to keep livestock away from it when it is in seed.

Identification
Burdock is a biennial herb up to 2 m tall, with finely hairy stems when in flower. Leaves are heart-shaped to ovate, up to 450 mm long, covered with a fine down on their under surfaces and with solid stalks (see colour plate 1.3). The leaves on the flowering stalks are smaller and less downy underneath. Flower heads are spherical, 25–50 mm across, with red to purple florets: when ripe these develop into hooked burrs.

Lesser Burdock (*Arctium minus*) is a very similar plant which is a little smaller. It differs in that it has hollow leaf stems and smaller flower heads 13–20 mm across. It is reported to be growing as a weed in parts of Australia.

Variety
There may be some scope for breeding a variety with a shorter stockier root. I found that seed saved from a few selected plants with shorter thicker roots tended to produce plants with shorter roots, which were easier to dig.

Propagation

Method
Burdock is grown from seed.

Timing
Seed can be planted in autumn or spring. Autumn sowing is preferable, as it gives the plants a longer growing period. This requires clean ground: in winter the Burdock seedlings go completely dormant and can be lost under a growth of weeds. If winter weed control is likely to be a problem, then spring sowing is preferable.

Planting
Direct sowing is preferable, but young seedlings can be transplanted with care.

Layout and Spacing
Burdock should be thinned to, or set out on, 75–150 mm centres in rows 900–1200 mm apart.

From the point of view of harvesting, it will be a little easier if there are two rows, or an even number of rows, beside each other.

Compost

Burdock needs a heavy dressing of compost at a rate of around 5 kg/m².

Weed Control

Weed control is mainly inter-row cultivation and hoeing around the plants while they are small. Once they have developed good-sized leaves, they shade out most weeds and then an occasional hoeing around the edges of the plants is all that is needed.

Rotation

Burdock fits in to a number of different rotations. If grown for seed there will usually be a lot of self-sown seedlings the next season, though these can be easily controlled with cultivation.

When grown for the root, it leaves the ground pretty clean and deeply worked: though this may tend to bring up weed seeds that have been lying dormant deep in the soil.

Irrigation

Burdock needs around 25–40 mm of water per week through summer.

Pests and Diseases

Grasshoppers

Grasshoppers can be a problem in Burdock, but if the plants get off to an early start and are growing vigorously, they can usually stand up to feeding a few grasshoppers. Late started plants and those that are lacking vigour may suffer more.

Harvesting Root

Timing

Dig Burdock in late autumn or winter, when the tops have died down or when they have nearly done so.

Digging will be easier if the subsoil is quite moist. This loosens the deep roots' grip on the soil, so it is best to wait until there have been some good soaking rains. Then after a few days of fine weather, the surface will be dry enough to work, but the subsoil will be still wet and loose.

Tools

A good narrow spade with a strong blade and handle are required for digging the roots.

Technique

If the growth is good, the roots will go down as far as a metre. This can make digging Burdock appear quite a formidable undertaking, requiring a strong back and a weak brain. However, by approaching the task systematically, it is not really difficult.

The method I use is to progressively work along the row, starting with a hole at one end, into which the first root is pulled sideways. This creates a space for pulling the next root into and so on down the row, in a sort of underground domino effect.

To begin, dig a hole in the line of the row just next to the first root and go down to about two-thirds of its depth. The spade is then driven down the other side of this root as deeply as possible and levered forward to lean the root towards the hole.

If this is successful, it should be possible to grasp the root and lift it out of the ground. If it still won't budge, try spearing the spade on either side of it. Your grip on the

Burdock root.

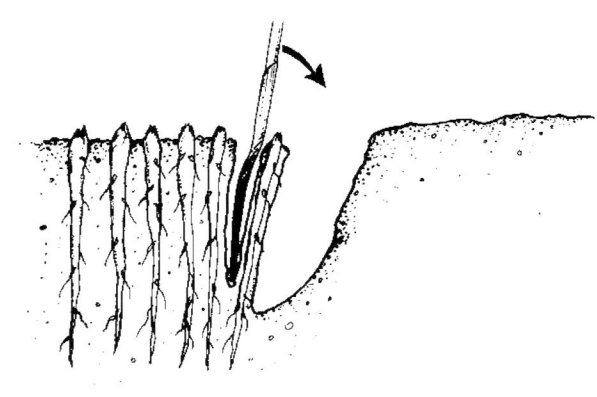

Diagram showing the digging technique used for Burdock.

root can be aided by wrapping a piece of old Burdock leaf around it.

Sometimes the root snaps off and you may have to dive after it with a trowel or else abandon the lower part.

When the root, or as much of it as possible, has been removed, you can then enlarge the hole created and work towards the next root, and so on progressively along the row.

In most soils it will be important to pile the topsoil and subsoil separately as you dig your way along the row. After a short distance, you can fill the trench in behind you as you dig, putting the subsoil into the bottom and the topsoil on top.

At the end of the row, you will be left with a hole, which will need to be filled with the soil that was stacked to the side in the beginning of the row. But if you have two parallel rows of Burdock, you can start digging the second row from this opposite end and put the first soil from it into the hole at the end of the first trench. The second trench will then be finished beside the beginning of the first, where there will be a pile of soil waiting to fill in the final hole.

It is worth going to a bit of trouble with all this, in order to avoid ending up with your garden looking like the aftermath of a First World War battlefield, with a thick layer of infertile subsoil on the surface and much of your valuable topsoil buried half a metre underground.

Mechanical Harvesting

Mechanical harvesting is a possibility, but the deep roots of Burdock would present something of a challenge. In a deep sandy alluvial soil, it might be feasible with specially designed equipment.

Post-harvest Treatment

A thorough cultivation after harvest will help to mix in any subsoil remaining on the surface. As Burdock will only regenerate from its crown, there is no problem with pieces of root remaining in the soil after harvest.

Drying

Burdock roots are easy to wash, as they have few branches. If the leaf buds have started to develop, they should be cut off, but usually no trimming is required apart from removing any remains of the previous season's leaf bases.

To facilitate drying, the roots should be chopped into round sections about 25 mm thick.

Check dryness by breaking some of the larger pieces in half and feeling the centres. If they are completely dry, these should be quite hard (though sometimes large roots are a little pithy in the centre).

Temperature

Burdock root should be dried at a temperature of 35–45°C.

Processing

Burdock is used to a moderate extent as a tea (or decoction really) and it is also manufactured into fluid extract.

Tea Grade

Tea grade should be milled down to pieces no larger than 6 mm across, though usually the grower is not required to do this.

When milled, Burdock root can sometimes look deceptively as though it is full of the webs of moth larvae, but this is really the fine hairs on the budding leaves in the crown clinging to each other and to pieces of root. I have even had customers return it, thinking the herb was infested with moths.

The problem could be avoided by trimming off the crowns of any roots that have started to swell, or by harvesting earlier – in late autumn or early winter – for as spring approaches the buds swell, so there is a higher proportion of this clinging hair among the milled dried root.

Manufacturing Grade

For manufacturing grade, the root can be sent in the form it comes from dryer.

Storage

Freeze treatment to control moth larvae is necessary before sale or storage.

Yields

At 'Twin Creeks' our yields were around 0.3–0.5 kg/m^2 dried root per annum. Mrs Grieve (*A Modern Herbal*) mentions yields of 1500–2000 lbs/acre (0.17–0.23 kg/m^2).

Harvesting Seed

Burdock will go to seed if allowed to grow for two full seasons. Some precautions need to be taken to prevent the seed spreading, in particular do not allow animals access to the ripening burrs or they may hook onto the animals' coats and be carried far and wide.

The plants die after forming seed, so the roots cannot be harvested as well.

Timing
Burdock will go to seed in its second full season and this will be ready to harvest in late summer–early autumn. The tops should be cut when the seed is just starting to fall.

Tools
On a small scale, the tops are best cut with a reaping hook.

Technique
As Burdock seed ripens unevenly, the tops need to be cured for 2 weeks or so to allow the undeveloped seeds to fill out. It is best cured in the drying shed to prevent losses to vermin, or to risk seed being spread by birds and animals.

When the seed heads are thoroughly dry, the seeds can be separated by rubbing the seed heads very firmly against a $2^1/_2$ or 3 dent screen to break them up and force the seeds through. They can then be cleaned by sifting and winnowing.

Mechanical Harvesting
Harvesting with a header is a possibility, but there may be problems with timing, because the seed ripens unevenly and starts to fall when it is ripe. It may need drying after harvest.

Post-harvest Treatment
After harvest, go over the patch and gather up any burrs which have been missed or dropped, to prevent them from being spread elsewhere. Do not graze the stubble with livestock, though chooks might help clean up any seed on the ground.

Storage
Freeze treatment to control moth larvae is necessary before sale or storage.

Harvesting Leaf
I haven't ever harvested Burdock leaf. Because the stems and midribs are quite thick and fleshy, it may be necessary to chop the leaves before drying to ensure the fleshy parts dry adequately.

Marketing
There is a moderate demand for Burdock root and occasionally a requirement for seed or leaf.

Price
Price to growers for premium quality, organically grown Burdock root is currently around $24/kg for 5-star tea grade.

Price to growers for manufacturing-grade, organically grown Burdock root is currently around $12–15/kg.

Current trade price is around $10–12/kg.

Calendula
Calendula officinalis L ASTERACEAE

Parts Used: **Flower**

Other Names: Marigold, English Marigold, Pot Marigold.

I prefer to use the name Calendula for this herb. It is commonly also known as Marigold, but this can lead to confusion with several *Tagetes* species, which are also known as Marigold but do not have the same qualities and are somewhat toxic.

Calendula is a much valued medicinal herb with a range of uses, mainly associated with its anti-inflammatory and antiseptic qualities, but it also has an action on the gall bladder.

The bright orange and yellow petals also enjoy some use as a culinary herb, mainly for their attractive colour in salads and other dishes. Orange Calendula flowers are a common ingredient of laying mixes. They give a nice dark colour to the yolks when the hens are locked up in cages and unable to obtain the green leaves that are part of their natural diet.

Origin
Calendula is a native of southern Europe.

GROWING CALENDULA
Calendula is easy to grow and its orange flowers are an attractive addition to the garden. It bears flowers over a long period and it is easy to maintain.

Calendula is a good crop to consider for new ground or for growers just starting out, as it does not require an initial bare fallow and it is into full production a few months after planting.

Calendula (*Calendula officinalis*).

Climate

Calendula prefers a cool climate with a mild summer. With adequate moisture it can be grown in hotter conditions, but the plants don't live as long nor yield as well.

The plant itself is quite frost hardy and will survive the winter in most parts of southern Australia. In very cold conditions, it will not flower much and frost may damage the flowers.

Site

Exposure
Calendula likes full sun and is tolerant of cold winds.

Drainage
Good drainage is necessary for optimum growth and survival through winter.

Soil Type
Calendula will grow in a range of soils.

Fertility
A moderate level of fertility is preferred. If soil is too fertile, Calendula produces a lush growth of leaves with fewer flowers.

Initial Weed Control
Calendula holds its own fairly well against weeds, so it is a good crop to put in where the weed situation is not ideal.

Neighbouring Plants
No particular requirements.

Containment
Inevitably Calendula will self-sow as some seed always forms and falls to the ground, but this is not a great problem as it does not carry far and the seedlings are easily killed with cultivation.

Identification

Calendula is an annual herb, though it will live a bit longer than a year in good conditions. It has a bushy form to about 800 mm. Leaves are oblong to lanceolate, 50–150 mm long. the flower heads 60–80 mm across, petals yellow to orange (see colour plate 1.5).

Calendula, with its simple undivided leaf, can be distinguished from French Marigold (*Tagetes patula*) which has a pinnate leaf.

Variety

The deep orange flowers are regarded as having the best medicinal qualities. Some authorities prefer the single-flowered varieties. These have a central button and several rows of petals.

Double-flowered varieties exist, but as they are a step away from the wild form, it is best to stick with single-flowered varieties. However, if your market requirement is for petals only, then a double-flowered variety may be a better proposition, as the petals weigh almost nothing, so you will need a lot of them to make it worth while.

Some varieties have brown centres, but when dried these don't look as attractive as the orange-centred flowers.

Some seed is a mixture of colours and forms, so it may be necessary to select for the desired deep orange colour. This can be done by planting rather thickly and culling plants as they begin to show their true colours. After a couple of generations of selecting, very few yellow flowers will show up.

The varieties Orange King and Nagasatu Orange both have good colour and flowers with multiple rows of petals.

Propagation

Method
Calendula is grown from seed.

Timing
Plant in autumn or early spring.

Planting
Calendula can be direct sown, but as it transplants easily there are advantages to establishing it in a nursery bed or in seed-trays and transplanting so it gets a good start on the weeds. Often there is a good supply of seedlings from self-sown seed.

Layout and Spacing
Calendula should be set out on, or thinned to, 300–400 mm centres in rows about 900 mm apart.

Closer spacing tends to give heavier initial yields but the bushes will be healthier and live longer if they are given more space.

Compost

If the soil is not sufficiently fertile, Calendula may need a moderate quantity of compost at the rate of up to 2 kg/m^2, but if fertility is reasonably good, it may not need any, or just a very thin application, especially if it is planted in rotation with crops that get a heavy dressing of compost.

Weed Control

Weed control is quite easily maintained with regular inter-row cultivation and hoeing around the plants. Once they are well established, they hold their own well, but an under-storey of weeds tends to develop because it is too hard to hoe among the bushy fully grown plants. As only the flowers are harvested, some weeds can be tolerated among the crop.

Pests and Diseases

Mildew

Mildew usually shows up in the summer of the second season as the plants start to age and lose vigour. Young plants are not normally affected.

Rotation

Calendula is a crop that can be used to follow others that tend to self-sow or become weedy.

As it is difficult to maintain thorough weed control in the crop once the bushes have filled out, and some seed inevitably falls to the ground, Calendula tends to leave a legacy of weeds for following crops.

Irrigation

Calendula needs 25–40 mm per week through summer, depending on the climate.

Harvesting

Timing

The flowers can be picked from when they open until they are a few days old. Once they begin to fade they are too old.

Frequency of picking will vary according to the season, from every 2–3 days to once a week or so.

Newly planted Calendula begins flowering in late spring and continues virtually throughout the year, but the greatest yields are through summer. The plants will often continue to flower into their second season, but as they lose their vigour their flowers usually become smaller and yellower and mildew weakens them, so they tend to die off as summer comes on.

Tools

For the best quality, Calendula flowers should be picked by hand. They can be picked with a harvesting comb, but this will inevitably gather some buds and over-mature heads.

A basket of a convenient size and shape will be needed for picking into: this can be set on the ground or attached to the waist in front.

For harvesting Calendula by combing, a flower comb with the same spaced teeth as for Chamomile (3.5–4.5 mm between the teeth) is needed.

Technique

While picking, watch out for caterpillars feeding on the flowers.

It is a good idea to save some flowers on the best plants for seed.

Hand-picking

Pluck the flowers so they break off just below the head: either clasp them between index and middle finger or between index finger and thumb (a long thumbnail may help). Hand-picking produces the best quality, as long as you don't mind spending the time and getting your hands covered with sticky sap (it washes off).

Combing

Alternatively, Calendula flowers can be combed off. This is faster but picks up some buds and over-mature flowers as well, which will lower the quality somewhat. Combing will probably not produce premium grade, but if you are supplying manufacturers who want a lower priced product, it may be an option.

Post-harvest Treatment

Go over the bushes and trim off any old flowers that are not wanted, as these will suppress the formation of new flowers (this might be done during the picking operation).

Drying

Calendula is reluctant to dry and needs very good conditions to complete the drying process. If the herb is too slow in drying, it fades badly, with a resulting loss in value.

For the best quality, optimum drying conditions are required. To ensure this, artificial heat may be necessary, or should at least be available as a back-up for when prevailing weather conditions are unsuitable.

Avoid overloading the screens: the flowers should only be about one layer thick. Some growers lay them all face down to ensure even drying.

Judging the dryness of Calendula flowers can be quite deceptive as the petals can feel very dry while the centres are still moist inside.

To check them, the flower stalks should break with a brittle snap when they are dry and the centre of the flowers should be hard and not pliable.

Calendula is often misjudged and put into storage in an inadequately dried state, containing enough hidden moisture to cause the whole bag to go mouldy. Until you are confident of your ability to assess the moisture content of Calendula, it should be re-checked after it has been in storage for a week or so. If the petals no longer feel crisp and dry, it should be put back on the drying screens for a spell.

If your market requires the petals alone, it would not be necessary to wait until the centres are dry. In fact it may be easier to rub when the petals are dry enough to fall off, but the centres are still pliable and less inclined to break up.

As Calendula's orange petals show up rather glaringly in other herbs, it is best to dry them in a spot where the wind won't blow them onto other screens. A few petals can also fall through shadecloth screens and onto herbs drying below.

If the screens are to be used for drying other herbs after the Calendula, they should be carefully cleaned to remove any remaining petals, as they tend to cling to the screen.

Temperature

Calendula should be dried at a temperature of not more than 35°C.

Processing

For whole flowers, there is no processing necessary after drying.

If petals alone are required, the heads could be rubbed. I have never done this, but I think a 2½ or 3 dent screen would be suitable. Winnowing may be necessary to remove any trash.

Tea Grade

The flowers are normally used whole for tea grade. Only the best quality bright orange flowers should be offered. Sometimes petals alone are required for blends.

Culinary Grade

For culinary use, petals alone are required.

Manufacturing Grade

Whole flowers are used for manufacturing. Quality requirements may not be as stringent, depending on the manufacturer.

Storage

Freeze treatment to control moth larvae is necessary before sale or storage. Calendula needs to be stored in a dark place to reduce fading, and even in the dark it will slowly lose its colour, so it will not store for as long as most other herbs.

Yields

Yields are around 0.1–0.25 kg/m^2 dried flower heads per annum.

Marketing

There is a moderate demand for Calendula as tea grade and for manufacturing.

Price

Price to growers for premium quality, organically grown Calendula is currently around $22–30/kg for 5-star tea grade.

Price to growers for manufacturing-grade, organically grown Calendula is currently around $15–22/kg.

Current trade price is around $7/kg.

Caraway

Carum carvi L   APIACEAE

Parts Used: Seed

Other Name: Caraway Seed.

Caraway is widely used as a culinary, but it also has some use medicinally as a digestive and respiratory herb.

GROWING CARAWAY

Caraway is a seed crop. It can be grown on a small scale but, like most other seed crops, this will not be a financial proposition. It lends itself to a broad-acre scale using conventional grain harvesting equipment.

The main problem is that Caraway is a biennial, so it has to be kept free of weeds over the winter, as weed seeds are difficult to remove from the harvested crop.

My experience with this species is by no means vast. I grew a small trial plot in Tasmania in the early 1980s. In eastern Canada, it occasionally used to grow wild in the hay-fields where we would harvest a bit for our own use.

Caraway will grow in a range of climates, but seems to prefer cooler climates with a mild summer and a high rainfall.

Initial Weed Control

Caraway needs very clean ground as it has to go through a winter, with associated weed problems. If there are many weeds present, a thorough bare fallow may be necessary prior to planting.

Because the foliage is so similar, it should not be grown in locations where Poison Hemlock is growing as a weed.

Identification

Caraway is a biennial with a rosette of Carrot-like feathery leaves the first season, followed by a tall slender flowering stalk, 200–1000 mm, in the second season with numerous small white flowers in umbels. The fruit (commonly referred to as seed) is 3–5 mm long, oblong and ribbed.

There are a number of species from the same family with similar leaves and flowers in umbels. The most notable is Poison Hemlock, but that can be identified by the purple blotches on its stems and its rank aroma when crushed.

Variety

Apparently there are a number of varieties of Caraway cultivated in different regions of Europe.

Propagation

Timing

If planted in spring, Caraway will not go to seed until the following season. Planting in autumn is supposed to result in it flowering the following summer, but this would have to be trialled first. It may need to be planted in early autumn so it gets a long enough period of growth to trigger it into flowering after the interruption of winter. Autumn planting may also affect yields.

Weed Control

Regular inter-row cultivation and hoeing around the plants are essential to maintain good weed control, as Caraway does not shade the weeds out very much.

Keeping the crop clean is important, as weed seeds are difficult to separate from Caraway.

Rotation

Caraway should follow crops that leave the ground fairly clean, as weeds in Caraway can be a problem.

In Europe a cover crop of an annual seed crop, such as Coriander, is sometimes sown with the Caraway. After the annual crop is harvested, the Caraway continues to grow. Here this system would depend on adequate moisture being available.

Harvesting

Timing

The seed should be harvested when it is ripe and before it begins to fall. It seems to hold in the heads reasonably well, so it can be allowed to ripen in the field, provided you take care to avoid losses from shattering when handling it.

For more detail on this herb see 'Information Charts'.

Caraway *(Carum carvi)*.

Catnip

Nepeta cataria L LAMIACEAE

Parts Used: Leaf & Flower (Aerial Parts)

Other Names: Catnep, Catmint.

Catnip gets its name from the way many cats relish it to the point of intoxication.

It is used in herbal medicine as a mild sedative, as a digestive herb and for colds and flu, especially for treating children.

Catnip has some usage as a herbal beverage and as stuffing in toy mice for cats to play with.

GROWING CATNIP

Catnip does not neatly fall into this group of herbs, as it is rather variable in its habits, in some situations it may last 3 or 4 years, but often it lasts only 2 years.

In its prime it is quite a large attractive plant with its soft blue-green foliage, but it declines quite rapidly and if you are not ready with some replacement plants, there can be a lot of gaps in the rows.

If it were to be regarded as a perennial, it would more or less fall into the woody perennial category. The stems do become almost woody and its focus of regrowth is above ground level.

Catnip seems to live longer in a cool climate with mild summers, but it will do quite well in hotter regions if it gets plenty of moisture. In Tasmania it typically lives 3–4 years, whereas in north-east Victoria only the only occasional plant lasts beyond 2 years.

Neighbouring Plants

Catnip can grow quite tall and bushy, so it should be given plenty of room as it can overshadow smaller herbs.

Identification

Catnip is a short-lived perennial, with square branching stems up to 1000 mm tall (see colour plate 1.6). Leaves are ovate, coarsely toothed, blue-green with whitish undersurfaces, pubescent, 30–70 mm long. Flowers are normally white, about 6 mm long, in whorls at the ends of the branchlets and on axillary spikes.

There are two smaller related species referred to as Catmint (a name sometimes also used for Catnip). *Nepeta* x *faassenii* is a similar plant, but only grows to 600 mm and has mauve-blue flowers. *Nepeta musinii* is shorter again with a trailing habit. It grows to 300 mm and also has grey leaves and mauve-blue flowers (perhaps it should be called Mousenip).

Catnip (*Nepeta cataria*) is the preferred species and is the only one which should be grown commercially.

Catnip (*Nepeta cataria*).

Propagation

Catnip is easy to grow from seed and where the plant is short-lived, this is the preferred method. In locations here it lives longer, young vigorous plants can also be divided, in a similar manner to that used for woody perennials.

As Catnip can decline very rapidly and unpredictably, it is good to have a supply of replacement plants on hand. Often in spring quite a few self-sown seedlings can be found around an established bed of Catnip, and these can be used to fill in the gaps.

Rotation

Catnip can be replanted continually in the same bed. As the plants die off erratically, the easiest way to maintain production is to replant the gaps. The young plants grow quite quickly to fill in the spaces.

Pests and Diseases

Cats

Some cats love to roll and sleep on Catnip, even until it dies, though many seem quite oblivious and ignore it. If the plot is large enough, Catnip should be able to accommodate a few cats and still provide a good yield.

Harvesting

Technique

Enough leaf should remain on the plants to enable them to regenerate, so usually the upper two-thirds to three-quarters is harvested.

Processing

Catnip is used as a tea and to some extent as a fluid extract.

Tea Grade

Rub through a 2½ or 3 dent screen and re-screen or winnow until the herb is free of stem.

Manufacturing Grade

First Screening

Catnip aerial parts usually contain a large proportion of stem – this can be well over 50%, particularly with tall plants which are over-mature – so for a better quality manufacturing grade, it can be rubbed once through a 2½ dent screen to remove the bulk of the stems. This first screening will contain more than 90% leaf and will produce an excellent quality fluid extract. The question is whether manufacturers are prepared to pay a price for it that will be adequate for the grower.

Much of the Catnip currently being used in the trade is of shocking quality, being typically two-thirds or more stalk. This may well be a case of a herb which has gone out of general usage because the quality generally available is so poor and ineffective.

I have been told by one herbalist that he found our premium grade Catnip leaf and flower to give results he has never been able to achieve with Catnip from other sources.

The *British Herbal Pharmacopoeia* (1983) specifies 'leaves and flowering tops of *Nepeta cataria*', which is in contrast to the 'aerial parts' or 'whole herb' it specifies for many other herbs.

Aerial Parts

If your purchaser insists on a lower-priced product, you could try to avoid the larger stems in harvesting so there is a better proportion of leaf in the aerial parts.

Another alternative would be to use the less stalky aerial parts of the second harvest for manufacturing grade (the first harvest could be made into tea grade by removing the stalks).

Aerial parts should be put through the chaff-cutter. Before chopping, make sure the upper parts of the stems are thoroughly dry.

For more detail on this herb see 'Information Charts'.

Celery

Apium graveolens L APIACEAE

Parts Used: Seed, (Leaf)

Other Names: Wild Celery, Smallage.

Celery is commonly thought of as a vegetable, but the leaf is used as a culinary herb and the seed has important medicinal properties, being used in the treatment of rheumatism and arthritis.

Origin

Celery is a native of southern Europe.

GROWING CELERY SEED

Growing Celery for seed is quite a different story to growing it as a vegetable. First, it is the wild form that should be grown. It would appear that the right climate is necessary, too. I did some trials growing it in southern Tasmania and north-east Victoria.

In Tasmania the plant often seems to behave as an annual, which is contrary to all the descriptions of Celery, as it is supposed to be biennial. This may have something to do with the local climate, which is very changeable in spring, with alternating warm and cold spells. Ordinary garden Celery is often difficult to grow in these conditions, which seem to trigger flowering, the plants being confused when it turns cold and then going to flower when it gets warm again. I did obtain some seed from this trial, but the yields were not very good.

In north-east Victoria the results were dismal, as the wild form Celery did not like the heat at all.

Climate

Celery seems to require a mild stable climate.

Identification

Celery is a biennial (annual under some conditions), with a branching flowering stem up to 1000 mm tall. Leaves are pinnate with stems of varying fleshiness, 100–150 mm long with deeply toothed, fan-shaped leaflets. The flowers are greenish-white, small, and in stalkless umbels.

Variety

I am unsure about Celery varieties, but being a variable species, there may well be a variety that is more suitable for seed production.

Weed Control

Good weed control in the early stages of development is essential, as the seedlings are small and early growth is slow.

Celery *(Apium graveolens)*

Harvesting

Timing

Timing of harvest is a bit tricky, as Celery flowers sequentially, so it will still be flowering and forming new seed when the first formed seeds are ripe and starting to fall. Generally with plants of this nature, the best yields are obtained by harvesting when the first seeds start to fall.

Technique

As many of the seeds will be still unripe, the stems should be cut fairly low.

Mechanical harvesting would be a bit of a challenge with this crop, because of the uneven ripening and the sprawling heads. The seed would probably need to be dried after harvest. Soil contamination could a difficult problem to address.

For more detail on this herb see 'Information Charts'.

Chamomile, German

Chamomilla recutita (L) Rauschert
ASTERACEAE

Parts Used: Flower

Other Names: Chamomile, Wild Chamomile, *Matricaria recutita*, *M. chamomilla*.

German Chamomile is the common Chamomile widely known as a pleasant herbal tea and for its mildly sedative properties. It is one medicinal herb that has survived in the popular consciousness and even Peter Rabbit was given some to settle him down after his close encounter with Mr MacGregor. It also has anti-inflammatory and carminative actions.

This species is more widely known as *Matricaria* but in 1974 a botanical reclassification resulted in the division of that genus into two seperate genera, *Matricaria* and *Chamomilla*, German Chamomile falling into the latter.

I am not sure how it warrants the name German Chamomile, perhaps because it commonly grows wild in that country, or perhaps (because it was found to be unsuitable for Corgis to run on) to distinguish it from the 'true blue-blood' English or Roman Chamomile.

Much of the literature seems to indicate that Roman Chamomile is the preferred species, but this is not the case. To most people's tastes, German Chamomile has a much nicer flavour, and it is this species which is almost always used even for medicinal purposes.

From a grower's point of view, the German Chamomile is much easier to harvest and gives better yields.

For detail on Roman Chamomile see Chapter 16, 'Expanding Clump Herbs'.

Origin

Native of Europe and northern Asia.

GROWING GERMAN CHAMOMILE

There is often some confusion between this species and Roman or English Chamomile *Chamaemelum nobile* (syn. *Anthemis nobilis*). Many books regarded as authoritative fail to clarify the picture and some even add to the confusion by illustrating with the wrong species.

Unfortunately this has not been helped at all by recent changes in their botanical names.

So the first thing to be done is to make sure you understand which Chamomile you are going to grow and that you can identify it.

German Chamomile is an easy plant to grow but, in my experience, good production is only achieved in mild conditions. Its yields in Tasmania are significantly higher than those in north-east Victoria where the hotter, drier climate seems to reduce flowering and the lifespan of the plant.

German Chamomile is a good possibility to consider as a crop when getting started because it is in full flower 2–3 months after planting. Once established, it holds its own well against weeds, and harvesting is usually over by the time they reach a problem stage. It does take a certain dexterity and patience to harvest large quantities by hand, though.

Climate

German Chamomile prefers a climate with a mild summer. It may be possible to grow it as a winter crop in hotter regions: it is grown commercially in Malawi, Sudan and Egypt as well as in central Europe and Argentina.

Site

Exposure

German Chamomile prefers full sun, and while it will grow in a windy situation, growth and flower production will be better with a bit of protection and the plants will be less likely to get blown over.

Drainage

It will not tolerate any waterlogging, but it may be able to be grown as a summer crop in sites subject to excess water in winter, provided it doesn't get prolonged periods of heavy rainfall or overwatering from irrigation.

Soil Type

Light to medium loams are preferred.

Fertility

A moderate level of fertility is required: enough to produce good growth without an excess of leaf. In soils which are more fertile, the plants will be lusher and leafier, but they will produce fewer flowers and the weight of the leaves will cause the plants to flop around and sprawl, resulting in a more tangly growth, which is difficult to harvest with the comb.

Initial Weed Control

While good weed control prior to planting is helpful, it is possible to cope with fairly weedy conditions with German Chamomile.

Neighbouring Plants

German Chamomile can be a problem for nearby crops as a lot of seed falls to the ground and some of it inevitably carries onto neighbouring crops. The seedlings are easy to kill with cultivation, but more will keep germinating.

If a lot of weeds are allowed to grow among the crop, these can be a problem for adjacent areas too.

Containment

While German Chamomile will not survive here without irrigation and cultivation, it is probably best to keep it in an area separate from meadow-crops and others where it may cause problems.

Any following crops on the same ground or adjacent areas will have to be of a nature that can cope with repeated massive germinations of Chamomile from fallen seed, which remains viable in the soil for a number of years. It is best if the following crops can be cultivated to kill any Chamomile before it goes to seed, as it will gradually find its way around the garden and become a weed.

Identification

German Chamomile is a short-lived annual up to 1000 mm tall in some conditions but commonly 400–700 mm, with erect branching stems. The leaves are mildly aromatic, 2–3 times pinnate, with fine filiform segments. The sweetly aromatic flower heads are yellow-centred with white petals, up to 25 mm across, arising singly on stalks from the apices of the branchlets (see colour plate 2.1).

German Chamomile's flower head has a conical hollow receptacle (the inside of the yellow central head), when mature, which can be seen by splitting the flower head down the middle with a sharp knife. This distinguishes it from Roman Chamomile, which has a receptacle filled with pith.

There are a few plants, including some weeds, which bear a flower rather similar to German Chamomile, but they lack the characteristic aroma and/or the hollow conical

Comparison of cross-sections of German (top) and Roman (bottom) Chamomile flowers.

receptacle. A specimen was once sent to me from some people on the upper Murray in Victoria who had 'several acres of flats with Chamomile growing wild ready for harvest'. The flowers did look like German Chamomile, but they didn't smell too good and they differed in that the seeds were much larger: about 2 mm long and 1 mm across. It turned out to be Stinking Mayweed (*Anthemis cotula*).

Variety

There is quite a lot of variation within the species and some strains are more suitable than others from a harvesting point of view. The main factors to consider in this are: density of flowers, size of flowers, and ease of combing. This last depends on the flowers breaking easily from the stalks and the plant being of an erect form that is free of tangles.

It is probably worth selecting seed from several of your best plants and developing a strain suited to your conditions and easy to harvest.

Another factor is the essential oil content of the flowers. There are selected strains with a higher essential oil content available overseas.

The height of plants is partly genetic, but also greatly influenced by the time of planting. German Chamomile plants which start early in the season will grow much taller than those started later.

A tall form of Chamomile with large flowers has been available here for a few years now. Its growth is more upright and taller than the 'standard' variety and it tends to live somewhat longer.

German Chamomile seems to cross-pollinate easily, so if you are growing different varieties it may be difficult to keep them separate.

Propagation

Method

German Chamomile is grown from seed: usually in seed trays or nursery beds, but some growers direct sow it.

As the seed is quite small, the surface needs to be to be continually moist for successful germination. Temperatures should not be too high. Good germination can be achieved at 10–20°C. For details on starting very small seeds like Chamomile, see 'Growing Herbs from Seed' in Chapter 5, 'Propagation and Planting'.

Timing

In a mild climate like Tasmania, seed can be started almost any time of year, so a series of sequential plantings will provide staggered harvests, enabling you to harvest almost year round if you wish to do so. However, crops harvested in the warmer part of the year give the best yields in cool climates.

An early winter sowing will be ready to set out in late winter or very early spring, and will come into full production mid-to-late spring. This is usually the best yielding crop. An early spring sowing will come into its peak just as the first crop is declining. Sowings can be staged every 6–8 weeks until about February.

A February sowing will be ready to transplant in mid-autumn, and will possibly flower lightly through a mild winter, and then peak early to mid-spring.

This planting schedule is based on growing in Tasmania. In hotter regions, the schedule would need to be varied so the crop's growth and flowering coincides with mild weather conditions.

Alternatively, you can put all your eggs in one basket and go for a big harvest at the optimum time. The advantage of staggered planting and harvesting is that one or two people who are good at harvesting can be kept going for 2–3 hours a day. This is about as long as the wrists, back and concentration can sustain and will keep a small drying facility operating to capacity throughout much of the year.

For a big harvest coming all at once, there are added costs: more pickers, more combs (or a mechanical harvester!) and a bigger drying facility, which may be under-utilised for the rest of the season.

Planting

Planting Seeds

German Chamomile seed is quite small and needs to be sown in seed trays following the method outlined for small seeds under 'Growing Herbs from Seed' in Chapter 5, 'Propagation and Planting'.

With adequate moisture, the seeds will will germinate freely in cool to mildly warm conditions, but if it is hot, the seed trays should be placed in a cool spot, asotherwise it won't germinate.

Self-sown Seedlings

Alternatively, a source of seedlings may be available from an area where German Chamomile has grown previously as there often is a heavy germination of self-sown seed, especially in the wetter part of the year. These can be used quite satisfactorily, though I think over a period of time the harvesting characteristics would deteriorate, because this seed will have come from flowers which have evaded the harvesting comb.

If you do use self-sown seedlings, it is probably wise to select some seed from good bushes and use it to establish some beds and then take your self-sown seedlings from those beds, so you are improving your stock, rather than letting it run out.

Transplanting

It is best to transplant German Chamomile seedlings when they are 25–50 mm tall. More advanced seedlings will not do as well.

Separating out individual seedlings from a dense stand damages their roots excessively. Instead, little clumps of seedlings can be planted out with less transplanting shock and these will do just as well or better than single plants.

A single plant will send out a large number of stems from the base, but if several plants are growing together, they just send up fewer stems each and the overall result is about the same.

A tray full of seedlings ready to transplant can be divided into little squares, about 20 mm across, by cutting lines up and down and across it with a knife, so each square contains one to six seedlings. Then the little squares, each with a clump of seedlings on them, can be planted out. The size of the squares can vary with the density of the seedlings.

Direct Seeding

Direct seeding is another possibility, but one I have not tried myself, though I know of other growers doing it. As the seed is very small and will only germinate if it is close to the surface, direct seeding will only work for late winter or early spring sowings when the surface of the soil can be relied upon to remain moist for sufficient time to ensure good germination.

Unfortunately these same conditions will encourage the germination of any weed seed in the soil and, as the young Chamomile seedlings are very small, unless prior weed control has been very thorough, weeds can be a major problem.

For direct sowing of small seeds like German Chamomile, a stale seedbed bare fallow technique needs to be followed (see under 'Bare Fallow' in Chapter 4, 'Weed Management and Control').

Support

Support is sometimes suggested as a means of preventing the crop being flattened by wind or rain. In the past I haven't worried about it too much and have just put up with the hassle of harvesting the fallen tangled growth.

After last season, though, when several growers had a lot of Chamomile flattened in a series of mini-cyclones which struck Tasmania, some of us have been thinking about trying some type of support to keep the plants upright so harvesting will be easier.

This might consist of wires or string-lines running along the rows, and attached to stakes. They would have to be erected when the crop was fairly advanced but not yet flowering, and when the weeds were well under control, as weeding would be difficult once wires or strings were in place.

Wires or strings running the length of the rows beside the plants should help keep them from falling sideways, but cross-ties may be needed to prevent the whole row going down domino-style.

At this stage it is a matter of conjecture as to how effective or worth while such a support system would be.

Layout and Spacing

German Chamomile should be set out on 225–300 mm centres in rows about 600 mm apart, though this can vary a little according to your system. Sometimes I make the inter-row space alternate between 500 mm and 700 mm. This way the Chamomile is laid out in two-row beds and the wider spaces are for walking in during cultivation and harvesting. If your arms are long and you don't mind stretching a bit, then three-row beds may be possible, but I find that little extra stretch a bit more of a strain as I get older and my back has less endurance.

If your site is subject to possible waterlogging, it may be advisable to raise the beds a little to give a margin of safety in the event of heavy rain or overwatering, but remember that excessively raised beds will dry out more, be harder to water and more difficult to manage for good weed control.

Self-sown Crops

After a crop of German Chamomile there is commonly a heavy growth of self-sown seedlings which can be used for a second crop. There are two ways of approaching this: the seedlings can be allowed to develop as and where they come up, or they can be cut back into rows and thinned to the same sort of layout as for transplanted seedlings.

If weeds are sparse and not likely to cause much problem for growth and harvesting, then the first system is probably easier. A dense growth as thick as the hairs on a dog's back will still produce a harvestable crop of flowers. Due to the competition between plants, the yields will probably be lower and the plants shorter-lived, but the slender straight growth is usually easier to harvest as the comb can be swept through the flowers with less tangling.

Alternatively, if higher yields are required or if weeds need to be controlled, then the self-sown plot can be cultivated in strips, leaving narrow rows of Chamomile every 600 mm, and the remaining plants can then be weeded and thinned.

It is important this thinning procedure be undertaken early in the seedlings' development so they have the opportunity to utilise the extra space created. If they are too advanced, their pattern of growth will already have been determined by the crowding in their early stages and they will not fill out much. The thinning operation should be carried out before the seedlings have grown more than 50 mm high.

Compost

Unless your soil is rather poor, it is best not to put much compost onto German Chamomile, as it will stimulate excessive leaf growth, reduce flowering and make harvesting more difficult. In poorer soils, a light dressing of compost (1–2 kg/m^2) may improve yields, but be cautious with it.

If the Chamomile follows a crop that has been heavily composted, results are often good as excess nitrogen will have been consumed by the previous crop.

Weed Control

While transplanted Chamomile comes away strongly, it still needs good regular inter-row cultivation and hoeing around the plants or clumps until they have filled out. This gives them a good start against the weeds, so the flowers will be clear of them.

If weed control is inadequate, there will be a lot of various weed flowers, seed heads and leaves among the Chamomile flowers, which can make harvesting difficult. While 100% weed control is not necessary and the odd weed can be tolerated and coped with, the better the weed control, the easier harvesting will be. In addition, any weeds allowed to establish in the Chamomile will leave a legacy of seeds or runners, causing problems in subsequent crops.

Nevertheless, German Chamomile is a relatively easy crop to manage from the point of view of weeds, even when the ground is not very clean. If the seedlings are transplanted into cultivated ground, they can be given a good start against weed germination and regeneration. Cultivation can be maintained for a few weeks to keep the weeds down until the plants have filled out the space around them. The harvest begins shortly after this, and is usually over before weed development catches up with the declining Chamomile plants.

Rather than let the plot sit there growing weeds, after harvest is finished it should be slashed and cultivated ready for the next crop or a period of fallow.

Rotation

It is normally possible to get a second crop of Chamomile in the same plot from self-sown seed. This is usually somewhat lower yielding, but as it involves less work in establishing the crop, it can be good value.

Chamomile seedlings will keep coming up thickly for years, but it is not good practice to take more than two crops in sequence: yields decline and pests and diseases are likely to show up. It needs to be rotated with other dissimilar crops, preferably ones that are not too closely related (though a number of other Asteraceae seem to have no problems following Chamomile). The most suitable are large plants grown on wide spacings, which allow plenty of room for cultivation around them to control the myriad of Chamomile seedlings which keep coming up.

Examples of good following crops are mostly from the 'Perennial Crown', 'Woody Perennial' and 'Annual, Biennial and Short-lived Perennial groups, such as Marshmallow, Elecampane, Broad-leaf Echinacea, Globe Artichoke, Motherwort, Catnip, Rosemary, Angelica, Burdock and Chicory, as well as most of the 'Tree and Shrub' group.

Avoid following German Chamomile with spreading herbs and low-growing plants that are closely spaced, as Chamomile seedlings coming up among them can be a real problem to control for a number of years afterwards.

Quite a few vegetable crops, such as potatoes, silver beet, corn, cabbages etc., would be also be suitable to follow German Chamomile in a rotation. After 3 years or so in other crops, German Chamomile could be planted again on the same ground.

Irrigation

German Chamomile needs 25–40 mm of water per week through summer. For best flower production, growth needs to be sustained by adequate moisture: avoid subjecting the plants to stress. Care should be taken, however, to avoid waterlogging through overwatering, as this can cause severe damage to the plants and greatly reduce their yields, if it doesn't kill them outright. On a well-drained soil this should not be a problem, but if there is a clay subsoil it can be disastrous.

As the Chamomile fills out and approaches flowering, care needs to be taken not to knock the bushes around while moving hoses or sprinklers, as this can create awful tangles in the flowers, making them difficult to harvest. It is best to have sufficient set sprinklers for this crop so moving them is not necessary.

When the Chamomile is heavily in flower, the additional weight of water from irrigation or rain can collapse the bushes, especially if it is windy. If possible, try to harvest before irrigation or expected rain. This may not be easy to fit into your schedule, especially if the harvesting is staggered and some Chamomile is being harvested each day.

Pests and Diseases

When it is growing in good conditions, German Chamomile is not much subject to pests and diseases. If you find the plants are suffering from dieback or mildew, it usually points to some problem with waterlogging.

Rabbits sometimes show a fondness for German Chamomile and will nibble the young plants off at a critical stage, as will geese.

Vegetable Weevil

I did have a problem with vegetable weevil at one stage. The larvae of this are little grubs which live in the soil, coming out at night to strip the foliage. One year a large area was affected, wiping out a few beds of my Chamomile.

Department of Agriculture information indicated that the vegetable weevil was rather intractable, necessitating Malathion in heavy doses and 'even that might not do the job'.

Naturally that course of action was of no interest to me, so I just watched and waited. It soon became clear that only those plots in German Chamomile for the third successive year were severely affected and I found that all I had to do was to rotate it with other crops after the second year. Since adopting that rotation, I have had no further problems with it.

The vegetable weevil seems to attack a few other Asteraceae species, particularly Capeweed, though not Elecampane.

Beetles

Though not really a pest, a few species of small beetle are often attracted to the flowers, and picking them out can slow down harvesting. Usually they are not so plentiful as to cause much of a problem, just a little pause now and again to shake a couple off a flower or pick one out of the comb.

I did observe once that they were particularly attracted to wet flowers under the spray of operating sprinklers. This might suggest a suitable control measure: if beetles are too thick during harvesting, by turning the sprinklers onto some flowering Chamomile adjacent to the section being harvested, it might be possible to lure the beetles away.

Harvesting

Timing

Stage of Development

Timing of the harvest is a matter of judgement. German Chamomile flowers sequentially or, in other words, over a period of 2 months or more it continues to form new flowers as the older ones mature. The aim is to harvest when most of the flowers are within the optimum range of maturity.

The Chamomile flower is actually a composite head of small individual florets, those around the circumference bearing white petals. The flowering process begins with the outside florets opening and enlarging. Flowering progressively moves towards the centre of the head, which begins as a flat button-shape and gradually swells to a rounded cone as flowering progresses. This takes about a week: more or less, according to prevailing conditions.

Stages of development of a German Chamomile flower.

German Chamomile *(Chamomilla recutita)* ready to harvest.

When flowering is completed, the petals remain folded back and then fall off. However, when assessing the stage of maturity of your German Chamomile, you need to be aware that the petals normally fold back at night as they become damp, or during the day if they are moist or wet. They open out again when they dry out.

The optimum stage to harvest, from the point of view of quality, is when flowering has progressed about halfway up the flower head. It is not practical, however, to get all the flowers at this exact stage. If you harvest as soon as the oldest flowers have reached this stage, your comb will also gather a large proportion of immature flowers and buds, so your yield will be lower and less potent.

On the other hand if you harvest too late, there will be an excessive proportion of over-mature flowers that are starting to fall apart and fade. These older flowers will shatter when dry, with a consequent loss of quality in the dried herb.

A general guide is to harvest when a small proportion, not more than 5%, are close to the completion of flowering and their petals are beginning to fold back, while most of the flowers are in early to mid-flowering, though they do tend to shatter.

With German Chamomile a small proportion of older flowers will not greatly affect the appearance of the final product, because they don't discolour much.

Another factor to consider is whether there is a big proportion of large buds just about to open, in which case you might delay picking a day or two to get them at a better stage. On the other hand, if flowering is very dense and the weight of the flowers is beginning to collapse the bushes, it is best to harvest before this happens, as collapsing results in a tangled mass, which is difficult to harvest.

Time of Year and Frequency of Harvest

German Chamomile will flower year round if the winter is mild enough, though the flowering is much heavier in the spring and summer, provided conditions don't get too hot. Early autumn flowering can be productive, too. Plants which start growth in winter to early spring will begin flowering in October–November. Usually these will produce three good harvests 10–14 days apart. After that the bushes go into decline and start to die off: there may be two or three further light harvests worth taking, if weed flowers and seedheads aren't too great a problem.

Plants that begin flowering in late autumn may flower over a much longer period, but harvests in the colder part of the year will be much lighter and less frequent.

Time of Day

Usually it is best to harvest Chamomile flowers in the morning, once the dew has dried off them. At this time of day the heads seem to snap off more easily. Later in the day, if the weather is warm, the plants will wilt slightly and the stems will become a little more leathery and tougher. This makes the flowers harder to pluck and more will come away with stems or branchlets attached. Chamomile flowers generate heat quite rapidly and in the heat of the day even a bushel basket can get very hot in half an hour or so.

However, if the situation necessitates it, you can harvest later in the day, provided you take precautions to prevent heating. Normally with German Chamomile it won't matter a great deal if drying doesn't commence until the following day, provided it then progresses reasonably quickly.

Tools

Flower Comb

On this scale of production, a combing device makes harvesting a viable proposition. Some growers do pick by hand and this may be preferable if the ultimate in quality is required for special uses or for those who like their biodynamic preparations to be absolutely free of stalk. However, for most purposes, a very good quality can be harvested with the comb.

For details on the design and construction of this implement see the section on 'Harvesting Flowers' in Chapter 11, 'Harvesting'.

For the average German Chamomile flower, the optimum space between the teeth of the comb is 3.5 mm, but for the large-flowered variety, this could be 4.5 mm. The 3.5 mm spacing will work on the larger flowers, but it tends to gather more buds. As the bushes get old, the flowers get smaller and some may slip through the teeth of the comb. It can be handy to have two different combs for these situations, or one comb with interchangeable sets of teeth of different spacings.

Hedge Trimmer

Some growers looking for an alternative to combing have adopted the use of an electric hedge trimmer, the aim being to cut the flowers off below the heads and above the foliage. I don't know if this is quicker, but the quality of a sample of their final product I have seen was poor, with very long stalks on most of the flowers, a lot of leaf, and even pieces of soil included.

Mechanical Harvesting: Mechanical harvesting is possible. To be viable, production would need to be on a large scale and it is doubtful whether a premium quality could be produced (see 'Mechanical Harvesting' under 'Harvesting Flowers' in Chapter 11).

Technique

The aim is to pluck the Chamomile flowers with stalks less than 25 mm long. This is something of an ideal and a small proportion of stalks up to 50 mm long is acceptable. Try to avoid longer stalks and, as far as possible, leaves. Inevitably some will be picked up by the comb, no matter how careful you are, but they can be picked out when the flowers are spread on the screens. The harvesting of leafy young plants that are just beginning to flower can be particularly difficult because the tangly leaf keeps getting caught in the comb.

Keep your eye out for insects, particularly beetles, revelling in rapture among the flowers, and gently shake them off or pick them out, so they don't end up hidden in your final product, dried and dead, to reappear floating around in someone's cup of tea.

Insects are not such a problem in an open drying facility as they will use their wings to regain their freedom when it suits them, but in a closed situation like a cabinet dryer, there may be no escape for them. Most customers don't appreciate insects in their herbs and they may not be amused when you assure them it is a clear indication of the herb's pesticide-free status.

As German Chamomile plants age and go into decline, the flowers often become smaller. This can make harvesting difficult as the flowers tend to slip through the comb. Failing a comb with closer spaced teeth, this problem can be partly overcome by harvesting as early as possible in the morning when the flowers are more turgid and pluck more easily.

If German Chamomile is in rows 600 mm apart, then by working along every second inter-row, you can comfortably harvest a row on each side. While it is physically possible to reach the flowers from every third inter-row, I find it too much of a strain on my back to be continually leaning so far.

The most challenging situation is combing a patch of Chamomile that has been flattened by wind or rain. The first harvest after this happens can be rather discouraging and you usually have to just skim the bushes as best as you can, allowing a lot of flowers too deeply buried to remain where they are.

The plants will sort themselves out a bit for subsequent pickings, in that new shoots will rise up vertically from the fallen plants, so the next lot of flowers will be more accessible.

If a prolonged spell of rain occurs during the flowering period, some of the older flowers will start to rot on the bushes. They form into little brown lumps and remain attached to the stalks, which means some of them get picked up by the comb and go in with the flowers. The problem is that in appearance and texture they very much resemble rat droppings, so they might cause a bit of a commotion somewhere along the line. They should be carefully picked out (fortunately they don't occur very often).

In good going I have harvested up to 1 kg dried Chamomile flowers in an hour's picking, but my average would be less than 0.5 kg/hr. I usually quit when production drops much below 0.25 kg/hr.

For detail and illustrations of combing technique, see 'Harvesting Flowers' in Chapter 11, 'Harvesting'.

Post-harvest Treatment

After the final harvest, slash and cultivate the plot in preparation for the next crop.

Drying

German Chamomile is relatively easy to dry, not being as touchy as some other flowers, but nevertheless it needs good conditions for optimum quality. Excessive delays or re-moistening during the drying process will result in the yellow centres discolouring and the white petals darkening, while the green of the stems and any leaf present will turn towards brown. There will also be more shattering as a result of flower heads continuing to mature rather than drying.

In order to obtain a premium price to make Chamomile production by these methods viable, it is worth setting up a drying system that will ensure a high quality dried product.

Combing Chamomile flowers.

Chamomile can be dried with ambient air if good drying conditions prevail in the warmer part of the season, but in the event of inclement weather, a back-up system with artificial heat will be needed to avoid loss of quality. As the main flowering season begins in mid-spring, when ambient air drying can hardly be relied upon and can extend into late autumn or even winter, some artificial heat will be essential in most situations.

Poor drying conditions will result in an inferior product, which will attract a severely discounted price.

When spreading the flowers on the screens, they should be carefully picked over for long stalks, weed heads and other extraneous matter. If drying in an enclosed cabinet dryer, any beetles should be removed, as they will be unable to escape.

If drying in ambient air, the flowers should be spread fairly thinly on the screen, no more than one and a half layers of flowers thick, or thinner if conditions are not ideal.

In a cabinet dryer with warm air forced up through the screens, the flowers can be safely loaded on somewhat more thickly – two to three layers thick – but be careful not to overload, or drying will be slowed and deterioration will result.

As some flowers inevitably shatter, fine particles and white petals will fall through the screen onto whatever is below. This can be a problem if you are drying Chamomile with other herbs, as a small amount of this fine light-coloured material shows up badly in leaf herbs. It is best to keep the Chamomile below the other herbs on the drying racks, or use a finer woven material on the drying screens. I have a number of screens covered with fine shower-cloth in addition to the usual shadecloth, and these serve well for this purpose.

Chamomile can be a bit deceptive when you are assessing it for dryness, as the smaller flowers and buds are the slowest to dry. A handful of flowers can feel superficially quite dry, but there can be enough moisture still in the centres of the smaller flowers to cause problems in storage.

These smaller flowers and buds should be carefully checked on several parts of the screen, crushing them between finger and thumb. If they shatter easily and completely, or feel very hard, then they are dry, but if they flatten limply or feel leathery or only partly shatter, then they need further drying.

No matter how confident you are of your ability to judge the dryness of the flowers, you should recheck it after it has been in storage for a week or so and has had the opportunity for any hidden moisture to spread around.

Before emptying the screens of dried flowers, pick over them again for any stalks and trash that may have become revealed during drying.

Temperature

Chamomile should be dried at a temperature of not more than 35°C but, if possible, temperatures should be maintained around 30–35°C for optimum drying. A solar drying system that maintains these temperatures for 10–15 hours a day should be adequate, provided prevailing humidity is low enough and a back-up heat source is available.

Processing

Tea Grade

Whole flowers are preferred for tea grade, so no further processing will be required, provided the flowers have been adequately cleaned of stem, leaf and trash during the picking and drying.

An alternative is to produce crushed Chamomile, by rubbing the flowers through a fine screen, about 8 dent, crushing them and sifting out the stem and trash. I have never resorted to this but some growers do, or even put it through a hammer-mill. This produces an inferior low-priced product, as it lacks the attractiveness of whole flowers and loses flavour owing to being broken into small particles.

Some producers find it tempting to include additional stem and leaf in this sort of preparation, as their presence is not obvious, but this results in poor flavour, as the leaf and stem have a different and weaker essential oil content.

A good quality crushed Chamomile would have potential for tea-bag production, but this is getting into a different ballgame.

Crushed Chamomile is a bit of a problem for normal tea-making as the fine particles tend to pass through the average strainer and end up floating around in the cup.

Manufacturing Grade

The presence of a certain amount of stem, leaf and other material may be acceptable for manufacturing grade Chamomile. *British Herbal Pharmacopoeia* (1983 edition) allows up to 8% foreign organic matter, which seems a lot, but was probably established to accommodate the normal poor quality trade product being used by the herb industry. I would expect a discerning manufacturer to set a lower limit.

Storage

Freeze treatment to control moth larvae is necessary before sale or storage, as Chamomile is very attractive to moths. It is probably the most susceptible of all herbs, so infestation is almost inevitable if some sort of control is not practised.

Yields

My average yields in Tasmania have been in the order of 0.1–0.15 kg/m² tea grade per crop, but a good crop could go somewhat higher than this. In less favourable climates, the yields would be rather lower.

Marketing

Chamomile is second only to Peppermint in popularity as a herb tea and is widely used medicinally, too. There is a large market for a high quality, organically grown, whole-flower Chamomile, as the quality of the product generally available in the trade is rather poor, and even that from several organically grown sources is rather disappointing.

I have twice found cigarette butts in organic Chamomile from North Africa, so evidently there is not a lot of attention paid to quality assurance. They must be paying better wages these days though: about 7 years ago I found a butt rolled out of newspaper in some imported organic Chamomile, whereas the recent discovery in a jar in a health food shop was a tailor-made filter tip.

Price

Premium price to growers is currently $45–50/kg for 5-star tea-grade German Chamomile.

Manufacturers are offering $10–18/kg for organically grown Chamomile, but hand-harvesting for that sort of price would not be viable.

A lot of Chamomile is used for tea-bag production and the market for the herb in this form may seem promising, but this is a packaging game, with many brightly displayed brands competing for the customer's attention. They are all sadly lacking in the quality of the herb inside the fancy packaging.

I have never had a decent cup of Chamomile tea made with a tea bag: even the organically grown brands are dismal compared with the real thing.

It would seem there could be a market for a high quality Chamomile in tea-bag form, but this would involve careful development of the product and sophisticated marketing.

Personally I have reservations about this sort of marketing as mostly what is being sold is the packaging (produced to the detriment of our forests) and the tea-bagged product can never be as good as the whole herb. I feel it is preferable to promote the use of tea infusers as a more ecologically sound alternative, which gives a much better flavoured cup of tea and is also cheaper for the consumer.

A healthy stand of German Chamomile nearly ready to harvest. Sequential plantings will stagger harvests and spread them over a longer period.

Chicory

Cichorium intybus L ASTERACEAE

Parts Used: **Root**

Other Name: Succory.

The roasted roots of Chicory have traditionally been used as a substitute for Coffee or mixed with it. Some even regard it as being able to reduce Coffee's harmful over-stimulating effects. Chicory has weak tonic, diuretic and laxative effects.

The young leaves are also used as a vegetable in some cultures: this form is known as Whitloof. The Endive is also of the same or a closely related species, depending on which botanist you follow.

Origin

Chicory is a native of Europe and temperate Asia. It has become naturalised and grows wild in many temperate parts of the world, including Australia.

GROWING CHICORY

Technically Chicory is a perennial crown herb but it must be harvested at the end of its first season and can't be propagated vegetatively. So from the herb grower's point of view, it is treated as an annual, and for that reason I have included it here.

Chicory is an easy crop to grow and is suited to a range of climates, but it is important to get the right variety as it is a very variable plant. Once quite significant volumes were grown in Victoria, on Phillip Island and nearby areas, but this industry has died out.

Lighter soils are preferred.

Identification

Chicory forms a rosette of leaves and a flowering stem up to 1200 mm tall, which normally does not appear until the second season. The basal leaves are 100–300 mm long, slightly rough to the touch, irregularly toothed or lobed, similar to Dandelion in shape but without the backward curving aspect of the teeth. The stem leaves are smaller. Flowers are normally blue (but white and pink forms do exist), 30–40 mm across, solitary or in clusters of two or three, and borne along the upper part of the stalks.

Chicory is a perennial. Some forms are active in winter, while others go dormant. Chicory can be confused with Dandelion if the flowers are not present. Chicory's leaves are somewhat rough to the touch, while Dandelion's are quite smooth.

Variety

There is a great variation in Chicory and the roots of many varieties are quite small. This is one case where the wild form is not preferred as its roots tend to be quite small and thin, though if you specifically want Chicory for medicinal use, the wild form may be better.

Most people, however, drink Chicory because they want a Coffee-like flavour without the problems of over-stimulation and addiction. So the preference is for varieties producing a large root, which is easier to harvest and to roast.

Giant Magdeburg is a traditional Chicory root variety.

Planting

Direct sowing is preferred as the seed is relatively large and is easy to get started.

Transplanting can set Chicory back and result in shorter, deformed roots, as the taproot gets damaged in the process.

Rotation

As it is direct sown, Chicory needs to follow a crop that leaves the ground pretty clean and with a good seedbed.

Being fairly easy with respect to weed control, Chicory can leave the ground clean for following crops, provided you have been diligent with your cultivation.

Harvesting

Timing

Harvest in late autumn or winter, at the end of the first season. Older plants which have reached the flowering stage will have woody roots.

Tools

On a small scale, harvesting will require a garden fork or a spade if your soil is a bit heavy. On a larger scale, mechanical harvesting is feasible as the roots are large and uniform, with few branches.

Technique

As long as the soil is not too heavy, digging is fairly simple, the large roots not having tenacious side branches nor growing too deep. Don't bother with the small pieces that break off: they are a nuisance to wash and roast, and they will not regenerate if left in the soil.

Washing is fairly easy as the roots don't have many branches.

Processing

Chicory root is normally used roasted. The grower may be required to do this.

It is important to have your Chicory thoroughly dry before roasting. Pieces still moist inside will roast unevenly, as the

Chicory (*Cichorium intybus*).

centres will still be drying while the outside will be darkening and burning.

Chicory root is fairly easy to roast in large pieces. Temperatures should be relatively low and care should be taken not to roast it for too long. Most commercial Chicory is over-roasted to the point of charring, which results in a very dark coloured beverage, but an inferior flavour: it might as well be burnt sawdust. If the pieces are roasted until they are just light brown inside, they make a superior beverage: somewhat lighter in colour but much richer in flavour.

The roasted pieces will have to be milled. Particle size varies from around 4 mm for a coarse percolation grade to 1 mm for a Turkish Coffee style semi-instant grade.

Roasted Chicory is mostly used in blends with Coffee or combined with roasted grains to make a coffee substitute. There is some market for it on its own as well. The particle size will depend on your market requirement.

Storage

Freeze treatment to control moth larvae is necessary before storage of unroasted Chicory. The unroasted root can be stored for some years before roasting, but roasted Chicory should not be held for too long as it will slowly lose its flavour and go stale: it also absorbs moisture from the air very readily, so it needs to be kept well sealed.

For more detail on this herb see 'Information Charts'.

Coriander

Coriandrum sativum L APIACEAE

Parts Used: Seed (i.e. Fruit), Leaf

Other Name: Cilantro (for dried Coriander leaf).

Coriander seed is widely used as a culinary herb and the dried leaf, known as Cilantro, is becoming popular as well. The seed has properties as a digestive herb.

GROWING CORIANDER

Coriander can be grown for its seed (technically a fruit), or harvested at an immature stage for its leaf. It is easy to grow, provided it gets an early start.

Coriander can be grown in almost any part of temperate Australia. For seed production on a commercial scale, regions suited to growing other grains and seeds are probably best. The growing of Coriander for leaf is suited to cooler moister regions.

Identification

Coriander is an erect annual, growing to 900 mm, with bright green aromatic pinnate leaves. The leaflets on the lower part of the plant are round to oval and slightly lobed. The leaflets become progressively narrower and more divided further up the stem, the uppermost being quite linear. The flowers are in short-stalked umbels, whitish-mauve to very pale pink (see colour plate 2.4). As the plant matures, the leaves turn a reddish colour and the fruits form: these are about 5 mm across, round, green at first, then turning reddish brown as they ripen.

Variety

There is a Cilantro variety used for leaf production.

Harvesting

Timing

Harvest the fruits when they are fully formed and dry. They usually hold on the plants fairly well.

The leaves must be harvested early. They rapidly deteriorate as flowering commences.

Processing

Coriander seed will need screening and winnowing to make it suitable for culinary use. The dried leaf needs to be rubbed to culinary grade.

Culinary Grade

Fruit should be free of stem and other trash.

Leaf should be rubbed through an 8 dent screen, until it is free of stem.

For more detail on this herb see 'Information Charts'.

Coriander (*Coriandrum sativum*).

Dill

Anethum graveolens L APIACEAE

Parts Used: Seed (i.e. Fruit), Leaf

Other Name: *Peucedanum graveolens.*

Dill seed and Dill leaf (sometimes referred to as Dill weed) are used as culinary herbs. Dill seed is used as a digestive and to stimulate milk flow.

GROWING DILL

Dill is a fast growing short-lived annual, which will grow in a range of climates, though it is rather delicate. In southern Australia it would be normally grown as a spring and summer crop.

Identification

Dill is an aromatic annual plant up to 1000 mm, with fine feathery leaves (that is, deeply divided into thread like segments), blue-green, up to 500 mm long. Flowers are yellow, numerous, small with the petals rolled inwards, in terminal umbels. The fruits (commonly referred to as seeds), are flattened, 3–4 mm long, and brown when mature.

In general appearance, Dill is similar to Fennel, but the plants can be easily distinguished. Fennel has multiple stems filled with pith, while Dill normally has only the one stem and this is hollow.

Variety

There are varieties of Dill selected for leaf production: Tetra is one of these. The common variety goes to seed very quickly.

*Dill (**Anethum graveolens**).*

Pests and Diseases

Dill seems lack the resilience of most other herbs and suffers badly if conditions are not suited to it. If attacked by pests or disease it gives up easily.

Aphids

This is one herb I found could not withstand aphids: it would be killed before ladybird numbers could build up to control them.

The aphids were probably due to some kind of stress on the plant, possibly overcrowding, fluctuating weather conditions or the use of chicken manure.

Harvesting

Timing

Leaf should be harvested as or before it starts to flower.

For seed, the plants should be cut when the first seeds formed have turned brown.

Processing

Culinary Grade

Leaf should be rubbed through an 8 dent screen. Seed needs to be screened and winnowed to remove any chaff or trash.

For more detail on this herb see 'Information Charts'.

Echinacea, Broad-leaf

Echinacea purpurea (L) Moench.
ASTERACEAE

Parts Used: Root, Seed, Aerial Parts, Whole Plant

Other Names: Purple Coneflower, *Rudbeckia purpurea.*

Broad-leaf Echinacea (*Echinacea purpurea*) has similar properties to Narrow-leaf Echinacea (*E. angustifolia*), being used as a stimulant for the immune system.

Origin

Broad-leaf Echinacea is native of north-eastern North America.

GROWING BROAD-LEAF ECHINACEA

Broad-leaf Echinacea is the easiest of the Echinaceas to grow, being more adaptable, faster growing and easier to get started. It is quite a magnificent plant when it is in its prime, with its vibrant purple cone-flowers and dark green foliage. It is sometimes grown for the cut-flower market.

This species comes from a more humid region than the other Echinaceas and consequently its form is somewhat different. It is a much leafier plant, taller but with a shallower, less extensive root system. It lends itself to cultivation, responding vigorously to the available moisture and nutrients in this situation.

It also differs in that commonly the fresh whole plant is used, in tincture form, though there is an increasing usage of the dried root as an alternative to Narrow-leaf Echinacea root. The dried aerial parts are also used to some extent, though while they may offer a cheaper alternative, there is some doubt as to their potency.

The seed of Echinacea is regarded by some authorities as being the most potent part of the plant: this can be confirmed by chewing just a few of them and feeling the effect on your tongue.

While Broad-leaf Echinacea is a perennial and can be propagated by division, best results will be had if it is treated as a biennial and sown from seed. The plants usually decline after their second year, though they may linger on for a while. Plants raised from divisions are not as vigorous as seedlings.

Much of the Broad-leaf Echinacea grown in Australia is used fresh by manufacturers, but if you are intending to supply it in this form, it is important to ensure you have a definite market for the product before you plant. Buyers may be difficult to line up at short notice when it is ready to harvest.

Climate

Broad-leaf Echinacea will grow in a wide range of climates.

Site

Exposure

Broad-leaf Echinacea thrives in full sun and will tolerate a fair bit of exposure to wind.

Drainage

It prefers good drainage, but is nowhere near as sensitive to excess soil moisture as the other Echinaceas.

Soil Type

It will grow in a wide range of soils, but for root production a lighter soil is preferable as it is easier to wash from the finer roots.

Fertility

Broad-leaf Echinacea needs fertile soil.

Initial Weed Control

For direct seeding it is important to have clean ground; weeds can be a problem as the early growth of the germinating seedlings is rather slow. If strong seedlings are being transplanted, initial weed control is not as critical.

Neighbouring Plants

Broad-leaf Echinacea is a strong-growing plant not greatly affected by neighbouring plants, once it is established. At the seedling stage, however, it is very vulnerable to snails and slugs, so it needs to be located well away from other herbs that may be harbouring these.

Growing it in proximity to other Echinaceas can lead to hybridisation occurring, but this only affects the seed.

Identification

Broad-leaf Echinacea grows up to 1500 mm, normally 1000 mm, with robust stems and dark green broadly lanceolate leaves, roughly toothed, up to 300 mm long with a central midrib and branching veins. Flowers have pink to purple petals about 30–40 mm long, at right angles to the stem or slightly drooping (see colour plate 3.2). The dark red centre of the flower is spiny, taking the form of a rounded cone as it matures. The root system is shallow and

A stand of Broad-leaf Echinacea at the optimum stage for harvesting aerial parts or the whole plant.

fibrous, spreading laterally, the roots are light brown on the outside, grey inside. The tops die down to ground level in winter.

Broad-leaf Echinacea differs from the other species of Echinacea with its broad leaves with branching veins, while they have narrower leaves with parallel veins. It also grows somewhat taller in good conditions, while its shallow, laterally spreading, much-branched root system contrasts with the deep roots with few branches of the other Echinaceas.

Variety

There may have been some selection of varieties of this herb, as it is grown as a cut flower. My initial seed was obtained from a flower seed merchant, but it seems to grow well and to be quite potent. It is often referred to as *Rudbeckia purpurea* in the flower trade but do not confuse it with the other *Rudbeckia* species, which have no medicinal value.

Occasionally plants show up which are intermediate in form between Broad-leaf Echinacea and Narrow-leaf Echinacea: these may have some breeding potential.

Propagation

Method

Seed is the preferred method of propagation for Broad-leaf Echinacea. While this species can be grown from crown divisions, the plants which arise from them lack the vigour of seedling plants.

Broad-leaf Echinacea seed does not require cold stratification and will germinate using normal procedures, though sometimes pre-soaking the seed for 24 hours is advised to improve percentage germination.

Timing

Seed is best sown in early autumn, so there are sufficiently advanced seedlings available for transplanting in early spring. If they can get off to an early start like this, the majority will flower before the end of the next autumn, but if they get off to a late start, they probably won't flower until the following season.

In cooler climates they will need to be sown earlier than this to ensure a long enough growth period.

Planting

Start the seeds in trays or nursery beds and transplant them early spring. This way you don't have to keep track of small dormant seedlings among the inevitable growth of winter weeds.

Direct seeding is a possibility. If it can be done it might gain a few weeks over transplanted stock. Irregular germination can be a problem and weeds would have to be under good control as the initial growth is rather slow.

Layout and Spacing

Plant on 300 mm centres in rows 600–900 mm apart. For ease of harvesting, plant in blocks two or more rows wide.

Compost

Broad-leaf Echinacea needs a good dressing of compost 3–5 kg/m^2, or 30–50 t/ha, worked in beside the plants in early spring.

Weed Control

Weed control can be rather demanding while the seedlings are small, as early growth is slow and they are liable to be overshadowed if the weeds are neglected. Once the plants get bigger and form a solid rosette of leaves, they hold their own very well and just need regular inter-row cultivation with a bit of hoeing around the plants.

The advantage of transplanting is that it enables the period of intensive weed control in the early seedling stage to be done in a concentrated area in the nursery bed or seedling tray, which is much easier than trying to keep long rows of small sparsely planted seedlings free of weeds.

Rotation

Provided you have strong seedlings to put in, Broad-leaf Echinacea can be used to follow problem crops, such as German Chamomile, in a rotation.

Irrigation

Broad-leaf Echinacea needs around 25 mm of water per week through summer.

Management

Broad Leaf Echinacea usually grows vigorously for two seasons, but then it starts to decline. It steadily goes down hill and after a few years it usually dies. While the crowns do produce daughter plants, these seem to retain whatever degree of senility their parent had and so they do not give rise to very productive plants.

The best approach is to grow this species from seed each year and treat it as a biennial.

If the crop is being grown primarily for root production, the plants should probably not be allowed to go to seed, as the root systems can die back after this as the plants decline.

Pests and Diseases

Generally Broad-leaf Echinacea is not subject to pests and diseases, though seedlings are vulnerable to snails and slugs.

Snails and Slugs

Snails and slugs can wreak havoc on young seedlings, especially when they have just been transplanted: being under stress, the seedlings must be more attractive to them. To avoid this problem try to locate your Echinacea bed away from any snail or slug havens, including other crops that tend to harbour them.

Harvesting Fresh Aerial Parts and Fresh Whole Plant

Timing

The optimum time for harvesting the aerial part of the plant is when it is at one-twelfth flowering, that is, when it is just coming into flower, with buds developing and one in twelve (or so) flowers open. This is when the plant's vitality is at its greatest. The active constituents in the aerial parts are then at their highest but they rapidly decline as flowering progresses.

Crude Taste Test for Assessing Potency of Echinacea

The level of isobutylamide – one of the active constituents in an Echinacea product – can be roughly ascertained by a taste test, if you are up to it. When you chew a piece of the plant for a short period, a strange tingling sensation should develop in your tongue, its intensity depending on the concentration of isobutylamide. With a really good sample, you should find yourself wishing you had never tasted the horrible stuff, while your face contorts, your eyes water and your tongue feels like nothing on earth.

Herbalists often delight in sampling a few drops of an extract, rather enjoying the sensation of its potency and comparing it with other vintages.

Co-ordination

It is important to co-ordinate the timing of the harvest with your manufacturer to ensure the crop arrives in good condition and they are organised to process it straight away. If it is being shipped, it should be harvested early in the week so it doesn't end up sitting in a depot somewhere over a weekend.

Uneven Maturing

A stand of Broad-leaf Echinacea never matures evenly. It will need to be gone over at least three times, at intervals of 2 weeks or so, to harvest the plants as close to the optimum stage as possible. This can be a source of annoyance to manufacturers, who like to make their batches as large as possible or who may have trouble fitting in with the timing of your crop.

If the plants have been established early enough, harvest should begin in the late summer of their first full season. In cooler regions the plants may not flower until later in autumn or perhaps the next season.

Even in warm regions, a proportion of the crop will carry through to the second full season before it flowers: this could be taken as a late spring harvest, or could be left for the harvest of root or seed.

Number of Harvests per Season

If the first harvest of aerial parts is early enough and the roots are not harvested, a second harvest is usually possible later in the season.

Tools

A reaping hook and garden fork, harvesting sheets and woven polypropylene bags are required for harvesting.

Technique

Go along the rows first with the reaping hook and cut off the tops of plants that are at the right stage. If these are laid carefully on the sheet, aligned together with the butts all at the same end, then the next stage will be much easier.

Some manufacturers like to include the fresh root with the tops to make a more potent whole-plant tincture. If this is the case, it is best to cut the tops off first and then dig the roots, keeping them separate, so the soil doesn't get all over the leaves. This simplifies the operation and makes washing easier. The two can be recombined when packing for shipment, or later at the milling stage.

When the roots are to be included, follow along with the garden fork and dig up the roots of all the harvested plants, shaking most of the soil off before washing. When shaking, take care not to shake soil onto the leaves of other plants.

The lower leaves of the tops may need washing if they have splashed soil adhering to them and there may be an odd yellow leaf to strip away. Avoid using the full blast of the jet when washing, as this can bruise the leaves. After washing, allow them to dry for only as long as it takes the surface moisture to dry off, otherwise they will begin to wilt.

The roots will need a thorough washing and also just enough drying to remove surface moisture: this usually only takes a short while in the sun.

The freshly harvested herb should be kept cool and rushed off to the manufacturer as quickly as possible: ideally they should be only a short distance away but in times of shortage we have shipped fresh Broad-leaf Echinacea from Victoria to Queensland by road, in plastic-lined cartons containing not more than 10 kg each (larger masses could begin to compost). Of course this was late in the season when temperatures were down: in hot weather there would be too much deterioration.

Manufacturers sometimes blend two or more harvests of fresh aerial parts with one harvest of fresh roots. This gives them a bigger volume per plant, but the resulting product would probably be less strong as the crown and root seem to be the most potent parts of the fresh plant.

Post-harvest Treatment

After harvesting, just level off the holes where the plants have been dug out. This species cannot regenerate from the roots, only from crown divisions (that is, daughter plants) and seed.

Drying Aerial Parts

There is currently a market for dried Broad-leaf Echinacea aerial parts as it is the cheapest Echinacea product available, but there are some doubts as to its effectiveness.

As it usually fails the crude taste test, it would appear the isobutylamide is largely lost in drying. An analysis of a sample of dried aerial parts, sent to a manufacturer by one grower, failed to detect any active constituents (see *The Growing of Echinacea in Australia* by Clifton Ellyett).

The lack of this active constituent in the dried aerial parts may be why they have traditionally always been used fresh. The iso-butylamide content and that of other active constituents appears to be retained in the dried root of this species and Narrow-leaf Echinacea.

There is some evidence that fine chopping the plant material before drying causes a greater loss of active constituents in Echinacea.

This is an area where some research is needed to determine whether it is possible to dry Broad-leaf Echinacea aerial parts in a manner that retains the active constituents, and then to ascertain the therapeutic effectiveness of these products when compared to other Echinacea products.

Sometimes Echinacea dried whole plant is offered, but it seems that it may only be the root component of this which is active.

Temperature

Dry at temperatures of up to 35°C.

Yields

Broad-leaf Echinacea yields 2.0–3.0 kg/m^2 fresh aerial parts per annum, and 0.5-0.75 kg/m^2 dried aerial parts per annum.

Harvesting Seed

There are two purposes for harvesting seed: for medicinal use and for propagation. While seed for medicinal use can be harvested from all Echinacea plants, if it is intended for propagation, it should only be taken from well-formed early-maturing plants that could not have crossed with other Echinaceas.

Timing

The seed must be allowed to ripen as long as possible on the plants, but not for so long that it begins to fall. Seed which ripens too late in the autumn may not be viable and won't be useful medicinally.

Tools

A reaping hook or shears and a bucket are required for harvesting.

Technique

Cut the heads off near the top of the stalk and spread them out on screens to dry in the drying shed.

Drying Seed

It is best if the drying of seed does not proceed too quickly, so any immature heads have the opportunity to finish ripening.

Temperature

Dry at temperatures of up to about 30°C.

Processing

To remove the seeds from the heads, put on some thick gloves and rub them firmly against a 2$^1/_2$ dent screen to break the seeds off and push them through the screen.

The seed will be mixed with a lot of bracts and other trash and will need to be sifted and winnowed to get it clean. Check a few seeds to see if they contain kernels. A good winnowing set-up will separate the empty seeds, as they are lighter.

Manufacturing Grade

For manufacturing grade, the seed needs to be fairly clean of all trash and contain a high percentage of full seeds.

Storage

Seed should be freeze treated before sale or storage to prevent moth infestation.

Yields

No yield figures are available for seed, but they would be quite low.

Harvesting Root

Timing

The root can be harvested either when the plant is beginning to flower, for inclusion with the fresh aerial parts, or in late autumn or winter.

If autumn or winter harvest is intended, the plants should probably not be allowed to go to seed. The root system can largely die off as the plants decline afterwards.

Digging should be undertaken when the soil is dry enough to be friable, so it can be shaken easily from the roots, which are fairly fine.

Tools

A garden fork will be required for harvesting the roots. On a large scale it should be possible to harvest mechanically, as the root systems are fairly shallow

Technique

Digging is quite straightforward, as the root system is fairly shallow and not extensive. In fact for the size of the tops of the plant, the roots are surprisingly small.

The tops should be trimmed back, but the base of the stem, which is usually slightly swollen and a reddish colour, should be included with the root (a taste test will indicate this part is actually quite potent).

Washing needs to be thorough, as the finer roots tend to cling to the soil and other debris. Roots of older plants may need breaking up if they are multi-crowned, but roots of younger plants with single crowns can often be washed adequately without breaking them up much, if you have a strong jet of water.

Drying Root

Some of the larger crowns may need to be split or coarsely chopped if they are thicker than your thumb, but avoid fine chopping before drying as this may cause a loss of active constituents.

The roots dry easily, but check the crowns for complete dryness: they should be quite hard in the centre.

Temperature

Broad-leaf Echinacea root should be dried at temperatures of up to 35°C.

Processing

Most of this herb is used in the manufacture of liquid herbal medicines. If it is to be used as a tea grade, it will need milling down to particles of 6 mm or less, but this may not have to be done by the grower.

Manufacturing Grade

For manufacturing grade, the root can be left as it comes from the dryer.

Storage

Freeze treatment is needed to control moth larvae before sale or storage.

Yields

Dried root yields have been up to around 0.2 kg/m² per annum.

Marketing

Fresh Broad-leaf Echinacea whole plant is one herb where manufacturers have to pay a realistic price that reflects the cost of production. Because it has to be used in the fresh state, they are dependent on local growers in order to obtain a continued supply.

However, because large-scale production of this herb is quite straightforward and it does not require drying, an oversupply situation can easily occur.

Shaking soil from Broad-leaf Echinacea roots.

The dried root of this species is not held in as high esteem as that of Narrow-leaf Echinacea (*E. angustifolia*), however, with the increasing price and decreasing availability of *E. angustifolia*, the demand for *E. purpurea* root is quite substantial and increasing.

Echinacea seed is not commonly used by manufacturers in this country, so growers should ensure they have a market for it at a realistic price before committing themselves to production.

Price

Manufacturers' prices for organically grown, dried Broad-leaf Echinacea root (*E. purpurea*) are in the range of $30–35/kg.

Organically grown, fresh whole plant brings $4–6/kg, while dried aerial parts are around $8/kg and the dried whole plant $12–15/kg.

Cleaned seed, organically grown, brings around A$100–150/kg overseas, but there does not seem to be much demand for it here.

Mullein, Great

Verbascum thapsus L
SCROPHULARIACEAE

Parts Used: Leaf, Flower

Other Names: Woolly Mullein, Mullein, Aaron's Rod.

Great Mullein is a respiratory herb and is also used in cosmetics. It is dealt with in more detail in Chapter 21, 'Wildcrafting and Weed Harvesting'.

GROWING GREAT MULLEIN

As harvestable sources of Great Mullein may not be easy to find close at hand, in some situations it may be necessary to consider growing it as a crop.

There are locations in southern Australia where large volumes are growing wild, so large-scale cultivation of Great Mullein is probably not viable.

Great Mullein will grow in a wide range of climates, but avoid growing it in a dusty location, as the woolly leaves will gather dust which doesn't ever wash off and will end up in the finished product.

Containment

Great Mullein is a declared noxious weed in Victoria.

At one stage a few Mullein plants found their own way into my herb garden in Tasmania and, rather enjoying the appearance of their large woolly rosettes of leaves and towering seed heads, I allowed them to grow among the rows. For some time they didn't seem to be a problem, but then one spring I suddenly found I had a massive infestation of some weed seedlings I didn't recognise.

It was a while before I figured out where they had come from and I realised the importance of harvesting the Mullein before seed is formed and removing the flowering heads from the garden. If seed is required, some heads could be allowed to cure indoors, or a cloth bag could be tied over them after enough flowers have set.

Variety

For general usage, leaf and aerial parts, the common species growing wild here, *Verbascum thapsus*, is suitable for the production of leaf, but if flowers alone are required, this species will not be a viable proposition unless you are getting a very good price. Its small flowers emerge sequentially and rather sparsely on the heads, over an extended period of time. A related European species, *Verbascum thapsiforme*, with larger flowers, 30–60 mm across, is used if flowers are required, but seed would need to be located overseas and imported.

For more detail on this herb see 'Information Charts' and Chapter 21 'Wildcrafting and Weed Harvesting'.

Oats

Avena sativa L   POACEAE

Parts Used: Seed, Green Plant

Oats are widely consumed as a food. This species is also used as a herb for its wonderful stimulating effect, which probably has something to do with why it has come to be traditionally eaten at breakfast. Horse owners will be quite familiar with the high spirits horses get into when fed on Oats. And we also have the expression 'feeling his oats' (always used in the masculine).

The green plant, fresh or dried, has in recent years been shown to have a valuable antinarcotic effect and is used in treating drug addiction, including Tobacco addiction.

GROWING OATS

Oats are a traditional crop in our culture and different varieties are available to suit most regions of southern Australia.

Oats require cool to mild growing conditions with plenty of moisture. They can be grown as a winter crop in most regions and as a summer crop in colder regions.

Identification

Oats are annual grasses, growing to 1300 mm tall, with green leaves 150–300 mm long and 4–10 mm wide. Flowers are pendulous spikelets arranged in open terminal panicles, 2–3 per spikelet. The grain is about 10 mm long and normally enclosed in a hull when it falls from the spikelet.

There are a number of species of wild oats which are similar in appearance and habit. Generally they can be distinguished by the long awns attached to the grains.

Variety

There is a great range of varieties of Oats. In order to avoid disease problems, you need to find one suited to your region and the time of year you are planting as there are spring Oats and winter Oats. It is best to enquire locally for information on good varieties, as quality depends on having a healthy crop.

Rotation

Oats is a valuable green manure crop as it grows well during the colder part of the year and provides a good bulk of organic matter to work back into the soil in spring in preparation for summer crops, so it is good to plant in areas that might otherwise sit bare all winter. However, it can take a few cultivations to kill it out, so it is best not to put it in areas such as inter-row spaces between perennial herbs where it might cause problems.

Oats (*Avena sativa*).

Pests and Diseases

Being a widely cultivated and selectively bred species, Oats have quite a number of pests and diseases, but most of these will not be a problem under organic or biodynamic management, provided you have a variety suited to your situation.

Smut

Smut is the worst problem I have had with Oats. It is a fungal disease which infects the heads, turning them into black powdery masses. Even if there are only a few affected, they show up badly in the finished product.

Some varieties are more susceptible than others, so smut can be reduced or prevented by finding a resistant variety. If it does show up, the affected heads can be removed ahead of the harvesting provided there are not too many of them.

Rust

Rust is another fungal disease, but it mostly affects the leaves and stems, turning them a rusty red colour. Usually it strikes plants that are under stress, for example if there is excess moisture or overcrowding. It works progressively up the stem, so it can often be avoided when harvesting the green plant by cutting above it.

Harvesting

Timing

Seed

On a small scale, Oats grown for seed should be cut when the grain is at the 'hard dough' stage, firm but still easily dented with the fingernail. It should be cured in stooks for about 2 weeks before threshing the seed out.

Harvesting on a broad-acre scale with a header, the crop is allowed to fully mature standing in the field.

Green Plant

In harvesting the green plant, the optimum time is when the seed is in the 'milk stage', yielding a milky liquid when crushed. The green plant can be harvested earlier than this stage, but not later, as the leaf soon deteriorates as the plant approaches maturity.

Technique

Green Plant

With the green plant, the aim is to harvest the upper part of the plant, avoiding excessive stalk and discoloured leaf. Keep an eye out for plants affected by smut or rust and avoid tall weeds.

Smut-affected heads should be removed before cutting. Rust can usually be avoided by cutting above affected leaves.

Drying

Oat plants harvested at the green stage are somewhat slow to dry owing to the juiciness of the stem. This can be overcome by feeding the freshly harvested tops through the chaff-cutter, before drying on screens. If drying conditions are good, they will not bruise and the result will be a brighter, greener product, drying in a fraction of the time.

When checking for dryness, it is the stem which needs to be watched, particularly at the nodes. It can be tested by crushing a piece of the stem between the front teeth. If a sort of carrot-like crunching is noticed, the stem still contains moisture, but if the piece crushes quietly, it is dry.

It is very easy for hidden moisture to remain undetected in dried green plant Oats not chopped before drying, as the moist stems are concealed by the very dry outside leaves. This problem is largely overcome by chaff-cutting prior to drying.

For more detail on this herb see 'Information Charts'.

Parsley

Petroselinum crispum (Mill.) Nyman
APIACEAE

Parts Used: **Leaf, Root**

Other Names: *Apium petroselinum, Carum petroselinum, Petroselinum sativum.*

Parsley leaf needs no introduction as a culinary herb, but this plant also has valuable medicinal properties. The root is generally used, as a diuretic, a digestive herb and for some disorders of the reproductive system.

Origin

Parsley is a native of the eastern Mediterranean.

GROWING PARSLEY

Parsley is quite easy to grow and the same plants can be harvested for both leaf and root: for dried leaf over summer and then the root in winter. Another possibility is to grow Parsley for the fresh herb trade and to harvest its roots and dry them at the end of the season.

Climate

Parsley is adapted to a wide range of climates.

Site

Exposure

Parsley prefers full sun, and it will tolerate a windy site.

Drainage

Good drainage is essential as Parsley will not survive water-logging.

Soil Type

Parsley is suited to a wide range of soil types.

Fertility

Parsley needs a fertile humus-rich soil.

Initial Weed Control

Good initial weed control prior to planting is advisable, as weeds can be difficult to control among young Parsley and its early development is fairly slow.

Neighbouring Plants

As Parsley foliage is quite fine, it tends to hold anything that falls on it, so it needs to be sited away from plants that are inclined to drop leaves or petals. Deciduous trees can be a particular problem and can ruin a crop if removing the fallen leaves from the Parsley becomes too time consuming.

Identification

Parsley is an annual or biennial plant, up to 1000 mm tall. the leaves are bipinnate or tripinnate, the leaflets 10–20 mm, dark green, mildly aromatic, wedge-shaped,

Parsley (*Petroselinum crispum*).

lobed and crispate or curled in some varieties. Flowers are around 2 mm across, creamish-yellow, in numerous compound umbels. The fruit is ovoid and slightly flattened, about 2.5 mm long.

Variety

For general purpose leaf and root production, the ordinary triple-curled variety seems to be the best. The actual curliness of the leaves is not important for the dried product, but this variety is consistently biennial.

Italian or Plain-leaf Parsley may have a bit more flavour, but it goes to seed early in the season and its roots become too woody once it flowers.

Hamburg or Turnip-rooted Parsley lacks the medicinal properties of the small-rooted varieties.

Propagation

Method

Parsley is grown from seed: direct sown or transplanted.

Timing

Spring is the best time for planting.

Planting

The seed tends to be slow to germinate, but this can be overcome by soaking in warm water for 24 hours before planting. Direct sowing is possible if the ground is clean enough, otherwise it should be started in seed trays or a nursery bed.

Seedlings should be transplanted at an early stage as advanced plants do not transplant well. They can be planted singly, or in clumps of two or three together.

Layout and Spacing

Parsley should be set out on 300 mm centres in rows 600 mm apart. For harvesting with the scythe, these should be in blocks at least four rows wide.

Compost

Parsley needs a good dressing of compost at the rate of 4–5 kg/m².

Weed Control

Thorough weed control during Parsley's early development is essential for good growth. Weeds can be a problem if they get a hold in the clumps.

Once established, Parsley is easily maintained with regular inter-row cultivation, hoeing and hand-weeding around the plants.

Rotation

Parsley should follow a crop that leaves the ground relatively clean.

Irrigation

Parsley needs 25–40 mm of water per week through summer.

Pests and Diseases

Generally Parsley is not bothered by much in the way of pests or diseases. If plants die, this has most likely been caused by excess moisture.

Harvesting Leaf

Timing

Harvest leaf when the clumps have grown to a good size, before the lower leaves get too old.

Two or three harvests are normally possible during the first season, and an early spring harvest in the second, before the Parsley goes to seed.

Tools

A harvesting scythe or reaping hook is required.

Technique

Parsley is something of an exception in this group because annual varieties will regenerate strongly from the base, so they can be cut down to ground level. Just be careful not to damage the crown and watch out for leaves which have been drooping in the soil, as the curls tends to hold soil particles.

Post-harvest Treatment

Trim off any old unwanted leaf after harvest.

Drying Leaf

Whole Parsley leaf is very slow to dry and if left to its own devices in ambient air, it will turn quite yellow. This is because it behaves differently from most other plants as its leaves do not shut themselves off as they start to wilt. Instead, they continue to draw moisture from the stems and slowly linger and die.

The solution to this problem is to chop the Parsley up into lengths of about 25 mm before drying, either by hand or with the chaff-cutter. This way much of the leaf is separated from the stems and the stems themselves can dry through their cut ends. Fortunately Parsley leaf is not inclined to bruise, but it does fade rapidly in the light, so it should be dried in a well-shaded spot.

Conditions need to be good so drying progresses reasonably quickly.

Temperature

Parsley leaf should be dried at a temperature of 25–35°C.

Processing Leaf

Parsley leaf is primarily used as a culinary herb.

Culinary Grade

Rub culinary grade through an 8 dent screen, repeating the process until it is free of stem. If this is done when the leaf is brittle dry but there is still enough moisture in the pieces of stem to make them pliable, they will be less inclined to go through the screen.

Storage

Freeze treatment to control moth larvae is necessary before sale or storage. Parsley fades very quickly in the light, so it should be stored in complete darkness.

Yields

Around 0.3 kg/m² culinary-grade dried Parsley leaf per annum.

Harvesting Root

Timing

The root is harvested at the end of its first season, in late autumn or winter when growth is slowed, as Parsley doesn't go dormant in our climate. Plants which are starting to flower will have woody roots unsuitable for harvesting.

Tools

A garden fork is needed to dig the roots.

Technique

It is best to trim the tops back to the crown before digging, as this saves handling afterwards. Digging is quite easy, as the roots are relatively shallow. If the plants are growing in clumps, they should be pulled apart before washing.

Washing is usually straightforward, as the root systems are fairly open and don't hold much soil.

Post-harvest Treatment

After harvest, rake over the ground to level it in preparation for the next crop.

Drying Root

Parsley root usually does not need much chopping before drying. The branches are generally not very thick, but the crowns may need to be split or quartered.

Parsley root dries easily but check that the thickest parts are dry right through before taking them off the screen.

Temperature

Parsley root should be dried at a temperature of close to but not more than 35°C, as it contains volatile oils.

Processing Root

Usage of the root is primarily as a medicinal herb, sometimes as a tea, but mostly as a liquid extract.

Tea Grade

Tea grade will need to be milled to particles of up to 6 mm, but the grower is not normally required to do this.

Manufacturing Grade

Roots can be sent off in the form they come from the dryer.

Storage

Freeze treatment to control moth larvae is necessary before sale or storage.

Yields

Parsley yields around 0.25 kg/m² dried root per annum.

Marketing

Dried Parsley leaf is in little demand in the organic trade, probably because fresh Parsley is usually preferred for culinary use.

There is a small demand for Parsley root for manufacturing.

Price

Price to growers for premium quality, organically grown Parsley leaf is currently $24/kg for 5-star culinary grade.

Price to growers for manufacturing-grade, organically grown Parsley root is currently $10–20/kg.

Current trade price is around $10/kg for Parsley leaf.

21

Wildcrafting and Weed Harvesting

Broom, English – *Cytisus scoparius*
Centaury – *Erythraea centaurium*
Chickweed – *Stellaria media*
Cleavers – *Galium aperine*
Couch Grass, English – *Elymus repens*
Dandelion – *Taraxacum officinale*
Dock, Yellow – *Rumex crispus*
Elder, Black – *Sambucus nigra*
Eucalyptus – *Eucalyptus globulus*
Fennel – *Foeniculum vulgare*
Hawthorn – *Crataegus monogyna, C. laevigata*
Horehound, White – *Marrubium vulgare*
Melissa Balm – *Melissa officinalis*
Mountain Pepper – *Tasmannia lanceolata*
Mullein, Great – *Verbascum thapsus*

Nettle, Lesser – *Urtica urens*
Pennyroyal – *Mentha pulegium*
Periwinkle, Greater – *Vinca major*
Plantain, Greater – *Plantago major*
Plantain, Narrow-leaf – *Plantago lanceolata*
Poke – *Phytolacca decandra*
Red Clover – *Trifolium pratense*
Rose, Dog – *Rosa canina*
Rose, Sweet Briar – *R. rubiginosa*
Shepherd's Purse – *Capsella bursa-pastoris*
Sorrel, Sheep – *Rumex acetosella, R. angiocarpa*
St John's Wort – *Hypericum perforatum*
Variegated Thistle – *Silybum marianum*
Yarrow – *Achillea officinalis*

The term 'wildcrafting' refers to the careful harvesting of plants found growing wild. It is a craft, because it is done with careful attention to the quality of the herb and is carried out in a manner that ensures the plant populations are not depleted and – in the case of larger species – not excessively damaged. There must also be minimal damage to the ecosystem and the environment, and no contamination of the product being harvested.

In Australia, where the majority of herbs that can be wild harvested are introduced species, some of the ecological considerations may be somewhat different.

An important aspect of wildcrafting in this country is often the need to prevent proliferation of the species involved if it is a noxious weed capable of invading bush or farming land.

Another related area is the harvesting of farm weeds: weeds of cultivation and pasture. This cannot really be termed wildcrafting because it is an adjunct to a farming system, but some of the same principles and problems apply.

The herbs covered in this chapter do not include all possibilities, but they do give a representative coverage of herbs growing wild in southern Australia which may be available for harvesting.

Guidelines

Guidelines for wildcrafting and weed harvesting in Australia are still in the process of evolution and development. Wildcrafting certification standards and procedures are needed. A 'Wild Harvest' certification category has been established by Biological Farmers of Australia and is being considered by other certifying organisations.

Wild harvest certification already exists for products such as honey and Tea-tree oil, and it could be used to cover many wildcrafted herbs. The costs and procedures are basically the same as for organic certification.

The main concern in wild harvest certification is whether the land has any history of chemical fertiliser usage or whether there is any possible source of contamination. For instance, grazing animals may bring chemical residues into the vicinity of wild plants.

Wild harvest certification is obtainable for herbs wildcrafted on crown land, provided permission has been obtained from the appropriate government department and the location is free from contamination. For private land it may be possible, but it might be more difficult to establish that there is no past or present contamination, and that there will be none in the future, if the land does not belong to the person doing the harvesting.

Parallel production is another problem. Because of the possibility of substitution (deliberate or accidental), certifying bodies are unwilling to certify a product from an organic or wild-harvested source if there is parallel production of the same product from an uncertifiable source.

Introduced species can be given wild harvest certification, provided the environment they are growing in meets requirements. However, species which are farm weeds occurring on cultivated or grazing land, such as Chickweed or Shepherd's Purse, may not be eligible for this category because they are essentially weeds associated with a farming system. If this system doesn't qualify for organic certification, then using the wild harvest category, or referring to these herbs as wildcrafted, could be regarded as bypassing certification procedures. The best solution to this problem is to find sources of these farm weeds growing on certified organic and biodynamic properties.

There are a number of grey areas like this, which need to be sorted out. To ensure credibility, it is important that standards for wildcrafted herbs be high and verifiable.

In order for a herb to be labelled as wildcrafted, it must comply with the following requirements: freedom from contamination, maintenance of the plant's population, minimal environmental and social impact, correct identification, and minimal adulteration. Each of these requirements is elaborated below.

Freedom from Contamination

The herb must be growing in a location where there is no likelihood of it being contaminated with herbicides and other agricultural or roadside spraying. If there is any evidence of spraying, harvest at least 30 m away from affected areas, and only then if you can be sure which areas have been sprayed.

Areas where chemical fertilisers have been used should also be avoided. It is usually best to ask somebody who knows the location about its fertiliser and pesticide history.

Avoid harvesting beside roads, as lead fallout from vehicle exhaust can be a source of contamination. Almost all of the lead falls out in the first 20 m. Some guidelines specify staying 15 m away from roadsides, but 30 m would be safer for high traffic roads. This distance will avoid most of the road dust as well.

Avoid high tension electric wires (may cause mutations), sites that are downstream or downwind from mines, industry, intensive chemical agriculture, weed control spraying or other sources of pollution.

Major urban areas are suspect at the best of times because of the intensity of pollution-generating activity and the widespread fallout, which affects everything growing in and around them. This also applies to industrial areas and coal-fired power plants.

Buffer Zones

Buffer zones need to be adequate to avoid contamination. For certification at least 30 m from localised contamination sources is usually regarded as sufficient, but for major sources of contamination, such as urban centres, much greater distances would be necessary.

Maintenance of the Plant's Population

Introduced Species

Maintaining populations is not a big issue with the majority of species listed here as they are introduced plants. Many are weeds quite capable of surviving despite major human efforts to get rid of them. Some are such a problem that they are a threat to native ecosystems, so total harvesting could even be justified on ecological grounds.

With non-noxious introduced species, the wildcrafter needs to consider that they may wish to harvest them again in subsequent years. Where whole plants or the sole reproductive parts of plants are removed, care must be taken to ensure at least one-third of the plants remain to maintain the population.

This does not apply to noxious weeds, which are going to be there next time no matter what you do. Indeed, one major concern with these is the possibility of inadvertently introducing them to new areas as a result of harvesting. Care must be taken not to disperse viable seed or other reproductive parts.

Native Species

When it comes to wildcrafting native species, it is important to consider carefully the impact of harvesting. The wildcrafter needs to develop an understanding of the growth habits and life cycle of the plant being harvested, and also the ecology of its habitat. This knowledge is needed to assess just how much harvesting and disturbance can be endured without the plant's population and ecology being significantly affected. After harvesting, wildcrafters should also monitor regeneration and the impact of their activity on the growth of the plants, their numbers and their habitat.

Fortunately, the two native species listed are not currently threatened by harvesting for herbal use. However, supplies of them are not unlimited and it is conceivable that in the future, increased harvesting of some of these or other native species could result in depletion if strict wildcrafting guidelines are not followed.

This has already happened overseas where, in places like North America, a number of species of herbs are threatened by overharvesting. The term wildcrafting originated in that country and there is a movement to respond to the need for more careful and responsible utilisation of native flora. Included at the end of this section is a set of 'Wildcrafting Guidelines' and a 'Wildcrafter's Criteria Sheet' put out by the Rocky Mountains Herbalist Coalition in the USA.

Minimal Environmental and Social Impact

Wildcrafters must take care not to damage fragile environments, such as steep hillsides, alpine areas and river banks.

On the social impact side, it only makes sense to get permission first, whether it be on crown land or private property. You don't want to be accused of stealing or some other crime: people seen carrying bundles of herbage out of the bush in isolated locations are liable to raise suspicions.

When asking permission, take the opportunity to find out a bit about the fertiliser and pesticide history of the location. Your contact may have valuable information on the best place to harvest and how to get access. Most people are only too happy to allow you to harvest a few weeds, and I don't recall ever having permission refused.

On crown land a modest royalty may be charged and locations where harvesting is allowed may be specified.

Impact on aesthetic qualities is another consideration, particularly when you don't own the land. Careful harvesting can reduce this impact as will tidying up afterwards. Harvesting should be avoided in the vicinity of walking trails, camping and picnic areas or scenic places.

Consideration should also be had for places of spiritual significance, particularly Aboriginal sacred sites. Harvesting should not be carried out in their vicinity.

Correct Identification

Correct identification is very important because there are too many instances of errors in identification, occasionally with tragic results. As it is often not easy to positively identify a herb from its dried processed form, a botanical specimen should be taken for identification, at least for an initial shipment of a wild-harvested herb.

This specimen should include leaves and flowers and, if possible, fruits, and, if relevant, the root. The specimen should be pressed, mounted and suitably packed so it won't be damaged in transit. It can be sent to the National Herbarium in major state capital cities for positive identification, or it can be supplied to your buyer for assurance.

It may be advisable to do both, as with some species positive identification is not easy and mistakes can be made. Herb buyers do not always have the resources to thoroughly assess specimens sent to them. There has been at least one instance of a manufacturer accepting a specimen and when a quantity was sent in they realised it was the wrong species and rejected the shipment.

You also need to be aware of any other species growing in the vicinity which might be confused with the species you are harvesting. Some species grow in mixed populations with other related species and in some instances hybrids occur.

Minimal Adulteration

As many of these species grow in mixed stands with other plants, it is easy to pick up a lot of foreign material if they are harvested without due care.

If the herb has not been harvested previously in that location, there may be a lot of dead material amongst it from the previous year's growth. Sometimes this can be avoided in harvesting, but it may need to be removed afterwards. For above ground parts, it may be possible to mow off wild stands before the growing season starts, but take care you are not also damaging native species.

Maximum acceptable levels of foreign material in wildharvested herbs will depend on the buyer's requirements. For the herb to be of good quality, foreign and dead material should be virtually absent.

Potential and Logistics of Wildcrafting and Weed Harvesting

Wildcrafting and weed harvesting may offer some potential in many regions of southern Australia, but before you jump to the conclusion that there is a lucrative bounty out there just waiting to be reaped, you need to consider what is involved.

Drying Facilities

Most of these plants need to be dried and processed in the same way as cultivated herbs, so you will need all the same drying and processing equipment. A drying and processing operation may be able to rely solely on wildcrafting and weed harvesting for its production if there is sufficient quantity and/or a range of species available to ensure adequate utilisation of the facility.

Wild harvesting is also a possibility as a sideline for herb growers who have a drying and processing facility and sufficient space and time to be able to accommodate additional herbs within their drying schedule.

Time

For some of these plants, particularly those harvested at the flowering stage, timing of harvest is critical. This can be demanding if you are harvesting some distance from where you live, as you may have to make several trips to keep an eye on their development and to make sure you are there at the right time.

For many species, harvesting in the wild is more time consuming than harvesting cultivated crops, because the plants are smaller, mixed with other species and scattered about so you have to spend time looking for them. Often they are growing in steep and difficult terrain some distance from a road. Previously unharvested stands of some species may contain a lot of dead material, which has to be avoided or removed.

The critical thing is to find a patch worth harvesting and be there when it is ready.

English Broom must be harvested when in flower, October to December.

Marketing

Marketing of wild-harvested species can have its own particular problems. Some manufacturers have a tendency to over order as their supply can be erratic. Yields depend on uncontrollable factors, such as the weather, and often people don't come up with the goods.

To ensure continuity of supply, manufacturers will order from a number of wildcrafters in different areas. In this way they usually get enough to supply them in a lean year, but it means that in a good year there can be an oversupply situation. If there is no contract, the price may drop or the wildcrafter may be unable to sell some or all of their product.

Some species are growing wild in huge volumes, so a massive oversupply situation could easily develop if the local market were flooded. A few species such as St John's Wort could offer export potential if large-scale harvesting and drying could provide economies of scale and viability at trade prices.

Financial Viability

Wildcrafting and weed harvesting are not without their problems and down sides. Part of the skill involved is in being able to assess what is worth while doing and what isn't. Sometimes one can do quite well out of it, but on other occasions it can involve a lot of effort for a meagre return. Much depends on the volumes obtainable and the rates of harvesting, which can be variable and difficult to estimate at first. Distance can be a major factor in costs.

Prices here tend to reflect overseas prices rather than cost of local production, so the wildcrafter needs to make an assessment as to whether the species available in their vicinity are worth harvesting at the prices offered.

On the other hand, there is something quite exhilarating about harvesting the Earth's freely given produce and I have spent many very enjoyable days wildcrafting. It has given me opportunities to explore the countryside, appreciate a change of scenery and spend a few quiet hours in the bush. Thus wildcrafting can bring something more than mere financial gain.

Guidelines Used in Other Countries

Following is an example of documentation and guidelines used for wildcrafted herbs in the USA. The sample guidelines were compiled by the Rocky Mountains Herbalists Coalition and reprinted here with their kind permission.

Rocky Mountains Herbalists Coalition Wildcrafting Ethics and Guidelines

WILDCRAFTED HERB: Wildcrafted means the plant has grown wild in nature without cultivation and has been harvested from unpolluted areas with regard to ecological balance without threatening the full survival of the plant species, as distinguished from pillaging. The essence of wildcrafting is harvesting plants in a manner that increases their number and health. This is one aspect of Planetary Stewardship.

1. NEVER GATHER AN ENDANGERED SPECIES OR THREATENED SPECIES, check your local herbarium or botanical garden for a list of these plants.
2. Taste but don't swallow a plant you don't know. I.D. positively before harvesting. Use identification keys or a voucher specimen when necessary.
3. Give thanks, acknowledge connection with all life, share your appreciation.
4. Leave grandparent plants at the top of a hill to seed downslope. Work your way up.

SITE SELECTION:

1. Obtaining permission: On BLM (Bureau of Land Management) land a free-use permit may be obtained for a minimal charge if you are collecting small amounts. Both the US Forest Service and BLM will tell you there is no picking: a) in or near campgrounds or picnic areas, b) any closer than 200 feet from trails and c) on the roadsides.
2. Stay away from downwind pollution, roadsides (at least 50 feet), high tension electric wires (may cause mutations), fertilizers in lawns and public parks, downstream from mining or agribusiness, around parking lots and possible sprayed areas. (Some BLM & Forest Service districts use routine spraying, this applies to private land as well, you may need to ask about herbicides and pesticides.)
3. Use discretion with fragile environments as one irresponsible wildcrafter can easily destroy a rocky hillside or riverbank environment.

GARDENING & PROPAGATION TECHNIQUES: This will insure minimal impact and increase harvest yields as well as continue to provide food for wildlife. Do not harvest the same stand year after year but tend the area as necessary. 'Gardening' techniques that apply include: thinning, root division, top pinching and preserving a wide selection of grandparent plants to seed and guard young plants.

1. Be aware of erosion factors. If digging roots, replant or scatter seeds and cover holes. Be mindful of hillside stands, replace foliage and dirt around harvested area. Gathering foliage from nearby unharvested plants and spreading it around may be necessary. Wearing hard-soled shoes may cause delicate hillside ecosystems irreparable damage.
2. If harvesting leaf, don't pull the roots. Flower pruning of certain plants will increase root yields as well as foliage.
3. Make seasonal observations of wildcrafted areas. Be mindful of your harvested stands and check different growth cycles. This will determine your real impact on the ecosystem.

VOUCHER SPECIMENS

Either a voucher specimen and/or a local botanist's verification signature on the completed Wildcrafter's Criteria Form should be provided with shipment of an herb.

A voucher specimen is a pressed plant specimen used to positively identify a plant. A sample of the whole plant should be provided including the leaves, flowers, fruits and root. From this reference a trained botanist inspects the characteristics of the various parts and verifies the plant's identity. The presence of the flower is essential.

Wildcrafters often rely on their own extensive knowledge in identifying plants. But when plants are used as medicines it is important that the I.D. be verifiable at least to the level of species and on record for future reference. With some genera of herbs the exact species or variety may make very little difference in therapeutic action. While with others constituents vary in type and concentration making an important difference.

A positive identification assures everyone involved that it is in fact the herb specified and lays a foundation for high quality herb standards and ethics.

(A list of botanical institutions available to do verification in the USA and Canada was included.)

ROCKY MOUNTAINS HERBALISTS COALITION
WILDCRAFTER'S CRITERIA SHEET

Used in Batch #

Botanical name: _____ Date collected: _____

Common name: _____ Gathered by: _____

Send voucher specimen or verification approved by: (see attached)

Name: _____ Authorization: _____

Describe location & ecological niche: (including information such as the neighboring plants, description of the area, slope direction, what is upstream if gathered streamside, etc.)

_____
_____
_____
_____

Note the last time you harvested in this area: _____ Approx % of species collected: _____

What kind of effect is harvesting having on the local environment & plant population: _____
_____

Optional section or by request of the buyer:

Time of day harvested:_____Temperature: _____ _____Moon phase: _____

Weather conditions: _____Farmed or logged? _____

Gathered on _____ Public or _____Private land, with permission? __

Distance from road or highway: (describe) _____and/or power lines: _____

History of fertilizers: _____ History of herbicides/pesticides: _____

Handling Methods:

Describe cleaning method: *(ex. shaking, peeling, washing, etc.)* _____
_____

Describe drying method & plant form: *(ex. shade, shed, racks, dehydrator, oven, kiln, whole, cut, sliced, etc)*
_____
_____Drying temp.: _____

The undersigned guarantees that these herbs were harvested in ecologically sound ways in appropriate unpolluted areas with regard for the integrity of the plant population and the natural environment in which they are found.

Signed: _____ Date: _____

Compiled by the Rocky Mountain Herbalists Coalition
c/- Feather Jones, Salina Star Rt., Gold Hill, Boulder, Colorado 80302 USA

Broom, English

Cytisus scoparius (L) Link FABACEAE

Parts Used: Aerial Parts

Other Names: Scotch Broom, Common Broom, Broom, *Sarothamnus scoparius, Spartium scoparius, Genista scoparius*.

Status in Australia: NOXIOUS WEED

English Broom is used medicinally for certain heart conditions. The seeds were traditionally used as a Coffee substitute; buds and flowers can be pickled and eaten like capers.

There is some concern about the safety of using English Broom. It is said to contain toxic substances, though it does have a long tradition of use in herbal medicine. Often substances which are toxic in isolation are modified in their action by other substances contained in a plant. This is the case with many plants used in herbal medicine.

Occurrence

English Broom is widespread in the tablelands of New South Wales and moderate to high rainfall areas of Victoria, South Australia and Tasmania. It usually invades disturbed bushland, roadsides and neglected areas. Where established, it tends to dominate local vegetation.

Identification

English Broom is an erect shrub up to 3 m tall. Stems are green to brownish green, with 5 prominent ribs, with numerous straight erect branchlets. Leaves consist of 3 leaflets (occasionally 1 leaflet on new growth) which are up to 20 mm long, softly hairy, ovate to lanceolate. Flowers are bright yellow, sometimes with red markings, pea-like, 20–25 mm long, occurring singly or in pairs in the leaf axils, mostly towards the ends of stems and branchlets. Fruit is a brown or black flattened pod about 50 mm long and 10 mm wide.

English Broom loses many of its leaves in winter. It can be distinguished from other Brooms by its straight, erect, spineless green stems and branchlets, and by its flowers being single or in pairs (other Brooms have them in clusters).

Precautions

If Broom is harvested in the early flowering stage, there is no risk of spreading its seeds.

Harvesting

Timing

Harvest English Broom when in flower, October–December, depending on the region.

Tools

A reaping hook is required.

Technique

Harvest the flowering tops. This means cutting the stems off where the flowers commence, so the harvested herb consists of green stems, leaves and flowers.

Yields and Market

Large volumes of this herb could be harvested in some regions, but as its usage in Australia is quite limited, the feasibility of this would depend on locating a viable overseas market at a price that makes it worth while.

For more detail on this herb see 'Information Charts'.

English Broom (*Cytisus scoparius*).

Centaury

Centaurium erythraea Rafn.
GENTIANACEAE

Parts Used: Aerial Parts

Other Names: Century, Lesser Centaury, *Erythraea centaurium*.

Status in Australia: INTRODUCED WEED

Centaury is used medicinally to stimulate appetite and help digestion. It is an ingredient of some bitter liqueurs.

Occurrence

Centaury is a smallish docile weed found in many parts of southern Australia. It tends to be in sparse stands mixed with grass and weeds.

Identification

Centaury is an annual or biennial herb with a basal rosette of leaves. The stems are erect, up to 500 mm tall. The leaves are green with 3–7 prominent veins; in the basal rosette they are elliptical, 20–50 mm long, 8–20 mm broad, while the stem leaves are smaller, opposite and more linear. The flowers are pink to pale red, about 10 mm long, clustered in corymbose cymes at the top of the stems.

The plants die off after flowering and setting seed.

There is another introduced species, *C. pulchella*, which is much smaller, up to 200 mm tall and lacking the basal rosette of leaves.

A native species, *C. australe* is described by Winifred Curtis in *Flora of Tasmania* as having an 'inflorescence cymose but with leafy branches which resemble one-sided spikes or racemes', while *C. erythraea* has flowers in equal-sided cymes, 'the lower branches with internodes much longer than the subtending leaves'.

Harvesting

Being short-lived, Centaury depends on forming sufficient seed to survive. Harvested plants probably won't regenerate and form seed, so sufficient numbers need to be left to ensure regeneration. It tends to be fairly sparse at the best of times, so having to leave one-third of the plants may make it a non-viable proposition.

Cultivation

In Europe, Centaury is grown in cultivation, but I have no experience with this. Seed would need to be collected from wild plants and sown in autumn.

Being very bitter, Centaury is not touched by livestock, rabbits, wallabies, possums etc. It might lend itself to being grown in a semi-cultivated situation, relying on some help from the voracious appetites of furry friends to keep the weeds down.

However, as the market for it is quite small, growers would need a firm commitment from buyers before producing significant quantities.

Pests and Diseases

Centaury often develops a fungal infection, which attacks the leaves and turns them whitish. It usually appears in the mid-to-late flowering stage, rapidly moving through the plants and soon making them unfit for harvesting.

To avoid this problem, they should be harvested in early flowering stage, before the fungus develops.

Drying

Being quite slow to dry, Centaury needs good drying conditions.

For more detail on this herb see 'Information Charts'.

Centaury (*Centaurium erythraea*).

Chickweed
Stellaria media (L) Vill.
CARYOPHYLLACEAE

Parts Used: Aerial Parts

Status in Australia: INTRODUCED WEED

Chickweed is a common garden weed and is used herbally for skin conditions and rheumatism. It also has some tradition as a salad herb and, of course, for feeding poultry and cage birds. It is commonly used as a succus, the juice of the fresh plant, pressed out and stabilised with alcohol. The dried herb is also used to some extent.

Origin

Chickweed is a native of Europe, but is now distributed worldwide.

Occurrence

In southern Australia, Chickweed has adapted to the climate by germinating in autumn and making its main growth through the cool moist part of the year before setting seed and dying off when conditions become dry in late spring. In moister regions it may continue through summer.

Chickweed is a farm weed, almost always associated with cultivation. If the situation where it is growing does not qualify for organic certification, then calling it 'wildcrafted' is rather questionable as this could be a way of bypassing organic standards.

Identification

Chickweed is an annual herb, erect when small, but prostrate or climbing when larger. The stems are much branched, up to 400 mm long, with a single line of hairs running along them. The leaves are ovate with pointed tips, opposite, up to 25 mm long. The flowers are small, white, about 6 mm across, with 5 bifid petals – so they appear to have 10 petals – on long stems in the leaf axils or terminal.

The plants set seed and die off in dry conditions.

It is important to be able to distinguish Chickweed from Scarlet Pimpernel (*Anagallis arvensis*), which has a similar form and habits and can be toxic.

Scarlet Pimpernel is easily identified by its red (occasionally blue) flowers, leaves which are darker green and shinier, and fruit which is a spherical capsule. If in doubt, a little taste of the leaves (which won't kill you) will settle any argument, because Scarlet Pimpernel is quite bitter.

Mouse Ear Chickweed (*Cerastium fontanum*) is superficially similar to Chickweed, but is hairy and more erect.

There are about 120 species of *Stellaria*, a few of which are native. Chickweed can be distinguished from most other species found in Australia by the single line of hairs along the stems. *Stellaria pallida*, Lesser Chickweed, is very similar but is smaller and paler, with flowers lacking petals.

Management

Maintaining production is not usually a problem. Growers are usually happy to get rid of Chickweed as it tends to smother dormant herbs so they have trouble finding daylight when they emerge. It starts to set seed very early in spring, and often has dropped plenty of seed by the time you harvest it.

If you are keen to keep it coming in future years, leave around 10% of the plants.

Chickweed does best in soil cultivated and fertilised over summer, but is occasionally found in undisturbed places, though not growing as vigorously.

Chickweed (*Stellaria media*).

Harvesting

Timing

Harvesting can be carried out any time after flowering has commenced and must be carried out before the plants start to turn yellow in spring. When harvesting for fresh plant use, timing must be co-ordinated with your buyer so when the herb arrives they are able to process it straight away.

Harvesting for drying must be done when good drying conditions are available.

Tools

Use the reaping hook, or just pull the Chickweed tops up with bare hands. Normally the stems break off easily.

Technique

The aim is to gather mostly green leafy material and leave old yellow growth behind, taking care not to pick up soil, as Chickweed sometimes strikes roots along its stems where it touches the ground. Also take care not to collect other debris, such as fallen leaves, dead plants etc.

Processing Fresh Plant

Processing of the fresh plant needs to be done by a licensed manufacturer in licensed premises or by a practitioner for their own practice. The herb needs to arrive fresh, in full vitality, for milling and pressing immediately.

Drying

Chickweed needs good conditions for drying, otherwise it will fade and deteriorate. It has a very high water content.

Temperature

Chickweed should be dried at a temperature of not more than 45°C.

Processing

Dried Chickweed is predominantly required as aerial parts for the manufacture of liquid extract.

Aerial Parts

Put the dried herb through the chaff-cutter after checking that it is thoroughly dry.

Storage

Freeze treatment to control moth larvae is necessary before sale or storage.

Yields

Because growth is quite variable, harvesting Chickweed can range from a tedious time-consuming operation to one that proceeds quickly. Before committing yourself, do a trial on a representative patch and estimate whether harvesting is going to be worth while.

Marketing

The market for dried Chickweed is very small, but there is some demand for the fresh herb.

Price

Price to growers for manufacturing-grade, dried Chickweed is currently $10–22/kg.

Current trade price is around $10–15/kg.

Cleavers

Galium aparine L   RUBIACEAE

Parts Used: Aerial Parts
Other Names: Clivers, Goosegrass.
Status in Australia: INTRODUCED WEED

Cleavers gets its name from its strong tendency to cleave or stick fast to passers-by. The plant has a multitude of little velcro-like hooks all over which latch onto clothes or fur to carry it to new places.

Highly regarded in herbal medicine, Cleavers is used as a lymphatic tonic with alterative and diuretic properties and for treating skin conditions. The juice of the fresh plant is generally used, preserved with alcohol to make a succus.

The seeds are said to make an excellent coffee substitute when roasted.

Origin

Native of Europe.

Occurrence

Cleavers is widespread in southern Australia, mostly on unmaintained land, such as roadsides and other areas where it is not subject to mowing or grazing. In warmer regions it prefers locations with partial shade, where it makes its growth while there is plenty of moisture in early spring, and matures by early summer.

Cleavers tends to grow in great masses, climbing over other vegetation or anything else it can get a grip on, or else, lacking the strength to hold itself up, it sprawls over the ground.

Cleavers *(Galium aparine)*.

Identification

Cleavers is an annual herb with quadrangular straggling stems up to 1500 mm long. The stems and leaves are rough with small recurved spines, so the plant has a rather sticky feeling. Leaves are in whorls of 6–8 around the stem, oblanceolate, 10–40 mm long. Flowers are about 2 mm in diameter, white, in groups of 2–5 on stalks which extend from the axils of the leaves. The fruits are produced in pairs and are also covered in clinging spiny hairs.

The plants normally germinate in autumn or very early spring, and mature and die off by late summer: earlier in hotter regions.

There are a number of related species with leaves in whorls, but Cleavers can be distinguished by its clinging nature and longer leaves in whorls of 6–8. While some other species may cling, these all have whorls of 4–6 leaves not longer than 15 mm.

Wildcrafting/Weed Harvesting

As Cleavers tends to be fairly prolific in its established locations, management for continued growth is not difficult and it is only necessary to leave a small proportion, 10–20% of plants, to ensure good regeneration.

Problems

Cleavers' clinging nature can be a problem in harvesting as this tends to bring parts of other plants and assorted debris with it. Some sites may not be suitable for harvesting because of this. It is easier when Cleavers is climbing on things that are sufficiently robust to stay behind when it is pulled away.

Harvesting is unlikely to spread Cleavers to new locations, as this is carried out well before the seed is formed.

Harvesting

Timing

Cleavers should be harvested just before or at the commencement of flowering. The plants start to deteriorate fairly rapidly after this stage. For making a succus, it is important to get it at an early stage or there won't be much juice in the plant.

Tools

A reaping hook or similar implement is needed for harvesting.

Technique

The plants should be cut just above the level of dead leaves, or where they can be pulled free, trying not to cut stems of other plants. Carefully pull the Cleavers away from whatever they are clinging to. Make sure you are not also collecting other plants, sticks, dead leaves, barbed wire, goanna skins or whatever else is caught up in the growth.

Leave 10–20% of plants unharvested, scattered evenly throughout the site to foster a good seeding for the next season.

Post-harvest Treatment

If there are obstacles and a lot of debris around the site causing problems in harvesting, these could be tidied up after harvest.

Processing Fresh Cleavers

Processing of the fresh plant must be done by a licensed manufacturer in a licensed premises or else by a practitioner for use in their own practice.

The milling and pressing of the juice must be done as soon as possible after harvesting, as Cleavers dehydrates rapidly. If you are supplying it to a manufacturer, you will need to have an arrangement for getting it to them quickly and at a time when they are able to process it straight away.

Drying

Cleavers dries easily.

Temperature

It should be dried at a temperature of not more than 45°C.

Processing Dried Cleavers

Tea grade is required in small quantities, but dried Cleavers mostly goes into liquid extract using the aerial parts.

Tea Grade

The dried herb can be rubbed through a $2^{1}/_{2}$ or 3 dent screen to break it up. As the stems are green and soft, they can be worked through the screen and included.

Aerial Parts

Put the dried aerial parts through the chaff-cutter, keeping an eye out for debris that may have found its way past scrutiny in harvesting.

Storage

Freeze treatment to control moth larvae is necessary before sale or storage.

Yields

In suitable locations Cleavers can be harvested in good volumes at a reasonable rate.

Marketing

There is a moderate demand from manufacturers.

Price

Price for manufacturing-grade, wildcrafted dried Cleavers is currently $10–15/kg.

Current trade price is around $10/kg.

Couch Grass, English

Elymus repens (L) Gould
POACEAE

Parts Used: Rhizome
Other Names: Couch, Twitch, Quitch, *Triticum repens*.
Status in Australia: INTRODUCED WEED

English Couch Grass is only too frequent in cultivated areas in the cooler parts of southern Australia. The opportunity to harvest some of it may arise in the process of getting rid of it. You might wish to maintain it if you establish a market for it.

It is important not to confuse it with another species often called Couch Grass (*Cynodon dactylon*), which is much more common in many regions and is often mistakenly thought to be the species used in herbal medicine.

For more detail on this herb see 'Information Charts' and Chapter 15, 'Spreading Herbs'.

Dandelion

Taraxacum officinale Weber ASTERACEAE

Parts Used: Root, Leaf
Status in Australia: INTRODUCED WEED

Dandelion is dealt with in full detail, including positive identification, in Chapter 16, 'Expanding Clump Herbs', but is included here because it is a common weed and there may be opportunities to harvest it. It is important to distinguish true Dandelion from a number of rather similar species that commonly occur as weeds throughout southern Australia.

Occurrence

Dandelion is widespread, especially in cooler areas. In hotter, drier regions it tends to be found growing in semi-shade.

Most wild Dandelions are too sparse or their roots are too small to consider harvesting on any scale. Occasionally one comes across good stands of big plants in alpine and sub-alpine areas, but most of these are in National Parks where harvesting would not be permitted (unless perhaps they were being removed as part of a native revegetation program).

Other occurrences tend to be in urban areas unsuitable for harvesting because of contamination.

Much of the 'wildcrafted' Dandelion available on the market is of disappointing quality. Some consists of small ugly discoloured roots, which look as if they have been dug out of manicured suburban lawns. Other batches have a bland flavour lacking the characteristic mild bitterness and mild sweetness of Dandelion: this could be a result of wrong identification or of harvesting too early in the season.

For more detail on this herb see 'Information Charts' and Chapter 16, 'Expanding Clump Herbs'.

Dock, Yellow

Rumex crispus L POLYGONACEAE

Parts Used: Root

Other Names: Curled Dock.

Status in Australia: INTRODUCED WEED (Declared Noxious Western Australia, Secondary Weed Tasmania)

The name Yellow Dock is a bit misleading as the root is no more yellow than other species of Dock, though the tops sometimes do have a yellower look about them.

Medicinal use is for its action on the gall bladder, as a mild purgative and for skin complaints.

There is also some tradition of use of the leaves as a vegetable, but the water needs to be changed twice during cooking to remove oxalic acid.

Origin

Yellow Dock is a native of Europe and northern Asia.

Occurrence

In southern Australia, Yellow Dock is widespread throughout agricultural regions. It prefers fertile soils that are subject to some waterlogging, and is often found growing in dense stands. It commonly occurs as a farm weed in pasture and cultivated areas.

Identification

Yellow Dock grows up to 1500 mm tall, though usually 500–1200 mm. The plant forms a rosette of leaves at first and then sends up a thick main stem, which is usually unbranched below the inflorescence. Leaves are green, lanceolate, 200–400 mm long, 80–120 mm wide, with a pointed apex and narrowing at the base with a long stalk, the margins typically wavy and undulating. The stem leaves are alternate, smaller and narrower. Flowers are greenish-red, small, in dense clusters along panicles at the top of the main stem (see colour plate 3.1). The fruit is a small 3-sided nut with smooth-edged wings.

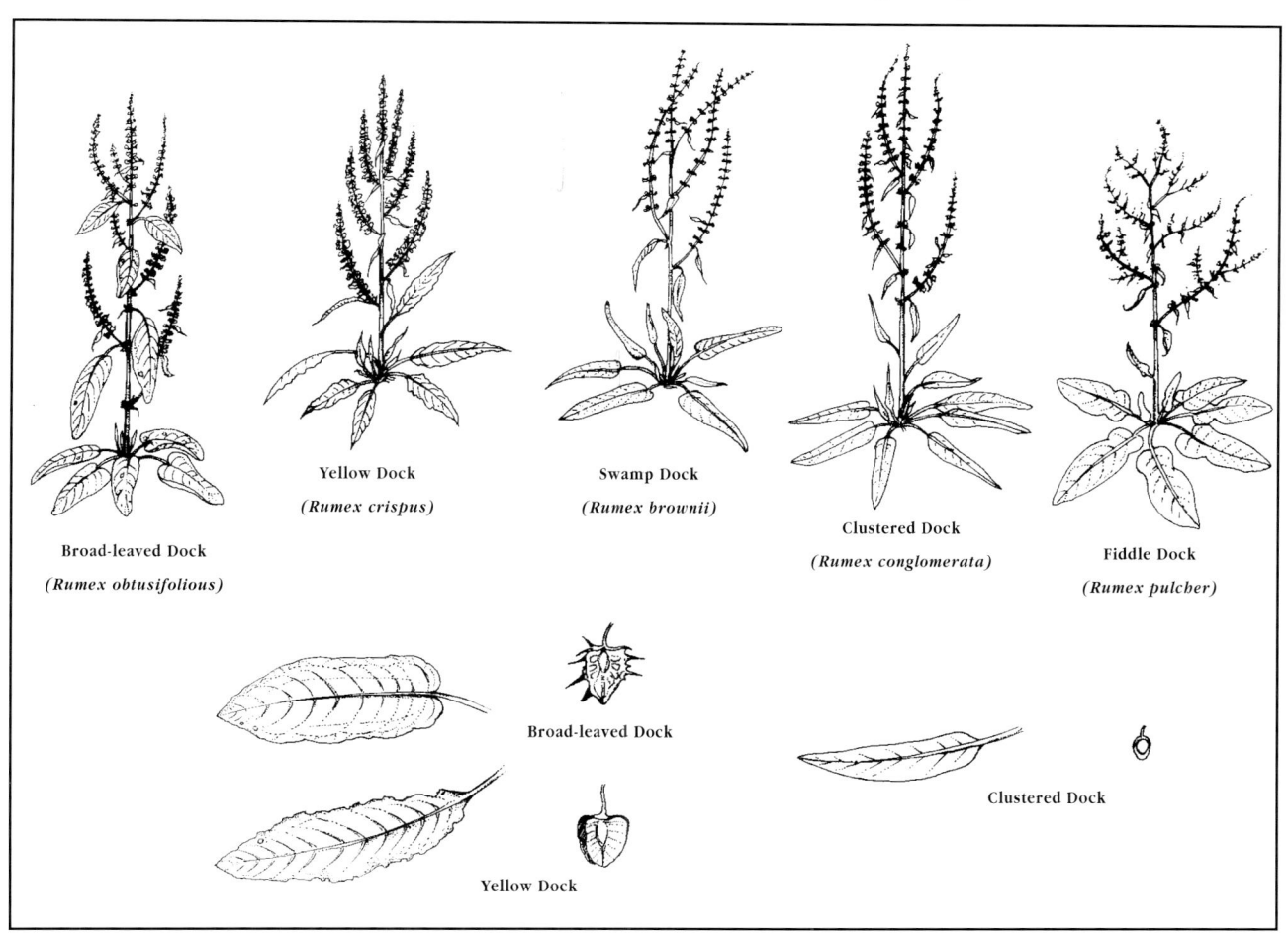

Distinguishing between Yellow Dock and other species of Dock.

As there are a number of other species of Dock commonly found in southern Australia and they often grow in mixed stands, it is important to be able to distinguish Yellow Dock.

The leaves are the first thing to look at. Other Docks generally have fairly straight edges to their leaves, while Yellow Dock leaves have undulating to frizzy edges, but this characteristic can be variable.

A distinguishing point is the base of the leaves at the rosette stage (Note: Stem leaves are different). The angle the leaf margin makes where it meets the stalk varies according to the species. In Yellow Dock it forms an angle of 30–90°, the leaf typically tapering into the stem. Broad-leaved Dock forms an angle of 90° or greater, the base of the leaf being cordate (like an ace of spades).

Another characteristic is in the wings of the seed (that is, fruit). Yellow Dock has wings about twice the width of the seed with clearly smooth edges. Other species have little barbs or hooks on their wings, or else they have smooth-edged wings that are shorter than the width of the seed.

It is common to find Yellow Dock growing together with Broad-leaved Dock (*Rumex obtusifolius*). If the above distinguishing characteristics are observed, it is not hard to separate them most of the time. However, there is often the odd plant with intermediate characteristics, probably the result of hybridisation between the two species.

The name of this species may give the impression that only Yellow Dock's roots are yellow, while other species' roots are more white. This is not the case as most Docks have similar coloured roots. The darker yellow roots are, in fact, older roots, while younger fast-growing roots are a lighter colour.

This can be confirmed by looking at the roots of an older plant that has put on a recent spurt of growth. The older part of its roots will be dark yellow, while young growth will be lighter coloured.

Before harvesting any quantity of Yellow Dock, it is important to get confirmation from your buyer that the Yellow Dock root you can supply will be acceptable. A representative sample should be sent beforehand.

Management

Problems

Small Roots

It can be difficult finding plants big enough to be worth digging. Often a grazing paddock choked with Dock will turn out to have mostly miserable little woody roots, which take ages to dig, are hard work to trim, and probably not a lot of value medicinally as they are mostly wood.

The best root development seems to be in plants that have not yet gone to seed. Mowing can sometimes leave a bit of a mulch which encourages good root growth. Ground that has been disturbed or cultivated is often quite good for finding young vigorous plants with a big rosette of leaves and a large root system. Places where there has been additional moisture during summer can be good too.

Mixed Stands

Where Yellow Dock is harvested from mixed stands and other species of Dock are left to grow, they will tend to dominate the site. In a season or two there will not much Yellow Dock to harvest. If the Yellow Dock production is to be sustainable, these other species need to be dug out at the same time and disposed of where they won't cause problems.

Cultivation of Yellow Dock

Cultivation of Yellow Dock is a definite possibility if sufficient supplies of good wild Yellow Dock are not available and the price is good enough. It can be done without creating a big weed problem by growing it from seed and digging it carefully. Seed-grown Yellow Dock doesn't normally go to seed in its first season.

One-year-old seed-grown plants will yield good-sized roots. Yellow Dock will also grow from root or crown divisions, but these will go straight to flower and most of their energy will flow into seed development, reducing root size. Cutting off these flower stems seems to make no difference.

Harvesting

Timing

Harvest Yellow Dock in autumn or winter.

Tools

A robust garden fork is required for digging up the Dock roots.

Technique

Loosen the roots with the fork and pull them out by the leaves. The tops need to be trimmed off after pulling. I like to get around with a wheelbarrow, tossing the roots into it as they are dug and then, working from the barrow with a thick board laid across it, chopping the tops off the crowns with a stout knife or a cleaver.

An alternative could be to plough and cultivate a Yellow Dock-infested patch with spring-tined implements, which will bring Dock roots up to the surface where they can be picked up.

This might be more efficient labour-wise, but a significant proportion of the roots would be left in the ground and some damage would occur. This method would not be suitable for mixed stands of Docks, as the other species could not be separated.

The soil adhering to Dock roots should not be allowed to dry out, as this can make washing them very difficult, particularly with the heavier soils Yellow Dock often grows in.

Post-harvest Treatment

Fill in any holes you have made during harvest and scatter any piles of trimmings. You may want to gather some Yellow Dock seed and sprinkle it onto the exposed soil to help regenerate it, but this should only be done in locations where the presence of Dock is not going to cause any problems and is acceptable to the landholder.

Drying

Yellow Dock needs to be washed before drying. This is not difficult, as the roots have a fairly open formation, though if the plant has been growing in heavy clay soil, it may take some removing.

Any pieces of root thicker than your thumb should be split or chopped to speed up drying.

When drying appears to be complete, check the centre of the larger pieces to ensure they are thoroughly dry.

Temperature

Yellow Dock Root should be dried at a temperature of 35–45°C.

Processing

Yellow Dock is used to some extent as a decoction, but most of it is used in fluid extract form.

Tea Grade

Tea grade will need to be milled down to pieces of up to 6 mm so it is suitable for making a decoction. Normally the grower is not required to do this. But if you do, take care not to get a nose full of the dust, as this can be quite irritating for some people.

Manufacturing Grade

The roots can be sent to the manufacturer in large pieces as they come from the dryer.

Storage

Freeze treatment to control moth larvae is necessary before sale or storage.

Yields

In a good stand of Yellow Dock, you can probably make wages but you won't make a fortune.

Marketing

There is a small demand for Yellow Dock as a dried herb and a moderate demand for it for manufacturing.

Price

Price to growers for premium quality, organically grown Yellow Dock is currently around $15-20/kg for 5-star tea grade.

Price for manufacturing-grade, wildcrafted Yellow Dock is currently around $10–15/kg.

Current trade price is around $13–14/kg.

Elder, Black

Sambucus nigra L   CAPRIFOLIACEAE

Parts Used: Flowers, (Berries, Leaf)

Other Names: Elder, Common Elder.

Status in Australia: INTRODUCED WEED

The Elder has a tradition of usage going back to prehistoric times. In herbal medicine the flowers are used mainly to relieve catarrh in the treatment of colds and influenza. The berries are also used for the same purpose and for rheumatism. The bark and the leaves also have their own medicinal effects, but they are not much used.

The flowers are still occasionally used as a culinary herb. The berries are used in wine-making and in jams and chutneys.

Mrs Grieve devotes 11 pages to Elder in *A Modern Herbal*, with much detail on traditional beliefs surrounding it. The plant was felt to be very powerful and many magical qualities were attributed to it. Even today many people hold a special regard for it.

The root system of the Elder tree is felt to have a special beneficial influence on compost heaps.

Origin

Black Elder is a native of Europe, North Africa and western Asia.

Occurrence

In southern Australia, Black Elder occurs sporadically in cooler regions where it has escaped from cultivation by means of birds eating the berries and dropping the seeds. In Victoria, its best growth is in high rainfall areas on alluvial soil near creeks or around the edge of cleared land. In Tasmania, it is less frequent, perhaps owing to the denser population of possums that tend to nibble on its leaves.

To be worth harvesting, there need to be a number of good trees near each other, which often happens because they spread locally with the help of birds.

Identification

Black Elder is a shrub or small tree up to 10 m tall, with hollow woody stems. Usually it has many stems or trunks arising from the base. The leaves are dull green, pinnate with 5–9 ovate-lanceolate leaflets, with toothed edges, 30–90 mm long. Flowers are 5-petalled, creamy white, 5 mm across, in flat-topped inflorescences up to 200 mm in diameter, borne all over the bushes (see colour plate 3.4). The flowers have a characteristic fragrance, described by Mrs Grieve as 'scarcely pleasant'. The fruit is a purple-black berry, 6–8 mm in diameter.

Black Elder sheds its leaves in winter.

White Elderberry (*Sambucus australasica*) is a native species that occurs in rainforests and moist scrub along the coast and ranges of Victoria, New South Wales and Queensland. It is a small tree up to 10 m tall, with creamy yellow flowers in terminal panicles. It can be distinguished by its leaves which are divided into only 3–5 narrower leaflets with prominent veins and by its yellow berries.

In somewhat drier forests you may come across the smaller Native Elder (*Sambucus gaudichaudiana*), which is found in all the eastern states and South Australia. It is a diminutive shrub up to 2 m tall, with similar leaves. The flowers are distinctly different, being 4-petalled, the petals curved inwards so the flower looks spherical. The flowers have a highly volatile aroma, resembling nail polish.

The leaves and flowers of the Rowan *(Pyrus aucuparia)* are superficially similar, but the flowers lack the characteristic odour and the leaflets are narrower. The Rowan is more tree-like in form, while the Elder is like a large bush.

Variety

In cultivation you may come across other species or varieties. There seems to be a cultivated variety, or it may be a different species that bears flowers continuously, but these are odourless. This species/variety would not be acceptable.

Wildcrafting/Weed Harvesting

Black Elder bushes are quite long-lived, so there is no threat to their survival by continued harvesting of the flowers. There are usually a few flowers missed or unreachable that go on to form seed.

Indeed, in places where Black Elder is invasive, it may be a good thing to prevent seed formation by picking all the flowers.

Problems

The main problem with wildcrafting is the time it takes to harvest the flowers. You really need to find good bushes with plenty of big flower heads for it to be worth while, unless you just enjoy the picking for its own sake.

Cultivation

In the case of Black Elder, cultivation is definitely worth considering and is dealt with in Chapter 19, 'Trees and Shrubs'.

Harvesting Flowers

Timing

Flowering begins late October–December, depending on the region, and continues for 3–4 weeks. Picking should

Black Elder *(Sambucus nigra)* in flower.

begin when a few of the first-opened flowers are just about to wither and fall. This will ensure there are sufficient opened flowers to make a picking worth while.

Picking can be repeated after about a week, so 3–4 pickings are possible.

Tools

The best thing for nipping off the flower heads is a long thumbnail, sharpened so that it can cut the flower stems by squeezing against the forefinger. A guitar plectrum might be a possible substitute if your thumbnail is too short.

A ladder may be useful to reach higher flowers, but I usually prefer to use a thick wire hook to pull the branches down within reach. This is shaped a bit like the hook used for catching chooks by the leg and is made out of 3 mm (8 gauge) wire. A large loop at the other end is to used as a handle for pulling the branch down. One boot can be hooked into this loop to hold the branch, leaving both hands free for picking. But don't get too ambitious or Mother Elder may take her revenge and spring the branch back up, leaving you hanging upside down by one leg!

A large basket that is light enough to be carried by a belt going behind the neck or over the shoulder is also needed. With this, both hands are then free for picking and the flowers can be dropped in as they are picked.

Technique

The flower heads should be nipped off at the base of the umbel, above any leaves. Only open or partially open flower heads should be picked, and advanced flower heads that have started to shrivel and lose their petals should be left.

Try to avoid breaking the branches when you pull them down to pick the flowers, being sure to apologise to Mother Elder for any damage to the tree.

The flowers are subject to rapid deterioration, so care must be taken to prevent them from heating. They need to be spread out to dry within a few hours of picking.

Drying Flowers

Elder flowers need to dry rapidly to retain good colour, otherwise they will turn a rather sad-looking murky brown. Good quality dried Elder flower is a pale yellow colour.

Temperature

Elder flowers should be dried at a temperature not over 35°C.

Processing Flowers

Elder flower is used quite a lot as a tea and also as a fluid extract.

Tea Grade

For tea grade the flowers can be rubbed forcefully through a $2^{1}/_{2}$ or 3 dent screen without under-screens, the aim being to break up the smaller stems and push them through the screen. Coarser pieces of stem that don't break up easily should be discarded. The herb should be rescreened at least once to ensure that it is adequate for tea grade.

The pollen dust which rises up from dried Elder flowers during processing can have quite a powerful effect on some people, setting their eyes and noses streaming, so make sure you have adequate protection.

Manufacturing Grade

For manufacturing, the flowers should be rubbed just once through the screen, simply to reduce their bulk.

Storage

Freeze treatment to control moth larvae is necessary before sale or storage.

Yields

In my experience, a person is doing fairly well to pick enough to make 0.5 kg dried flowers per hour, not including travelling time.

Picking Elder flowers: showing the basket with the belt worn around the neck and the use of a branch hook for pulling down high branches.

Harvesting Leaves

Elder leaves should be harvested at flowering or shortly after.

Harvesting Berries

There may be some market for dried Elder berries, but I have never explored harvesting these as most people prefer to use the flowers and my trees always had all the flowers stripped so they never formed any fruit.

You would need to find additional trees to harvest berries from, or refrain from harvesting the flowers.

Birds and possums may offer a fair bit of competition too.

Marketing

There is a moderate demand for Elder flower as a herb tea, and some significant use of it by manufacturers.

Price

Price to growers for premium quality, organically grown or wildcrafted Elder flower is currently around $40/kg for 5-star tea grade.

Price to growers for manufacturing-grade, wildcrafted Elder Flower is currently around $10–20/kg.

Current trade price is around $11/kg.

Eucalyptus: Blue Gum

Eucalyptus globulus Labill.

Eucalyptus globulus var. *globulus*

Eucalyptus globulus var. *bicostata*

MYRTACEAE

Parts Used: Leaf

Other Names: Southern or Tasmanian Blue Gum, Eurabbie, *E. bicostata*

Status in Australia: NATIVE SPECIES

This species of *Eucalyptus* is one of the few Australian plants to be adopted into the realms of Western herbal medicine.

It is mainly used for its benefits to the respiratory system, and while it is predominantly used as an essential oil for inhalation, there is some usage of the dried leaf and the tincture.

A Blue Gum leaf floated in the billy when making tea out in the bush gives it a traditional Australian flavour, having one of the most pleasant flavours of the many species of *Eucalyptus*.

Occurrence

Blue Gum is native to the eastern half of Tasmania and cooler parts of Victoria, with some isolated occurrences in the tablelands of New South Wales.

Identification

Eucalyptus globulus is actually a complex of species, which includes the Tasmanian Blue Gum (*E. globulus* var. *globulus*) and the common Blue Gum or Eurabbie of Victoria (*E. globulus* var. *bicostata*).

They are trees reaching up to 80 m tall with white to grey deciduous bark on the trunk and branches, and thicker darker bark at the base. Juvenile leaves are opposite, with square stems, glaucous blue, ovate up to 200 mm long. Mature leaves are alternate, green, lanceolate, up to 300 mm long (the longest leaved species in its area of distribution). Flowers are single with fruits 15–30 mm across (*E. globulus* var. *globulus*) or in threes with fruits 14–20 mm across (*E. globulus* var. *bicostata*).

Eucalyptus – Tasmanian Blue Gum – (*Eucalyptus globulus* var. *globulus*).

They are easily distinguished from other Eucalypts by their long leaves and large fruits (other longish-leaved species have rough bark or small fruits). Do not confuse this species with the Sydney Blue Gum (*E. saligna*) or the South Australian Blue Gum (*E. leucoxylon*).

Wildcrafting

Being a native species, some consideration needs to be given to the impact of harvesting. Because the trees are so tall, felling them is usually the only way of getting the leaves.

As long as the dried herb is only used in small volumes, harvesting an occasional tree will not have a great impact. Sufficient leaves can usually be harvested from trees being cut for other purposes, nevertheless, replanting should be carried out.

This species does coppice, but not vigorously.

Harvesting

Timing

Blue Gum leaves can be harvested any time after the spring growth has hardened off. This will be after January or February, depending on the region. Harvesting is possible until the commencement of spring growth around September.

If leaves are taken before they have hardened sufficiently, they will turn black during drying.

Tools

Some means of felling the tree will be needed when this is necessary, and secateurs will be needed for cutting small branchlets.

Technique

Only mature leaves should be harvested (juvenile leaves have a different appearance and aroma). Small branchlets with leaves attached should be cut, so they can be easily handled in the drying and processing operations. Discard leaves with blotches or discoloration, but insect holes are inevitable and acceptable.

Post-harvest Treatment

Clean up the site and make sure you use the rest of the tree, even if just for firewood. Plant replacement trees.

Drying

Eucalyptus leaves dry relatively easily. If drying space is limited, branches can be hooked over a wire slung across the drying room.

Temperature

Eucalyptus should be dried at a temperature of not more than 35°C.

Processing

When the leaves are dry, the branches can be whacked and rubbed against each other in the 'boat' or other large tray, until all the leaves are knocked off.

Tea Grade

For tea use, the leaves should be broken up into smaller fragments, but as they are too tough to be broken up by rubbing, this is a hammermill job and the grower is not normally required to do it.

Manufacturing Grade

Whole leaves are required for manufacturing.

Storage

Freeze treatment to control moth larvae is not really necessary with Eucalyptus.

Yields

Harvesting progresses quite quickly.

Marketing

The market for *Eucalyptus* leaf is very limited at present: there is a small usage of dried leaf in teas and blends but very little use of fluid extract.

Price

Price to growers for premium quality, wildcrafted *Eucalyptus* leaf is currently around $14/kg for 5-star tea grade. Trade price overseas is around $6/kg (Australia actually imports some dried *Eucalyptus* leaf).

Fennel

Foeniculum vulgare Mill.   APIACEAE

Parts Used: Seed

Status in Australia: INTRODUCED WEED (Declared noxious in Tasmania and Victoria.)

Fennel has been widely used as a culinary herb and as a vegetable for thousands of years. It is used medicinally for digestive and respiratory disorders.

Prometheus is said to have carried fire – which he gave to the human race – in a hollowed out Fennel stem.

Origin
Fennel is native to southern Europe and western Asia.

Occurrence
In Australia, Fennel is widespread in the southern states in regions of moderate rainfall or in drier regions where additional moisture is available. It is mostly a weed of roadsides and other neglected areas, where it occurs in fairly solid stands. Fennel usually does not encroach onto agricultural land, apparently being controlled by grazing, mowing and cultivation.

Identification
Fennel grows to 2500 mm tall, with thick blue-green pith-filled stems. Leaves are alternate, very deeply divided into thread-like segments, up to 450 mm long, with a strong aniseed-like aroma. The flowers are yellow, small, in compound umbels to 150 mm across, 15–20 flowers per small umbel (see colour plate 3.5). The fruit is grey to brown or yellowish brown, 3–6 mm long.

Fennel (Foeniculum vulgare).

The plant is a perennial and dies down to the ground after going to seed.

Variety
There is some considerable variation in the species, with a number of cultivated varieties, some selected for their swollen stalk bases, others for their edible stalks or their decorative foliage.

I don't know how much variation there is in the seed of wild Fennel in Australia, but the flavour of some from Tasmania and from South Australia is very good and definitely superior to much of the imported Fennel.

Harvesting Wild Stands
While Fennel is a common weed, it is not very easy to find stands of it that are away from roadsides and other contaminated places.

Being prolific in its growth and a perennial, there is not much concern about maintaining population as only the seed is harvested and never so thoroughly that none falls on the ground.

Some care should be taken, though, not to accidentally distribute seeds to new sites as a result of your harvesting. Be careful where you do your screening and winnowing.

Cultivation of Fennel
Cultivation is a definite possibility if sufficient wild growing seed is not available. Fennel is grown as an essential oil crop in Tasmania.

Harvesting

Timing
Around February or March in warmer regions, the seed will be ready to fall. At this stage it can be harvested and threshed in the field.

In cooler, moister regions, it may not be ready to fall until May. It is best not to wait that long, as it may deteriorate in moist conditions. The top part of the plant should be harvested when the seeds have filled and are at the hard dough stage.

Tools
A reaping hook is useful for cutting off the stems. A flail or other suitable implement for threshing out the seeds and a large sheet are needed.

On a larger scale, harvesting with a header should be quite feasible, provided a pure enough stand exists on smooth enough ground.

Technique

There are basically two methods that can be used for small-scale harvesting, depending on the prevailing conditions.

In warmer regions where Fennel matures in late summer or early autumn, the seed can be threshed out in the field. A large sheet should be placed on the ground for threshing on. The tops can be cut off, laid on the sheet and threshed with the flail.

In cooler regions where the Fennel matures late in the season, the tops should be cut with some of the stems included, so the sap can continue to flow into the seeds as they ripen. They should then be placed on screens, where they can cure and dry out, but not too quickly.

Drying

Seed harvested by the second method should be allowed to dry slowly so the sap in the stems can move into the seeds, completing the ripening process. Being very attractive to vermin, precautions should be taken to prevent rats and mice getting access to the seed.

When the Fennel has dried adequately, it can be threshed in the manner described above.

After threshing, the seed should be checked for dryness. This is best assessed by leaving the seed in a bag for a few days and then plunging a hand into it. If it feels at all damp, it should be spread out thinly on screens for a few days of further drying.

Temperature

Fennel should be dried at a temperature of not more than 35°C.

Processing

For the small-scale herb producer, the main problem with Fennel, as with other seed crops, is the cleaning: separating all the dust and chaff from the seed. It can be done on a small scale with improvised equipment, but this is quite labour intensive and only feasible for small volumes. For detail on this see 'Harvesting Seed' in Chapter 11, 'Harvesting'.

Alternatively, you may be able to get access to some seed-cleaning equipment. (But be aware of certification requirements, as it is possible for contamination to occur from residues left from previous use for cleaning pesticide-treated crops.)

The cleaned seed is used both for tea and for manufacturing fluid extract.

Storage

Freeze treatment to control moth larvae is necessary before sale or storage. Make sure it is protected from vermin.

Yields

Harvesting by the first method progresses reasonably quickly in good conditions and can be a viable proposition. The second method is more time consuming and is only worth while if you have need for a small quantity for clinical use or to complement a range of herbs for a localised market.

Marketing

Fennel is fairly popular as a herb tea, and is also used in moderate amounts as a fluid extract.

Price

Price to growers for premium quality, wildcrafted or organically grown Fennel seed is currently around $10/kg for 5-star tea grade.

Price to growers for manufacturing-grade, wildcrafted Fennel Seed is currently around $4–6/kg.

Current trade price is around $2/kg.

Hawthorn
Crataegus monogyna Jacq.
Crataegus laevigata DC
ROSACEAE

Parts Used: **Leaf & Flower, Fruit**

Other Names: May, Whitethorn, *C. oxyacanthoides* (= *C. laevigata*).

Status in Australia: PLANTED AS A HEDGE, WEED (Declared Noxious in Victoria and South Australia.)

Hawthorn berries are widely used as a cardiac tonic and for circulatory problems. Hawthorn leaf and flower is used for the same purpose. The berries can be used in jams as they have a high pectin content. They also make a delightful liqueur, while the flowers make a very pleasant wine.

Origin

Hawthorn is a native of Europe, North Africa and west Asia.

Occurrence

Hawthorn is widespread throughout the moderate and high

rainfall areas of Victoria and south-eastern New South Wales, throughout Tasmania and in parts of South Australia, particularly the Adelaide Hills.

It has been widely planted as a hedge on farms. In some regions it does not spread much, but in others it is quite invasive.

Identification

The two species are very similar, differing mainly in the number of seeds per fruit. They are erect shrubs or small trees, up to 9 m tall, but more commonly 4–6 m. They have many spreading branches with thorns 5–25 mm long and grey bark, generally smooth but roughened towards the base. Leaves are alternate and somewhat variable, mostly triangular to ovate, deeply lobed and toothed at the tips, 15–50 mm long. Flowers are white (pink forms exist), 8–12 mm across, with 5 petals, in clusters of 5–12 at the ends of branches and branchlets, with a strong, but not exactly sweet scent. Fruit is a berry about 8 mm across, which turns red as it ripens (see colour plate 3.7). The flesh is yellowish, the flavour rather boring. *C. monogyna* has 1 seed per fruit, while *C. laevigata* has 2 or 3 seeds per fruit.

A related species, Azzarola, *C. sinaica*, occurs in the Adelaide Hills. It is a larger tree, up to 10 m tall, with white downy hairs on young branches and flower stalks, fewer spines, larger flowers (15 mm across), larger leaves, wedge-shaped, 3–5 lobed, and larger fruit, 10–25 mm across, with an apple flavour. As it not recognised as an alternative species, it should not be harvested.

The other two species are both equally acceptable, but *C. monogyna* is the more common.

Management

Harvesting Hawthorn berries is not likely to cause it to spread, but in regions where it is invasive, care should be taken not to introduce it to new sites.

Harvesting leaf and flower is somewhat destructive as the easiest method is to cut moderate-sized branches. Hawthorn will stand heavy pruning, but where trees provide valuable shelter, this may not be appropriate.

Problems

Flowering, and particularly fruiting, is erratic and in some locations big stands of Hawthorn never seem to bear much fruit. Harvesting is much quicker on trees that are well-loaded with berries, so if you want to harvest any quantity, it is worth scouting around for the best locations.

Unfortunately the best yielding trees are often growing along roadsides or as hedges on farms, where they are likely to be affected by contamination from vehicle exhaust or agricultural practices.

The best option would be to find Hawthorn growing on certified organic or biodynamic farms or in ungrazed bushland.

Harvesting Leaf & Flower

Timing

Leaves and flowers should be harvested when the flowers are mostly open, but before they start to fade. This is in October or November, depending on the region. The optimum harvest period is relatively short, only about a week, so keep an eye on them.

Tools

Large pruning loppers and a small bow saw will be required for harvesting larger volumes. Alternatively, gloves and basket will be needed for hand stripping.

Technique

For harvesting larger volumes to supply manufacturers etc.,

Hawthorn (*Crataegus monogyna*) in flower.

the only practical method is to cut whole branches that are well laden with flowers and take these back to the drying facility.

Select them carefully for a good proportion of flowers, as flower density varies greatly from tree to tree and even on different parts of the same tree. There is generally no point in cutting off thick branches, as the pieces need to be small enough to easily load onto the screens and to fit in your rubbing screen.

Some consideration should be given to the effect your pruning will have on the appearance of the trees and the shelter they provide.

An alternative technique, which yields a higher proportion of flowers but is more time consuming, is to hand-strip the flowers from the trees, wearing gloves of course to avoid shredding your hands on the spines. With this method you can selectively strip the densest flowering portions, but you still get some leaves, as it would be very slow work to avoid them entirely.

Drying Leaf & Flower

Hawthorn leaves and flowers dry quite easily but they should be dried quickly to retain good colour in the flowers.

Temperature

Hawthorn leaf and flower should be dried at a temperature of not more than 35°C.

Processing Leaf & Flower

There may be a small requirement for leaf and flower for tea use, but mostly it is used in liquid extract form.

Tea Grade

When thoroughly dry, the leaves and flowers fall off easily. The branches can then be vigorously rubbed against each other over a 2½ or 3 dent screen. Once the leaves and flowers have been knocked off, they can be rubbed through the screen (with under-screens in place) and rescreened or winnowed until they are free of stalk.

Leaf and flower harvested by hand-stripping will still need to be rubbed and rescreened or winnowed to remove any twigs.

Manufacturing Grade

For manufacturing a first screening is required. Follow the same procedure as for tea grade, but don't worry about under-screens and just rub the herb through the screen once: a small amount of stem is acceptable in manufacturing grade.

Harvesting Berries

Timing

Harvesting should commence as soon as the berries have turned red. At this stage their level of active constituents is highest and, being firmer, they are less susceptible to damage during harvesting and processing. They are normally ready in March–April, depending on the region.

In warmer regions, an early harvest has the advantage of being able to be dried or partially dried in ambient conditions.

Tools

Harvesting berries calls for gloves, a basket that can be suspended in front of you by means of straps around the neck or over the shoulder (or an apple picking bag), and possibly a ladder or some means of getting to well-laden branches beyond normal reach.

Technique

Strip the berries with gloved hands into the basket suspended in front of you. Don't worry about getting some leaves and stalks as well, as these can be separated easily enough by rubbing when they are dry.

Keep a sack handy so you can empty your basket when it gets too heavy. For the lower branches you can give your

Hawthorn (*Crataegus monogyna*) tree in fruit

neck a rest by placing the basket on the ground and stripping the berries into it.

The berries will hold okay in bags for quite a few hours, as they don't seem to heat, so you can pick all day and then take your booty home to spread on the drying screens.

Drying

Hawthorn berries need good drying conditions. If you are relying on ambient air, they will need to be up next to a corrugated iron roof to get enough heat to dry. This will only work in warm early-autumn weather.

In cooler regions, where the berries are harvested later in the season, they will need to be dried with artificial heat. If your heated drying capacity is limited, the rest of the crop should be spread on screens, placing them in the best positions possible. After some initial ambient air drying, they can go through the dryer as space becomes available.

Be careful when handling screens loaded with Hawthorn berries: they very easily roll off if tipped and it is disheartening to watch from the top of the ladder as a screenful of berries pours onto the floor.

When thoroughly dry, Hawthorn berries should be hard enough to resist denting with a thumbnail: they should chip rather than dent. If the berries still feel a bit like sultanas, they are not quite dry.

Temperature

Hawthorn berries should be dried at a temperature of 30–40°C, but not more than 45°C.

Processing

Once thoroughly dry, the berries should first be rubbed through a 2 or 2½ dent screen (3 dent will probably be too small and too many berries will be held back). This is to separate larger pieces of twig and stalk, break up the leaves, and make the berries and leaves flow.

They should then be rubbed over a finer screen with spaces just big enough to hold all the berries back. This will require a 4 or 5 dent screen. I have usually used a 5 dent screen, with apertures of 4 mm, but if the berries are consistently large, then screen with 5 mm openings.

The aim of this second rubbing operation is to break up all the leaves and push them through the screen, along with as much of the stalks and twigs as will go. There will still be some twigs left: these will need to be picked out by hand. The thoroughness of cleaning depends on the buyer's requirements, but usually it is not necessary that berries be totally picked clean, as for most purposes they are going to be milled anyway.

I usually pick out the coarser twigs and anything with lichen on it, so the finished product looks quite presentable and is at least 99.5% berries. The small thin berry stalks are quite acceptable and the odd small twig is okay, too. If you have large quantities of berries to clean it may be worth getting access to seed-cleaning equipment, but make sure there is no possibility of contaminating your product with residues (chemical or other) from previous usage.

Tea Grade

For most purposes, the berries need to be milled to enable better extraction, but the grower is not normally required to do this.

Sometimes Hawthorn berries are used whole in blends and for this purpose their appearance is quite important, as they make an attractive contrast to the green of the other herbs in the blend. If intended for this purpose, the berries need to be unmarked and free of small twigs.

Manufacturing Grade

The dried berries can be sent off for manufacturing as they come from processing.

Storage

Freeze treatment to control moth larvae is necessary before sale or storage.

Hawthorn flower is particularly vulnerable to moth infestation, as it is dried when moth numbers peak and they are very attracted to flowers.

Yields

Hawthorn berry harvesting progresses reasonably quickly in a good stand of well-laden trees. Harvesting leaves and flowers is a bit slower.

Marketing

There is quite a large usage of Hawthorn berries, mostly in manufacturing, but a small amount goes for tea use. The market for Hawthorn leaf and flower is limited and mainly in manufacturing.

Price

Price to growers for premium quality, wildcrafted or organic Hawthorn is currently around $14/kg for 5-star grade berries.

Price to growers for manufacturing-grade, wildcrafted or organic Hawthorn is currently $10–13/kg for berries and $15–17/kg for leaf and flower.

Current trade price is around $8/kg for Hawthorn berries.

Horehound, White

Marrubium vulgare L LAMIACEAE

Parts Used: **Leaf & Flower, Aerial Parts**
Other Name: Horehound.
Status in Australia: INTRODUCED NOXIOUS WEED

White Horehound has been used since ancient times as a remedy for coughs and sore throats and as an expectorant. It is extremely bitter, which often seems to make it so effective that just the mention of the possibility of Horehound tea is enough to give instant relief from the symptoms, especially in children!

It is also used in the manufacture of Horehound candy and Horehound beer.

Origin

White Horehound is a native of Europe, central and western Asia and North Africa.

Occurrence

It has spread throughout the sheep-grazing regions of southern Australia because its burrs catch in wool. It commonly occurs around stockyards, sheep camps, fence lines, buildings etc., but can become very invasive in conditions which favour it, taking over large areas of grazing land. It is a major cause of vegetable fault in wool.

White Horehound is essentially a farm weed, but sometimes can be found in disturbed bushland.

Identification

White Horehound is woody at the base with much-branched, woolly, quadrangular stems, up to 750 mm tall. Leaves are opposite, grey-green, woolly, wrinkled, ovate, bluntly toothed, up to 70 mm long, faintly aromatic. Flowers are white, 6–10 mm long, in dense whorls in the leaf axils.

The plant is perennial and tends to die back after flowering and setting seed, but it always has some green leaf on it. It is winter-active.

Its white flowers and lighter-coloured appearance distinguish it from Black Horehound, which lacks woolly stems, has a darker appearance, pink flowers, and is not found growing wild.

Harvesting

Vast stands of Horehound exist in many parts of southern Australia, but in other regions there are just odd plants here and there, too few and too scattered to justify collection on any scale.

To qualify for wildcrafted status, White Horehound must be growing away from farming activity, though it may be possible to find some growing on a certified organic property. Some harvestable stands occur in bushland.

Care should be taken to ensure you are not inadvertently facilitating its further spread by means of burrs adhering to clothes or harvesting sheets.

Problems

The old dead stems tend to stay on the plants for a long time. If possible, the bushes should be cut back some time before harvesting for the first time so there is clean new growth of good quality. Even though it appears to be a woody perennial, White Horehound can be cut back quite hard, but don't take it down to ground level.

If the bushes are harvested regularly, dead material will not be a problem.

Timing

Harvest White Horehound when in flower and when there is enough growth to make it worth while. This can occur at almost any time of year as it depends on seasonal conditions, particularly moisture. It should be borne in mind that White Horehound needs very good drying conditions, so harvest when these are available.

White Horehound (*Marrubium vulgare*).

Tools

Tough growth may need to be cut with a reaping hook, though harvesting with the catching scythe is usually possible. One wildcrafter cuts it with a brush-cutter and then picks up the pieces.

Technique

Cut the plants just above the level of dead and discoloured leaf.

Post-harvest Treatment

If there is a lot of leggy growth left after harvest, it is best to cut this back so the next harvest will be easier and cleaner.

Drying

White Horehound needs very good conditions for drying, as it is quite slow to dry, especially the stems. Pre-chopping might be a possibility for manufacturing grade where the stalks are included, but this should be tested first to see if it bruises the leaf.

Herb intended for manufacturing should be totally dry: all the stems should snap when bent sharply.

Temperature

The temperature at which it is dried should not be more than 35°C.

Processing

White Horehound is used to a small extent as a tea, but more for manufacturing into a fluid extract, for which aerial parts are used.

Tea Grade

The herb should be rubbed through a $2^1/_2$ or 3 dent screen, with under-screens in place and rescreened or winnowed until free of stem.

Aerial Parts

After carefully checking that the stems are thoroughly dry, White Horehound should be fed through the chaff-cutter. This is a herb that commonly arrives from inexperienced harvesters in a damp condition because of moisture still in the stems, even though the leaves may have felt very dry when bagged.

Storage

Freeze treatment to control moth larvae is necessary before sale or storage.

Yields

Getting a good return from White Horehound depends very much on having a good stand of it where harvesting can proceed quickly.

Marketing

There is a moderate demand for this herb for manufacturing and a small demand for tea grade.

Price

Price to growers for premium quality, wildcrafted or organically grown White Horehound leaf and flower is about $18/kg for 5-star tea grade.

Price to growers for manufacturing-grade, wildcrafted or organically grown White Horehound aerial parts is currently around $8–10/kg.

Melissa Balm

Melissa officinalis L   LAMIACEAE

Parts Used: **Leaf & Flower, Aerial Parts**
Other Names: Lemon Balm, Balm, Melissa.
Status in Australia: INTRODUCED WEED

Melissa Balm is dealt with in detail as a cultivated plant in Chapter 16, 'Expanding Clump Herbs', but in some locations wildcrafting is a possibility.

Occurrence

Melissa Balm grows as a weed in parts of Victoria and Tasmania, and possibly elsewhere. In north-east Victoria, it is sometimes found growing beside creeks, where it often looks healthier than cultivated Melissa Balm, which frequently suffers from mites.

Harvesting

Harvesting aerial parts of Melissa Balm will not deplete it. It is not a potential problem weed as the habitat suited to it is limited.

Problems

The first time Melissa Balm is harvested there is often a lot of old stalk among it from previous growth, and this will lower its quality. To prevent this problem, the plants should be trimmed in autumn or very early spring before new growth starts.

For more detail on this herb see 'Information Charts' and Chapter 16, 'Expanding Clump Herbs'.

Mountain Pepper

Tasmannia lanceolata (Poir.) Sm.
WINTERACEAE

Parts Used: **Leaf, Fruit, (Bark)**

Other Name: *Drimys lanceolata.*

Status in Australia: NATIVE SPECIES

My first introduction to Mountain Pepper went something like this: 'Have a taste of this leaf, Greg, but you have to chew it really hard to get the flavour out of it'.

For about 15 seconds there was a very mild aromatic flavour a bit like nutmeg, while I chewed vigorously. Then suddenly my mouth was on fire, much to the amusement of my companions.

Since then I have learned to be a little more cautious when tasting it. Because its piercing peppery flavour is released by the action of enzymes in the saliva, there is a short delay in its onset. A gentle initial nibble is recommended until you get an idea of its intensity.

The berries and the bark have a similar flavour, but are perhaps a little more aromatic. It is becoming a popular condiment in Australian bush food cuisine and there is some interest overseas in using essential oils extracted from Mountain Pepper as a flavouring.

There are other species of *Tasmannia* in New South Wales and Victoria that have similar qualities.

Related species from South America and New Zealand have traditional medicinal uses, so there may be traditional Aboriginal uses as well.

Origin

Mountain Pepper is endemic to Australia, with related species in the Pacific and South America.

Occurrence

Mountain Pepper occurs in subalpine to lower mountain regions of Victoria, New South Wales, and Tasmania, extending to lower altitudes in East Gippsland and Tasmania, where it reaches its best development, usually on south-facing slopes. In high rainfall areas in Tasmania, it is occasionally found in good stands in old pastures where rainforest has been cleared. Stock are, understandably, not partial to it, nor are native animals.

Identification

Mountain Pepper is evergreen and ranges from a small shrub to a small tree, 0.2–8.0 m tall, much branched, with reddish brown bark and bright red young stems. The leaves are quite variable in size, depending on the locality. They are elliptical to oblanceolate, quite thick, 15–130 mm long, and slightly aromatic with the characteristic peppery taste. The flowers are white or cream, 2–9 petalled, in terminal clusters, with male and female flowers on separate plants. Fruit is a dark purple or black berry (see colour plate 5.3). Flowers and fruit are usually not prolific, though in some locations good crops of berries occur in some years.

A little taste of the leaf will confirm identification.

Variety

For wild harvesting, only the larger forms with their larger leaves and fruit will be a realistic proposition. There is considerable regional variation in the flavour, so for some purposes harvesting has to be selective.

Wildcrafting

Being a native plant, harvesting of any quantity of Mountain Pepper could only be justified in locations where there are good stands.

Leaf harvesting can be carried out as a light pruning about every 3 years to allow the bushes sufficient time to recover.

If bark is required, it can be obtained from the twigs left over from the leaf harvest. This apparently has the same properties as the bark of the main trunk. This avoids having to harvest the bark by methods that would destroy or damage valuable leaf-producing trees, reduce populations, and affect local ecosystems.

The wild food industry is very concerned about how wild foods are harvested. Any unsustainable or destructive harvesting practices will result in blacklisting of the persons responsible and their products.

Mountain Pepper (*Tasmannia lanceolata*) foliage.

Cultivation

'Wild cultivation' is worth considering if significant quantities are required. Seed or cuttings could be taken from productive trees to raise young trees for planting in good locations in the bush.

Farm planting would be worth while if you have a suitable site, but this is a long-term project. It prefers a southern or south-eastern aspect.

Mountain Pepper can be grown from cuttings. Semi-softwood cuttings taken in February and started with bottom heat and misting have reportedly been successful. They can be a bit touchy when transplanting, so disturbance of the roots should be minimised.

Harvesting

Timing

Leaf can be harvested after the new growth has hardened off: probably from January or February to June.

Fruit is harvested in autumn.

Tools

Secateurs will be required for harvesting the leaf.

Technique

Leaf

Leaves should be harvested sparingly so as not to excessively damage the bushes. A light pruning which removes only the new growth will harvest the better quality leaf, with sufficient remaining on the tree for regeneration if this is done once every 3 years or so.

Fruit

The berries can be picked by hand. Leaf and other trash can be easily removed by screening after drying in a manner similar to Hawthorn berries.

Leave some fruit on the bushes or, better still, propagate from the more productive bushes and plant them in the vicinity and in other suitable places (some non-fruiting male plants will be required as well).

Bark

It looks as if it will be relatively straightforward to develop a technique using milling equipment to remove the bark from the twigs remaining after the leaf has been processed. At this stage there is no particular demand for the bark.

Drying

All parts of the Mountain Pepper dry easily.

Temperature

Mountain Pepper should be dried at a temperature of not more than 35°C.

Processing

The fruit and the leaves are best presented whole.

Leaf

Leaves are separated from the stems when they are dry enough to easily break off by stirring them around together in a large tray, similar to the method described for Bay leaves and for Eucalyptus. For premium grade, whole leaf is required. Broken leaf can be used for milling, but brings a lower price.

Fruit

Premium-grade berries should be whole and have all the little peduncles or fruit stalks removed. The bulk of the cleaning can be done with screens, but there will be a few that need to be removed by hand.

Milling grade can include some broken berries and a few peduncles still attached to the berries.

Storage

Freeze treatment to control moth larvae is probably best undertaken as a precaution, though it seems unlikely they would attack it.

Marketing

There seems to be a growing market for Mountain Pepper, but what its capacity will be is unclear at this stage.

Price

Prices quoted for Mountain Pepper seem to vary enormously. It may be a while before a stable price pattern emerges, as it is a new product.

Mullein, Great

Verbascum thapsus L SCROPHULARIACEAE

Parts Used: Leaf, Flower

Other Names: Woolly Mullein, Aaron's Rod.

Status in Australia: INTRODUCED WEED (Declared noxious in Victoria.)

Mullein is mainly used as a respiratory herb, but it is also used in cosmetics.

Origin

Great Mullein is a native of Europe and Asia.

Occurrence

Sporadic through southern Australia, Great Mullein prefers poor soils and stony sites, often on steep ground or on old mine tailings. The best stands I know of are in the Southern Tablelands of New South Wales in 'flogged-out' sheep country, but there are occasional good stands of it elsewhere.

Many of the sites where it grows would not qualify for wildcrafted status as they are on grazing land or contaminated mine sites, but it may be possible to locate suitable stands, either in bushland or on organic farms.

Identification

Great Mullein is a biennial herb with a thick erect woolly flowering stem, growing up to 2.5 m tall. The leaves are grey-green, woolly, ovate. Basal leaves are 150–450 mm long, while the stem leaves are progressively smaller towards the top. Flowers are yellow, 15–30 mm across, on a thick woolly spike, one or more at the top of the stem.

The plants are biennial, forming a large rosette in the first season, flowering in the second season.

Most other species of Mullein are easily separated from Great Mullein by their lack of woolliness.

A similar species, *V. thapsiforme* has larger flowers, 30–60 mm across, but does not seem to have been introduced to Australia.

Management

Management of Great Mullein depends on whether you want to maintain the population or get rid of it.

If sustainable harvesting is the aim, some flowering plants should be left to maintain the population because it regenerates from seed.

Alternatively, if you are just harvesting the leaves, they can be stripped from the stems and the flowers left to mature and form seed.

If it is growing in a situation where it is not wanted, the whole aerial parts should be carefully removed. Take care with the seed head as it produces large amounts of viable seed.

Problems

Being biennial, the density of a stand can vary greatly from year to year, according to the germination and growing conditions of the previous seasons.

Harvesting

Timing

Great Mullein flowers are harvested as they open.

The leaf is best harvested in early flowering but as it holds quite well on the plants, there is some flexibility.

Leaf can be harvested from the first year rosettes as well, providing they are large enough.

Tools

A reaping hook or similar implement will be suitable for harvesting.

Technique

Cut the healthy leaves from the stem with the reaping hook or cut the whole stem near the base and strip the leaves off later.

Flowers need to be picked by hand, but with this species the flowers would need to bring a good price as they are so small and only open a few at a time.

Great Mullein (*Verbascum thapsus*).

Drying

Whole leaves of Mullein are very reluctant to dry as they are covered with wool and have a fleshy midrib. To get them to dry fast enough, they need to be chopped into pieces about 50 mm long. Chopping it into even shorter pieces with a chaff cutter may be okay provided it doesn't bruise and discolour: do a test run first.

Herb intended for manufacturing use should be checked very carefully for dryness. The pieces should shatter when crushed and the midribs should be thoroughly dry.

Temperature

Mullein should be dried at a temperature of not more than 35°C.

Processing

There is a small demand for Mullein as a tea, and some demand for manufacturing fluid extract.

Tea Grade

Once dry, the herb should be rubbed vigorously through a 2 or $2^1/_2$ dent screen. I rather think 3 dent would be too small, but if it is all you have, then try it. Under-screens are just a nuisance for a woolly herb like this, so don't bother with them.

There is very little to screen out with Mullein. It is just a matter of ensuring that it is thoroughly dry and breaking the leaf up to make it suitable for tea grade.

Manufacturing Grade

Check that the herb is thoroughly dry. If it has been pre-chopped, you should be able to bag it up as is, but if it is too bulky, then do a first screening, throwing in any pieces of midrib, provided they are totally dry.

Check the herb again for dampness after it has been in the bag for a few days.

Storage

Freeze treatment to control moth larvae is necessary before sale or storage.

Yields

I don't have any experience with handling significant volumes of Mullein, but the viability of harvesting would depend mostly on finding a good accessible stand that was not too far away.

Marketing

There is a moderate demand for Mullein leaf for manufacturing.

Price

Price for premium quality, wildcrafted or organically grown Mullein leaf is currently around $15–20/kg for 5-star tea grade.

Price for wildcrafted or organically grown manufacturing-grade Mullein leaf is currently around $8–12/kg.

Current trade price is around $9–10/kg.

For more information on growing Great Mullein see Information Charts and Chapter 20, 'Annuals, Biennials and Short-lived Perennials'.

Nettle, Lesser
Urtica urens L   URTICACEAE

Parts Used: Leaf, Aerial Parts
Other Names: Stinging Nettle, Dwarf Nettle, Small Nettle.
Status in Australia: INTRODUCED WEED

While this species of Nettle is sometimes used in place of Greater Nettle (*Urtica dioica*), it is not the preferred species for use in herbal medicine, though it is the species used for making homeopathic Nettle preparations.

From a botanical and a production point of view, Lesser Nettle is significantly different from Greater Nettle. Nevertheless, it is quite suitable for general beverage use. Being lower in fibre, the leaves are less stringy and it makes a nicer Nettle soup.

Origin

Lesser Nettle is a native of Europe.

Occurrence

In southern Australia Lesser Nettle is widespread and mainly found in disturbed areas where there is a high level of nutrients, particularly nitrogen, such as stockyards, poultry runs, stock camps and house gardens.

Being an annual winter-active plant, it is most likely to be found during winter and spring, usually dying off in early summer and starting from seed in autumn.

Identification

Lesser Nettle is an annual herb, covered in stinging hairs, with erect branched stems, 100–600 mm tall. The leaf is ovate to broadly elliptical, 15–40 mm long, with deep regular teeth along the margins. The flowers are in short clusters in the leaf axils.

This species should be distinguished from the Native Nettle (*U. incisa*), which is the only other species likely to be found growing wild. Lesser Nettle is an annual and has small simple hairs among the stinging hairs on the stems and underside of the young leaves. It also has broader, rounder leaves.

Native Nettle is a perennial, winter-active, usually found in high rainfall regions in moist gullies or around rainforest margins and has a spreading growth with white underground runners. It has rather variable triangular pointed leaves. Its stems are smooth between the fairly large stinging hairs.

Greater Nettle (*U. dioica*) is perennial, winter-dormant, tall with flowers near the tops of the stems. It forms underground rhizomes and these and the roots are a characteristic yellow colour. It is not found wild but it is grown by the occasional grower.

A few plants mimic the Lesser Nettle to avoid being eaten by grazing animals and are sometimes found growing in mixed stands with it, so you need to keep your eye out for them when harvesting.

Nettle-leaved Goosefoot or Sowbane (*Chenopodium murale*) is such a plant, and grows as a weed throughout southern Australia. At first glance it can easily be mistaken for Lesser Nettle, but can be distinguished by its irregularly toothed margins and the lack of stinging hairs.

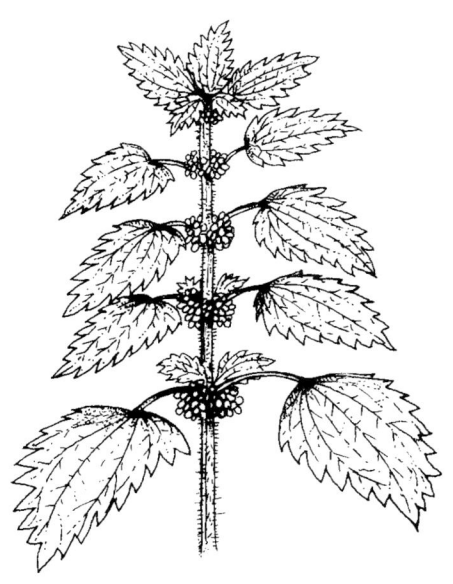

Lesser Nettle (*Urtica urens*).

Management

Lesser Nettle is a weed of farms and cultivated situations, only occurring in very fertile situations where there is regular disturbance or cultivation. It cannot really be classed as wildcrafted, but it may be possible to find harvestable stands on farms with organic certification.

As it starts forming seeds shortly after germinating, there is not much likelihood of depleting Lesser Nettle by harvesting, because by that stage it has already dropped plenty of seed.

Harvested material will contain a lot of viable seed, which falls from the herb in drying and processing. This can be a problem if it finds its way into your garden.

The plant is very specific in its habitat and simply will not thrive without a high level of nitrogen. In a fertile cultivated situation, though, it can be difficult to eradicate as it begins flowering very early in its development. Viable seeds can be formed in the axils of the first pair of true leaves.

Problems

Take care to avoid old sheep dip sites and other places around stockyards on farms where chemicals may have been used. Because it usually grows in places where there is a high concentration of stock and their manure, be careful you don't accidentally pick up any manure when harvesting, and watch out for splattered leaves.

Make sure you are well protected from being stung: wear gloves, long sleeves and trousers. Even so, some stings are inevitable. Dock leaves rubbed on vigorously are the traditional antidote. Another good treatment is the juice or tea of the Nettle itself.

The plants are often heavily infested with aphids in late spring and early summer. This seems to be a symptom of moisture stress because as conditions get warmer and drier, the plants start to decline.

Harvesting

Timing

Harvesting of Lesser Nettle should be carried out when the plants are large enough for it to be worth while. This depends very much on the season. If the autumn break comes early enough to allow for good growth, a late autumn or winter harvest may be possible if you have the facilities to dry it. Otherwise the first cut will be around September–October.

The plants will usually recover sufficiently to allow a second harvest 4–6 weeks later. In wetter springs a third harvest may be possible, and if summer is wet and cool enough, growth may continue through summer, but late regrowth is not usually prolific and can be heavily infested with aphids.

Tools

If it is a thick stand and the site is free of obstacles, the catching scythe can be used for harvesting, otherwise use the reaping hook and a basket. Gloves are definitely recommended.

Technique

Harvesting is fairly straightforward. Being an annual, some leaf should be left on the bushes if you anticipate being able to take a subsequent harvest.

Avoid old and discoloured leaf.

Drying

Lesser Nettle dries easily and is not as susceptible to darkening as Greater Nettle. If it is harvested between late autumn and mid-spring, some artificial heat may be needed, at least to finish it.

Temperature

Lesser Nettle should be dried at a temperature of not more than 45°C.

Processing

Lesser Nettle leaf is predominantly used as a tea.

CAUTION: Dried Nettle stingers are very irritating to the eyes and respiratory system, so suitable protective equipment should be worn when handling the dried plant, including respirator and goggles. Particular care should be taken when filling and tying bags, as it is very easy to get a puff of dried stingers right in your face.

Tea Grade

Rub tea grade through a 2$\frac{1}{2}$ or 3 dent screen, with underscreens in place and rescreen or winnow until free of stem.

The dried stingers may need to be sifted out, as for Greater Nettle.

Storage

Freeze treatment to control moth larvae is necessary before sale or storage.

Yields

A good stand of Lesser Nettle can be well worth while harvesting.

Marketing

Greater Nettle is the preferred species in general usage, but in some situations Lesser Nettle may be acceptable. Requirements should be checked with buyers beforehand to ensure that the harvest is going to be marketable.

Price

Current trade price is around $9/kg.

Pennyroyal

Mentha pulegium L   LAMIACEAE

Parts Used: **Leaf & Flower, Aerial Parts**

Other Names: Pudding Grass, *Pulegium vulgare.*

Status in Australia: INTRODUCED WEED (Declared noxious in Western Australia.)

Pennyroyal is dealt with in detail in Chapter 15, 'Spreading Herbs'.

Occurrence

Pennyroyal occurs in scattered locations throughout southern Australia, but is most abundant in south-west Western Australia and south-west Victoria, where it even has a town named after it. Its usual occurrence is in pastures where it can become dominant because stock don't like it.

Harvesting Wild Pennyroyal

As Pennyroyal is a weed of pastures and part of a farming system, it does not qualify for wildcrafted status. It may be possible to find it growing on an organic farm.

Problems

The main problem is that it is a short-growing plant and there is almost always at least some grass amongst it. Only a very small proportion of grass will make the dried herb look badly contaminated.

Still, this may be acceptable for some purposes and in a large stand you may find there are patches pure enough to harvest for a higher quality product.

For more detail on this herb see 'Information Charts' and Chapter 15, 'Spreading Herbs'.

Periwinkle, Greater
Vinca major L APOCYNACEAE

Parts Used: Aerial Parts

Status in Australia: INTRODUCED WEED

Periwinkle is used medicinally as an astringent.

Occurrence
Periwinkle is often found growing wild around houses, old ruins and on roadsides in southern Australia

Identification
Greater Periwinkle is a leafy vine that spreads over the ground, with stems 600–1000 mm long. Leaves are opposite, broadly ovate, shiny dark green, 50–90 mm long. Flowers are single in the leaf axils, bluish purple, 30–50 mm across. Fruit are follicles 40–50 mm long containing 1–2 seeds, but seldom forming in cooler climates.

Harvesting
Finding an occurrence away from contamination of roadsides or house sites may be difficult.

Where it gets established, it often forms a pure stand because it grows over other vegetation, smothering it out.

Technique
Harvest just the green leafy material. Avoid going down far into the mass of vines, as you will pick up old dead stems.

Drying
Drying is rather slow, so it needs good conditions.

Marketing
The market for Periwinkle is quite limited, so make sure you have a definite commitment from a buyer before harvesting.

For more detail on this herb see 'Information Charts'.

Greater Periwinkle (*Vinca major*).

Plantain
Plantago spp.   PLANTAGINACEAE

Plantago lanceolata L: **Narrow leaf Plantain**

Plantago major L: **Greater Plantain**

Parts Used: Leaf

Other Names: Greater Plantain: Broad-leaf Plantain, Rat-tail Plantain, White Man's Foot.

Narrow-leaf Plantain: Ribwort Plantain.

Status in Australia: INTRODUCED WEEDS

Greater Plantain has been dealt with in more detail in Chapter 17, 'Perennial Crown Herbs'. Narrow-leaf Plantain is recognised as having similar uses, mainly as a demulcent and respiratory herb.

Occurrence
Greater Plantain is often found in places that are waterlogged in winter or where traffic and trampling makes life difficult for other plants. In warmer regions it is usually found in shade or semi-shade, but in cooler areas it often grows in full sun.

Narrow-leaf Plantain is much more common, being better adapted to our climate. It is more likely to be found growing wild as large plants in a harvestable stand, though it is seldom pure. It is common in run-down pastures and is quite happy growing in full sun in most areas.

Identification
Greater Plantain has already been described in Chapter 17. Narrow-leaf Plantain grows in a rosette with the leaves more or less erect. Leaves are lanceolate or ovate-lanceolate, smooth-edged or slightly toothed, 30–200 mm long, with 3–7 distinct longitudinal veins. Flowers are in a cylindrical or ovoid spike, much shorter than the length of their stalks, which stand taller than the leaves. The flower-stalk has conspicuous longitudinal furrows in it.

The plants are perennial and retain some leaf through winter.

Narrow-leaf Plantain can be distinguished from other species of Plantain by its short fat flower heads and the flower-stalks with their conspicuous longitudinal furrows.

Harvesting
Although Greater Plantain is not uncommon, I can't say I have ever seen a location where it could be feasibly be harvested wild. Usually it occurs as small plants mixed with grasses, which would make harvesting very tedious.

One is more likely to find harvestable stands of the related Narrow-leaf Plantain (*Plantago lanceolata*), which may be acceptable to some buyers: it is regarded by some as equivalent to Greater Plantain. Harvesting, drying and processing is similar to that described for Greater Plantain.

Yields

Yields will depend on how good a stand you can find, but it is probably still going to be fairly slow work because of the problems with harvesting a low-growing plant from a mixed stand.

Marketing

Demand for both these Plantains is quite small

Price

Price for organically grown, manufacturing-grade Plantain is currently $6–18/kg for both species.

Current trade price is around $10/kg for Greater Plantain.

For more detail on these herbs see 'Information Charts' and Chapter 17, 'Perennial Crown Herbs'.

Narrow-leaf Plantain (*Plantago lanceolata*).

Poke

Phytolacca decandra L
PHYTOLACCACEAE

Parts Used: Root

Other Names: Pokeweed, Pocan, Red Ink Plant, *P. americana*.

Status in Australia: INTRODUCED WEED

Poke root is used in small doses in treating respiratory and other infections and for its action on the lymphatics.

While parts of this plant are quite toxic, there is a widespread tradition in North America of using it as a wild food, particularly the young leaves, after special cooking treatments to remove the toxins.

Origin

Poke is a native of North America, where it grows in disturbed and recently cleared areas.

Occurrence

Phytolacca decandra can be found growing wild in some places in Australia. I have seen it at Thora, near Bellingen, where it grows as a weed, and it has been recorded in other locations on the north coast of New South Wales.

Much of the Poke in Australia is *P. octandra*, a similar species, which is regarded by most herbalists as inferior.

Wildcrafting/Weed Harvesting

Whether Poke can be classified as wildcrafted will depend on the site where it is growing. It is often found in cultivated areas or other farm situations.

The root can get to quite a large size in the wild, so digging it out can be a major operation.

It will regrow from pieces of crown, so if you want to ensure continued production, these should be replanted when you harvest it, provided this is acceptable to the landholder.

At one stage Poke was classed as a noxious weed in some states, but it no longer holds that distinction.

Harvesting the wild plant is not likely to cause it to spread. If suitable quality and volume can be obtained this way, it will obviate the need to grow Poke in cultivated situations and risk it spreading into new areas.

Identification

Poke has thick, hollow purplish stems and grows 1.2–2.7 m tall. Leaves are alternate, ovate-lanceolate or oblong, 100–300 mm long, with an unpleasant smell. Flowers are white or sometimes pinkish, 7.5 mm wide on many-flowered lateral racemes that droop as the berries develop. Berries are purple-black when mature, consisting of 10 segments (like a pie divided into ten pieces), each containing one seed.

The plant dies down to ground level in autumn and has a large taproot. A cross-section of the root shows concentric circles of woody fibres.

Phytolacca octandra is a related species, which can be distinguished by its berries having 8 segments. This species is found wild in many parts of Australia and tends to be more common than *P. decandra*.

Harvesting

Timing

Poke root can be harvested from autumn to late winter. It is best to leave it until last, after all the other root crops have been harvested and dried, so there is no chance of getting it mixed up with them.

Tools

A garden fork will be needed for harvesting, or if the roots are well developed, a good digging spade and perhaps a crowbar may be necessary.

Technique

Gloves should be worn, as the juice can cause skin irritation. Poke can develop a rather large root system, with substantial laterals extending in several directions and a good-sized taproot. This can make harvesting a major operation, particularly if the root is several years old. It may be necessary to dig down deeply beside the taproot to get it out.

If the plants are dug every year or so, their roots will not get as big.

Post-harvest Treatment

After harvest, the ground surface should be levelled off and restored. The crowns can be cut off the harvested roots, divided and replanted if this is acceptable to the landholder.

Drying

The root needs to be washed, which is fairly straightforward, and then chopped.

CAUTION: Some protective measures need to be taken in these operations, as Poke juice can cause irritation and even burns. It is rather insidious because the effects don't show up for some hours, by which time it is too late to think about washing it off.

Be particularly careful of your eyes: it is probably best to wear goggles while chopping the root. Even a splash of water from washing the roots can cause eye irritation and should be washed out immediately.

All roots thicker than your thumb should be chopped into sections about 25 mm long to facilitate drying. Do this by hand, though, as using the chaff-cutter could easily result in contamination of the next herb to be processed if pieces of Poke remain somewhere in the machine.

For the same reason, Poke should not be dried with other herbs in the dryer at the same time and some care should be taken in cleaning the dryer out after the Poke has been through. If a piece of root accidentally finds its way into somebody's cup of tea, they might get an unpleasant surprise.

Drying should be straightforward, but check the insides of the larger pieces to ensure they are completely dry.

Temperature

Poke Root should be dried at a temperature of 35–45°C.

Processing

Poke is not normally sold as a tea herb: it is used to make liquid extracts and tinctures, for which the herb can be sent to the manufacturer in the form it comes from the dryer.

Storage

Freeze treatment is probably not necessary, as Poke root should be toxic to any moth larvae.

Poke (*Phytolacca decandra*).

Yields

Yields will depend on how good the stand is and how hard it is to dig the roots out.

Marketing

The market for this crop is limited and it should only be sold to manufacturers or herbalists who are aware of the appropriate dosage and its manner of usage. It is really not a safe herb to sell in health food shops, as people may not be aware that it must only be used in very small doses.

Price

Price to growers for manufacturing-grade, organically grown Poke is currently around $12–16/kg.

Current trade price is around $11/kg.

Red Clover

Trifolium pratense L   FABACEAE

Parts Used: Flowers

Other Name: Cow Grass (for variety Hamua).

Status in Australia: CULTIVATED FOR PASTURE AND HAY

Red Clover is covered in detail in Chapter 17, 'Perennial Crown Herbs'.

Occurrence

Red Clover is often found growing wild in the cooler, moister parts of southern Australia and sometimes good stands can be found in pastures and hay paddocks where it has been sown down, especially in Tasmania.

It is essentially a plant associated with farming systems, so it could not be classified as wildcrafted, but it may be possible to find harvestable stands growing on certified organic or biodynamic farms.

Harvesting

Red Clover can sometimes be found flowering heavily in regrowth after cutting hay, if the paddock is left shut up. Plants growing wild on unused land are usually too scattered and too mixed with grass to be harvestable.

Problems

Finding a suitable stand without a history of chemicals may be difficult.

Timing is very critical with Red Clover flowers as they need to be harvested shortly after they begin opening, and then must be dried rapidly. For premium quality they should be harvested every few days.

For more detail on this herb see 'Information Charts' and Chapter 17, 'Perennial Crown Herbs'.

Rose, Wild

Rosa spp.
Rosa canina L: Dog Rose
Rosa rubiginosa L: Sweet Briar Rose
ROSACEAE

Parts Used: Fruit ('Hips')

Other Name: Dog Rose: Dog Briar. Sweet Briar Rose: Eglantine, *Rosa eglanteria*.

Status in Australia: Dog Rose: INTRODUCED WEED (Declared noxious in South Australia). Sweet Briar Rose: INTRODUCED WEED (Declared noxious in Victoria, and in parts of New South Wales and South Australia. It is a declared secondary weed in Tasmania.)

These two species are rather similar in their general usage, but while they are both suitable for making herbal tea, for some reason Dog Rose is the preferred medicinal species. This may have something to do with the nature of the fruit, as Dog Rose hips are easier to pick, being free of prickles, and they dry much more quickly, which may result in higher retention of Vitamin C. Dog Rose is used for its medicinal properties as a nutritive and as a mild laxative and diuretic. They both can be used for making Rose hip syrup or as a puree, with the seeds and hairs removed.

Origin

Both species are natives of Europe and western Asia.

Occurrence

In Australia, Sweet Briar is the more widespread of the two species, being common in the tablelands and western slopes of New South Wales, most of Victoria and Tasmania, and the moister parts of South Australia and Western Australia. It occurs in pastures and on unused land, particularly roadsides.

Dog Rose is much less frequent. It tends to be found around old settlements, as it has been widely used as a rootstock for ornamental roses. It occurs in scattered locations in inland New South Wales, central and western Victoria, around Adelaide and in a few other locations in South Australia and northern Tasmania.

Identification

Dog Rose (*Rosa canina*) is a climbing and trailing shrub up to 5 m tall, but usually 2–3 m (see colour plate 6.3). The stems have scattered large backward-curving thorns. Leaves are alternate, pinnate, with 5–7 leaflets, each 20–40 mm long, ovate to elliptic, with toothed margins. Flowers are pink or white, 25–50 mm across, 5-petalled, in small groups at the end of branches. Fruit is a fleshy hip, scarlet when ripe, around 15 mm long. When mature, the sepals (or remains of the flower) fall off, leaving a small scar at the end of the hip.

Sweet Briar (*Rosa rubiginosa*) is similar to the above, but its growth is shorter, being up to 3 m tall, commonly 1.5–2 m. The stems have a variety of different sized thorns and prickles on them. The leaves have many glandular hairs on their undersides, which emit a sweet fragrance when crushed. The fruit is a little larger than that of Dog Rose (up to 20 mm long), has little prickles on the stalks and sometimes on the hip itself. The sepals remain attached to the hip when ripe (see colour plate 6.6). This latter characteristic is an important distinguishing feature, as it is most likely to be noticed when harvesting.

Occasionally the two species occur together in mixed stands. Dog Rose hips ripen earlier and tend to dry somewhat on the bushes, while Sweet Briar hips ferment inside their skins if they remain on the bush.

Preferred Species

As mentioned above, Dog Rose is the preferred species. If it is not available, Sweet Briar may be acceptable for some purposes, but generally not for medicinal use.

Wildcrafting/Weed Harvesting

Where they occur on grazing land, these Roses probably cannot be classed as wildcrafted.

Harvesting usually is not destructive, though with Sweet Briar most people would rather see less of it, so landholders may have no problems with destructive harvesting methods.

Problems

Yields of fruit are erratic and tend to be better where the plants are growing in the open and where there is no stock grazing. Unfortunately the best yielding bushes are often found alongside roads where they are subject to council spraying and the lead fallout from traffic.

Cultivation

The cultivation of Dog Rose is worth consideration in areas where it not regarded as noxious (for detail on this see Chapter 19, 'Trees and Shrubs').

Harvesting

Timing

The hips are ready as soon as they have attained their full colour and can be harvested over a period of up to 2 months, from around mid-March to late May (depending on the region), until they start to shrivel (Dog Rose) or turn soft inside (Sweet Briar). Earlier harvested hips will be of better quality.

It is important that harvesting be timed with the availability of drying space, as they need to be dried promptly at relatively high temperatures.

Dog Rose (*Rosa canina*).

Sweet Briar Rose (*Rosa rubiginosa*).

Tools

Tools for harvesting depend on the technique used. Dog Rose hips are smooth and can be safely harvested with bare hands. Sweet Briar hips are a bit hazardous to pick this way, as you inevitably get your fingers laced with prickles. I did endure this for a few seasons, but one year my hands became badly infected, so I tried combing the hips, which has turned out to be much quicker.

A Chamomile comb can be used, but the spacing of the teeth is a little close. A spacing of 4–5 mm between the teeth is preferable for Sweet Briar hips and should work for Dog Rose hips.

If you are constructing a comb specifically for picking Rose hips, it would be worth using some heavier 2.85 mm (11 gauge) high tensile wire for the teeth as they take something of a beating combing the tough bushes (for more detail on making a comb see 'Making a Flower Comb' in Chapter 11, 'Harvesting').

An alternative method of harvesting Rose hips involves cutting the whole canes and threshing them (see below).

Technique

Hand-picking

Hand-picking is fairly straightforward, but not recommended for Sweet Briar, on account of the prickles on its stalks and fruit.

Combing

Combing involves running the comb through the bushes to pick up the hips while the leaves and stems mostly slip through the comb. You soon learn how much you can thrust the comb into the bush as it tends to snag. Mostly it is best to comb off the hips that are towards the outside of the bush or on the long branches, and leave those in the middle as they aren't worth the hassle.

Sweet Briar Rose on grazing land. It is more widespread than Dog Rose.

A certain amount of leaf and stem will also be gathered in the combing operation, but this is easily separated once the hips are dry.

Cutting and Threshing

Cutting whole canes and threshing the hips off them is a possibility if you aren't going to be coming back again for a while, or if the Sweet Briar needs to be cleared anyway. They will reshoot, but it will be a few years before they are back in full production.

The canes can be cut a bit above ground level. They are then whacked against rabbit wire screen (the heavier mesh with holes 30–50 mm across) or reinforcing mesh placed across the back of a ute or a large tray until all the hips have been knocked off.

This method tends to collect a lot more trash than combing and too much of it may interfere with drying. Some could be eliminated by shaking and skimming, or by winnowing.

Drying

Rose hips need good conditions for drying: this is particularly so with Sweet Briar.

Sweet Briar

The problem with Sweet Briar hips is that their skins are virtually impermeable, so they need relatively high temperatures to dry quickly enough, otherwise they ferment.

Sweet Briar hips need artificial heat, and even so will take around two weeks to thoroughly dry. People often have difficulty the first time they try to dry Sweet Briar hips, because this fruit is so much more reluctant to dry than other herbs. They need a steady flow of warm dry air.

One method to reduce fuel costs is to place the hips close to the heat on the bottom shelves in the cabinet dryer. They will then be in the best drying spot, but because they give off moisture so slowly, the air passing over them will still have almost its full drying capacity. This air can be used for drying other crops: often there are leaf crops in late autumn that need some heat to finish them.

Dog Rose

The hips of Dog Rose are less of a problem as they seem to have a more permeable skin. In warmer regions, it may be possible to dry Dog Rose hips close to the roof in an ambient air drying shed if they are picked early enough.

Loading

Be careful in loading your screens with Rose hips as they are heavy and inclined to roll off if tipped.

Assessing

To test for dryness, the hips should shatter when squeezed tightly between thumb and forefinger.

CAUTION: Rose hips contain fine hairs that are quite irritating, especially when dry. Avoid getting them in your eyes, nose, mouth and lungs.

Temperature

Rose hips should be dried at a temperature of 35–45°C. For Sweet Briar this should be continual, or at least for most of the time.

Processing

The dried hips will probably need to be cleaned of leaf and other debris. This can be done by rubbing them over a $2^1/_2$ or 3 dent screen, the aim being to shatter the leaves so they will go through the screen, while the hips are held back. Pieces of stem should be removed by hand, but the sepals (remains of the flowers) and the thin fruit stalks can remain on the hips.

Sweet Briar is used as a beverage herb, while Dog Rose is used as a beverage and for manufacturing herbal medicines.

Tea Grade

For tea grade the hips should be milled. Some purchasers like the seeds and hairs to be removed as well, but the grower/wildcrafter is not normally required to do either of these operations.

Manufacturing Grade

The whole hips can be sent to the manufacturer just as they come from the cleaning process.

Storage

Freeze treatment to control moth larvae is necessary before sale or storage.

Yields

In a good stand, harvesting can progress quite quickly.

Marketing

There is a moderate demand for Rose hip. Both species are used for general tea purposes, but for manufacturing, Dog Rose is required.

Price

Price for premium quality, wildcrafted or organically grown product is currently around $17/kg for Dog Rose hip and $13/kg for Sweet Briar Rose hip.

Price for manufacturing-grade wildcrafted or organically grown Dog Rose hip for medicinal purposes is currently around $10–15/kg.

Current trade price is around $9/kg for both species.

Shepherd's Purse

Capsella bursa-pastoris (L) Medic.
BRASSICACEAE

Parts Used: **Aerial Parts**

Status in Australia: INTRODUCED WEED

Shepherd's Purse gains its name from a rather euphemistic reference to the shape of the fruiting body, likening it to the purse traditionally made by shepherds from a ram's scrotum.

It is used as a uterine stimulant, diuretic and astringent. In many countries, it is also eaten as a vegetable in spring.

Occurrence

Throughout Australia, Shepherd's Purse occurs as a weed on farms. It is a common weed of cultivated crops, but is usually found growing in mixed stands with other weeds. Being associated with farming systems, it cannot really be given wildcrafted status, but locating some on certified organic properties is a possibility.

Shepherd's Purse (*Capsella bursa-pastoris*).

Identification

Shepherd's Purse is an annual or biennial herb that forms a basal rosette and then sends up a flower stem 50–500 mm tall.

Leaves are very variable, even on the one plant, being oblanceolate, entire, toothed or almost pinnate, up to 100 mm long, shorter on the stems.

The flowers are white, 2–4 mm across, 4-petalled, in loose racemes at the top of the stems.

The fruit are heart shaped, and 6–9 mm long.

Shepherd's Purse is difficult to identify by its leaves, as they are so variable and resemble a number of other weeds, however, the erect flowering stems, bearing the distinctive heart (or ram scrotum) shaped fruits make identification easy.

Problems

Finding a harvestable stand of Shepherd's Purse seems to be the main problem. It is very widespread, so possibly conditions somewhere in southern Australia suit its development sufficiently to give rise to pure stands of large plants, but I have only ever seen scattered plants in mixed stands, and usually these have been quite small.

It may be worth looking in grain-growing areas, as it is an annual weed of cultivation.

Shepherd's Purse is often affected by a white rust.

Harvesting

Timing

Harvest when in the early flowering stage.

Technique

I found it frustrating to harvest small scattered plants. The leafy parts were very close to the ground and there was not much anyway.

If there were larger plants growing close together, harvesting might be easier and more productive.

Shepherd's Purse is sometimes used in the fresh state for manufacturing a fresh plant tincture.

For more detail on this herb see 'Information Charts'.

Sorrel, Sheep
Rumex acetosella L
R. angiocarpus Murb.
POLYGONACEAE

Parts Used: Aerial Parts
Other Names: Sorrel, *Acetosella vulgaris*, *A. angiocarpa*.
Status in Australia: INTRODUCED WEED

Sheep Sorrel has suddenly become the flavour of the month in the herb trade, being used in herbal medicine as supportive therapy for cancer patients.

It has a long tradition of use as a food and was among the contents of the stomach of an Iron Age man, buried for around 2000 years and found almost perfectly preserved in a peat bog in Denmark.

It is still used in soups today (but there is no guarantee that it will do for you what it did for the Bog Man!).

Origin

Sheep Sorrel is a native of Europe.

Occurrence

Widespread throughout southern Australia, Sheep Sorrel is a frequent weed of cultivated crops and poor pastures. It can sometimes smother other vegetation and be a worrisome weed in some herb crops.

It has the reputation of being an indicator of acid soils, but this is only partially correct. It will thrive in non-acid soils if its competition is reduced by occasional cultivations, because it regenerates more rapidly than most weeds. In acid soils it is often able to dominate because its growth is not greatly affected by the acid conditions that stunt the growth of other plants.

It grows in warmer regions, but its best development is in mild to cold regions: it can be a problem weed in alpine areas.

Identification

Sheep Sorrel is a perennial herb with branching flowering stems growing up to 600 mm tall but usually 100–400 mm. The plant forms a loose rosette, with leaves 40–200 mm long, ovate to lanceolate. The leaf typically has lobes at the base giving it a rounded arrowhead shape. The stem leaves are progressively shorter towards the top of the stem. The flowers are in whorls towards the ends of the branches, 2–4 mm across. Male and female flowers are on separate plants.

The plants are winter active and spread rampantly by means of its extensive yellow roots.

There are two very similar species of Sheep Sorrel, which are distinguished only by differences in the seed. *R. acetosella* is more common, with a nut that is free of flower segments. *R. angiocarpus* has thin silvery-grey flower segments adhering to the nut.

Harvesting

Sheep Sorrel harvested on farms can't really be classed as wildcrafted, as it is a product of a farming system.

The main difficulty is to find stands that are pure enough and tall enough to be worth harvesting. It may be necessary to manage it with an intermittent cultivation regime to maintain a sufficiently pure stand.

When harvesting and drying take care that seed does not find its way into cultivated areas where it will cause problems.

Timing

Harvest Sheep Sorrel when it starts to flower. This will be from mid-spring onwards. Don't let the plants become too advanced as they will start to deteriorate, becoming mostly stem.

Sheep Sorrel (*Rumex acetosella*).

Tools

If the stand is pure enough and the ground is level enough, the catching scythe can be used for harvesting, otherwise the reaping hook will be preferable.

Technique

Because Sheep Sorrel is a low-growing plant, care should be taken not to pick up soil and debris in harvesting. Avoid including other plants in the harvest.

Post-harvest Treatment

Mow off all the growth to ground level to ensure a good quality regrowth.

Drying

Sheep Sorrel leaf is rather fleshy, with a high water content, so it needs good conditions for drying. If it dries too slowly, it will turn brown.

Pre-chopping before it wilts will help speed up drying, and provided drying conditions are good, it won't cause bruising.

Make sure the leaves and stems are all thoroughly dry before processing.

Temperature

Sheep Sorrel should be dried at a temperature of not more than 45°C.

Processing

Aerial parts are required.

Manufacturing Grade

The dried aerial parts should be fed through the chaff-cutter. Check the herb a few days after bagging to make sure it still feels dry, in case there was some hidden moisture amongst it.

Storage

Freeze treatment to control moth larvae is necessary before sale or storage.

Yields

Feasibility of harvesting will depend on being able to find a good stand of Sheep Sorrel. Being mostly water, it loses a lot of weight in drying, so large volumes will be required.

Marketing

At present there appears to be a good demand for Sheep Sorrel, but this is a situation where the market could be volatile.

Price

Prices offered for organically grown, manufacturing-grade Sheep Sorrel vary considerably and are in the range of $10–30/kg.

Current trade price is around $21/kg.

St John's Wort

Hypericum perforatum L
HYPERICACEAE

Parts Used: **Leaf & Flower, Aerial Parts**

Other Name: *H. perforatum* var. *angustifolium*.

Status in Australia: INTRODUCED WEED (Declared noxious in New South Wales, Victoria, Western Australia. Declared secondary and prohibited weed in Tasmania.)

This herb has a long history of use in religious ceremonies from well before the Christian era. Being in full flower and at its greatest potency at the summer solstice, it featured in midsummer festivals where it was burned to drive off evil spirits.

With the spread of Christianity, the midsummer festival became St John's Day, and so this herb became associated with the saint. The red pigment produced when crushing the flowers was said to represent the blood of St John when he was beheaded.

St John's Wort has valuable medicinal properties as a healing herb, a tonic and an antidepressant. The herb grown in Australia seems to have a greater potency than that from Europe.

Its leaves were once used as a salad herb.

Origin

St John's Wort is native to the temperate zones of Europe and western Asia, but is now widely naturalised in the other parts of the world.

Occurrence

St John's Wort is a very prolific noxious weed in many parts of the country. It probably escaped from medicinal herb gardens during the gold-rush days and has since taken over vast areas of the inland slopes and tablelands of the Great Dividing Range in New South Wales and Victoria. It is also a problem weed in parts of South Australia and Western Australia.

St John's Wort is not very palatable to livestock and can cause photosensitisation (a chronic sunburn condition) if they eat too much of it. It mostly invades rough pasture by means of its rhizomes and by seed that is carried on the coats of livestock and native animals. It is capable of invading open bushland, due to its early spring development and its ability to both grow in shade and to withstand hot dry conditions over summer.

In Tasmania and in coastal areas on the mainland, it does not seem to be such an invasive weed.

Identification

St John's Wort grows 300–1200 mm tall, the round reddish stems branching near the top. The leaves are opposite, ovate to linear, smooth-edged, 15–30 mm long, with many translucent oil glands. The flowers are bright yellow, 20–30 mm across, with 5 petals, in many flowered cymes at the ends of the branchlets (see colour plate 7.3).

The plant is perennial, dying back to ground level after seeding, but continues growth as leafy surface runners, which spread rapidly. It is usually found growing in a stand, many of which are quite extensive.

The form of St John's Wort found in Australia has leaves narrower and stiffer than the form commonly found in Europe. It is sometimes referred to as *Hypericum perforatum* var. *angustifolium*.

There are a number of other species of *Hypericum*, including several native species.

St Peter's Wort (*H. tetrapteron*) is a very similar introduced species and also has oil glands in the leaves. It differs in

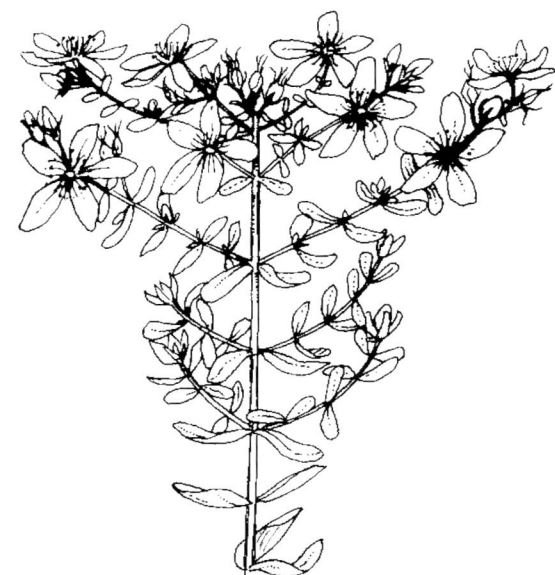

St John's Wort *(Hypericum perforatum).*

that it has square stems and smaller flowers, 10–20 mm across. It has a limited occurrence in the Dandenong Ranges east of Melbourne, mainly along creek banks and in moist or swampy sites.

Wildcrafting/Weed Harvesting

Some herb growers with available drying space may wish to take advantage of St John's Wort growing wild in their vicinity and go out and harvest some of it. In doing so, a few points need to be borne in mind: namely, contamination, the incidence of dead material, maintenance of the stand, and containment of further spread. Each of these is discussed below.

Contamination

Avoid roadsides and farmland (apart from organic farms), as these areas are likely to be subject to spraying to eradicate St John's Wort. Crown land is usually the best place to wildcraft, but watch out for token sprayed areas: these are usually next to tracks where there is access and where the public will see that the authorities are attempting to contain this weed's spread. Give these situations at least a 30 m buffer zone.

Dead Material

When harvesting a patch of St John's Wort for the first time, there will be a lot of old dead stems and seed heads still standing from the previous season. If these get mixed in with the harvest, they will reduce its quality. Avoiding them can greatly slow your harvesting and prevent the use of the scythe. If you can get on to these stands by early spring and mow them or otherwise knock these old dead plants down, your subsequent crop will be much cleaner and easier to harvest.

Maintenance

Wildcrafters don't have to worry about depleting stocks of this herb. Harvesting does not seem to diminish its stranglehold on the places where it grows and there will be just as much there next year. You may want to just go over the area you have been taking it from and cut down any plants you have missed, so they aren't still standing in the middle of your patch next year.

Containment

It is important to ensure you are not helping St John's Wort spread by rhizomes or seeds. As long as the herb is harvested in the early flowering stage, there will be no seeds in it. Old seed heads from the previous year, which may be in the harvest, don't seem to be a problem. Over a number of years harvesting, we have never noticed any St John's Wort coming up around locations where it was dried.

Biological Controls

Attempts are being made to introduce insects that will eat St John's Wort, but so far these have not been very successful. There is a small dark blue beetle seen on the leaves sometimes, but it doesn't seem to do much damage.

Harvesting

Timing

St John's Wort should be harvested just as the first flowers are opening. Depending on altitude, this is late November to late December in north-east Victoria. At this stage the active constituents are at their highest, but they decline as the plant matures.

Tools

The catching scythe is the most efficient for harvesting, but on steep and stony sites, the reaping hook may be more appropriate if the stands are thick enough.

Technique

Cut the plant just below the level of the good quality leaves. As St John's Wort can sometimes grow quite tall with a high proportion of stem, it is important when harvesting the herb for aerial parts to make sure the plant is not being cut too low down the stem, resulting in an excessively stalky final product.

Take care the herb doesn't build up heat in the harvested bundles, as St John's Wort can turn dark if it gets too hot. Avoid handling the herb when it is wilted.

Sometimes St John's Wort is harvested for fresh plant use in making an oil infusion. For this purpose it should be harvested at the same early flowering stage, but just gather the top 100–150 mm to get a high proportion of leaf and flower.

CAUTION: Handling the fresh herb and breathing the vapours from it for several hours can make you rather intoxicated, to such a degree that your driving ability can be impaired and you may be placing yourself and others at risk.

While it is in the fresh state, St John's Wort is quite a potent herb which can strongly affect people who are working with it for extended periods of time: 3 or 4 hours harvesting is enough, especially if you have to drive home with it in inside the vehicle.

Harvesting with a reaping hook seems to be worse (more skin contact perhaps) and the effects seem to be cumulative. I found that when I went out and harvested St John's Wort several days in a row this feeling of inebriation progressed to a state of melancholic depression, as I proceeded to get an overdose.

To avoid these effects, the best approach is to harvest the best spots quickly and get home and spread the herb out. Taking breaks during the work and having something to eat, especially protein, helps reduce the effects. If it is possible to harvest with the scythe, this will also help, as the herb is further away from the nose and it is not being handled.

If you have any distance to travel, St John's Wort should not be inside the vehicle: use a roof rack or a ute.

If you are doing more than a day's harvest, wait a day or two before going out and harvesting more. Once the herb is dry, it no longer seems to have this intoxicating effect, so handling it during processing is no problem.

Drying

St John's Wort is a bit tricky to dry. While the leaves dry relatively easily, the flowers and the stems can be a problem. Any flowers that have advanced to the stage where they have started to form a fruiting body need a close check before declaring the herb dry enough. This immature fruit is somewhat juicy and holds on to its moisture for quite a while.

The thicker stems are also quite slow to dry. Being fairly woody, they may snap as if they are dry, when in fact they still have too much moisture in them. The test is to scrape your thumbnail along some of the thick stems: if you are able to scrape the outside layer off, it is still too moist. The bark of the stems will be quite hard when they are dry.

If St John's Wort is subjected to excessive remoistening, heating or bruising, it will turn a dark reddish brown colour. Much of the St John's Wort produced overseas looks like this, due to improper handling and drying.

Temperature

St John's Wort should be dried at a temperature of not more than 35°C.

Processing

St John's Wort is mostly required as dried aerial parts for manufacturing liquid medicines and ointments, but a certain proportion is sold as tea grade.

Tea Grade

Tea grade is simply rubbed through a 2 1/2 or 3 dent screen It is not necessary to wait until all stalks and flowers are completely dry before rubbing, because any material that is not dry enough will not go through the screen.

Aerial Parts

Before putting St John's Wort through the chaff cutter, make sure it is thoroughly dry, as outlined above under 'Drying'.

Storage

Freeze treatment to control moth larvae is necessary before sale or storage.

Yields

If you can get on to a good patch of St John's Wort, harvesting can be worth while, but the problems listed above can limit production.

Marketing

Marketing of St John's Wort can be a bit tricky. Only a small quantity is used as tea grade, the rest is purchased by manufacturers. With so much of it growing wild, an oversupply situation that lowers the price can easily develop. Harvesting it is something of a speculative venture, but St John's Wort is used in large volumes in the trade.

Price

Price to wildcrafters for premium quality St John's Wort is currently around $16/kg for 5-star tea grade.

Price to wildcrafters for manufacturing-grade St John's Wort is currently $5–13/kg.

Current trade price is around $9–10/kg.

Variegated Thistle

Silybum marianum (L) Gaertn.
ASTERACEAE

Parts Used: Seed, Leaf

Other Names: Milk Thistle, St Mary's Thistle, Cabbage Thistle, Gundagai Thistle, *Carduus marianus*.

Status in Australia: INTRODUCED WEED (Declared noxious in Victoria, South Australia and Western Australia. Declared secondary weed in Tasmania.)

Variegated Thistle is highly esteemed in herbal medicine for its action on the liver. It also has been widely used as a vegetable. The young leaves, peeled stems, flower-heads and roots can all be cooked and eaten (this food use is reflected in one of the Australian names for it: Cabbage Thistle).

Origin

Variegated Thistle is native to southern Europe and western Asia.

Occurrence

Variegated Thistle is thought to have been introduced as a medicinal plant early last century, but it rapidly spread throughout southern Australia. It is mainly a weed of fertile soils. In some regions its occurrence is localised around stock camps and yards, but it can form dense infestations over large areas, usually as a weed of cultivation or pasture.

Identification

Variegated Thistle grows up to 2500 mm tall, commonly 900–1800 mm. The longitudinally grooved stems are usually much branched at the base. The leaves are large and broad, dark green with pronounced milky variegation, shiny above, duller below, spiny, deeply divided with undulating margins, up to 600 mm long in the rosette, shorter on the stem.

Florets are purple, in large heads up to 130 mm across including spines, and solitary at the ends of the branches. Seed is black or brown, 6–8 mm long.

The plants are annual or biennial and die off after setting seed.

Variegated Thistle is fairly easy to distinguish from other thistles by its large broad leaves with their milky markings. Its flower heads are larger than any other thistle, except for Artichoke Thistle, which has distinctly different leaves.

Management

There are no worries about depleting this species. No matter how you harvest the seed, a good proportion will find its way to the ground. In view of its noxious nature, it would be hard to justify any cultivation of Variegated Thistle, though it is grown as a crop overseas and has been grown in the past in Tasmania.

Some care should be taken not to introduce it to new locations, though the birds are doing a good job of this already.

Problems

Timing is critical for mechanical harvesting, as there is only a short period between the heads being dry and the seeds all floating away on the breeze.

In some regions rosellas and other parrots are very quick to attack the heads as soon as they are mature.

Harvesting Leaves

Variegated Thistle leaves are preferably used fresh and should be harvested when the plant is in flower. They are sometimes used to make a fresh plant tincture, but there is not a lot of demand for this part of the herb.

Harvesting Seed

Timing

Harvest is from late November to February, depending on the region and the season.

Hand Harvesting

Variegated Thistle can be taken when the seed has just started to shed, but if birds are a problem it is best to cut the heads somewhat earlier.

Birds, particularly those of the parrot family, sometimes rip into the heads as soon as the seeds have formed, leaving very few behind. This is more likely to be a problem in places where the stands of Variegated Thistle are small and localised. In this situation the heads should be cut as soon as the seeds start to turn dark.

Mechanical Harvesting

When using a header, Variegated Thistle should be harvested as soon as the majority of heads are mature, but before too much seed is shed. This timing needs to be fairly precise, as most of the seed can be lost in the space of a few days.

Tools

On a small scale, the heads can be cut with a reaping hook and threshed with a flail.

Large-scale harvesting can be done with an ordinary grain header.

Variegated Thistle (*Silybum marianum*).

Technique

On a small scale, the heads are harvested with a reaping hook, wearing gloves, of course. A basket with a tight enough weave to hold any falling seeds can be used as a container.

Drying

Heads that have been cut a little before maturity need to be left in a place where they can cure, but not dry out too quickly, so the sap can continue to flow into the seeds. A drying screen may be suitable, as long as conditions are not too hot and dry, in which case the heads should be placed in small heaps to slow down their drying. (This is the same principle as stooking grain.)

Vermin are attracted to any seed, so your control measures need to be very effective

Temperature

The curing process should take place at a temperature of not more than 30°C.

Processing

Once dry, the heads can be laid out on a large sheet and threshed with the flail. The resulting pile of broken bits of thistle then needs sifting and winnowing.

This can be a spiky job, which I found gave a new meaning to the old tongue twister: 'Theo the thistle sifter, while sifting a sift full of unsifted thistles, thrust three thousand thistles through the thick of his thumb'.

I had always thought Theo must have been rather daft to be sifting thistles, but in fact he was pursuing a most honourable trade, harvesting medicinal herbs.

By the time I had my thistle seed threshed and cleaned and I had spiked my thumb more than a few times, I found myself identifying quite strongly with Theo.

For more detail on small-scale threshing, sifting and winnowing, see 'Harvesting Seed' in Chapter 11, 'Harvesting'.

Tea Grade

Variegated Thistle should be milled for tea use, but the wholesaler usually does this.

Manufacturing Grade

For manufacturing, the whole seeds – cleaned of any trash – are required.

Storage

Freeze treatment to control moth larvae is necessary before sale or storage.

Yields

Small-scale hand-harvesting is only worth while if you are getting a premium price for the product, or if you require small quantities for a particular purpose. Yields from mechanical harvesting are rather variable and depend on the density of the stand, the timing of harvest, and how much seed has set.

Marketing

There is a market for quite large quantities of Variegated Thistle among manufacturers.

Price

Price for premium quality, organically grown Variegated Thistle seed is currently around $4–6/kg.

Price for manufacturing grade Variegated Thistle seed is currently around $3–4/kg.

Yarrow

Achillea millefolium L ASTERACEAE

Parts Used: Leaf & Flower, Aerial Parts

Other Name: Milfoil.

Status in Australia: INTRODUCED WEED

Yarrow is described in detail in Chapter 15, 'Spreading Herbs'.

Occurrence

In the cooler parts of southern Australia, Yarrow is found as a weed along roadsides and on unused land. It is very palatable to stock and does not usually survive in pastures.

Harvesting

Sometimes Yarrow can be found in stands that might be harvestable, but there is usually grass mixed with it. As it is a perennial which spreads by rhizomes, harvesting will not deplete it.

Problems

The best stands are usually on roadsides where it will be contaminated with lead fallout from traffic and possibly contaminated by spraying.

Where Yarrow is growing wild mixed with grass, it will require patient harvesting with a reaping hook.

For more detail on this herb see 'Information Charts' and Chapter 15, 'Spreading Herbs'.

22

Information Charts

Notes on Using the Information Charts

Basic information about the various herbs covered in this book has been gathered into the two charts on the following pages for easy reference. As space does not allow lengthy coverage of important aspects, the reader should also refer to the individual entries, which cover major herbs in full detail and the more important aspects of minor herbs.

INFORMATION CHART 1: GROWING

Explanation of Terms Used

Growth Type

Each herb is classified according to the growth types described in Chapter 6.

Climate Range

The climatic range refers to the herb's preferred climate, specifically the summer temperature range. Of course there will be some flexibility, as individual sites can result in some modification of microclimate, or some annual crops can be grown at a cooler or warmer times of the year. For conversion to Fahrenheit see page 556.

Cool: Cool climate includes the higher altitudes of Tasmania (above about 300 m) and some of the highest altitude regions of Victoria and New South Wales (above 1000–1300 m), where summer temperatures seldom rise much above 25°C.

Mild: Mild climate refers to the lower altitudes of Tasmania, some of the coastal, tableland and highland regions of Victoria, New South Wales, South Australia and the south-west tip of Western Australia, where summer temperatures don't often reach 35°C.

Warm: Warm climate generally refers to inland slopes, foothills and other regions with some moderation of summer heat due to altitude or ocean influence. Summer temperatures are normally in the mid-30s (°C).

Hot: Hot climate refers to the inland plains where summer temperatures are regularly over 40°C and more.

Sun

The amount of exposure to sun the species enjoys is noted. Naturally this will vary according to the climate. A species that does well in full sun in a cool climate may need partial shade in a warmer region.

Full: Full sun most of the day.

Part: This can be sun for part of the day, preferably morning, and shade for the rest of the day, or it can mean a 40–50% shadecloth covering during summer.

Wind Tolerance

Species vary greatly as to their tolerance of wind, though few actually require it.

Tolerant: Will tolerate an exposed windy site without too adverse an effect on growth or quality.

Moderate: Will tolerate some exposure to wind, but will suffer significantly in extreme conditions.

Needs shelter: Needs good protection from the wind for optimum growth.

Soil Type

Soil type refers to preferred topsoil texture. In some cases a crop will grow in heavier soils, but a light to medium soil is recommended to facilitate weed control.

Deep: Soil that is easily dug down to a depth of 500–1000 mm, usually for harvesting deep roots.

Drainage Required

This refers to how good the subsoil drainage needs to be to avoid damage or loss of crop from waterlogging.

Propagation

Method

Methods of propagation available, in order of normal preference. A method noted in brackets may be unreliable, difficult or otherwise unsatisfactory.

Time

Optimum time of planting will vary according to climate and site.

Rows Apart

Recommended distance between rows, in millimetres, except for larger spacings which are in metres (as noted).

Plants Apart

Recommended spacing between plants within the rows.

Continuous row

This refers to the placing of rhizomes end-to-end along the row.

Compost

Approximate compost requirements. These figures are intended as a relative guide only, as soil fertility varies so much. For metric conversion tables see page 556.

Total Water

Approximate weekly water requirement for optimum growth in summer, in terms of rainfall plus irrigation. This will vary according to climate, site, soil, stage of development and size of crop.

Weed Control

An indication of how demanding weed control will be in growing a crop.

Minimal

A crop which can be managed with minimal attention to weed control, usually by mowing, and perhaps in conjunction with some mulching close to the plant, as with trees.

Minimal once estab(lished)

As above, except that it may need more attention in the early stages.

Regular

Regular inter-row cultivation and hoeing around the plants will be needed to maintain adequate weed control. In some cases, other techniques will supplement it or be substituted for it.

Regular easy

As above, but because of the dense or dominant nature of the herb, weed control is generally easy to maintain.

Regular until estab(lished)

As for regular, but the need for weed control will diminish as the crop gets established.

Intensive

These crops require special techniques and/or greater amounts of time and labour to maintain adequate weed control.

Information Chart 1: *Growing*

Species	Growth Type	Climate Range	Sun	Wind Tolerance	Soil Type	Drainage	Propagation Method	Propagation Time	Rows Apart	Plants Apart	Compost kg/m²	Total Water mm/week	Weed Control	Comments
Agrimony *Agrimonia eupatoria*	Perennial Crown	cool–warm	full, part	tolerant of cold wind	wide range	good	division, seed	spring or autumn	900mm	150–300mm	2–3	25–50	regular easy	
Alfalfa *Medicago sativa*	Perennial Crown	cool–hot	full	moderate	light–medium	very good	direct seed	spring	broadcast or drills	1–2 g/m²	1–2	12–25	minimal once estab.	Good source of compost material.
Angelica *Angelica archangelica*	Biennial/Triennial	cool–mild	full, part	moderate	light–medium	good	seed	summer–autumn	900–1200mm	450–600mm	4–6	25–50	regular easy	Best results with fresh seed.
Aniseed *Pimpinella anisum*	Annual	mild–hot	full	needs shelter	wide range	good	direct seed	spring	300–450mm or drills	40–75mm	1–2	20–30	good prior control	Vulnerable to weeds.
Artichoke, Globe *Cynara scolymus*	Perennial Crown	cool–warm	full	moderate	light–medium	good	seed, division	spring or autumn	900–1200mm	600mm	3–4	20–30	regular easy	
Balm of Gilead *Populus candicans*	Tree	cool–warm	full, part	tolerant	wide range	poor–good	cutting, sucker	spring	4–6m	2–6m	1–2	15–25	minimal	
Basil *Ocimum basilicum*	Annual	warm–hot	full	needs shelter	wide range	good	seed	spring	300–450mm	100–150mm	4–5	25–50	regular	Frost sensitive.
Bay *Laurus nobilis*	Tree	cool–hot	full	moderate	avoid heavy	good	seed, cutting	spring or autumn	4–6m	2–6m	1–2	0–25	minimal once estab.	Plant as separate trees or as a hedge.
Bergamot *Monarda didyma*	Expanding Clump	cool–mild	full, part	needs shelter	light	very good	division, seed	spring or autumn	900mm	150–300mm	2–4	25–50	regular	Stems break off easily at the base.
Blessed Thistle *Cnicus benedictus*	Annual	cool–mild	full	tolerant	light–medium	good	direct seed	spring or autumn	600–900mm	150–300mm	1–3	20–30	regular	Winter crop in warmer regions.
Broom, English *Cytisus scoparius*	Shrub	cool–warm	full	tolerant	wide range	good	seed	spring or autumn					noxious weed	Wildcrafted or harvested weed only.
Burdock *Arctium lappa*	Annual	cool–warm	full	tolerant	light, deep	good	seed	spring or autumn	900–1200mm	75–150mm	4–6	25–40	regular easy	Needs deep easy-to-dig soil. Can be direct sown.
Calendula *Calendula officinalis*	Annual	cool–warm	full	tolerant	wide range	good	seed	spring or autumn	900mm	300–400mm	0–2	25–40	regular easy	Can be direct sown.
Caraway *Carum carvi*	Biennial	cool–mild	full	moderate	wide range	good	direct seed	spring	300–600mm or drills	40–75mm or drills	2–3	20–30	regular	
Cascara Sagrada *Rhamnus purshiana*	Tree	cool–warm	full, part	moderate	light–medium	good	cutting, seed	spring or autumn	2–3m	1–2m	2–3	12–25	minimal once estab.	
Catnip *Nepeta cataria*	Short-lived Perennial	cool–warm	full	tolerant	wide range	good	seed (division)	spring or autumn	900mm	300mm	3–4	20–30	regular easy	Longer-lived in cooler climates

Information Chart 1: *Growing*

Species	Growth Type	Climate Range	Sun	Wind Tolerance	Soil Type	Drainage	Propagation Method	Propagation Time	Rows Apart	Plants Apart	Compost kg/m²	Total Water mm/week	Weed control	Comments
Celandine, Greater *Chelidonium majus*	Perennial Crown	cool–mild	full, part	needs shelter	wide range	good	division, seed	spring or autumn	900mm	300mm	2–3	25–50	regular	Yellow latex can irritate skin.
Celery *Apium graveolens*	Annual/Biennial	cool–warm	full	needs shelter	light–medium	good	seed	spring	300–600mm or drills	100–200mm	0–2	20–30	regular to intensive	Might be direct sown.
Centaury *Erythraea centaurium*	Annual	cool–warm	full	tolerant	wide range	good	seed							Grows wild, but good stands rare. Cultivation possible.
Chamomile, German *Chamomilla recutita*	Annual	cool–mild	full	needs shelter	light–medium	good	seed	flexible	500–700mm	225–300mm	0–2	25–40	regular	Plant at 6–8 week intervals. Self-sows heavily.
Chamomile, Roman *Chamaemelum nobile*	Expanding Clump	cool–warm	full, part	moderate	light–medium	good	division, seed	spring or autumn	900mm	150–300mm	1–2	25–40	regular easy	
Chaste Tree *Vitex agnus-castus*	Tree/Shrub	mild–warm	full, part	moderate	light–medium	good	seed, cutting	spring or autumn	4m	2–3m	1–3	20–30	regular until estab.	
Chickweed *Stellaria media*	Annual	cool–mild	full, part	tolerant of cold wind	wide range	poor–good	seed							Harvested weed. Winter growth.
Chicory *Cichorium intybus*	Short-lived Perennial	cool–mild	full	tolerant	light, deep	good	direct seed	spring	600mm	150mm	4–5	20–30	regular easy	
Chives *Allium schoenoprasum*	Perennial Crown	cool–warm	full	tolerant	wide range	good	division, seed	spring or autumn	450–600mm	150–300mm	4–5	25–50	regular	Weeding must be thorough to control grass in clumps.
Cleavers *Galium aparine*	Annual	cool–mild	full, part	needs shelter	wide range	good	seed							Harvested weed. Winter growth in warm regions.
Coltsfoot *Tussilago farfara*	Spreading	cool–mild	full, part	needs shelter	wide range	poor–good	rhizome	spring or autumn	600–900mm	continuous row	4–5	25–50	regular to intensive	Needs shade in warmer regions.
Comfrey, English *Symphytum officinale*	Expanding Clump	cool–hot	full	moderate	wide range	poor–good	root/crown division	spring or autumn	900mm	300mm	4–5	25–50	regular easy	Hard to get rid of.
Coriander *Coriandrum sativum*	Annual	cool–warm	full	needs shelter	light–medium	good	direct seed	spring	300–450mm	20–50mm	0–3	20–30	regular	The higher compost rate is needed for leaf production.
Couchgrass, English *Elymus repens*	Spreading	cool–mild	full	tolerant	light–medium	good	rhizome (seed)	spring or autumn	900mm	continuous row	3–5	20–30	intensive	Problem weed: good containment needed.
Crampbark *Viburnum opulus*	Tree/Shrub	cool–mild	full, part	tolerant of cold wind	light–medium	good	cutting, seed	spring or autumn	2–3m	1–2m	2–3	12–25	minimal once estab.	Also propagate by layering.
Dandelion *Taraxacum officinale*	Expanding Clump	cool–warm	full, part	tolerant of cold wind	light–medium	good	seed, root div.	spring or autumn	600mm	300mm	1–3	25–50	regular to intensive	Avoid excessive fertilisation.

Information Chart 1: *Growing*

Species	Growth Type	Climate Range	Sun	Wind Tolerance	Soil Type	Drainage	Propagation Method	Propagation Time	Rows Apart	Plants Apart	Compost kg/m²	Total Water mm/week	Weed control	Comments
Dill *Anethum graveolens*	Annual	mild–warm	full	needs shelter	wide range	good	direct seed	spring to autumn	300–450mm	20–50mm	1–4	25–50	regular	Aphids can be a problem. Plants are delicate.
Dock, Yellow *Rumex crispus*	Expanding Clump	cooler warm	full	tolerant	wide range	poor–good	seed (root div.)	spring or autumn	600mm	150–300mm	3–4	20–30	regular	Harvested weed. Cultivation may be worth while.
Echinacea, Broad-leaf *Echinacea purpurea*	Short-lived Perennial	cool–warm	full	tolerant	light–medium	good	seed (division)	spring or autumn	600–900mm	300mm	3–5	20–30	regular	Slow early growth.
Echinacea, Narrow-leaf *Echinacea angustifolia*	Perennial Crown	mild–hot	full	tolerant	light, deep	very good	seed (division)	spring or autumn	600–900mm	200–300mm	2–3	10–25	regular	Can be a difficult crop. Slow growing.
Elder, Black *Sambucus nigra*	Tree	cool–mild	full, part	moderate	wide range	poor–good	cutting; seed:	winter spring	4–5m	2–4m	3–4	20–30	minimal	Wildcrafted/harvested weed or planted tree.
Elecampane *Inula helenium*	Perennial Crown	cool–warm	full	moderate	wide range	poor–good	crown div., (seed)	spring	900mm	300–450mm	3–4	25–40	regular easy	
Eucalyptus (Blue Gum) *Eucalyptus globulus*	Tree	cool–mild	full	tolerant	wide range	poor–good	seed	spring	6m	4–6m	0–1	0–25	minimal	Usually wildcrafted.
Fennel *Foeniculum vulgare*	Perennial Crown	cool–hot	full	tolerant	wide range	good	direct seed	spring	300–600mm	50–100mm	0–2	10–25	regular	Harvested weed. Easy to grow.
Feverfew *Tanacetum parthenium*	Woody Perennial	cool–warm	full	tolerant	wide range	good	seed, cutting	spring or autumn	900mm	300mm	2–3	20–30	regular	Early growth not woody.
Figwort *Scrophularia nodosa*	Perennial Crown	cool–warm	full	tolerant of cold wind	light–medium	good	division, seed	spring or autumn	900mm	300mm	3–4	25–50	regular easy	
Garlic *Allium sativum*	Perennial Crown	cool–warm	full	tolerant	light–medium	good	division	autumn	300–450mm	100–150mm	1–2	10–25	regular to intensive	Winter crop. Needs dry conditions towards maturity.
Ginkgo *Ginkgo biloba*	Tree	cool–warm	full range	moderate	wide	good	seed, cutting	spring	6m	6m	0–2	12–25	minimal	Locate where falling leaves can be gathered.
Ginseng *Panax spp.*	Perennial Crown	cool–mild	80% shade	needs shelter	light–medium	very good	direct seed	autumn	600mm	300mm	0	20–30	regular	Very slow difficult crop. 18 months seed dormancy.
Gipsywort *Lycopus virginicus*	Spreading	cool–mild	full, part	needs shelter	light–medium	poor–good	rhizome, seed	spring or autumn	900mm	continuous row	3–4	25–50	regular to intensive	Spreads to form a meadow-crop.
Goat's Rue *Galega officinalis*	Perennial Crown	cool–warm	full	moderate	wide range	good	division, seed	spring or autumn	900mm	300–450mm	2–3	25–50	regular easy	
Golden Seal *Hydrastis canadensis*	Spreading	cool–mild	80% shade	needs shelter	light–medium	very good	rhizome, root div.	autumn	600mm	300mm	0–1	20–30	intensive	Very slow growing difficult crop.

Information Chart 1: *Growing*

Species	Growth Type	Climate Range	Sun	Wind Tolerance	Soil Type	Drainage	Propagation Method	Propagation Time	Rows Apart	Plants Apart	Compost kg/m²	Total Water mm/week	Weed control	Comments
Ground Ivy *Glechoma hederacea*	Spreading	cool–mild	full, part	needs shelter	light–medium	good	stolon	spring	600–900mm	150mm	3–4	25–40	intensive	Spreads to form a meadow crop.
Gum Plant *Grindelia robusta*	Woody Perennial	cool–hot	full	tolerant	light–medium	good	seed, division	spring or autumn	900mm	300mm	1–2	20–30	regular	Early growth not woody.
Hawthorn *Crataegus monogyna*	Tree	cool–warm	full	tolerant	wide range	poor–good	cutting, (seed)	autumn	4–6m or hedge	1–4	1–2	10–25	minimal	Seed needs stratification. Wildcrafted/harvested weed.
Hops *Humulus lupulus*	Expanding Clump	cool–warm	full	needs shelter	wide range	good	cutting, sucker	spring or autumn	1.2–3.0m	900–1200mm	4–7	25–50	minimal once estab.	Needs provision for vines to climb.
Horehound, Black *Ballota nigra*	Perennial Crown	cool–mild	full	tolerant	light–medium	good	division, seed	spring or autumn	900mm	225–450mm	2–3	25–50	regular easy	
Horehound, White *Marrubium vulgare*	Perennial Crown	cool–hot	full	tolerant	wide range	good	seed, division						Noxious weed	Caution: burrs attach to cloth and fleece.
Horse Chestnut *Aesculus hippocastanum*	Tree	cool–mild	full	needs shelter	wide range	good	seed	autumn	8–10m	8–10m	2–3	12–25	minimal	Plant seed early in autumn.
Horseradish *Armoracia rusticana*	Spreading	cool–mild	full	moderate	wide range	poor–good	root div.	spring or autumn	900mm	300mm	4–5	25–50	regular	Spreads to form a meadow-crop
Horsetail, Field *Equisetum arvense*	Spreading	cool–mild	part (full)	needs shelter	light–medium	poor–good	rhizome	spring	900–1200mm	continuous row	2–3	25–50	intensive	May need shade. Spreads: good containment essential.
Hyssop *Hyssopus officinalis*	Woody Perennial	cool–warm	full	tolerant	light–medium	very good	division, seed	spring or autumn	1800mm	150–300mm	2–3	20–30	regular easy	
Lady's Mantle *Alchemilla vulgaris*	Perennial Crown	cool–mild	full, part	moderate	wide range	good	division, seed	spring or autumn	900mm	300mm	3–4	25–50	regular	
Lavender, English *Lavandula angustifolia*	Woody Perennial	cool–warm	full	moderate	light–medium	good	cutting, layering	spring or autumn	1200–2000mm	750–1000mm	2–3	10–30	regular easy	Will not flower in humid sub-tropical regions.
Lemon Grass *Cymbopogon citratus*	Perennial Crown	warm–hot	full	moderate	wide range	good	division	spring	900mm	300–450mm	4–5	25–50	regular easy	Frost sensitive.
Lemon Thyme *Thymus x citriodorus*	Woody Perennial	cool–warm	full	tolerant	light–medium	good	division, layering	spring or autumn	600mm	150mm	2–3	20–30	regular until established	Develops into a solid stand.
Lemon Verbena *Aloysia triphylla*	Tree	cool–warm	full	needs shelter	wide range	very good	cutting	variable	4m	2–3m	2–3	20–30	regular until established	Propagation can be difficult.
Licorice *Glycyrrhiza glabra*	Spreading	mild–hot	full	moderate	light, deep	good	rhizome, (seed)	spring	900–1200	300mm	5–7	20–30	regular	Spreads to form a meadow-crop.

Information Chart 1: *Growing*

Species	Growth Type	Climate Range	Sun	Wind Tolerance	Soil Type	Drainage	Propagation Method	Propagation Time	Rows Apart mm	Plants Apart mm	Compost kg/m²	Total Water mm/week	Weed control	Comments
Linden *Tilia spp.*	Tree	cool–mild	full	moderate	wide range	good	seed, cutting	late summ. autumn	10–20m	10–20m	1–2m	12–25	minimal	
Marjoram, Sweet *Origanum majorana*	Woody Perennial	cool–mild	full	moderate	light–medium	good	division, seed	spring or autumn	600mm	225–300mm	3–4	20–30	regular	
Marshmallow *Althaea officinalis*	Perennial Crown	cool–mild	full	tolerant	wide range	poor–good	division, seed	spring	900mm	450–600mm	3–4	25–50	regular easy	
Meadowsweet *Filipendula ulmaria*	Perennial Crown	cool–mild	full, part	moderate	wide range	good	division, seed	spring or autumn	900mm	150–300mm	3–4	25–50	regular	
Melissa Balm *Melissa officinalis*	Expanding Clump	cool–mild	full, part	needs shelter	wide range	poor–good	division, seed	spring or autumn	900mm	150mm	2–4	25–50	regular	Suffers from two-spotted mite if under stress.
Motherwort *Leonurus cardiaca*	Perennial Crown	cool–warm	full	tolerant	wide range	good	seed, division	spring or autumn	900mm	300mm	2–3	25–40	regular easy	
Mountain Pepper *Tasmannia lanceolata*	Tree/Shrub	cool	full, part	moderate	wide range	good	seed, cutting	spring summer	3–4m	2–3m	0–1	25–40	minimal	Wildcrafted native species; will grow in cultivation
Mugwort *Artemisia vulgare*	Expanding Clump	cool–hot	full	tolerant	wide range	good	division, rhizome	spring or autumn	900–1200mm	150–300mm	2–3	25–45	regular easy	Grows very tall.
Mullein, Great *Verbascum thapsus*	Biennial	cool–warm	full	tolerant	light–medium	good	seed	spring or autumn	600–900mm	300–450mm	0–3	12–25	regular easy	Harvested weed. Can be cultivated.
Nettle, Greater *Urtica dioica*	Expanding Clump	cool–warm	full	tolerant	light–medium	good	rhizome, (seed)	spring or autumn	900–1200mm	continuous row	4–6	25–50	regular	Expands to form a meadow-crop.
Nettle, Lesser *Urtica urens*	Annual	cool–mild	full	tolerant	light–medium	good	seed				very fertile			Harvested weed. Not recommended for cultivation
Oats *Avena sativa*	Annual	cool–mild	full	tolerant	wide range	fairly good	direct seed	autumn	broadcast or drilled	7–11g/m²	0–2	12–25	prior cultivation	Winter crop in most regions.
Oregano *Origanum vulgare*	Expanding Clump	cool–hot	full	tolerant	light–medium	good	division, seed	spring or autumn	600–900mm	150–300mm	2–3	20–30	regular	Transplant with care.
Parsley *Petroselinum crispum*	Biennial/Annual	cool–warm	full	tolerant	light–medium	good	seed	spring	600mm	300mm	4–5	25–40	regular	Curly variety biennial, Italian variety annual.
Pasque Flower *Anemone pulsatilla*	Perennial Crown	cool–mild	full	tolerant	light–medium	good	division, (seed)	spring or autumn	600–900mm	150–300mm	2–3	25–40	regular	
Passionflower *Passiflora incarnata*	Spreading	mild–hot	full	needs shelter	wide range	good	shoot, seed	spring	1.5–2.0m	300–600mm	3–4	20–30	minimal once estab.	Mulching works well for weed control.

Information Chart 1: *Growing*

Species	Growth Type	Climate Range	Sun	Wind Tolerance	Soil Type	Drainage	Propagation Method	Propagation Time	Rows Apart	Plants Apart	Compost kg/m²	Total Water mm/week	Weed control	Comments
Pennyroyal *Mentha pulegium*	Spreading	cool–warm	full	tolerant	wide range	wet sites	stolon, (seed)	spring or autumn	900mm	continuous row	3–4	25–50	regular to intensive	
Peppermint *Mentha x piperita*	Spreading	cool–hot	full	tolerant	light–medium	fairly good	rhizome	spring or autumn	900–1200mm	continuous row	5–7	25–50	intensive	Needs good weed control. Subject to rust.
Periwinkle *Vinca major*	Spreading	mild–warm	full	moderate	wide range	good	stolon							Harvested weed.
Plantain, Greater *Plantago major*	Perennial Crown	cool–mild	full, part	needs shelter	wide range	poor–good	seed	spring or autumn	450mm	150mm	3–4	25–50	regular to intensive	Self-sows thickly.
Plantain, Narrow–leaf *Plantago lanceolata*	Perennial Crown	cool–warm	full	tolerant	wide range	fairly good	seed							Harvested weed.
Poke *Phytolacca decandra*	Perennial Crown	warm–hot	full	moderate	light, deep	good	crown div., seed	spring	900mm	450mm	3–4	25–40	regular easy	Caution: Toxic. Can become a problem weed.
Raspberry *Rubus idaeus*	Expanding Clump	cool–mild	full, part	needs shelter	light–medium	good	division	winter	3–4m or 1.2m	450–600mm	3–4	25–50	regular until estab.	Rows can be 1.2 m to establish a meadow-crop.
Red Clover *Trifolium pratense*	Perennial Crown	cool–warm	full	tolerant	wide range	good	seed	spring or autumn	broadcast	1g m²	0–3	20–30	minimal	Good soil builder and source of compost material.
Rose, Dog *Rosa canina*	Shrub	cool–warm	full	tolerant	wide range	good	cutting, (seed)	autumn	3–4m	2–3m	2–3	12–25	regular until estab.	Occasionally found wild. Preferred species of Rose hip.
Rose, Sweet Briar *Rosa rubiginosa*	Shrub	cool–warm	full	tolerant	wide range	good	seed							Wildcrafted/Harvested weed.
Rosemary *Rosmarinus officinalis*	Woody Perennial	cool–hot	full	tolerant	light–medium	good	cutting, (seed)	autumn	1800mm	600mm	1–2	10–20	regular until estab.	
Rue *Ruta graveolens*	Woody Perennial	mild–hot	full	tolerant	wide range	good	seed, cutting	spring or autumn	900mm	300–450mm	2–3	12–25	regular until estab.	
Sage *Salvia officinalis*	Woody Perennial	cool–hot	full	tolerant	wide range	good	seed, division	spring or autumn	600–900mm	150–300mm	2–3	20–30	regular	Can also be grown from cuttings and layering.
Scullcap *Scutellaria lateriflora*	Expanding Clump	cool–warm	full	needs shelter	light–medium	very good	rhizome, seed	spring or autumn	900mm	300mm	2–3	25–40	regular	Stems break easily. Transplant with care.
Shepherd's Purse *Capsella bursa-pastoris*	Annual	cool–warm	full	tolerant	wide range	good	seed	spring or autumn						Harvested weed of cultivation.
Sorrel, Sheep *Rumex acetosella R. angiocarpus*	Spreading	cool–warm	full	tolerant	light–medium	poor–good	seed, root div.	spring or autumn	600 mm	150 mm	2–3	20–30	intensive	Harvested weed. Could be cultivated.

Information Chart 1: *Growing*

Species	Growth Type	Climate Range	Sun	Wind Tolerance	Soil Type	Drainage	Propagation Method	Propagation Time	Rows Apart	Plants Apart	Compost kg/m²	Total Water mm/week	Weed control	Comments
Spearmint *Mentha spicata*	Spreading	cool–hot	full	tolerant	light–medium	poor–good	rhizome	spring or autumn	900 mm	continuous row	5–7	25–50	intensive	Needs good weed control. Subject to rust.
St John's Wort *Hypericum perforatum*	Spreading	cool–hot	full	tolerant	wide range	poor–good	stolon	spring or autumn	900mm	continuous row	2–3	12–25	regular	Harvested weed: noxious. Can be very invasive.
Stoneroot *Collinsonia canadensis*	Perennial Crown	cool–mild	full	needs shelter	wide range	good	division, seed	spring	600–900 mm	300 mm	2–3	25–40	regular	Stems break easily.
Tansy *Tanacetum vulgare*	Spreading	cool–hot	full	tolerant	wide range	good	rhizome, seed	spring or autumn	900–1200 mm	continuous row	1–3	20–30	regular easy	
Tarragon, French *Artemisia dracunculus*	Expanding Clump	cool–mild	full	tolerant	light–medium	good	rhizome, division	spring or autumn	750–900 mm	150–300 mm	3–4	25–40	regular to intensive	
Thyme *Thymus vulgaris*	Woody Perennial	cool–warm	full	tolerant	light–medium	good	div., layer, seed	spring or autumn	600 mm	150 mm	2–3	20–30	regular until estab.	Develops into a solid stand
Valerian *Valeriana officinalis*	Expanding Clump	cool–mild	full, part	needs shelter	light	poor–good	rhizome, seed	spring or autumn	900 mm	150–300 mm	2–3	25–50	regular	Soil must wash easily and be free of small stones.
Variegated Thistle *Silybum marianum*	Biennial	mild–hot	full	tolerant	wide range	good	seed				fertile			Harvested weed.
Vervain *Verbena officinalis*	Perennial Crown	mild–hot	full	tolerant	light–medium	good	seed, division	spring or autumn	900 mm	150–300 mm	2–3	25–40	regular	Self-sows.
Violet, Sweet *Viola odorata*	Expanding Clump	cool–mild	part	needs shelter	light–medium	good	stolon	spring or autumn	600–900 mm	continuous row	2–3	25–50	regular	May need shade.
Willow *Salix* spp.	Tree	cool–warm	full	moderate	wide range	poor–good	cutting	winter	3–6 m	3–6 m	0–2	20–30	minimal	
Witch Hazel *Hamamelis virginiana*	Tree/Shrub	cool–mild	full	needs shelter	wide range	good	cutting (seed)	autumn	3 m	2–3 m	2–3	20–30	minimal once estab.	
Wood Betony *Stachys officinalis*	Perennial Crown	cool–mild	full	moderate	wide range	good	division, seed	spring or autumn	600–900 mm	150–300 mm	3–4	25–40	regular	
Wormwood *Artemisia absinthium*	Perennial Crown	cool–hot	full	moderate	wide range	good	division, seed	spring or autumn	900 mm	450 mm	2–3	20–30	regular easy	
Yarrow *Achillea millefolium*	Spreading	cool–warm	full	tolerant	wide range	fairly good	rhizome, seed	spring or autumn	900–1200 mm	continuous row	3–4	25–40	regular	Needs rejuvenation every 2–3 years.

INFORMATION CHART 2: HARVESTING, DRYING, PRICES AND MARKETING

Explanation of Terms Used

Part Used

The part(s) of the plant commonly used. Where more than one part is listed, the details applicable to each part are listed on the same line.

A.P.

Aerial Parts, or the whole of the above-ground plant. Often the lower leafless stalky part of the plant is excluded. Aerial parts are normally coarsely chopped for a manufacturing grade.

A.P. Tea Gde

Aerial Parts Tea Grade: in a few cases the stem is included in tea grade, for which it needs to be finely chopped or milled.

Bark

The live inner bark of the tree or shrub.

Cones: In the case of Hops, it is the cone-like strobile or fruiting body that is harvested.

Drd. Grn. Plnt

Dried Green Plant: in the case of Oats aerial parts, this term is used to distinguish it from straw or seed.

Flower

The flowers or flower heads, at the early blossom stage.

Fruit

The fruit or berries, usually at an early stage of maturity.

Fr.Whl.Plnt.

Fresh Whole Plant: the aerial parts plus the roots are harvested and used in the fresh state.

Leaf

Leaf alone is required, though usually this is separated from the stems after drying. Sometimes the leaf includes a small proportion of flowers as well.

Lf & Flwr

Leaf and flower: when the herb normally contains the flowers as well as leaf.

Lf (& Flwr): As for Lf & Flwr, except that some harvests may not contain much, if any, flower.

Root

The below ground parts of the plant: crown, roots and rhizomes, if present.

Seed

The ripe seed. Sometimes these are technically fruits.

Harvest

The normal range of harvesting times. This may vary somewhat according to region, season and time of planting.

Drying

Generally an upper temperature limit is specified. For optimum drying, the temperature should be somewhere approaching this limit. For conversion to Fahrenheit see page 556.

Yield kg/m²/ann

This is the annual dry weight yield, expressed as kilograms per square metre in production (this includes the cultivated area around the plants). To convert to tonnes per hectare simply multiply by 10. For metric conversion tables see page 556.

Where several harvests are taken during the season, the figures given are a total annual yield.

The yields listed are an indication of what can reasonably be expected from a good crop once it is established. The figures are generally based on the experience of local organic growers. Many perennial crops will take one or two seasons to reach full production. Actual yields will vary in response to a number of different factors.

Yields and rates of harvest for wildcrafted herbs and harvested weeds are too variable to list here.

(averaged)

Where a root crop takes two or more seasons to reach a harvestable stage, the figure listed is the total yield divided by the number of years taken to achieve it. This gives an annual return that can be compared with the annual yields of other crops.

Trade Herbs

Price/kg

This is in Australian dollars and represents the 1995 price charged by importers for non-organic herbs generally sourced from overseas. Bear in mind that these are spot prices and can fluctuate somewhat. They give an indication of the prices that wholesalers pay for poor to mediocre quality herbs.

Certified Organically Grown and Wildcrafted Herbs

Certified organic and wildcrafted herbs are dealt with in two categories:

Manufacturing Grade

This category refers to a level of quality and processing suitable for manufacturing liquid medicines and other products.

Premium Tea & Culinary Grade

This category refers to the highest quality possible, with maximum colour aroma and flavour, processed to a form suitable for tea or for culinary use.

Volume

As many manufacturers and wholesalers are very secretive about how much and what quality of herb they use, it is not possible to give more than a rubbery indication of market volumes.

The overall quantity of herbs used in the organic manufacturing sector is currently significantly greater than that used in the Premium Tea and Culinary sectors.

Consequently they are not directly comparable: a 'moderate' in the Manufacturing column might represent several times the volume of a 'moderate' in the Premium Grade column (though an increased availability of premium grade could see this sector increase to a size comparable to the manufacturing sector).

Price/kg

These are prices to growers and vary according to quality, availability and demand. The higher prices of the premium grade market may only be obtainable for a limited volume of some herbs from wholesalers who specialise in this level of quality. Any additional amount of it may have to be sold at lower manufacturing-grade prices.

Gross Return

This is a range of figures in Australian dollars that gives an indication of the return possible for organically grown herbs per square metre of production (to convert to return per hectare, multiply by 10 000). The lower end of the range represents the return obtained from the lower yield and sold at the lower price listed, while the higher end of the range represents the higher yield, sold at the highest price obtainable.

The actual gross return obtained will depend on who the grower is selling to and the quality of the herb. As mentioned above, the higher end of the return scale may only be obtainable for a limited volume of a particular herb.

Where a market exists for both aerial parts and tea-grade leaf of a herb, separate figures allow a realistic comparison of returns to be made.

Note: To save space, the symbol '~' has been used to denote an approximate figure.

For metric conversion tables see page 556.

Information Chart 2: Harvesting, Drying, Prices and Marketing

Species	Part Used	Harvest	Drying	Yield kg/m²/annum	Trade Herbs Price/kg	Manufacturing Grade Volume	Manufacturing Grade Price/kg $	Manufacturing Grade Gross Return $/m²/annum	Premium Tea & Culinary Grade Volume	Premium Tea & Culinary Grade Price/kg $	Premium Tea & Culinary Grade Gross Return $/m²/annum	Comments
Agrimony *Agrimonia eupatoria*	Aerial Parts Lf & Flwr	Nov.–Mar. "	to 35°C "	0.4–0.5 0.25–0.35	9–14	small–mod.	12–15	4.80–7.50	small	20–24	5.00–8.40	
Alfalfa *Medicago sativa*	Aerial Parts Lf & Flwr	Oct.–May "	to 45°C "	0.8–1.0 0.5–0.7	7	fairly large	3–8	2.40–8.00	moderate	13–16	6.50–11.20	
Angelica *Angelica archangelica*	Root	Jun.–Aug.	to 35°C	0.2–0.3	15	moderate	15–17	3.00–5.10	small	25–35	5.00–10.50	
Aniseed *Pimpinella anisum*	Seed	variable	to 35°C	around 0.1	3–5	very small	3–5	0.30–0.50	small	4–7	0.40–0.70	
Artichoke, Globe *Cynara scolymus*	Leaf	spring	to 35°C	quite low	9	moderate	13–18					Needs chopping before drying.
Balm of Gilead *Populus candicans*	Winter Bud	Jun.–Aug.	to 35°C			very small	12–18					
Basil *Ocimum basilicum*	Leaf	Dec.–Apr.	to 35°C	0.2–0.3	3–4				moderate	20–24	4.00–7.20	
Bay *Laurus nobilis*	Leaf	Jan.–Aug.	to 35°C	10–20 kg/ mature tree	4				moderate	~30	variable	
Bergamot *Monarda didyma*	Leaf (& Flwr)	Nov.–May	to 45°C	0.2–0.4					small	24	4.80–9.60	
Blessed Thistle *Cnicus benedictus*	Aerial Parts	Oct.–Nov.	to 45°C	0.3–0.45	8–9	small	12–15	3.60–6.75				
Broom, English *Cytisus scoparius*	Aerial Parts	spring	to 45°C			very small	10					Wildcrafted or harvested weed.
Burdock *Arctium lappa*	Root Seed	Jun.–Aug. Feb.–Apr.	to 45°C to 30°C	0.3–0.5	10–12	moderate	12–15	3.60–7.50	small	24	7.20–12.00	Digging can be challenging.
Calendula *Calendula officinalis*	Flower	spring to autumn	to 35°C	0.1–0.25	7	moderate	15–22	1.50–5.50	moderate	22–30	5.50–7.50	
Caraway *Carum carvi*	Seed	late spring–summer	to 35°C	0.05–0.2 (in Europe)	4–6		6–8	0.30–1.60	potential	10–15	0.50–3.00	
Cascara Sagrada *Rhamnus purshiana*	Bark	early spring	to 35°C		13	small–mod.	12–17					Age dried bark at least 6 months.
Catnip *Nepeta cataria*	Aerial Parts Lf & Flwr	Dec.–May "	to 35°C "	0.4–0.6 0.3–0.4	13–14	small "	7–8 11–15	2.80–4.80 3.30–6.00	very small	24	7.20–9.60	

Information Chart 2: Harvesting, Drying, Prices and Marketing

Species	Part Used	Harvest	Drying	Yield kg/m²/annum	Trade Herbs Price/kg	Certified Organically Grown or Wildcrafted Herbs							Comments
						Manufacturing Grade			Premium Tea & Culinary Grade			
						Volume	Price/kg $	Gross Return $/$/m²/annum	Volume	Price/kg $	Gross Return $/m²/annum	
Celandine, Greater *Chelidonium majus*	Aerial Parts	Dec.–May	to 35°C	0.25–0.35	10	small	11–20	2.75–7.00				
Celery *Apium graveolens*	Seed	variable	to 35°C		1–3	moderate	7–10					
Centaury *Centaurium erythraea*	Aerial Parts	Dec.–Jan.	to 35°C		13–17	very small	13–17					Wildcrafted or harvested weed.
Chamomile, German *Chamomilla recutita*	Flower	Oct.–May	to 35°C	0.1–0.2	10–12	large	10–18	1.00–3.60	large	10–50	4.00–10.00	
Chamomile, Roman *Chamaemelum nobile*	Flower	Dec.–Jan.	to 35°C	quite low	30–35	very little used						
Chaste Tree *Vitex agnus-castus*	Fruit	May	to 35°C		8–9	large	5–10					Yields in trials have been quite low.
Chickweed *Stellaria media*	A.P. dried A.P. fresh	winter, spring "	to 45°C		10–15	very small small–mod.	10–22					Weed of cultivation
Chicory *Cichorium intybus*	Root		to 45°C	0.4–0.5	6–8				small	15–20	6.00–10.00	Unroasted price.
Chives *Allium schoenoprasum*	Leaf	spring to autumn	to 35°C		12–14				very small	24		Chop finely before drying.
Cleavers *Galium aparine*	Aerial Parts	spring	to 35°C		10	moderate	10–15					Wildcrafted or harvested weed.
Coltsfoot *Tussilago farfara*	Leaf	Dec.–May	to 35°C	0.5–0.4	10				very small	25		Scheduled: sale restricted.
Comfrey *Symphytum officinale*	Aerial Parts Root	Nov.–May winter	to 35°C to 45°C	around 0.5 0.4–0.5	11 7	small	10	around 5.00				Scheduled: external use only.
Coriander *Coriandrum sativum*	Seed Leaf	variable "	to 35°C "	0.1–0.2	2				small small	uncertain 24		
Couch Grass, English *Cynodon dactylon*	Rhizome	winter	to 45°C		10	moderate	8–15		small	about 20		Harvestable weed.
Crampbark *Viburnum opulus*	Bark	early spring	to 45°C		17	moderate	15–35					
Dandelion *Taraxacum officinale*	Leaf Root	Nov.–May winter	to 45°C "	0.1–0.2 0.2–0.3	9 12–13	moderate very large	10–15 10–17	1.00–3.00 2.00–5.10	small large	30 30	3.00–6.00 6.00–9.00	

Information Chart 2: Harvesting, Drying, Prices and Marketing

Species	Part Used	Harvest	Drying	Yield kg/m²/annum	Trade Herbs Price/kg	Manufacturing Grade Volume	Price/kg $	Gross Return $/m²/annum	Premium Tea & Culinary Grade Volume	Price/kg $	Gross Return $/m²/annum	Comments
Dill *Anethum graveolens*	Seed	variable	to 35°C		2	very small	2–5		small	4–6		
	Leaf	"	"						small	24	around 4.80	
Dock *Rumex crispus*	Root	winter	to 45°C	around 0.2	13–14	moderate	10–15		small	15–20		Harvested weed.
Echinacea, Broad–leaf *Echinacea purpurea*	Aerial Parts	early flwr	to 35°C	0.4–0.7		moderate	~8	~3.20–5.60	small	12–15	6.00–11.25	
	Root	winter	"	around 0.2		very large	30–35	~6.00–7.00	moderate	30–35	~6.00–7.00	
	Fr. Whl. Plnt.	early flwr		1.5–2.5 fresh wt.		large	4–6	6.00–15.00				
Echinacea, Narrow–leaf *Echinacea angustifolia*	Root	winter	to 35°C	0.05–0.1 (averaged)		very large	45–80	2.25–8.00	small	60–80	3.00–8.00	Slow growing, 2 yrs to harvest. Difficult to harvest.
Elder, Black *Sambucus nigra*	Flower	Oct.–Jan.	to 35°C		11	moderate	10–20		small–mod.	40–45		Some wildcrafted.
Elecampane *Inula helenium*	Root	winter	to 35°C	0.3–0.4	9–11	moderate	10–15	3.00–6.00	very small	24	7.20–9.60	
Eucalyptus (Blue Gum) *Eucalyptus globulus*	Leaf	Jan.–Aug.	to 35°C		6 (overseas)				very small	14		Wildcrafted.
Fennel *Foeniculum vulgare*	Seed	Feb.–May	to 35°C	around 0.5	2	mod–large	4–10		small–mod.	~10		Harvested weed.
Feverfew *Tanacetum parthenium*	Aerial Parts	Nov.–May	to 35°C	0.4–0.8	22	moderate	5–20	2.00–16.00	small	24	7.20–12.00	Market can be volatile.
	Lf & Flwr	"		0.3–0.5								
Figwort *Scrophularia nodosa*	Aerial Parts	Dec.–May	to 35°C			small	10–16	6.00–8.00				
Garlic *Allium sativum*	granules	Nov.–Dec.	to 35°C	~0.2 dried			~8 (fresh)		small	20–25	~4.00–5.00	Dried Garlic needs special facilities. Also used fresh.
	corm	"								~8 (fresh)		
Ginkgo *Ginkgo biloba*	Leaf	autumn	to 35°C		17	moderate	9–18					Harvest falling leaves.
Ginseng *Panax spp.*	Root	winter	to 35°C	0–0.015 (averaged)	50–150	moderate	50–150	0.00–2.25	moderate	150–400	0.00–6.00	Difficult crop with low success rate. 5–7 yrs to harvest.
Gipsywort *Lycopus virginicus*	Aerial Parts	Nov.–May	to 35°C	0.3–0.6	18	small	11–15	3.30–9.00				
Goat's Rue *Galega officinalis*	Aerial Parts	Dec.–May	to 45°C	0.4–0.55	9	small	10–15	4.00–8.25				
Golden Seal *Hydrastis canadensis*	Root	winter	to 35°C	0–0.05 (averaged)	250	large	60–150	0.00–7.50				Difficult crop with a low success rate. 4 years to harvest.

Information Chart 2: Harvesting, Drying, Prices and Marketing

Species	Part Used	Harvest	Drying	Yield kg/m²/annum	Trade Herbs Price/kg	Manufacturing Grade Volume	Manufacturing Grade Price/kg $	Manufacturing Grade Gross Return $/m²/annum	Premium Tea & Culinary Grade Volume	Premium Tea & Culinary Grade Price/kg $	Premium Tea & Culinary Grade Gross Return $/m²/annum	Comments
Ground Ivy *Glechoma hederacea*	Aerial Parts	spring–autumn	to 35°C	low		small	12–15					
Gum Plant *Grindelia robusta*	Aerial Parts	Dec.–May	to 35°C	0.4–0.7		small–mod.	11–16	4.40–11.20				Very slow drying.
Hawthorn *Crataegus spp.*	Lf & Flwr / Fruit	Oct.–Nov. / Mar.–May	to 35°C / to 45°C		8	small–mod. / moderate	15–17 / 10–13		small	~14		Wildcrafted or harvested weed.
Hops *Humulus lupulus*	Cones	Feb.–April	to 35°C	around 0.1	8–10	small–mod.	8–14	~.80–1.40	small	17	1.70	
Horehound, Black *Ballota nigra*	Aerial Parts	Dec.–May	to 35°C	0.3–0.5		small	8–18	2.40–9.00				
Horehound, White *Marrubium vulgare*	Aerial Parts / Lf & Flwr	variable	to 35°C		8	moderate	8–10		small	18		
Horse Chestnut *Aesculus hippocastanum*	Seed	Mar.–Apr.	to 35°C		7	moderate						
Horseradish *Armoracia rusticana*	Root	winter	to 35°C	0.2–0.3	8	small	10–15	2.00–4.50				Yields are better in a cooler climate.
Horsetail, Field *Equisetum arvense*	Aerial Parts	Nov.–May	to 35°C	0.3–0.4	9	moderate	12–16	3.60–6.40	moderate	25	7.50–10.00	
Hyssop *Hyssopus officinalis*	Aerial Parts / Lf & Flwr	Nov.–May	to 35°C	0.4–0.65 / 0.3–0.45	11–12	small–mod.	10–16	4.00–10.40	moderate	24	7.20–10.80	
Lady's Mantle *Alchemilla vulgaris*	Aerial Parts / Lf & Flwr	Dec.–May	to 35°C	0.3–0.45 / 0.3–0.45		small–mod.	13–17	3.90–7.60	very small	24	7.20–10.80	
Lavender, English *Lavandula angustifolia*	Flower	Dec.–Apr.	to 35°C	0.02–0.05	12	very small	10–20	0.20–1.00	small	20–30	1.00–1.50	
Lemon Grass *Cymbopogon citratus*	Leaf	Nov.–May	to 35°C	0.3–0.5	8				large	12–16	3.60–8.00	Higher yields in sub-tropical regions.
Lemon Thyme *Thymus x citriodorus*	Lf & Flwr	Nov.–May	to 35°C	0.2–0.35					moderate	20–24	4.00–8.40	
Lemon Verbena *Aloysia triphylla*	Lf & Flwr	Jan.–May	to 35°C	around 0.2	11				small	34	around 6.80	
Licorice *Glycyrrhiza glabra*	Root	winter	to 35°C	0.25–0.3 (averaged)	9	large	12–18	3.00–5.40	moderate	25	6.25–7.50	3–4 yrs to harvest.

Information Chart 2: Harvesting, Drying, Prices and Marketing

Species	Part Used	Harvest	Drying	Yield kg/m²/annum	Trade Herbs Price/kg	Manufacturing Grade Volume	Manufacturing Grade Price/kg $	Manufacturing Grade Gross Return $/m²/annum	Premium Tea & Culinary Grade Volume	Premium Tea & Culinary Grade Price/kg $	Premium Tea & Culinary Grade Gross Return $/m²/annum	Comments
Linden *Tilia* spp.	Flwr	Dec.	to 35°C		16	moderate	14–18		moderate	40		
Marjoram, Sweet *Origanum majorana*	Leaf (& flower)	late spring autumn	to 35°C	0.2–0.3	3				small	24	4.00–7.20	
Marshmallow *Althaea officinalis*	Lf & Flwr / Root	Jan.–Apr. / winter	to 35°C / 35–45°C	0.1–0.2 / 0.3–0.4	12–13	small-mod. / moderate	10–15 / 10–18	1.00–3.00 / 3.00–7.20	very small / small	24 / 24	2.40–4.80 / 7.20–9.60	
Meadowsweet *Filipendula ulmaria*	Aerial Parts / A.P. Tea Gde	Dec.–May / "	to 35°C / "	0.25–0.35	8	mod.-large	10–18	2.50–6.30	moderate	24	6.00–8.40	
Melissa Balm *Melissa officinalis*	Aerial Parts / Leaf	Nov.–May / "	to 35°C / "	0.3–0.45 / 0.2–0.3	9–12	moderate	12–15	3.60–6.75	moderate	24	4.80–7.20	
Motherwort *Leonurus cardiaca*	Aerial Parts / Leaf	Dec.–May / "	to 35°C / "	0.25–0.5 / 0.2–0.35	8	small-mod.	10–16	2.50–8.00	very small	24	4.80–7.20	
Mountain Pepper *Tasmannia lanceolata*	Leaf / Fruit	Feb–Aug. / May–Jun.	to 35°C / "									Wildcrafted. New product with some potential.
Mugwort *Artemisia vulgaris*	Aerial Parts / Leaf	Jan.–Apr. / "	to 35°C / "	0.55–0.75 / 0.3–0.4	8	small	12–15	6.60–11.25	very small	24	7.20–9.60	Higher yield first season.
Mullein, Great *Verbascum thapsus*	Leaf / Flower	summer / "	to 35°C / "		9–10	moderate	8–12		very small		15–20	Wildcrafted/Harvested weed. *V. thapsoides* has larger flowers
Nettle, Greater *Urtica dioica*	Aerial Parts / Leaf	Nov.–May / "	to 45°C / "	0.35–0.5 / 0.25–0.4	9	moderate	9–13 / 10–16	3.15–6.50 / 2.50–6.40	large	24	6.00–9.60	Preferred species.
Nettle, Lesser *Urtica urens*	Leaf	winter to spring	to 45°C		9							Harvested weed. Limited market.
Oats *Avena sativa*	Drd. Grn. Plnt. / Seed	Sept.–May / winter	to 35°C / "	around 0.3 / around 0.25	10	small / moderate	5–8 / 0.40–0.60	~1.50–2.40 / 0.06–0.21	small	16	~4.80	Green plant harvested and dried at the milk stage.
Oregano *Origanum vulgare*	Leaf (& Flwr)	Dec.–May	to 35°C	0.3–0.4	6				small-mod.	24	7.20–9.60	
Parsley *Petroselinum crispus*	Leaf / Root	Sept.–May / winter	to 35°C / "		10	small	10–20	~2.50–5.00	very small	24	~7.20	Chop leaf before drying
Pasque Flower *Anemone pulsatilla*	Aerial Parts	Oct.–May	to 35°C	0.3–0.5	16	small	15–19	4.50–9.50				
Passionflower *Passiflora incarnata*	Aerial Parts / Leaf	Mar.–Apr. / "	to 35°C / "	0.35–0.5 / 0.3–0.4	10–12	large	16–20	5.60–10.00	small	24	7.20–9.60	

Information Chart 2: Harvesting, Drying, Prices and Marketing

Species	Part Used	Harvest	Drying	Yield kg/m²/annum	Trade Herbs Price/kg	Certified Organically Grown or Wildcrafted Herbs					Comments
						Manufacturing Grade			Premium Tea & Culinary Grade			
						Volume	Price/kg $	Gross Return $/m²/annum	Volume	Price/kg $	Gross Return $/m²/annum	
Pennyroyal *Mentha pulegium*	Aerial Parts Lf & Flwr	Dec.–May "	to 35°C "	0.25–0.4 0.2–0.3	9	very small	11–15	2.75–6.00	very small	24	4.80–7.20	Grown as crop or harvested weed.
Peppermint *Mentha x piperita*	Aerial Parts Leaf	late Nov.–May "	to 35°C "	0.3–0.6 0.25–0.45	6–9	large	5–12 10–15	1.50–7.20 2.50–6.75	very large	24–26	6.00–11.70	
Periwinkle *Vinca major*	Aerial Parts	flowering Oct.–Feb.	to 45°C			small	10–16					Harvested weed.
Plantain, Greater *Plantago major*	Leaf	Dec.–Apr.	to 45°C	0.2–0.35	10	small	6–18	1.20–6.30				
Plantain, Narrow-leaf *Plantago lanceolata*	Leaf	spring–summer	to 45°C			small	6–18					
Poke *Phytolacca decandra*	Root	winter	to 45°C	0.4–0.5	11	moderate	12–16	4.80–8.00				Caution: Toxic. Juice is caustic.
Raspberry *Rubus idaeus*	Leaf	Nov.–Apr.	to 35°C	0.2–0.3	7–8	moderate	8–16	1.60–4.80	large	24	4.80–7.20	
Red Clover *Trifolium pratense*	Flower Lf & Flwr	Jan.–May "	to 35°C "	0.05–0.15 around 0.3	9–10	mod.–large	10–15	~3.00–4.50	mod.–large	30–50	1.50–7.50	
Rose, Dog *Rosa canina*	Fruit	Mar.–May	to 45°C		9	moderate	10–15		moderate	17		Wildcrafted/harvested weed.
Rose, Sweet Briar *Rosa rubiginosa*	Fruit	Mar.–May	to 45°C		9				moderate	13		Wildcrafted/harvested weed.
Rosemary *Rosmarinus officinalis*	Leaf	Jan.–Mar.	to 35°C	around 0.2	5	small	10–13	~2.00–2.60	small	24	~4.80	
Rue *Ruta graveolens*	Aerial Parts	Dec.–May	to 35°C	0.4–0.6	9	very small	8–15	3.20–9.00				Chop before drying.
Sage *Salvia officinalis*	Aerial Parts Leaf (& Flwr)	Nov.–May "	to 35°C "	0.3–0.5 0.2–0.35	5–9	moderate	12 12–16	3.60–6.00 2.40–5.60	small	24	4.80–8.40	
Scullcap *Scutellaria lateriflora*	Aerial Parts Leaf	Nov.–May "	to 35°C "	around 0.4 around 0.3		large	12–20	~4.80–8.00	small	24	~7.20	
Shepherd's Purse *Capsella bursa–pastoris*	Aerial Parts	in flower	to 35°C		8	small	11–15					Harvested weed. Also used fresh.
Sorrel, Sheep *Rumex acetosella, R. angiocarpus*	Aerial Parts	in flower	to 45°C		21	moderate	10–30					Harvested weed.

Information Chart 2: Harvesting, Drying, Prices and Marketing

Species	Part Used	Harvest	Drying	Yield kg/m²/annum	Trade Herbs Price/kg	Certified Organically Grown or Wildcrafted Herbs							Comments
						Manufacturing Grade				Premium Tea & Culinary Grade				
						Volume	Price/kg $	Gross Return $/m²/annum		Volume	Price/kg $	Gross Return $/m²/annum	
Spearmint *Mentha spicata*	Leaf	Nov.–May	to 35°C	0.25–0.5	3–9					large	24	6.00–12.00	
St John's Wort *Hypericum perforatum*	Aerial Parts Lf & Flwr	Nov.–Jan. "	to 35°C "		9–10	very large	5–13			small	16		Wildcrafted/Harvested weed.
Stoneroot *Collinsonia canadensis*	Root	winter	35–45°C	around 0.1 (averaged)		very small	18	around 1.80					2 years to harvest.
Tansy *Tanacetum vulgare*	Aerial Parts	Dec.–May	to 35°C	around 0.5	8	very small	8–16	~4.00–8.00					
Tarragon, French *Artemisia dracunculus*	Leaf	Nov.–May	to 35°C	0.15–0.3	22					small	24	3.60–7.20	
Thyme *Thymus vulgaris*	Leaf (& Flwr)	Oct.–May	to 35°C	0.2–0.35	4–5	moderate	12–15	2.40–5.25		moderate	24–26	4.80–9.10	
Valerian *Valeriana officinalis*	Root	winter	to 35°C	0.2–0.4	13–16	large	15–18	3.00–7.20		small	37	7.40–14.80	Cleaning roots is main challenge.
Variegated Thistle *Silybum marianum*	Seed	seed ripe Dec.–Feb.	to 35°C			very large	3–4			small	4–6		Harvested weed.
Vervain *Verbena officinalis*	Aerial Parts A.P. Tea Gde	Oct.–May "	to 35°C "	0.5–0.7 0.5–0.7	10–11	moderate	12–14	6.00–9.80		very small	20	10–14.00	
Violet, Sweet *Viola odorata*	Leaf	Nov.–May	to 35°C	0.2–0.3		small	12–25	2.40–7.50					
Willow *Salix* spp.	Bark	spring to early summer	to 45°C		9	moderate	10–15						
Witch Hazel *Hamamelis virginiana*	Leaf Bark	Jan.–Mar. early spring	to 35°C		8	small	12–15						
Wood Betony *Stachys officinalis*	Aerial Parts	Dec.–May	to 35°C	0.3–0.5	9–10	small	12–15	3.60–7.50					
Wormwood *Artemisia absinthium*	Aerial Parts Leaf & Flwr	Jan.–May "	to 35°C "	0.45–0.60 0.25–0.35	9	variable "	8–11 16	3.60–6.60 4.00–5.60					Market can be volatile.
Yarrow *Achillea millefolium*	Aerial Parts Leaf (& Flwr)	Dec.–May "	to 35°C "	0.4–0.5 0.3–0.35	7–8	mod.-large	10–15	4.00–7.50		small	24	7.20–8.40	

Appendixes

Appendix 1
Leaf Shapes and Arrangements

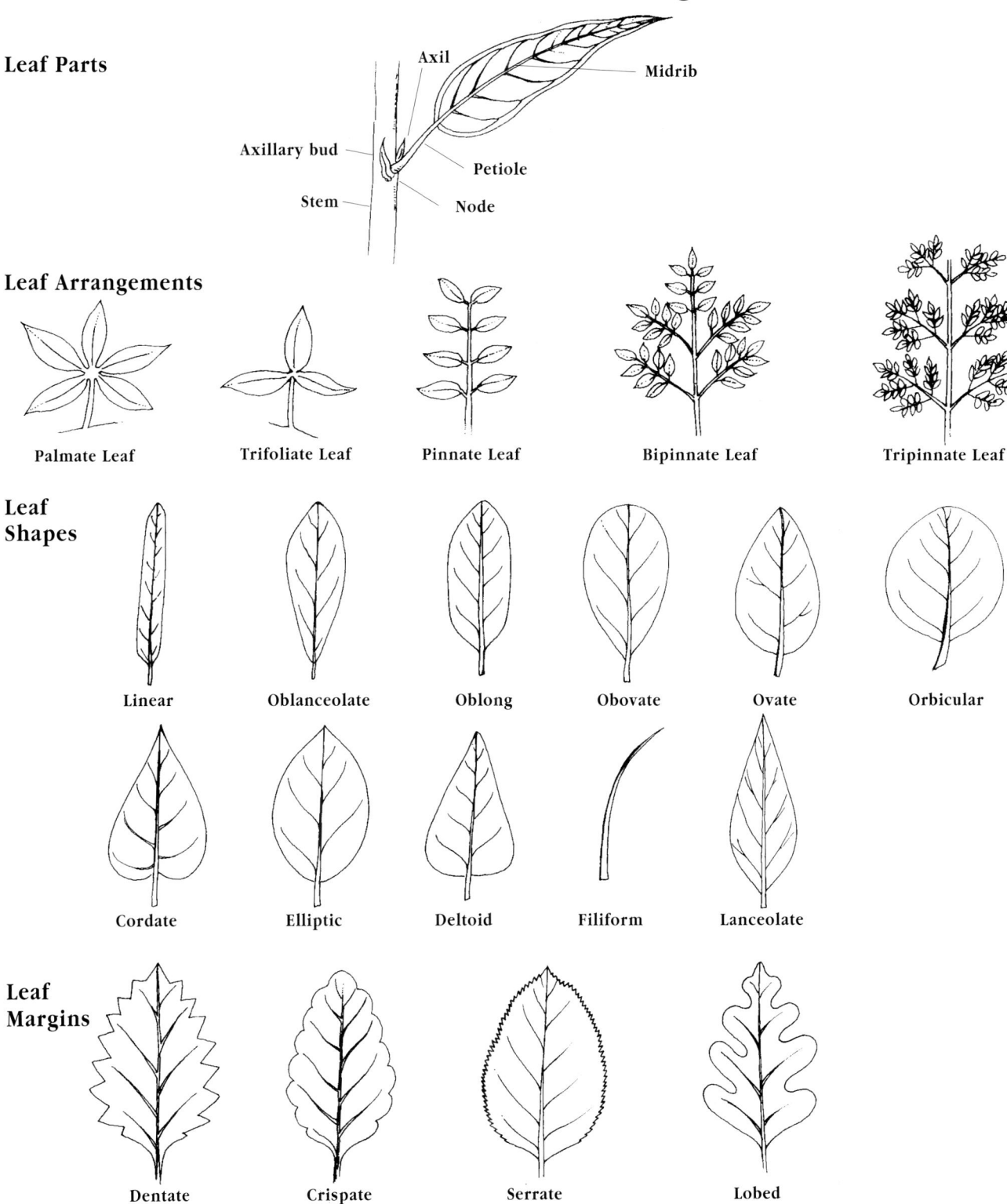

Appendix 2
Flower Parts and Arrangements

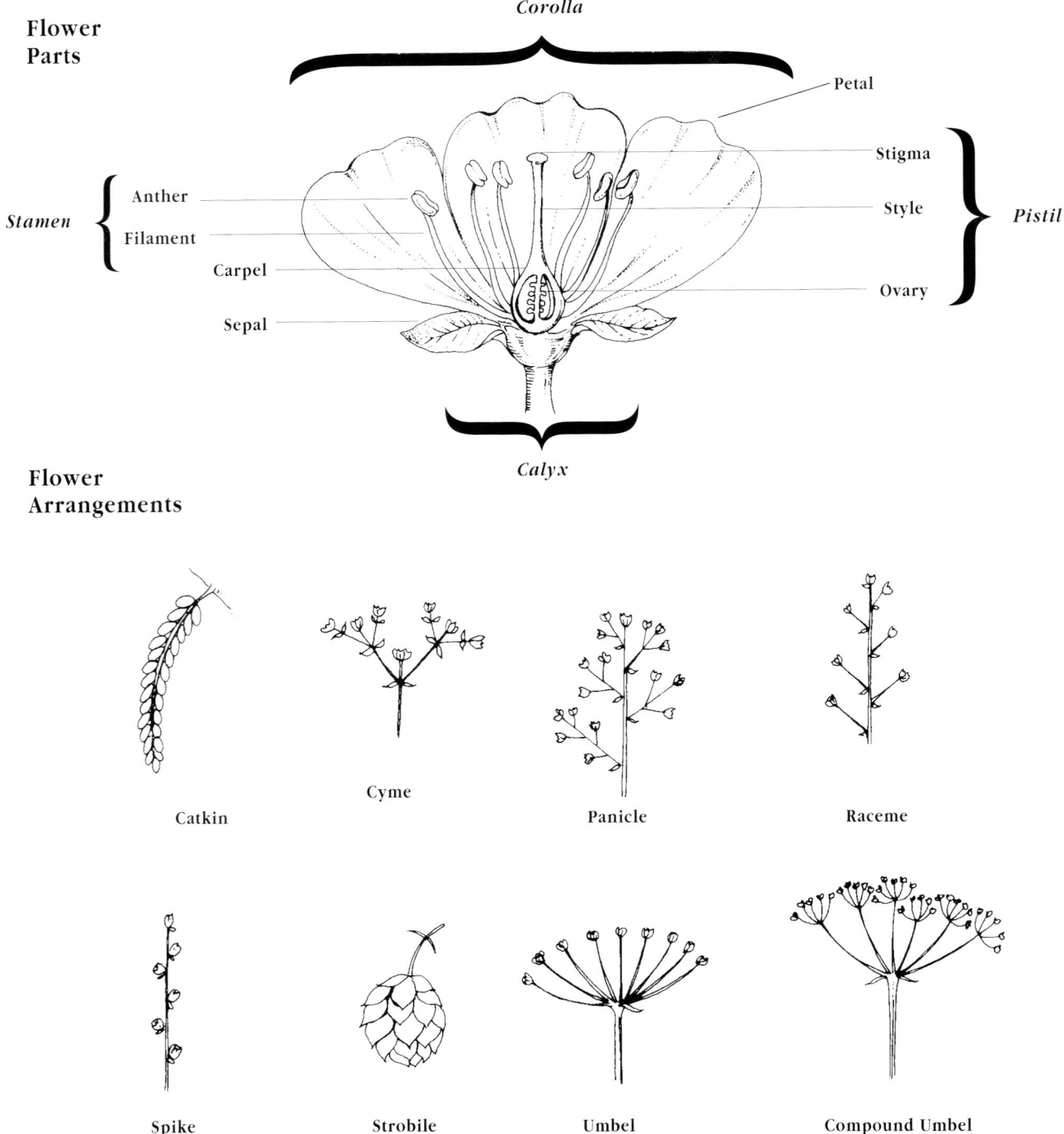

Appendix 3
Suppliers, Buyers and Organisations

Books
The Green Book Company
PO Box 5255, Burnley, VIC 3121
ph 03 9427 8866 or 1800 646 533
(international +61 3 9427 8866)
fax 03 9427 9066 (+61 3 9427 9066)
sales@greenbook.com.au
www.greenbook.com.au
Horticultural and agricultural books.

Buyers and Distributors of Organic Herbs

Bronzewing Herbal Teas
134 Maudsleys Rd, TAS 7109
ph/fax 03 6266 4308
(+61 3 6266 4308)
bronzewing@iprimus.com.au
Produces premium quality herbs – does not purchase.

Highland Herbs Tasmania
690 Gulf Rd, Liffey, TAS 7302
ph 03 6397 3461 (+61 3 6397 3461)
herbstas@vision.net.au
Produces and purchases premium quality herbs

Mediherb Pty Ltd
PO Box 713, Warwick, QLD 4370
ph 07 4661 0770 (+61 7 4661 0770)
fax 07 4661 0799 (+61 7 4661 0799)
www.mediherb.com.au
Purchases in minimum 20 kg lots.

Southern Light Herbs
PO Box 227, Maldon, VIC 3463
ph 03 5475 2763 (+61 3 5475 2763)
fax 03 5475 1477 (+61 3 5475 1477)
Winner of Australian Small Business of the Year Award 2001. Produces and purchases premium quality herbs, including 1st screenings.

UK

The Organic Herb Trading Company
Milverton, Somerset, TA4 1NF,
United Kingdom
ph +44 (0)18 2340 1205
fax +44 (0)18 2340 1001
www.organicherbtrading.com

USA

Pacific Botanicals
4840 Fish Hatchery Rd.
Grants Pass, Oregon 97527
ph +1 541 479 7777
fax +1 541 479 7780
www.pacificbotanicals.com

See also: Listings in *Australian Herb Industry Resource Guide* by Kim and Michael Fletcher (see below under **Further Resources**) and listings under Health Foods and under Organic Products in the Yellow Pages.

Herb Plants and Vegetative Propagation Material

See: Listings under Nurseries and under Growers in *Australian Herb Industry Resource Guide* by Kim and Michael Fletcher (see next page under **Further Resources**), under Herbs in the Yellow Pages and www.nurseriesonline.com.au/PAGES/Herbs.htm

Herb Seeds

Austral Herbs & Seeds
The Shamrock, Niangala, NSW 2354
ph 02 6769 2263 (+61 2 6769 2263)
fax 02 6769 2325 (+61 2 6769 2325)
www.australherbs.com.au

Eden Seeds
M.S.905, Lower Beechmont, Qld 4211
ph 07 5533 1107 (+61 7 5533 1107)
fax 07 5533 1108 (+61 7 5533 1108)
www.edenseeds.com.au

Eshcol Springs Pty Ltd
agents for Richters Herb Seeds, Canada
PO Box 61 Gingin, WA 6053
ph 08 9575 7522 (+61 8 9575 7522)
fax 08 9575 7622 (+61 8 9575 7622)

Green Patch Organic Seeds
PO Box 1285, Taree, NSW 2430
www.greenpatchseeds.com.au
ph/fax 02 6551 4240
(+61 2 6551 4240)

Kings Herb Seeds
PO Box 975, Penrith, NSW 2751
ph 02 4776 1493 (+61 2 4776 1493)
kirstenole@ozemail.com.au

Pleasance Herb Seeds
11935 Summerland Way, Casino, NSW 2470
ph/fax 02 6663 3390
(+61 2 6663 3390)
www.pleasanceherbs.com.au

Royston Petrie Seeds Pty Ltd
77 Kenthurst Rd (PO Box 77),
Kenthurst, NSW 2156
ph 02 9654 1186 (+61 2 9654 1186)
fax 02 9654 2658 (+61 2 9654 2658)
roseed@bigpond.com

NORTH AMERICA

Horizon Herbs
PO Box 69 Williams,
Oregon 97544 USA
ph +1 (541) 846 6704,
fax +1 (541) 846-6233
www.horizonherbs.com

Johnny's Selected Seeds
RR1 Box 2580
Albion, Maine 04910-9731
USA
ph +1 207 437 4395
www.johnnyseeds.com
Organic seeds and gardening tools

Richters Herb Seeds
Goodwood, Ontario, L0C 1A0 Canada
www.richters.com
(see Australian agent Eshcol Springs above)

See also: Listings under Seeds in *Australian Herb Industry Resource Guide* by Kim and Michael Fletcher (see next page under **Further Resources**).

Herb Photography

Elemental Vision Photographics
7 McPherson Rd,
Warrandyte, VIC 3113
ph/fax 03 9844 4443
(+61 3 9844 4443)
www.evphotographics.com.au
Photographer Jenny Grinlington did the cover photo of this book and has a large range of superb photos of herbs and other plants.

Organic and Biodynamic Certification Organisations

Australian Certified Organic/ Biological Farmers of Australia
PO Box 530,
Chermside, QLD 4032
ph 07 3350 5716 (+61 7 3350 5716)
fax 07 3350 5996 (+61 7 3350 5996)
www.bfa.com.au

Biodynamic Research Institute
Powelltown, VIC 3797
ph 03 5966 7333 (+61 3 5966 7333)

National Association for Sustainable Agriculture in Australia
PO Box 768, Stirling, SA 5152
ph 08 8370 8455 (+61 8 8370 8455)
fax 08 8370 8381 (+61 8 8370 8381)
www.nasaa.com.au

Organic Herb Growers of Australia
PO Box 6171
South Lismore, NSW 2480
ph 02 6622 0100 (+61 2 6622 0100)
www.organicherbs.org

Tasmanian Organic-Dynamic Producers (TOP)
1333 Russell Road
Lonnavale TAS 7109
ph 03 6266 0330 (+61 3 6266 0330)

Other Countries

For information on organic certification in other countries see International Federation of Organic Agriculture Movements website: www.ifoam.org

Tools and Equipment

Gameco Group of Companies
65a Chifley Drive, Preston VIC 3072
ph 03 9480 3633 (+61 3 9480 3633)
fax 03 9416 9728 (+61 3 9416 9728)
www.gameco.com.au
Flaming equipment

Gundaroo Tiller
Allsun Farm, Gundaroo, NSW 2620
ph 02 6236 8173 (+61 2 6236 8173)
www.allsun.com.au
Hoes, wheel hoe, broad-fork etc

Hollander Imports
87 Brooker Highway
Hobart, TAS 7000
ph 03 6234 5111 (+61 3 6234 5111)
Machinery and garden tools.

UK

Littlemesters
23 Newsham Road
Norton Hammer
Sheffield S8 9EA U.K.
ph/fax +44 (0)11 4220 0516
www.littlemesters.com
Rivetted scythes, garden tools

James Cookson (Stockport) Ltd
Cookson Armstrong House
Swallow Street
Stockport, Cheshire SK1 3LG U.K.
ph +44 (0)16 1480 2388
fax +44 (0)16 1480 1968
www.cooksons.com
Rivetted scythes, garden tools

Ladbrooke Ltd
The Grange
Exhall near Alcester
Warwickshire B49 6EA U.K.
ph +44 (0)17 8977 3898
fax +44 (0)17 8977 3898
www.ladbrooke.co.uk
Manufacturers of soil blocking tools.

The Lazy Dog Tool Company
Hill Top Farm
Spaunton
Appleton Le Moors
North Yorkshire YO62 6TR U.K.
ph/fax: +44 (0)1 751 417 351
www.lazydogtoolco.co.uk
Innovative tools for organic weed control

USA

Peaceful Valley Farm Supply
P.O. Box 2209
Grass Valley, California 95945 USA
ph +1 530 272 4769
fax +1 530 272 4794
www.groworganic.com
Ships USA but not internationally

Scythe Supply
496 Shore Road
Perry, Maine 04667 USA
ph +1 207 853 4750
www.scythesupply.com

The Scythe Source
202 Linden Avenue
Towson, Maryland 21286 USA
ph +1 410 404 6827
www.scythesource.com

See also: Listings under Equipment in *Australian Herb Industry Resource Guide* by Kim and Michael Fletcher (see next page under **Further Resources**) *Johnny's Selected Seeds* under Herb Seeds

Woven Wire Screen
Lockers Pty Ltd
45 Keys Rd, Moorabbin, VIC 3189
ph 03 9555 7744 (+61 3 9555 7744)
fax 03 9553 3675 (+61 3 9553 3675)

Richardson Pacific
330 Ballarat Rd, Braybrook, VIC 3019
www.richpac.com.au
ph 03 9313 2222 (+61 3 9313 2222)

Wire Mesh Industries Pty Ltd
7-9 Rhodes St, West Ryde, NSW 2114
ph 02 9809 0900 (+61 2 9809 0900)
fax 02 9807 4073 (+61 2 9807 4073)
www.wiremesh.com.au

See also: Listings under Wire Products in the Yellow Pages.

Further Resources

Australian Herb Industry Resource Guide with New Zealand Supplement by Kim and Michael Fletcher.

Available from the authors or The Green Book Company (see under **Books** above).
PO Box 203,
Launceston, TAS 7250
ph 03 6330 1493 (+61 3 6330 1493)
fax 03 6330 1498 (+61 3 6330 1498)
admin@focusonherbs.com.au

This invaluable little book lists virtually every business, service, grower association or other organisation of any significance in all aspects of the herb industry in Australia and New Zealand. Updated every 2 years.

Bibliography

Auld, B.A., & Medd R.W., *Weeds: An Illustrated Guide to the Weeds of Australia*, Inkata Press.

Bio-Dynamic Farming & Gardening Association Of New Zealand, *Bio-dynamics – New Directions – New Zealand*. A compilation of studies and stories of a number of different biodynamic farms, very practical and down to earth.

Bremness, Lesley, *The Complete Book of Herbs*, R D Press, 1988.

British Herbal Medicine Association, *British Herbal Pharmacopoeia*, B.H.M.A., 1983.

Bureau of Flora and Fauna, Canberra, *Flora of Australia*, Australian Government Printing Service, 1982.

Caplan, Basil (ed.), *The Complete Manual of Organic Gardening*, Headline Book Publishing 1992.

Coleman, Eliot, *The New Organic Grower, A Master's Manual of Tools and Techniques for the Home and Market Gardener*, Chelsea Green, Vermont, 1989. A valuable source of information on organic gardening on a small commercial scale, based on the authors many years of experience and innovation.

Cribb, A.B. & J.W., *Wild Food in Australia*, Fontana Collins, 1976.

Crochmal, A. & C., *A Guide to the Medicinal Plants of the United States*, Quadrangle/New York Times Book Co., 1980.

Curtis W.M. & Morris D.L., *The Student's Flora of Tasmania*, Government Printer, Tasmania, 1975.

Edlin, Herbert L. & Nimmo, Maurice, *The World of Trees*, Orbis Publishing, London.

Ellyett, Clifton D., *How Heat Pump Drying Works*, Self-published: PO Box 84, Ourimbah, NSW 2258. This little booklet gives a clear, easy to understand explanation of the principle and operation of a heat pump dehumidifier with some details on suppliers and costs.

Ellyett, Clifton D., *The Growing of Echinacea in Australia*, Self-published: PO Box 84 Ourimbah, NSW 2258.

Fletcher, Kim, *A Modern Australian Herbal*, Viking – Penguin Books.

Fletcher, Kim, *Australian Herb Industry Resource Guide*, Focus On Herbs Consultancy & Information Service, PO Box 203, Launceston, TAS 7250. Updated every 2 years, this valuable book lists virtually every business of any significance in all aspects of the herb industry.

Flöck, Hans, *Medicinal Plants and Their Uses*, W. Foulsham & Co., 1976.

Forman, Terry, *Bio-Dynamics: A Modern Sustainable Agriculture*. Terry has a wonderful ability to explain bio-dynamics in ways that make sense and inspire.

Foster, Steven, 'Herb Quality – A Tempest In Your Teacup?' in *Whole Life Times*, July–August 1985 (Published In USA).

Grieve, Mrs. M., *A Modern Herbal*, Dover, 1971. First published in 1931, this book gives a fascinating perspective on herbs, folklore and traditional medicinal usage. It also has quite a lot of useful information on herb growing scattered through it.

Harrison, Richmond E., *Handbook of Trees and Shrubs*, 5th Edition, A. H. & A. W. Reed.

Hoffman, David, *The New Holistic Herbal*, Element Books, 1990. One of the best books on herbal medicine written for the lay person.

Hyde-Wyatt, B.H. & Morris, D.I., *Tasmanian Weed Handbook*, Tasmanian Department Agriculture, 1975.

Jacobs, Betty, *Growing and Using Herbs Successfully*, Garden Way Publishing, Vermont, 1981.

Johnson, Hugh, *Hugh Johnson's Encyclopaedia of Trees*, Mitchell Beazley, 1979.

Keats, Brian, & Pearson, Sue, *Antipodean Astro Calendar*. Self-published each year: 117 Sunset Drive, Mittagong, NSW 2575. An excellent resource for biodynamic growers.

Lamp, Charles & Collet, Frank, *Field Guide to Weeds in Australia*, Inkata Press, 1989.

Lowenfeld, Claire, *Herb Gardening*, Faber & Faber, 1964.

Macoboy, Stirling, *What Tree Is That?*, Landsdowne Press.

Macoboy, Stirling, *What Shrub Is That?*, Weldon Publishing.

Mathews, F. Schuyler, *Field Book of American Trees and Shrubs*, G. Putnam's Sons, 1915.

Miller, Richard Alan, *The Potential of Herbs as a Cash Crop*, Acres, USA, 1985. This book is about growing herbs on a broad-acre scale using organic methods. A number of glaring inadequacies leave the reader, perhaps, a little more ignorant than when they picked the book up. The 'organic' methods he advocates involve the use of herbicides for weed control and the book is riddled with other misinformation and inaccuracies, including some incredibly low figures for costs of production. The reader is left with the impression that producing a crop of herbs is no more difficult than producing a crop of hay. Unfortunately these far-fetched figures and other misinformation put forward by Miller are being used in this country as a basis for promoting the production of herbs.

Minnich, J Hunt, M., et al., *The Rodale Guide to Composting*, Rodale Press, 1979.

Parsons, W.T. & Cuthbertson, E.G., *Noxious Weeds of Australia*, Inkata Press, 1992.

Peterson R.T. & M. Mckenny, *A Field Guide to Wildflowers of North East And North Central North America*, Houghton Mifflin, 1968.

Proctor, Peter, *Grasp the Nettle, making biodynamic farming work*. Random House New Zealand, 1997. An esay-to-read and practical outline of biodynamics. Peter Proctor has 30 years experience in biodynamic management and as a field advisor both in New Zealand and internationally.

Small, Ernest, *Culinary Herbs*, NRC Research Press, Ottawa, 1997.

Steiner, Rudolf, *Agriculture*. A translation of his original lectures, heavy going in parts, but it is interesting and relevant to go to the source.

Storl, *Culture and Horticulture*. An outline of biodynamic philosophy that most people will find quite readable.

Stuart, Malcolm, (ed.), *The Encyclopaedia of Herbs and Herbalism*, Orbis, London, 1979.

The Revolutionary Health Committee of Hunan Province, *A Barefoot Doctor's Manual*, Second Back Row Press (Cloudburst Press, 1977).

Tutin, T.G. et al, *Flora Europaea*, Cambridge University Press, Cambridge, 1964-80. Usher, George, *A Dictionary of Plants Used by Man*, Constable, London, 1974.

Weiss, Gaea & Shandor, *Growing & Using Healing Herbs*, Rodale Press, USA. This book has good illustrations.

Woodward, Penny, *An Australian Herbal*, Hyland House, 1986. A useful source of information for growing herbs on a home-garden scale in Australian conditions.

Glossary

ACTIVE CONSTITUENTS

Chemical substances contained in a herb and responsible, partly or wholly, for its therapeutic effect.

ADAPTOGEN

A herb which helps the body's processes to return normal when subjected to stress.

AERIAL PARTS

Those parts of the plant above ground. In practice this often excludes the lower portion which is predominantly stem.

ALKALOID

A group of physiologically active, basic organic compounds which occur in plants.

ALTERATIVE

A herb that helps increase the health and vitality of the body.

ALTERNATE

Arranged singly at the nodes, alternating from one side of the stem to the other.

AMBIENT AIR

Unheated air, the temperature and humidity of which depend on prevailing weather conditions.

ANALGESIC

A herb that reduces pain.

ANNUAL

A plant which goes through its life cycle, from germination to death, within one year.

ANTHROPOSOPHIC MEDICINE

Anthroposophy is a system of beliefs and practices based on the teachings of Rudolph Steiner (see 'Biodynamic').

ANTI–EMETIC

A herb that reduces nausea.

ANTI–INFLAMMATORY

A herb that reduces inflammation.

ANTISPASMODIC

A herb that eases spasms and cramps.

ANTITUSSIVE

A herb that helps relieve coughing.

AROMATIC

A herb with a fragrant odour due to volatile components.

ASTRINGENT

A herb that causes tissues to shrink, tighten or dry out, reducing haemorrhage, secretion or diarrhoea.

AXIL

The upper surface of the junction between a leaf stalk and the main stem (see illustration of 'Leaf Arrangements' below).

AXILLARY

Forming in the axil, where the leaf stalk joins the main stem.

BARE FALLOW

Frequently repeated cultivation of an area of ground to destroy weeds before planting. To be effective, this usually needs to be carried out for at least 3 months.

BASAL LEAVES

The leaves growing at the base of a plant.

BIODYNAMIC

A system of organic farming and gardening based on the principles and methods indicated by Rudolf Steiner, including the use of horn manure (500) and other preparations.

BIPINNATE

Doubly pinnate. The compound leaf consisting of leaflets arranged pinnately along secondary ribs, which are attached to a central midrib. Many Wattles have typical bi-pinnate leaves (see illustration in Appendix 1).

BIENNIAL

A plant that goes through its life cycle, from germination to death, in two growing seasons, normally only flowering in the second season.

BIFID
Deeply divided into two lobes.

BRACT
A much reduced or scale-like leaf.

BRUISING
The darkening, usually of leaf or flower herbs, as a result of damage to or fermentation of the tissues before or during the drying process.

CALYX
The outside whorl of flower parts, or outside covering of the flower.

CANOPY
A more or less contiguous layer formed by the leafy tops of a stand of plants.

CARMINATIVE
A herb that helps relieve colic and flatulence.

CATKIN
A cylindrical inflorescence, usually hanging.

CHOOK
Australian colloquial term for a domestic hen.

COMPOSITE
A cluster of small flowers or florets fused together into one flower head.

COMPOST
An organic fertiliser consisting of a proportionate mixture of plant and animal products, which have been through a decomposition process together, usually involving the generation of heat. The resulting material should contain a balanced level of nutrients and a good ratio of carbon to nitrogen.

COMPOUND LEAF
A leaf made up of several or many separate leaflets (see illustration of 'Leaf Arrangements' below).

CONTAINMENT
A system for restricting the spread of invasive plants.

COPPICE
The development of new shoots from the stump of a tree which has been cut down.

CORDATE
Heart-shaped.

CORM
A swollen underground stem that contains reserves for one season only, sometimes dividing into segments that each give rise to a new plant or a new corm forming above or beside the parent.

COROLLA
The petals of a flower.

COTYLEDON
The leaf or leaves contained within the seed, usually differing in shape from the true leaves. They may emerge above the ground on germination, or may stay within the seed-coat.

CRISPATE
With a marked undulating or wavy surface and/or edge.

CULINARY GRADE
A dried herb processed to a standard suitable for use in cooking.

CULTIVAR
A cultivated variety selected from the wild or developed through breeding.

CURING
A slow drying process. In the case of seed crops cut before full maturity, the seed continues filling out and maturing as it draws on the sap in the rest of the plant.

CUTTING
A piece of stem or leaf lacking roots, or root lacking leaves, taken from a parent plant and capable of giving rise to a new plant.

CYME
A broad cluster of flowers, in an inverted cone shape, with the oldest flowers in the centre.

DECIDUOUS
Losing leaves in autumn.

DECOCTION
A dilute extraction of a herb obtained by simmering it in water for about 15 minutes.

DELTOID
Roughly triangular in shape.

DEMULCENT
A herb that soothes inflamed membrane or skin.

DENT
A measurement of mesh size, denoting number of squares per inch or 25 mm.

DENTATE
With coarse or sharp teeth along the leaf margin.

DESICCATION
Complete drying or loss of moisture.

DIAPHORETIC
A herb that promotes perspiration.

DICOTYLEDON
Member of a group of flowering plants that have two cotyledons or seed leaves.

DIOECIOUS
Bearing male and female flowers on separate plants.

DIPLOID
Having the normal complement of paired chromosomes.

DISTILLATION
The process of vaporising a liquid and re-condensing it by cooling. This is used to separate components with different boiling points.

DIURETIC
A herb that increases the volume of urine.

DIVISION
A piece separated from a parent plant and possessing both roots and leaves (or buds ready to produce these) so it will give rise a new plant.

DORMANCY
The suspension of growth, as in winter, or the delayed germination of a seed.

DRILLS
Closely spaced rows 100–200 mm apart, usually sown with a seed drill or combine where inter-row cultivation is not envisaged.

DRUPE
Fleshy fruit with the seed(s) enclosed in a stone.

ELLIPTIC
In the shape of an ellipse, widest at the middle with narrower rounded ends.

EMMENAGOGUE
A herb that helps stimulate menstrual flow.

ENTIRE
Not lobed or toothed.

ENZYME
An organic catalyst essential in the functioning of a biochemical reaction.

ESSENTIAL OIL
An aromatic oil of plant origin, which evaporates or distils readily.

EXPECTORANT
A herb that assists the removal of excess mucus from the lungs and air passages.

EXTENSIVE ROOT
A long horizontal root which extends from the plant and gives rise to new plants along its length.

EXTRACT
A concentrated liquid obtained by treating plant material with a solvent or a mixture of solvents to dissolve its active constituents. In herbal medicine usually a mixture of alcohol and water is used.

FILIFORM
Thin and thread-like.

FINES
The small flour-like particles in a milled or rubbed product.

FLORET
A small flower, sometimes without petals, which forms part of a composite flower.

FLUID EXTRACT
See 'Extract'.

FOLLICLE
A dry fruiting body, which splits to release the seed.

FRIABLE
Having a fine crumbly texture.

GALACTAGOGUE
A herb that promotes the flow of milk.

GLABROUS
Lacking hairs.

GLAUCOUS
Covered with a bluish white waxy bloom.

HARROW
An implement for cultivating ploughed ground to break up the soil, control weeds and prepare it for planting.

HEEL
A small portion of stem attached to the base of a shoot when it is broken away from the main stem.

HERB
1. (botanical) A non-woody plant. 2. A plant or part of a plant used for medicine, aroma, flavour etc.

HERBICIDE
A substance, usually a synthetic chemical, used to kill plants.

HIP
Fleshy fruit of the Rose.

HOMEOPATHY
A system of medicine based on potentised infinitesimally minute quantities of substances, which are used to treat symptoms identical to the symptoms that would be produced if a healthy person were given a large amount of the same substance.

HUMUS
Broken down organic matter in the soil.

HYBRID
The result of crossing two parents of different genetic make-up.

INFLORESCENCE
An arrangement of flowers.

INFUSION
A dilute extraction of a herb obtained by pouring boiling water onto it and allowing it to stand for a few minutes.

INTERNODE
The part of the stem between two nodes.

LANCEOLATE
Shaped like the head of a spear, with the widest part at about one-third of the length and tapering to a point.

LATEX
A thick milky fluid that exudes from injuries of some plants.

LINEAR
Narrow with parallel sides.

LIQUID EXTRACT   See 'Extract'.

LOAM
A soil type intermediate between clay and sand.

MANUFACTURING GRADE
A dried herb processed to a standard and form suitable for the manufacturer's requirements, who will process it further in the production of medicines.

METABOLISM
The biochemical processes in living organisms involved in the building up and breaking down of chemical substances.

MIDRIB
The principle and central vein of a leaf.

MOISTURE STRESS
A state of stress in a plant caused by lack of moisture.

MONOCOTYLEDON
Member of a group of flowering plants having one cotyledon or seed leaf.

MONOECIOUS
Separate male and female flowers occurring on the same plant.

MUCILAGE
A slimy substance consisting of gum and water, found in some plants.

MULCH
A layer of material placed on the surface of the soil to suppress weeds, reduce water loss etc.

NERVINE
A herb that helps allay nervous disorders.

NODE
The point on a stem, often appearing as a joint, from which a leaf or leaves (and sometimes roots) arise.

NOXIOUS WEED
A weed with particularly undesirable toxic or invasive characteristics. A weed is usually only declared noxious if there is some prospect of eradicating it or preventing its spread.

NUTRITIVE
A herb used for its nourishment or food value.

OBLANCEOLATE
Similar in shape to lanceolate, except that the broadest part is about two-thirds along its length.

OBLONG
Having parallel sides and rounded ends, the length two to three times greater than the width.

OBOVATE
Egg-shaped with the broadest part towards the tip.

OCTOPLOID
Having four times the normal complement of paired chromosomes.

OPPOSITE
Arranged in pairs at the nodes, on opposite sides of the stem.

ORBICULAR
Circular in outline.

ORGANIC
A system of farming and gardening based on the building up of fertility and plant and animal vitality which avoids the use of synthetic inputs, such as chemical fertilisers or pesticides.

ORGANOLEPTIC
With the use of normal human senses, as in assessing the quality or identification of a herb by its appearance, smell, taste and feel.

OVAL
In the shape of a somewhat flattened circle.

OVATE
Egg-shaped with the broadest part nearer the base (2-dimensional).

OVOID
A solid the shape of a hen egg (3-dimensional).

OXYTOCIC
A herb that stimulates uterine contractions.

PALMATE
Divided into leaflets attached at a central point, or into lobes that extend like a hand.

PANICLE
A branched arrangement of flowers in racemes. Sometimes used to refer to any loosely branched arrangement of flowers.

PAPPUS
Fluffy down attached to the seed to aid its dispersion by wind.

PECTIN
A constituent of some fruits used as a demulcent or thickener.

PERCOLATION
The liquid extraction of soluble constituents by passing suitable solvents down a column of finely milled herb.

PERENNIAL
A plant that lives for several to many seasons, going through a number of flowering and seeding cycles in that period.

PESTICIDE
A substance used to kill pests, usually insects and other invertebrates. Sometimes the term is used more broadly to include all substances used to kill unwanted living things, including weeds and disease organisms.

PEDICEL
The stalk of an individual fruit or flower.

PEDUNCLE
The stalk of a multiple arrangement of fruits or flowers.

PETIOLE
The stalk of a leaf.

pH
A scale used to describe acidity-alkalinity. A pH of 7 is neutral, figures above this are increasingly alkaline, while figures below 7 are increasingly acid. For most crops a desirable pH is between 5 and 7, but as organic matter has some buffering effect, pH is not so critical in an organic system of management.

PINNATE
A compound leaf with leaflets arranged along both sides of a central stem, like a feather.

PINNATIFID
A leaf cut into lobes in a pinnate style, but the divisions not reaching the midrib.

PLOUGH
An implement used to carry out the initial breaking of soil established in grass or a crop.

PREMIUM GRADE
A standard of quality virtually as high as it is possible to attain.

PRIMING
Treatment of seed prior to planting by soaking in water for 24 hours or so.

PROPAGATION
Multiplication of plants by seed or vegetative means.

PROSTRATE
Trailing along the ground or low-growing.

PUBESCENT
Softly hairy.

PURGATIVE
A herb with a strong laxative effect.

RACEME
An unbranched arrangement of flowers along a stem, with the oldest flowers at the base and the youngest at the tip.

REAPING HOOK
A small sickle-shaped harvesting tool with a saw-toothed blade.

RECEPTACLE
The base from which the parts of a flower arise.

REJUVENATION
Cultivation treatment to re-invigorate a stand of a perennial herb whose production has declined.

REGENERATIVE WEED
A weed capable of regenerating from pieces of root, rhizome or stem left in the soil.

REHYDRATION
See 'Remoistening'.

REMOISTENING
The re-absorption of water from moist air by dry or partially dry plant material.

RHIZOME
An elongated underground stem growing more or less horizontally, giving rise to roots and leaf shoots at the nodes and/or at the tip.

ROOTSTOCK
Fleshy central core of a root system.

ROSETTE
A clustered circular arrangement of leaves attached to a central crown and close to the ground.

RUBBING
A processing method of working dried herb through a wire screen to break up the leaves and separate unwanted stem.

RUNNER
1. Any horizontal stem or root, above or below ground, which extends from the parent plant and gives rise to new plants.

2. Sometimes the term is used more specifically to refer to a stem that trails over the ground surface, taking root at the nodes and/or the tip.

SCHEDULED
Officially listed on one of the Poisons Schedules.

SECONDARY WEED
In some states, a category of undesirable weeds second to noxious weeds.

SEDATIVE
A herb that helps reduce tension or induce sleep.

SELF-SOWING
The propagation or spreading of a herb without the grower's assistance by means of seed that falls to the ground.

SEPALS
The usually leaf-like outside parts or covering of a flower.

SEQUENTIAL
One after the other over an extended period, rather than all at once.

SERRATED
A leaf margin with saw-like teeth.

SESSILE
Without a stalk.

SPATHE
A large bract more or less enclosing a cluster of flowers.

SPIKE
An arrangement of stalkless flowers along a single stem, with the oldest flowers at the base.

Glossary

SPIKELET
A small secondary spike, as typically found in grasses.

SPORE
A minute reproductive cell, usually airborne, formed by primitive plants such as mosses, ferns, horsetails etc.

STOLON
An above-ground stem that bends over or grows horizontally and forms roots at the nodes and/or the tip.

STRATIFICATION
A seed treatment, which usually involves subjecting it to a period of moist cold, just above or below freezing, depending on the species.

STROBILE
A cone-like fruiting body.

SUCCUS
The pressed out juice of a fresh plant, preserved by the addition of alcohol.

SYMBIOSIS
A relationship between two organisms, which is of benefit to both.

STERILE
Unable to produce viable seed.

SWG
Standard Wire Gauge.

TAPROOT
A central root, usually enlarged and containing food reserves.

TEA GRADE
A dried herb processed to a particle size suitable for use in making tea.

TENDRIL
A thin, wiry appendage on a plant enabling it to attach itself and climb up other plants or objects.

TERMINAL
Occurring at the end of a main stem or branch.

TETRAPLOID
Having twice the normal complement of paired chromosomes.

TINCTURE
An extraction of the soluble constituents of a herb produced by soaking it for an extended period in a mixture of alcohol and water.

TOMENTOSE
Covered with a dense, more or less matted layer of fine soft hairs, usually giving a whitish appearance.

TONIC
A herb that has restorative powers and helps invigorate the body.

TRADE HERBS
Herbs of mainly conventionally grown origin handled by the global herb trade.

TRADE QUALITY
The poor to mediocre level of quality usually found in trade herbs.

TRANSPIRATION
The loss of water from the leaves of living plants.

TRIENNIAL
A plant that goes through its life cycle, from germination to death, in 3 years, normally only flowering in its third season.

TRIFOLIATE
Compound leaf consisting of three leaflets.

TRIPINNATE
Triply pinnate: the compound leaf consisting of leaflets arranged pinnately along tertiary ribs, which are attached to secondary ribs attached to a central midrib (see illustration of 'Leaf Arrangements' below).

TUBER
A swollen part of a stem or root containing food reserves.

UMBEL
A flat-topped, umbrella-shaped or spherical arrangement of flowers whose stalks are attached at a single central point.

VARIEGATED
Marked with patches of a different, usually lighter, colour.

VASCULAR BUNDLES
Conductive tissue in plants consisting of minuscule tubes.

VEGETATIVE PROPAGATION
The production of new plants from parts of an existing plant other than seeds.

VERMIFUGE
A herb used to help expel or destroy intestinal worms.

VOLATILE OIL
An oil, usually aromatic, which evaporates or distils readily.

VOLUNTEER
A plant or crop that has self-sown.

VULNERARY
A herb used to treat or help heal wounds.

WATERLOGGING
The retention of excess water in the root zone of the soil, causing the exclusion of air and damage to susceptible plants.

WELL-DRAINED
A soil structure that allows excess water to flow down through it rather than retaining it within the root zone.

WILDCRAFTING
The harvesting of plants growing wild in an environment not subject to farming or other human activity that might cause any contamination or otherwise adversely affect them. Harvesting is carried out with special attention to the quality of the product, the ecological impact of harvesting and the maintenance of natural populations of the species.

WINTER-ACTIVE
Continues to grow during winter instead of going dormant.

General Index

Major entries are in bold type
For references to *specific herbs* see **Index of Individual Herb Crops** p553

5-star assessment of quality 216

Accounts for shipment 224
acidity of soil 19
active constituents 3, 8, 191, 206, **216**, 535
    see also medicinal herbs; therapeutic herbs
adulteration 535
    see also contamination; foreign material
aerial parts
    see also leaf herbs
    assessing dryness **170–171**, 172, 211, 215
    definition 244, 535
    drying 160, **166–171**, 179, 181–182
    drying costs 194, 238
    field curing 132, 190
    harvesting 57, 58, 65–66, 67, 68, 70, 93, **119–133**
    heated air drying 166, 179, 181–182
    inadequate drying 153–154, 211, 215
    manufacturing grade productivity 239
    marketing **228**, 240, 519, 520, 521–527
    particle size *Plate 8.7*, 197–198
    percentage of stem 237–238
    price comparisons 237–238, 521–527
    processing costs 238, 239
    processing with chaff cutter *Plate 8.7*, 202
air exchange in drying 175–176
air supply options for drying 155
airflow *see* drying – airflow
alluvial soils 17
ambient air drying *see* drying – ambient air
animal manures
    composting 26, **72, 73, 74**, 85, 233
    contamination of herbs 45, 72, 215, **233**
    weed seeds in 26, 44, 73, 74
annual operating costs 236, 237
annuals 60, 69–70, 421–461, 535
    retaining leaf at harvest 70, 133
    weed control 69, 422
aphids 101–102, 233, 283
'Appendix B' 233–234
area in crops 5, 21, 556
aromatic herbs **13**, 151, 175, 191, 192, 203, 535
    see also essential oil levels
assessing moisture *see* drying; soil
assessing quality 4, 197–198, 214–216
associations 532
Australian herb market *see* marketing
autumn
    planting 47, 56, 249
    compost application 84
    slug damage 103
    watering 94
    weed growth 39

Back care 38
back-up drying system 162, **166**, 170, 174–179

bare fallow **27–29**, 40, 47, 250, 535
bark crops
    assessing dryness 184
    chopping for drying 184
    drying 160, 184
    harvesting 68, 147–149, 402–403
    marketing 229
    processing 198, 214
    tannin content 184
bark in compost 72
batch number 219–220
batch separation 218
beetles on Chamomile flowers 444
bending and back strain 38
biennials 69–70, 421–461, 535
    post-harvest treatment 133
    weed control 24, 422
biodynamic
    associations 118
    calendar 113–116, 118
    certification 532
    compost 78, 102, 108–113
    definition 7, 105, 535
    farm as an organism 105, 117–118
    further reading 118
    harvesting times 116, 134
    moon and planet rhythms 113–116, 118
    peppering 116–117
    pest and disease control 116–117
    preparations
        compost preps (Preparations 502–507) 78, 108–112
        horn manure (Preparation 500) 105–107
        horn silica (Preparation 501) 107–108
        Horsetail tea (Preparation 508) 116
    weed control 116–117
black plastic
    for mulching 41–42, 65, 251
    over greenhouse for drying 194
black spot 412
Blackberries 20
blood and bone in compost 73
blueberry rake 135
botanical names 47–48, 243–244
boxes for packing and shipping 217, 223
Bracken 20, 27, 72
broad-acre operations 5, 21, 63
    drying 191–193 *see also* field curing
    harvesting 126, 136–137, 143–144, 147
broad-fork 32, 33, 140, 532
broadcasting seed 53
bruising of herb **120, 151**, 156, 169, 192, 212, 216, 536
buds, assessing dryness 171, 172, 173
buffer zone
    containment of invasive herbs 29–30, 31, 60, **248**
    spraydrift avoidance 464

buffer zone (*continued*)
    weed management **29–30**, 31, 33, 43, 61
bunches for drying 155, 190–191
buttercups 18
buyers of Australian grown herbs 227–228, 532

Cabbage white butterfly 260
cabinet dryer 176–188
    see also heated air drying
    bark herbs 184
    building 176–179, 184–185
    'dry chimney' 180–181
    finishing partially dried herbs 179
    flower herbs 182
    freshly harvested herbs 179–182
    fruit herbs 182
    leaf herbs 177, 181–182
    loading 179, 180–181, 183
    operating 179–184, 185
    root herbs 176, 177, 182–183
    solar heat options 185–188
    vermin control 178
    wood-fired 184–185
Capeweed 38, 104
capital outlay 11, 235–236
carbon dioxide fumigation 154, 223
carbon:nitrogen ratio in compost 72, 73, 74, 85
case hardening and drying roots 183
catching scythe *see* scythe
catching tray for rubbing herbs 200, 201
central drying facility 10, 194–195
certification, organic 8, 22, 532
    acceptable residue levels 215
    animal manures 26
    expense 22
    foliar feeding 85
    labelling requirements 221
    shared drying facility 195
    snail bait 102
    soil test 20
chaff-cutting 201–203 *see also* chopping
    bark herbs 184, 214
    before drying 169, 182–183, 184, 209, **212**
    manufacturing grade *Plate 8.7*, 211–212
    tea grade 209, 212
Chamomile flowers (Preparation 503) 109
chemical residues 4, 8, **19–20**, 42, 71, 144, **215**, 464
chicken manure 17
    causing aphids on crops 101–102
    compost ingredient 72
    contamination of herbs 215, 233
choice of crops *see* crop choice
chopping *see also* chaffcutting
    bark herbs 184, 214
    leaf herbs *see* chaffcutting
    root herbs 182–183
circulatory airflow drying 176, 188–190

543

clay 17, 18–19
climate 13–16, 509, 511–517
    see also specific herbs; humidity
  evaluating 15
  frost 15
  microclimate 13, **15–16**, 58–59, 401, 509, 511–517
  rainfall **14–15**, 87, 95, 510
  temperate 5, 8
  temperature **13–14**, 15, 16, 95, 509, 511–517
  wind **15**, 16, 89, 94, 509, 511–517
cloches 55
cloning 48–49
coal heat for drying 174
cockchafers 329
Coleman Gung-hoe 35, 36, 37, 531
comb
  flower harvesting 135–136, 137–139, 445, 446
  fruit harvesting 144
combine harvester for seed harvesting 145, 146, 147
commercial organic fertilisers 7, 85
companion planting 60
complementary crops 21
compost 8, 11, **71–86**, 245, 510, 511–517, 536
    see also specific herbs
  aeration 78–79
  alternatives to making compost 85–86
  animal manures 26, 72, 73, 74, 85, 233
  application 26, **80, 83–4**, 510, 511–517
  biodynamic 78, 102, 108–113
  building heap 74–75, 76–79
  carbon:nitrogen ratio 72, 73, 74, 85
  contamination of harvest 83, 84
  desirable characteristics 71
  factors for success in making 78–79
  forage harvester 11, 75–76
  gathering materials 74–76
  hot composting 43, 44, 71, **78, 79–80**, 81, 82, 233
  immature 80
  ingredients 22, 44, **71–76**, 84–85
  materials 22, 44, **71–76**, 84–85
  moisture content 78, 79
  monitoring 79
  pH level of soil 19
  preparations (biodynamic) 78, 108–113
  problem crops 84
  rates of application 80, 510, 511–517, 556
  requirements of crops 80, 83–84, 510, **511–517**
  snakes in compost heaps 79
  spreading 26, **80, 83–84**, 510, 511–517
  sustainable production 84–85
  temperature of heap 43, 44, 71, 78, **79**, 80, 81, 82, 233
  tools and equipment 75, 76–77, 83
  turning 79–80, 81–82, 233
  unturned 79
  weed control 43, 44, 71, 79, 80
composting *see* compost; heating of harvested material
compressing herbs when packing 218
concealed moisture when drying 153–154, 171, 172, **211**, 215
containment beds 29–30, 31, 60, **247–248**, 536
contamination
  acceptable levels 71, 213, 215

contamination (*continued*)
  animal manures in herb 45, 72, 215, 233
  chemical 4, 8, 19–20, 42, 71, 144, **215**, 464
  compost in herb 83, 84
  during drying 154–155, 168, 190
  foreign plant material 23, 42, 132, **154–155**, 156, 191, 215, 465
  heavy metal 19, 71, 262
  insects 139–140, 233
  microbial 143, 215, 233 *see also* animal manures
  quality standards 215
  removing from herb 103, 132–133, 154–155, 156, 209, **213–214**, 233
  residues in soil 19–20, 42, 144
  snails and slugs 103
  soil in herb 132–133, 143, 154, **213–214**, 218, 233
  tree leaves 59, 401
  vermin 154, 178, 191, 194, 222
  visibility 156, 218
  weeds in harvest 23, 132, 154, 156, 191, 465
  wild harvest problems 464, 465
contract harvesting 144, 147
cool storage and insects 223
coppicing and bark harvesting 148
cosmic influences and biodynamics 113–116
costs *see* expenditure; production costings
cow manure in compost 73
crop choice 9, 21
    see also growth types; trial plot
  comparisons of crops 509–528
  diversity 9, 21, 98–99
  marketing considerations 9, 21, 228
  weed control considerations 23–25 *see also* weed control
crop flexibility 232
crop rotation 60–61, 99, 245, 421
crops, individual 509–527, 553–555
crown division 49, 50, 67, 69, 335
culinary grade herbs 197, 230
  acceptable stem 208
  assessing dryness 170–171, 172, 215
  definition 197, 536
  marketing 10, 230, 520, 521–527
  particle size *Plate 8.5*, 197
  processing *Plate 8.5*, 197, 198–201, 205–211, 212–214
  quality 23, 214–216, 230
  rescreening 206, 208
  rubbing 205–211
cultivation tools and equipment 22, **26**, 27, 30, **32–37**, 531
cultivation and management of herbs 8 *see also specific herbs*; growth types; propagation; soil; weed control
curing *see also* drying
  field curing leaf crops 132, 155, 190
  seed crops 145, 146–147
cuttings for propagation **49**, 50, **51**, 536
Czech Chamomile harvester 136–137

Dam capacity 87
Dandelion flowers (Preparation 506) 110–111
Dandelion root rot *Plate 2.7*, 305, 308
decoction herbs *Plate 8.6*, 198, 536
dehumidifier for drying 155, 175, **193**, 194
dense planting and weed control 32

dent (mesh size) 199–200, 537
deterioration during harvesting and drying
  bruising of herb **120**, **151**, 156, 169, 192, 212, 216
  composting *see* heating of harvested herb
  fading **151**, 161, 173, 190, 193, 216
  floor drying 190–192
  flowers **133–134**, 136, 138, 140, **171, 173**
  fruit 144–145
  handling wilted herb 120, 151 *see also* bruising of herb
  heating of harvested material
    aerial parts **120**, 127, 151, 156, 191, 192
    flowers **134**, 135, 136, 138, 140
  inadequate drying **153–154**, 171, 172, 211, 215
  reducing deterioration 156
  remoistening 152, **153**, 166, 170, 171
  roots 142, 151
  slow drying 151, 171
  sweating 165, 191
  temperature of drying **151–152**, **175**, 245, 519, 520–527
digging roots 141–142, 143–144
direct burner for drying 155, 174, 192
direct planting 51, 53
  stale seed-bed 27, 28
disc harrows 26, 27
diseases *see* pests and diseases
distances between rows and plants
    see also specific herbs
  general layout **31–32**, 61–62, 94, 244–245
  growth types 249–250, 297, 335, 381
  individual crops 509–510, 511–517
  metric conversion 62, 556
distribution, doing your own 227–228
diversity of crops 9, 21, 98–99
division as propagation method **49**, 50, 51, 537
Dock (as a weed) 18, 20
dormancy 537
  perennial crown herbs 67
  seed dormancy 55
  spreading herbs 41, 65
  weed proliferation in winter 8, 14, 23, **40–41**, 65, 250–251
  winter dormancy mulch **41–42**, 65, 251
drainage of soil **17–19**, 57, 509, 511–517, 542
  planning layout 59
  raised beds 56–58, 250
drawknife 148, 149
drilling seed 53, 537
drums for storage and shipping 217, 223
'dry chimney' 180–181
drying 9–10, 21–22, **151–195**, 245, 519, 521–527
  aerial parts 160, **166–171**, 179, 181–182 *see also* leaf herbs
  airflow **152**, 160, 162, 163, 169–170, **175–176**
  cabinet dryer **178–181**, 182, 183, 184, 185
  circulatory airflow 176, 188–190
  relative humidity 152, 169–170, 175
  solar heat systems 185–188
  suspended floor dryer 191, 191–192
  vertical airflow 176–188, 191–193
  ambient air drying 10, 120, 155, 156, **160–173**, 194, 535
  advantages and disadvantages 161
  airflow 162, 163, 169–170

General Index

drying (*continued*)
    heated back-up system 162, **166**, 170, 174–179
    monitoring 170–171
    relative humidity 14–15, **152–153**, 160, 166, 169–170
    remoistening **153**, 170, 171, 210–211
    solar heat gain 155, 160, 161–162
    unfavourable weather 169–170
    ventilation 162, 163, 169–170
  assessing dryness 170–171, **172**, 173, 205, 211, **215**
  bark herbs 160, 184
  bruising of herb **120**, **151**, 156, 169, 192, 212, 536
  bunches 9, 155, 190–191
  cabinet dryer 176–188
  central drying facility 10, 194–195
  choice of system 155–156
  comparisons of drying costs 194, 238–240
  composting *see* heating of harvested material
  contamination **154–155**, 156, 168, 190, 191
  cost comparisons of drying 194, 238–240
  dehumidifier 155, 175, **193**, 194
  designing a system 155–156
  dryness levels for safe storage 154, 215
  fading **151**, 161, 173, 190, 193, 216
  field curing **132**, 145, 146–147, 155, 190
  finishing partially dried herbs 179
  floor drying 155, 190, 191–193
  flower herbs 160, 171–173, 182
  for quality **151–155**, 215–216 *see also* deterioration
  freeze drying 194
  fruit herbs 160, 173, 182
  greenhouse with black plastic cover 194
  handling herb 120, 151
  harvest timing and drying space 120
  heat pump 155, 175, 193, 194
  heated air drying 174–190, 191–192, 194
    see also cabinet dryer
    airflow **175–176**, 182, 183, 184, 185, 188–190, 191, 192–193
    back-up system for ambient air drying 162, **166**, 170, 174–179
    circulatory airflow 188–190
    direct burner 155, 174, 192
    electric heat 174, 177, 194
    gas heat 174–175, 194
    heat pump 155, 175, 193, 194
    heat source 155, **174–175**, 177, 178, 180, 184, 185–187, 188, 189, 191–192, 194
    insulation **176**, **177–178**, 184, 189, 191
    materials for chamber 176, 177
    solar heat 155, 160, 161–162, 175, **185–188**, 194
    temperature control 175, 185, 188
    wood heat 174, 177, **184–185**, 194
  holding time 127, 140, 151 *see also* heating of harvested herb
  humidity 14–15, 151, **152–153**, 160, 166, 169–170, 175
  in bunches 9, 155, 190–191
  inadequate drying **153–154**, 171, 172, 173, 211, 215
  inside floor 155, 191–193
  intermixing 154–155
  large scale 10, 132, 190, 191–192

drying (*continued*)
  leaf herbs 159, 160, **167–171**, 179, 181–182
  loading herbs **166–169**, 173, 179, 180–181, 183, 189, 192
  mice and rats, controlling 154, **163–164**, 178, 191, 194, 209
  moisture migration 154 *see also* inadequate drying
  monitoring 155, 170–171, 173, 180, 182, 183, 184
  moth infestation 154, **222–223**, 448
  number of screens required 159–160
  outside 132, 155, 190
  portable screens *see* screens, portable
  preliminary chopping **169**, 182–183, 184, 209, **212**
  quality and drying 151–155, 215 *see also* deterioration
  rats and mice, controlling 154, **163–164**, 178, 191, 194, 209
  recirculation of used air 193
  recording system 155, 221–222
  rehydration *see* remoistening
  relative humidity 14–15, 151, **152–153**, 160, 166, 169–170, 175
  remoistening **153**, 170, 171, 210–211
  root herbs 154, 160, 172, 176, 177, **182–183**, 189
  running costs 194
  screen shelves 155
  screens, portable 155, 156–160
    advantages 156
    cost and lifespan 160
    loading **166–169**, 173, 179, 180–181, 183, 189
    making 156–159
    number required 159–160
    placement 164–166
  seed herbs 145, 146–147
  selecting a system 155–156
  shed design and structure 160–165
  slug-infested herb 103
  small scale 9, 10, 190
  snail-infested herb 103
  soil contamination 132–133, 143, 154, 213–214, 218, 233
  solar heat 155, 160, 161–162, 175, **185–188**, 194
  sunlight exposure 132, 151, 190
  sun drying under cover 193–194
  suspended screen floor 155, 191–193, 194
  systems, selecting and designing 155–156
  temperature **151–153**, **175**, 185, 188, 245, 519, 521–527
  testing for dryness 170–171, **172**, 173, 205, 211, **215**
  time (length of time) 151–152
  turning herb during drying 151, 191, 192
  ventilation 162, 163, 169–170 *see also* airflow
  vermin control 154, **163–164**, 178, 191, 194, 209
  wild harvested herbs 465–466
  wilting 120, 151, 192 *see also* bruising of herb
duplex soils 18–19
dust mask **204**, 205, 212, 237 *see also* protective equipment

Dutch hoe 35, 36, 39

Ear muffs 204
earthworms 17
economics of herb growing 235–240
effectiveness of medicinal herbs v, 7, 8, 206, 216, 229
electric heat for drying 174, 177, 194
environmental impact when wildcrafting 465, 467
equipment *see* tools and equipment; protective equipment
erosion control 19
  following cultivation 29
  green manuring 26
  layout 58
essential oil levels 537
  climatic effect 13
  drying temperature 151, 175, 245
  hammermilling 203
  harvest timing 119–120
  suspended floor drying 191, 192
ethylene oxide fumigation 4, 223, 233
expanding clump herbs 66–67, 133, 297–333
expenditure
  annual operating costs 236–237
  capital outlay 11, 235–236
  production costs compared 237–240
export markets 11, 226
exposure 15–16, 509, 511–517
extensive roots 40, 49, 65

Fabric-type mulches 42
fading **151**, 161, 173, 190, 193, 216
family names of herbs 244
fan *see* airflow
farm as an organism 105, 117–118
feed-out shed as compost source 74–75
fences 21, 30, 31, 61
fertilisers, organic 7, 85–86 *see also* compost
fertility of soil **17**, **71**, 80, 510, 511–517 *see also* compost
field curing 132, 145, 146–147, 155, 190
field ripening seed 145, 146–147
filling bags 218–219
filters for irrigation system 92
financial viability of herb growing 235–238
fine-toothed harrows 26
fines, removing 213–214
first screening in processing leaf 206, 207
  holding first screening 210
fish waste in compost 73
flail 145, 146
flaming
  rust control **281–282**, 291–292, 531
  weed control 44–45
flat profile beds 56–57
flea beetle 283
flood irrigation 88
flood waters as contamination source 20
flower arrangements and shapes 530
flower comb **135–136**
  combing technique 137–139, 445, 446
flower herbs
  assessing dryness 171, 172, **173**, 215
  definition 244
  deterioration 133–134, 136, 138, 140, 171, 173
  drying area needed 160
  drying in ambient air 160, 171–173

545

flower herbs (*continued*)
    drying in cabinet dryer 182
    fading 173
    hand picking 139
    harvesting 133–140
    inadequate drying **153–154**, 171, 172, **173**
    marketing 229
    moths 222–223
    overmature flowers 134, 140
    prices for 229, 521–526
    processing 198, 214
    productivity in harvesting and drying 240
flower shapes and arrangements 530
focus of regeneration *see* regeneration
foliar feeding 85
forage harvester
    compost making 11, 72, 75–76
    harvesting herbs 127
    trimming root tops 142
foreign material *see* contamination
forks *see also* broad-fork
    for compost handling 75, 77, 83
    for digging roots 32, 140, 141
form of herb and labelling 219
fossil fuels 174–175
freeze drying 155, 194
freezing for insect control 154, 221, 223
freight costs 223–224
fresh herb crop prices 229
frost 15
fruit herbs
    artificial heat for drying 160, 173, 182
    assessing dryness 171, 172, 173, 215
    cabinet dryer 182
    circulation drying 189
    cleaning 198, 214
    definition 244
    deterioration 144–145
    drying area needed 160
    drying in ambient air 160, 173
    harvesting 144–145
    processing 198, 214
fumigation
    carbon dioxide 154, 223
    methyl bromide, ethylene oxide 4, 223, 233
fungal control 116 *see also* rust; pests and diseases
furrow irrigation 88–89

Gas heat for drying 174–175
genetic diversity growing from seed 51
germination of dormant seed 55
gloves 204
goggles 204
gooseneck hoe 34, 36
grain cradle 121, 145
grass
    as soil builder 26, 46
    composting 72, 73
grasshoppers 21, 61, 98, **100–101**, 233
    on individual crops 255, 283, 289, 299, 305, 317, 322, 329, 372, 424, 430
grazing animals as weed control 45
green manures 26, 61, 85
Green Pharm Health Products 3–4
greenhouse 55, 56
    covered in black plastic for drying 194
gross returns 236–237, 520, 521–527

growing requirements of herb crops 509–519 *see also specific herbs*
group marketing 242
grower associations 532
grower network 235
growers and *Therapeutic Goods Act* 232–235
growers' options if prices fall 232
growth stage in harvesting 119, 134, 140, 144, 145, 147
growth types **65–70**, 243, 509, 511–517
    see also harvesting; propagation methods; weed control
    annuals, biennials, short-lived perennials 69–70, 421–461
    expanding clump 66–67, 297–333
    perennial crown 67, 335–380
    spreading 65–66, 247–296
    trees and shrubs 68–69, 401–420
    woody perennials 67–68, 381–399
guidelines for wildcrafting 463–465, 466–467

Habits of growth *see* growth types
hammermill 198, **203**, 209, 211, 212
hand cultivation tools 22, 33–7, 531
hand-digging roots 143, 144
hand hoes *see* hoes
hand-picking flowers 139
hand threshing 145, 146, 147
handling harvested herb 120, 151
hard pan 26, 32, 33
hardening off before transplanting 56
harrowing and soil preparation 26, 27, 538
harvesting 9, **119–149**, 245
    see also specific herbs; specific plant parts
    aerial parts 65–66, 67, 68, 70, **119–133**
        equipment 22, 120–127
        field curing 132
        heating of harvested herb 120, 127, 151
        height to cut 65–66, 67, 68, 70, 123, 126, 132
        large scale mechanical 126–127
        layout of crop 57, 58
        post-harvest treatment **133**, 252–253, 336
        regeneration **65**, 66, 67, 68, 69, 70, 132, 133
        technique 127–133
        timing 93, **119–120**, 127, 466, 519, 521–527
    annuals, biennials, short-lived perennials 70, 422
    bark 68, **147–149**, 402–403
    biodynamic calendar 116, 118
    bruising of herb **120**, **151**, 169, 212, 536
    contamination of harvest
        animal manures 45, 72, 215, **233**
        compost 83, 131
        foreign plant material 23, 42, 131, 132, **154–155**, 215, 465
        insects 139–140, 233
        microbial 143, 215, 233 *see also* animal manures
        pesticides 464
        removing 103, 132–133, 154–155, 156, 209, **213–214**, 233
        slugs and snails 103
        soil 131, 132, 143, 154, **213–214**, 233
    contract harvesting 144, 147

harvesting (*continued*)
    equipment 22, 120–127, 135–137, 140–141, 144, 145–146, 147–148
    expanding clump herbs 67, 133
    flat profile beds 56–57
    flowers 133–140
        combing **135–139**, 445, 446
        deterioration **133–134**, 136, 138, **140**, 171, 173
        drying capacity 134, 160
        equipment 135–137
        hand picking 139
        heating of harvested flowers 134, 135, 136, **138**, 140
        insects 139–140
        leaf inclusion 137–138, 139
        mechanical harvesting 136–137
        other methods of harvesting 139
        over-mature flowers 134, 140
        rate of harvest 138–139
        technique 137–140
        timing **134–135**, 466, 519, 521–527
    forest trees 68
    fruit 68, 144–145
    handling of wilted herb 120, 151
    heating of harvested material
        aerial parts **120**, 127, 151
        flowers 134, 135, 136, **138**, 140
    irrigation cycle 93, 120
    layout of crops for ease of harvest 58
    leaf *see* aerial parts; trees and shrubs
    leaf and flower *see* aerial parts; trees and shrubs
    mechanical 22, 126–127, 136–137, 143–144, 145, 146, 147
    perennial crown herbs 67, 335–336
    post-harvest treatment 39, **133**, 251, 252–253, 336
    quality considerations 119 *see also* deterioration
    raised beds 57, 58
    records of harvest 155, 221–222
    roots 66, 67, 140–144
    seed 145–147
    slugs and snails 103
    soil depletion 71
    spreading herbs 65–66, 255
    stage of growth 119, 134, 140, 144, 145, 147
    technique 127–132, 137–140, 141–144, 144–145, 146–147, 148–149
    timing 93, 116, **119–120**, 127, **134–135**, **140**, **144**, **145**, **147**, 466, 519, 521–527
    tools *see* harvesting – equipment
    trees and shrubs 68, 402–403
    weed contamination **23**, 132, 154, 156, 191, 465
    wilting 120, 151 *see also* bruising of herb
    woody perennials 68, 123, 126, 381
    yields 246, 519, 521–527 *see also specific herbs*
hay in compost 72, 73, 75
header for seed harvesting 145, 146, 147
heat bank for drying 175
heat pump dehumidifier 155, 175, 193
heat source *see* drying – heated air
heated back up system 162, **166**, 170, 174–179
heating of harvested material
heavy metals 19, 71, 262

General Index

heavy soil **17**, 57, 103, 509
hedge trimmer for harvesting 123, 126
height of harvest 65–66, 67, 68, 70, 123, 126, 132
herb
    fever 11–12
    grower associations 532
    growth habits *see* growth types
    marc in compost 72
    parts used 244, 519, 521–527
    propagation material suppliers 531
    seed suppliers 531
    teas 10, 23 *see also* tea grade herbs
herbal medicines **v**, 4, 10, 23, 154, 214
    see also therapeutic herbs
    active constituents 3, 8, 10, 191, 206, **216**, 535
    effectiveness of **v**, 7, 8, 206, **216**, 229
    quality of raw materials **v**, 3, 10, 23, 206, **216**, 228, 229
herbs, individual species *see*
    Index of Individual Herb Crops 553–555
    Information Charts 509–527
herbs that become weeds 29–30, 43, 464
hoes **33–37**, 38, 39 *see also* rotary hoe
    vine hoe 141
    wheel hoe 1, **33–34**, 36, 39, 531
holding time before drying 127, 140, 151
homogeneity of finished product 218
hormone rooting powder 49
horn manure (Preparation 500) 105–107
horn silica (Preparation 501) 107–108
horse manure in compost 73
Horsetail tea (Preparation 508) 116
hot-water weeding 45
humidity 14–15, 151, **152–153**, 160, 166, 169–170, 175, 205, 209
humus 71, 109, 538
hydroponics 8

Identification 47–8, 215, 244
    see also specific herbs
    during drying 155
    flower shapes and arrangements 530
    leaf shapes and arrangements 529
    plants grown from seed 51
    varieties, good and inferior 48, 51, 216
    vegetative propagation material 48
    wild harvested herbs 465
imported herbs 3, 4, 10, 11, 13, 190, **225–226**
    see also trade herbs
inadequate drying **153–154**, 171, 172, 173, 211, 215
income 11–12, **236–237**, 520, 521–527
infusion herbs *see* tea grade herbs
insects *see also* pests and diseases
    control in storage 154, 221, 222–223
    on crops when harvested 139–140, 233
inside floor drying 155, 191–193
insulation of drying chamber **176**, **177–178**, 184, 189, 191
intermixing when drying 154–155
invasive herbs *see* spreading herbs
irradiation 4, 223, 233
irrigation 8, 11, 19, **87–95**, 245, 510, 511–517
    see also specific herbs; watering
    automatic control 92, 94
    beeswax for sealing fittings 92
    drip 90–91, 95

irrigation (*continued*)
    flood irrigation 88
    frequency 95
    furrow irrigation 88–89
    harvesting timing 93, 120
    layout plan 57, 59, 91–93
    leaf burn 89, 94
    management 91–95, 279
    moisture stress 56, **91**, 98, 538
    mulches 43
    newly planted crops 56
    night watering 94
    overwatering 90–91, 94, 95
    pipes 91–92
    pressure variation 92
    rates 87, 95, 510, 511–517
    setting up 91–93
    soil moisture 56, 91, 93, 94
    sprinkler shadow 59, 402
    sprinklers 59, **89–90**, 91, **92–95**
    storage 87
    systems 88–93
    tall plants 59, 402
    teflon tape alternative 92
    timing 93–5, 120
    trickle 90–91, 95
    water quality 20, 88
    water supply 20, 87–88
    wind effects 89, 94

Joint marketing 228

Kikuyu 20

Labelling 155, 219–221, 224
ladybirds 101
lawn-mower for mowing stubble 133
layering 50, 51, 67
layout 29–32, 56–63, 244–245
    see also specific herbs
    companion planting 60
    containment 29–30, 31, 60, **247–248**, 536
    drainage 17–19, **57–58**, 59, 509, 511–517
    erosion control 19, 58
    expanding clump herbs 66, 297
    flat profile beds 56–57
    harvesting considerations 57, 58
    irrigation considerations 57, 59, 91–93
    low-profile raised beds 58, 250
    microclimate 15–16, 58–59, 401
    perennial crown herbs 67, 335
    pest and disease management 61, 99, 102
    proximity to trees 59, 401, 402
    raised beds 56, 57–58, 250
    shade 15, 16, 58, **62**, 401, 509, 511–517
    soil types **17–19**, 57, 59, 509, 511–517
    spacing **30–32**, **61–62**, 94, 244–245
        for growth types 249–250, 297, 335, 381
        for individual crops 509–510, 511–517
    spreading herbs 65, 249–250
    trees and shrubs 59, 401–403
    vehicle access 63
    weed control considerations 29–32, 58
    woody perennials 68, 381
leaf and flower, definition 244
leaf arrangements and shapes 529
leaf blight 260
leaf burn 89, 94

leaf damage and plant growth 97–98
leaf herbs
    see also aerial parts; harvesting
    assessing dryness 170–171, 172, 215
    chopping before drying 169, 212
    definition 244
    drying in ambient air 159, 160, 167–171
    drying in cabinet dryer 177, 181–182
    heated air drying 166, 179, 181–182
    particle size *Plates 8.5, 8.6, 8.7*, 197–198
    price comparisons 228, 237–238, 519, 520, 521–527
    processing *Plates 8.3, 8.4, 8.5*, 198, 199–214
    soil contamination 132–133, 154, 209, **213–214**, 233
    stem content 206–209
    stripping 209, 212–213
leaf shapes and arrangements 529
leaf spot 305
leafminer 255
legal requirements when packaging 221, 232–234
legumes 46, 84
length of storage of dried herbs 10, 218
life cycle interruption of pests and diseases 98, 104
light soil **17**, 18–19, 57, 94
liquid manures 85
livestock and weed control 44, 45
loading drying screens **166–169**, 173, 179, 180–181, 183, 189
locusts 100
longevity of crops 8, 69, 252, 336, 381 *see also* rejuvenation
lopping and bark harvesting 148, 403
low-profile raised beds 58, 250
LP gas
    drying with 174–175, 177
    flaming for rust control 281–282, 531
    flaming for weed control 44–45

Mail order retailing 228
making compost *see* compost
management of crops *see* growth types; *specific herbs*
manufacturers and herb market 227, 228, 229
manufacturing grade herbs
    aerial parts and leaf crops 228
    bark crops 229
    chaff-cutting *Plate 8.7*, 169, 182–183, 184, **211–212**
    consumption and demand 229, 520, 521–527
    definition 10, 538
    flower crops 229
    fresh herb crops 229
    marketing 10–11, **228–229**, 236, 246, 520, 521–527
    particle size *Plate 8.7*, 198
    price 11, **228–229**, 236, 520, **521–527**
    quality 228
    root crops 228
    seed crops 228–229
    stem content 171, 198, 208, **237–238**
market projections 241
marketing 7–8, 9, 10–11, **225–242**, 246, 519–527
    see also specific herbs; price
    boom and bust 231–232
    buyers of Australian grown herbs 227–228, 532

547

marketing (*continued*)
  comparison of gross returns 236–238
  crop choice 9, 21, 228
  culinary herbs 230, 520, 521–527
  distribution 227–228
  establishing a market 240–241
  expenditure 11, 235–237
  export markets 11, 226
  financial viability 235–238
  flexibility 232, 240
  future trends 231–232, 241
  group marketing 242
  grower network 235
  import replacement 226
  imported herbs 3, 4, 10, 11, **225–226**, 519, 521–527
  joint marketing 228
  mail order retailing 228
  maintaining a market 241
  manufacturers 227, 228, 229
  manufacturing grade 10–11, **228–229**, 236, 246, 520, 521–527
  organic herbs 7, **226–232**, 520, 521–527
  'other use' categories 230
  overview of market 225–228
  premium grade 10–11, **226**, 227, **229–230**, 231, 237, 246, 520, 521–527
  price stability 230–232
  quality of product 5, 7–8, 10–11, **225–227**, **228–231**
  second-grade organic 229, 231
  tea herbs 10–11, **229–230**, 520, 521–527
  Therapeutic Goods Act 1991 229, 232–235
  trade herbs 10, 11, **225–226**, 519, 521–527
  value adding 228, 232
  wholesalers 227, 532
  wild-harvested herbs 466
marsupials 61, 154, 164
meadow-crops
  compost spreading 83, 84
  direct seeding 30, 53
  expanding clump herbs 66
  grasshopper damage 101
  perennial crown herbs 67
  rejuvenation 251–252
  trickle irrigation 90
  weed management and control 46, 251
measuring spacing 61–62, 556
mechanical harvesting 22
  flowers 136–137
  leaf and aerial parts 126–127
  roots 143–144
  seeds 145, 146, 147
mechanical washing 144
mechanised rubbing equipment 204
medicinal herbs *see* herbal medicines; therapeutic herbs
mesh screen for rubbing 199–200, 531
methyl bromide fumigation 4, 223, 233
mice and rats 154, **163–164**, 178, 191, 194, 209, 222
microbial contamination 45, 72, 143, 215, **233**
  see also animal manures; compost
microclimate 13, **15–16**, 58–59, 401, 509, 511–517
mildew 434, 444
milling root herbs *Plate 8.6*, 198, 214
minor herbs 243

Mint rust *Plate 6.5*, 98, 116, **280–283**, 284, 291–292
mites *see* red-legged earth mite; two-spotted mite
moisture migration in dried herb 154 *see also* inadequate drying
moisture stress 56, **91**, 98, 538
monoculture 9, 98–99
moon and planet rhythms 113–116, 118
mosaic culture 9, 99
moth, warehouse 154, 222–223, 448
mower 22, 133
mowing
  buffer zone 29, **30**, 31, 60, 248
  stubble post-harvest 133, 252, 336
  weed control 24, 29, 30, 31, **45–46**
mulches **41–43**, 44, 251, 538
  fertilising value 86
  organic 42–43
  permanent plastic 41
  problems and cautions 41, 42–43, 44
  winter dormancy 41–42, 65, 251
multiplication plots 22, 47

Names of plants 47–48, 219, 243–244
native species, wildcrafting 1, 464–465, 467
natural pesticides 99, 539
natural resistance to pests and diseases 97
net returns 236–237
nettle butterfly 317
Nettle (Preparation 504) 109
networking 235, 532
'next bin syndrome' 48
nitrogen and aphids 101–102, 283
nitrogen:carbon ratio in compost 72, 73, 74
nitrogen starvation 42, 73
nodes (lunar) 113
nodes (stem) 49, 249, 538
nomenclature 47–48, 219, 243–244
noxious weeds 43, 464, 538
nursery beds 51, 54
nutrients 71, 85, 86 *see also* compost

Oak bark (Preparation 505) 109–110
oil heat for drying 155, 174, 191–192
old stock 218
operating costs 236–237
orchards as weed source 20
organic certification *see* certification, organic
organic farming methods 7–8, 105, 539
organic fertilisers 7, 8, 44, **85** *see also* compost
organic herbs *see* manufacturing grade herbs; marketing; premium grade organic; second grade organic
organic matter 17, 29, **71**, 85 *see also* humus
outside floor drying 155, 190
over-mature flowers 134, 140
oversupply and prices 230–231
overwatering 90–91, 94, 95

Packing and storage
  containers 10, 216–218
  dryness levels for safe storage 154, 215
  ease of inspection 218
  filling bags 218–219
  insect and vermin control 154, 222–223
  labelling 216, 219–221, 224
  legal requirements 221, 232–234

packing and storage (*continued*)
  length of storage 10, 218
  moisture migration 154
  permeable packaging materials 217
  plastic bags 217, 218–219
  records 221–222
  separation of batches 218
  shipping 223–224
  soil contamination visibility 218
pan formation 26, 32, 33
particle size *Plates 8.5, 8.6, 8.7*, 197–198, 216
parts of herb used 244, 519, 521–527
passive floor drying 191
pastures as weed source 20
patch stripping and bark harvesting 148, 403
Paterson's Curse (*Echium lycopsis*) 43
paths and weed control 30, 31
pear slug 412
peeling bark 147–149, 403
peppering for weeds and pests 99, 116–117
perennial crown herbs 67, 133, 335–380
perennials 8, 539
  see also growth types; expanding clump herbs; perennial crown herbs; spreading herbs; woody perennials
  and pests 100, 102
  autumn planting 47
  compost application 80, 83
  rotation 60
  short-lived **69–70**, 421–422, 436–438, 451–457
  waterlogging 18
permeable packaging materials 217
pesticide residues 4, 8, **19–20**, 42, 71, 144, **215**, 464, 539 *see also* contamination
pesticides, natural 99, 539
pests and diseases 21, 61, **97–104**, 233, 245
  see also specific herbs
  and timing of harvest 120
  aphids 101, 233, 283
  biodynamic control 99, 116–117
  control strategies 98–104, 233
  factors reducing 97
  fungal control 116 *see also* rust
  grasshoppers 21, 61, 98, **100–101**, 233
    on individual crops 255, 283, 289, 299, 305, 317, 322, 329, 372, 424, 430
  layout of crops 61, 99, 102
  management strategies 97–104
  mites
    red-legged earth mite 21, 104
    two-spotted mite (red spider) 103–104
  moth, warehouse 154, 222–223, 448
  natural controls 97
  natural resistance 97
  peppering for 99, 116–117
  plant vigour 97, 101, 104
  polyculture 9, 21, 98–99
  predators 97, 98, 101, 103, 104
  red spider 98, **103–104**, 311, 312, 365
  rust *Plate 6.5*, 98, 116, **280–283**, 284, 289, 291–292, 420, 458
  slugs 21, 42, 99, **103**, 283, 454
  snails 21, 42, 61, 68, 99, **102–103**, 255, 283, 348, 454
  stress 98, 100
  vegetable weevil 444
  warehouse moth 154, 222–223, 448
  weevils 222, 444

General Index

pH of soil 19, 539
planet rhythms 113, 116, 118
plant suppliers 531
plant vigour and pests 97, 101, 104
planting out 56–6
    see also layout; propagation
    spacing **30–32**, **61–62**, 94, 244–245
        for growth types 249–250, 297, 335, 381
        for individual crops 509–510, 511–517
    supportive measures 56
    timing 47, 56, 509, 511–517
    weed growth 39
plastic bags for packing and storage 217, 218–219
plastic mulch 41–42, 65, 251
ploughing and soil preparation 25–26, 540
pollarding and bark harvesting 148, 403
pollution of site 19–20 *see also* contamination
polyculture 9, 21, 98–99
pooks 145, 146–147
portable screens *see* drying screens
possums 21, 61, 154, 164
post-harvest treatment 39, **133**, 251, 252–253, 336
pots for propagation 51, 56
poultry and pest control 101, 102, 103
preliminary chopping **169**, **182–183**, 184, 209, 212 *see also* chaff-cutting
premium-grade organic herbs 216, 540
    marketing 10–11, **226**, 227, **229–230**, 231, 237, 246, 520, 521–527
Preparation 500 (horn manure) 105–107
Preparation 501 (horn silica) 107–108
Preparation 502 (Yarrow flowers) 108–109
Preparation 503 (Chamomile flowers) 109
Preparation 504 (Nettle) 109
Preparation 505 (Oak Bark) 109–110
Preparation 506 (Dandelion flowers) 110–111
Preparation 507 (Valerian flowers) 111–112
Preparation 508 (Horsetail tea) 116
price *see also* specific herbs
    comparisons 10–11, 228, 229, 230, **237–238**, 519, 520, **521–527**
    culinary herbs 10, **230**, 520, 521–527
    future trends 231–232
    manufacturing herbs 10–11, **228–229**, 236, 246, 520, **521–527**
    organically grown herbs 10, 226, **228–230**, 231, 236–237, 520, **521–527**
    premium grade 10, 226, 229, 231, 237, 246, 520, **521–527**
    stability 230–232
    tea herbs 10, **229–230**, 520, **521–527**
    trade herbs 10, **225–226**, 236, 246, 519, **521–527**
    wildcrafted herbs **466**, 520, 521–527
priming seeds 55
processing 10, 197–216, 245
    see also specific herbs
    aerial parts 201–203, 209, 211–213 *see also* leaf herbs
    bark herbs 214
    categories of herb usage 197–198
    chaff-cutting 169, 184, **201–203**, 209, **211–212**, 214
    concealed moisture 153–154, 171, 172, **211**, 215
    costings 238–240
    culinary grade *Plate 8.5*, 197, 198–201, **205–211**, 212–214

processing (*continued*)
    dryness 170–171, 172, 173, 211, 215
    equipment 21–22, 198, **199–204**, 531
    facilities 21–22, 198
    first screening 206, 207, 210
    flexibility in marketing 240
    flower herbs 198, 214
    fruit herbs 198, 214
    hammermilling 198, **203**, 209, 212
    leaf herbs *Plates 8.3, 8.4, 8.5*, 198, **199–214**
    manufacturing grade *Plate 8.7*, 198, 208, **211–212**
    mechanised rubbing 204
    mesh screen 199–200, 531
    packing 216–221, 223
    particle size *Plates 8.5, 8.6, 8.7*, **197–198**, 216
    protective equipment **204**, 205, 212, 237, 265, 302, 317, 494
    quality assessment 214–216
    root herbs 198, 214
    rubbing **199–201**, **204–211**, 214, 531, 540
    rubbing winnower 201
    screening *see* rubbing
    shared facility 10, 195
    soil contamination 209, **213–214**, 218, 233
    storage 216–218, 222–223
    stripping leaves 209, 212–213
    tea grade 197–198, **199–211**, 212–213
    winnowing 103, 145, 146, 201, **209**, 210, 214
    yields 246, 519, 521–527 *see also* specific herbs
production costs 7–8, 236–237, 238–240
productivity (harvesting, drying, processing) 238–240
propagation 47–55, 244–245, 540
    see also specific herbs; layout; planting out
    and compost 83–84
    direct planting 28, 51, 53
    facilities 55
    growth types
        annuals, biennials, short-lived perennials 69
        expanding clump herbs 66, 297
        perennial crown herbs 67, 335
        spreading herbs 65, 248–250
        trees and shrubs 68
        woody perennials 67, 381
    identification 47–48
    material 22, 47, 49, 51, 531
    methods 48–55, 509, 511–517
        cuttings **49**, 50, 51, 67, 68, 536
        division **49**, 50, 67, 537
        layering 50, **51**, 67
        rhizomes **49**, 50, 51, 249–250, 540
        runners **49**, 50, 65, 66, 249
        seed 48, **51–55**, 67, 68, 69, 249, 297, 335, 531
        vegetative propagation **48–51**, 297, 531, 542
    nursery beds 51, 54
    snails in propagation material 102
    timing **56**, 244, 509, **511–517**
    transplanting 54, 56
    trial plot 47
    weed control aspects 44
protective equipment **204**, 205, 212, 237, 265, 302, 317, 494

pruning and bark harvesting 148
Purslane 38

Quality 4, 214–216 *see also* deterioration
    5-star assessment system 216
    acceptable levels of contamination 71, 213, **215**
    active constituents 3, 8, 191, 206, **216**, 535
    assessment 4, 197–198, 214–216
    correct species and identification 47–48, 215
    drying for quality 151–155, 215–216
    harvesting for quality 119
    marketing 5, 7–8, 10–11, **225–231**
    raw materials for herbal medicines v, 7, 8, 206, **216**, 228, 229
    standards 206, 208, **214–216**, 228
    therapeutic effectiveness v, 7, 8, 206, **216**, 229
    Therapeutic Goods Act 1991 232–235
quarantine 43–44, 99

Rabbits 61, 359, 444
radiation 4, 223, 233
rainfall **14–15**, 87, 95, 510
raised beds 56, 57–58, 250
raspberry rust 289
rats and mice 154, **163–164**, 178, 191, 194, 209, 222
reaper and binder 127, 146
reaping hook **121**, **127–128**, 145, 540
recirculation of used air for drying 193
records 155, 221–222
red-legged earth mite 21, 104
red spider 98, **103–104**, 311, 312, 365
regeneration
    after harvesting 65, 132, **133**, 148–149
    annuals, biennials, short-lived perennials 69, 70, 133
    expanding clump herbs 66, 133
    focus of regeneration **65**, 66, 67, 68, 69, 133
    perennial crown herbs 67, 133, 336
    spreading herbs 65–66, 133, 252–253
    trees and shrubs 68, 148–149, 403
    woody perennials 67, 68, 133, 381
regenerative weeds 27, 28, 37, **40**, 41, 251, 540
rehydration *see* remoistening
rejuvenation 251–252, 297
relative humidity 14–15, 151, **152–153**, 160, 166, 169–170, 175, 205, 209
remoistening 540
    and deterioration 153, 166, 170, 171
    and holding first screening 210–211
    reducing 170, 171
removing contamination *see* contamination
removing fines 213–214
reservoir of weed seeds in soil 20, 25, 27, 28, 43
residues *see* chemical residues; contamination; pesticide residues
respirator *see* protective equipment
returns, financial 236–237, 520, 521–527
rhizomes **49**, 50, 51, 249, 540
Rocky Mountains Herbalists Coalition 467–468
root crops
    chopping before drying 182–183
    contract harvesting 144
    definition 244
    deterioration 142, 151

root crops (*continued*)
   digging 141–142, 143–144
   drying
      ambient air drying 160
      assessing dryness 172, 183
      cabinet dryer 176, 177, 182–183
      case hardening 183
      chopping before drying 182–183
      circulation drying 189
      drying capacity needed 160
      inadequate drying 153–154, 172, 215
   harvesting 66–67, 140–144
   marketing 228
   mechanical harvesting 143–144
   milling *Plate 8.6*, 198, 214
   price for 228, 520, 521–529
   processing 198, 214
   productivity in harvesting and drying 239–240
   rubbing 214
   soil contamination 143, 214
   storing 142, 143
   trimming 142
   washing 141, 142–143, 144
root division 49, 50, 51
root rot *Plate 2.7*, 305, 348
rotary hoe 11, 22, 36
   buffer zone maintenance 30
   cultivating 22, 26, 27, **32–33**
   hard pan formation 26, 32, 33
   rejuvenating herb plots 252
rotary thresher 204
rotation **60–61**, 99, 245, 421
row crops 30–32
   direct planting 51, 53
   expanding clump herbs 66
   flood irrigation 88
   maintaining soil organic matter 85
   spacing **30–32**, **61–62**, 94, 244–245
      for growth types 249–250, 297, 335, 381
      for individual crops 509–510, **511–517**
   spreading herbs 249–250, 252
   string lines 30–31, 252
   weed control 30–32, 33–34
rubbing 540
   air humidity 205
   dryness of leaf 205
   equipment for leaf herbs 199–201, 207, 531
   fruit herbs 214
   mechanised 204
   protective equipment 204, 205, 237
   root herbs 214
   rubbing winnower 201
   screen 199–200, 531
   technique
      action 205, 207
      first screening 206, 207, 210
      holding first screening 210
      loading tray 205
      rescreening 206, 207, 208–209
      stem elimination 205, 206, 208–209
      time management 209–211
   under-screens 199, **200–201**, 206, 207
runners **49**, 50, 51, 65, 66, 249, 540
rushes 18
rust on crops *Plate 6.5*, 98, 116, **280–283**, 284, 289, 291–292, 420, 458

Salinity 88, 90

sawdust in compost 72, 73, **74**, 75
scale of operations 5, 21, 63
screen shelves 155
screening *see* rubbing
screens, drying *see* drying screens
screwdriver 35, 36
scythe 3, 58, 121, **122–125**, 126, **128–132**, 145
   for mowing stubble 133
   making catching scythe 122–125
   technique 128–132
seagrass in compost 74
seaweed in compost 72, 74
second grade organic herbs 216, 229, 231
sedges 18
seed crops **145–147**, 154, 198, 228–229, 244
seed, growing from 48, **51–55**, 531
seed suppliers 531
seeds, weed *see* weed seeds
selective grazing and weed control 45
shade drying 151, **155–190**, 191–193, 194
shade for crops 15–16, 55, 56, 58, **62**, 401, 509, **511–517**
shadecloth 62, 157, 158, 159
shadehouse 55, 56, 62
shallow-rooted plants 88, 89, 90
shared central drying facility 10, 194–195
sharpening hoe blades 37
sharpening scythe 129
shears for harvesting 121
sheep manure in compost 74
Sheep Sorrel 20, 27, 28, 33, **40**
shipping 223–224
short-lived perennials **69–70**, 421–422, 436–438, 451–457
shrubs *see* trees and shrubs
sickle 121, 127–128
sidereal rhythm 115
sifting to remove fines 213–214
site selection **13–21**, 244, 509, 510, 511–517 *see also specific herbs; layout*
slasher 22, 133, 142
slope 19, 29, 58
slugs 21, 42, 99, **103**, 283, 454
small herb gardens 21, 63
small seeds, special procedures 51–53
smut 458
snails 21, **102–103**, 255, 283, 348, 454
   bait 102
   control 102–103
   layout to reduce snails 61
   mulches harbouring 42
   perennials harbouring 61, 68, 99
   timing of harvest 120
Sneeboer gooseneck hoe 34, 36
soil 17–19
   auger 18
   blocks 54, 56
   contamination (residues in soil) 19–20, 42, 144
   contamination (soil in herb)
      acceptable limits 213
      microbial content 143, 233
      removing 132–133, 154–155, 209, **213–214**, 233
      root herbs 143, 214
      sources of 89, 132, 143, **213**
      visibility when packing 218
   cultivation 22, **25–29**, 31, **32–39**, 40

soil (*continued*)
   depth of watering 93
   drainage 17–19
   duplex soils 18–19
   earthworms 17
   erosion 19, 26, 29, **58**
   fertility 8, **17**, **71**, 80, 510, 511–517
   hard-pan 26, 32, 33
   heavy **17**, 57, 103, 509
   humus levels 17, 71, 538
   moisture 56, 91, **93**, 94
   organic matter 17, 29, **71**, 85
   pesticide residues **19–20**, 42, 71, 144
   pH 19, 539
   planting out 56
   preparation 25–29
   splash 89, 132–133, 213
   stones 17
   temperature 42–43, 89
   test and certification 20
   texture 17, 509, 511–517
   types **17–19**, 59, 509, 511–517
   water-holding capacity 94
solar
   cabinet dryer 177, 185–188
   chimney 186, 187
   collector 185–187
   efficiency of drying shed 160, 161–162
   heat for drying 155, 175
   panel 185, 187
   roof collector 186, 187
   tunnel 185, 186, 187–188
Sorrel 20, 27, 28, 33, **40**
Southern Light Herbs 3, 4, 532
spacing of plants **30–32**, **61–62**, 94, 244–245
   see also specific herbs; growth types
   for growth types 249–250, 297, 335, 381
   for individual crops 509–510, 511–517
spade 140–141
specific herbs *see*
   Index of Individual Herb Crops 553–555
   Information Charts 509–527
spreading herbs 65–66, 247–296
   autumn planting 249
   black plastic dormancy mulch 41–42, 65, 251
   containment 60, 65, 247–248
   cultivation and management 65, 247–253
   definition 65, 247
   dormancy 41, 65, 251
   harvesting 65–66
   invasive growth 60, 65, 247–248
   longevity 252
   planting 28, 248–250
   post-harvest treatment 133, 252–253
   propagation 65, 248–250
   regeneration 65, 133, 252–253
   rejuvenation 251–252
   stale seedbed 28
   weed control 24, 28, 40–41, 65, 247, **250–251**, 252, 277–279
spring
   planting 56
   slug problems 103
   watering 94
   weed growth 39
spring-tined harrows 26, 27
sprinkler irrigation 59, **89–90**, 91, **92–95**
sprinkler shadow 59, 402

General Index

stage of growth for harvesting 119, 134, 140, 144, 145, 147
stale seed bed 27, 28
stalks *see* stems
standards 197–198, 215–216 *see also* quality
Steiner, Rudolf 105, 106, 107, 108, 116, 118
stems
    acceptable limits in processed herb 197–198, **206**, **208**, 209, 216
    aerial parts 171, 198, 237–238
    assessing dryness 171, 172
    culinary herbs 197
    fleshy 171
    fruit herbs 214
    leaf herbs 171, 197, 205–209
    pliable 171, 205
    woody 171
stirrup hoe 35
stolons **49**, 50, 51, 541
stones 17
stooks 145, 146, 147
storage *see* packing and storage
stratification of seeds 55, 541
straw in compost 74, 75
stressed plants 15
    and pests 98, 100
    moisture stress 56, **91**, 98, 538
stripping leaves 209, 212–213
stubble, mowing 133, 252, 336
suitability of crops 9, 21, 23–25, 47, **509–517** *see also* crop choice
sun drying under cover 193–194
sun requirements of crops 15–16, 509, 511–517
sunlight and fading of herb 132, 151, 190
suppliers
    seeds and propagation material 531
    tools and equipment 37, 200, 531
supply and demand 230–231
surface evaporation 95
suspended floor drying with forced air 155, 191–193
sweating and deterioration 154, 156, 191 *see also* heating of harvested material

Tannin content of bark herbs 184
tea grade herbs 541 *see also* tea herbs
    acceptable stem content 197, **206**, **208**, 209, 216
    assessing dryness 170–171, 172
    chaff cutting 212
    consumption and demand 229–230
    leaf particle size 197–198
    marketing 10–11, 229–230
    price 10, **229–230**, 520, **521–527**
    processing 197–198, **199–211**, 212–213
    quality 197, 214–216, 229
    rescreening 206, 208–209
    rubbing 205–209
teflon tape, beeswax as an alternative 92
temperature (climatic) **13–14**, 15, 16, 95, 556
    essential oil levels 13
    humidity 175
    individual crop preferences 509, 511–517
    winter weed growth 14
temperature (drying) **151–153**, 175, 245, 519, **521–527**, 556
Therapeutic Goods Act 1991 4, 229, **232–235**
    'Appendix B' 233–234
    effect on growers 232–235

Therapeutic Goods Act 1991 (*continued*)
    effect on markets 234–235
    microbial levels 45, 72, 215, **233**
therapeutic effectiveness **v**, 7, 8, 206, **216**, 229
therapeutic herbs **v**, 4, 8, 10, 45, 206, **216**, 229, **232–235** *see also* medicinal herbs
threshing 145, 146, 147
tickle weeding 45
tillers 26
tools and equipment
    compost 75, 76–77, 83
    cultivation 22, **26**, 27, 30, **32–37**, 531
    harvesting 22, 120–127, 135–137, 140–141, 144, 145–146, 147–148
    weed control 32–37, 531
toxic residues 4, 8, **19–20**, 42, 71, 144, **215**, 464
tractor 11, **22**, 75, 76, 77, 235, 236
trade herbs
    Australian market 11, 225–226
    definition 10, 541
    price 10, **225–226**, 236, 246, 519, **521–527**
    quality 216
transplanting 54, 56
trays for propagation 51, 52, 53
trees and shrubs 68, 401–420
    see also specific herbs
    effects on other crops 15, 16, **59**, **401**, 402
    harvesting 68–69, 148–149, 402–403
    influence on microclimate 15, 16, 59, **401**
    irrigation 59, 402
    layout 59, 401–403
    picking tree flowers 139
    planning 59, 403
    propagation 68, 401
    weed control 24, 45, 68, **402**
trial plot 47, 100, 240–241
trickle irrigation 90–91, 95
trimming harvested roots 142
tubs for containing invasive herbs 247
turning compost 79–80, 81–82, 233
tweaker 35, 36
'Twin Creeks' 4, 58, 59
two-spotted mite 98, **103–104**, 311, 312, 365

Under-screens 199, **200–201**, 206, 207
unharvested crops, mowing off 133
urine in compost 74
used air recirculation in drying 193

Valerian flowers (Preparation 507) 111–112
value adding 228, 232
vegetable weevil 444
vegetative propagation **48–51**, 248, 297, 542
    advantages and disadvantages 48–49
    suppliers of material 532
vehicle access 63
ventilation of drying shed 162, 163, 169–170 *see also* air exchange; airflow
vermin control 154, **163–164**, 178, 191, 194, 209, 222
vertical airflow systems 176–188, 191–193
vine hoe 141
volatile constituents *see* essential oils
volcanic soils 17

Warehouse moth 154, 222–223, 448
washing
    removing soil contamination 154–155, 213, 233

washing (*continued*)
    root crops 141, 142–143, 144
water supply 20, 87–88 *see also* irrigation
watering *see also* irrigation; water supply
    and crop spacing 59, 94, 402
    and grasshopper control 101
    plant requirements 91, 95, 510, **511–517**
    planting out 56
    raised beds 57
    slugs 103
    small seeds 52
    total water requirements 510, 511–517
waterlogging **17–18**, 19, 59, 509, 542 *see also* drainage
weather
    drying 169
    harvesting 120
    planting out 56
weed *see also* weed control; weeds
    competition 23
    consciousness 25, 46
    contamination of harvest 23, 132, 154, 156, 191, 465
    control *see* weed control (below)
    definition 23
    harvesting 243, 463–508 *see also* wildcrafting
    legacy from previous land use 20
    management *see* weed control
    seeds
        from weeds going to seed 24, 26, 28, 30, 37, 38, 39, 40, 61, 85
        germination of 27, 28, 39, 94
        in animal manures 26, 44, 73, 74
        in compost 44, **71**, 79, 80
        in mulches 42, 44
        in propagation material 44
        in soil 20, 25, 27, 28, 43, 44
    survival mechanisms 28, 38–39
weed control 18, **23–46**, 510–517
    see also specific herbs; growth types
    animal manures, seeds in 26, 44, 73, 74
    annuals, biennials and short-lived perennials 69, 422
    autumn plantings 47, 56, 249
    bare fallow **27–29**, 40, 47, 535
    biodynamic 116–117
    buffer zones **29–30**, 31, 33, 43, 61
    composting system 43, 44, **71**, 79, 80
    critical times for action 39
    crop choice 23–25
    crop types and requirements 23–24
    dense planting 32
    disposal of chopped weeds 38–39
    during dormancy 8, 14, 23, **40–42**, 65, 250–251
    equipment 22, **26**, 27, **32–37**
    expanding clump herbs 66, 297
    flame weeding 44–45, 531
    germination of weed seeds 27, 28, 39, 94
    green manure crops 26, 61, 85
    herbs that become weeds 29–30, 43
    hot-water weeding 45
    implements 22, **26**, 27, **32–37**
    initial bare fallow 27–29
    invasive weeds *see* regenerative weeds
    killing weeds thoroughly 39
    laissez-faire approaches 23
    large perennial root weeds 20

weed control (*continued*)
- layout of crops 29–32, 58
- livestock 44, 45
- making it a priority 25, 46
- meadow-crops 46, 251
- mowing 24, 29, 30, 31, **45–46**
- mulching **41–43**, 44, 251, 538
- newly planted crops 39
- noxious weeds 43, 464, 538
- overwatering and weed problems 94
- paths harbouring weeds 30, 31
- peppering 116–117
- perennial crown herbs 68, 336
- plan of action 25, 37–38, 46
- post-harvest 39, 251
- priorities 25, 46
- problem weeds 40–41
- quarantining 43–44
- raised beds 57–58
- rapid regrowth crops 24, 46
- regenerative weeds 20, 27, 28, 37, **40**, 41, 251, 540
- rejuvenation 251–252
- requirements of different crop types 23–24
- row crops 30–32, 33–34
- selective grazing 45
- soil preparation 25–27
- sources of weeds 20, 26, 42, **43–44**, 71, 73, 74
- spacing of plants 30–32
- spreading herbs 24, 28, 40–41, 65, 247, **250–251**, 252, 277–279
- spring plantings 56, 249
- stale seedbed 27, 28
- systematic approach 25, 37–38, 46

weed control (*continued*)
- tickle weeding 45
- tools 22, 32–37
- trees and shrubs 24, 45, 68, **402**
- trickle irrigation and weed growth 90
- winter-active weeds 8, 14, 23, **40–41**, 65, 251, 542
- winter dormancy mulch 41–42, 65, 251
- woody perennials 68, 381

weeding claw 35, 36

weeds 25–46
- *see also* weed; weed control
- benefits of 46
- controlling *see* weed control
- disposal of chopped weeds 38–39
- herbs that become weeds 29–30, 43
- invasive *see* regenerative weeds
- killing thoroughly 39
- large perennial root 20
- noxious 43, 464, 538
- problem 40–41
- regenerative 20, 27, 28, 37, **40**, 41, 251, 540
- sources of 20, 26, 42, **43–44**, 71, 73, 74
- spreading *see* regenerative weeds
- toxic 23
- winter-active 8, 14, 23, **40–41**, 65, 251, 540

weevils 222, 444 *see also* warehouse moth

weight when labelling 221

wheel hoe 1, **33–34**, 36, 39, 531

wholesalers 227, 532

wild-harvesting *see* weed harvesting; wildcrafting

wildcrafting 243, 463–508
- certification 463–464
- definition 463, 542

wildcrafting (*continued*)
- drying 465–466
- guidelines 463–465, 466–468
- logistics 465–466

wilting of harvested herb **120**, 151, 169, 192 *see also* bruising of herb

wind **15**, 16, 509, 511–517
- and drying process 169
- and sprinkler irrigation 89, 94
- and trickle irrigation 95

wingless grasshopper *see* grasshoppers

winnower 22, **201**, 209, **210**

winnowing
- seed crops 145, 146
- to remove foreign material 103, 209, 214
- to remove stem 201, 209, 210

winter-active herbs 41, 542

winter-active weeds 8, 14, 23, **40–41**, 65, 251, 542

winter dormancy mulch 41–42, 65, 251

Winter Grass (Poa annua) 14, 40–41, 43–44

wire mesh for rubbing 199–200, 531

wood ash in compost 74

wood heat for drying 174, 177, **184–185**, 194

woody perennials 67–68, 381–399
- compost spreading 84–85
- harvesting 68, 123, 126, 381
- propagation 67, 381
- rejuvenation 381
- retaining leaf at harvest 68, 133, 381, 390
- snails 61, 248

wool bales 217–218

Yarrow flowers (Preparation 502) 108–109

yields 246, 519, 521–527 *see also specific herbs*

Index of Individual Herb Crops

Major entries are in bold type.

Achillea millefolium (Yarrow) Plate 7.6, 65, 108-9, 237, 240, **294-6, 508, 517, 527**

Aesculus hippocastanum (Horse Chestnut) **413, 514, 524**

Agnus Castus *see* Chaste Tree

Agrimony *(Agrimonia eupatoria)* 100, **336-7, 511, 521**

Agropyron repens see Elymus repens

Alchemilla mollis see Alchemilla vulgaris

Alchemilla vulgaris (Lady's Mantle) Plate 4.5, 67, 206, 238, **358, 514, 524**

Alfalfa *(Medicago sativa)* Plate 1.1, 18, 24, 29, 44, 46, 53, 67, 84, 85, 131-2, 151, 154, 161, 209, 222, 237, 240, **337-41, 511, 521**

Allium sativum (Garlic) 4, 67, **351-4, 513, 523**

Allium schoenoprasum (Chives) 151, 169, **343-5, 512, 522**

Aloysia triphylla (Lemon Verbena) 209, 212, **414-15, 514, 524**

Althaea officinalis (Marshmallow) Plate 4.6, 60, 67, 69, 142, 183, 239, **361-3, 515, 525**

American Ginseng *see* Ginseng

Anemone pulsatilla (Pasque Flower) Plate 5.2, **367, 515, 525**

Anethum graveolens (Dill) 101, **451, 513, 523**

Angelica *(Angelica archangelica)* Plate 1.2, 70, **422-4, 511, 521**

Anise *see* Aniseed

Aniseed *(Pimpinella anisum)* **424-5, 511, 521**

Anthemis nobilis see Chamaemelum nobile

Apium graveolens (Celery) **438-9, 512, 522**

Arctium lappa (Burdock) Plate 1.3, 69, 140, 142, 222, 239, **428-32, 511, 521**

Armoracia rusticana (Horseradish) 65, **259-61, 514, 524**

Artemesia absinthium (Wormwood) 67, 100, **379-80, 517, 527**

Artemesia vulgaris (Mugwort) 30, 209, 212, 237, **313-14, 515, 525**

Artemisia dracunculus (Tarragon, French) 48, 197, **323-6, 517, 527**

Artichoke, Globe *(Cynara scolymus)* 169, **341-2, 511, 521**

Avena sativa (Oats) 61, 70, 130, 169, 171, 209, 212, **457-9, 515, 525**

Ballota nigra (Horehound, Black) **357, 514, 524**

Balm of Gilead Poplar *(Populus candicans)* **403-4, 511, 521**

Balm *see* Melissa Balm

Basil *(Ocimum basilicum)* 15, 55, 69, 153, **425-7, 511, 521**

Bay *(Laurus nobilis)* 197, 209, 212, 222, **404-7, 511, 521**

Bergamot *(Monarda didyma)* Plate 1.4, **298-300, 511, 521**

Betony, Wood *see* Wood Betony

Black Horehound *see* Horehound, Black

Blessed Thistle *(Cnicus benedictus)* **427-8, 511, 521**

Blue Gum *see* Eucalyptus

Briar, Sweet *see* Rose, Sweet Briar

Broad-leaf Echinacea *see* Echinacea, Broad-leaf

Broom, English *(Cytisus scoparius)* **469, 511, 521**

Bugleweed *see* Gipsywort

Burdock *(Arctium lappa)* Plate 1.3, 69, 140, 142, 222, 239, **428-32, 511, 521**

Calendula *(Calendula officinalis)* Plate 1.5, 69, 140, 152, 239, 240, **432-5, 511, 521**

Camomile *see* Chamomile

Capsella bursa-pastoris (Shepherd's Purse) **501-2, 516, 526**

Caraway *(Carvum carvi)* 69, **435-6, 511, 521**

Carduus benedictus see Cnicus benedictus

Carduus marianus see Silybum marianum

Carum petroselinum see Petroselinum crispum

Carum carvi (Caraway) 69, **435-6, 511, 521**

Cascara Sagrada *(Rhamnus purshiana)* **407, 511, 521**

Catmint *see* Catnip

Catnip *(Nepeta cataria)* Plate 1.6, 69, 70, **436-8, 511, 521**

Celandine, Greater *(Chelidonium majus)* 103, **342, 512, 522**

Celery *(Apium graveolens)* **438-9, 512, 522**

Centaury *(Centaurium erythraea)* **470, 512, 522**

Chamaemelum nobile (Chamomile, Roman) 30, 60, 61, 109, 136, 140, 239, 240, 248, **300-1, 512, 522**

Chamomile, English *see* Chamomile, Roman

Chamomile, German *(Chamomilla recutita)* Plate 2.1, 69, 80, 109, 136, 300, 301, **439-48, 512, 522**

Chamomile, Roman *(Chamaemelum nobile)* 30, 60, 61, 109, 136, 140, 239, 240, 248, **300-1, 512, 522**

Chamomilla recutita (Chamomile, German) Plates 2.1 & 8.2, 69, 80, 109, 136, 300, 301, **439-48, 512, 522**

Chaste Tree *(Vitex agnus-castus)* Plate 2.3, 44, 173, **408, 512, 522**

Chelidonium majus (Celandine, Greater) 103, **342, 512, 522**

Chickweed *(Stellaria media)* **471-2, 512, 522**

Chicory *(Cichorium intybus)* 239, **448-50, 512, 522**

Chives *(Allium schoenoprasum)* 151, 169, **343-5, 512, 522**

Chrysanthemum parthenium see Tanacetum parthenium

Cichorium intybus (Chicory) 239, **448-50, 512, 522**

Cilantro *see* Coriander

Cleavers *(Galium aparine)* **472-4, 512, 522**

Clivers *see* Cleavers

Clover, Red *see* Red Clover

Cnicus benedictus (Blessed Thistle) **427-8, 511, 521**

Cochlearia armoracia see Armoracia rusticana

Collinsonia canadensis (Stoneroot) 239, **375-6, 517, 527**

Coltsfoot *(Tussilago farfara)* 15, 29-30, 41, 60, 65, 152, 206, 238, **253-6, 512, 522**

Comfrey, English *(Symphytum officinale)* Plate 2.2, 30, 66, 141, 204, 239, **301-2, 512, 522**

Coneflower *see* Echinacea

Coriander *(Coriandrum sativum)* Plate 2.4, **450, 512, 522**

Couchgrass, English *(Elymus repens)* 248, **256-7, 474, 512, 522**

Crampbark *(Viburnum opulus)* Plate 2.5, **409-10, 512, 522**

Crataegus laevigata (Hawthorn) 69, 145, 173, 148, **412, 483-6, 514, 524**

Crataegus monogyna (Hawthorn) Plate 3.7, 69, 145, 173, 148, 214, **412, 483-6, 514, 524**

Crataegus oxyacanthoides see Crataegus laevigata

Curled Dock *see* Dock, Yellow

Cymbopogon citratus (Lemon Grass) 15, 169, 212, **358-60, 514, 524**

Cynara scolymus (Artichoke, Globe) 169, **341-2, 511, 521**

Cytisus scoparius (Broom, English) **469, 511, 521**

Dandelion *(Taraxacum officinale)* Plates 2.6 & 2.7, 30, 60, 66, 110-11, 142, 152, 182, 239, 248, **302-8, 474, 512, 522**

Dill *(Anethum graveolens)* 101, **451, 513, 523**

Dock, Curled *see* Dock, Yellow

Dock, Yellow *(Rumex crispus)* Plate 3.1, 18, 20, 35, 239, **475-7, 513, 523**

553

Dog Rose *see* Rose, Dog

Drimys lanceolata see *Tasmannia lanceolata*

Echinacea angustifolia (Echinacea, Narrow-leaf) Plate 3.3, 140, 239, **345-9, 513, 523**

Echinacea, Broad-leaf *(Echinacea purpurea)* Plate 3.2, 69, 345, 346-7, 348, **451-7, 513, 523**

Echinacea, Narrow-leaf *(Echinacea angustifolia)* Plate 3.3, 140, 239, **345-9, 513, 523**

Elder, Black *(Sambucus nigra)* Plate 3.4, **410, 477-80, 513, 523**

Elecampane *(Inula helenium)* 60, 67, 69, 141, 142, 179, 239, **349-50, 513, 523**

Elymus repens (Couchgrass, English) 248, **256-7, 474, 512, 522**

English Broom *see* Broom, English

English Chamomile *see* Chamomile, Roman

English Comfrey *see* Comfrey, English

English Couch Grass *see* Couch Grass, English

English Lavender *see* Lavender, English

Equisetum arvense (Horsetail, Field) Plate 4.2, 29-30, 41, 44, 60, 65, 116, **261-5, 514, 524**

Erythraea centaurium see *Centaurium erythraea*

Eucalyptus (Blue Gum) *(Eucalyptus globulus)* 59, **480-1, 513, 523**

Fennel *(Foeniculum vulgare)* Plate 3.5, **350, 481-3, 513, 523**

Feverfew *(Tanacetum parthenium)* **382-4, 513, 523**

Field Horsetail *see* Horsetail, Field

Figwort *(Scropularia nodosa)* 67, **350-1, 513, 523**

Filipendula ulmaria (Meadowsweet) 15, 44, 103, 238, **364-6, 515, 525**

Foeniculum vulgare (Fennel) Plate 3.5, **350, 481-3, 513, 523**

French Tarragon *see* Tarragon, French

Galega officinalis (Goat's Rue) **356, 513, 523**

Galium aparine (Cleavers) **472-4, 512, 522**

Garlic *(Allium sativum)* 4, 67, **351-4, 513, 523**

German Chamomile *see* Chamomile, German

Gilly-over-the-Ground *see* Ground Ivy

Ginkgo *(Ginkgo biloba)* 119, **410-11, 513, 523**

Ginseng *(Panax* spp.) Plate 3.6, **354-5, 513, 523**

Gipsywort *(Lycopus virginicus)* **257, 513, 523**

Glechoma hederacea (Ground Ivy) **258-9, 514, 524**

Globe Artichoke *see* Artichoke, Globe

Glycyrrhiza glabra (Licorice) 140, **265-9, 514, 524**

Goat's Rue *(Galega officinalis)* **356, 513, 523**

Golden Seal *(Hydrastis canadensis)* **258, 513, 523**

Great Mullein *see* Mullein, Great

Greater Celandine *see* Celandine, Greater

Greater Nettle *see* Nettle, Greater

Greater Plantain *see* Plantain, Greater

Grindelia camporum see *Grindelia robusta*

Grindelia robusta (Gum Plant) **384, 514, 524**

Ground Ivy *(Glechoma hederacea)* **258-9, 514, 524**

Guelder Rose *see* Crampbark

Gum Plant *(Grindelia robusta)* **384, 514, 524**

Gum, Blue *see* Eucalyptus

Hamamelis virginiana (Witch Hazel) **420, 517, 527**

Hawthorn *(Crataegus* spp.) Plate 3.7, 69, 145, 173, 148, 214, **412, 483-6, 514, 524**

Holy Thistle *see* Blessed Thistle

Hops *(Humulus lupulus)* Plate 4.1, 20, 103, 174, 191, **309-12, 514, 524**

Horehound, Black *(Ballota nigra)* **357, 514, 524**

Horehound, White *(Marrubium vulgare)* 45, 237, **487-8, 514, 524**

Horse Chestnut *(Aesculus hippocastanum)* **413, 514, 524**

Horseradish *(Armoracia rusticana)* 65, **259-61, 514, 524**

Horsetail, Field *(Equisetum arvense)* Plate 4.2, 29-30, 41, 44, 60, 65, 116, **261-5, 514, 524**

Humulus lupulus (Hops) Plate 4.1, 20, 103, 174, 191, **309-12, 514, 524**

Hydrastis canadensis (Golden Seal) **258, 513, 523**

Hypericum perforatum (St John's Wort) Plate 7.3, 43, 171, 237, 248, **292-3, 504-6, 517, 527**

Hyssop *(Hyssopus officinalis)* Plate 4.3, 67, 68, 123, 237, 240, **385-7, 514, 524**

Inula helenium (Elecampane) 60, 67, 69, 141, 142, 179, 239, **349-50, 513, 523**

Korean Ginseng *see* Ginseng

Lady's Mantle *(Alchemilla vulgaris)* Plate 4.5, 67, 206, 238, **358, 514, 524**

Laurel *see* Bay

Laurus nobilis (Bay) 197, 209, 212, 222, **404-7, 511, 521**

Lavender, English *(Lavandula angustifolia)* **387, 514, 524**

Lemon Balm *see* Melissa Balm

Lemon Grass *(Cymbopogon citratus)* 15, 169, 212, **358-60, 514, 524**

Lemon Thyme *(Thymus* x *citriodorus)* Plate 4.4, 84, **388-90, 514, 524**

Lemon Verbena *(Aloysia triphylla)* 209, 212, **414-15, 514, 524**

Leonurus cardiaca (Motherwort) 237, **366, 515, 525**

Lesser Nettle *see* Nettle, Lesser

Licorice *(Glycyrrhiza glabra)* 140, **265-9, 514, 524**

Lime *see* Linden

Linden *(Tilea* spp.) **416-18, 515, 525**

Lippia citriodora see *Aloysia triphylla*

Liquorice *see* Licorice

Lucerne *see* Alfalfa

Lycopus virginicus (Gipsywort) **257, 513, 523**

Marigold *see* Calendula

Marjoram, Sweet *(Origanum majorana)* 319, 320, **391, 515, 525**

Marrubium vulgare (Horehound, White) 45, 237, **487-8, 514, 524**

Marshmallow *(Althaea officinalis)* Plate 4.6, 60, 67, 69, 142, 183, 239, **361-3, 515, 525**

Matricaria chamomilla see *Chamomilla recutita*

Matricaria recutita see *Chamomilla recutita*

Meadowsweet *(Filipendula ulmaria)* 15, 44, 103, 238, **364-6, 515, 525**

Medicago sativa (Alfalfa) Plate 1.1, 18, 24, 29, 44, 46, 53, 67, 84, 85, 131-2, 151, 154, 161, 209, 222, 237, 240, **337-41, 511, 521**

Melissa Balm *(Melissa officinalis)* Plate 5.1, 15, 66, 103, 104, 237, **312-13, 488, 515, 525**

Mentha pulegium (Pennyroyal) 41, **272-3, 494, 516, 526**

Mentha spicata (Spearmint) Plates 7.1 & 8.4, 15, 24, 25, 41, 60, 65, 66, 102, 208, 209, **291-2, 517, 527**

Mentha x *piperita* (Peppermint) Plates 5.5, 6.5 & 8.3, 15, 17, 24, 25, 41, 45, 48, 60, 65, 84, 100, 102, 119, 171, 209, 237, 240, **273-86, 516, 526**

Milk Thistle *see* Variegated Thistle

Mint *see* Peppermint; Spearmint

Monarda didyma (Bergamot) Plate 1.4, **298-300, 511, 521**

Motherwort *(Leonurus cardiaca)* 237, **366, 515, 525**

Mountain Pepper *(Tasmannia lanceolata)* Plate 5.3, **489-90, 515, 525**

Mugwort *(Artemesia vulgaris)* 30, 209, 212, 237, **313-14, 515, 525**

Mullein, Great *(Verbascum thapsus)* 169, **457, 490-2, 515, 525**

Narrow-leaf Echinacea *see* Echinacea, Narrow-leaf

Narrow-leaf Plantain *see* Plantain, Narrow-leaf

Nepeta cataria (Catnip) Plate 1.6, 69, 70, **436-8, 511, 521**

Nettle, Greater *(Urtica dioica)* Plate 5.6, 17, 41, 66,73, 80, 84, 109, 154, 204, 222, 240, **314-18, 515, 525**

Nettle, Lesser *(Urtica urens)* 17, 28, 109, 204, 237, 314, 315, **492-4, 515, 525**

Oats *(Avena sativa)* 61, 70, 130, 169, 171, 209, 212, **457-9, 515, 525**

Ocimum basilicum (Basil) 15, 55, 69, 153, **425-7, 511, 521**

Oregano *(Origanum vulgare)* 66, **319-21, 515, 525**

Origanum majorana (Marjoram, Sweet) 319, 320, **391, 515, 525**

Origanum vulgare (Oregano) 66, **319-21, 515, 525**

Index of Individual Herb Crops

Panax spp. (Ginseng) Plate 3.6, **354-5, 513, 523**

Parsley *(Petroselinum crispum)* 59, 69, 151, 169, 212, 219, 239, **459-61, 515, 525**

Pasque Flower *(Anemone pulsatilla)* Plate 5.2, **367, 515, 525**

Passionflower *(Passiflora incarnata)* Plate 5.4, 65, **269-72, 515, 525**

Pennyroyal *(Mentha pulegium)* 41, **272-3, 494, 516, 526**

Pepper, Mountain *see* Mountain Pepper

Peppermint *(Mentha x piperita)* Plates 5.5, 6.5 & 8.3, 15, 17, 24, 25, 41, 45, 48, 60, 65, 84, 100, 102, 119, 171, 209, 237, 240, **273-86, 516, 526**

Periwinkle, Greater *(Vinca major)* **495, 516, 526**

Petroselinum crispum (Parsley) 59, 69, 151, 169, 212, 219, 239, **459-61, 515, 525**

Phytolacca americana see Phytolacca decandra

Phytolacca decandra (Poke) 43, **368-9, 496-8, 516, 526**

Pimpinella anisum (Aniseed) **424-5, 511, 521**

Plantago lanceolata (Plantain, Narrow-leaf) 368, **495-6, 516, 526**

Plantain, Greater *(Plantago major)* 60, **367-8, 495-6, 516, 526**

Plantain, Narrow-leaf *(Plantago lanceolata)* 368, **495-6, 516, 526**

Poke *(Phytolacca decandra)* 43, **368-9, 496-8, 516, 526**

Populus candicans (Balm of Gilead Poplar) **403-4, 511, 521**

Populus gileadensis see Populus candicans

Pulsatilla see Pasque Flower

Purple Coneflower *see* Echinacea

Raspberry *(Rubus idaeus)* Plate 6.1, 45, 206, 209, **286-90, 516, 526**

Red Clover *(Trifolium pratense)* Plates 6.2 & 8.1, 46, 53, 67, 86, 127, 136, 239, 240, **369-75, 498, 516, 526**

Rhamnus purshiana (Cascara Sagrada) **407, 511, 521**

Ribwort *see* Plantain, Narrow-leaf

Roman Chamomile *see* Chamomile, Roman

Rosa canina (Rose, Dog) Plate 6.3, 135, 173, 214, **418-19, 498-501, 516, 526**

Rosa rubiginosa (Rose, Sweet Briar) Plate 6.6, 173, 182, **498-501, 516, 526**

Rosa spp. (Rose, Wild) **498-501**

Rose Hip *see* Rose, Dog *and* Rose, Sweet Briar

Rose, Dog *(Rosa canina)* Plate 6.3, 135, 173, 214, **418-19, 498-501, 516, 526**

Rose, Sweet Briar *(Rosa rubiginosa)* Plate 6.6, 173, 182, **498-501, 516, 526**

Rose, Wild *(Rosa* spp.) **498-501**

Rosemary *(Rosmarinus officinalis)* 30, 67, **392-4, 516, 526**

Rubus idaeus (Raspberry) Plate 6.1, 45, 206, 209, **286-90, 516, 526**

Rudbeckia purpurea see Echinacea, Broad-leaf

Rue *(Ruta graveolens)* Plate 6.4, 30, 169, 209, **394-5, 516, 526**

Rumex crispus (Dock, Yellow) Plate 3.1, 18, 20, 35, 239, **475-7, 513, 523**

Rumex spp. (Sorrel, Sheep) 4, 20, 27, 28, 33, 251, **502-4, 516, 526**

Ruta graveolens (Rue) Plate 6.4, 30, 169, 209, **394-5, 516, 526**

Sage *(Salvia officinalis)* Plate 6.7, 4, 67, 198, 209, **395-7, 516, 526**

Salix alba (Willow) 59, **419-20, 517, 527**

Salvia officinalis (Sage) Plate 6.7, 4, 67, 198, 209, **395-7, 516, 526**

Sambucus nigra (Elder, Black) Plate 3.4, **410, 477-80, 513, 523**

Sarothamnus scoparius see Cytisus scoparius

Scropularia nodosa (Figwort) 67, **350-1, 513, 523**

Scullcap *(Scutellaria lateriflora)* **321-3, 516, 526**

Sheep Sorrel *see* Sorrel, Sheep

Shepherd's Purse *(Capsella bursa-pastoris)* **501-2, 516, 526**

Silybum marianum (Variegated Thistle) **506-8, 517, 527**

Skullcap *see* Scullcap

Snowball Tree *see* Crampbark

Sorrel, Sheep *(Rumex* spp.) 4, 20, 27, 28, 33, 40, 251, **502-4, 516, 526**

Spearmint *(Mentha spicata)* Plates 7.1 & 8.4, 15, 24, 25, 41, 60, 65, 66, 102, 208, 209, **291-2, 517, 527**

St John's Wort *(Hypericum perforatum)* Plate 7.3, 43, 171, 237, 248, **292-3, 504-6, 517, 527**

St Mary's Thistle *see* Variegated Thistle

Stachys betonica see Stachys officinalis

Stachys officinalis (Wood Betony) Plate 7.4, **378-9, 517, 527**

Stellaria media (Chickweed) **471-2, 512, 522**

Stoneroot *(Collinsonia canadensis)* 239, **375-6, 517, 527**

Sweet Basil *see* Basil

Sweet Briar *see* Rose, Sweet Briar

Sweet Marjoram *see* Marjoram, Sweet

Symphytum officinale (Comfrey, English) Plate 2.2, 30, 66, 141, 204, 239, **301-2, 512, 522**

Tanacetum parthenium (Feverfew) **382-4, 513, 523**

Tansy *(Tanacetum vulgare)* 293, **517, 527**

Taraxacum officinale (Dandelion) Plate 2.6 & 2.7, 30, 60, 66, 110-11, 142, 152, 182, 239, 248, **302-8, 474, 512, 522**

Tarragon, French *(Artemisia dracunculus)* 48, 197, **323-6, 517, 527**

Tasmannia lanceolata (Mountain Pepper) Plate 5.3, **489-90, 515, 525**

Thistle, Blessed *see* Blessed Thistle

Thistle, Holy *see* Blessed Thistle

Thistle, Milk *see* Variegated Thistle

Thistle, St Mary's *see* Variegated Thistle

Thistle, Variegated *see* Variegated Thistle

Thyme *(Thymus vulgaris)* Plate 7.5, 59, 67, 68, 84, 123, 198, 208, 209, **397-9, 517, 527**

Thyme, Lemon *see* Lemon Thyme

Thymus vulgaris (Thyme) Plate 7.5, 59, 67, 68, 84, 123, 198, 208, 209, **397-9, 517, 527**

Thymus x citriodorus (Lemon Thyme) Plate 4.4, 84, **388-90, 514, 524**

Tilea spp. (Linden) **416-18, 515, 525**

Trifolium pratense (Red Clover) Plates 6.2 & 8.1, 46, 53, 67, 86, 127, 136, 239, 240, **369-75, 498, 516, 526**

Tussilago farfara (Coltsfoot) 15, 29-30, 41, 60, 65, 152, 206, 238, **253-6, 512, 522**

Twitch *see* Couchgrass, English

Urtica dioica (Nettle, Greater) Plate 5.6, 17, 41, 66, 73, 80, 84, 109, 154, 204, 222, 240, **314-18, 515, 525**

Urtica urens (Nettle, Lesser) 17, 28, 109, 204, 237, 314, 315, **492-4, 515, 525**

Valerian *(Valeriana officinalis)* Plate 7.2, 17, 41, 60, 111-12, 141, 143, 174, 179, 214, 239, **326-32, 517, 527**

Variegated Thistle *(Silybum marianum)* **506-8, 517, 527**

Verbascum thapsus (Mullein, Great) 169, **457, 490-2, 515, 525**

Vervain *(Verbena officinalis)* 60, 67, 169, 209, 238, **376-8, 517, 527**

Viburnum opulus (Crampbark) Plate 2.5, **409-10, 512, 522**

Vinca major (Periwinkle, Greater) **495, 516, 526**

Violet, Sweet *(Viola odorata)* 15, **332-3, 517, 527**

Vitex agnus-castus (Chaste Tree) Plate 2.3, 44, 173, **408, 512, 522**

White Horehound *see* Horehound, White

Wild Rose *see* Rose, Wild

Willow *(Salix alba)* 59, **419-20, 517, 527**

Witch Hazel *(Hamamelis virginiana)* **420, 517, 527**

Wood Betony *(Stachys officinalis)* Plate 7.4, **378-9, 517, 527**

Wormwood *(Artemesia absinthium)* 67, 100, **379-80, 517, 527**

Yarrow *(Achillea millefolium)* Plate 7.6, 65, 108-9, 237, 240, **294-6, 508, 517, 527**

Yellow Dock *see* Dock, Yellow

Metric Conversion Tables

Although Australia has now officially converted to metric there are other countries, particularly U.S.A, that have retained traditional systems of measurements for length, weight, volume and temperature. Even in Australia many people on the land still think and work in the imperial measurement systems and many metric dimensions in common use are simply metric equivalents of previous imperial dimensions.

These tables include a number of rounded-off equivalents which, while not being highly accurate, are handy in converting from one system to another for most practical purposes on the farm. (See also: **An ancient but convenient measuring system that is always at hand**, on page 62.) In these tables the symbol '≈' indicates 'approximately equal', while '=' indicates 'exactly equal'.

Length
1 millimetre (mm) ≈ .04 inches
1 inch ≈ 25 mm
100 mm ≈ 4 inches
1 foot = 12 inches ≈ 300 mm
600 mm ≈ 2 feet

1 yard = 3 feet ≈ 900 mm
1200 mm ≈ 4 feet
1 centimetre (cm) = 10mm ≈ 0.4 inches
1 metre (m) = 100cm = 1000mm ≈ 40 inches
1 chain = 22 yards = 66 feet ≈ 20 metres

Area
1 square metre (m²) ≈ 1.2 square yards ≈ 11 square feet
20 metres x 20 metres = 400 m² ≈ 484 square yards = 1 square chain = 1/10 acre
4000 m² = 0.4 hectare (ha) ≈ 1 acre
1 hectare = 10,000 m² ≈ 2.5 acres

Weight
1 gram (g) ≈ 0.035 ounces (oz)
1 ounce ≈ 28 g
1 kilogram (kg) = 1000 g ≈ 2.2 pounds (lbs)
50 kg ≈ 1 hundredweight (cwt) = 112 lbs
1 tonne (metric) = 1000 kg ≈ 1 long ton (British or imperial) ≈ 1.1 short tons (U.S., Canadian)

Weight/Area
1 kg/m² = 10 tonnes/hectare ≈ 0.2 pounds/square foot ≈ 4 long tons/acre ≈ 4.5 short tons/acre
1 pound/square foot ≈ 5kg/m²
1 tonne/hectare ≈ 0.4 long tons/acre ≈ 0.45 short tons/acre
1 long ton/acre ≈ 2.5 tonnes/hectare
1 short ton/acre ≈ 2.25 tonnes/hectare

Volume (liquid)
1 litre (L) = 1.76 pints (British, Canadian, or imperial) = 2.11 pints (U.S.)
1 gallon (British, Canadian, or imperial) ≈ 4.5 litres
1 gallon (U.S.) ≈ 3.8 litres
200 litres ≈ 44 gallons (British, Canadian, or imperial) = 53 gallons (U.S.)
1 cubic metre (m³) = 1000 L ≈ 1.3 cubic yards
1 megalitre (ML) = 1,000,000 L = 1000 m³ ≈ 35,000 cubic feet ≈ 0.8 acre feet
    ≈ 220,000 gallons (British, Canadian, or imperial) ≈ 260,000 gallons (U.S.)

Temperature
Converting between Celsius or Centigrade (C°) and Fahrenheit (F°):
Temperature C° = (Temperature F° - 32) x 5 ÷ 9   Temperature F° = (Temperature C° x 9 ÷ 5) + 32

-40° C = -40° F   0° C = 32° F   20° C = 68° F   40° C = 104° F
-30° C = -22° F   5° C = 41° F   25° C = 77° F   45° C = 113° F
-20° C = -4° F   10° C = 50° F   30° C = 86° F   50° C = 122° F
-10° C = 14° F   15° C = 59° F   35° C = 95° F   100° C = 212° F